New Mathematical Monographs

For information about Cambridge University Press mathematics publications visit http://publishing.cambridge.org/stm/mathematics/

Representation Theory of Finite Reductive Groups

At the crossroads of representation theory, algebraic geometry and finite group theory, this book blends together many of the main concerns of modern algebra, synthesizing the past 25 years of research, with full proofs of some of the most remarkable achievements in the area.

Cabanes and Enguehard follow three main themes: first, applications of étale cohomology, leading, via notions of twisted induction, unipotent characters and Lusztig's approach to the Jordan decomposition of characters, to the proof of the recent Bonnafé–Rouquier theorems. The second is a straightforward and simplified account of the Dipper–James theorems relating irreducible characters and modular representations, while introducing modular Hecke and Schur algebras. The final the is local representation theory. One of the main results here is the authors' version Fong–Srinivasan theorems showing the relations between twisted induction and of modular representations.

Throughout, the text is illustrated by many examples; background is provid several introductory chapters on basic results, and appendices on algebraic ge and derived categories. The result is an essential introduction for graduate s reference for all algebraists.

MARC CABANES is Chargé de Recherche at Université Paris 7

MICHEL ENGUEHARD is Professeur Emérite, Université Paris 8.

Representation Theory of Finite Reductive Groups

MARC CABANES, MICHEL ENGUEHARD

Université Paris

CAMBRIDGE
UNIVERSITY PRESS

CAMBRIDGE UNIVERSITY PRESS
Cambridge, New York, Melbourne, Madrid, Cape Town, Singapore,
São Paulo, Delhi, Dubai, Tokyo, Mexico City

Cambridge University Press
The Edinburgh Building, Cambridge CB2 8RU, UK

Published in the United States of America by Cambridge University Press, New York

www.cambridge.org
Information on this title: www.cambridge.org/9780521825177

First published 2004

A catalogue record for this publication is available from the British Library

Library of Congress Cataloguing in Publication Data
Cabanes, Marc, 1959–
Representation theory of finite reductive groups / Marc Cabanes, Michel Enguehard.
p. cm. – (New mathematical monographs)
Includes bibliographical references and index.
ISBN 0 521 82517 2
1. Finite groups. 2. Representations of groups. I. Title. II. Series.
QA177.C33 2004
512'.2–dc21 2003055189

ISBN 978-0-521-82517-7 Hardback

Contents

Preface

This book is an introduction to the study of representations of a special class of finite groups, called finite reductive groups. These are the groups of rational points over a finite field in reductive groups. According to the classification of finite simple groups, the alternating groups and the finite reductive groups yield all finite non-abelian simple groups, apart from 26 "sporadic" groups.

Representation theory, when applied to a given finite group G, traditionally refers to the program of study defined by R. Brauer. Once the ordinary characters of G are determined, this consists of expressing the Brauer characters as linear combinations of ordinary characters, thus providing the "decomposition" matrix and Cartan matrix of group algebras of the form $k[G]$ where k is some algebraically closed field of prime characteristic ℓ. One may add to the above a whole array of problems:

- blocks of $k[G]$ and induced partitions of characters,
- relations with ℓ-subgroups,
- computation of invariants controlling the isomorphism type of these blocks,
- checking of finiteness conjectures on blocks,
- study of certain indecomposable modules,
- further information about the category $k[G]-$**mod** and its derived category $D(k[G])$.

In the case of finite reductive groups, only parts of this program have been completed but, importantly, more specific questions or conjectures have arisen. For this reason, the present book may not match Brauer's program on all points. It will generally follow the directions suggested by the results obtained during the last 25 years in this area.

Before describing the content of the book, we shall outline very briefly the history of the subject.

The subject. Finite simple groups are organized in three stages of mounting complexity, plus the 26 sporadic groups. First are the cyclic groups of prime order. Second are the symmetric groups (or, better, their derived subgroups) whose representation theory has been fairly well known since the 1930s. Then there are the finite reductive groups, each associated with a power q of a prime p, a dimension n, and a geometry in dimension n taken in a list similar to the one for Dynkin "ADE" diagrams. A little before this classification was complete, Deligne–Lusztig's paper [DeLu76] on representations of finite reductive groups appeared. It introduced to the subject the powerful methods of étale cohomology, primarily devised in the 1960s and 1970s by Grothendieck and his team in their re-foundation of algebraic geometry and proof of the Weil conjectures. Deligne–Lusztig's paper set the framework in which most subsequent studies of representations over the complex field of finite reductive groups took place, mainly by Lusztig himself [Lu84] with contributions by Asai, Shoji and others.

The *modular* study of these representations was initiated by the papers of Fong–Srinivasan [FoSr82], [FoSr89], giving the partition of complex characters induced by the blocks of the group algebras over a field of characteristic ℓ ($\ell \neq$ the characteristic p of the field of definition of the finite reductive group). Meanwhile, Dipper [Dip85a–b] produced striking results about the decomposition numbers (relating irreducible representations over the complex field and over finite fields of order ℓ^a) for finite linear groups $\mathrm{GL}_n(p^b)$, emphasizing the rôle played by analogues in characteristic ℓ of concepts previously studied only over the complex field, such as Hecke algebras and cuspidal representations. These works opened a new field of research with numerous contributions by teams in Paris (Broué, Michel, Puig, Rouquier, and the present authors) and Germany (Dipper, Geck, Hiss, Malle), producing several new results on blocks of modular representations, Deligne–Lusztig varieties, non-connected reductive groups, and giving new impulse to adjacent (or larger) fields such as derived categories for finite group representations, cyclotomic Hecke algebras, non-connected reductive groups, quasi-hereditary rings or braid groups. Dipper's work was fully rewritten and generalized in a series of papers by James and himself, linking with James' study of modular representations of symmetric groups, thus generalizing the latter to Hecke algebras of type **A** or **B**, and introducing the so-called q-Schur algebra.

In 1988 and 1994, Broué published a set of conjectures postulating that most correspondences in Lusztig theory should be consequences of Morita or derived equivalences of integral group algebras. One of them was recently proved by Bonnafé–Rouquier in [BoRo03]. It asserts that the so-called "Jordan decomposition" of characters ([Lu84]; see [DiMi91] 13.23) is induced by a Morita equivalence between group algebras over an ℓ-adic coefficient ring.

Their proof consists of a clever use and generalization of Deligne–Lusztig's most significant results, in particular a vanishing theorem for étale sheaves supplemented by the construction of Galois coverings for certain subvarieties in the smooth compactification of Deligne–Lusztig varieties.

The book. Our aim is to gather the main theorems around Bonnafé–Rouquier's contribution and the account of Deligne–Lusztig's methods that it requires. This makes a core of six chapters (7–12). After establishing the main algebraic-geometric properties of the relevant varieties, we expound Deligne–Lusztig theory and Bonnafé–Rouquier theorems. The methods are a balanced mix of module theory and sheaf theory. We use systematically the notions and methods of derived categories.

In contrast to this high-flying sophistication, our Part I gathers most of the properties that can be proved by forgetting about algebraic groups and working within the framework of finite BN-pairs, or "Tits systems," a framework common to finite, algebraic or p-adic reductive groups. (There are not even BN-pairs in Chapter 1 but finite groups possessing a set of subquotients satisfying certain properties.) This, however, allows us to prove several substantial results, such as the determination of simple modules in natural characteristic ([Ri69], [Gre78], [Tin79], [Tin80]), the results about the independence of Harish-Chandra induction in relation to parabolic subgroups in transversal characteristics ([HowLeh94], [DipDu93]), or the theorem asserting that Alvis–Curtis–Deligne–Lusztig duality of characters induces an auto-equivalence of the derived category (transversal characteristics again, [CaRi01]). Chapter 5 on blocks is a model of what will be done in Part V, while Chapter 4 gives a flavor of sheaf theory and derived categories, topics that are at the heart of Part II.

Apart from in Part I, the finite groups we consider are built from (affine connected) reductive $\mathbf{F}$-groups $\mathbf{G}$, where $\mathbf{F}$ is an algebraically closed field of non-zero characteristic (we refer to the books [Borel], [Hum81], [Springer]). When $\mathbf{G}$, as a variety, is defined over a finite subfield of $\mathbf{F}$ and $F:\mathbf{G}\to\mathbf{G}$ is the associated Frobenius endomorphism (think of applying a Frobenius field automorphism to the matrix entries in $\mathrm{GL}_n(\mathbf{F})$), the finite group $\mathbf{G}^F$ of fixed points is specifically what we call a **finite reductive group**.

Parts III to V of the book give proofs for the main theorems on modular aspects of character theory of finite reductive groups, i.e. the type of theorems that started the subject, historically speaking. Just as characters are a handy computational tool for approaching representations of finite groups over commutative rings, these theorems should be considered as hints of what the categories $\mathcal{O}\mathbf{G}^F$**−mod** should look like ($\mathcal{O}$ is a complete discrete valuation ring), either absolutely, or relative to $\mathcal{O}\mathbf{L}^F$**−mod** or $\mathcal{O}W$**−mod** categories for certain F-stable Levi subgroups $\mathbf{L}$ or Weyl groups W (see Chapter 23). The results in Parts III to V are thus less complete than the ones in Part II.

The version we prove of Fong–Srinivasan theorems (Theorem 22.9) is our generalization [CaEn94], introducing and using polynomial orders for tori, and "e-generalized" Harish-Chandra theory [BrMaMi93]. This allows us to check Broué's "abelian defect" conjecture when $e = 1$ (Theorem 23.12). As for decomposition numbers, we prove the version of Gruber–Hiss [GruHi97], giving the relation between decomposition numbers for the unipotent blocks of $\mathbf{G}^F$ and the decomposition numbers of q-Schur algebras (see Theorem 20.1). The framework is an extended "linear" case which comprises (finite) general linear groups, and classical groups with the condition that both ℓ and the order of q mod. ℓ are odd.

Chapter 16 gives a full proof of a theorem of Lusztig [Lu88] about the restriction of characters from $\mathbf{G}^F$ to $[\mathbf{G}, \mathbf{G}]^F$. This checking consists mainly in a quite involved combinatorial analysis of conjugacy classes in spin groups.

The general philosophy of the book is that proofs use only results that have previously appeared in book form.

Instead of giving constant references to the same set of books in certain places, especially in Part II, we have provided this information in three appendices at the end of the book. The first gathers the basic knowledge of derived categories and derived functors. The second does the same for the part of algebraic geometry relevant to this book. The third is about étale cohomology. Subsections and results within the appendices are referenced using A1, A2, and A3 (i.e. A2.12 etc.).

Historical notes, indicating authors of theorems and giving references for further reading, are gathered at the end of each chapter.

We thank Cédric Bonnafé and Raphaël Rouquier for having provided early preprints of their work, along with valuable suggestions and references.

To conclude, we should say that there are surely many books to be written on neighboring subjects. For instance, we have not included Asai–Shoji's determination of the $\mathrm{R}_{\mathbf{L}}^{\mathbf{G}}$ functor, which is a crucial step in the definition of generic blocks [BrMaMi93]; see also [Cr95]. Character sheaves, Kazhdan–Lusztig cells, or intersection cohomology are also fundamental tools for several aspects of representations of finite reductive groups.

Terminology

Most of our terminology belongs to the folklore of algebra, especially the group theoretic branch of it, and is outlined below.

The **cardinality** of a finite set S is denoted by $|S|$. When G is a group and H is a subgroup, the **index** of H in G, i.e. the cardinality of G/H, when finite, is denoted by $|G : H|$.

The unit of groups is generally denoted by 1.

Group actions and modules are on the left unless otherwise stated.

If G acts on the set S, we denote by S^G the subset of fixed points $\{s \in S \mid gs = s \text{ for all } g \in G\}$.

The subgroup of a group G generated by a subset S is denoted by $<S>$. In a group G, the action by **conjugation** is sometimes denoted exponentially, that is ${}^h g = hgh^{-1}$ and $g^h = h^{-1}gh$. The **center** of G is denoted by $\mathrm{Z}(G)$. If S is a subset of G, we denote its **centralizer** by $\mathrm{C}_G(S) := \{g \in G \mid gs = sg \text{ for all } s \in S\}$. We denote its **normalizer** by $\mathrm{N}_G(S) := \{g \in G \mid gSg^{-1} = S\}$. The notation $H \lhd G$ means that H is a **normal** subgroup of G. The notation $G = K \rtimes H$ means that G is a semi-direct product of its subgroups K and H, with H acting on K.

For $g, h \in G$, we denote their **commutator** by $[g, h] = ghg^{-1}h^{-1}$. If H, H' are subgroups of G, one denotes by $[H, H']$ the subgroup of G generated by the commutators $[h, h']$ for $h \in H$, $h' \in H'$.

If π is a set of primes, we denote by π' its complementary set in the set of all primes. If $n \geq 1$ is an integer, we denote by n_π the biggest divisor of n which is a product of powers of elements of π. If G is a group, we denote by G_π the set of elements of finite order n satisfying $n = n_\pi$. We call them the π-elements of G. A **π-group** is any group of finite order n such that $n_\pi = n$. The π'-elements of G are sometimes called the π-regular elements of G. Any element of finite order $g \in G$ is written uniquely as $g = g_\pi g_{\pi'} = g_\pi g_{\pi'}$ where $g_\pi \in G_\pi$ and $g_{\pi'} \in G_{\pi'}$. We then call g_π the π-part of g.

If $n \geq 1$ is an integer, we denote by $\phi_n(x) \in \mathbb{Z}[x]$ the nth **cyclotomic polynomial**, defined recursively by $x^n - 1 = \Pi_d \phi_d(x)$ where the product is over divisors $d \geq 1$ of n.

Let A be a (unital) ring. We denote by $J(A)$ its **Jacobson radical.** If M is an A-module, we denote the head of M by $\mathrm{hd}(M) = M/J(A).M$. We denote by $\mathrm{soc}(M)$ the sum of the simple submodules of M (this notion is considered only when this sum is non-empty, which is ensured with Artin rings, for instance finite-dimensional algebras over a field).

If $n \geq 1$ is an integer, we denote by $\mathrm{Mat}_n(A)$ the ring of $n \times n$ **matrices** with coefficients in A (generally for a commutative A). We denote the **transposition** of matrices by $X \mapsto {}^tX$ for $X \in \mathrm{Mat}_n(A)$.

We denote by $A^\times$ the group of **invertible elements** of A, sometimes called **units**. We denote by A^{opp} the **opposite ring**.

We denote by $A-\mathbf{Mod}$ (resp. $A-\mathbf{mod}$) the category of A-modules (resp. of finitely generated A-modules). Note that we use the sign $\in$ for objects in categories, so $M \in A-\mathbf{mod}$ means that M is a finitely generated A-module.

When $M \in A-\mathbf{mod}$, we denote by $\mathrm{GL}_A(M)$ the group of automorphisms of M. For a field $\mathbb{F}$ and an integer $n \geq 1$, we abbreviate $\mathrm{GL}_n(\mathbb{F}) = \mathrm{GL}_{\mathbb{F}}(\mathbb{F}^n)$.

If A, B are two rings, an **A-B-bimodule** M is an $A \times B^{\mathrm{opp}}$-module, that is the datum of structures of left A-module and right B-module on M such that $a(mb) = (am)b$ for all $a \in A$, $b \in B$ and $m \in M$. Recall that $M \otimes_B -$ then induces a functor $B-\mathbf{Mod} \to A-\mathbf{Mod}$. When $A = B$, we just say **A-bimodule**. The category of finitely generated A-B-bimodules is denoted by $A-\mathbf{mod}-B$.

If C is a commutative ring and G is a group, we denote by CG (sometimes $C[G]$) the associated group ring, or **group algebra**, consisting of finite linear combinations $\sum_{g \in G} c_g g$ of elements of G with coefficients in C endowed with the C-bilinear multiplication extending the law of G. The **trivial module** for this ring is C with the elements of G acting by Id_C. This CG-module is sometimes denoted by 1.

The commutative ring C is sometimes omitted from the notation. For instance, if H is a subgroup of G the **restriction** of a CG-module M to the subalgebra CH is denoted by $\mathrm{Res}_H^G M$. In the same situation, CG is considered as a CG-CH-bimodule, so we have the **induction** functor Ind_H^G defined by tensor product $CG \otimes_{CH} -$.

Let $\mathcal{O}$ be a complete local principal ideal domain (i.e. a complete discrete valuation ring) with field of fractions K and residue field $k = \mathcal{O}/J(\mathcal{O})$. Let A be an $\mathcal{O}$-algebra which is $\mathcal{O}$-free of finite rank over $\mathcal{O}$. Then $\mathcal{O}$ is said to be a **splitting system** for A if $A \otimes_{\mathcal{O}} K$ and $A \otimes_{\mathcal{O}} k/J(A \otimes_{\mathcal{O}} k)$ are products of matrix algebras over the fields K and k, respectively. Note that this implies that

$A \otimes_{\mathcal{O}} K$ is semi-simple. For group algebras $\mathcal{O}G$ (G a finite group) and their blocks, this is ensured by the fact that $\mathcal{O}$ has characteristic zero.

If G is a finite group and ℓ is a prime, a triple $(\mathcal{O}, K, k)$ is called an **ℓ-modular splitting system** for G if $\mathcal{O}$ is a complete discrete valuation ring containing the $|G|$th roots of 1, free of finite rank over $\mathbb{Z}_\ell$, K denoting its field of fractions (a finite extension of $\mathbb{Q}_\ell$) and k its residue field (with $|k|$ finite, a power of ℓ). Then $\mathcal{O}$ is a splitting system for $\mathcal{O}G$, i.e. KG (resp. $kG/J(kG)$) is split semi-simple over K (resp. k); see [NaTs89] §3.6. Note that if $(\mathcal{O}, K, k)$ is an ℓ-modular splitting system for G, it is one for all its subgroups.

Let G be a finite group. We denote by $\mathrm{Irr}(G)$ the set of **irreducible characters** of G, i.e. trace maps $G \to \mathbb{C}$ corresponding with simple $\mathbb{C}G$-modules. **Generalized characters** are $\mathbb{Z}$-linear combinations of elements of $\mathrm{Irr}(G)$. They are considered as elements of $\mathrm{CF}(G, \mathbb{C})$, the space of central functions $G \to \mathbb{C}$ of which $\mathrm{Irr}(G)$ is a base. Since finite-dimensional $\mathbb{C}G$-modules may be realized over $\mathbb{Q}[\omega]$ for ω a $|G|$th root of 1, classically $\mathrm{Irr}(G)$ is identified with the trace maps of simple KG-modules for any field K of characteristic zero containing a $|G|$th root of 1. They form a basis of $\mathrm{CF}(G, K)$ (central functions $G \to K$).

Classically we consider on $\mathrm{CF}(G, K)$ the "scalar product" $\langle f, f'\rangle_G := |G|^{-1}\sum_{g\in G} f(g)f'(g^{-1})$ for which $\mathrm{Irr}(G)$ is orthonormal.

PART I

Representing finite BN-pairs

This first part is an elementary introduction to the remainder of the book.

Instead of finite reductive groups $G := \mathbf{G}^F$ defined as the fixed points under a Frobenius endomorphism $F: \mathbf{G} \to \mathbf{G}$ in an algebraic group, we consider finite groups G endowed with a split BN-pair. This is defined by the presence in G of subgroups B, N satisfying certain properties (see Chapter 2 for precise definitions). Part I gathers many of the results that can be proved about representations of G within this axiomatic framework. Though some results are quite recent, this should not mislead the reader into the idea that finite reductive groups can be studied without reference to reductive groups and algebraic varieties.

However, many important ideas are evoked in this part. We shall comment on Harish-Chandra induction, cuspidality, Hecke algebras, the Steinberg module, the duality functor, and derived categories.

The six chapters are almost self-contained. We assume only basic knowledge of module theory (see, for instance, the first chapter of [Ben91a]). We also recall some elementary results on BN-pairs (see, for instance, [Asch86] §43, [Bour68] §IV).

A group with a split BN-pair of characteristic p is assumed to have *parabolic subgroups* decomposed as $P = U_P \rtimes L$, the so-called Levi decomposition, where L is also a finite group with a split BN-pair. A leading rôle is played by the G-L-bimodules

$$RG.e(U_P)$$

where $R = \mathbb{Z}[p^{-1}]$ and $e(U_P) = |U_P|^{-1} \sum_{u \in U_P} u$, and their R-duals

$$e(U_P).RG.$$

A first natural question is to ask whether this bimodule depends on P and not just on L. We also study "Harish-Chandra induced" modules $RG.e(U_P) \otimes_{RL} M$ for M simple kL-modules (k is a field where $p \neq 0$) such that (L, M) is minimal with regard to this induction process. It is important to study the law of the

"Hecke algebra" $\mathrm{End}_{kG}(RG.e(U_P) \otimes_{RL} M)$ and show that it behaves a lot like a group algebra.

The RG-bimodules

$$RG.e(U_P) \otimes_{RL} e(U_P).RG$$

allow us to build a bounded complex defining an equivalence

$$D^b(RG\text{−}\mathbf{mod}) \to D^b(RG\text{−}\mathbf{mod})$$

within the derived category of the category of finitely generated RG-modules.

We see that the main ingredients in this module theory are in fact permutation modules. In finite group theory, these are often used as a first step towards the study of the full module category, or, more importantly, through the functors they define. In our context of groups $G := \mathbf{G}^F$, we may see the G-sets G/U_P as 0-dimensional versions of the Deligne–Lusztig varieties defined in $\mathbf{G}$.

1
Cuspidality in finite groups

The main functors in representation theory of finite groups are the restriction to subgroups and its adjoint, called induction.

We focus attention on a slight variant. Instead of subgroups, we consider subquotients $V \lhd P$ of a finite group G. It is natural to consider the fixed point functor $\mathrm{Res}^G_{(P,V)}$ which associates with a G-module M the subspace M^V of its restriction to P consisting of fixed points under the action of V. This is a P/V-module. When the coefficient ring (we denote it by Λ) is such that the order of V is invertible in Λ, the adjoint of

$$\mathrm{Res}^G_{(P,V)}\colon \Lambda G\mathbf{-mod} \to \Lambda P/V\mathbf{-mod}$$

is a kind of induction, sometimes called Harish-Chandra induction,

$$\mathrm{Ind}^G_{(P,V)}\colon \Lambda P/V\mathbf{-mod} \to \Lambda G\mathbf{-mod}$$

which first makes the given $\Lambda P/V$-module into a V-trivial ΛP-module, then induces it from P to G.

The usual Mackey formula, which computes $\mathrm{Res}^G_P \mathrm{Ind}^G_{P'}$, is then replaced by a formula where certain non-symmetric intersections $(P, V) \cap\!\downarrow (P', V') := ((P \cap P')V', (V \cap P')V')$ occur. This leads naturally to a notion of $\cap\!\downarrow$-stable Λ-regular sets $\mathcal{L}$ of subquotients of a given finite group. For such a set of subquotients, an $\mathcal{L}$-cuspidal triple is (P, V, M), where $(P, V) \in \mathcal{L}$ and M is a V-trivial ΛP-module such that $\mathrm{Res}^P_{(P',V')} M = 0$ for all $(P', V') \in \mathcal{L}$ such that $V \subseteq V' \subseteq P' \subseteq P$ and $(P', V') \neq (P, V)$.

The case of a simple M above (for $\Lambda = K$ a field) has remarkable properties. The induced module $\mathrm{Ind}^G_{(P,V)} M$ is very similar to a projective ΛG-module. The indecomposable summands of $\mathrm{Ind}^G_{(P,V)} M$ have a unique simple quotient, and a unique simple submodule, each determining the direct summand that yields it.

A key fact explaining this phenomenon is the property of the endomorphism algebra $\mathrm{End}_{KG}(\mathrm{Ind}^G_{(P,V)} M)$ of being a *self-injective algebra*. This last property seems to be intimately related with cuspidality of M. These endomorphism algebras are what we call Hecke algebras.

Self-injectivity is a property Hecke algebras share with group algebras. To prove self-injectivity, we define a basis of the Hecke algebra. The invertibility of these basis elements is related to the following quite natural question. Assume (P, V) and (P', V') are subquotients such that $P \cap P'$ covers both quotients P/V and P'/V' and makes them isomorphic. Then, the V-trivial P-modules and the V'-trivial P'-modules can be identified. The "independence" question is as follows. Are $\mathrm{Ind}_{(P,V)}^G$ and $\mathrm{Ind}_{(P',V')}^G$ tranformed into one another by this identification? A positive answer is shown to be implied by the invertibility of the basis elements mentioned above.

1.1. Subquotients and associated restrictions

Let G be a finite group. A subquotient of G is a pair (P, V) of subgroups of G with $V \lhd P$.

Definition 1.1. *When $V \lhd P$ and $V' \lhd P'$ are subgroups of G, let $(P, V) \cap\!\downarrow (P', V') = ((P \cap P').V'), (V \cap P').V')$.*

One denotes $(P, V) \leq (P', V')$ if and only if $V' \subseteq V \subseteq P \subseteq P'$.

One denotes (P, V)—(P', V') if and only if $(P, V) \cap\!\downarrow (P', V') = (P', V')$ and $(P', V') \cap\!\downarrow (P, V) = (P, V)$.

Proposition 1.2. *Keep the above notation. (i) If (P, V)—(P', V'), then $V \cap P' = V' \cap P = V \cap V'$ and $P/V \cong P'/V' \cong P \cap P'/V \cap V'$.*

(ii) $((P, V) \cap\!\downarrow (P', V'))$—$((P', V') \cap\!\downarrow (P, V))$.

(iii) (P, V)—(P', V') if and only if $(P, V) \cap\!\downarrow (P', V') = (P', V')$ and $|P/V| = |P'/V'|$.

Proof. (i), (ii) are easy from the definitions.

(iii) The "only if" is clear from (i). Assume now that $(P, V) \cap\!\downarrow (P', V') = (P', V')$ and $|P/V| = |P'/V'|$. Then $(P \cap P').V' = P'$ and $V \cap P' \subseteq V'$. Then P'/V' is a quotient of $(P \cap P')/(V \cap P')$ by reduction mod. V'. But $(P \cap P')/(V \cap P')$ is a subgroup of P/V. Since $|P/V| = |P'/V'|$, all those quotients coincide, so $P \cap P' \cap V' = V \cap P'$ and $(P \cap P').V = P$. This gives $(P', V') \cap\!\downarrow (P, V) = (P, V)$ and therefore (P, V)—(P', V'). □

Notation 1.3. When $V \lhd P$ are subgroups of G, let $\Lambda P/V$−**mod** be the category of finitely generated ΛP-modules having V in their kernel (we sometimes call them (P, V)-modules).

Let

$$\mathrm{Res}^G_{(P,V)}\colon \Lambda G-\mathbf{mod}\to \Lambda P/V-\mathbf{mod}$$

be the functor defined by $\mathrm{Res}^G_{(P,V)}(M)=M^V$ (fixed points under the action of V) as ΛP-module.

Definition 1.4. *When G is a finite group, Λ is a commutative ring, and V, V' are two subgroups of G whose orders are invertible in Λ, let $e(V)=|V|^{-1}\sum_{u\in V}u\in\Lambda G$. If VV' is a subgroup, then $e(V)e(V')=e(VV')$. In particular $e(V)$ is an idempotent.*

Proof. Clear.

Proposition 1.5. *Let Λ be a commutative ring. Let $V\lhd P$ and $V'\lhd P'$ in G with $|V|$ and $|V'|$ invertible in Λ. Let $L\subseteq P$ be a subgroup such that $P=LV$. Let N be a $\Lambda P/V$-module identified with a $\Lambda L/(L\cap V)$-module. Denote $e=e(V)$.*

(i) ΛGe is a G-L-bimodule and $\Lambda Ge\otimes_{\Lambda L}N\cong \mathrm{Ind}^G_{(P,V)}N$ by $ge\otimes m\mapsto g\otimes m$ for $g\in G$, $m\in N$.

(ii) If M is a ΛG-module, then $\mathrm{Res}^G_{(P,V)}M=eM$. If moreover $(P',V')\le(P,V)$, then $\mathrm{Res}^P_{(P',V')}\circ\mathrm{Res}^G_{(P,V)}=\mathrm{Res}^G_{(P',V')}$.

*(iii) Ind^G_P and $\mathrm{Res}^G_{(P,V)}$ induce exact functors preserving projectivity of modules, and adjoint to each other between $\Lambda P/V-$**mod** and $\Lambda G-$**mod**.*

(iv) If N' is a $\Lambda P'/V'$-module, we have

$$\mathrm{Hom}_{\Lambda G}(\mathrm{Ind}^G_P N,\mathrm{Ind}^G_{P'}N')$$
$$\cong\bigoplus_{PgP'\subseteq G}\mathrm{Hom}_{\Lambda(P\cap {}^gP')}\big(\mathrm{Res}^P_{{}^g(P',V')\cap\downarrow(P,V)}N,\ \mathrm{Res}^{{}^gP'}_{(P,V)\cap\downarrow{}^g(P',V')}{}^gN'\big)$$

as Λ-modules.

Proof. (i) One has clearly $\Lambda G\otimes_{\Lambda P}\Lambda Pe\cong\Lambda Ge$ by $g\otimes pe\mapsto gpe$ ($g\in G$, $p\in P$). So one may assume $G=P$. Then one has to check $\Lambda Pe\otimes_{\Lambda L}N\cong N$ by $pe\otimes m\mapsto pem$. This is clear, the reverse map being $m\mapsto e\otimes m$ since $P=LV$.

(ii) It is clear that the elements of eM are V-fixed. But an element fixed by any $u\in V$ is fixed by e. So $eM=\mathrm{Res}^G_{(P,V)}M$ as subspaces of $\mathrm{Res}^G_P M$. The composition formula comes from $e(V').e=e$.

(iii) The right ΛP-module ΛGe is projective (as direct summand of ΛG, which is free), so $\Lambda Ge\otimes$ is exact. The image of $\Lambda P/V$ is a projective ΛG-module. So $\Lambda Ge\otimes$ sends projective $\Lambda P/V$-modules to projective ΛG-modules. Similarly, $e\Lambda G$ is a projective right ΛG-module and a projective left $\Lambda P/V$-module, so $\mathrm{Res}^G_{(P,V)}\colon\Lambda G-\mathbf{mod}\to\Lambda P/V-\mathbf{mod}$ is exact and preserves projectives. Concerning adjunction, the classical adjunction between induction and restriction gives $\mathrm{Hom}_{\Lambda G}(\mathrm{Ind}_P^G(N), M) \cong \mathrm{Hom}_{\Lambda P}(N, \mathrm{Res}_P^G(M))$ and $\mathrm{Hom}_{\Lambda G}(M, \mathrm{Ind}_P^G(N)) \cong \mathrm{Hom}_{\Lambda P}(\mathrm{Res}_P^G(M), N)$ for all ΛP-modules N and ΛG-modules M (see [Ben91a] §3.3). When, moreover, N is V-trivial one may replace the Res_P^G by fixed points under V since $eN = N$, $(1-e)N = 0$ and therefore for all ΛP-modules N', $\mathrm{Hom}_{\Lambda P}(N', N) = \mathrm{Hom}_{\Lambda P}(eN', N)$ and $\mathrm{Hom}_{\Lambda P}(N, N') = \mathrm{Hom}_{\Lambda P}(N, eN')$.

(iv) Note first that the expression $\mathrm{Hom}_{\Lambda(P\cap {}^gP')}(\mathrm{Res}^P_{{}^g(P',V')\cap\downarrow(P,V)}N, \mathrm{Res}^{{}^gP'}_{(P,V)\cap\downarrow {}^g(P',V')}{}^gN')$ makes sense since $P\cap {}^gP'$ is a subgroup of the first terms in both ${}^g(P',V')\cap\downarrow(P,V)$ and $(P,V)\cap\downarrow {}^g(P',V')$. The Mackey formula and adjunction between induction and restriction give $\mathrm{Hom}_{\Lambda G}(\mathrm{Ind}_P^G N, \mathrm{Ind}_{P'}^G N') \cong \oplus_{PgP'\subseteq G}\mathrm{Hom}_{\Lambda(P\cap {}^gP')}(\mathrm{Res}^P_{P\cap {}^gP'}N, \mathrm{Res}^{{}^gP'}_{P\cap {}^gP'}{}^gN')$. Now, we may replace the second term with its fixed points under $V\cap {}^gP'$ (hence $(V\cap {}^gP').{}^gV'$) since $N = e(V\cap {}^gP')N$. Similarly, we may replace the first term with its fixed points under $P\cap {}^gV'$ (hence $(P\cap {}^gV').V$) since $(1-e(P\cap {}^gV')){}^gN' = 0$. □

1.2. Cuspidality and induction

We fix Λ a commutative ring.

Definition 1.6. *A G-stable set $\mathcal{L}$ of subquotients (P, V) is said to be Λ-regular if and only if, for all $(P,V)\in\mathcal{L}$, V is of order invertible in Λ. One says that $\mathcal{L}$ is $\cap\downarrow$-stable if and only if $\mathcal{L}$ is G-stable, $(G,\{1\})\in\mathcal{L}$, and, for all (P,V), $(P',V')\in\mathcal{L}$, one has $(P,V)\cap\downarrow(P',V')\in\mathcal{L}$.*

For the remainder of the chapter, we assume that G is a finite group, and $\mathcal{L}$ is a Λ-regular, $\cap\downarrow$-stable set of subquotients of G.

Example 1.7. (i) When $V \lhd P$ are subgroups of the finite group G, and $|V|$ is invertible in Λ, it is easy to show that there is a minimal Λ-regular, $\cap\downarrow$-stable set of subquotients $\mathcal{L}_{(P,V)}$ such that $(P,V)\in\mathcal{L}_{(P,V)}$. It consists of finite $\cap\downarrow$-intersections (with arbitrary hierarchy of parenthesis) of G-conjugates of (P,V).

(ii) Let $\mathbb{F}$ be a finite field, let $G := \mathrm{GL}_2(\mathbb{F})$ be the group of invertible matrices $\begin{pmatrix} a & c \\ b & d \end{pmatrix}$ $(a,b,c,d\in\mathbb{F}, ad-bc\neq 0)$. Let B be the subgroup defined by $c = 0$, let $U\lhd B$ be defined by $a = d = 1, c = 0$. Then the pair $(G,\{1\})$ along with the G-conjugates of (B,U) is $\cap\downarrow$-stable. This is easily checked by the equality $G = B\cup BwB$ for $w = \begin{pmatrix} 0 & 1 \\ 1 & 0 \end{pmatrix}$, and the (more obvious) fact that $B = (B\cap B^w)U$ with $B\cap U^w = \{1\}$.

The system just defined is Λ-regular as long as the characteristic of $\mathbb{F}$ is invertible in Λ.

(iii) In the next chapter, we introduce more generally the notion of groups with split BN-pairs of characteristic p (p is a prime). These groups G have a subgroup B which is a semi-direct product $U.T$ where U is a subgroup consisting of all p-elements of B. The subgroups of G containing B are called **parabolic subgroups**. They decompose as semi-direct products $P = V.L$, where V is the largest normal p-subgroup of P. The set of G-conjugates of pairs (P, V) is $\cap\!\downarrow$-stable (see Theorem 2.27(ii)) and of course Λ-regular for any field of characteristic not equal to p.

The reader familiar with these groups may assume in what follows that our system $\mathcal{L}$ corresponds with this example. The notation, however, is the same.

Definition 1.8. *A ΛG-module M is said to be $\mathcal{L}$-cuspidal if and only if $(G, \{1\}) \in \mathcal{L}$, and $\mathrm{Res}^G_{(P,V)} M = 0$ for each $(P, V) \in \mathcal{L}$ such that $(P, V) \neq (G, \{1\})$. This clearly implies that, if some pair $(P, \{1\})$ is in $\mathcal{L}$, then $P = G$. When $(P, V) \in \mathcal{L}$ and M is a $\Lambda P/V$-module, M is said to be $\mathcal{L}$-cuspidal if and only if it is $\mathcal{L}_{P/V}$-cuspidal for $\mathcal{L}_{P/V} = \{(P'/V, V'/V) \mid \mathcal{L} \ni (P', V') \leq (P, V)\}$ (this implies that the only pair $(P', V) \leq (P, V)$ in $\mathcal{L}$ is (P, V)).*

Remark 1.9. If $\mathcal{L}$ is $\cap\!\downarrow$-stable, Λ-regular, and $(P, V) \in \mathcal{L}$ is such that there is a cuspidal $\Lambda P/V$-module, then the condition

$$(P', V) \leq (P, V) \text{ implies } P' = P$$

is a strong constraint. Applying it to the pairs $(P, V)^g \cap\!\downarrow (P, V)$, one gets that, if $g \in G$ is such that $V^g \cap P \subseteq V$, then $P = (P \cap P^g).V$.

Notation 1.10. A cuspidal triple in G relative to $\mathcal{L}$ and Λ is any triple $\tau = (P, V, M)$ where $(P, V) \in \mathcal{L}$ and M is an $\mathcal{L}$-cuspidal $\Lambda P/V$-module.

When $\tau' = (P', V', M')$ is another cuspidal triple, one denotes $\tau — \tau'$ if and only if $(P, V)—(P', V')$ and $\mathrm{Res}^P_{P\cap P'} M \cong \mathrm{Res}^{P'}_{P\cap P'} M'$.

Let $I^G_\tau = \mathrm{Ind}^G_P M$. Then $I^G_\tau = I^G_{{}^g\tau}$ for all $g \in G$.

When $\Lambda = k$ is a field, let $\mathbf{cusp}_k(\mathcal{L})$ be the set of all triples $\tau = (P, V, S)$ such that (P, V) is in $\mathcal{L}$ and S is a **simple** cuspidal kP/V-module (one for each isomorphism type).

Proposition 1.11. *Assume that $\Lambda = k$ is a field. Let M be a simple kG-module. Then*

(a) there exists a simple cuspidal triple
$\tau' \in \mathbf{cusp}_k(\mathcal{L})$ *such that* $M \subseteq \mathrm{soc}(I^G_{\tau'})$,

(a′) there exists a simple cuspidal triple
$\tau \in \mathbf{cusp}_k(\mathcal{L})$ *such that* $M \subseteq \mathrm{hd}(I^G_\tau)$.

Proof. Let $(P, V) \in \mathcal{L}$ of minimal $|P/V|$ be such that $\mathrm{Res}^G_{(P,V)} M \neq 0$ (recall $(G, \{1\}) \in \mathcal{L}$). Let S be a simple component of the head of $\mathrm{Res}^G_{(P,V)} M$. There is a surjection $\mathrm{Res}^G_{(P,V)} M \to S$. For every $(P', V') \leq (P, V)$, one has a surjection $\mathrm{Res}^G_{(P',V')} M \to \mathrm{Res}^P_{(P',V')} S$ since $\mathrm{Res}^P_{(P',V')}$ is exact. Then $\mathrm{Res}^P_{(P',V')} S = 0$ when $|P'/V'| < |P/V|$ by the choice of (P, V). So $(P, V, S) \in \mathbf{cusp}_k(\mathcal{L})$. This gives (a). One would get (a′) with $\tau' = (P, V, S')$ by considering S' a simple component of $\mathrm{soc}(\mathrm{Res}^G_{(P,V)} M)$. □

Remark. Assume (P, V)—(P', V') in $\mathcal{L}$. For each V-trivial kP-module M there is a unique V'-trivial kP'-module M' with the same restriction to $P \cap P'$ as M. This clearly defines an isomorphism between $k[P/V]$−**mod** and $k[P'/V']$−**mod**. Then M is simple if and only if M' is so. Similarly one checks that (P, V, M) is cuspidal if and only if (P', V', M') is so. Indeed, if $M'^{V''} \neq 0$ for $(P'', V'') \leq (P', V')$ in $\mathcal{L}$, then $M'^{P\cap V''} = M^{P\cap V''} \neq 0$ and therefore $M^{(P\cap V'').V} \neq 0$. But $(P'', V'') \cap\!\downarrow (P, V) \in \mathcal{L}$ and $(P, V, M) \in \mathbf{cusp}_k(\mathcal{L})$, so $(P'', V'') \cap\!\downarrow (P, V) = (P, V)$. Along with $(P'', V'') \leq (P', V')$—(P, V), this clearly implies $(P'', V'') = (P', V')$.

1.3. Morphisms and an invariance theorem

In the following, Λ is a commutative ring, G is a finite group and $\mathcal{L}$ is a Λ-regular, $\cap\!\downarrow$-stable set of subquotients of G.

Definition 1.12. *When $\tau = (P, V, M)$, $\tau' = (P', V', M')$ are cuspidal triples (see Notation 1.10), and $g \in G$ is such that τ—${}^g\tau'$, choose an isomorphism $\theta_{g,\tau,\tau'}\colon \mathrm{Res}^P_{P\cap {}^gP'} M \cong \mathrm{Res}^{{}^gP'}_{P\cap {}^gP'} {}^gM'$. It can be seen as a linear isomorphism $\theta_{g,\tau,\tau'}\colon M \to M'$ such that $\theta_{g,\tau,\tau'}(x.m) = x^g.\theta_{g,\tau,\tau'}(m)$ for all $m \in M$ and $x \in P \cap {}^gP'$. Assume $\theta_{1,\tau,\tau} = \mathrm{Id}_M$.*

When τ—${}^g\tau'$, define $a_{g,\tau,\tau'}\colon I^G_\tau \to I^G_{\tau'}$ by $I^G_\tau = \Lambda G \otimes_{\Lambda P} M$ and

$$a_{g,\tau,\tau'}(1 \otimes_{\Lambda P} m) = e(V)g \otimes_{\Lambda P'} \theta_{g,\tau,\tau'}(m). \tag{1}$$

Proposition 1.13. *Assume τ—${}^g\tau'$. Let L be a subgroup such that $P = LV$, ${}^gP' = L.{}^gV'$ (for instance $L = P \cap {}^gP'$). Denote $e = e(V)$, $e' = e(V')$.*

The equation (1) defines a unique ΛG-morphism $I^G_\tau \to I^G_{\tau'}$. Through the identifications $I^G_\tau \cong \Lambda Ge \otimes_L M$ and $I^G_{\tau'} \cong I^G_{{}^g\tau'} \cong \Lambda G{}^ge' \otimes_L {}^gM'$ (see Proposition 1.5(i)) the map $a_{g,\tau,\tau'}$ identifies with the map $\mu \otimes_L \theta_{g,\tau,\tau'}$: $\Lambda Ge \otimes_L M \to \Lambda G{}^ge' \otimes_L {}^gM'$ where $\mu\colon \Lambda Ge \to \Lambda G{}^ge'$ is multiplication by ${}^ge'$ on the right.

Proof. Since the morphism μ is clearly a morphism of G-L-bimodules, it suffices to check the second statement to have that $a_{g,\tau,\tau'}$ is well-defined and ΛG-linear.

Let us recall that, by Proposition 1.5(i), $\Lambda G \otimes_{\Lambda P} M \cong \Lambda Ge \otimes_L M$ by $x \otimes m \mapsto xe \otimes m$ for $x \in G, m \in M$. Similarly $I_{\tau'}^G \cong I_{^g\tau'}^G \cong \Lambda G^g e' \otimes_L {}^g M'$ by $x(^g e') \otimes m' \mapsto xg \otimes m'$ for $x \in G$ and $m' \in M'$. Now the map $a_{g,\tau,\tau'}$ would send the element corresponding with $xe \otimes m$ to the one corresponding with $x.e.^g e' \otimes \theta_{g,\tau,\tau'}(m)$. This is clearly the image by $\mu \otimes_L \theta_{g,\tau,\tau'}$. □

Theorem 1.14. *Assume $\Lambda = k$ is a field. Assume that for each τ, τ' in $\mathbf{cusp}_k(\mathcal{L})$ and τ—τ', the map $a_{1,\tau,\tau'}\colon I_\tau^G \to I_{\tau'}^G$ is an isomorphism.*

Then, whenever (P, V)—(P', V') in $\mathcal{L}$, the map $kGe(V) \to kGe(V')$ defined by $x \mapsto xe(V')$ is an isomorphism (and therefore $|V| = |V'|$).

We give a homological lemma used here in a special and quite elementary case, but stated also for future reference.

Lemma 1.15. *Let A be a finite-dimensional algebra over a field. Let $X = (\ldots \xrightarrow{d_{i+1}} X_i \xrightarrow{d_i} X_{i-1} \xrightarrow{d_{i-1}} \ldots)$ be a bounded complex of projective right A-modules. That is, the d_i are A-linear maps, $d_i \circ d_{i+1} = 0$ for any i, and $X_i = 0$ except for a finite number of i's.*

Let $\mathcal{M}$ be a set of (left) A-modules such that any simple A-module is in some $\mathrm{hd}(M)$ for $M \in \mathcal{M}$. Then X is exact (that is $\mathrm{Ker}(d_i) = d_{i+1}(X_{i+1})$ for all i) if and only if

$$X \otimes_A M = (\ldots \xrightarrow{d_{i+1}\otimes_A M} X_i \otimes_A M \xrightarrow{d_i \otimes_A M} X_{i-1} \otimes_A M \xrightarrow{d_{i-1}\otimes_A M} \ldots)$$

is exact for any $M \in \mathcal{M}$.

Proof of Lemma 1.15. We use the classical notation $\mathrm{H}_i(X)$ for the quotient $\mathrm{Ker}(d_i)/d_{i+1}(X_{i+1})$.

Suppose X is not exact, i.e. $X \otimes_A A$ is not exact. Let i_0 be the maximal element in $\{i \mid \mathrm{H}_{i-1}(X \otimes_A M) = 0$ for all M in $A-\mathbf{mod}\}$.

By the projectivity of the X_i's, any extension $0 \to M_3 \to M_2 \to M_1 \to 0$ gives rise to an exact sequence of complexes $0 \to X \otimes_A M_3 \to X \otimes_A M_2 \to X \otimes_A M_1 \to 0$ and then, by the homology long exact sequence (see [Ben91a] 2.3.7), to the exact sequence

$$\begin{aligned} \mathrm{H}_{i_0}(X \otimes_A M_3) \to \mathrm{H}_{i_0}(X \otimes_A M_2) \to \mathrm{H}_{i_0}(X \otimes_A M_1) \to 0 \\ = \mathrm{H}_{i_0-1}(X \otimes_A M_3). \end{aligned}$$

Suppose first $\mathrm{H}_{i_0}(X \otimes_A M_2) \neq 0$, then either $\mathrm{H}_{i_0}(X \otimes_A M_1)$ or $\mathrm{H}_{i_0}(X \otimes_A M_3) \neq 0$. This allows us to assume that there is a simple A-module S such that $\mathrm{H}_{i_0}(X \otimes_A S) \neq 0$. Now, there is an extension with $M_1 = S$ and $M_2 \in \mathcal{M}$. Then $\mathrm{H}_{i_0}(X \otimes_A M_2) = 0$ by the hypothesis, and the above exact sequence gives $\mathrm{H}_{i_0}(X \otimes_A S) = 0$, a contradiction. □

Proof of Theorem 1.14. We apply the above lemma for

$$X = (\ldots \to 0 \ldots \to 0 \to X_1 = kGe(V) \xrightarrow{\mu} X_0$$
$$= kGe(V') \to 0 \ldots \to 0 \ldots)$$

where $\mu(b) = b.e(V')$ and $A = kG'$ for $G' = P \cap P'/V \cap V'$ (isomorphic to both P/V and P'/V', by Proposition 1.2(i)). This tells us that μ is an isomorphism if and only if $\mu \otimes M$ is an isomorphism for each M in a set $\mathcal{M}$ of kG'-modules satisfying the condition given by Lemma 1.15.

The kG'-modules can be considered as restrictions to $P \cap P'$ of V-trivial kP-modules (resp. V'-trivial kP'-modules). By Proposition 1.11, a set $\mathcal{M}$ may be taken to be the set of induced modules $M_i = \mathrm{Res}^P_{P\cap P'}\mathrm{Ind}^P_{(P_i,V_i)}N_i$ for $(P_i, V_i, N_i) \in \mathbf{cusp}_k(\mathcal{L})$ and $(P_i, V_i) \leq (P, V)$. Let $\tau = (P_0, V_0, N_0)$ be such a triple. It is easily checked that $(P', V') \cap\!\!\downarrow (P_0, V_0) = (P_0, V_0)$ (see also Exercise 3). Then Proposition 1.2(ii) implies (P_0, V_0)—$((P_0, V_0) \cap\!\!\downarrow (P', V')) = ((P_0 \cap P').V', (V_0 \cap P').V')$. So let N_0' be the $(P_0, V_0) \cap\!\!\downarrow (P', V')$-module defined by having the same restriction to $P_0 \cap P'$ as N_0.

Denote $\tau' = ((P_0, V_0) \cap\!\!\downarrow (P', V'), N_0')$—$\tau$. Now, recalling that $\theta_{1,\tau,\tau'} = \mathrm{Id}$, Definition 1.12 implies that $a_{1,\tau,\tau'}$ is the map defined by

$$(1) \qquad kGe(V_0) \otimes_{P_0\cap P'} N_0 \to kGe((V_0 \cap P').V') \otimes_{P_0\cap P'} N_0',$$
$$x \otimes n \mapsto xe((V_0 \cap P').V') \otimes n$$

for $x \in kGe(V_0)$, $n \in N_0$ (see also Proposition 1.13).

The hypothesis of our theorem tells us that map (1) is an isomorphism.

Thanks to Lemma 1.15, we just have to check that $\mu \otimes M_0$ is an isomorphism, where $M_0 = \mathrm{Res}^P_{P\cap P'}\mathrm{Ind}^P_{(P_0,V_0)}N_0 = \mathrm{Ind}^{P\cap P'}_{(P_0\cap P', V_0\cap P')}N_0 \cong k(P \cap P').e(V_0 \cap P') \otimes_{P_0\cap P'} N_0$ (see Proposition 1.5(i)). With this description of M_0, $\mu \otimes M_0$ is the map

$$(2) \qquad kGe(V) \otimes_{P\cap P'} k(P \cap P')e(V_0 \cap P') \otimes_{P_0\cap P'} N_0 \to$$
$$kGe(V) \otimes_{P\cap P'} k(P \cap P')e(V_0 \cap P') \otimes_{P_0\cap P'} N_0,$$
$$y \otimes y' \otimes n \mapsto ye(V') \otimes y' \otimes n$$

where $y \in kGe(V)$, $y' \in k(P \cap P')e(V_0 \cap P')$, $n \in N_0$.

Note that $e(V)e(V_0 \cap P') = e(V_0)$ by Definition 1.4 and since $V.(V_0 \cap P') = V_0 \cap (V.(P \cap P')) = V_0 \cap P = V_0$. Through the trivial G-$(P_0 \cap P')$-bimodule isomorphisms

$$kGe(V) \otimes_{P\cap P'} k(P \cap P')e(V_0 \cap P') \to kGe(V_0), \quad y \otimes y' \mapsto yy',$$
$$kGe(V') \otimes_{P\cap P'} k(P \cap P')e(V_0 \cap P') \to kGe(V'.V_0 \cap P'),$$
$$z \otimes z' \mapsto zz',$$

(for $y \in kGe(V)$, $y' \in k(P \cap P')e(V_0 \cap P')$, $z \in kGe(V')$, $z' \in k(P \cap P')e(V_0 \cap P')$), map (2) becomes the map

$$kGe(V_0) \otimes_{P_0 \cap P'} N_0 \to kGe(V'.V_0 \cap P') \otimes_{P_0 \cap P'} N_0',$$
$$yy' \otimes n \mapsto ye(V')y' \otimes n$$

with the same notation as above. But this is map (1) since we may take $y = ge(V)$ ($g \in G$), $y' = e(V_0 \cap P')$, and we have $yy' = ge(V)e(V_0 \cap P') = ge(V_0)$. So map (2) is an isomorphism. □

1.4. Endomorphism algebras of induced cuspidal modules

We keep the hypotheses of §1.3 above. We show that, under certain assumptions, the elements introduced in Definition 1.12 give a basis of the endomorphism algebra of the induced module I_τ^G.

Proposition 1.16. *Let $\tau = (P, V, M)$, $\tau' = (P', V', M')$ be two cuspidal triples (see Notation 1.10).*

(i) The set of $g \in G$ such that τ —${}^g\tau'$ is a union of double cosets in $P\backslash G/P'$.

(ii) Assume Λ is a subring of the algebraically closed field K and $M \otimes K$, $M' \otimes K$ are simple. If τ —${}^g\tau'$, then the maps $a_{g,\tau,\tau'}$ and $a_{1,\tau,{}^g\tau'}$ differ by an isomorphism $I_{\tau'}^G \to I_{{}^g\tau'}^G$.

Proof. (i) is trivial. (ii) is clear from the definitions. □

We now study the modules $\mathrm{Ind}_{(P,V)}^G M$ for cuspidal triples (P, V, M) (see Notation 1.10). We are mainly interested in the case where Λ is a field and M is absolutely simple, but we may also need a slight variant where Λ is not a field. When $\mathcal{C}$ is a set of cuspidal triples, one defines the following.

Condition 1.17. *Either*

(a) Λ is a splitting field for the group algebra ΛG and $\mathcal{C} \subseteq$ $\mathbf{cusp}_\Lambda(\mathcal{L})$,

or

(b) Λ is a principal ideal domain, subring of a splitting field K of KG, $\mathcal{C}$ has a single element (P, V, M) such that $(P, V, M \otimes K) \in \mathbf{cusp}_K(\mathcal{L})$, and, whenever $g \in G$ satisfies $(P, V, M \otimes K)$—$(P, V, M \otimes K)^g$, then (P, V, M)—$(P, V, M)^g$.

Proposition 1.18. *Let $\mathcal{C}$ be a set of cuspidal triples satisfying the above Condition 1.17.*

Consider the $a_{g,\tau,\tau'}$ of Definition 1.12 as endomorphisms of $I := \bigoplus_{\tau\in\mathcal{C}} I_\tau^G$.

(i) One may define a linear form $f{:}\,\mathrm{End}_{\Lambda G}(I) \to \Lambda$ by $f(\mathrm{Hom}_{\Lambda G}(I_\tau^G, I_{\tau'}^G)) = 0$ when $\tau \neq \tau'$, $x(1 \otimes m) \in f(x)(1 \otimes m) + \bigoplus_{PgP \neq P} \Lambda PgP \otimes M$ when $(P, V, M) \in \mathcal{C}$, $m \in M$ and $x \in \mathrm{End}_{\Lambda G}(\mathrm{Ind}_P^G M)$.

Then

(ii) $f(a_{g',\sigma,\sigma'}a_{g,\tau,\tau'}) \neq 0$ *only if* $P'g'P \ni g^{-1}$, *and* $(\tau,\tau') = (\sigma',\sigma)$;

(iii) $|V'|.f(a_{g^{-1},\tau',\tau}a_{g,\tau,\tau'}) = |V|.f(a_{g,\tau,\tau'}a_{g^{-1},\tau',\tau}) = \lambda|{}^gV'\cap V|$, *where* $\theta_{g,\tau,\tau'}\theta_{g^{-1},\tau',\tau} = \lambda\mathrm{Id}_M$ *for* $\lambda \in \Lambda^\times$.

Proof. (i) It suffices to check that, if $(P,V,M)\in\mathcal{C}$, $x\in\mathrm{End}_{\Lambda G}(\Lambda G\otimes_P M)$ and $m \in M$, then $x(1\otimes m)\in f(x)(1\otimes m)+\bigoplus_{PgP\neq P}\Lambda PgP\otimes M$ for a unique scalar $f(x)\in\Lambda$. The restriction to P of $\mathrm{Ind}_P^G(M) = \Lambda G\otimes_P M$ is the direct sum of ΛP-submodules $\Lambda PgP\otimes M$ associated to the double cosets PgP. Then there is $\theta\in\mathrm{End}_{\Lambda P}(M)$ such that $x(1\otimes m)\in 1\otimes\theta(m)+\bigoplus_{PgP\neq P}\Lambda PgP\otimes M$ for all $m\in M$. But $\mathrm{End}_{\Lambda P}(M)=\Lambda$ by hypothesis, hence our claim.

(ii)–(iii) The elements of $\mathrm{End}_{\Lambda G}(I)$ are in the matrix form $(x_{\tau,\tau'})_{\tau,\tau'\in\mathcal{C}}$ with $x_{\tau,\tau'}\in\mathrm{Hom}_{\Lambda G}(I_\tau^G, I_{\tau'}^G)$. The linear form satisfies $f((x_{\tau,\tau'})_{\tau,\tau'}) = \sum_\tau f(x_{\tau,\tau})$.

One clearly has $f(a_{g',\sigma,\sigma'}a_{g,\tau,\tau'}) = 0$ when $(\sigma,\sigma')\neq(\tau',\tau)$.

Denote $\tau = (P,V,M)$, $\tau' = (P',V',M')$. Let $m\in M$. One has $a_{g',\tau',\tau}a_{g,\tau,\tau'}(1\otimes m) = a_{g',\tau',\tau}(|V|^{-1}\sum_{u\in V}ug\otimes\theta_{g,\tau,\tau'}(m)) = |V|^{-1}|V'|^{-1}\sum_{u\in V,\,u'\in V'}(ugu'g'\otimes\theta_{g',\tau',\tau}\theta_{g,\tau,\tau'}(m))$. Using the direct sum decomposition $\Lambda G\otimes_P M = \bigoplus_{PhP}\Lambda PhP\otimes M$, the projection on $1\otimes M$ is non-zero only if $VgV'g'\cap P\neq\emptyset$. Thus (ii).

If in addition $g'=g^{-1}$, then $VgV'g^{-1}\cap P = V$ by Proposition 1.2(i). Thus $a_{g^{-1},\tau',\tau}a_{g,\tau,\tau'}(1\otimes m)\in 1\otimes m' + \bigoplus_{PxP\neq P}\Lambda Px\otimes M$ where $m' = |V'|^{-1}|{}^gV'\cap V|\theta_{g^{-1},\tau',\tau}\theta_{g,\tau,\tau'}(m)$. So, for all $m\in M$, $f(a_{g^{-1},\tau,\tau'}a_{g,\tau',\tau})(1\otimes m) = |V'|^{-1}|{}^gV'\cap V|(1\otimes\theta_{g^{-1},\tau',\tau}\theta_{g,\tau,\tau'}(m))$. This gives us that $\theta_{g^{-1},\tau',\tau}\theta_{g,\tau,\tau'}$ is a scalar, necessarily invertible, denoted by λ and therefore $f(a_{g^{-1},\tau,\tau'}a_{g,\tau',\tau}) = \lambda.|V'|^{-1}|{}^gV'\cap V|\in\Lambda^\times$. Changing (g,τ,τ') into (g^{-1},τ',τ) gives the same λ since, if $\theta_{g^{-1},\tau',\tau}\theta_{g,\tau,\tau'} = \lambda\mathrm{Id}_M$, then $\theta_{g^{-1},\tau',\tau} = \lambda(\theta_{g,\tau,\tau'})^{-1}$ and $\theta_{g,\tau,\tau'}\theta_{g^{-1},\tau',\tau} = \lambda\mathrm{Id}_{M'}$. Thus (iii). □

Let us recall the following notions (see [Ben91a] §1.6, [NaTs89] §2.8, [Thévenaz] §6).

Definition 1.19. *Let* Λ *be a principal ideal domain and let* A *be a* Λ*-free finitely generated* Λ*-algebra.*

A is said to be a symmetric algebra if and only if there exists $f\colon A\to\Lambda$ *a* Λ*-linear map such that* $f(ab)=f(ba)$ *for all* $a,b\in A$, *and* $a\mapsto(b\mapsto f(ab))$ *induces an isomorphism* $A\to\mathrm{Hom}_\Lambda(A,\Lambda)$.

One says that A is a Frobenius algebra (see [Ben91a] §1.6) if and only if $\Lambda = K$ *is a field, and there exists* $f\colon A\to K$ *a* K*-linear map such that, for all* $a\in A\setminus\{0\}$, $f(aA)\neq\{0\}$ *and* $f(Aa)\neq\{0\}$.

Note that A is symmetric (resp. Frobenius) if and only if the opposite algebra A^{opp} is symmetric (resp. Frobenius). Note also that, when $\Lambda = K$ is a field, any symmetric algebra is Frobenius.

Theorem 1.20. *Let $\mathcal{C}$ be a set of cuspidal triples satisfying Condition 1.17.*

(i) Let $\tau = (P, V, M)$, $\tau' = (P', V', M') \in \mathcal{C}$. Take a representative in each double coset PgP' such that τ—${}^g\tau'$. Then the corresponding $a_{g,\tau,\tau'}$ form a Λ-basis of $\mathrm{Hom}_{\Lambda G}(I_\tau^G, I_{\tau'}^G)$.

(ii) In case (b) of Condition 1.17 (which implies $\mathcal{C} = \{\tau_0\}$), the endomorphism algebra $\mathrm{End}_{\Lambda G}(I_{\tau_0}^G)$ is a symmetric algebra.

(iii) In case (a) of Condition 1.17 (which implies Λ is a field), the Λ-algebra $\mathrm{End}_{\Lambda G}(\bigoplus_{\tau\in\mathcal{C}} I_\tau^G)$ is a Frobenius algebra.

(iv) In case (a) of Condition 1.17 and if $\mathcal{L}$ has the additional property that any relation (P, V)—(P', V') in $\mathcal{L}$ implies $|V| = |V'|$, then $\mathrm{End}_{\Lambda G}(\bigoplus_{\tau\in\mathcal{C}} I_\tau^G)$ is a symmetric algebra.

Proof of Theorem 1.20. Consider the $a_{g,\tau,\tau'}$ above as endomorphisms of $I = \bigoplus_{\tau\in\mathcal{C}} I_\tau^G$. Denote $E := \mathrm{End}_{\Lambda G}(I)$. Having chosen representatives $g \in G$ for each pair $\tau, \tau' \in \mathcal{C}$, denote by $\mathcal{T}$ the resulting set of triples (g, τ, τ').

Lemma 1.21. *$E \cong \Lambda^{\mathcal{T}}$ as a Λ-module.*

Proof of Lemma 1.21. By Proposition 1.5(iv), one has

$$\begin{aligned}&\mathrm{Hom}_{\Lambda G}(I_\tau^G, I_{\tau'}^G)\\ &\quad\cong \oplus_{PgP'\subseteq G}\mathrm{Hom}_{\Lambda(P\cap {}^gP')}\left(\mathrm{Res}^P_{{}^g(P',V')\,\cap\!\downarrow\,(P,V)}M,\ \mathrm{Res}^{{}^gP'}_{(P,V)\,\cap\!\downarrow\,{}^g(P',V')}{}^gM'\right)\end{aligned}$$

where the summand is zero unless (P, V)—${}^g(P', V')$ by cuspidality of M and M'. By Condition 1.17 on $\mathcal{C}$, the corresponding summand is isomorphic to Λ if τ—${}^g\tau'$, zero otherwise. □

Returning to the proof of Theorem 1.20, take f as in Proposition 1.18.

Let K_0 be the field of factions of Λ. Let $x = \sum_{(g,\tau,\tau')\in\mathcal{T}} \lambda_{g,\tau,\tau'} a_{g,\tau,\tau'} \in E \otimes K_0$ be a linear combination of the $a_{g,\tau,\tau'}$'s with coefficients in K_0. Proposition 1.18(ii) and (iii) yield

$$\lambda_{g,\tau,\tau'} = f(xa_{g^{-1},\tau',\tau})f(a_{g,\tau,\tau'}a_{g^{-1},\tau',\tau})^{-1} \tag{1}$$

(where f denotes also the extension of f to $E \otimes K_0$). This implies at once that the $a_{g,\tau,\tau'}$'s for $(g, \tau, \tau') \in \mathcal{T}$ are K_0-linearly independent. Then the $a_{g,\tau,\tau'}$'s for $(g, \tau, \tau') \in \mathcal{T}$ are a K_0-basis of $E \otimes K_0$ by Lemma 1.21. But (1) above and Proposition 1.18(iii) show that any $x \in E$ is a combination of the $(a_{g,\tau,\tau'})_{(g,\tau,\tau')\in\mathcal{T}}$ with coefficients in Λ. Thus (i) is proved.

The $a_{g,\tau,\tau'}$'s for $(g, \tau, \tau') \in \mathcal{T}$ and the $a_{g^{-1},\tau',\tau}$'s for $(g, \tau, \tau') \in \mathcal{T}$ are both bases of E by (i). The formula in (1) also implies that f induces an isomorphism between E and $\mathrm{Hom}(E, \Lambda)$, the basis dual to $(a_{g,\tau,\tau'})_{(g,\tau,\tau')\in\mathcal{T}}$ being

$$(f(a_{g^{-1},\tau',\tau}a_{g,\tau,\tau'})^{-1}a_{g^{-1},\tau',\tau})_{(g,\tau,\tau')\in\mathcal{T}}.$$

This gives (iii). When, moreover, $\mathcal{C}$ has a single element, Proposition 1.18(ii)–(iii) for $\tau = \tau'$ (hence $V = V'$) gives $f(aa') = f(a'a)$ for all basis elements, hence for every $a, a' \in E$. This implies our (ii). A similar result holds if in $\mathcal{L}$ the relation (P, V)—(P', V') implies $|V| = |V'|$, whence (iv) is proved. □

Remark 1.22. The linear form f gives the coefficient on $\mathrm{Id}_{I_\tau^G} = a_{1,\tau,\tau}$ (see Definition 1.12) in the basis of Theorem 1.20(i).

Proposition 1.23. *Let H be a subgroup of G and let M be a ΛH-module. Then the subalgebra of $\mathrm{End}_{\Lambda G}(\mathrm{Ind}_H^G M)$ consisting of $f\colon \mathrm{Ind}_H^G M \to \mathrm{Ind}_H^G M$ such that $f(1 \otimes M) \subseteq 1 \otimes M$ is isomorphic to $\mathrm{End}_{\Lambda H}(M)$ by the restriction map $f \mapsto f_{|1\otimes M}$.*

Let $(P', V') \subseteq (P, V)$ be in $\mathcal{L}$, and let $\tau = (P', V', N)$ be a cuspidal triple satisfying Condition 1.17. Then the injection above sends $a_{g,\tau,\tau} \in \mathrm{End}_{\Lambda P}(I_\tau^P)$ to the element denoted the same in $\mathrm{End}_{\Lambda G}(I_\tau^G)$.

Proof. Writing $\mathrm{Ind}_H^G M = \Lambda G \otimes_H M = \bigoplus_{HgH\in H\backslash G/H} \Lambda HgH \otimes M$ as a ΛH-module, the summand M_H for $g \in H$ is isomorphic to M. Let E be the subalgebra of $\mathrm{End}_{\Lambda G}(\mathrm{Ind}_H^G M)$ of endomorphisms x such that $xM_H \subseteq M_H$. To show that E is isomorphic to $\mathrm{End}_{\Lambda H}(M)$, it suffices to show that every $y \in \mathrm{End}_{\Lambda H}(M)$ extends to a unique $\bar{y} \in E$. The uniqueness is ensured by the fact that M_H generates $\mathrm{Ind}_H^G M$ as a ΛG-module. The existence is just the functoriality of $\mathrm{Ind}_H^G = \Lambda G \otimes_H$. One takes $\bar{y} = \Lambda G \otimes_H y$, defined by $\bar{y}(g \otimes m) = g \otimes y(m)$ for $m \in M$, $g \in G$. This gives our first claim.

For the second, let us recall the defining relation for $a_{g,\tau,\tau}$: $a_{g,\tau,\tau}(1 \otimes_{P'} n) = e(U')g \otimes_{P'} \theta_g(n)$ for any $n \in N$. It is clear that $a_{g,\tau,\tau}$ stabilizes $M = \Lambda P \otimes_{P'} N$ when $g \in P$ and coincides with the element denoted $a_{g,\tau,\tau}$ in $\mathrm{End}_{\Lambda P}(I_\tau^P)$. □

1.5. Self-injective endomorphism rings and an equivalence of categories

In the following, k is a field and A is a finite-dimensional k-algebra. One denotes by A–**mod** (resp. **mod**–A) the category of finitely generated left (resp. right)

A-modules. One has the contravariant functor $M \mapsto M^\vee = \mathrm{Hom}_k(M, k)$ between them.

Notation 1.24. Let Y be a finite-dimensional A-module and let $E := \mathrm{End}_A(Y)$. Let H be the functor from A–**mod** to **mod**$-E$ defined by $H(V) = \mathrm{Hom}_A(Y, V)$, where E acts on $H(V)$ by composition on the right.

Let $A - \mathbf{mod}_Y$ be the full subcategory of $A-\mathbf{mod}$ whose objects are the A-modules V such that there exist $l \geq 1$ and $e \in \mathrm{End}_A(Y^l)$ with $V \cong e(Y^l)$.

Theorem 1.25. *Let Y be a finitely generated A-module. Let $E := \mathrm{End}_A(Y)$ and let $H = \mathrm{Hom}_A(Y, -)$ be as above. Assume that E is Frobenius (see Definition 1.19). Then*

*(i) H is an equivalence of additive categories from $A-\mathbf{mod}_Y$ to **mod**-E. Assume moreover that all simple A-modules are in $A-\mathbf{mod}_Y$. Then*

(ii) if M is in $A-\mathbf{mod}_Y$ then it is simple if and only if $H(M)$ is simple. This induces a bijection between the simple left A-modules and the simple right E-modules.

(iii) If Y' is an indecomposable direct summand of Y, then $\mathrm{soc}(Y')$, $\mathrm{hd}(Y')$ are simple, and $H(\mathrm{soc}(Y')) = \mathrm{soc}(H(Y'))$, $H(\mathrm{hd}(Y')) = \mathrm{hd}(H(Y'))$.

(iv) If Y', Y'' are indecomposable direct summands of Y, then $\mathrm{soc}(Y') \cong \mathrm{soc}(Y'')$ (and $\mathrm{hd}(Y') \cong \mathrm{hd}(Y'')$) if and only if $Y' \cong Y''$.

Over a Frobenius algebra, projective modules and injective modules coincide (see [Ben91a] 1.6.2(ii)). Considering injective hulls, we get the following.

Lemma 1.26. *If E is a Frobenius algebra, then every finitely generated E-module embeds into a free module E^l for some integer l.*

We shall use the following notation.

Notation. If $M \subseteq H(V)$, we denote $M.Y := \sum_{m \in M} m(Y) \subseteq V$.

Assume that E is Frobenius.

Lemma 1.27. *Let V be in A–**mod**.*

(i) $H(Y) = E_E$ (E considered as right E-module) and, if l is an integer ≥ 1, $H(\mathrm{Hom}_A(Y^l, V)) = \mathrm{Hom}_E((E_E)^l, H(V))$.

(ii) If $V = e(Y^l)$ for $e \in \mathrm{End}_A(Y^l)$, then $H(V).Y = V$.

(iii) Let $l \geq 1$ and $M \subseteq H(Y^l) = (E_E)^l$ be a right E-submodule. Then $M.Y$ is in $A-\mathbf{mod}_Y$ and $H(M.Y) \cong M$, the latter being induced by the image by H of the inclusion $M.Y \subseteq Y^l$.

Proof of Lemma 1.27. (i) is straightforward.

(ii) Writing $V = e(Y^l)$ for $e \in \mathrm{End}_A(Y^l)$, $H(V)$ clearly contains e composed with all the coordinate maps $Y \to Y^l$, hence $H(V).Y = V$.

(iii) One has $M.Y \subseteq Y^l$ and $M \subseteq H(M.Y) \subseteq (E_E)^l$ as right E-modules. The sum $M.Y = \sum_{m \in M} mY$ may be turned into a finite sum since M is finite dimensional, so $M.Y$ is a sub-A-module of some finite power of Y. Therefore $M.Y$ is in $A-\mathbf{mod}_Y$.

Let us assume $H(M.Y)/M \neq 0$. Then there exists a right E-module N such that $M \subset N \subseteq H(M.Y) \subseteq (E_E)^l$ and N/M is simple. By Lemma 1.26, N/M injects into some $(E_E)^m$. So there is a non-zero map $f: N \to E_E$ such that $f(M) = 0$. By the self-injectivity mentioned above ([Ben91a] 1.6.2), the module E_E is injective, so f extends into $\widehat{f}: (E_E)^l \to E_E$. But then $\widehat{f}$ is in the form $\widehat{f} = H(e)$ where $e \in \mathrm{Hom}_A(Y^l, Y)$ (Lemma 1.27(i)). The hypothesis on f implies $e(M.Y) = 0$, $e(N.Y) \neq 0$. But $M.Y \subseteq N.Y \subseteq H(M.Y).Y \subseteq M.Y$, so $N.Y = M.Y$, a contradiction. □

Proof of Theorem 1.25. (i) Let M be a right E-module, then M is a submodule of some $(E_E)^l$ by Lemma 1.26. Then Lemma 1.27(iii) applies, so one gets $M = H(V)$ for $V = M.Y$, which is in $A-\mathbf{mod}_Y$.

It remains to check that H is faithful and full. Let V, V' be A-modules in $A-\mathbf{mod}_Y$; one must check that H induces an isomorphism of vector spaces between $\mathrm{Hom}_A(V, V')$ and $\mathrm{Hom}_E(H(V), H(V'))$.

Obviously H is linear. If $f \in \mathrm{Hom}_A(V, V')$ is in its kernel, then $f(H(V).Y) = 0$ by definition of H, but $H(V).Y = V$ by Lemma 1.27(ii), so $f(V) = 0$ and $f = 0$.

In order to check surjectivity, one may assume that $V = e(Y^l)$, $V' = e'(Y^l)$ for $e, e' \in \mathrm{End}_A(Y^l)$. Then $H(V)$ and $H(V')$ are submodules of $(E_E)^l$. Let $g \in \mathrm{Hom}_E(H(V), H(V'))$. By injectivity of $(E_E)^l$, g extends to $\widehat{g} \in \mathrm{Hom}_E((E_E)^l, (E_E)^l)$ which is $H(\mathrm{Hom}_A(Y^l, Y^l))$ by Lemma 1.27(i). Then $\widehat{g} = H(\widehat{f})$ for $\widehat{f} \in \mathrm{End}_A(Y^l)$. We have $\widehat{f}(V) \subseteq V'$ since $\widehat{f}(V) = \widehat{f}(H(V).Y) = (\widehat{g}.H(V)).Y = (g.H(V)).Y \subseteq H(V').Y = V'$. Therefore $g = H(f)$, where $f: V \to V'$ is the restriction of $\widehat{f}$.

This completes the proof of (i).

Assume now that all simple A-modules are in $A-\mathbf{mod}_Y$.

(ii) Take V in A–$\mathbf{mod}_Y$ and assume that $H(V)$ is simple. One may assume that there is some l such that $V \subseteq Y^l$ and $V = H(V).Y$ by Lemma 1.27(ii). Let X be a simple submodule of V. Then X occurs in $\mathrm{hd}(Y)$, so $H(X) \neq 0$. But $H(X) \subseteq H(V)$ so $H(X) = H(V)$ and therefore $X = H(X).Y = H(V).Y = V$, so V is simple.

Conversely, assume that V is a simple A-module. By the hypothesis on Y, V is a submodule of Y. Let S be a simple submodule of $H(V)$, then $0 \neq S.Y \subseteq H(V).Y = V$ by Lemma 1.27(ii). Then $S.Y = V$ and $S = H(S.Y) = H(V)$ by Lemma 1.27(iii).

The equivalence of (i) then implies (ii).

(iii) Let Y' be an indecomposable direct summand of Y, then $H(Y')$ is a (projective) indecomposable direct summand of E_E, so its head and socle are simple (see [Ben91a] 1.6).

Now it is clear by (ii) above that $H(\mathrm{soc}(Y'))$ is a non-zero semi-simple submodule of $\mathrm{soc}(H(Y'))$, whence the first claimed equality.

By what we have just checked, $\mathrm{soc}(Y')$ is simple. We may now apply this to $Y^\vee$ since $\mathrm{End}_A(Y^\vee) = \mathrm{End}_A(Y)^{\mathrm{opp}}$. We obtain that the indecomposable direct summands of Y have simple heads. To check the second equality of (iii), note that we have a non-zero element in $\mathrm{Hom}_A(Y', \mathrm{hd}(Y'))$ while both modules are in $A-\mathbf{mod}_Y$, so by the equivalence of (i), there is a non-zero element in $\mathrm{Hom}_E(H(Y'), H(\mathrm{hd}(Y')))$. The first module has simple head while the second is simple by (ii) and what we have just said. So we have $\mathrm{hd}(H(Y')) = H(\mathrm{hd}(Y'))$ as claimed.

(iv) This follows from (iii) and the fact that this is true for indecomposable direct summands of E_E. □

1.6. Structure of induced cuspidal modules and series

We take again a finite group G, k a field such that $kG/J(kG)$ is split (i.e. a product of matrix algebras over k), and $\mathcal{L}$ a k-regular $\cap\!\downarrow$-stable set of subquotients of G. This allows us to consider the set $\mathbf{cusp}_k(\mathcal{L}))$ of cuspidal triples (see Notation 1.10).

Theorem 1.28. *For each cuspidal triple (P, V, M) where $(P, V) \in \mathcal{L}$ and M is a simple cuspidal kP/V-module, the induced module* $\mathrm{Ind}_P^G M$ *can be written as a direct sum $\bigoplus_i Y_i$ where*

(a) each Y_i is indecomposable,

(b) $\mathrm{soc}(Y_i) \cong \mathrm{soc}(Y_j)$ *if and only if $Y_i \cong Y_j$,*

(b′) $\mathrm{hd}(Y_i) \cong \mathrm{hd}(Y_j)$ *if and only if $Y_i \cong Y_j$.*

(c) If moreover $\mathcal{L}$ has the property that any relation (P, V)—(P', V') implies $|V| = |V'|$, then $\mathrm{soc}(Y_i) \cong \mathrm{hd}(Y_i)$ *for all i.*

Proof. Let $Y = \bigoplus_\tau \mathrm{Ind}_P^G S$ where τ ranges over $\mathbf{cusp}_k(\mathcal{L}))$. Theorem 1.20 tells us that $\mathrm{End}_{kG}(Y)$ is Frobenius. The $H(Y_i)$ are the indecomposable projective E-modules. Any simple kG-module occurs in both $\mathrm{hd}(Y)$ and $\mathrm{soc}(Y)$ by Proposition 1.11. We may now apply Theorem 1.25 to the module Y.

When the condition of (c) is satisfied, Theorem 1.20(iv) tells us that E is symmetric. This implies that $\mathrm{hd}(H(Y_i)) \cong \mathrm{soc}(H(Y_i))$ (see [Ben91a] 1.6.3), whence (b) by Theorem 1.25(iv). □

Notation 1.29. When $\tau, \tau' \in \mathbf{cusp}_k(\mathcal{L})$, we write $\tau —_G \tau'$ if and only if there exists $g \in G$ such that $\tau —{}^g\tau'$.

When $\tau \in \mathbf{cusp}_k(\mathcal{L})$, denote by $\mathcal{E}(kG, \tau)$ the set of simple components of $\mathrm{hd}(I_\tau^G)$.

One has $\mathcal{E}(kG, \tau) = \mathcal{E}(kG, {}^g\tau)$ for all $g \in G$.

Theorem 1.30. *Assume that $\mathcal{L}$ has the property that any relation $(P, V)—(P', V')$ in $\mathcal{L}$ implies $|V| = |V'|$.*

(i) $\bigcup_{\tau\in\mathbf{cusp}_k(\mathcal{L})} \mathcal{E}(kG, \tau)$ gives all simple kG-modules.

(ii) If $\mathcal{E}(kG, \tau) \cap \mathcal{E}(kG, \tau') \neq \emptyset$, then $\tau —_G \tau'$.

(iii) Assume $\mathcal{L}$ satisfies the hypotheses of Theorem 1.14. Then $—_G$ is an equivalence relation on $\mathbf{cusp}_k(\mathcal{L})$, and the union in (i) is a partition of the simple kG-modules indexed by the quotient $\mathbf{cusp}_k(\mathcal{L})/—_G$.

(iv) If S' is a simple composition factor of some I_τ^G ($\tau = (P, V, N) \in \mathbf{cusp}_k(\mathcal{L})$), then $S' \in \mathcal{E}(kG, \tau')$ where $\tau' = (P', V', N')$ and $(P', V') \cap\!\downarrow (P, V) = (P, V)$ (and therefore $|P'/V'| \geq |P/V|$). If moreover $|P'/V'| = |P/V|$, then $\tau —_G \tau'$.

Proof. (i) is clear from Proposition 1.11.

(ii) Since the head and socle of each I_τ^G yield the same simple kG-modules thanks to Theorem 1.28(c) above, $\mathcal{E}(kG, \tau) \cap \mathcal{E}(kG, \tau') \neq \emptyset$ implies that there is a non-zero morphism $I_\tau^G \to I_{\tau'}^G$. Then $\tau —_G \tau'$ by Theorem 1.20(i).

(iii) When the hypotheses of Theorem 1.14 are satisfied, $I_\tau^G \cong I_{\tau'}^G$ whenever $\tau —_G \tau'$. Since the converse is true (see Notation 1.10), there is an equivalence. Therefore $—_G$ is an equivalence relation. Then we also have $\mathcal{E}(kG, \tau) = \mathcal{E}(kG, \tau')$ as long as $\tau —_G \tau'$, so the union in (i) is a partition.

(iv) This won't be used. We leave it as an exercise (hint: consider a projective cover of S'). □

Exercises

1. Find a counterexample in a commutative group showing that — is not transitive. Find one with minimal cardinality of G.
2. Let a, b be subquotients of a finite group.
   (a) Show that $a \cap\!\downarrow (a \cap\!\downarrow b) = a \cap\!\downarrow b$ and $a \cap\!\downarrow (b \cap\!\downarrow a) = (a \cap\!\downarrow b) \cap\!\downarrow a = (b \cap\!\downarrow a) \cap\!\downarrow a = b \cap\!\downarrow a$. More generally, when $b' \leq b$, relate $b' \cap\!\downarrow (b \cap\!\downarrow a)$, $b \cap\!\downarrow (b' \cap\!\downarrow a)$, $(b \cap\!\downarrow a) \cap\!\downarrow b'$, $b \cap\!\downarrow (a \cap\!\downarrow b')$, and $(a \cap\!\downarrow b) \cap\!\downarrow b'$ with $b' \cap\!\downarrow a$ and $a \cap\!\downarrow b'$.

(b) Show that $\cap\!\!\downarrow$ induces a structure of an (associative) monoid on $\mathcal{L}_{a,b} = \{a, b, a \cap\!\!\downarrow b, b \cap\!\!\downarrow b\}$.

(c) Let M be the monoid generated by two generators x, y, subject to the relations $x^2 = x$, $y^2 = y$, $xyx = yx$ and $yxy = xy$. Show that $\mathbb{Z}[M] \cong \mathbb{Z} \times \mathbb{Z} \times U$ where U is the ring of upper triangular matrices in $\mathrm{Mat}_2(\mathbb{Z})$.

3. Let a, a', b, b' be subquotients of a finite group. If $a \cap\!\!\downarrow b = b$, $a' \geq a$, and $b \geq b'$, show that $a' \cap\!\!\downarrow b' = b'$.
4. Let $\mathcal{L}$ be a set of subquotients of a finite group G. Show that $a \cap\!\!\downarrow b = b \cap\!\!\downarrow a$ for all $a, b \in \mathcal{L}$, if and only if there is a subgroup $H \subseteq G$ such that, for all $(P, V) \in \mathcal{L}$, $V \subseteq H \subseteq P$.
5. Show that there are groups G with subgroups U, V such that UVU is a subgroup but $e(UVU) \neq e(U)e(V)e(U)$.
6. Prove a Mackey formula implying Proposition 1.5(iv),

$$\mathrm{Res}^G_{(P',V')}\mathrm{Ind}^G_P N$$
$$\cong \bigoplus_{P'gP \subseteq G} \mathrm{Ind}^{P'}_{{}^g(P,V) \cap\downarrow (P',V')} i_g \mathrm{Res}^{{}^gP}_{(P',V') \cap\downarrow {}^g(P,V)} {}^gN,$$

where i_g is the functor making a $P' \cap {}^gV$-trivial $P' \cap {}^gP$-module into a ${}^gV$-trivial $(P' \cap {}^gP){}^gV$-module.

You may use the following steps in relation to P'-P-bimodules.

(a) $e(V')\Lambda P'Pe(V) \cong \Lambda P'e(V'.(V \cap P')) \otimes_{P \cap P'} e((V' \cap P).V)\Lambda P$.

(b) If $g \in G$, then $e(V')\Lambda P'gPe(V) \cong \Lambda P'e(V'.({}^gV \cap P')) \otimes_{{}^gP \cap P'} e((V' \cap {}^gP).{}^gV)\Lambda {}^gP \otimes_{{}^gP} \Lambda gP$.

(c) Decompose $e(V')\Lambda G \otimes_G \Lambda Ge(V) = e(V')\Lambda Ge(V)$.

7. Show Theorem 1.20(i) more directly, without using the linear form f or the rank argument.
8. If $\tau = (P, V, S) \in \mathbf{cusp}_k(\mathcal{L})$, show that $\mathrm{N}_G(V) \subseteq \mathrm{N}_G(P)$.
9. Let $\mathcal{L}$ be a $\cap\!\!\downarrow$-stable set of subquotients (P, V) $(V \lhd P \subseteq G)$.

(a) Assume a—a' in $\mathcal{L}$. Show that $x \mapsto x \cap\!\!\downarrow a'$ and $x' \mapsto x' \cap\!\!\downarrow a$ induce inverse isomorphisms between the intervals $\{x \in \mathcal{L} \mid x \leq a\}$ and $\{x' \in \mathcal{L} \mid x' \leq a'\}$ in $\mathcal{L}$. Show that x—$x \cap\!\!\downarrow a$ (resp. x'—$x' \cap\!\!\downarrow a'$) for all $x \leq a$ (resp. $x' \leq a'$).

(b) If $a = (P, V)$—$a' = (P', V')$, define a set $\mathcal{L}_{P \cap P'}$ of subquotients of $P \cap P'$ in bijection with the above intervals. Show that $\mathbf{cusp}_k(\mathcal{L}_{P \cap P'})$ injects into $\mathbf{cusp}_k(\mathcal{L})$ in two ways. Apply this to the proof of Theorem 1.14.

(c) See which simplification of that proof can be obtaind by assuming the existence of a subgroup L such that $P = LV$, $P' = LV'$ are semi-direct products.

10. Assume $\mathcal{L}$ is a $\cap\downarrow$-stable Λ-regular set of subquotients of G. Assume that, for every relation (P, V)—(P', V') in $\mathcal{L}$ with $|P| = |P'|$, the map $x \mapsto xe(V')$ is an isomorphism from $\Lambda Ge(V)$ to $\Lambda Ge(V')$. Show that, for all (P, V), (Q, W) in $\mathcal{L}$, one has $\Lambda Ge(V)e(W) = \Lambda Ge(V \cap Q)e(W) = \Lambda Ge(W)e(V \cap Q)$.
11. Show a converse of Theorem 1.14.
12. Let A be a finite-dimensional k-algebra. Let $i, j \in A$ be two idempotents such that $i + j = 1$. Show that A is symmetric if and only if the following conditions are satisfied:
    (a) iAi and jAj are symmetric for forms f_i, f_j such that $f_i(iajbi) = f_j(jbiaj)$ for all $a, b \in A$,
    (b) for all $0 \neq x \in iAj$, $xAi \neq 0$ and for all $0 \neq y \in jAi$, $yAj \neq 0$.

    Application (see Theorem 1.20): if $Y = Y_1 \oplus \ldots \oplus Y_n$ is a sum of A-modules, show that $\mathrm{End}_A(Y)$ is a symmetric algebra if and only if each algebra $\mathrm{End}_A(Y_i)$ is symmetric for a form f_i such that, for all $x_{i,j} \in \mathrm{Hom}_A(Y_j, Y_i)$ and $y_{j,i} \in \mathrm{Hom}_A(Y_i, Y_j)$, one has $f_i(x_{i,j}y_{j,i}) = f_j(y_{j,i}x_{i,j})$ and, if $x_{i,j} \neq 0$, there is a $y_{j,i}$ such that $x_{i,j}y_{j,i} \neq 0$.
13. Check Theorem 1.25 assuming that both E and E^{opp} are self-injective instead of Frobenius. Recall that a ring is said to be self-injective when the (projective) regular module is also injective.
14. Prove Theorem 1.25 assuming that only E^{opp} is self-injective.

    Hint: only the proofs of (iii) and (iv) need some adaptation. Let Y' be an indecomposable direct summand of Y. Show that $\mathrm{hd}(Y')$ is simple using the following steps. Show that $\mathrm{End}_A(\mathrm{hd}(Y')) \cong \mathrm{End}_E(H(Y'), H(\mathrm{hd}(Y')))$. Then use a decomposition of $\mathrm{hd}(Y')$ as a sum of simple A-modules and its image by H.
15. Let Y be an A-module such that $\mathrm{End}_A(Y)$ is Frobenius and the semi-simple A-modules $\mathrm{soc}(Y)$ and $\mathrm{hd}(Y)$ have the same simple components (possibly with different multiplicities). Prove a version of Theorem 1.25 where the simple A-modules are replaced by the ones occurring in $\mathrm{soc}(Y)$.

Notes

Modular versions of Harish-Chandra induction (see, for instance, [DiMi91] §6 for the characteristic zero version) were used by Dipper [Dip85a], [Dip85b]. The general definition for BN-pairs is due to Hiss [Hi93] and was quickly followed by Dipper–Du [DipDu93] who partly axiomatized it and gave a proof of the independence with regard to (P, U) of the Harish-Chandra induction (our

Theorem 1.14). §1.5 comes from [Gre78] and [Ca90] (see also Chapter 6 below). The application to generalized Harish-Chandra theory is due to Linckelmann and Geck–Hiss; see [GeHi97] and [Geck01]. Exercise 14 is due to Linckelmann (see [Geck01] 2.10).

For a more general approach to Harish-Chandra induction and restriction, see [Bouc96]. For more general category equivalences induced by the functor $\mathrm{Hom}_A(Y, -)$, see [Ara98] and [Aus74].

2
Finite BN-pairs

The aim of this chapter is to give a description of a $\cap\downarrow$-stable set of subquotients (see the introduction to Chapter 1) present in many finite simple groups. The axiomatic setup of BN-pairs (see [Asch86] §43, [Bour68] §IV, [CuRe87] §65) has been devised to cover the so-called Chevalley groups and check their simplicity. Such a G has subgroups B, N such that $B \cap N \lhd N$ and the quotient group $N/B \cap N$ is generated by a subset S such that the unions $P_s := B \cup BsB$ are *subgroups* of G for any $s \in S$. More generally, the subgroups of G containing B (*standard parabolic* subgroups) are in bijection with subsets of S:

$$I \subseteq S \mapsto P_I.$$

Under certain additional hypotheses, defining a notion of a *split BN-pair of characteristic* p (p a prime), each P_I has a semi-direct product decomposition, called a Levi decomposition,

$$P_I = U_I L_I,$$

where U_I is the biggest normal p-subgroup of P_I and L_I is a group with a split BN-pair given by the subgroups $B \cap L_I$, $N \cap L_I$ and the set I.

Among other classical properties, we show that the set $\mathcal{L}$ of G-conjugates of subquotients (P_I, U_I) (I ranging over the subsets of S) is $\cap\downarrow$-stable.

The approach we follow uses systematically the reflection representation of the group $W = N/B \cap N$ and the associated finite set of so-called *roots*. The set Φ of roots α has a very rich structure which gives us a lot of information about W and the structure of G itself (*root subgroups*). While the axiomatic study of BN-pairs involves many (elementary) computations on double cosets BwB, once the notion of root subgroups B_α is introduced, the description of subgroups of B of the form $B \cap B^{w_1} \cap B^{w_2} \cap \dots$ for $w_1, w_2, \dots \in W$ and of double cosets BwB is much easier.

We develop several examples.

The theory of cuspidal simple modules and their induced modules then applies to finite groups with a split BN-pair. Lusztig has classified the cuspidal triples (P_I, U_I, M) (see Definition 1.8) over fields of characteristic 0 (see [Lu84]). Chapters 19 and 20 give the first steps towards a classification of the cuspidal triples over fields of non-zero characteristic (see Theorem 19.20 for groups $\mathrm{GL}_n(\mathbb{F}_q)$).

2.1. Coxeter groups and root systems

In the present section, (W, S) is a Coxeter system in the usual sense (see [Asch86] §29, [Bour68] §IV, [CuRe87] §64.B, [Hum90] §5). We consider it as acting on a real vector space $\mathbb{R}\Delta$ with a basis Δ in bijection with S ($\delta \mapsto s_\delta$) and a symmetric form such that $\langle\delta, \delta'\rangle = -\cos(\pi/|<s_\delta s_{\delta'}>|)$. Then W acts faithfully on $\mathbb{R}\Delta$ by a morphism which sends s_δ to the orthogonal reflection through δ.

Defining $\Phi := \{w(\delta) \mid w \in W,\ \delta \in \Delta\}$ ("the root system of W"), each element of Φ is a linear combination of elements of Δ ("simple roots") with coefficients either all ≥ 0 or all ≤ 0. This gives the corresponding partition $\Phi = \Phi^+ \cup \Phi^-$ (see [CuRe87] 64.18 and its proof, [Hum90] §5.3).

We use subsets of Δ as subsets of S and denote accordingly W_I the subgroup of W generated by the elements of S corresponding with elements of I, $\Phi_I = \Phi \cap \mathbb{R}I$, for $I \subseteq \Delta$.

We use diagrams to represent the set Δ of simple roots. These are graphs, where, in the examples given below, a simple (resp. double) link between two elements means an angle of $2\pi/3$ (resp. $3\pi/4$). This also means that the product of the two corresponding reflections is of order 3 (resp. 4). There is no link when the angle is $\pi/2$ (commuting reflections).

In this notion of "root system," each root could be replaced by the half-line it defines. This notion is well adapted to the classification and study of Coxeter groups.

In the classical notion of (finite, crystallographic) root systems (see [Bour68] or A2.4 below), roots are indeed elements of a $\mathbb{Z}$-lattice $X(\mathbf{T})$ and root lengths may be $\neq 1$ (this is the notion that will be used from Chapter 7 on to describe algebraic reductive groups).

Example 2.1. (i) **Coxeter group of type** A_{n-1} (see [Bour68] Planche I).

It is easy to see that the symmetric group on n letters $\mathfrak{S}_n$ is a Coxeter group for the subset of generators $\{s_i := (i, i+1) \mid i = 1, \ldots, n-1\}$. Let $E = \mathbb{R}e_1 \oplus \ldots \oplus \mathbb{R}e_n$ be the n-dimensional euclidean space where the e_i are orthogonal of norm $\sqrt{2}/2$. The reflection representation of $\mathfrak{S}_n$ is given by the hyperplane orthogonal to $e_1 + \cdots + e_n$, where $\mathfrak{S}_n$ acts on E by permutation of the e_i's, with $\Delta_{\mathrm{A}} = \{\delta_i := e_{i+1} - e_i \mid i = 1, \ldots, n-1\}$, represented by the following diagram

(A_{n-1})

$\Phi = \{e_i - e_j \mid i \neq j,\ 1 \leq i, j \leq n-1\}$ and $\Phi^+ = \{e_i - e_j \mid 1 \leq j < i \leq n-1\}$.

(ii) **Coxeter group of type** BC_n (see [Bour68] Planche II, III). In E above, take the basis $\Delta_{\mathrm{BC}} = \{\delta_0\} \cup \Delta_{\mathrm{A}}$ with $\delta_0 = \sqrt{2}e_1$.

(BC_n)

The corresponding reflections generate the matrix group $W(\mathrm{BC}_n)$ of permutation matrices with ± 1 instead of just 1's. Denoting by s'_i the reflection of vector e_i, every element in the Coxeter group of type BC can be written in a unique way as

$$s'_{i_1} \ldots s'_{i_k} w$$

for $w \in \mathfrak{S}_n$ and $1 \leq i_1 < \ldots < i_k \leq n$.

(iii) **Coxeter group of type** D_n (see [Bour68] Planche IV). This time, take $\Delta_{\mathrm{D}} = \{\delta'_0\} \cup \Delta_{\mathrm{A}}$ with $\delta'_0 = e_1 + e_2$.

(D_n)

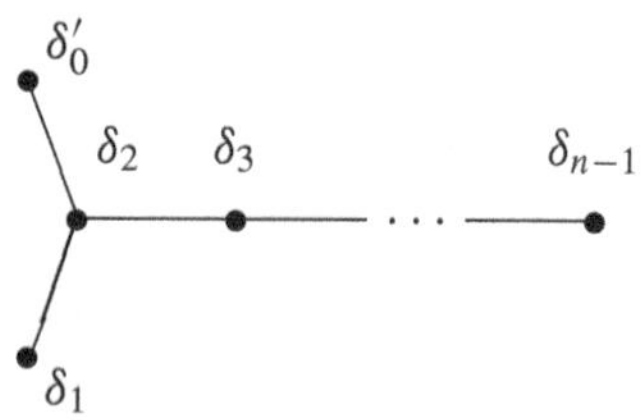

The corresponding group $W(\mathrm{D}_n)$ is the subgroup of $W(\mathrm{BC}_n)$ of matrices with an even number of -1's. This corresponds to the condition that k is even in the decomposition above.

Definition 2.2. *When $w \in W$, denote $\Phi_w = \Phi^+ \cap w^{-1}(\Phi^-)$. If $I \subseteq \Delta$, let $D_I = \{w \in W \mid w(I) \subseteq \Phi^+\}$. If $I,\ J \subseteq \Delta$, let $D_{IJ} = D_J \cap (D_I)^{-1}$.*

Proposition 2.3. *(i) If $\delta \in \Delta$, then $\delta \in \Phi_w$ if and only if $l(ws_\delta) = l(w) - 1$.*

(ii) Φ_w is finite of cardinality $l(w)$.

(iii) If $v,\ w \in W$, then $l(vw) = l(v) + l(w) \Leftrightarrow \Phi_w \subseteq \Phi_{vw} \Leftrightarrow \Phi^+ \subseteq v^{-1}(\Phi^+) \cup w(\Phi^+) \Leftrightarrow \Phi_{vw} = \Phi_w \dot{\cup} w^{-1}(\Phi_v) \Leftrightarrow \Phi_v \cap \Phi_{w^{-1}} = \emptyset$.

Proof. (i) and (ii) are standard ([Hum90] §5.6, [Stein68a] (22) p. 270). (iii) is left as an exercise. More generally, if Φ'_w denotes the set of lines corresponding to elements of Φ_w, one has $\Phi'_{vw} = \Phi'_w \dot{+} w^{-1}(\Phi'_v)$ (boolean sum) for any $v, w \in W$. □

Proposition 2.4. *If $I,\ J \subseteq \Delta$, then every double coset in $W_I\backslash W/W_J$ contains an element of minimal length, which is in D_{IJ}. This induces a bijection $W_I\backslash W/W_J \leftrightarrow D_{IJ}$. (The letter D is for distinguished representatives.)*

Proof. If w is of minimal length in $W_I w W_J$, it is of minimal length in wW_J and $W_I w$, whence $w(\delta) \in \Phi^+$ if $\delta \in J$ and $w^{-1}(\delta) \in \Phi^+$ if $\delta \in I$, thanks to Proposition 2.3(i). □

Example 2.5. Let $\mathfrak{S}_{n-1} \subseteq \mathfrak{S}_n$ be the inclusion corresponding to permutations of n letters fixing the last one. In the reflection representation of Example 2.1 above, this corresponds to the subset $\Delta' = \Delta \setminus \{e_n - e_{n-1}\}$. When $i < j$, let $s_{i,j}$ be the cycle of order $j - i + 1$ equal to $(i, \ldots, j)$. When $i > j$, let $s_{i,j} = (s_{j,i})^{-1}$. Checking images of the elements of Δ', it is easy to see that $s_{i,n} \in D_{\emptyset,\Delta'}$ for each $i = 1, \ldots, n$. Then $s_{n,i} \in D_{\Delta',\emptyset}$. Moreover, if $w \in \mathfrak{S}_n$, it is clear that $s_{n,w(n)}w$ and $ws_{w^{-1}(n),n}$ fix n, hence are in $\mathfrak{S}_{n-1}$. By Proposition 2.4 above, this implies that

$$\{s_{i,n} \mid 1 \leq i \leq n\} = D_{\emptyset,\Delta'} \quad \text{and} \quad \{s_{n,i} \mid 1 \leq i \leq n\} = D_{\Delta',\emptyset}.$$

For any $w \in \mathfrak{S}_n$, one gets $w \in s_{w(n),n}\mathfrak{S}_{n-1}$ and $w \in \mathfrak{S}_{n-1}s_{n,w^{-1}(n)}$.

Theorem 2.6. *If $I,\ J \subseteq \Delta$ and $w \in D_{IJ}$, let $K = I \cap w(J) \subseteq \Delta$. Then $W_I \cap {}^wW_J = W_K$, $\Phi_I \cap w(\Phi_J) = \Phi_K$, and $\Phi_I^+ \cap w(\Phi_J^+) = \Phi_K^+$.*

Lemma 2.7. *Given the same hypotheses as for Theorem 2.6, we have $wJ \cap \Phi_I^+ \subseteq I$.*

Proof of Lemma 2.7. Let $\delta \in J$ be such that $w(\delta) \in \Phi_I^+$. Let us write $w(\delta) = \sum_{\delta' \in I} \lambda_{\delta'}\delta'$ with $\lambda_{\delta'} \geq 0$. Then $\delta = \sum_{\delta' \in I} \lambda_{\delta'} w^{-1}(\delta') \in \Delta$ with each $w^{-1}(\delta')$ a positive root since $w^{-1} \in D_I$. One of the $\lambda_{\delta'}$ is non-zero, say $\lambda_{\delta_0} > 0$. Then $w^{-1}(\delta_0)$ must be proportional to δ, hence equal to it. That is, $\delta_0 = w(\delta) \in I$. □

Proof of Theorem 2.6. The inclusions $W_K \subseteq W_I \cap {}^wW_J$ and $\Phi_K \subseteq \Phi_I \cap w(\Phi_J)$ are clear.

Conversely, let $x \in W_I \cap {}^wW_J$ and let us check that $x \in W_K$. We use induction on the length of x. If $x = 1$ this is clear. Otherwise let $y = w^{-1}xw \in W_J \setminus 1$

and $\delta \in J$ be such that $y(\delta) \in \Phi_J^-$, or equivalently $y = y's_\delta$ with $l(y') = l(y) - 1$. Since $w \in D_J$, one has $xw(\delta) = wy(\delta) \in \Phi^-$. One has $w(\delta) \in \Phi^+$ since $w \in D_J$. Then $w(\delta) \in \Phi_x \subseteq \Phi_I^+$. Now Lemma 2.7 implies $w(\delta) \in I$. Denoting $\delta' = w(\delta)$, one has $\delta' \in I \cap w(J)$. Moreover, $x(\delta') \in \Phi^-$, so $x = x's_{\delta'}$ with $l(x') = l(x) - 1$. One may then apply the induction hypothesis to $x' = wy'w^{-1}$.

Now let α be an element of $\Phi_I^+ \cap w(\Phi_J)$. Let t be the element of W corresponding to the reflection associated with α in the geometric representation. Now $t \in W_I \cap {}^wW_J$. Then $t \in W_K$ by what we have just proved, and therefore $\Phi_t \subseteq \Phi_K^+$. But $t(\alpha) = -\alpha$, so $\alpha \in \Phi_K^+$ as claimed. Thus $\Phi_I^+ \cap w(\Phi_J) = \Phi_K^+$. Then, making the union with its opposite, we get $\Phi_I \cap w(\Phi_J) = \Phi_K$. The equality $\Phi_I^+ \cap w(\Phi_J^+) = \Phi_K^+$ also follows since $w(\Phi_J^+) \subseteq \Phi^+$. □

We assume W is finite. Then the form $\langle -, - \rangle$ on $\mathbb{R}\Delta$ is positive definite (see [CuRe87] 64.28(ii), [Bour68] §V.4.8, [Hum90] §6.4). Moreover, W has a unique element of maximal length, characterized by several equivalent conditions, among which is the fact that it sends Δ to $-\Delta$ (see [Hum90] §1.8).

Notation 2.8. If $I \subseteq \Delta$, one denotes by w_I the element of maximal length in W_I. If $\delta \in \Delta \setminus I$, let $v(\delta, I) = w_{I\cup\{\delta\}}w_I$.

Example 2.9. (i) In $\mathfrak{S}_n$ (see Example 2.1(i)), the element of maximal length is defined by $w_0(i) = n + 1 - i$ (it is easily checked that this element makes negative all $\delta_i = e_{i+1} - e_i \in \Delta_{\mathrm{A}}$). It is easily checked that $v(\delta_{n-1}, \Delta_{\mathrm{A}} \setminus \{\delta_{n-1}\}) = s_{n,1}$ (cycle of order n, see Example 2.5).

(ii) For the Coxeter group of type BC_n, w_0 is $-\mathrm{Id}_E$ in the geometric representation (see Example 2.1(ii)). One has $v(\delta_{n-1}, \Delta_{\mathrm{BC}} \setminus \{\delta_{n-1}\}) = s'_n$.

(iii) In the geometric representation of D_n, one gets $w_0 = s'_1 \dots s'_n = -\mathrm{Id}_E$ if n is even, $w_0 = s'_2 \dots s'_n = -s'_1$ if n is odd. One has $v(\delta_{n-1}, \Delta_{\mathrm{D}} \setminus \{\delta_{n-1}\}) = s'_1 s'_n$.

Proposition 2.10. *Let $I \subseteq \Delta$, $\delta \in \Delta \setminus I$.*

(i) We have $\Phi_{w_I} = \Phi_I^+$, $\Phi_{v(\delta,I)} = \Phi_{I\cup\{\delta\}}^+ \setminus \Phi_I^+$, and $v(\delta, I)(I) \subseteq I \cup \{\delta\}$.

(ii) Let $w \in W$ satisfy $w(I) \subseteq \Delta$. Then $l(w.v(\delta, I)^{-1}) = l(w) - l(v(\delta, I))$ if and only if $w(\delta) \in \Phi^-$. Otherwise, $l(w.v(\delta, I)^{-1}) = l(w) + l(v(\delta, I))$.

Proof. (i) We have $w_I(I) = -I$, so $w_I(\Phi_I^+) = \Phi_I^-$ and $v(\delta, I)(I) = -w_{I\cup\{\delta\}}(I) \subseteq -w_{I\cup\{\delta\}}(I \cup \{\delta\}) = I \cup \{\delta\}$. But $l(w_I) = \Phi_I^+$, hence $\Phi_{w_I} = \Phi_I^+$. This also gives $\Phi_{v(\delta,I)} = \Phi_{\{\delta\}\cup I}^+ \setminus \Phi_I^+$ since $l(w_{\{\delta\}\cup I}) = l(w_I) + l(v(\delta, I))$.

(ii) Let now w be such that $w(I) \subseteq \Delta$ and $w(\delta) \in \Phi^-$. To show that $l(w.v(\delta, I)^{-1}) = l(w) - l(v(\delta, I))$, as a result of Proposition 2.3(iii), it is enough to show that $\Phi_w \supseteq \Phi_{v(\delta,I)}$. Let $\alpha \in \Phi_{v(\delta,I)}$, i.e. $\alpha \in \Phi_{\{\delta\}\cup I}^+ \setminus \Phi_I^+$ thanks to (i) above. Let us write $\alpha = \lambda_\delta\delta + \sum_{\delta'\in I} \lambda_{\delta'}\delta'$ with $\lambda_\delta > 0$ and $\lambda_{\delta'} \geq 0$ for

$\delta' \in I$. Then $w(\alpha) = \lambda_\delta w(\delta) + \sum_{\delta' \in I} \lambda_{\delta'} w(\delta')$. If we had $w(\alpha) \in \Phi^+$, since $w(\delta) \in \Phi^-$ and $w(I) \in \Delta$, the non-zero coefficients in $w(\delta)$ would be for elements of $w(I)$. So $w(\delta) \in \Phi_{w(I)}$, or equivalently $\delta \in \Phi_I$. But $\delta \in \Delta \setminus I$, a contradiction.

It remains to show that, if $w(I) \subseteq \Delta$ and $w(\delta) \in \Phi^+$, then $l(w.v(\delta, I)^{-1}) = l(w) + l(v(\delta, I))$. We apply the implication we have just proved with $w.v(\delta, I)$ instead of w, so we just have to check that $w.v(\delta, I)(\delta) \in \Phi^-$. We have seen that $\delta \in \Phi_{v(\delta, I)}$, so $v(\delta, I)(\delta) \in \Phi^+_{\delta \cup I}$. Then its image by w is in Φ^+ since w sends both I and δ into Φ^+ by hypothesis. □

Theorem 2.11. *Assume as above that W is finite. Let $w \in W$ and $\Delta_1 \subseteq \Delta$ be such that $w(\Delta_1) \subseteq \Delta$. Then there exist $\Delta_1, \ldots, \Delta_k$ subsets of Δ, and a sequence $\delta_1 \in \Delta \setminus \Delta_1, \ldots, \delta_k \in \Delta \setminus \Delta_k$, such that, for all $1 \leq j \leq k-1$, $v(\delta_j, \Delta_j)(\Delta_j) = \Delta_{j+1}$ and $w = v(\delta_k, \Delta_k) \ldots v(\delta_1, \Delta_1)$ with $l(w) = l(v(\delta_k, \Delta_k)) + \cdots + l(v(\delta_1, \Delta_1))$.*

If moreover $\Delta \setminus \Delta_1$ has a single element δ_1, then $w = 1$ or $v(\delta_1, \Delta_1)$.

Proof. The first point is by induction on the length of w. Everything is clear when $w = 1$. Otherwise, there is $\delta \in \Delta$ such that $w(\delta_1) \in \Phi^-$. Then $\delta_1 \notin \Delta_1$ and Proposition 2.10(ii) allows us to write $w = w_1 v(\delta_1, \Delta_1)$ with lengths adding. Letting $\Delta_2 = v(\delta_1, \Delta_1)(\Delta_1)$, one may clearly apply the induction hypothesis to Δ_2 and w_1.

We also prove the second point by induction on the length. If $w = 1$, we are done. Otherwise, one has $w = w_1 v(\delta_1, \Delta_1)$ with lengths adding and w_1 satisfying the same conditions as w for $\Delta_2 = v(\delta_1, \Delta_1)(\Delta_1) = -w_0(\Delta_1)$. The induction implies $w' = 1$ or $v(-w_0(\delta_1), \Delta_2)$. But the latter is $w_0 v(\delta_1, \Delta_1) w_0 = v(\delta_1, \Delta_1)^{-1}$. We get our claim. □

2.2. BN-pairs

We now define BN-pairs.

Definition 2.12. *A BN-pair (or Tits system) consists of the data of a group G, two subgroups B, N and a subset S of the quotient $N/B \cap N$ such that, denoting $T := B \cap N$ and $W := N/T$:*

(TS1) *$T \lhd N$ (W is therefore a quotient group), W is generated by S and $\forall s \in S$, $s^2 = 1$.*

(TS2) *$\forall s \in S$, $\forall w \in W$, $sBw \subseteq BwB \cup BswB$.*

(TS3) *$B \cup N$ generates G.*

(TS4) *$\forall s \in S$, $sBs \neq B$.*

Remark 2.13. The notation Bw is unambiguous since w is a class mod. T and $T \subseteq B$. Similarly, if X is a subgroup of B normalized by T, the notation X^w makes sense (and is widely used in what follows).

The hypothesis (TS4) implies that all the elements of S are of order 2.

(TS2) implies

(TS2′) $\forall s \in S, \forall w \in W, wBs \subseteq BwB \cup BwsB$.

For the next two theorems, we refer to [Bour68] §IV, [Cart85] §2, [CuRe87] §65.

Theorem 2.14. (Bruhat decomposition) *If G, B, N, S is a Tits system, the subsets $(BwB)_{w \in W}$ are distinct and form a partition of G.*

Definition 2.15. *If G, B, N, S is a Tits system, and $I \subseteq S$ is a subset, let $W_I = \langle I \rangle$, and N_I be the subgroup of N containing T such that $N_I/T = W_I$. Let $P_I := BN_IB = \bigcup_{w \in W_I} BwB$.*

Theorem 2.16. *Let (G, B, N, S) be a group with a BN-pair.*

(i) W is a Coxeter group with regard to S.

(ii) The P_I defined above are subgroups of G ("parabolic" subgroups) and $N \cap P_I = N_I$. If P is a subgroup of G containing B, we have $P = P_J$ for $J := \{s \in S | s \subseteq P\}$.

(iii) If P_I is a parabolic subgroup, (P_I, B, N_I, I) is a BN-pair.

(iv) If I, J are subsets of S, then $P_I\backslash G/P_J \cong W_I\backslash W/W_J$.

Example 2.17. (see [Cart72b] §11.3, §14.5, [DiMi91] §15) Let $\mathbb{F}$ be a field, let $n \geq 1$ be an integer.

(i) Let $\mathrm{GL}_n(\mathbb{F})$ be the group of invertible elements in the ring $\mathrm{Mat}_n(\mathbb{F})$ of $n \times n$ matrices with coefficients in $\mathbb{F}$. Let U (resp. T, resp. W) be the subgroup of upper triangular unipotent (resp. invertible diagonal, resp. permutation) matrices. Let $B = UT$ (upper triangular matrices in $\mathrm{GL}_n(\mathbb{F})$), $N = TW$ and S be the set of elements of W corresponding to the transpositions $(i, i+1)$ for $i = 1, \ldots, n-1$.

Then $B \cap N = T$ and (B, N, S) makes a BN-pair for $\mathrm{GL}_n(\mathbb{F})$ (see Exercise 1). The associated Coxeter system (W, S) corresponds to Example 2.1(i), i.e. type A_{n-1}.

A slight adaptation of the above allows us to show a similar result for $\mathrm{SL}_n(\mathbb{F})$, the group of matrices of determinant 1.

(ii) Assume now that $\mathbb{F}$ has an automorphism $\lambda \mapsto \overline{\lambda}$ of order 2. This extends as $g \mapsto \overline{g}$ for $g \in \mathrm{GL}_n(\mathbb{F})$. Let $w_0 \in \mathrm{GL}_n(\mathbb{F})$ be the permutation matrix corresponding to $i \mapsto n+1-i$ for $i = 1, \ldots, n$. Denote by

$$\sigma : \mathrm{GL}_n(\mathbb{F}) \to \mathrm{GL}_n(\mathbb{F})$$

the group automorphism defined by $\sigma(g) = w_0.{}^t\overline{g}^{-1}.w_0$.

Let $\mathrm{GU}_n(\mathbb{F})$ be the group of fixed points in $\mathrm{GL}_n(\mathbb{F})$, i.e. $g \in \mathrm{GL}_n(\mathbb{F})$ satisfying $g.w_0.{}^t\overline{g} = w_0$. Let B^σ, T^σ, W^σ be the subgroups of $\mathrm{GU}_n(\mathbb{F})$ consisting of fixed points under σ in the corresponding subgroups of $\mathrm{GL}_n(\mathbb{F})$. Let $m := [\frac{n}{2}]$ be the biggest integer $\leq \frac{n}{2}$. Then it is easily checked that W^σ is isomorphic to $(\mathbb{Z}/2\mathbb{Z})^m \rtimes \mathfrak{S}_m$ and that it is generated by the set S_0 of permutation matrices corresponding to the following elements of $\mathfrak{S}_n$: $(i, i+1)(n+1-i, n-i)$ for $2i < n$, plus an element equal to $(m, m+1)$ when $n = 2m$ is even, and equal to $(m, m+2)$ when $n = 2m+1$ is odd.

This makes a Coxeter system of type BC_m.

From the fact that B, N, S of (i) above make a BN-pair, one may prove that B^σ, N^σ, and S_0 are a BN-pair for $\mathrm{GU}_n(\mathbb{F})$ (see Exercise 2).

When $\mathbb{F}$ is finite and $\lambda \mapsto \overline{\lambda}$ is non-trivial, $|\mathbb{F}|$ is a square q^2 and $\overline{\lambda} = \lambda^q$ for all $\lambda \in \mathbb{F}$. Then, it is also easily checked that the above group $\mathrm{GU}_n(\mathbb{F})$ is isomorphic to the group of matrices satisfying $a.{}^t\overline{a} = \mathrm{Id}_n$ (a more classical definition of unitary groups). For that, it suffices to find an element $g_0 \in \mathrm{GL}_n(\mathbb{F})$ such that $g_0.{}^t\overline{g_0} = w_0$. This reduces to dimension 2 where one takes $g_0 = \begin{pmatrix} 1 & \varepsilon \\ \varepsilon(\varepsilon-\eta)^{-1} & \varepsilon\eta(\varepsilon-\eta)^{-1} \end{pmatrix}$ for $\varepsilon \neq \eta$ in $\mathbb{F}$ satisfying $\varepsilon^{q+1} = \eta^{q+1} = -1$. Then $a \mapsto a^{g_0}$ is the isomorphism sought.

(iii) Let $\mathbb{F}$ be a field of characteristic $\neq 2$, $m \geq 2$ an integer. Recall $w_0 \in \mathrm{SL}_{2m}(\mathbb{F})$, the permutation matrix associated with the permutation $i \mapsto 2m - i + 1$.

Let $\mathrm{SO}^+_{2m}(\mathbb{F})$ denote the subgroup of $\mathrm{SL}_{2m}(\mathbb{F})$ consisting of matrices satisfying ${}^tg.w_0.g = w_0$. This is the special orthogonal group associated with the symmetric bilinear form on $\mathbb{F}^{2m}$ of maximal Witt index (hence the $+$ in SO^+). Let B' (resp. T') be its subgroups of upper triangular, resp. diagonal, matrices. Let S' be the set of permutation matrices corresponding to the following elements of $\mathfrak{S}_{2m}$: $(i, i+1)(2m-i, 2m-i+1)$ for $i = 1, \ldots, m-1$, and $(m-1, m+1)(m, m+2)$. Clearly $S' \subseteq \mathrm{SO}^+_{2m}(\mathbb{F})$ and it generates the centralizer W' of w_0 in the group of permutation matrices in $\mathrm{SL}_{2m}(\mathbb{F})$. Along with S', this makes a Coxeter system of type D_m (note that the embedding of W' in W^σ above corresponds with the embedding of type D_m in type BC_m suggested in Example 2.1).

Using a method similar to (ii) above (see Exercise 2), one may check that B', T', W', and S' make a BN-pair of type D_m for $\mathrm{SO}^+_{2m}(\mathbb{F})$.

2.3. Root subgroups

We keep the same notation as in §2.2. G is a group with a BN-pair and finite W. We show how to associate certain subgroups of G with the roots of W.

Assume $B \cap B^{w_0} = T$.

Definition 2.18. *If $w \in W$, $\delta \in \Delta$, let $B_w = B \cap B^{w_0 w}$, $B_\delta = B_{s_\delta}$.*

Theorem 2.19. *Let G be a group with a BN-pair and finite W. Assume $B \cap B^{w_0} = T$. Let $w \in W$, $s \in S$, $\delta \in \Delta$.*

(i) If $l(ws) = l(w) + 1$, then $B \cap B^{ws} \subseteq B \cap B^s$.

(ii) $B = B_s(B \cap B^s) = (B \cap B^s)B_s$.

(iii) ${}^w B_\delta$ depends only on $w(\delta)$. We write ${}^w B_\delta = B_{w(\delta)}$.

(iv) There is a sequence $\alpha_1, \ldots, \alpha_N$ giving all the elements of Φ^+ with no repetition, such that $B = B_{\alpha_1} \ldots B_{\alpha_N}$. The corresponding decomposition of the elements of B is unique up to elements of T.

(v) If $\delta, \delta' \in \Delta$ are such that $w_0(\delta) = -\delta'$, then $P_\delta \cap P_{\delta'}{}^{w_0} = B_\delta \cup B_\delta s_\delta B_\delta$.

Proof. (i), (ii), (iii), (iv) are classic (see [Cart85] §2.5, [CuRe87] 69.2). They can be deduced in a fairly elementary way from the axioms of the BN-pair (see [Asch86] Exercise 10, p. 227).

(v) Let us show first

(v′) $B \cap (P_{\delta'})^{w_0} = B_\delta$.

Using (ii), one has $P_{\delta'} = B \cup Bs_{\delta'}B = B \cup B_{\delta'}s_{\delta'}B$. Therefore $P_{\delta'} = s_{\delta'}P_{\delta'} = s_{\delta'}B \cup s_{\delta'}B_{\delta'}s_{\delta'}B = s_{\delta'}B \cup B_{-\delta'}B$ and $(P_{\delta'})^{w_0} = s_\delta B^{w_0} \cup B_\delta B^{w_0}$ by the definition of δ' from δ and (iii). Now $B \cap s_\delta B^{w_0} = (Bw_0 \cap s_\delta w_0 B)w_0 = \emptyset$ by Theorem 2.14 (Bruhat decomposition). So $B \cap (P_{\delta'})^{w_0} = B \cap B_\delta B^{w_0} = B_\delta$ since $B \cap B^{w_0} = T$. This is (v′).

Let us write $P_\delta = B \cup Bs_\delta B_\delta$, again by (ii). We have s_δ, $B_\delta \subseteq (P_{\delta'})^{w_0}$, so (v′) implies that $P_\delta \cap (P_{\delta'})^{w_0} = B_\delta \cup B_\delta s_\delta B_\delta$ as claimed. □

Definition 2.20. *The BN-pair (G, B, N, S) is said to be* **split of characteristic** *p if and only if G is finite, $B \cap B^{w_0} = T$, and there is a semi-direct product decomposition $B = UT$ where $U \lhd B$ is a p-group and T is a commutative group of order prime to p. The BN-pair is said to be* **strongly split** *when moreover, for all $I \subseteq S$, $U_I := U \cap U^{w_I}$ is normal in U.*

When $\alpha \in \Phi$, let X_α be the set of p-elements of B_α (see Definition 2.18).

Note that $B_\alpha = X_\alpha.T$ (semi-direct product). In the following, $(G, B = UT, N, S)$ is a split BN-pair of characteristic p.

Theorem 2.21. *(i) U is a Sylow p-subgroup of G and G has no normal p-subgroup $\neq \{1\}$.*

(ii) ${}^w X_\delta$ depends only on $w(\delta)$, so we can write ${}^w X_\delta = X_{w(\delta)}$. It is not equal to $\{1\}$.

(iii) There is a sequence $\alpha_1, \ldots, \alpha_N$ giving all the elements of Φ^+ with no repetition, such that $U = X_{\alpha_1} \ldots X_{\alpha_N}$ with uniqueness of the decompositions (i.e. $|U| = \Pi_{\alpha\in\Phi^+}|X_\alpha|$).

Proof. (i) Since U is a Sylow p-subgroup of B, it suffices to check that $\mathrm{N}_G(U) = B$ to have that U is a Sylow p-subgroup of G. We have $\mathrm{N}_G(U) \supseteq B$, so $\mathrm{N}_G(U) = P_I$ for some $I \subseteq S$. Then, if $s \in I$, we have $U^s = U$ and therefore $B^s = B$. This contradicts (TS4).

There is no normal p-subgroup $\neq \{1\}$ in G since such a subgroup would be in $U \cap U^{w_0} \subseteq B \cap B^{w_0} = T$, a group of order prime to p.

(ii) is a consequence of Theorem 2.19(iii). The group X_δ is non-trivial since $X_\delta = \{1\}$ would imply $B_\delta = T$, and therefore $B = B \cap B^{s_\delta}$ by Theorem 2.19(ii), contradicting (TS4).

(iii) is a consequence of Theorem 2.19(iv). □

Lemma 2.22. *Let $\alpha_1, \ldots, \alpha_m$ be m distinct positive roots and, for every i, let $x_i \in X_{\alpha_i} \setminus \{1\}$. Let $w \in W$, then $x_1x_2\ldots x_m \in U^w$ if and only if $w(\alpha_i)$ is a positive root for every i.*

Proof. The "if" is clear since ${}^w(X_\alpha) = X_{w(\alpha)} \subseteq U$ when $w(\alpha)$ is positive (Theorem 2.21(iii)).

We prove the converse by induction on the length of w. If $w = 1$, this is clear. If $w = s_\delta$ with $\delta \in \Delta$, one must check only that δ is none of the α_i's. Suppose on the contrary that $\delta = \alpha_{i_0}$. Then in the product $x_1x_2\ldots x_m$, all the terms on the left of x_{i_0} are in U^{s_δ} (by the "if" above), and the same is true for the ones on the right. So $x_{i_0} \in U \cap U^{s_\delta}$, while $x_{i_0} \in X_\delta = U \cap U^{w_0 s_\delta}$. But $B^{s_\delta} \cap B^{w_0 s_\delta} = T^{s_\delta} = T$, so $x_{i_0} = 1$, contradicting the hypothesis.

For an arbitrary w of length ≥ 1, write $w = w's$ with $l(w) = l(w') + 1$. We have $U \cap U^w \subseteq U \cap U^s$ as a result of Theorem 2.19(i) so, by the case just treated, $s(\alpha_i)$ is positive for all i. Now defining $\alpha'_i = s(\alpha_i)$, $x'_i = (x_i)^g$ for $g \in N$ a representative of s, we have $x'_1 \ldots x'_m \in U^{w'}$ and the induction hypothesis gives our claim. □

Theorem 2.23. *Let $(G, B = UT, N, S)$ be a split BN-pair of characteristic p (see Definition 2.20).*

If A is a subset of W containing 1, denote $\Psi_A = \bigcap_{a \in A} a^{-1}(\Phi^+)$ and $U_A = \bigcap_{a \in A} U^a$.

(i) Let $\alpha_1, \alpha_2, \ldots, \alpha_N$ be a list of the positive roots such that $U = X_{\alpha_1} \ldots X_{\alpha_N}$ (see Theorem 2.21(iii)). Let A be a subset of W containing 1. Then, denoting $\Psi_A = \{\alpha_{i_1}, \alpha_{i_2}, \ldots, \alpha_{i_m}\}$ with $1 \leq i_1 < \ldots < i_m \leq N$, one has

$$U_A = X_{\alpha_{i_1}} \ldots X_{\alpha_{i_m}} = \langle X_\alpha \,; \alpha \in \Psi_A \rangle$$

and

$$\Psi_A = \{\alpha \in \Phi^+ \mid X_\alpha \subseteq U_A\}.$$

(ii) If A, A', A'' are subsets of W such that $1 \in A \cap A' \cap A''$ and $\Psi_A \subseteq \Psi_{A'} \cup \Psi_{A''}$, then $U_A = (U_A \cap U_{A'}).(U_A \cap U_{A''}) \subseteq U_{A'}.U_{A''}$.

Proof. (i) Repeated application of Lemma 2.22.

(ii) One has clearly, for arbitrary subsets of W containing 1, $U_A \cap U_{A'} = U_{A \cup A'}$ and $\Psi_A \cap \Psi_{A'} = \Psi_{A \cup A'}$. By (i), one has $|U_A| = \Pi_{\alpha \in \Psi_A} |X_\alpha|$. Therefore $(U_A \cap U_{A'}).(U_A \cap U_{A''})$ has cardinality $|U_{A \cup A'}|.|U_{A \cup A''}|.|U_{A \cup A' \cup A''}|^{-1} = \Pi_{\alpha \in \Psi_{A \cup A'} \cup \Psi_{A \cup A''}} |X_\alpha|$. We have $\Psi_{A \cup A'} \cup \Psi_{A \cup A''} = \Psi_A \cap (\Psi_{A'} \cup \Psi_{A''}) = \Psi_A$ by hypothesis. So $|(U_A \cap U_{A'}).(U_A \cap U_{A''})| = |U_A|$. This implies $(U_A \cap U_{A'}).(U_A \cap U_{A''}) = U_A$ since the inclusion $(U_A \cap U_{A'}).(U_A \cap U_{A''}) \subseteq U_A$ is clear. This gives (ii). □

Remark. The sets Ψ_A of Theorem 2.23 coincide with the intersections of Φ^+ with convex cones (see Exercise 3).

When the BN-pair is strongly split, i.e. $U \cap U^{w_I} \lhd U$ for all I, the root subgroups X_α satisfy a commutator formula (see Exercise 5).

2.4. Levi decompositions

We now assume that G has a strongly split BN-pair of characteristic p (see Definition 2.20).

Definition 2.24. *Let $I \subseteq \Delta$. Let N_I be the inverse image of W_I in N, and recall that $U_I = U \cap U^{w_I}$. Let $L_I = \langle B_{w_I}, N_I \rangle$; this is called the Levi subgroup associated with I.*

Denote $\mathcal{L} = \{(P_I, U_I)^g \mid I \subseteq \Delta \text{ , } g \in G\}$.

Proposition 2.25. *L_I has a strongly split BN-pair of characteristic p given by (B_{w_I}, N_I, I). One has a semi-direct product decomposition $P_I = U_I \rtimes L_I$, and U_I is the largest normal p-subgroup of P_I. So $\mathcal{L}$ above is a set of subquotients.*

Proof. Let us check first that L_I has a split BN-pair. The axioms (TS1) and (TS3) are clear. (TS4) and $B_{w_I} \cap (B_{w_I})^{w_I} = T$ both follow using Theorem 2.23(i). One has $B_{w_I} = (U \cap B_{w_I})T$, a semi-direct product.

It remains to check (TS2). Let $s_\delta \in S$ correspond with $\delta \in I$ and let $w \in W_I$. Using Theorem 2.23(ii), one has $U = U_\delta X_\delta$ and therefore $B_{w_I} = T(B_{w_I} \cap U) = T(B_{w_I} \cap U_\delta)X_\delta$. One has $s_\delta(\Phi_I^+ \setminus \delta) = \Phi_I^+ \setminus \delta$, so Theorem 2.23(i) implies that s_δ normalizes $B_{w_I} \cap U_\delta$. If $w^{-1}(\delta)$ is positive, it is an element of $\Phi_I^+ = \Phi_{w_I}$, so $s_\delta B_{w_I} w \subseteq T(B_{w_I} \cap U_\delta)s_\delta w B_{w_I} \subseteq B_{w_I} s_\delta w B_{w_I}$. If $w^{-1}(\delta)$ is negative, one may apply the preceding case to $s_\delta w$, so it suffices to show

$s_\delta B_{w_I} s_\delta \subseteq B_{w_I} \cup B_{w_I} s_\delta B_{w_I}$. Now using the same decomposition of B_{w_I} as before, one has $s_\delta B_{w_I} s_\delta \subseteq s_\delta B_\delta s_\delta B_{w_I}$. Theorem 2.19(v) told us that $B_\delta \cup B_\delta s_\delta B_\delta$ is a group, so $s_\delta B_\delta s_\delta \subseteq B_\delta \cup B_\delta s_\delta B_\delta$. Thus we have our claim.

We must show that the BN-pair of L_I is strongly split. We have seen that $B_{w_I} = X_I T$, a semi-direct product where $X_I = U \cap B_{w_I} = U \cap U^{x_0 w_I}$. So, given $J \subseteq I$, we must check $X_I \cap (X_I)^{w_J} \lhd X_I$. We have $X_I = U \cap U^{w_0 w_I}$ and $X_I \cap (X_I)^{w_J} = U \cap U^{w_0 w_I} \cap U^{w_J} \cap U^{w_0 w_I w_J}$. Knowing that $U \cap U^{w_J} \lhd U$ by the strongly split condition satisfied in G, it suffices to check that $U \cap U^{w_0 w_I} \cap U^{w_J} \cap U^{w_0 w_I w_J} = U \cap U^{w_0 w_I} \cap U^{w_J}$. By Theorem 2.23(i), this may be checked at the level of the corresponding subsets of Φ^+. This follows from $\Phi^+ \cap w_J(\Phi^+) = \Phi^+ \setminus \Phi_J^+$, $\Phi^+ \cap w_I w_0(\Phi^+) = \Phi_I^+$ and the fact that $\Phi_I^+ \setminus \Phi_J^+$ is made negative by $w_I w_J$ (all this follows from Proposition 2.10(i)).

The strongly split condition gives $U_I \lhd U$. But U_I is clearly normalized by N_I (use Theorem 2.23 (i)). Then $U_I \lhd P_I$. Now L_I has no non-trivial normal p-subgroup, so $U_I \cap L_I = \{1\}$. To check that $U_I L_I = P_I$ it suffices to check $B = U_I B_{w_I}$. This is clear by Theorem 2.23(ii). □

Definition 2.26. *When $I \subseteq \Delta$, let W^I be the subgroup $\{w \in W \mid wI = I\}$.*

Theorem 2.27. *Let $(G, B = UT, N, S)$ be a strongly split BN-pair of characteristic p (see Definition 2.20). Let $I, J \subseteq \Delta$, $g \in G$.*

(i) If $d \in D_{IJ}$, then ${}^d(P_J, U_J) \cap\!\!\downarrow (P_I, U_I) = (P_K, U_K)$ for $K = I \cap dJ$.

(ii) $\mathcal{L}$ (see Definition 2.24) is k-regular and $\cap\!\!\downarrow$-stable for all fields k of characteristic $\ell \neq p$.

(iii) (P_I, U_I)—${}^g(P_J, U_J)$ if and only if $g \in P_I d P_J$ where $d \in W$ satisfies $dJ = I$. This induces a bijection between $P_I \backslash \{g \in G \mid (P_I, U_I)$—${}^g(P_J, U_J)\}/P_J$ and $\{d \in W \mid dJ = I\}$. When (P_I, U_I)—${}^g(P_J, U_J)$, one has $P_I = U_I \rtimes L$ and ${}^g P_J = {}^g U_J \rtimes L$ for an L which is a P_I-conjugate of L_I.

(iv) $\{g \in G \mid (P_I, U_I)$—$(P_I, U_I)^g\} = P_I \mathrm{N}_G(L_I) P_I = P_I W^I P_I$ (see Definition 2.26).

(v) Any relation (P, V)—(P', V') in $\mathcal{L}$ implies $|P| = |P'|$ and $|V| = |V'|$.

Proof. (i) Let $(P, V) = (({}^d P_J \cap P_I)U_I, ({}^d U_J \cap P_I)U_I) = {}^d(P_J, U_J) \cap\!\!\downarrow (P_I, U_I)$.

Let us show first that $P \supseteq P_K$. We have $T \subseteq P_I \cap {}^d P_J \subseteq P$. If $\alpha \in \Phi_I^+$, then $X_\alpha \in P_I$ and $d^{-1}(\alpha) \in \Phi^+$ since $d \in D_I$, therefore $X_\alpha \subseteq P$. But $U_I \subseteq P$, therefore (Theorem 2.23(ii)) $B \subseteq P$. It now suffices to check $W_K \subseteq P$. But $W_K \subseteq W_I \cap {}^d W_J \subseteq P_I \cap {}^d P_J \subseteq P$.

Since $P \subseteq P_I$, there exists a subset K' such that $K \subseteq K' \subseteq I$ and $U_I.(P_I \cap {}^d P_J) = P_{K'}$.

Let us show that $X_\alpha \subseteq V$ for all $\alpha \in \Phi^+ \setminus \Phi_K$. If $\alpha \in \Phi^+ \setminus \Phi_I$, then $X_\alpha \subseteq U_I \subseteq V$. If $\alpha \in \Phi_I^+ \setminus \Phi_K$, $X_\alpha \subseteq P_I$. It remains to show that $X_\alpha^d \subseteq U_J$, which in turn comes from $d^{-1}(\alpha) \notin \Phi_J$. If $\alpha \in d(\Phi_J)$, then $\alpha \in \Phi_I \cap d(\Phi_J) = \Phi_K$ (Theorem 2.6). This contradicts the hypothesis. Therefore $U_K \subseteq V$.

Now, since V is a normal p-subgroup in $P = P_{K'}$, one has $V \subseteq U_{K'}$, whence $U_K \subseteq U_{K'}$ and therefore $U_K \cap X_\alpha = \{1\}$ for all $\alpha \in \Phi_{K'}$ (Theorem 2.23(i)). Then $\Phi^+ \setminus \Phi_K^+ \subseteq \Phi^+ \setminus \Phi_{K'}^+$, i.e. $K' \subseteq K$. We already had the reverse inclusion, whence the equality.

(ii) We have $U_I \lhd P_I$ and $|U_I|$ is a power of p. It remains to check that $(P_I, U_I)^g \cap\!\!\downarrow (P_J, U_J)^h$ is in $\mathcal{L}$. This reduces to (i) recalling that $G = P_I D_{IJ} P_J \ni gh^{-1}$.

(iii) is clear by (i) and (ii). We take $L = {}^x(L_I)$, where $x \in P_I$ is such that $g \in xdP_J$ and $d \in W$ satisfies $dJ = I$.

(iv) We clearly have $\{g \in G \mid (P_I, U_I)—(P_I, U_I)^g\} \supseteq P_I \mathrm{N}_G(L_I) P_I \supseteq P_I W^I P_I$. Now, if $(P_I, U_I)—(P_I, U_I)^g$, let us write $g \in P_I d P_I$ for $d \in D_{II}$ satisfying $(P_I, U_I)—(P_I, U_I)^d$. By (i), this means $dI = I$, i.e. $d \in W^I$ as stated.

(v) follows from (iii). □

Example 2.28. Let us give some examples of subgroups W^I (see Definition 2.26). We use the notation of Example 2.1.

(i) For type A_{n-1} ($n \geq 2$), $\Delta = \{\delta_1, \ldots, \delta_{n-1}\}$, let $I = \{\delta_1, \ldots, \delta_i\}$ ($i \geq 1$).

It is easy to see that any element $w \in W^I$ must correspond to a permutation which increases on the set $\{1, \ldots, i+1\}$ but is also such that $\{1, \ldots, i+1\}$ is preserved. So W^I coincides with permutations fixing all elements of this set, i.e. $W^I = \langle s_{i+2}, \ldots, s_{n-1} \rangle = W_{I'}$ where $I' = \{\delta_{i+2}, \ldots, \delta_{n-1}\}$.

(ii) For type BC_n, $\Delta_{\mathrm{BC}} = \{\delta_0, \delta_1, \ldots, \delta_{n-1}\}$, let $I = \{\delta_0, \delta_1, \ldots, \delta_{i-1}\}$ ($i \geq 1$). An element in W^I must permute the basis elements e_j in a fashion similar to the case above, with trivial signs on $e_1, \ldots, e_i$. So we get $W^I = \langle s'_{i+1}, s_{i+1}, \ldots, s_{n-1} \rangle$, which is a Coxeter group of type BC_{n-i} represented in the space generated by $\Delta^I = \{e_{i+1}, \delta_{i+1}, \ldots, \delta_{n-1}\}$.

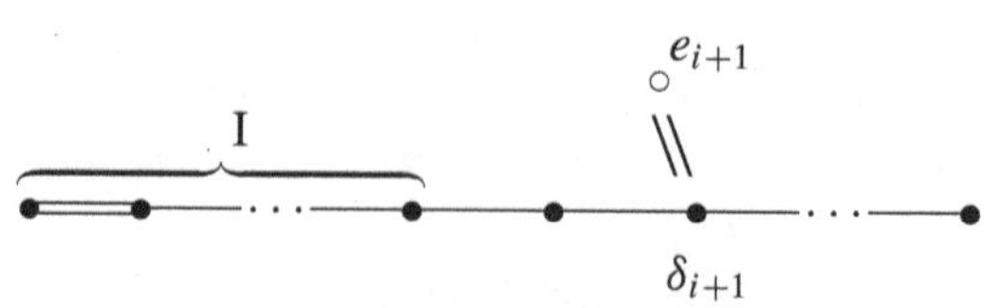

(iii) For type D_n, $\Delta_{\mathrm{D}} = \{\delta'_0, \delta_1, \ldots, \delta_{n-1}\}$ ($n \geq 4$), let $I = \{\delta'_0\} \cup \{\delta_1, \ldots, \delta_{i-1}\}$ ($i \geq 1$).

If $I = \{\delta'_0\}$, one finds $W^I = <s_1> \times <s'_3 s'_4, s_3, \ldots, s_{n-1}>$, which is of type $A_1 \times D_{n-2}$ with simple roots $\{\delta_1\} \cup \{e_4 - e_3, \delta_3, \delta_4, \ldots, \delta_{n-1}\}$. When $i \geq 2$, one finds $W^I = <s'_1 s'_{i+1}, s_{i+1}, \ldots, s_{n-1}>$ isomorphic with the Coxeter group of type BC_{n-i} through its action on the space generated by $\Delta^I = \{e_{i+1}, \delta_{i+1}, \ldots, \delta_{n-1}\}$.

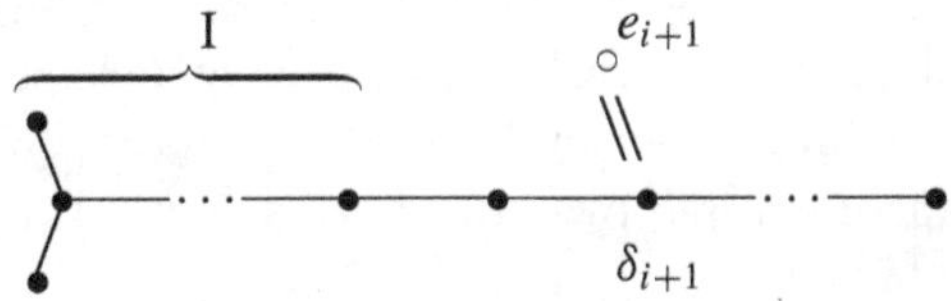

2.5. Other properties of split BN-pairs

In the following, G is a finite group with a strongly split BN-pair $(G, B = UT, N, S)$ (see Definition 2.20). We state (and prove) the following results for future reference.

Proposition 2.29. *Let $I \subseteq \Delta$. Then the following hold.*

(i) $N_G(U_I) = P_I$ and U_I is the largest normal p-subgroup of P_I.

(ii) If $g \in G$ is such that ${}^g U_I \subseteq U$, then $g \in P_I$ and ${}^g U_I = U_I$. If moreover ${}^g U_I = U_J$ for some $J \subseteq \Delta$, then $I = J$.

Proof. (i) $N_G(U_I)$ contains P_I, so $N_G(U_I)$ is a parabolic subgroup P_J with $J \supseteq I$. Assume $\delta \in \Delta \setminus I$ is such that $(U_I)^{s_\delta} = U_I$. Since $X_\delta \subseteq U_I$, we have $X_{-\delta} = (X_\delta)^{s_\delta} \subseteq U_I \subseteq U$, a contradiction. So $J = I$.

The second statement of (i) is in Proposition 2.25.

(ii) Write $g \in BwB$ (Bruhat decomposition). Since B normalizes U_I and U, one gets ${}^w U_I \subseteq U$. By Lemma 2.22, the inclusion $U_I \subseteq U^w$ implies that $\Phi_w \subseteq \Phi_I$. This implies $w \in W_I$ (use Proposition 2.3(i)) and therefore $g \in P_I$.

When $U_I = U_J$, the normalizer gives $P_I = P_J$ and therefore $I = J$. □

Lemma 2.30. *Assume that W is of irreducible type (i.e. there is no partition of Δ into two non-empty orthogonal subsets). Let I be a subset of Δ such that $C_G(U_I) \cap Bw_I B \neq \emptyset$. Then $I = \emptyset$ or Δ.*

Proof. Let us take $g \in C_G(U_I) \cap Bw_I B$. Then g can be written as

$$g = un_I u'$$

with $u, u' \in U$ and $n_I \in N$ such that $n_I T = w_I$. Denoting $X_I := U \cap U^{w_0 w_I} = <X_\alpha \mid \alpha \in \Phi_I^+>$, we have $U = X_I U_I$ (see Theorem 2.23(i) and Proposition 2.25). So we may assume $u \in X_I$.

Let us show that $w_I(\delta) = \delta$ for all $\delta \in \Delta \setminus I$. Assume $w_I(\delta) \neq \delta$. Then $w_I(\delta) \in \Phi^+ \setminus \{\delta\}$ and therefore $(X_\delta)^{n_I} \subseteq U_\delta$. Take $x \in X_\delta$, $x \neq 1$. We have $u \in X_I \subseteq U_\delta \lhd U$, so ${}^u x \in xU_\delta$. But ${}^u x \in {}^u U_I = U_I$, so it is centralized by $g = un_I u'$. We get ${}^u x = x^{n_I u'} \in (X_\delta)^{n_I u'} \subseteq (U_\delta)^{u'} = U_\delta$. This contradicts ${}^u x \in xU_\delta$ since $U_\delta \cap X_\delta = \{1\}$. So $w_I(\delta) = \delta$.

Suppose that $I \subset \Delta$ is a proper non-empty subset. Let $\delta \in \Delta \setminus I$ and $\delta' \in I$. Then $\langle \delta', \delta \rangle \leq 0$ since it is the scalar product of two elements of Δ. However, $\langle \delta', w_I(\delta) \rangle = \langle w_I(\delta'), \delta \rangle$ but $w_I(I) = -I$, so $-\langle w_I(\delta'), \delta \rangle \leq 0$. So we get $\Delta = I \cup \Delta \setminus I$, a partition into two orthogonal subsets. This contradicts the irreducibility of W. □

Theorem 2.31. *Assume that W is of irreducible type (i.e. there is no partition of Δ into two non-empty orthogonal subsets) with $|W| \neq 2$. Then $\mathrm{C}_G(U) = \mathrm{Z}(G)\mathrm{Z}(U)$.*

Proof. By Proposition 2.29(i), we have $\mathrm{N}_G(U) = B$, so $\mathrm{C}_G(U) \subseteq B$. Our statement reduces to checking that $\mathrm{C}_B(U) \subseteq \mathrm{Z}(G)\mathrm{Z}(U)$. Note that $\mathrm{C}_B(U)$ contains $\mathrm{Z}(U)$ as a central subgroup and that $\mathrm{C}_B(U)/\mathrm{Z}(U) = \mathrm{C}_B(U)/\mathrm{C}_U(U)$ injects in $B/U \cong T$, hence is a commutative p'-group. Then $\mathrm{C}_B(U) = \mathrm{Z}(U) \times A$ where A is a commutative p'-subgroup (take a Hall subgroup). Let ℓ be a prime $\neq p$. Then A_ℓ is a Sylow ℓ-subgroup of $\mathrm{C}_B(U)$, while T_ℓ is a Sylow ℓ-subgroup of B. By Sylow theorems, there is $b \in B$ such that ${}^b A_\ell \subseteq T$. But B normalizes $\mathrm{C}_B(U)$, so ${}^b A_\ell \subseteq T \cap \mathrm{C}_B(U) = \mathrm{C}_T(U)$. So, to get our theorem, it suffices to check

$$\mathrm{C}_T(U) \subseteq \mathrm{Z}(G).$$

Now take $\delta \in \Delta$. Denote $Z_\delta = \mathrm{C}_{L_\delta}(U_\delta)$. We know that X_δ normalizes U_δ, hence Z_δ, while U_δ centralizes Z_δ. Then $U = X_\delta U_\delta$ normalizes Z_δ.

Using Lemma 2.30 for $I = \{\delta\} \neq \Delta$, along with $P_\delta = B \cup Bs_\delta B$, we get $\mathrm{C}_{P_\delta}(U_\delta) \subseteq B$. Since P_δ normalizes $\mathrm{C}_{P_\delta}(U_\delta)$, we get $\mathrm{C}_{P_\delta}(U_\delta) \subseteq B \cap B^{s_\delta} = TU_\delta$. Hence $Z_\delta \subseteq TU_\delta \cap L_\delta \subseteq T$. Hence $Z_\delta = \mathrm{C}_T(U_\delta)$ is normalized by U, and also normalizes U. So Z_δ commutes with U since $U \cap T = \{1\}$. Then

$$Z_\delta = \mathrm{C}_T(U).$$

Since $Z_\delta = \mathrm{C}_{L_\delta}(U_\delta)$ is normalized by T and s_δ, this implies that $\mathrm{C}_T(U)$ is normalized by T and any s_δ for $\delta \in \Delta$. Then $\mathrm{C}_T(U)$ is normalized by N. But $\mathrm{C}_G(\mathrm{C}_T(U)) \supseteq TU = B$, hence is a parabolic subgroup: $\mathrm{C}_G(\mathrm{C}_T(U)) = P_I$ for some $I \subseteq \Delta$. By Proposition 2.29(i), $\mathrm{N}_G(P_I) = \mathrm{N}_G(U_I) = P_I$, so the fact that N normalizes $\mathrm{C}_T(U)$ (and therefore $\mathrm{C}_G(\mathrm{C}_T(U))$) implies that $I = \Delta$. Then $\mathrm{C}_T(U) \subseteq \mathrm{Z}(G)$. □

Exercises

1. Let $G = \mathrm{GL}_n(\mathbb{F})$, $B = UT$, $N = TW$, S be as in Example 2.17(i). Then (W, S) is a Coxeter system of type A_{n-1} with root system denoted by Φ and simple roots $\Delta_\mathrm{A} \subseteq \Phi$ (see Example 2.1(i)). When $\delta \in \Delta_\mathrm{A}$, one associates $s_\delta \in S$.
   (a) Define and describe B_δ and X_δ for $\delta \in \Delta_\mathrm{A}$ (see Definitions 2.18 and 2.20). Show that $B = (B \cap B^{s_\delta}).X_\delta$, a semi-direct product.
   (b) Describe ${}^wX_\delta$ for all $\delta \in \Delta_\mathrm{A}$ and $w \in W$. Show that it depends only on $w(\delta) \in \Phi$. Denote ${}^wX_\delta = X_{w(\delta)}$. Check that $X_\alpha \subseteq B$ when $\alpha \in \Phi^+$.
   (c) Show that, if $\delta \in \Delta_\mathrm{A}$ and $w \in W$ are such that $w(\delta) \in \Phi^+$, then $wBs_\delta \subseteq Bws_\delta B$ (use (a)).
   (d) Show that $G = B \cup Bs_\delta B$ when $n = 2$ and $\Delta_\mathrm{A} = \{\delta\}$. Deduce that $G = \langle B, N \rangle$ for all n.
   (e) Show that $(B = UT, N, S)$ is a strongly split BN-pair for $\mathrm{GL}_n(\mathbb{F})$. Describe the Levi decompositions $P_I = U_I.L_I$ of the standard parabolic subgroups for $I \subseteq S$.
2. (a) Let (W, S) be a Coxeter system with finite W. Let $\sigma\colon W \to W$ be a group automorphism such that $\sigma S = S$. Show that the group of fixed points W^σ is generated by the w_I for $I \subseteq S$ among the orbits of σ on S (use the induced action of σ on roots and apply Proposition 2.3).
   (b) Let $(G, B = UT, N, S)$ be a finite group with a strongly split BN-pair such that the extension $N \to N/T$ splits, i.e. there is a subgroup $W \subseteq N$ such that $N = T.W$ is a semi-direct product (so S is considered as a subset of N). Let $\sigma\colon G \to G$ be a group automorphism such that U, T and S are σ-stable.

   Show that the group of fixed points G^σ is endowed with the strongly split BN-pair $B^\sigma = U^\sigma.T^\sigma$, $N^\sigma = T^\sigma.W^\sigma$, $S_0 = \{w_I \mid I \in S/\langle\sigma\rangle\}$. *Hint*: prove and use a refined version of Bruhat decomposition where each $g \in G$ can be written uniquely as $g = unu'$ for $u \in U$, $n \in N$, $u' \in U \cap (U^{w_0})^n$.
   (c) Let $n \geq 1$. Let $\mathbb{F}$ be a field endowed with an involution $\lambda \mapsto \overline{\lambda}$ (possibly trivial). Let $G = \mathrm{GL}_n(\mathbb{F})$ and extend $\lambda \mapsto \overline{\lambda}$ to G. Let $w_0 \in G$ be the permutation matrix associated with $i \mapsto n + 1 - i$. Let $\sigma\colon G \to G$ be defined by $\sigma(g) = w_0.{}^\sigma\overline{g}^{-1}.w_0$. Use the above and the usual split BN-pair of G to show that G^σ has a split BN-pair of type $\mathrm{BC}_{[n/2]}$. Apply this to orthogonal groups for maximal Witt index and unitary groups (Example 2.17(ii)). Deduce also Example 2.17(iii).
   (d) Assume $n = 2m$ is even. Let $\varepsilon \in \mathrm{GL}_n(\mathbb{F})$ be the diagonal matrix with m first diagonal elements equal to 1, the others equal to -1. Let $J := w_0.\varepsilon$.

Use a slight adaptation of (c) above to check that the *symplectic group* $\mathrm{Sp}_{2m}(\mathbb{F})$, defined as the subgroup of matrices $g \in \mathrm{GL}_{2m}(\mathbb{F})$ satisfying ${}^t g.J.g = J$, has a split BN-pair of type BC_m.

(e) Show that $\mathrm{GU}_2(\mathbb{F}_{q^2})$ (see Example 2.17(ii)) is conjugated in $\mathrm{GL}_2(\mathbb{F}_{q^2})$ with $\mathrm{SL}_2(\mathbb{F}_q).Z$ where $Z \cong \mathbb{F}_{q^2}^\times$ is the group of matrices $\mathrm{diag}(a, \overline{a}^{-1})$ for $a \in \mathbb{F}_{q^2}^\times$. Show that $\mathrm{SO}_2^+(\mathbb{F}) \cong \mathbb{F}^\times$ (see Example 2.17(iii)) is the group of diagonal matrices $\mathrm{diag}(a, a^{-1})$ whenever $\mathbb{F}$ is a field of characteristic $\neq 2$.

3. Generalizing Theorem 2.23.

(a) Let E be a euclidean space endowed with a scalar product $\langle ., . \rangle$. If C and D are two closed convex cones of E such that $C \cap -C = D \cap -D = C \cap D = \{0\}$, show that there is $f \in E^\vee$ such that $f(C \setminus \{0\}) \subseteq]0, +\infty[$ and $f(D \setminus \{0\}) \subseteq]-\infty, 0[$.

(b) Let F be a finite subset of the unit sphere of E. Let $\mathcal{S}_{\frac{1}{2}}(F)$ denote the set of subsets $F' \subseteq F$ of the form $F' = \{x \in F \mid \langle x, a \rangle < 0\}$ for some $a \in E$ such that $F \cap a^\perp = \emptyset$. Let $\mathcal{S}(F)$ (*slices* of F) denote the set of intersections of any family of elements $\mathcal{S}_{\frac{1}{2}}(F)$.

Show that the slices of F are the sets of the form $F \cap C$ for C a closed convex cone of E such that $C \cap -C = \{0\}$. Show that this also coincides with the sets of the form $F \cap C$ for C an open convex cone of E such that $C \neq E$.

(c) We fix a set of simple roots $\Delta \subseteq \Phi$. If $A \subseteq W$ is a subset, denote $\Psi_A = \bigcap_{a \in A} a^{-1}(\Phi_\Delta^+)$. Show that $w \mapsto \Psi_{\{w\}}$ is a bijection between W and $\mathcal{S}_{\frac{1}{2}}(\Phi)$. Show that $\mathcal{S}(\Phi) = \{\Psi_A \mid A \subseteq W, A \neq \emptyset\}$.

(d) Let G be a finite group endowed with a split BN-pair of characteristic p and associated root system Φ. Generalize Theorem 2.23 to get an injective map

$$\Psi \mapsto U(\Psi) = <X_\alpha \mid \alpha \in \Psi>$$

from $\mathcal{S}(\Phi)$ into the set of p-subgroups of G.

Show that $U(\Psi) \cap U(\Psi') = U(\Psi \cap \Psi')$.

4. Let G be a group endowed with a BN-pair and finite W. Denote $\Phi \supseteq \Delta$ associated with the reflection representation of W.

Let $I \subseteq \Delta$ and let $J = -w_0(I)$. Show that $P_I \cap (P_J)^{w_0} = <B_{w_I}, N_I> = \bigcup_{w \in W_I} B_{w_I} w B_{w_I}$. Deduce that this group has a BN-pair (B_{w_I}, N_I). This allows L_I to be defined without assuming that the BN-pair of G is split. *Hint*: mimic the proof of Theorem 2.19(v); in particular, show that $B \cap (P_J)^{w_0} = B_{w_I}$.

5. Let G be a finite group endowed with a split BN-pair of characteristic p. We call the following hypothesis the *commutator formula*:

   (C) If $\alpha, \beta \in \Phi$ and $\alpha \neq \pm\beta$, then $[X_\alpha, X_\beta] \subseteq <X_\gamma \mid \gamma \in C_{\alpha,\beta}>$, where $C_{\alpha,\beta}$ is the set of roots $\gamma \in \Phi$ such that $\gamma = a\alpha + b\beta$ with $a > 0$ and $b > 0$.

   (a) Show that (C) implies that the BN-pair is strongly split.
   In the remainder of the exercise, we show a converse. So we assume that G has a strongly split BN-pair with associated root system Φ and let $\alpha \neq \pm\beta$ in Φ.
   (b) Show that $C_{\alpha,\beta} \cup \{\alpha, \beta\}$, $C_{\alpha,\beta} \cup \{\alpha\}$, $C_{\alpha,\beta} \cup \{\beta\}$, $C_{\alpha,\beta} \in \mathcal{S}(\Phi)$ (notation of Exercise 3(b)).
   (c) Show $<X_\alpha, X_\beta> \subseteq U(C_{\alpha,\beta} \cup \{\alpha, \beta\})$ (notation of Exercise 3(d)).
   If, moreover, $\alpha \in \Delta$, $\beta \in \Phi^+$, $\beta \neq \alpha$, show that $[X_\alpha, X_\beta] \subseteq U(C_{\alpha,\beta} \cup \{\beta\})$ (use $U_\alpha \lhd U$).
   (d) If $\alpha, \beta \in \Phi$ and $\alpha \neq \pm\beta$, show that there exists $w \in W$ such that $w(\alpha) \in \Delta$ and $w(\beta) \in \Phi^+$. *Hint*: Take w such that $w(\alpha) = \delta \in \Delta$, but if $w(\beta) \in \Phi^-$, try $w_0 s_\delta w$ instead of w.
   (e) Show $[X_\alpha, X_\beta] \subseteq U(C_{\alpha,\beta} \cup \{\beta\})$. Deduce (C).
   (f) Show that, if G is a finite group endowed with a split BN-pair of characteristic p defined by $B = UT$, N, S, then the following three conditions are equivalent:
   (A) the BN-pair is strongly split,
   (B) for any $s \in S$, $U \cap U^s \lhd U$,
   (C) the commutator formula is satisfied.
   Hint: note that in (c) we have used only the fact that $U_\delta \lhd U$ for any $\delta \in \Delta$.
   Show that, if any X_δ, $\delta \in \Delta$, has cardinality p, then the above conditions are satisfied.

6. Let G be a finite group with a split BN-pair given by $B = UT$, N, $T = T \cap B$, $W = N/T$ and associated root system $\Phi \supseteq \Delta$. Given $\delta \in \Delta$, denote by G_δ the group generated by X_δ and $X_{-\delta}$. Let $N_\delta = N \cap G_\delta$, $T_\delta = T \cap G_\delta$.
   (a) Show that $G = \bigcup_{n\in N} UnU$ (a disjoint union).
   Let $\delta \in \Delta$; show that there is $n_\delta \in N \cap X_\delta X_{-\delta} X_\delta$ such that $n_\delta T = s_\delta$ (see Proposition 6.3(i)). We assume n_δ has been chosen in that way for the remainder of the exercise.
   Check that G_δ has a split BN-pair $(X_\delta T_\delta, N_\delta, \{s_\delta\})$.

(b) Show that $[T, n_\delta] \subseteq T_\delta$ and that N_δ normalizes any subgroup of T containing T_δ. If $\delta_1, \delta_2, \ldots, \delta_k \in \Delta$ are such that $s_{\delta_1} \ldots s_{\delta_k} = 1$, show that $n_{\delta_1} \ldots n_{\delta_k} \in T_{\delta_1} \ldots T_{\delta_k}$. *Hint*: assume $\{\delta_1 \ldots \delta_k\} = \Delta$ and define a map $s_\delta \mapsto n_\delta T_{\delta_1} \ldots T_{\delta_k}$, checking that $(n_\delta n_{\delta'})^m \in T_\delta \cap T_{\delta'}$ when $\delta, \delta' \in \Delta$ and m is the order of $s_\delta s_{\delta'}$.

We want to show that there is a bijection $P' \mapsto (I, T')$ between the subgroups $P' \subseteq G$ containing U and the pairs (I, T') where $I \subseteq \Delta$ and T' is subgroup of T containing $<T_\delta \mid \delta \in I>$.

(c) If (I, T') is as above, define N' as the group generated by T' and the n_δ's for $\delta \in I$. Show that $P' := UN'U$ is a subgroup. Show that $P'T = P_I$ and $T' = N' \cap T$, $N' = P' \cap N$.

(d) Let P' be a subgroup of G containing U. Show that $P'T$ is a subgroup. Denote $I \subseteq \Delta$ such that $P'T = P_I$. Show that $G_\delta \subseteq P'$ for all $\delta \in I$. Show that $P' \cap N$ is generated by $P' \cap T$ and the n_δ's such that $\delta \in I$.

(e) Check the bijection stated above.

7. Show that — is transitive in the system $\mathcal{L}$ of Definition 2.24.

Notes

The notions of BN-pairs is due to Tits and originates in the simplicity proofs common to reductive groups and their finite analogues; see [Tits], Ch55. It also applies to p-adic groups. BN-pairs of rank 3 and irreducible W were classified by Tits (see [Tits]). He also introduces a more geometrical object, the "building" (see also [Brown], [Asch86] §43).

For finite groups, it has been proved that BN-pairs with an irreducible W of rank 2 are split and correspond to finite analogues of reductive groups [FoSe73]. Finite BN-pairs of rank 1 were classified by Suzuki and Hering–Kantor–Seitz (see [Pe00] for an improved treatment and the references). This is an important step in the classification of finite simple groups (see [Asch86], and the survey by Solomon [So95]).

The study of intersections $(P, V) \cap\!\downarrow (P', V')$ in finite reductive groups somehow started with Lemma 1 in Harish-Chandra's paper [HaCh70] (expanded by Springer [Sp70]). We follow the approach of Howlett [How80]. We have also used [Stein68a], [Ri69], and [FoSe73]. Exercise 5 is due to Genet [Gen02].

3

Modular Hecke algebras for finite BN-pairs

We now study the Hecke algebras introduced in Chapter 1, in the case of BN-pairs. Let G be a finite group with a (strongly) split BN-pair of characterisitic p. It is defined by subgroups B, N, $W := N/(B \cap N)$, $S \subseteq W$, giving rise to parabolic subgroups $B \subseteq P_I = U_I \rtimes L_I \subseteq G$ for each $I \subseteq S$ (see Chapter 2).

Let $\mathcal{H}(G, B)$ be the endomorphism algebra of the permutation representation on G-conjugates of B (over $\mathbb{Z}$ or any commutative ring). One finds for $\mathcal{H}$ the well-known law defined by generators $(a_w)_{w \in W}$ and relations

$$a_w a_{w'} = a_{ww'} \quad \text{if } l(ww') = l(w) + l(w')$$

and

$$(a_s)^2 = q_s a_s + (q_s - 1)$$

for $w, w' \in W$, $s \in S$, where $q_s = |B : B \cap B^s|$ is a power of p.

Let k be an algebraically closed field of characteristic $\neq p$. Let M be a simple cuspidal kL_I-module. We study the Hecke algebra $\mathrm{End}_{kG}(\mathrm{Ind}_{P_I}^G M)$ and find that the generators defined in Chapter 1 give rise to a presentation related to the above. The main difference is that W is replaced by a subgroup $W(I, M)$ which is not generated by a subset of S. See §19.4 and §20.2 for more precise descriptions.

A first result on the law of the Hecke algebra tells us that our generators are invertible. This invertibility implies that the "independence" theorem of Chapter 1 holds in these groups. As we have seen in Chapter 1, it may be important to consider the endomorphism ring not just of such an induced module but of a sum of induced cuspidal modules.

3.1. Hecke algebras in transversal characteristics

In the following sections, $(G, B = UT, N, S)$ is a strongly split BN-pair of characteristic p (see Definition 2.20).

Let k be a field of characteristic $\ell \neq p$. We apply certain notions and results introduced in Chapter 1 with $\mathcal{L}$ the $\cap\downarrow$-stable, k-regular set of pairs (P, V) introduced in Definition 2.24.

In this section we fix $I \subseteq \Delta$, and P_I, L_I, U_I the corresponding subgroups of G. Let M be a simple cuspidal kL_I-module.

If $n \in N$ is of class $w \in W$ mod. T and $wI \subseteq \Delta$, then ${}^nL_I = L_{wI}$ and ${}^nM$ is a cuspidal kL_{wI}-module. It can also be considered as a cuspidal $k(P_{wI}/U_{wI})$-module. Using the notation of §1.3, for $\tau := (P_I, U_I, M)$ and $\tau' = (P_{wI}, U_{wI}, {}^nM)$, one has $\tau \overset{n^{-1}}{—} \tau'$ and one may clearly take $\theta_{n^{-1},\tau,\tau'} = \mathrm{Id}_M$ (where ${}^nM$ is defined as the same underlying space as M with an action of ${}^nL_I$ which is that of L_I composed with conjugacy by n). One may now put forward the following.

Definition 3.1. *Let $W(I, M)$ be the subgroup of elements $w \in W$ such that $wI = I$ and any representative $n \in N$ of w satisfies $M \cong {}^nM$ as kL_I-module.*

Let $N_{I,\Delta}$ be the set of elements $n \in N$ such that their class w mod. T satisfies $wI \subseteq \Delta$. For such an n, define $b_{n,M} \in \mathrm{Hom}_{kG}(\mathrm{Ind}_{P_I}^G M, \mathrm{Ind}_{P_{wI}}^G {}^nM)$ as $b_{n,M} = a_{n^{-1},\tau,\tau'}$ with $\theta_{n^{-1},\tau,\tau'} = \mathrm{Id}_M$ in the notation of §1.3.

Let $\mathrm{ind}(w) = \prod_{\alpha\in\Phi_w} |X_\alpha| = |U : U \cap U^w| = |U \cap U^{w_0w}|$ (see Theorem 2.23(i)).

Lemma 3.2. *Assume $n'n$ and $n \in N_{I,\Delta}$, and $l(w'w) = l(w') + l(w)$. Then*

$$e(U_I)e(U_{wI})^w e(U_{w'wI})^{w'w} = e(U_I)e(U_{w'wI})^{w'w}$$

in $\mathbb{Z}[p^{-1}]G$.

Proof. Any $x \in U_I.{U_{w'wI}}^{w'w}$ satisfies $e(U_I)xe(U_{w'wI})^{w'w} = e(U_I)e(U_{w'wI})^{w'w}$. So it suffices to check that ${U_{wI}}^w \subseteq U_I.{U_{w'wI}}^{w'w}$ or equivalently $U_{wI} \subseteq {}^wU_I.{U_{w'wI}}^{w'}$. By Theorem 2.21(ii) and using $\Phi_{wI}^+ = w(\Phi_I^+) = w'^{-1}(\Phi_{w'wI}^+)$, it suffices to check that $\Phi^+ \setminus \Phi_{wI}^+ \subseteq (w\Phi^+ \setminus \Phi_{wI}^+) \cup (w'^{-1}\Phi^+ \setminus \Phi_{wI}^+)$. This is a consequence of the additivity of lengths (see Proposition 2.3(iii)). □

Theorem 3.3. *Let $R := \mathbb{Z}[p^{-1}]$. We let the same letter R denote the trivial RB-module.*

The R-algebra $\mathrm{End}_{RG}(\mathrm{Ind}_B^G R)$ has a presentation by generators $(a_s)_{s\in S}$ obeying the following relations for any $s, s' \in S$:

$(a_s)^2 = (\mathrm{ind}(s) - 1)a_s + \mathrm{ind}(s)$ *(quadratic relation), and*

$a_s a_{s'} \ldots = a_{s'} a_s \ldots$ *with* $|ss'|$ *terms on each side (braid relations).*

Another presentation is with generators $(a_w)_{w \in W}$ *subjected to the quadratic relations for* $w \in S$ *and the relations* $a_w a_{w'} = a_{ww'}$ *whenever* $w, w' \in W$ *satisfy* $l(ww') = l(w) + l(w')$.

In this second presentation, one has $\mathrm{End}_{RG}(\mathrm{Ind}_B^G R) = \bigoplus_{w \in W} R a_w$ *and no* a_w *is zero.*

Proof. It is clear that $\tau = (B, U, R)$ is a cuspidal triple satisfying Condition 1.17(b). In the notation of Definition 1.12, one may take $\theta_{g,\tau,\tau} = \mathrm{Id}_R$ and define $a_w \in \mathrm{End}_{RG}(\mathrm{Ind}_B^G R)$ for $w \in G$ by $a_w(1 \otimes 1) = \mathrm{ind}(w) e(U) w^{-1} \otimes 1$ (see Proposition 1.13). Theorem 1.20(i) implies that $\mathrm{End}_{RG}(\mathrm{Ind}_B^G R) = \bigoplus_{w \in W} R a_w$. Let us show that we have all the relations of the theorem.

Take $w, w' \in W$, and assume $l(ww') = l(w) + l(w')$. Then $a_w a_{w'}(1 \otimes 1) = \mathrm{ind}(w) a_w (e(U) w'^{-1} \otimes 1) = \mathrm{ind}(w) e(U) w'^{-1} a_w(1 \otimes 1) = \mathrm{ind}(w)\mathrm{ind}(w') e(U) w'^{-1} e(U) w^{-1} \otimes 1 = y \otimes 1$ where $y = \mathrm{ind}(ww') e(U) w'^{-1} e(U) w^{-1} e(U)$. By Lemma 3.2 with $I = \emptyset$, we have $y = \mathrm{ind}(ww') e(U)(ww')^{-1} e(U)$, so $a_w a_{w'}(1 \otimes 1) = \mathrm{ind}(ww') e(U)(ww')^{-1} e(U) \otimes 1 = \mathrm{ind}(ww') e(U)(ww')^{-1} \otimes 1 = a_{ww'}(1 \otimes 1)$. This implies $a_w a_{w'} = a_{ww'}$.

The braid relations are a special case since $ss' \ldots = s's \ldots$ ($|ss'|$ terms) are reduced expressions.

Take $s \in S$ corresponding to $\delta \in \Delta$. One has $U = X_\delta U_\delta$ (Theorem 2.23(i)) and s normalizes U_δ (Proposition 2.25), so $a_s(1 \otimes 1) = \mathrm{ind}(s) e(X_\delta) s \otimes 1 = \mathrm{ind}(s) s e(X_{-\delta}) \otimes 1$. Then $(a_s)^2(1 \otimes 1) = \mathrm{ind}(s) \sum_{x \in X_{-\delta}} e(X_\delta) x \otimes 1$. The summand for $x = 1$ gives $1 \otimes 1$. For other x, we argue in the split BN-pair L_δ. The Bruhat decomposition in this group gives $L_\delta = T X_\delta \cup X_\delta s T X_\delta \subseteq B \cup X_\delta s B$. Moreover $X_{-\delta} \cap B = \{1\}$ since $B \cap B^{w_0} = T$. So $X_{-\delta} \setminus \{1\} \subseteq X_\delta s B$. Then the summands for $x \neq 1$ give $\mathrm{ind}(s) - 1$ times $e(X_\delta) s \otimes 1$. Then $(a_s)^2 = (\mathrm{ind}(s) - 1) a_s + \mathrm{ind}(s)$ since they coincide on $1 \otimes 1$.

Let E_1 (resp. E_2) be the first (resp. the second) algebra defined by generators and relations in the theorem. With our notation, the evident map gives a surjection $E_2 \to \mathrm{End}_{RG}(\mathrm{Ind}_B^G R)$ since we have checked the defining relations. There exists a surjective morphism $E_1 \to E_2$, since, by the "Word Lemma" (see [Bour68] IV.1 Proposition 5), the expression $a_{s_1} \ldots a_{s_m} \in E_1$ when $s_1 \ldots s_m$ is a reduced expression, depends only on $s_1 \ldots s_m \in W$. We therefore have two surjections

$$E_1 \to E_2 \to \mathrm{End}_{RG}\big(\mathrm{Ind}_B^G R\big).$$

Since R is principal, it suffices to check now that E_1 is generated by $|W|$ elements as an R-module. We show that the elements of type $a_{s_1} \ldots a_{s_m} \in E_1$

with $l(s_1 \ldots s_m) = m$ generate E_1 by verifying that the module they generate is stable under right multiplication by the a_s's. If $l(s_1 \ldots s_m s) = m + 1$, the checking is trivial. Otherwise, the "Word Lemma" argument (see above) allows us to assume $s_m = s$. Then the quadratic relation gives $a_{s_1} \ldots a_{s_m} a_s = (\mathrm{ind}(s) - 1)a_{s_1} \ldots a_{s_m} + \mathrm{ind}(s)a_{s_1} \ldots a_{s_{m-1}}$, thus our claim. □

The above theorem makes natural the following definition of a Hecke algebra associated with a Coxeter system (W, S) and parameters q_s $(s \in S)$.

Definition 3.4. *Let R be any commutative ring.*

If (W, S) is a Coxeter system such that W is finite and if $(q_s)_{s \in S}$ is a family of elements of R such that $q_s = q_t$ whenever s and t are W-conjugate, one defines $\mathcal{H}_R((W, S), (q_s))$ as the R-algebra with generators a_s $(s \in S)$ obeying the relations

$$(a_s + 1)(a_s - q_s) = 0$$

and

$$a_s a_t a_s \ldots = a_t a_s a_t \ldots$$

($|st|$ terms on each side) for all $s, t \in S$.

If G is a finite group with a BN-pair defined by $B, N, S \subseteq W := N/(B \cap N)$, one defines $\mathcal{H}_R(G, B) := \mathcal{H}_R((W, S), (q_s))$ for $q_s = |B : B \cap B^s|$.

Remark 3.5. We will sometimes use abbreviations such as $\mathcal{H}_\Lambda(W, (q_s))$, for $\mathcal{H}_R((W, S), (q_s))$.

Using classical arguments, some similar to the proof of Theorem 3.3, one proves that $\mathcal{H}_R((W, S), (q_s))$ is R-free with basis a_w $(w \in W)$ satisfying $a_w a_{w'} = a_{ww'}$ when $l(ww') = l(w) + l(w')$ (see [Bour68] p. 55, [GePf00] 4.4.6, [Hum90] §7). This, along with the relations given in the definition about the a_s $(s \in S)$, may serve as another presentation.

Many properties follow, such as that $\mathcal{H}_R((W, S), (q_s)) \otimes_R R' \cong \mathcal{H}_{R'}((W, S), (q_s))$ and $\mathcal{H}_R((W_I, I), (q_s)) \subseteq \mathcal{H}_R((W, S), (q_s))$ when R' is a commutative R-algebra and I is a subset of S.

Those results will be used mainly in Chapters 18–20.

In the general case of the endomorphism ring of a kG-module induced from a cuspidal simple module of a Levi subgroup, we shall try to obtain similar presentations. We first prove a series of propositions about the composition of the $b_{n', {}^n M}$'s (see Definition 3.1). When $n, n', n_1, \ldots \in N$, their classes mod. T are denoted by $w, w', w_1, \ldots \in W$.

Definition 3.6. (see Definition 1.12) *If n is such that $w \in W(I, M)$, then choose $\theta_n \in \mathrm{End}_k M$ such that it induces an isomorphism of kL_I-modules ${}^nM \to M$ (i.e. $\theta_n(pm) = ({}^np)m$ for all $p \in L_I$). Let λ be the associated cocycle on the subgroup of N corresponding to $W(I, M)$, i.e. $\theta_n\theta_{n'} = \lambda(n, n')\theta_{nn'}$. Denote by $\widetilde{\theta_n} = \mathrm{Id} \otimes \theta_n$ the associated morphism $kGe(U_I) \otimes_{L_I} {}^nM \to kGe(U_I) \otimes_{L_I} M$. If $w'I \subseteq \Delta$, let ${}^{n'}\widetilde{\theta_n}$ be the map $1 \otimes_{kL_{w'I}} \theta_n : kGe(U_{w'I}) \otimes_{kL_{w'I}} {}^{n'n}M \to kGe(U_{w'I}) \otimes_{kL_{w'I}} {}^{n'}M$.*

Proposition 3.7. *(i) With the notation above, one may choose θ_n so that ${}^{n'}\widetilde{\theta_{nn'}} = {}^{n'}\widetilde{\theta_n}$ and $b_{n',M} \circ \widetilde{\theta_n} = {}^{n'}\widetilde{\theta_n} \circ b_{n',{}^nM}$.*

(ii) If $[W(I, M)]$ is a representative system of $W(I, M)$ in N, then $(\widetilde{\theta_n} \circ b_{n,M})_{n \in [W(I,M)]}$ is a k-basis for $\mathrm{End}_{kG}\mathrm{Ind}_{P_I}^G M$.

Proof. (i) The first equality is clear from the definition. The second follows from the definition of $b_{n,M} = a_{n^{-1},\tau,\tau'} : kGe(U_I) \otimes_{L_I} M \to kGe(U_{w'I})^{n'} \otimes_{L_I} M$ as $\mu \otimes \mathrm{Id}_M$ where μ is the right multiplication by $e(U_{w'I})^{n'}$ (Proposition 1.13).

(ii) See Theorem 1.20(i) and Theorem 2.27(iv). □

Proposition 3.8. *$\frac{\mathrm{ind}(w)\mathrm{ind}(w')}{\mathrm{ind}(ww')}$ is a power of p^2. It is 1 when $l(ww') = l(w) + l(w')$.*

Proof. Easy by Proposition 2.3 and induction on $l(w')$. □

Proposition 3.9. *(i) Assume $n'n$ and $n \in N_{I,\Delta}$, and $l(w'w) = l(w') + l(w)$. Then $b_{n',{}^nM}b_{n,M} = b_{n'n,M}$.*

(ii) If $\delta \in \Delta \setminus I$ and $n \in N$ is of class $w = v(\delta, I)$ (see Notation 2.8), then, denoting $\widetilde{M} := \mathrm{Ind}_{P_I}^G M$,

$$b_{n^{-1},{}^nM}b_{n,M} = \begin{cases} \mathrm{ind}(w)^{-1}\mathrm{Id}_{\widetilde{M}}, & \text{if } w \notin W(I, M); \\ \mathrm{ind}(w)^{-1}\mathrm{Id}_{\widetilde{M}} + \beta\widetilde{\theta_n} \circ b_{n,M} \text{ (with } \beta \in k), & \text{if } w \in W(I, M). \end{cases}$$

(iii) $b_{n,M}$ is an isomorphism for every $n \in N_{I,\Delta}$.

Proof. (i) By Definition 3.4, $b_{n,M} = a_{n^{-1},\tau,\tau'}$ with $\theta_{n^{-1},\tau,\tau'} = \mathrm{Id}_M$. Then Proposition 1.13 tells us that this identifies with the morphism $kGe(U_I) \otimes_{L_I} M \to kGe(U_{wI})^w \otimes_{L_I} M$ obtained by multiplying the left-hand side by $e(U_{wI})^w$ on the right. Now $b_{n',{}^nM}b_{n,M}$ consists in multiplying by $e(U_{wI})^we(U_{ww'I})^{ww'}$. This is the same as multiplying by $e(U_{ww'I})^{ww'}$ as a result of Lemma 3.2. Hence our claim.

(ii) Let $J = \{\delta\} \cup I$. The spaces $kG \otimes_{kP_I} M$ and $kG \otimes_{kP_{wI}} {}^nM$ have subspaces $kP_J \otimes_{kP_I} M$ and $kP_J \otimes_{kP_{wI}} {}^nM$ respectively. It is clear from the definition of these maps that $b_{n,M}$ sends the first into the second and $b_{n^{-1},{}^nM}$ the other way around. These are clearly kP_J-linear, so that $b_{n^{-1},{}^nM}b_{n,M} \in \mathrm{End}_{kP_J}(kP_J \otimes_{kP_I} M)$. Now, since U_J acts trivially on $kP_J \otimes_{kP_I} M$, this induced module can be considered as a kL_J-module induced from the cuspidal simple kL_I-module M. So, by Proposition 3.7(ii), $\mathrm{End}_{kP_J}(kP_J \otimes_{kP_I} M)$ has a basis indexed by $W(I, M) \cap W_J$ and consisting of the restrictions of the $b_{n',M}$'s such that $w'I = I$ and $M \cong {}^{n'}M$ (Proposition 1.23). By Theorem 2.11 (last statement), this group is $W(I, M) \cap \{1, w\}$. This gives the dichotomy of the Proposition, while the coefficient on $\mathrm{Id}_{\widetilde{M}}$ is given by Proposition 1.18(iii) with $\lambda = 1$ (recall that $\theta_{n^{-1},\tau,\tau'} = \mathrm{Id}_M$ and that the linear form of Proposition 1.18 gives the component on Id in the basis of Proposition 3.7(ii) since $\widetilde{\theta_n} \circ b_{n,M}(1 \otimes M) \subseteq kP_I nP_I \otimes M$). We find a coefficient whose inverse is $|U_I : (U_{wI})^n \cap U_I|$. This is $\mathrm{ind}(w) = |U : U^n \cap U|$ since $U^n \cap U \supseteq X_\alpha$ for each $\alpha \in \Phi_I^+$ and $U^n \cap U_I \subseteq (U_{wI})^n$.

(iii) The equality in (ii) reads $b' \circ b_{n,M} = \mathrm{Id}_M$ for some map b'. Then $b_{n,M}$ is injective, hence an isomorphism since $\mathrm{Ind}_{(P_I,U_I)}^G M$ and $\mathrm{Ind}_{(P_{wI},U_{wI})}^G {}^nM$ have the same dimension. This applies to n of type $v(\delta, I)$. For arbitrary $n \in N_{I,\Delta}$, one may use a decomposition of w as in Theorem 2.11. Then (i) turns this into a decomposition of $b_{n,M}$ as a product of isomorphisms of the type above. □

Theorem 3.10. *Let $(G, B = UT, N, S)$ be a strongly split BN-pair of characteristic p (see Definition 2.20).*

Let $R = \mathbb{Z}[p^{-1}]$. Assume (P, V)—(P', V') in $\mathcal{L}$ (see Theorem 2.27(ii)), then

$$RGe(V) \to RGe(V'),$$
$$x \mapsto xe(V')$$

is an isomorphism of G-$(P \cap P')$-bimodules.

Proof. The map $x \mapsto xe(V')$ from $RGe(V)$ to $RGe(V')$ is an R-linear map between two R-free modules of the same rank. R being a principal ideal domain, it suffices to prove that this map has no invariant divisible by ℓ, for every prime $\ell \neq p$. This is equivalent to proving that $x \mapsto xe(V')$ from $kGe(V)$ to $kGe(V')$ is an isomorphism for every algebraically closed field k of characteristic $\ell \neq p$ and every relation (P, V)—(P', V'). By Theorem 1.14, it suffices to check that $a_{1,\tau,\tau'}$ is an isomorphism for each relation τ—τ' in $\mathbf{cusp}_k(\mathcal{L})$. By Theorem 2.27(iii), one may assume $\tau = (P_I, U_I, M)$, $\tau' = ((P_{wI})^n, (U_{wI})^n, M)$ for I, $wI \subseteq \Delta$ where M is considered as kL_I-module. Then $a_{1,\tau,\tau'}$ differs from $b_{n,M}$ by an isomorphism as a result of Proposition 1.16(ii). By the above Proposition 3.9(iii), each $b_{n,M}$ is an isomorphism. This completes our proof. □

Notation 3.11. Keep $(G, B = UT, N, S)$ as a strongly split BN-pair of characteristic p (see Definition 2.20). Let L be a Levi subgroup of G. If Λ is a commutative ring where p is invertible, we denote by

$$\mathrm{R}_L^G : \Lambda L\mathbf{-mod} \to \Lambda G\mathbf{-mod}$$

and

$$^*\mathrm{R}_L^G : \Lambda G\mathbf{-mod} \to \Lambda L\mathbf{-mod}$$

the adjoint functors $\mathrm{Ind}_{(P,V)}^G$ and $\mathrm{Res}_{(P,V)}^G$, respectively, where $P = LV$ is a Levi decomposition. By the above theorem, it is not necessary to mention P and V.

3.2. Quotient root system and a presentation of the Hecke algebra

Let us recall some properties of root systems (see [Bour68] §IV, [Stein68a] Appendix, [Hum90] §1).

Proposition 3.12. *Let Φ be a finite subset of the unit sphere of a real euclidean space E. Let $W(\Phi)$ be the subgroup of the orthogonal group generated by the reflections through elements of Φ.* **Assume** $w\Phi = \Phi$ **for all** $w \in W(\Phi)$ *(Φ is then called a "root system"). Then,*

(i) for any positive cone C such that $C \cap -C = \emptyset$ and $\Phi \subseteq C \cup -C$, there is a unique linearly independent subset $\Delta \subseteq C \cap \Phi$ such that $\Phi \cap C$ is Φ_Δ^+, i.e. the set of elements of Φ which are combinations with coefficients all ≥ 0 of elements of Δ (such a set $\Phi \cap C$ is called a "positive system", and Δ is called a "set of simple roots" of Φ). Such C (and Δ) exist.

(ii) If Δ' is another set of simple roots of Φ, then there is $w \in W(\Phi)$ such that $\Delta' = w\Delta$.

(iii) If S is the set of reflections through elements of Δ, then $(W(\Phi), S)$ is a Coxeter system, and the length of an element w is the cardinality of $\Phi_\Delta^+ \setminus w^{-1}(\Phi_\Delta^-)$.

(iv) If I is a subset of Δ, the subgroup of $W(\Phi)$ generated by the reflections corresponding to elements of I equals $\{w \in W(\Phi) \mid (w-1)(I^\perp) = 0\}$.

Definition 3.13. *We take G, $\Phi \supseteq \Delta \supseteq I$, M a cuspidal kL_I-module as in §2.5 and §3.1. If $\alpha \in \Phi \setminus I$, we say that "$v(\alpha, I)$ is defined" if and only if there exists $w \in W$ such that $I \cup \{\alpha\} \subseteq w^{-1}\Delta$. We then write $v(\alpha, I) = v(w\alpha, wI)^w$. Let $\Omega(I, M)$ be the set of $\alpha \in \Phi \setminus I$ such that $v(\alpha, I)$ is defined, belongs to $W(I, M)$ and is an involution. Let $R(I, M)$ be the group generated by the $v(\alpha, I)$*

such that $\alpha \in \Omega(I, M)$. *Let* $C(I, M) = \{w \in W(I, M) \mid w(\Omega(I, M) \cap \Phi^+) = \Omega(I, M) \cap \Phi^+\}$.

Remark. The definition of $v(\alpha, I)$ above is clearly independent of w chosen such that $I \cup \{\alpha\} \subseteq w^{-1}\Delta$ since, if $J \subseteq \Delta$ and $w'J \subseteq \Delta$, then $w_J = (w_{w'J})^{w'}$. Note also that $v(\alpha, I)^2 = 1$ is equivalent to $v(\alpha, I)(I) = I$.

Proposition 3.14. *The group* $W(I, M)$ *stabilizes* $\mathbb{R}I$ *and acts faithfully on* $(\mathbb{R}I)^\perp$. *Let us identify* $W(I, M) \supseteq R(I, M)$, $C(I, M)$ *with subgroups of* $GL_{\mathbb{R}}(I^\perp)$. *Let* $\Omega' \subseteq I^\perp$ *be the orthogonal projection of* $\Omega(I, M)$. *Let* $\bar{\Omega}'$ *(resp.* $\bar{\Omega}'^+$*) be the set of quotients of elements of* Ω' *(resp. the orthogonal projections of* $\Phi^+ \cap \Omega(I, M)$*) by their norms.*

(i) $\bar{\Omega}'$ *is a root system in* $I^\perp$ *with positive system* $\bar{\Omega}'^+$.

Denote by $\Delta(I, M)$ *the associated set of simple roots.*

(ii) The image of $R(I, M)$ *is the Weyl group of the root system* $\bar{\Omega}'$.

(iii) $W(I, M) = R(I, M) \rtimes C(I, M)$.

Proof. The elements of $\Omega(I, M)$ are outside $\mathbb{R}I$ by definition, so $\bar{\Omega}'$ makes sense.

The group $W(I, M)$ stabilizes I so it stabilizes $I^\perp$. The kernel of the action of $W(I, M)$ on $I^\perp$ is $W_I \cap W(I, M)$, by Proposition 3.12(iv). One has $W_I \cap W(I, M) = \{1\}$ since a non-trivial element of W_I must send some element of I to a negative root (use Proposition 2.3(ii)).

Take $\alpha \in \Omega(I, M)$, $\alpha' \in \Omega'$ its projection on $I^\perp$. Let us show

(ii′) $v(\alpha, I)$ acts on $I^\perp$ by the reflection through α'.

The fact that $v(\alpha, I)(I) = I$ allows us to write $v(\alpha, I) = w'w_I$ where $w'I = -I$, $w'(\alpha) = -\alpha$ and w' fixes all the elements of $(I \cup \alpha)^\perp$ (assume I and α are in Δ). Then $v(\alpha, I)(\alpha) = -w_I(\alpha) \in -\alpha + \mathbb{R}I$, so $v(\alpha, I)$ acts on $I^\perp$ as the reflection through α'.

Now (ii′) implies that the image of $R(I, M)$ in the orthogonal group of $I^\perp$ is the group generated by the reflections associated with elements of Ω'. Moreover Ω' is stable under $R(I, M)$, so $\bar{\Omega}'$ satisfies the hypothesis of Proposition 3.12 in $I^\perp$, with associated $W(\bar{\Omega}')$ the restriction of $R(I, M)$ to $I^\perp$. Thus (ii) is proved.

The cone C generated by $\Phi^+ \cap \Omega(I, M)$ clearly satisfies $C \cap -C = \emptyset$ and $\Omega(I, M) \subseteq C \cup -C$ by the properties of Φ^+ itself. Then the normalized projections satisfy the same since $C \cap \mathbb{R}I = \emptyset$. Thus (i) is proved.

(iii) The whole group $W(I, M)$ acts faithfully on $\bar{\Omega}'$, so, by the transitivity of the Weyl group of $\bar{\Omega}'$ on its sets of simple roots (hence on its positive systems), one has $W(I, M) = R(I, M) \rtimes C$ where C is the stabilizer of $\bar{\Omega}'^+$. This stabilizer is $C(I, M)$ since, if an element of $\Omega(I, M) \cap \Phi^+$ is sent to Φ^-

by $w \in W(I, M)$, then its image in $\bar{\Omega}'^+$ is sent into $-\bar{\Omega}'^+$ since w stabilizes $\Omega(I, M)$. □

In the following, $(G, B = UT, N, S)$, k, $I \subseteq \Delta$, P_I, L_I, U_I, M are as in §2.5.

Proposition 3.15. *Let $n, n' \in N$ be such that $nn', n' \in N_{I,\Delta}$, and their classes mod. T satisfy $\Phi_w \cap \Phi_{w'^{-1}} \cap \Omega(w'I, {}^{n'}M) = \emptyset$. Then*

$$\left(\frac{\mathrm{ind}(w)\mathrm{ind}(w')}{\mathrm{ind}(ww')}\right)^{\frac{1}{2}} .b_{n,{}^{n'}M} b_{n',M} = b_{nn',M}$$

(where the quotient $\mathrm{ind}(w)\mathrm{ind}(w')/\mathrm{ind}(ww')$ *is a power of p^2).*

Proof. The proof is by induction on $l(w)$. If $w = 1$, it is clear. Otherwise, by Theorem 2.11, there is a decomposition $w = w_1 w_2$ with lengths adding and $w_1 = v(\delta, J)$ for $J = w_2 w'(I) \subseteq \Delta$ and $\delta \in \Delta \setminus J$. Let $n = n_1 n_2$ be a corresponding decomposition in N. Then, by Proposition 3.9(i), $b_{n,{}^{n'}M} = b_{n_1,{}^{n_2n'}M} b_{n_2,{}^{n'}M}$. The induction hypothesis applies to (n_2, n') replacing (n, n') since $\Phi_{w_2} \subseteq \Phi_w$ by Proposition 2.3(iii). Therefore

$$(1) \qquad b_{n,{}^{n'}M} b_{n',M} = \left(\frac{\mathrm{ind}(w_2 w')}{\mathrm{ind}(w_2)\mathrm{ind}(w')}\right)^{\frac{1}{2}} b_{n_1,{}^{n_2n'}M} b_{n_2n',M}.$$

If $l(w_1 w_2 w') = l(w_1) + l(w_2 w')$, one has

$$(2) \qquad b_{n_1,{}^{n_2n'}M} b_{n_2n',M} = b_{nn',M}$$

by Proposition 3.9(i). One then gets the present proposition by combining (1) and (2) since $\mathrm{ind}(ww') = \mathrm{ind}(w_1)\mathrm{ind}(w_2w')$ by the additivity of lengths.

If $l(w_1 w_2 w') \neq l(w_1) + l(w_2 w')$, then $l(w_1 w_2 w') = -l(w_1) + l(w_2 w')$ and $(w_2 w')^{-1}(\delta) \in \Phi^-$ by Proposition 2.10(ii). Proposition 3.9(i) gives

$$(3) \qquad b_{n_2n',M} = b_{n_1^{-1},{}^{n_1n_2n'}M} b_{n_1n_2n',M}.$$

Denote $\alpha = w_2^{-1}(\delta)$. One has $\alpha \in \Phi^+$ by the additivity in $v(\delta, J)w_2$ and Proposition 2.10(ii). Then $\alpha \in \Phi_{w'^{-1}}$. Also $\alpha \in w_2{}^{-1}\Phi_{w_1} \subseteq \Phi_w$ by Proposition 2.3(iii). Therefore, by our hypothesis, $\alpha \notin \Omega(w'I, {}^{n'}M)$.

But $(\alpha, w'I) = w_2^{-1}(\delta, J)$, so $v(\alpha, w'I)$ is defined and equals $v(\delta, J)^{w_2}$. The fact that $\alpha \notin \Omega(w'I, {}^{n'}M)$ means that $v(\alpha, w'I) \notin W(w'I, {}^{n'}M)$, or equivalently $v(\delta, J) \notin W(J, {}^{n_2n'}M)$. But now Proposition 3.9(ii) implies that $b_{n_1,{}^{n_2n'}M} b_{n_1^{-1},{}^{n_1n_2n'}M} = \mathrm{ind}(w_1)^{-1}.\mathrm{Id}_{\tilde{M}}$. Combining with (1) and (3) then gives our claim since $\mathrm{ind}(ww') = \mathrm{ind}(w_2w')/\mathrm{ind}(w_1)$. □

Theorem 3.16. *Let $(G, B = UT, N, S)$ be a strongly split BN-pair of characteristic p (see Definition 2.20). Let k be an algebraically closed field of characteristic $\neq p$, let $P_I \supseteq B$, and let M be a simple cuspidal kL_I-module.*

Choose a section map $W \to N$, $w \mapsto \dot{w}$. If $w \in W(I, M)$, choose $\theta_{\dot{w}}\colon M \to M$ a k-isomorphism such that $\theta_{\dot{w}}(x.m) = {}^{\dot{w}}x.\theta_{\dot{w}}(m)$ for all $x \in L_I$, $m \in M$. Assume $\theta_1 = \mathrm{Id}_M$. Define $\theta_{\dot{w}t}(m) = \theta_{\dot{w}}(tm)$ for all $t \in T$. Then $\theta_{\dot{w}}\theta_{\dot{w}'} = \lambda(w, w')\theta_{\dot{w}\dot{w}'}$ for a cocycle $\lambda\colon W(I, M) \times W(I, M) \to k^\times$. Assume that, if $w^2 = 1$, then $(\theta_{\dot{w}})^2$ acts as $(\dot{w})^2 \in T$.

The algebra $\mathrm{End}_{kG}(\mathrm{Ind}_{P_I}^G M)$ has a basis $(a_w)_{w \in W(I,M)}$ such that

- *$a_w a_{w'} = \lambda(w, w')a_{ww'}$ if $w' \in C(I, M)$, or $w \in C(I, M)$, or $w \in R(I, M)$ and $w' = v(\alpha, I)$ for $\alpha \in \Delta(I, M)$ and $w(\alpha) \in \Phi^+$,*
- *$(a_{v(\alpha,I)})^2 = c_\alpha a_{v(\alpha,I)} + 1$ where $c_\alpha \in k$.*

The above relations on the a_w ($w \in W(I, M)$) provide a presentation of $\mathrm{End}_{kG}(\mathrm{Ind}_{P_I}^G M)$.

Proof. If $n \in N$ is such that $w := nT \in W(I, M)$, one has chosen $\theta_n\colon M \to M$ a k-linear map such that $\theta_n(x.m) = nxn^{-1}\theta_n(m)$ for all $x \in L_I$, $m \in M$ (see Definition 3.6). This gives rise to a cocycle λ on the inverse image of $W(I, M)$ in N. Changing θ changes λ into some cohomologous cocycle. One may choose θ such that θ_t acts as t whenever $t \in T$, and $\theta_{nt} = \theta_n\theta_t$. Then the product $\widetilde{\theta_n} \circ b_{n,M}$ depends only on the class nT (note that, if $t \in T$, $b_{t^{-1},M} = \widetilde{\theta_t}$ is the morphism induced by the action of t on M).

Note that θ is just defined by the choice of the $\theta_{\dot{w}}$'s for $w \in W(I, M)$. We adjust this choice so that, if $w^2 = 1$, then $(\theta_{\dot{w}})^2$ acts as $(\dot{w})^2 \in T$ (divide $\theta_{\dot{w}}$ by some square root of $\lambda(\dot{w}, \dot{w})$).

One may check that composing λ with any section $w \mapsto \dot{w}$, one gets a cocycle (denoted by λ again) on $W(I, M)$ (see [Cart85] 10.3.3, or Exercise 10 below).

Since k is algebraically closed, one may choose a square root of p in $k^\times$ and define accordingly $(\mathrm{ind}(w))^{\frac{1}{2}}$ for each $w \in W$. Denote now

$$a_w = (\mathrm{ind}(w))^{\frac{1}{2}}\widetilde{\theta_{\dot{w}}} \circ b_{\dot{w},M}.$$

By Proposition 3.7(ii), this is a basis of the endomorphism algebra. It is clear from the description of $W(I, M)$ (Proposition 3.14) that the two formulae stated in the theorem allow us to compute any product of two basis elements, so those formulae give a presentation. Let us check them.

Assume w, $w' \in W(I, M)$ are as in the theorem. Then $\Phi_w \cap \Phi_{w'^{-1}} \cap \Omega(I, M) = \emptyset$ since each $x \in C(I, M)$ satisfies $\Phi_x \cap \Omega(I, M) = \Phi_{x^{-1}} \cap \Omega(I, M) = \emptyset$ and $\Phi_{v(\alpha,I)} \cap \Omega(I, M) = \{\alpha\}$. So $a_w a_{w'} = (\mathrm{ind}(w)\mathrm{ind}(w'))^{\frac{1}{2}}\widetilde{\theta_{\dot{w}}} \circ b_{\dot{w},M} \circ \widetilde{\theta_{\dot{w}'}} \circ b_{\dot{w}',M} = (\mathrm{ind}(w)\mathrm{ind}(w'))^{\frac{1}{2}}\widetilde{\theta_{\dot{w}}} \circ {}^{\dot{w}}\widetilde{\theta_{\dot{w}'}} \circ b_{\dot{w},{}^{\dot{w}'}M} \circ b_{\dot{w}',M} = (\mathrm{ind}(ww'))^{\frac{1}{2}}\widetilde{\theta_{\dot{w}}} \circ$

${}^{\dot w}\widetilde{\theta_{\dot w'}} \circ b_{\dot w\dot w',M}$ by Proposition 3.15. The map $\widetilde{\theta_{\dot w}} \circ {}^{\dot w}\widetilde{\theta_{\dot w'}}$ on $kGe(U_{wI}) \otimes_{kL_{wI}} M$ is $1 \otimes (\theta_{\dot w} \circ {}^{\dot w}\theta_{\dot w'})$, i.e. $\lambda(w, w')\widetilde{\theta_{\dot w\dot w'}}$. This gives $a_w a_{w'} = \lambda(w, w')a_{ww'}$ as claimed.

Let now $v = v(\alpha, I)$ with $\alpha \in \Delta(I, M)$, i.e. $v = u^{-1}v'u$ with $u \in W$, $v' := v(u\alpha, uI)$ for $u\alpha \cup uI \subseteq \Delta$ and $\Phi_v \cap \Omega(I, M) = \{\alpha\}$.

Lemma 3.17. *Let* $a'_{v'} = (\mathrm{ind}(v'))^{\frac{1}{2}}({}^{\dot u}\widetilde{\theta_{\dot v'}})b_{\dot v',{}^{\dot u}M}$. *Then* $b_{\dot u,M} \circ a_v = a'_{v'} \circ b_{\dot u,M}$.

In view of this Lemma and Proposition 3.9(iii), it now suffices to check that $(a'_{v'})^2 \in 1 + ka'_{v'}$. Replacing I with uI, M with ${}^{\dot u}M$ and the θ_n by the ${}^{\dot u}\theta_n$, we get the same cocycle and our claim reduces to showing that $(a_v)^2 \in 1 + ka_v$ as long as $v = v(\delta, I)$ for $\delta \in \Delta \cap \Omega(I, M)$. Using the definition of a_v, we have $(a_v)^2 = \mathrm{ind}(v)\widetilde{\theta_{\dot v}} b_{\dot v,M}\widetilde{\theta_{\dot v}} b_{\dot v,M} = \mathrm{ind}(v)\widetilde{\theta_{\dot v}}{}^{\dot v}\widetilde{\theta_{\dot v}}(b_{\dot v,M})^2$. But $(\theta_{\dot v})^2$ is the action of $\dot v^2 \in T$ on M, so $\widetilde{\theta_{\dot v}}{}^{\dot v}\widetilde{\theta_{\dot v}} = b_{\dot v^{-2},M}$. Then Proposition 3.15 and Proposition 3.9(ii) give $(a_v)^2 = \mathrm{ind}(v)b_{\dot v^{-1},M}b_{\dot v,M} \in 1 + ka_v$. □

Proof of Lemma 3.17. Proposition 2.10(ii) implies that $l(v(u\alpha, uI).u) = l(v(u\alpha, uI)) + l(u)$ and therefore $b_{\dot v',{}^{\dot u}M}b_{\dot u,M} = b_{\dot v'.\dot u,M}$ by Proposition 3.15 (or Proposition 3.9(i)). Multiplying this by ${}^{\dot u}\widetilde{\theta_{\dot v'}}$ on the left, one gets $a'_{v'}b_{\dot u,M} = \mathrm{ind}(v')^{\frac{1}{2}}{}^{\dot u}\widetilde{\theta_{\dot v'}}b_{\dot v'\dot u,M}$, i.e.

$$a'_{v'}b_{\dot u,M} = \mathrm{ind}(v')^{\frac{1}{2}}{}^{\dot u}\widetilde{\theta_{\dot v'}}b_{\dot u\dot v,M}b_{t,M}.$$

for $\dot u\dot v t = \dot v'\dot u$.

The equality $\Phi_v \cap \Omega(I, M) = \{\alpha\}$ implies that Proposition 3.15 may be used with $n = \dot u$ and $n' = \dot v$, thus giving $b_{\dot u\dot v,M} = \left(\frac{\mathrm{ind}(uv)}{\mathrm{ind}(u)\mathrm{ind}(v)}\right)^{-\frac{1}{2}} b_{\dot u,{}^{\dot v}M}b_{\dot v,M}$. Substituting in the equation above gives

$$a'_{v'}b_{\dot u,M} = \mathrm{ind}(v)^{\frac{1}{2}}{}^{\dot u}\widetilde{\theta_{\dot v'}}b_{\dot u,{}^{\dot v}M}b_{\dot v t,M}.$$

The equality $\dot v t = \dot u^{-1}\dot v'\dot u$ and Proposition 3.7(i) give ${}^{\dot u}\widetilde{\theta_{\dot v'}}b_{\dot u,{}^{\dot v}M} = {}^{\dot u}\widetilde{\theta_{\dot v t}}b_{\dot u,{}^{\dot v}M} = b_{\dot u,M}\widetilde{\theta_{\dot v t}}$, so the above equality becomes

$$a'_{v'}b_{\dot u,M} = \mathrm{ind}(v)^{\frac{1}{2}}b_{\dot u,M}\widetilde{\theta_{\dot v t}}b_{\dot v t,M}.$$

This gives our claim since a_v can be defined by taking the representative $\dot v t$ for v. □

Remark 3.18. Concerning the cocycle λ, it can be shown that λ is cohomologous to a cocycle which depends only on classes mod. $R'(I, M) := <v(\alpha, I); \alpha \in \Delta(I, M), c_\alpha \neq 0>$ (see [Cart85] §10).

Theorem 3.19. *Keep the hypotheses of Theorem 3.16.*

Assume that k is (algebraically closed) of characteristic zero. Let t be an indeterminate and let $A(t)$ be the $k[t]$-algebra defined by the generators a_w

($w \in W(I, M)$) and the following relations (where $c_\alpha \in k$ and λ are associated with I, M as in Theorem 3.16):

- $a_w a_{w'} = \lambda(w, w') a_{ww'}$ *if* $w' \in C(I, M)$, *or* $w \in C(I, M)$, *or* $w \in R(I, M)$, $w' = v(\alpha, I)$ *for* $\alpha \in \Delta(I, M)$ *and* $w(\alpha) \in \Phi^+$,
- $(a_{v(\alpha,I)})^2 = t.c_\alpha a_{v(\alpha,I)} + 1$.

Then

(i) the a_w *yield a* $k[t]$*-basis of* $A(t)$,

(ii) the specializations $A(1) \cong \mathrm{End}_{kG}(\mathrm{Ind}_{P_I}^G M)$ *and* $A(0) \cong k_\lambda(W(I, M))$ *are isomorphic.*

Proof. (i) The proof follows the standard lines. One considers a free $k[t]$-module $\bigoplus_w k[t].a_w$ with two families $(L_x)_x$ and $(R_x)_x$ of operators indexed by the set $X = C(I, M) \cup \{v(\alpha, I) \mid \alpha \in \Delta(I, M)\}$. They are defined by the expected outcome of multiplication on the left (resp. on the right) by the a_x. The main point is to show that $L_x R_y = R_y L_x$ for all $x, y \in X$. This is essentially a discussion on w, x, y to check that $L_x R_y(a_w) = R_y L_x(a_w)$. In all cases, the equality follows from the fact that it is satisfied when $t = 1$ in the law of $\mathrm{End}_{kG}(\mathrm{Ind}_{P_I}^G M)$.

(ii) It is clear that a presentation of $W(I, M)$ is obtained by the above relations with $t = 0$ and $\lambda = 1$. Therefore $A(0)$ is isomorphic with $k_\lambda(W(I, M))$ (recall that $\lambda(v(\alpha, I), v(\alpha, I)) = 1$; see Theorem 3.16). Then $A(t) \otimes_{k[t]} k(t)$ is separable and therefore all the semi-simple specializations of $A(t)$ are isomorphic (see [CuRe87] §68). □

Exercises

1. Define $\mathcal{C} := \{x \in E \mid \forall\delta \in \Delta\ (\delta, x) \geq 0\}$, $\mathcal{C}' := \{x \in E \mid \forall\delta \in \Delta\ (\delta, x) > 0\}$.
   (a) Show that $\mathcal{C}' \neq \emptyset$ and that $\bigcup_{w\in W(\Phi)} w(\mathcal{C}')$ is a disjoint union.
   (b) Show that $E = \bigcup_{w\in W(\Phi)} w(\mathcal{C})$ (if $v \in E$ take $d \in W(\Phi).v$, $d = \sum_{\delta\in\Delta} c_\delta \delta$ with maximal $\sum_\delta c_\delta$, then check that $d \in \mathcal{C}$).
   (c) Take $d \in \mathcal{C}$ and $w \in W(\Phi)$ such that $w(d) \in \mathcal{C}$. Show that $w = s_{\delta_1} s_{\delta_2} \ldots s_{\delta_l}$ where $\forall i\ \delta_i \in \Delta \cap d^\perp$.
   (d) If $d \in \mathcal{C}$, show that $\Phi_\Delta \cap d^\perp$ satisfies the hypothesis of Proposition 3.12 on Φ with $\Delta \cap d^\perp$ replacing Δ. Show that $\{w \in W(\Phi) \mid w(d) = d\}$ is generated by $\{s_\alpha \mid \alpha \in \Phi \cap d^\perp\}$ (use (c)).
   (e) If X is an arbitrary subset of E, show that $\{w \in W(\Phi) \mid \forall x \in X\ w(x) = x\}$ is generated by $\{\sigma_\alpha \mid \alpha \in \Phi \cap X^\perp\}$ (assume first that X is finite and

such that $X \cap \mathcal{C} \neq \emptyset$, using induction, (b) and (d)). Derive Proposition 3.12(iv).

2. Using the notation of Theorem 3.10, let $\ell \neq p$ be a prime, let (P, V)—(P', V') with common Levi L. Show that $e(V)$ and $e(V')$ are conjugate in $(\mathbb{Q}_\ell G)^L$ (apply Theorem 3.10, and see [Ben91a] 1.7.2). Show that (P, V)—(P', V') may occur without $e(V)$ and $e(V')$ being G-conjugate (cases where $dI \neq I$ in the notation of Theorem 2.27).
3. Prove a version of Proposition 3.14 where $W(I, M)$ is replaced with any subgroup X of $W^I = \{w \in W \mid wI = I\}$.
4. Use the notation of Definition 3.1. If $n \in N_{I,\Delta}$, and $m \in M$, show that

$$b_{n,M}(1 \otimes m) = \mathrm{ind}(w)^{-1} \sum_{u \in U \cap U^{w_0 w}} un^{-1} \otimes m.$$

5. (Howlett-Lehrer) Let $(G, B = UT, N, S)$ be a split BN-pair of characteristic p (see Definition 2.20) with associated Δ. Let J, K be subsets of Δ, let $w \in D_{KJ}$. Let $M = K \cap w(J)$, $M' = w^{-1}(M) = w^{-1}(K) \cap J$.
   (a) Show that $(U_M \cap X_K)^w \subseteq U_J$ and let ${}^w(U_{M'} \cap X_J) \subseteq U_K$ (use Theorem 2.23).
   (b) Let $n_w \in N$ such that $w = n_w T$. Show that $e(U_K)n_w e(U_J) = e(U_M)n_w e(U_{M'})$ in $\mathbb{Z}[p^{-1}]G$.
6. Let R be a principal ideal domain, A an R-free finitely generated R-algebra, $e, f \in A$ two idempotents. Assume $e \in Afe$, $f \in Aef$. Show that $Ae \cong Af$ by the map $x \mapsto xf$.
7. (Howlett-Lehrer's proof of Theorem 3.10) Let $R = \mathbb{Z}[p^{-1}]$. Let I be a subset of Δ and let $w \in W$ be such that $w(I) \subseteq \Delta$, $n \in N$ a representative of w. The goal is to show that $A = RG$, $e = e(U_I)$, $f = e(U_{wI})^n$ satisfy the hypotheses of Exercise 6. Denote $\mathcal{I} = RGe(U_{w(I)})ne(U_I)$. By symmetry, it suffices to check $e(U_I) \in \mathcal{I}$. One shows this by induction on $|I|$.
   (a) Show that one may assume $\Delta \setminus I = \{\delta\}$ and $w = v(\delta, I)$. (Use Theorem 2.11 and Lemma 2.2.) This is now assumed in what follows. Denote $\epsilon = efe = \sum_{u \in (U_{wI})^n} e(U_I)ue(U_I)$.
   (b) Show that each u in the sum above is in some coset $P_I w_u l_u U_I$ for a $w_u \in W_{I \cup \delta} \cap D_{II}$ and a $l_u \in L_I$. Then show that

$$e(U_I)n^{-1}une(U_I) \in RGe(U_I)n_u e(U_I)l_u$$

   where $n_u \in N$ has class w_u mod. T.
   (c) Assume $w_u(I) \neq I$. Denote $M = I \cap w_u^{-1}(I)$. Using Exercise 5 and the induction hypothesis, show that $e(U_I)ue(U_I) \in RGe(U_{w(M)})ne(U_M)l_u$. Use Exercise 5 again to get $e(U_I)ue(U_I) \in \mathcal{I}$, whenever $w_u(I) \neq I$.
   (d) Assume $w_u = w$. Show that $e(U_I)ue(U_I) \in \mathcal{I}$.

(e) If $w_u = 1$, show that $u \in U_I$ (use Theorem 2.27(i) on $\cap\downarrow$) and therefore $e(U_I)ue(U_I) \in \mathcal{I}$.

(f) Show that (c)–(d)–(e) above exhaust all possibilities for w_u (use Theorem 2.11) and therefore $\varepsilon \in |U_{w(I)} : U_{w(I)} \cap {}^w P_I|^{-1} e(U_I) + \mathcal{I}$. Complete the proof.

8. Let G be a group and A be a commutative group. Assume that any element of A is a square. Let $\lambda: G \times G \to A$ be a map such that $\lambda(x, y)\lambda(xy, z) = \lambda(x, yz)\lambda(y, z)$ for any $x, y, z \in G$ (that is, a 2-cocycle).

   Show that there is a map $f: G \to A$ such that the cocycle μ defined by $\mu(x, y) = f(x)f(y)\lambda(x, y)$, satisfies the following relations
   (i) $\mu(x, 1) = \mu(1, x) = 1$,
   (ii) $\mu(x, x^{-1}) = 1$, and
   (iii) $\mu(x, y) = \mu(y^{-1}, x^{-1})^{-1} = \mu(y^{-1}x^{-1}, x)$.

   Deduce that, if $x^2 = 1$, then $\mu(x, xy) = \mu(x, y)^{-1}$. Similarly $\mu(xy, y^{-1}xy) = \mu(y, y^{-1}xy)^{-1}$. Then $\mu(x, y)^2 = \mu(y, y^{-1}xy)^2$.

9. We use the notation of Theorem 3.16. Take $\alpha \in \Delta(I, M)$ and take $w \in R(I, M)$ such that $w(\alpha) \in \Delta$. Let $u = v(\alpha, I)$, $v = v(w(\alpha), I)$. Show that $c_\alpha = c_{w(\alpha)}$. Show that $a_w a_u = \varepsilon a_v a_w$ where $\varepsilon = \pm 1$. Deduce $\varepsilon = 1$ when $c_\alpha \neq 0$.

   Deduce Remark 3.18.

10. Let λ be a 2-cocycle (written additively) on a finite group G. Let $T \lhd G$ be such that $\lambda(T \times G) = \lambda(G \times T) = 0$. Show that, for any section $s: G/T \to G$, the map $\lambda \circ s$ is a 2-cocycle on G/T.

11. Find a common generalization for Theorem 3.3 and Theorem 3.16.

Notes

Hecke algebras were first defined as endomorphism algebras of induced modules $\mathrm{R}_L^G M$ where $L = T$ and $M = \mathbb{C}$ (see [Bour68] Exercises VI.22–27). Then Lusztig gave deep theorems on general L and cuspidal $\mathbb{C}L$-module M. For instance, in the notation of Theorem 3.16, $C(I, M) = \{1\}$ and the cocycle is trivial ([Lu84] §8; see also [Geck93b]).

Our exposition is based on [Lu76b] §5, [HowLeh80] and the adaptation by Geck–Hiss–Malle [GeHiMa96] to the modular case. We have also used the notes of a course given by François Digne. The reference for Exercise 7 is [HowLeh94].

4

The modular duality functor and derived category

Let G be a finite group endowed with a strongly split BN-pair of characteristic p, giving rise to parabolic subgroups $P_I = U_I \rtimes L_I$ for $I \subseteq S$ (see Chapter 2).

Let $R := \mathbb{Z}[p^{-1}]$. In this chapter, we introduce a bounded complex of RG-bimodules

$$D_{(G)}\colon (\ldots \to D^i_{(G)} \to D^{i+1}_{(G)} \to \ldots)$$

where $D^i_{(G)}$ is the direct sum of RG-bimodules $RGe(U_I) \otimes_{RP_I} e(U_I)RG$ for $|I| = i$. One considers the functor

$$D_{(G)} \otimes_{RG} -\colon C^b(RG-\mathbf{mod}) \to C^b(RG-\mathbf{mod})$$

within the category of bounded complexes of RG-modules. The main theorem in this chapter is that this functor induces an equivalence within the derived category $D^b(RG-\mathbf{mod})$.

Here, the derived category is the category obtained by inverting the complex morphisms $f\colon C \to C'$ inducing isomorphisms of cohomology groups $\mathrm{H}^i(f)\colon \mathrm{H}^i(C) \to \mathrm{H}^i(C')$ for all i. This is particularly well adapted to explain isometries of Grothendieck rings over fields of characteristic zero. Let, for instance, Λ be a complete discrete valuation ring with field of fractions K, let G be a finite group such that KG is split semi-simple, and let A, B be two summands (i.e. sums of blocks) of the group algebra ΛG. Then any equivalence

$$D^b(A-\mathbf{mod}) \to D^b(B-\mathbf{mod})$$

of the type described above, i.e. a tensor product functor and its adjoint as inverse (any equivalence $D^b(A-\mathbf{mod}) \to D^b(B-\mathbf{mod})$ of "triangulated" categories implies the existence of such a functor; see [KLRZ98] §9.2.2), gives the same equivalence for $A \otimes K$ and $B \otimes K$. But, between split semi-simple algebras, this can only be a bijection between the simple modules along with certain signs (see [KLRZ98] §9.2.3 or [GelMan94] §4.1.5). Then in the case of the functor $D_{(G)}\otimes -$ above, we obtain the permutation with signs of Irr(G) known as Alvis–Curtis duality (see [DiMi91] §8). The fact that this isometry of characters is produced by a derived equivalence over Λ implies that it preserves many invariants defined over ΛG, such as the partition of simple KG-modules induced by the blocks of ΛG (see [KLRZ98] §6.3).

Let us return to our $D_{(G)}\otimes_{RG} -\colon D^b(RG-\mathbf{mod})\to D^b(RG-\mathbf{mod})$. The main lemma states that, when k is a field of characteristic $\neq p$ and $M=\mathrm{ind}_{P_I}^G N$ for N a cuspidal kL_I-module, then $D_{(G)}\otimes_{RG} M$ has its cohomology $=0$ except in degree $|I|$ where it is isomorphic to M. The proof involves a study of the reflection representation of the Weyl group W of G and the triangulations of spheres of lower dimensions associated with the fundamental domain of W. This lemma implies that

$$\begin{aligned}&D_{(G)}\otimes_G D_{(G)}^\vee\otimes_G - ,\\ &D_{(G)}^\vee\otimes_G D_{(G)}\otimes_G -\colon D^b(kG-\mathbf{mod})\to D^b(kG-\mathbf{mod})\end{aligned}$$

coincides with the identity on those M. By an argument similar to the proof of the "invariance" Theorem 1.14, this gives our auto-equivalence.

In the case of duality, we show that the well-known property of commutation with Harish-Chandra induction of characters (see [DiMi91] 8.11) can be generalized as the equality

$$RGe(U_I)\otimes_{RL_I} D_{(L_I)} = D_{(G)}\otimes_{RG} RGe(U_I)$$

in $D^b(RG-\mathbf{mod}-RL_I)$, the derived category of R-modules acted on by G on the left and by L_I on the right.

It should be noted that all those results give also complete proofs of the corresponding statements in characteristic 0.

4.1. Homology

We give below some prerequisites about complexes and some classical ways to construct them. We refer to a few books on homological algebra; see also Appendix 1 for a more complete description of derived categories and sheaf cohomology.

4.1.1. Complexes and associated categories

(See [KLRZ98] §2, [Ben91a] §2, [God58] §1, [Weibel] §1, 10, [KaSch98] §1, [GelMan94].) The complexes we consider are mainly chain complexes

$$\dots C_i \xrightarrow{\partial_i} C_{i-1} \longrightarrow \dots$$

where the C_i's and the ∂_i's are objects and morphisms in a category $A-\mathbf{mod}$ for a ring A, satisfying $\partial_{i-1}\partial_i = 0$ for all i. They are always bounded in what follows, i.e. $C_i = 0$ except for finitely many i's. These complexes form a category $C^b(A-\mathbf{mod})$. This category is abelian (see [GelMan94] §2.2, [KaSch98]) in the sense that morphism sets are commutative groups; kernels and cokernels exist.

The homology of a complex is the sequence of A-modules $\mathrm{H}_i(C) = \mathrm{Ker}(\partial_i)/\partial_{i+1}(C_{i+1})$. This is a functor H from $C^b(A-\mathbf{mod})$ to the category of graded A-modules. A complex C such that $\mathrm{H}_i(C) = 0$ for all i is said to be **acyclic**. A morphism $f\colon C \to C'$ is said to be a **quasi-isomorphism** if and only if $\mathrm{H}(f)\colon \mathrm{H}(C) \to \mathrm{H}(C')$ is an isomorphism.

We shall come across many morphisms $C \xrightarrow{f} C'$ such that each $C_i \xrightarrow{f_i} C'_i$ is onto. Then we have an exact sequence $0 \to K \to C \xrightarrow{f} C' \to 0$ where $K_i = \mathrm{Ker} f_i$. A classical application of the homology long exact sequence tells us that f is a quasi-isomorphism if and only if K is acyclic ([Spanier] 4.5.5, [Bour80] §2 Corollaire 2).

The A-modules are considered as complexes with $C_i = 0$ except for $i = 0$, and $\partial_i = 0$ for all i. If n is an integer and C is a complex, one denotes by $C[n]$ the complex such that $C[n]_i = C_{i-n}$ (see [KaSch98] 1.3.2, [GelMan94] 4.2.2).

Tensor products of complexes are defined as the total complex associated with the usual bi-complex: if $C_i \xrightarrow{\partial_i} C_{i-1} \ldots$ and $C'_i \xrightarrow{\partial'_i} C'_{i-1} \ldots$ are complexes, then $(C \otimes_A C')_i = \oplus_{a+b=i} C_a \otimes_A C'_b$ with differential defined by $c \otimes c' \mapsto \partial_a(c) \otimes c' + (-1)^b c \otimes \partial'_b(c')$ (see [Weibel] 2.7.1). If C is a bounded complex of $A-\mathbf{mod}-B$, then $C \otimes_A -$ provides a functor from $C^b(A-\mathbf{mod})$ to $C^b(B-\mathbf{mod})$.

By a localization process which essentially consists of inverting the quasi-isomorphisms, one obtains a category called the derived category $D^b(A-\mathbf{mod})$ (see [KaSch98] §1.7, [Weibel] §10.4, [KLRZ98] §2.5 or Appendix 1 below).

Assume for simplification that A and B are symmetric algebras over a principal ideal domain Λ. Let $X \in C^b(A-\mathbf{mod}-B)$, $X' \in C^b(B-\mathbf{mod}-A)$ be complexes such that all terms are bi-projective (that is projective when restricted to A and B) and such that $X \otimes_B X' \cong A$ in $D^b(A-\mathbf{mod}-A)$ and $X' \otimes_A X \cong B$ in $D^b(B-\mathbf{mod}-B)$. Then $X' \otimes_A -$ and $X \otimes_B -$ induce inverse equivalences between $D^b(A-\mathbf{mod})$ and $D^b(B-\mathbf{mod})$ (see [KLRZ98] §9.2).

4.1.2. Simplicial schemes

We use a slight variant of what is usually called "simplicial complex" (see [Spanier] §3.1, [CuRe87] §66) or "schéma simplicial" ([God58] §3.2). An (augmented) simplicial scheme Σ is a set of finite subsets of a given set Σ_0 such that, if $\sigma \in \Sigma$ and $\sigma' \subseteq \sigma$, then $\sigma' \in \Sigma$. The elements of Σ are called **simplexes**. In particular, if $\Sigma \neq \emptyset$, then $\emptyset \in \Sigma$. We always assume that Σ_0 is the union of all simplexes.

The degree of σ is defined as its cardinality minus 1, denoted by $\deg(\sigma)$. The elements of Σ_0 are the ones of degree zero; they are called the **vertices**.

A simplicial scheme Σ is said to be **ordered** if it is endowed with a (partial) ordering of the vertices such that each simplex is totally ordered. Of course any total ordering of Σ_0 will do, and this is easy to choose when Σ_0 is finite (this is always the case below). If $\sigma \neq \emptyset$ is a simplex of such an ordered simplicial scheme, it is customary to list its elements in increasing order, $x_0 < x_1 < \ldots < x_{\deg(\sigma)}$. Then, if $0 \leq j \leq \deg(\sigma)$, we write $\sigma_j = \sigma \setminus \{x_j\}$.

One may define a topological vector space $\mathrm{Top}(\Sigma)$ associated with Σ (see [CuRe87] §66, [Spanier] §3.1, [God58] p. 39). The definition is as follows: let $\mathrm{Top}(\Sigma)$ be the set of maps $p\colon \Sigma_0 \to [0,1]$ such that $p^{-1}(]0,1])$ is a simplex and $\sum_{x\in\Sigma_0} p(x) = 1$. The topology on $\mathrm{Top}(\Sigma)$ is the usual topology on the set of almost constant maps on $[0,1]$. Conversely, Σ is said to be a triangulation of $\mathrm{Top}(\Sigma)$.

When Σ_0 is a finite subset of a real vector space and all simplexes are affinely free, $\mathrm{Top}(\Sigma)$ is easily described (see Exercise 3).

4.1.3. Coefficient systems

([God58] §3.5, [Ben91b] §7.1) Let Σ be a poset (for instance a simplicial scheme), let $\mathbf{C}$ be a category. A coefficient system on Σ with values in $\mathbf{C}$ is a collection of objects M_σ and morphisms $f^\sigma_{\sigma'}\colon M_\sigma \to M_{\sigma'}$ in $\mathbf{C}$, for all $\sigma' \subseteq \sigma$ in Σ, satisfying $f^\sigma_\sigma = \mathrm{Id}$, $f^{\sigma'}_{\sigma''} f^\sigma_{\sigma'} = f^\sigma_{\sigma''}$ when $\sigma'' \subseteq \sigma' \subseteq \sigma$. This can be seen as a functor from the category associated with the poset Σ to $\mathbf{C}$. In particular this can be composed with functors $\mathbf{C} \to \mathbf{C}'$. Having fixed Σ and $\mathbf{C}$, one has a category of coefficient systems on Σ with coefficients in $\mathbf{C}$; thus, for instance, a notion of isomorphic coefficient systems.

4.1.4. Associated homology complexes

If Σ is an ordered simplicial scheme and $(M_\sigma, f^\sigma_{\sigma'})$ is a coefficient system on it with coefficients in a module category $A-\mathbf{mod}$, one defines a complex $\mathrm{C}((M_\sigma, f^\sigma_{\sigma'}))$ (see [Spanier] §4.1, [CuRe87] §66, [God58] §§3.3, 3.5, [Ben91b] 7.3)

$$\ldots \xrightarrow{\partial_{i+1}} C_i \xrightarrow{\partial_i} C_{i-1} \ldots C_0 \xrightarrow{\partial_0} C_{-1} \to 0$$

where $C_i = \bigoplus_{\sigma,\deg(\sigma)=i} M_\sigma$ and ∂_i is the map defined on M_σ ($\deg(\sigma) = i$) by $\partial_i(m) = \sum_{j=0}^{i} (-1)^j f^\sigma_{\sigma_j}(m)$.

This gives a functor from coefficient systems on Σ with values in $A-\mathbf{mod}$ to the category of complexes $C(A-\mathbf{mod})$. The associated complex does not depend on the choice of the ordering on Σ_0 (see [Ben91b] 7.3 or Exercise 1 below).

A particular case is the constant coefficient system $M_\sigma = \mathbb{Z}$, $f^\sigma_{\sigma'} = \mathrm{Id}$.

The relation with the singular homology defined in topology is that the homology of the constant coefficient system on Σ is the "reduced singular homology" of Top(Σ) (see [CuRe87] §66, [Spanier] 4.6.8).

Note that a contractible topological space has reduced homology equal to 0 (in all degrees); see [Spanier] 4.4.4.

4.2. Fixed point coefficient system and cuspidality

Let Λ be a commutative ring. We take G a finite group and $\mathcal{L}$ a Λ-regular, $\cap\!\downarrow$-stable set of subquotients of G (see Definition 1.6).

Definition 4.1. *Let $\sigma = (P, V) \in \mathcal{L}$. One defines the G-G-bimodule*

$$(\Lambda G)_\sigma := \Lambda G \otimes_{\Lambda P} e(V)\Lambda G = \Lambda G e(V) \otimes_{\Lambda P} e(V)\Lambda G.$$

If moreover $\sigma \le \sigma' = (P', V')$, let

$$\phi^\sigma_{\sigma'} \colon (\Lambda G)_\sigma \to (\Lambda G)_{\sigma'}, \quad x \otimes_{\Lambda P} y \mapsto x \otimes_{\Lambda P'} y$$

for all $x \in \Lambda G$ and $y \in e(V)\Lambda G \subseteq e(V')\Lambda G$.

If M is a ΛG-module, then one defines a ΛG-module $M_\sigma = (\Lambda G)_\sigma \otimes_{\Lambda G} M = \mathrm{Ind}^G_P \mathrm{Res}^G_{(P,V)} M = \Lambda G \otimes_{\Lambda P} e(V)M$ and the maps $\phi^\sigma_{\sigma'} \otimes_{\Lambda G} M$ are just denoted by $\phi^\sigma_{\sigma'}$.

It is easily checked that $\phi^\sigma_{\sigma'}$ is a morphism of G-G-bimodules, and, if $\sigma'' \ge \sigma' \ge \sigma$, then $\phi^{\sigma'}_{\sigma''} \circ \phi^\sigma_{\sigma'} = \phi^\sigma_{\sigma''}$. Therefore

Proposition 4.2. *$(\Lambda G)_\mathcal{L} := ((\Lambda G)_\sigma, \phi^\sigma_{\sigma'})$ is a coefficient system on the poset $\mathcal{L}^{\mathrm{opp}}$ with coefficients in $\Lambda G-\mathbf{mod}-\Lambda G$. Each $(\Lambda G)_\sigma$ is bi-projective.*

If M is a ΛG-module, then $M_\mathcal{L} := ((M_\sigma)_\sigma, \phi^\sigma_{\sigma'})$ is a coefficient system on $\mathcal{L}^{\mathrm{opp}}$ with coefficients in $\Lambda G-\mathbf{mod}$.

Definition 4.3. *When $\sigma_i = (P_i, V_i) \in \mathcal{L}$ $(i = 0, 1, 2)$ with $\sigma_1 \le \sigma_2$, define $\mathcal{X}(\sigma_0)_{\sigma_i} := \{P_i g P_0 \mid g \in G\,,\ \sigma_i^g \cap\!\downarrow \sigma_0 = \sigma_0\}$. It is easily checked that the map $\psi^{\sigma_1}_{\sigma_2}$ defined by $\psi^{\sigma_1}_{\sigma_2}(P_1 g P_0) = P_2 g P_0$ sends $\mathcal{X}(\sigma_0)_{\sigma_1}$ into $\mathcal{X}(\sigma_0)_{\sigma_2}$ (see Exercise 1.3). Taking a fixed σ_0, one gets a coefficient system $\mathcal{X}(\sigma_0) = (\mathcal{X}(\sigma_0)_\sigma, \psi^\sigma_{\sigma'})$ on $\mathcal{L}^{\mathrm{opp}}$ with coefficients in the category of finite sets.*

Hypothesis 4.4. We take G a finite group, k a field and $\mathcal{L}$ a k-regular, $\cap\!\downarrow$-stable set of subquotients (P, V). We assume in addition that, whenever (P, V)—(P', V') in $\mathcal{L}$ (see Definition 1.1),

$$\begin{aligned} kGe(V) &\to kGe(V'), \\ x &\mapsto xe(V') \end{aligned}$$

is an isomorphism (and therefore $(|P|, |V|) = (|P'|, |V'|)$). Note that, as a result of Theorem 3.10, this is satisfied by the system of parabolic subgroups in strongly split BN-pairs.

We prove the following.

Theorem 4.5. *Let $\mathcal{L}$ be a subquotient system satisfying Hypothesis 4.4. Let $(P_0, V_0, N_0) \in \mathbf{cusp}_k(\mathcal{L})$ (see Notation 1.10). Denote $\sigma_0 = (P_0, V_0)$ and $M = \mathrm{Ind}_{P_0}^G N_0$.*

One has an isomorphism of coefficient systems on $\mathcal{L}^{\mathrm{opp}}$ with values in $kG{-}\mathbf{mod}$:

$$(kG)_{\mathcal{L}} \otimes_{kG} M \cong M \otimes_{\mathbb{Z}} \mathbb{Z}\mathcal{X}(\sigma_0).$$

Lemma 4.6. *If $\sigma := (P, V) \in \mathcal{L}$, then $e(V)g.kP_0 \otimes_{P_0} N_0 = 0$ unless $PgP_0 \in \mathcal{X}(\sigma_0)_\sigma$.*

Proof of Lemma 4.6. One has $e(V)g.kP_0 \otimes_{P_0} N_0 = e(V)g.e(V^g \cap P_0)kP_0 \otimes_{P_0} N_0 = e(V)g \otimes_{P_0} e(V^g \cap P_0)N_0$, so $e(V^g)kP_0 \otimes_{P_0} N_0 \neq 0$ implies $e(V^g \cap P_0)N_0 \neq 0$. Since V_0 acts trivially on N_0, one has $e((V^g \cap P_0).V_0)N_0 \neq 0$, i.e. $\mathrm{Res}^{P_0}_{(P,V)^g \cap\downarrow (P_0,V_0)} N_0 \neq 0$. By cuspidality (see Definition 1.8), this implies $(P, V)^g \cap\!\downarrow (P_0, V_0) = (P_0, V_0)$. □

Proposition 4.7. *If $\sigma := (P, V) \in \mathcal{L}$ and $PgP_0 \in \mathcal{X}(\sigma_0)_\sigma$, then the following is an isomorphism*

$$\begin{aligned} kG \otimes_{kP} kPe(V)g.kP_0e(V_0) &\to kGe(V_0), \\ x \otimes y &\mapsto xy. \end{aligned}$$

Proof of Proposition 4.7. The map is clearly defined (and is a morphism of G-P_0-bimodules). Multiplying by g^{-1} on the right allows us to assume $g = 1$.

Lemma 4.8. *If $(P, V) \cap\!\downarrow (P_0, V_0) = (P_0, V_0)$, then $(|(P \cap P_0).V|, |(P \cap V_0).V|) = (|P_0|, |V_0|)$, $e(V)kP_0e(V_0) = k(P \cap P_0)e(V)e(V_0)$ and $kGe(V)e(V_0) = kGe(V_0)$.*

Proof of Lemma 4.8. By the hypothesis, one has $P_0 = (P \cap P_0).V_0$, so $e(V)kP_0e(V_0) = e(V)k(P \cap P_0)k(V_0)e(V_0) = k(P \cap P_0)e(V)e(V_0)$. By Proposition 1.2(ii), one has $(P_0, V_0) \cap\!\downarrow (P, V) = ((P \cap P_0).V, (P \cap V_0).V)$— (P_0, V_0). Then Hypothesis 4.4 implies $kGe(V_0) = kGe((P \cap V_0).V)e(V_0)$. But this last expression is $kGe(V)e(V_0)$ by Definition 1.4. This also gives the equality of cardinalities (see Hypothesis 4.4 above). □

Let us check the surjectivity of the map defined in Proposition 4.7 (with $g = 1$). Its image is $kGe(V)kP_0e(V_0) = kGe(V)e(V_0) = kGe(V_0)$ by Lemma 4.8.

It remains to check that the dimension of $kG \otimes_{kP} kPe(V)kP_0e(V_0)$ is less than or equal to that of $kGe(V_0)$, i.e. $|G : V_0|$. By Lemma 4.8, one has $kG \otimes_{kP} kPe(V)kP_0e(V_0) = kG \otimes_{kP} kPe(V)e(V_0) = kG \otimes_{kP} kPe(V)$ $e(V_0 \cap P)e(V_0)$ and this is clearly equal to the subspace $\{x \otimes e(V_0) \mid x \in kGe(V.(V_0 \cap P))\}$, so its dimension is less than or equal to the dimension of $kGe(V.(V_0 \cap P))$, i.e. $|G : V(V_0 \cap P)|$. This last expression is indeed $|G : V_0|$ by Lemma 4.8. □

Proof of Theorem 4.5. We denote $\sigma = (P, V) \leq \sigma' = (P', V')$.

As a P-P_0-bimodule, $kG = \bigoplus_{PgP_0} kPgkP_0$, so $e(V)M = e(V)kG \otimes_{kP_0} N_0 = \bigoplus_{PgP_0} kPe(V)g.kP_0 \otimes_{kP_0} N_0$. By Lemma 4.6, this is also $\bigoplus_{PgP_0 \in \mathcal{X}_0} k$ $Pe(V)g.kP_0 \otimes_{kP_0} N_0$.

Then $M_\sigma = \bigoplus_{PgP_0 \in \mathcal{X}_0} kG \otimes_{kP} e(V)kPg.kP_0 \otimes_{kP_0} N_0$. By Proposition 4.7, each factor $kG \otimes_{kP} e(V)kPg.kP_0 \otimes_{kP_0} N_0$ is isomorphic with $M = kGe(V_0) \otimes_{kP_0} N_0$ by the map $x \otimes_{kP} e(V)y \otimes_{kP_0} n \mapsto xe(V)y \otimes_{kP_0} n$ for $x \in kG$, $y \in PgP_0$, $n \in N_0$. Then $M_\sigma \cong M \otimes_{\mathbb{Z}} \mathbb{Z}\mathcal{X}_0$ by the map

$$i\colon x \otimes_{kP} e(V)y \otimes_{kP_0} n \mapsto (xe(V)y \otimes_{kP_0} n) \otimes_{\mathbb{Z}} PyP_0$$

for $x \in kG$, $y \in \bigcup_{PgP_0 \in \mathcal{X}(\sigma_0)_\sigma} PgP_0$, $n \in N_0$.

Similarly, one gets an isomorphism $M_{\sigma'} \cong M \otimes_{\mathbb{Z}} \mathbb{Z}\mathcal{X}(\sigma_0)_{\sigma'}$ by the map

$$i'\colon x' \otimes_{kP'} e(V')y' \otimes_{kP_0} n \mapsto (x'e(V')y' \otimes_{kP_0} n) \otimes_{\mathbb{Z}} P'y'P_0$$

for $x' \in kG$, $y' \in \bigcup_{P'gP_0 \in \mathcal{X}(\sigma_0)_{\sigma'}} P'gP_0$, $n \in N_0$.

It suffices to check $i' \circ \phi^\sigma_{\sigma'} = (M \otimes_{\mathbb{Z}} \psi^\sigma_{\sigma'}) \circ i$. Taking $x \in kG$, $y \in PgP_0 \in \mathcal{X}(\sigma_0)_\sigma$, $n \in N_0$, one has $\psi^\sigma_{\sigma'} \circ i(x \otimes_{kP} e(V)y \otimes_{kP_0} n) = \psi^\sigma_{\sigma'}((xe(V)y \otimes_{kP_0} n) \otimes_{\mathbb{Z}} PyP_0) = (xe(V)y \otimes_{kP_0} n) \otimes_{\mathbb{Z}} P'yP_0$. But $\phi^\sigma_{\sigma'}(x \otimes_{kP} e(V)y \otimes_{kP_0} n) = x \otimes_{kP'} e(V)y \otimes_{kP_0} n$. This also equals $x \otimes_{kP'} e(V')e(V)y \otimes_{kP_0} n = |V|^{-1} \sum_{v \in V} x \otimes_{kP'} e(V')vy \otimes_{kP_0} n$ with $P'vyP_0 = P'yP_0 \in \mathcal{X}(\sigma_0)_{\sigma'}$ for each v. So the image under i' is $(|V|^{-1} \sum_{v \in V} xe(V')vy \otimes_{kP_0} n) \otimes_{\mathbb{Z}} P'yP_0 = (xe(V')e(V)y \otimes_{kP_0} n) \otimes_{\mathbb{Z}} P'yP_0 = (xe(V)y \otimes_{kP_0} n) \otimes_{\mathbb{Z}} P'yP_0$. This finishes our proof. □

Let $\mathcal{L}$ be a Λ-regular set of subquotients of a finite group G. Here is a construction devised to study more generally the tensor product $(\Lambda G)_{\mathcal{L}} \otimes_{\Lambda G} \Lambda Ge(V)$. We take some $(P_1, V_1) \in \mathcal{L}$. Then we have an associated system of subquotients of P_1: $]\leftarrow, (P_1, V_1)] = \{(P, V) \mid (P, V) \leq (P_1, V_1)\} \subseteq \mathcal{L}$ (see Definition 1.8), and we can define as in Definition 4.1 a coefficient system $(\Lambda P_1)_{]\leftarrow,(P_1,V_1)]}$ of ΛP_1-bimodules with regard to this subquotient system. We may even extend it by 0 to the whole of $\mathcal{L}$ (or more properly $\mathcal{L}^{\mathrm{opp}}$), thus leading to the following.

Definition 4.9. *If* $(P_1, V_1) \in \mathcal{L}$, *let* $(\Lambda P_1)_{\mathcal{L}}$ *be the coefficient system on* $\mathcal{L}^{\mathrm{opp}}$ *defined by* $(\Lambda P_1)_\sigma = \Lambda P_1 e(V) \otimes_P e(V)\Lambda P_1$ *if* $\sigma = (P, V) \leq (P_1, V_1)$, $(\Lambda P_1)_\sigma = 0$ *otherwise. The connecting map* $\varphi^{(P,V)}_{(P',V')}: (\Lambda P_1)_{(P,V)} \to (\Lambda P_1)_{(P',V')}$ *is defined by* $\varphi^{(P,V)}_{(P',V')}(x \otimes_P y) = x \otimes_{P'} y$ *if* $(P, V) \leq (P', V') \leq (P_1, V_1)$, $x \in \Lambda P_1 e(V)$, $y \in e(V)\Lambda P_1$.

Proposition 4.10. *One may define a surjective map of coefficient systems on* $\mathcal{L}^{\mathrm{opp}}$ *with coefficients in* ΛG–**mod**–ΛP_1,

$$(\Lambda G)_{\mathcal{L}} \otimes_G \Lambda Ge(V_1) \xrightarrow{\pi} \Lambda Ge(V_1) \otimes_{P_1} (\Lambda P_1)_{\mathcal{L}} \to 0.$$

Proof. We define the map π as follows.

Assume $(P, V) \leq (P_1, V_1)$. One has $(\Lambda G)_{(P,V)} \otimes_G \Lambda Ge(V_1) = \Lambda Ge(V) \otimes_P e(V)\Lambda G \otimes_G \Lambda Ge(V_1)$ and this coincides with the subspace $\Lambda Ge(V) \otimes_P e(V) \otimes_G \Lambda Ge(V_1)$. Letting $x \in \Lambda Ge(V)$, $y \in G$, one takes $\pi_{(P,V)}(x \otimes_P e(V) \otimes_G ye(V_1)) = x \otimes_{P_1} e(V) \otimes_P e(V)ye(V_1) \in \Lambda Ge(V_1) \otimes_{P_1} \Lambda P_1 e(V) \otimes_P e(V)\Lambda P_1 = \Lambda Ge(V_1) \otimes_{P_1} (\Lambda P_1)_{(P,V)}$ if $y \in P_1$, $\pi_{(P,V)}(x \otimes_P e(V) \otimes_G ye(V_1)) = 0$ otherwise.

If $(P, V) \not\leq (P_1, V_1)$, take $\pi_{(P,V)} = 0$.

Denote $\sigma = (P, V) \in \mathcal{L}$, $e = e(V)$, $e_1 = e(V_1)$.

The above does indeed give a well-defined surjective morphism π_σ of ΛG–**mod**–ΛP_1 since $e\Lambda G \otimes_G \Lambda Ge_1 \cong e\Lambda Ge_1 = \bigoplus_{PgP_1 \subseteq G} e\Lambda[PgP_1]e_1$ as a P-P_1-bimodule and since $\Lambda Ge_1 \otimes_{P_1} \Lambda P_1 e \otimes_P e\Lambda P_1$ clearly coincides with the subspace $\Lambda Ge_1 \Lambda P_1 e \otimes_{P_1} e \otimes_P e\Lambda P_1 = \Lambda Ge \otimes_{P_1} e \otimes_P e\Lambda P_1$.

Let $\sigma \leq \sigma' = (P', V') \in \mathcal{L}$, denote $e' = e(V')$. One must show that

$$\left(\Lambda Ge_1 \otimes_{P_1} \varphi^\sigma_{\sigma'}\right) \circ \pi_\sigma = \pi_{\sigma'} \circ \left(\phi^\sigma_{\sigma'} \otimes_G \Lambda Ge_1\right).$$

Since $(\Lambda P_1)_{\sigma'} = 0$ unless $\sigma' \leq (P_1, V_1)$, one may assume $\sigma \leq \sigma' \leq (P_1, V_1)$.

Let us take $x, y \in G$ so that the general element of a basis of $(\Lambda G)_\sigma \otimes_G \Lambda Ge_1 = \Lambda Ge \otimes_P e\Lambda G \otimes_G \Lambda Ge_1$ is $xe \otimes_P e \otimes_G ye_1$.

If $y \in P_1$, then the effects of the two compositions of maps above on this element are respectively

$$xe \otimes_P e \otimes_G ye_1 \mapsto xe \otimes_{P_1} e \otimes_P eye_1 \mapsto xe \otimes_{P_1} e \otimes_{P'} eye_1$$

and

$$xe \otimes_P e \otimes_G ye_1 \mapsto xe \otimes_{P'} e \otimes_G ye_1 = xee' \otimes_{P'} e'e \otimes_G ye_1$$
$$\mapsto xee' \otimes_{P_1} e' \otimes_{P'} e'eye_1 = xe \otimes_{P_1} e' \otimes_{P'} eye_1,$$

since $ey \in \Lambda P_1$. But $e \in \Lambda P'$, so $xe \otimes_{P_1} e' \otimes_{P'} eye_1 = xe \otimes_{P_1} e \otimes_{P'} eye_1$.

If $y \notin P_1$, then we get

$$xe \otimes_P e \otimes_G ye_1 \mapsto 0 \mapsto 0$$

and

$$xe \otimes_P e \otimes_G ye_1 \mapsto xe \otimes_{P'} e \otimes_G ye_1 = xee' \otimes_{P'} e' \otimes_G ye_1 \mapsto 0.$$

□

Assume, moreover, that $\Lambda = k$ is a field and that $\mathcal{L}$ satisfies Hypothesis 4.4. Let (P_0, V_0, N_0) be a cuspidal triple where $(P_0, V_0) \leq (P_1, V_1)$ and N_0 is a cuspidal $k[P_0/V_0]$-module. Denote $M = \mathrm{Ind}_{P_0}^{P_1} N_0$, $\widetilde{M} = \mathrm{Ind}_{P_1}^{G} M = \mathrm{Ind}_{P_0}^{G} N_0$, so that $(kG)_{\mathcal{L}} \otimes_G kGe(V_1) \otimes_{P_1} M = (kG)_{\mathcal{L}} \otimes_G \widetilde{M}$. Applying Theorem 4.5, we have isomorphisms $(kG)_{\mathcal{L}} \otimes_G \widetilde{M} \cong \widetilde{M} \otimes_{\mathbb{Z}} \mathbb{Z}\mathcal{X}(\sigma_0)^{(G)}$ and $(kP_1)_{\mathcal{L}} \otimes_{P_1} M \cong M \otimes_{\mathbb{Z}} \mathbb{Z}\mathcal{X}(\sigma_0)^{(P_1)}$ where $\sigma_0 = (P_0, V_0)$, and the exponent in $\mathcal{X}(\sigma_0)^{(P_1)}$ recalls the ambient group. Recall that $\mathcal{X}(\sigma_0)^{(P_1)}$ is a coefficient system defined on the poset $]\leftarrow, (P_1, V_1)]^{\mathrm{opp}} \subseteq \mathcal{L}^{\mathrm{opp}}$. We extend $\mathbb{Z}\mathcal{X}(\sigma_0)^{(P_1)}$ by zero to make it into a coefficient system on the whole of $\mathcal{L}^{\mathrm{opp}}$. Looking at the explicit definition of the isomorphism in Theorem 4.5 (see its proof) and of the map π of Proposition 4.10, we easily check the following.

Proposition 4.11. *Through the isomorphisms of Theorem 4.5, the map $\pi \otimes_{P_1} M$ of Proposition 4.10 identifies with $\widetilde{M} \otimes_{\mathbb{Z}} \theta$, where $\theta: \mathbb{Z}\mathcal{X}(\sigma_0)^{(G)} \to \mathbb{Z}\mathcal{X}(\sigma_0)^{(P_1)}$ is the map which sends PgP_0 satisfying $(P, V)^g \cap\!\downarrow (P_0, V_0) = (P_0, V_0)$, to PgP_0 if $(P, V) \leq (P_1, V_1)$ and $g \in P_1$, to 0 otherwise.*

4.3. The case of finite BN-pairs

We now take G a finite group with a strongly split BN-pair of characteristic p with subgroups $B = UT$, N, S (see Definition 2.20). Then the set of pairs

(P, V) for P a parabolic subgroup of G, and V its biggest normal p-subgroup, satisfies Hypothesis 4.4 for any field of characteristic $\neq p$ (Theorem 3.10).

Denote by Δ the set of simple roots of the root system associated with G (see §2.1), $n = |\Delta|$.

Definition 4.12. *Let Λ be a commutative ring where p is invertible. Let us define a coefficient system on the simplicial scheme $\mathcal{P}(\Delta)$ of all subsets of Δ by composing the map $\mathcal{P}(\Delta) \to \mathcal{L}^{\rm opp}$ defined by $I \mapsto (P_{\Delta\setminus I}, U_{\Delta\setminus I})$ with the coefficient system $(\Lambda G)_{\mathcal{L}}$ of Definition 4.1. One denotes by DC the associated complex of G-G-bimodules*

$$\begin{aligned}&\dots DC_n = 0 \longrightarrow DC_{n-1} = \Lambda G \otimes_{\Lambda B} e(U)\Lambda G \xrightarrow{\partial_{n-1}} \dots\\ &\dots DC_i = \bigoplus\nolimits_{I;|I|=n-i-1} \Lambda G \otimes_{\Lambda P_I} e(U_I)\Lambda G \xrightarrow{\partial_i} \dots DC_{-1}\\ &= \Lambda G \longrightarrow DC_{-2} = 0 \dots\end{aligned}$$

If M is a ΛG-module, denote $DC(M) = DC \otimes_{\Lambda G} M$, the complex of ΛG-modules with $DC(M)_{-1} = M$, $DC(M)_i = \bigoplus_{I\subseteq\Delta\,;\,|I|=n-i-1} M_{(P_I,U_I)}$ (see Definition 4.1) if $-1 \leq i \leq n-1$, $DC(M)_i = 0$ for other i.

If A is a Λ-free algebra and X is a complex of A-modules, we denote by $X^\vee = \mathrm{Hom}(X, \Lambda)$ its dual as a complex of right A-modules, with indices multiplied by -1 in order to get a chain complex like X.

Proposition 4.13. *If $\lambda = k$ is a field of characteristic $\neq p$ and $M = \mathrm{Ind}^G_{(P_{I_0},U_{I_0})} N_0$ for N_0 a simple cuspidal kL_{I_0}-module, then $DC \otimes_{kG} M \cong M[|\Delta \setminus I_0| - 1]$ and $DC^\vee \otimes_{kG} M \cong M[-|\Delta \setminus I_0| + 1]$ in $D^b(kG-\mathbf{mod})$.*

The proof consists essentially in a study of the coefficient system defined by the sets $\mathcal{X}(\sigma_0)_\sigma$ of Theorem 4.5. This is done by use of standard results on the geometric representation of Coxeter groups (see [Stein68a] Appendix, [Hum90] §5). We recall the euclidean structure on $\mathbb{R}\Delta$ and the realization of W in the associated orthogonal group (§2.1).

Definition 4.14. *Let $\mathcal{C} = \{x \in \mathbb{R}\Delta \mid \forall\delta \in \Delta\ (x, \delta) \geq 0\}$. If $I \subseteq \Delta$, let $\mathcal{C}_I = \mathcal{C} \cap I^\perp$.*

Lemma 4.15. *Let $I_0 \subseteq \Delta$, denote $\sigma_0 = (P_{I_0}, U_{I_0})$. Denote by $\mathcal{X}'(\sigma_0)$ the coefficient system on $\mathcal{P}(\Delta)$ obtained as in Definition 4.12 from the coefficient system $\mathcal{X}(\sigma_0)$ on the poset of parabolic subgroups of G.*

(i) If $\sigma = (P_I, U_I)$, then $\mathcal{X}(\sigma_0)_\sigma$ identifies with $\mathcal{Y}_I := \{w\mathcal{C}_I;\ w \in W,\ w\mathcal{C}_I \subseteq (I_0)^\perp\}$ by a map sending $w\mathcal{C}_I$ to $P_I w^{-1} P_{I_0}$. If $\sigma' = (P_{I'}, U_{I'})$ with $I \subseteq I'$, then $\psi^\sigma_{\sigma'}$ (see Theorem 4.5) corresponds to the map $\psi^I_{I'}$ sending $w\mathcal{C}_I$ to $w\mathcal{C}_{I'}$ (which is a subset of $w\mathcal{C}_I$).

(ii) The above identifies $\mathbb{Z}\mathcal{X}'(\sigma_0)$ with the constant coefficient system on a triangulation of the unit sphere of $(I_0)^\perp$ whose set of simplexes of degree d corresponds with the $P_I g P_{I_0} \in \mathcal{X}'(\sigma_0)$ such that $|\Delta \setminus I| = d + 1$.

(iii) Through the above identification, $\mathcal{X}'(\sigma_0) \setminus \{P_{I_0}\}$ identifies with a triangulation of a contractible topological space.

Proof of Lemma 4.15. (i) It clearly suffices to check the first statement of (i). To describe $\mathcal{X}(\sigma_0)_\sigma = \{P_I g P_{I_0} \mid (P_I, U_I)^g \cap\!\downarrow (P_{I_0}, U_{I_0}) = (P_{I_0}, U_{I_0})\}$, one may take $g \in D_{II_0}$ (see Proposition 2.4 and Theorem 2.16(iv)). Then Theorem 2.27(i) gives $\mathcal{X}(\sigma_0)_\sigma = \{P_I w P_{I_0} \mid wI_0 \subseteq I \subseteq w\Phi^+\}$ in bijection with the corresponding subset of $W_I \backslash W / W_{I_0}$. For those w, one has $W_I w W_{I_0} = W_I w$. However, the set of all cosets $W_I w$ ($I \subseteq \Delta$, $w \in W$) is in bijection with the set of all subsets $w^{-1}C_I$ by the obvious map (see [CuRe87] 66.24, [Bour68] V.4.6). So it only remains to check that, if $w \in W$ and $I \subseteq \Delta$, then $w^{-1}C_I \subseteq (I_0)^\perp$ if and only if $W_I w W_{I_0} \ni v$ such that $vI_0 \subseteq I \subseteq v\Phi^+$. The "if" is clear since $v^{-1}C_I \subseteq v^{-1}I^\perp \subseteq (I_0)^\perp$. For the "only if", one may take v of minimal length. Then $I \subseteq v\Phi^+$. The condition $v^{-1}C_I \subseteq (I_0)^\perp$ gives $vI_0 \subseteq \mathbb{R}I$ by taking orthogonals. But now Lemma 2.7 gives the remaining inclusion $vI_0 \subseteq I$.

(ii) By (i), $\mathrm{C}(\mathbb{Z}\mathcal{X}'(\sigma_0))$ is isomorphic to the complex associated with the coefficient system $(\mathcal{Y}_I, \psi_{I'}^I)$.

If Γ is a cone in $\mathbb{R}\Delta$, let Γ^{ex} denote the extremal points of its intersection with the unit sphere. Each $\Gamma \in \mathcal{Y}_I$ is generated as a cone by Γ^{ex}, so the coefficient system may be replaced by those finite sets and corresponding restrictions of maps ψ.

We now refer to the topological description of [Bour68] §V.3.3, [CuRe87] §66.B, or [Hum90] §1.15 for instance (see also Exercise 3.1). The set $\mathcal{C}^{\mathrm{ex}}$ is a basis of $\mathbb{R}\Delta$, and $\mathcal{C}$ is a fundamental domain for the finite group W. The faces of $\mathcal{C}$ are the $\mathcal{C}_I$'s, so the $(w\mathcal{C}_I)^{\mathrm{ex}}$ ($I \subseteq \Delta$) are elements of a triangulation of the unit sphere of $\mathbb{R}\Delta$ ([CuRe87] 66.28.(i); see also Exercise 3). The intersection with $(I_0)^\perp$ provides a triangulation of the unit sphere of $(I_0)^\perp$, since $(I_0)^\perp$ is the subspace generated by the face $\mathcal{C}_{I_0}$. Since Γ^{ex} for $\Gamma \in \mathcal{Y}_I$ has cardinality $|\Delta \setminus I|$, we get our second claim.

(iii) Through the above identification, $\mathcal{X}'(\sigma_0) \setminus \{P_{I_0}\}$ identifies with $\mathcal{Y}^{\mathrm{ex}} := \{\Gamma^{\mathrm{ex}} \mid \Gamma \in \mathcal{Y}_I,\ I \subseteq \Delta\}$ where we have deleted $(\mathcal{C}_{I_0})^{\mathrm{ex}}$, i.e. a simplex of highest dimension $m_0 = |\Delta \setminus I_0| - 1$ in our triangulation of the sphere $\mathbb{S}^{m_0}$. Since the convex hulls of the $w\mathcal{C}_I$'s intersect only on their boundaries (in $I^\perp$) by the property of a fundamental domain, the topological space associated with $\mathcal{Y}^{\mathrm{ex}} \setminus \{\mathcal{C}_{I_0}^{\mathrm{ex}}\}$ is $\mathbb{S}^{m_0} \setminus \mathcal{C}_0$, where $\mathcal{C}_0$ is the interior of $\mathcal{C}_{I_0}$ in $(I_0)^\perp$. The outcome is contractible (any point $x \in \mathbb{S}^{m_0} \cap \mathcal{C}_0$ defines a homeomorphism $\mathbb{S}^{m_0} \setminus \{x\} \cong \mathbb{R}^{m_0}$ sending $\mathbb{S}^{m_0} \setminus \mathcal{C}_0$ to a compact star-shaped set; see also Exercise 3(b)). $\square$

Proof of Proposition 4.13. We first check $DC \otimes_{kG} M \cong M[|\Delta \setminus I_0| - 1]$ in $D^b(kG\text{--}\mathbf{mod})$.

By its definition, $DC \otimes_{kG} M$ is the complex associated with the restriction to $\mathcal{L}_\Delta := \{(P_I, U_I) \mid I \subseteq \Delta\}$ of the coefficient system $M_\mathcal{L}$ where $\mathcal{L}$ is the poset defined in Definition 2.24. By Theorem 4.5, one has $M_\mathcal{L} \cong M \otimes_\mathbb{Z} \mathbb{Z}\mathcal{X}(\sigma_0)$. When restricted to $\mathcal{L}_\Delta$, we get $DC \otimes_{kG} M \cong M \otimes_\mathbb{Z} \mathrm{C}(\mathbb{Z}\mathcal{X}'(\sigma_0))$ where $\mathrm{C}(\mathbb{Z}\mathcal{X}'(\sigma_0))$ is the complex associated with the coefficient system $\mathbb{Z}\mathcal{X}'(\sigma_0)$ (see Lemma 4.15) on the simplicial scheme of subsets of Δ.

Then, to check our first claim, it suffices to check that we have a quasi-isomorphism $\mathrm{C}(\mathbb{Z}\mathcal{X}'(\sigma_0)) \cong \mathbb{Z}[|\Delta \setminus I_0| - 1]$. (This is formally stronger than saying that both have the same homology; see, however, Exercise 12.) One may define a map $\mathrm{C}(\mathbb{Z}\mathcal{X}'(\sigma_0)) \to \mathbb{Z}[|\Delta \setminus I_0| - 1]$ by sending P_{I_0} to a generator of $\mathbb{Z}$ at degree $|\Delta \setminus I_0| - 1$, all other elements of $\mathcal{X}'(\sigma_0)$ to 0 at the appropriate degree (checking that this is a map in $C^b(\mathbb{Z}\text{--}\mathbf{mod})$ is easy since the only choice is about the highest degree). This gives an exact sequence $0 \to \mathrm{C}(\mathbb{Z}(\mathcal{X}'(\sigma_0) \setminus \{P_{I_0}\})) \to \mathrm{C}(\mathbb{Z}\mathcal{X}'(\sigma_0)) \to \mathbb{Z}[|\Delta \setminus I_0| - 1] \to 0$. The second term is acyclic by Lemma 4.15(iii). Thus we have our first claim.

We now check the second isomorphism $DC^\vee \otimes_{kG} M \cong M[-|\Delta \setminus I_0| + 1]$.

For any kG-module M, denote by M^* the usual notion of duality on kG-modules (thus M^* is a left kG-module). This extends to complexes by $(C_i, \partial_i)^* = (C^*_{-i}, \partial^*_{-i})$.

To check that $DC^\vee \otimes_{kG} M \cong M[-|\Delta \setminus I_0| + 1]$, we deduce it from $DC \otimes_{kG} M \cong M[|\Delta \setminus I_0| - 1]$ and the following lemma.

Lemma 4.16. *For all Λ where p is invertible, and for all finitely generated ΛG-modules M, $DC^\vee \otimes_{\Lambda G} M \cong (DC \otimes_{\Lambda G} M^*)^*$ in $C^b(\Lambda G\text{--}\mathbf{mod})$.*

This gives our claim since $(\mathrm{Ind}_{P_{I_0}}^G N)^* \cong \mathrm{Ind}_{P_{I_0}}^G (N^*)$ and N^* is cuspidal when N is. □

Proof of Lemma 4.16. Let $x \mapsto x^\iota$ be the involution of group algebras over Λ induced by the inversion in the group. This gives a covariant functor $M \to M^\iota$ from $\Lambda G\text{--}\mathbf{mod}\text{--}\Lambda H$ to $\Lambda H\text{--}\mathbf{mod}\text{--}\Lambda G$. This extends to complexes. One clearly has $(L \otimes_{\Lambda G} M)^\iota \cong M^\iota \otimes_{\Lambda G} L^\iota$ for modules for which the tensor product makes sense.

Considering also the (contravariant) functor $M \to M^\vee$ relating the same categories, one clearly has $M^* = (M^\vee)^\iota = (M^\iota)^\vee$ for one-sided modules. One has $(L \otimes M)^\vee \cong M^\vee \otimes L^\vee$ by the evident map as long as L or M is projective on the side we consider to make this tensor product (see also [McLane63] V.4.3). This applies to complexes with the suitable renumbering for $M \mapsto M^\vee$ due to contravariance.

Our claim now follows once we check $DC \cong DC^\iota$ as complexes of bimodules. This in turn follows from the same property of the coefficient system $(\Lambda G)_{\mathcal{L}}$, having noted that $(\Lambda G)_{(P,V)} = \Lambda Ge(V) \otimes_{\Lambda P} (\Lambda Ge(V))^\iota$ and $\phi^{(P,V)}_{(P',V')} = j \otimes_{\Lambda P} j^\iota$ where j is the inclusion map of $\Lambda Ge(V)$ in $\Lambda Ge(V')$. □

4.4. Duality functor as a derived equivalence

We keep G a finite group with a strongly split BN-pair of characteristic p. Note first that the complex DC of Definition 4.12 is defined in an intrinsic way from the subgroup B since the subgroups of G containing B and their unipotent radicals make the whole poset used to define DC. Since the outcome would be the same with a G-conjugate of B and since B is the normalizer of a Sylow p-subgroup of G, one sees that DC is defined in an intrinsic way from the abstract structure of G.

Definition 4.17. *Let G be a finite group with a strongly split BN-pair of characteristic p (see Definition 2.20). Let n be the number of simple roots of its root system or of its set S (see Definition 2.12). Let Λ be a commutative ring where p is invertible. Let us denote by $D_{(G)}$ the cochain complex of bimodules in ΛG–**mod**–ΛG*

$$\begin{aligned}&\dots D^{-1}_{(G)} = 0 \longrightarrow D^0_{(G)} = \Lambda G \otimes_{\Lambda B} e(U)\Lambda G \xrightarrow{\partial^0} \dots\\ &\dots D^i_{(G)} = \bigoplus\nolimits_{I;|I|=i} \Lambda G \otimes_{\Lambda P_I} e(U_I)\Lambda G \xrightarrow{\partial^i} \dots D^n_{(G)}\\ &= \Lambda G \longrightarrow D^{n+1}_{(G)} = 0 \dots\end{aligned}$$

obtained from $DC[-|\Delta|+1]$ by taking the opposite of indices.

This is the version over $\mathbb{Z}[p^{-1}]$ tensored with Λ.

We prove the following.

Theorem 4.18. *Let G be a finite group with a strongly split BN-pair of characteristic p. Let Λ be a commutative ring where p is invertible. Then $D^\vee_{(G)} \otimes_{\Lambda G} D_{(G)} \cong D_{(G)} \otimes_{\Lambda G} D^\vee_{(G)} \cong \Lambda G$ in $D^b(\Lambda G$–**mod**–$\Lambda G)$.*

As recalled in §4.1 above, one gets the following.

Corollary 4.19. *$D_{(G)}$ induces an equivalence from $D^b(\Lambda G)$ into itself.*

Lemma 4.20. *Let X be a bounded complex of free $\mathbb{Z}[p^{-1}]$-modules. If $X \otimes k$ is acyclic for any field of characteristic $\neq p$, then X is acyclic.*

Proof of Lemma 4.20. This is a standard application of the universal coefficient theorem (see [Bour80] p. 98, [Weibel] 3.6.2) or of more elementary arguments (see Exercise 5). □

Proof of Theorem 4.18. We may also work with our initial DC (see Definition 4.12) instead of $D_{(G)}$.

Since the algebra ΛG is symmetric, if X is in $C^b(\Lambda G{-}\mathbf{mod}{-}\Lambda G)$ with all terms bi-projective, the functors $X \otimes -$ and $X^\vee \otimes -$ are adjoint to each other ([KLRZ98] 9.2.5) as functors on $C^b(\Lambda G{-}\mathbf{mod}{-}\Lambda G)$. Then we have a unit map $\eta \colon \Lambda G \to X^\vee \otimes X$ and a co-unit map $\varepsilon \colon X \otimes X^\vee \to \Lambda G$ of complexes of bimodules (see [McLane97] IV.Theorem 1) corresponding with the identity as element of the right-hand side in each isomorphism:

$$\mathrm{Hom}(X \otimes \Lambda G, X \otimes \Lambda G) \cong \mathrm{Hom}(\Lambda G, X^\vee \otimes X \otimes \Lambda G) \text{ and}$$
$$\mathrm{Hom}(X^\vee \otimes \Lambda G, X^\vee \otimes \Lambda G) \cong \mathrm{Hom}(X \otimes X^\vee \otimes \Lambda G, \Lambda G)$$

(all Hom being defined within $C^b(\Lambda G{-}\mathbf{mod}{-}\Lambda G)$). The fundamental property of adjunctions (see [McLane97] IV.(9)) implies, in this case of $C^b(\Lambda G{-}\mathbf{mod}{-}\Lambda G)$, that the map

$$X \otimes X^\vee \otimes X \xrightarrow{\varepsilon \otimes X} X$$

is split surjective, a section being given by $X \otimes \eta$.

In the case of $X = DC$, let us show that ε itself is onto. Its image $V \subseteq \Lambda G$ is a two-sided ideal. The surjectivity of $\varepsilon \otimes DC$ implies that $V \otimes DC = DC$. But one has $DC_{-1} \cong \Lambda G$ as bimodule, so $(V \otimes DC)_{-1} \cong V$ and a direct summand of ΛG. Therefore $V = \Lambda G$.

We now get an exact sequence in $C^b(\Lambda G{-}\mathbf{mod}{-}\Lambda G)$

$$0 \to Y \to DC^\vee \otimes DC \xrightarrow{\varepsilon} \Lambda G \to 0.$$

Moreover, by projectivity of ΛG, this exact sequence splits in each degree as a sequence of right ΛG-modules.

In order to check the first isomorphism of the theorem, it suffices to check that Y is acyclic.

We have $DC = DC_R \otimes_{\mathbb{Z}} \Lambda$, $DC^\vee = DC_R^\vee \otimes_{\mathbb{Z}} \Lambda$, $\varepsilon = \varepsilon_R \otimes_{\mathbb{Z}} \Lambda$, $Y = Y_R \otimes_{\mathbb{Z}} \Lambda$, etc. where DC_R, ε_R, Y_R are defined in the same way over $R = \mathbb{Z}[p^{-1}]$. It suffices to check that Y_R is acyclic. So Lemma 4.20 implies that we may assume that Λ is a field. If M is a ΛG-module, the above exact sequence becomes

$$0 \to Y \otimes M \to DC^\vee \otimes DC \otimes M \xrightarrow{\epsilon \otimes M} M \to 0$$

thanks to the splitting property mentioned above. By Theorem 1.30(i), one may apply Lemma 1.15 to the modules M of type $\mathrm{Ind}_{P_0}^G N_0$ for cuspidal N_0. It therefore suffices to check that $Y \otimes_G M$ is acyclic for those M. By Proposition 4.13, $M \cong (DC \otimes M)[m] \cong DC \otimes (M[m])$ for some $m \in \mathbb{Z}$, so $\varepsilon \otimes M \cong \varepsilon \otimes DC \otimes M[m]$ is a split surjection, a section being given by $DC \otimes \eta \otimes (M[m])$. Then $DC^\vee \otimes DC \otimes M \cong M \oplus (Y \otimes M)$ in $C^b(\Lambda G\mathbf{-mod})$. Proposition 4.13 again implies that $DC^\vee \otimes DC \otimes M$ has homology M. So $Y \otimes M$ is acyclic.

We get that $DC^\vee \otimes DC \cong \Lambda G$ for $\Lambda = \mathbb{Z}[p^{-1}]$, hence for every commutative ring where p is invertible. A similar proof would give $DC \otimes DC^\vee \cong \Lambda G$. □

4.5. A theorem of Curtis type

The following generalizes Proposition 4.13 above. The version in characteristic zero is well known (see [DiMi91] 8.11).

Theorem 4.21. *Let P be a parabolic subgroup of G, with Levi decomposition $P = L.U_P$. Let Λ be any commutative ring where p is invertible. Denote $\mathrm{R}_L^G := \Lambda G e(U_P)$ in $\Lambda G\mathbf{-mod}-\Lambda L$. Then*

$$D_{(G)} \otimes_{\Lambda G} \mathrm{R}_L^G \cong \mathrm{R}_L^G \otimes_{\Lambda L} D_{(L)}$$

(see Definition 4.17) in $D^b(\Lambda G\mathbf{-mod}-\Lambda L)$

Proof. We take $\Lambda = \mathbb{Z}[p^{-1}]$. We may choose P containing B, so P corresponds to a subset I_1 of Δ. Denote by $DC^{(L)}$ the same complex as in Definition 4.12 with regard to $L = L_{I_1}$.

Concerning DC, what we have to check amounts to

$$DC^{(G)} \otimes_{\Lambda G} \Lambda G e(U_P)[-|\Delta \setminus I_1|] \cong \Lambda G e(U_P) \otimes_{\Lambda L} DC^{(L)}$$

in the derived category $D^b(\Lambda G\mathbf{-mod}-\Lambda L)$.

If $I \subseteq \Delta$, let $\sigma_I = (P_I, U_I)$. Let $e = e(U_P)$ (recall $P = P_{I_1}$, so $U_P = U_{I_1}$). $DC^{(G)}$ is the complex associated with the coefficient system on $\mathcal{P}(\Delta)$ (subsets of Δ) obtained by composing $I \mapsto \sigma_{\Delta \setminus I}$ with the coefficient system $(\Lambda G)_{\mathcal{L}}$ (see Definition 4.12). But $DC^{(L)}$ is the complex associated with the system obtained by composing $I \mapsto \sigma_{I_1 \setminus I}$ from $\mathcal{P}(I_1)$ to $]\leftarrow, (P, U_P)] \subseteq \mathcal{L}$ with the coefficient system $(\Lambda P)_{]\leftarrow,(P,U_P)]}$. Had we taken the complex associated with $(\Lambda P)_{\mathcal{L}}$ (see Definition 4.9) composed with $I \mapsto \sigma_{\Delta \setminus I}$ from $\mathcal{P}(\Delta)$ to $\mathcal{L}$, we would have obtained $DC^{(L)}[|\Delta \setminus I_1|]$.

So the map of Proposition 4.10 gives a surjection in $C^b(\Lambda G-\mathbf{mod}-\Lambda L)$

$$DC^{(G)}\otimes_G \Lambda Ge \xrightarrow{\mathrm{C}(\pi)} (\Lambda Ge\otimes_L DC^{(L)})[|\Delta\setminus I_1|]\to 0.$$

(We may replace $P_1 := P$ by L in the tensor products of Proposition 4.10 since there $V_1 = U_P$ always acts trivially, see also Proposition 1.5(i)). We now consider the above map only as in $C^b(\mathbf{mod}-\Lambda L)$. The above surjection is split in each degree since the modules are all projective, so we have exact sequences

$$(S)\qquad 0\to Y\to DC^{(G)}\otimes_G \Lambda Ge \xrightarrow{\mathrm{C}(\pi)} \Lambda Ge\otimes_L DC^{(L)}[|\Delta\setminus I_1|]\to 0$$

in $C^b(\mathbf{mod}-\Lambda L)$ and

$$(E)\quad 0\to Y\otimes_L M\to DC^{(G)}\otimes_G \Lambda Ge\otimes_L M \xrightarrow{\mathrm{C}(\pi)\otimes M} \Lambda Ge\otimes_L DC^{(L)}\otimes_L M[|\Delta\setminus I_1|]\to 0$$

for any kL-module M where k is a field of characteristic $\neq p$.

Our claim reduces to checking that Y is acyclic. As in the proof of Theorem 4.18, Lemma 4.20 and Lemma 1.15 allow us to check only that $\mathrm{C}(\pi)\otimes M$ is a quasi-isomorphism for any kL-module in the form $M = \mathrm{Ind}_{P_0}^P N$ for $P_0 = P_{I_0}$ with $I_0 \subseteq I_1$ and $(P_{I_0}, U_{I_0}, N)\in \mathbf{cusp}_k(\mathcal{L})$.

By Proposition 4.11, $\mathrm{C}(\pi)\otimes_P M$ identifies with a map $(\mathrm{Ind}_P^G M)\otimes_{\mathbb{Z}} \mathrm{C}(\theta_P^G)$ where $\theta_P^G\colon \mathbb{Z}\mathcal{X}'(\sigma_0)^{(G)}\to \mathbb{Z}\mathcal{X}'(\sigma_0)^{(P)}$ is the map of coefficient systems on $\mathcal{P}(\Delta)^{\mathrm{opp}}$ sending $P_I gP_0\in \mathcal{X}'(\sigma_0)_I^{(G)}$ to $P_I gP_0\in \mathcal{X}'(\sigma_0)_I^{(P)}$ if $g\in P$ and $I\subseteq I_1$, to 0 otherwise (see the notation $\mathcal{X}'(\sigma_0)$ in Lemma 4.15).

Lemma 4.15(ii) tells us that $\mathrm{C}(\mathbb{Z}\mathcal{X}'(\sigma_0)^{(P)})$ is the (augmented) chain complex of singular homology of the sphere of $\mathbb{R}I_1\cap (I_0)^{\perp}$, up to a shift bringing its support into $[|\Delta\setminus I_0|-1, |\Delta\setminus I_1|]$ (the shift is due to the fact that $\mathbb{Z}\mathcal{X}'(\sigma_0)^{(P)}$ is made into a coefficient system on $\mathcal{P}(\Delta)$ like $\mathbb{Z}\mathcal{X}'(\sigma_0)^{(G)}$, instead of just $\mathcal{P}(I_1)$, by extending it by 0; see the paragraph before Proposition 4.11).

The homologies of $\mathrm{C}(\mathbb{Z}\mathcal{X}'(\sigma_0)^{(G)})$, $\mathrm{C}(\mathbb{Z}\mathcal{X}'(\sigma_0)^{(P_1)})$, and $\mathrm{C}(\mathbb{Z}\mathcal{X}'(\sigma_0)^{(P_0)})$ are all isomorphic to $\mathbb{Z}[|\Delta\setminus I_0|-1]$ by the well-known result on homology of spheres (see [Spanier] 4.6.6 and Exercise 4). We have two maps $\mathrm{H}(\mathrm{C}(\theta_{P_1}^G))$ and $\mathrm{H}(\mathrm{C}(\theta_{P_0}^{P_1}))$ between them. To show that the first is an isomorphism, it suffices to show that the composition is. It is clear from their definitions (see Proposition 4.11) that $\theta_{P_0}^{P_1}\circ\theta_{P_1}^G = \theta_{P_0}^G$, so the composition we must look at is in fact $\mathrm{H}(\mathrm{C}(\theta_{P_0}^G))$. The map $\theta_{P_0}^G$ annihilates every element of $\mathcal{X}'(\sigma_0)^{(G)}$ except P_0. So the kernel of $\theta_{P_0}^G$ is $\mathbb{Z}(\mathcal{X}'(\sigma_0)^{(G)}\setminus\{P_0\})$. The associated complex is acyclic by Lemma 4.15(iii).

This completes our proof. □

Exercises

1. Let Σ be a simplicial scheme and $(M_\sigma, f^\sigma_{\sigma'})$ be a coefficient system on it with values in the category of commutative groups. Show that the following defines the associated complex. Let $C_i = \bigoplus_{\sigma, \deg(\sigma)=i} \mathrm{Hom}_{\mathbb{Z}}(\wedge^{i+1}(\mathbb{Z}^\sigma), M_\sigma)$ and let d have component on $\sigma\sigma'$ ($\sigma' \subseteq \sigma$ with $\sigma \setminus \sigma' = \{\alpha\}$) the map $x \mapsto f^\sigma_{\sigma'} \circ x \circ r_\alpha$ where $r_\alpha \colon \wedge^i(\mathbb{Z}^{\sigma'}) \to \wedge^{i+1}(\mathbb{Z}^\sigma)$ is $\omega \mapsto \omega \wedge \alpha$.
2. Let Σ be a simplicial scheme such that Σ_0 is a finite subset of a finite-dimensional real vector space E where each $\sigma \in \Sigma$ is linearly independent. Let $c(\Sigma) := \bigcup_{\sigma \in \Sigma} c(\sigma)$ be the union of the convex hulls of the simplexes σ of Σ. Show that $\mathrm{Top}(\Sigma)$ is homeomorphic with $c(\Sigma)$ for the usual topology of E (define $p \mapsto \sum_{x \in \Sigma_0} p(x)x$).
3. Let B be a basis of a finite-dimensional euclidean space E.
   (a) Assume that the convex hull $c(B)$ of B is a fundamental domain for a finite subgroup W of the general linear group of E. Show that $\Sigma := \{wB' \mid B' \subseteq B,\ w \in W\}$ is a simplicial scheme such that $\mathrm{Top}(\Sigma)$ is homeomorphic with the unit sphere of E (define $U = \bigcup_{w \in W} c(wB)$ in the notation of Exercise 2, and check that the map associating the half-line generated by elements of U is a homeomorphism from U to the quotient of $E \setminus \{0\}$ by $\mathbb{R}^\times_+$). Apply this to the proof of Proposition 4.13.
   (b) Let S (resp. $c(S)$) be the unit sphere (resp. ball) of E, let C be an open convex cone of E, and denote the border of C by C'. Show that $S \setminus C$ is homeomorphic to $C' \cap c(S)$, hence contractible. *Hint*: choose $c_0 \in C \cap S$, define $x \mapsto f(x)$ for all $x \in S \setminus C$, by $c(\{x, c_0\}) \cap C' = \{f(x)\}$.
4. Show the classical results about the reduced singular homology of the spheres ([Spanier] 4.6.6) as a consequence of Lemma 4.15.
5. Let Λ be a principal ideal domain, let M be a free Λ-module of finite rank, and $\partial \in \mathrm{End}_\Lambda(M)$ such that $\partial^2 = 0$. Denote $\mathrm{H}(\partial) := \mathrm{Ker}(\partial)/\partial(M)$.

   Show that there is a basis of M where ∂ has matrix $\begin{pmatrix} 0 & 0 \\ a & 0 \end{pmatrix}$, where $a \in \mathrm{Mat}_{m,n}(\Lambda)$.

   Show that if Λ is a field and $\mathrm{H}(\partial) = \{0\}$, then a is square and invertible.

   Show that if Λ is no longer a field but $\mathrm{H}(\partial \otimes k) = \{0\}$ for any field $k = \Lambda/\mathfrak{M}$, then a is square and invertible and therefore $\mathrm{H}(\partial) = \{0\}$.

   Deduce Lemma 4.20.
6. Give an explicit contruction of $DC^\vee$ with $DC_i^\vee \cong DC_i$ and show the second statement of Proposition 4.13 in the same fashion as the first.
7. Count how many "dualities" we have used in this chapter.

8. Show that the isomorphisms of Proposition 4.13 are in fact homotopies.
9. Let A, B, C be three rings. Let $M \in A-\mathbf{mod}-B$, $N \in B-\mathbf{mod}-C$ be two bi-projective bimodules. Show that $M \otimes_B N$ is bi-projective.
10. Give an example of a short exact sequence of complexes of modules $0 \to X \to Y \to Z \to 0$ where Z is a module, $\mathrm{H}(Y) \cong Z$, but X is not acyclic (of course the isomorphism $\mathrm{H}(Y) \cong Z$ is not induced by the exact sequence).
11. Let A be a ring. If M is a finitely generated A-module and $i \in \mathbb{Z}$, define $M^{[i,i-1]} \in C^b(A-\mathbf{mod})$ by $M_i^{[i,i-1]} = M_{i-1}^{[i,i-1]} = M$, $M_j^{[i,i-1]} = 0$ elsewhere, $\partial_i = \mathrm{Id}_M$.
    (a) If C is a complex, show that $\mathrm{Hom}_{C^b(A-\mathbf{mod})}(M^{[i,i-1]}, C) \cong \mathrm{Hom}_A(M, C_i)$ and $\mathrm{Hom}_{C^b(A-\mathbf{mod})}(C, M^{[i,i-1]}) \cong \mathrm{Hom}_A(C_{i-1}, M))$.
    (b) Show that if M is a projective module, then $M^{[i,i-1]}$ is an indecomposable projective object of $C^b(A-\mathbf{mod})$.
    (c) Show that the projectives of $C^b(A-\mathbf{mod})$ are the acyclic complexes whose terms are projective A-modules. Deduce a projective cover of a given complex from projective covers of its terms.
12. (a) Assume that A is a ring, C an object of $C^b(A-\mathbf{mod})$ such that, for any i, C_i and $\mathrm{H}^i(C)$ are projective. Show that there is an isomorphism $C \cong \mathrm{H}(C) \oplus C'$ in $C^b(A-\mathbf{mod})$. Note that C' is null homotopic.
    (b) Find $D^b(A-\mathbf{mod})$ when A is semi-simple.
13. With the same hypotheses as Theorem 4.21, show that $e(U_P)\Lambda G \otimes_{\Lambda G} D_{(G)} \cong D_{(L)} \otimes_{\Lambda L} e(U_P)\Lambda G$.

Notes

The duality functor on ordinary characters was introduced around 1980 by Alvis, Curtis, Deligne–Lusztig and Kawanaka (see [CuRe87] §71, [DiMi91] §8, [Cart86] §8.2 and the references given there). The question of constructing an auto-equivalence of the derived category $D^b(\Lambda G-\mathbf{mod})$ inducing the duality functor on ordinary characters was raised by Broué [Bro88], and solved in [CaRi01]. There, Theorem 4.18 and Theorem 4.21 are conjectured to hold in the homotopy categories K^b, and thus to imply a "splendid equivalence" in the sense of Rickard [Rick96] (see also [KLRZ98] §9.2.5).

The construction of the complex, at least in the form $D_{(G)} \otimes M$, goes back to Curtis [Cu80a] and [Cu80b], and Deligne–Lusztig [DeLu82]. Related results are in [CuLe85]. Other results in natural characteristic have been obtained by Ronan–Smith (see [Ben91b] 7.5 and corresponding references).

"Character isometries" abound in finite group theory, especially in the classification of simple groups (see [Da71], [CuRe87] §14). They appear in the form of "bijections with signs" between sets of irreducible characters. Equivalences $D^b(A-\mathbf{mod}) \to D^b(B-\mathbf{mod})$ between blocks of group algebras are probably one of the best notions to investigate those isometries further. Stable categories are also relevant (see for instance [Rou01], [KLRZ98] §9). Broué's notion of a "perfect isometry" (see [Bro90a], and Exercise 9.5 below) stands as a first numerical test for a bijection of characters to be induced by a derived equivalence.

5

Local methods for the transversal characteristics

Let G be a finite group, ℓ a prime, and (Λ, K, k) an ℓ-modular splitting system for G. The algebra ΛG splits as a sum of its blocks, usually called the *ℓ-blocks of* G. The inclusion $\Lambda G \subseteq KG$ implies a partition $\mathrm{Irr}(G) = \bigcup_B \mathrm{Irr}(G, B)$ where B ranges over the blocks of ΛG (or equivalently of kG). The local methods introduced by R. Brauer associate ℓ-blocks of G with those of centralizers $\mathrm{C}_G(X)$ (with X an ℓ-subgroup), essentially by use of a ring morphism

$$\mathrm{Br}_X\colon \mathrm{Z}(kG) \to \mathrm{Z}(k\mathrm{C}_G(X)).$$

The word "local" comes from the fact that information about proper subgroups (centralizers and normalizers of ℓ-subgroups) provides information about G, and also that information about kG-modules implies results about characters over K (see [Al86])

In the case when $X = <x>$ is cyclic, Brauer's "second Main Theorem" shows that the above morphism relates well with the decomposition map

$$f \mapsto d^{x,G} f$$

associating with each central function $f\colon G \to K$ (for instance, a character) the central function on the ℓ'-elements of $\mathrm{C}_G(x)$ defined by $d^{x,G} f(y) = f(xy)$. Brauer's theory associates with each ℓ-block of G an ℓ-subgroup of G, its "defect group" (the underlying general philosophy being that the representation theory of kG should reduce to the study of certain ℓ-subgroups of G and their representations). As an illustration of those methods we give a proof of the first application, historically speaking: i.e. the partition of characters into ℓ-blocks for symmetric groups $\mathfrak{S}_n$ (proof by Brauer–Robinson of the so-called Nakayama Conjectures, see [Br47]).

We give another application. In order to simplify our exposition, take now $G = \mathrm{GL}_n(q)$ with $q \equiv 1$ mod. ℓ. Let $B = U \rtimes T$, $N = \mathrm{N}_G(T)$, $W = \mathfrak{S}_n$ be its usual BN-pair; see Example 2.17. We show that all characters occurring in $\mathrm{Ind}_B^G K$ (this includes the trivial character) are in a single ("principal") ℓ-block. This is essentially implied by commutation of the above decomposition map with Harish-Chandra restriction, a key property that will be used again in the generalization of Chapter 21. This also relates naturally to the Hecke algebra $\mathcal{H} := \mathrm{End}_{\Lambda G}(\mathrm{Ind}_B^G \Lambda)$. Since k is the residue field of a complete valuation ring Λ, we may define decomposition matrices for Λ-free finitely generated Λ-algebras (see [Ben91a] §1.9). We show that the decomposition matrix of $\mathcal{H}$ is a submatrix of the decomposition matrix of ΛG. This inclusion property will be studied to a greater extent in Chapters 19 and 20. In §23.3, we will show that this can be generalized into a Morita equivalence between principal ℓ-blocks of G and N.

5.1. Local methods and two main theorems of Brauer's

Let us recall briefly the notion of an ℓ-block of a finite group G.

Let (Λ, K, k) be an ℓ-modular splitting system for G. Denote by $\lambda \mapsto \overline{\lambda}$ the reduction mod. $J(\Lambda)$ of elements of Λ. This extends into a map $\Lambda G \to kG$.

The group algebra ΛG decomposes as a product of blocks

$$\Lambda G = B_1 \times \cdots \times B_m.$$

Similarly, the unit of the center $\mathrm{Z}(\Lambda G)$ decomposes as a sum of primitive idempotents

$$1 = b_1 + \cdots + b_m$$

with $B_i = \Lambda G.b_i$ for all i (see [Ben91a] §1.8, [NaTs89] §1.8.2).

As a result of the lifting of idempotents mod. $J(\Lambda)$ and since $\mathrm{Z}(kG) = \mathrm{Z}(\Lambda G)/J(\Lambda).\mathrm{Z}(\Lambda G)$, the blocks of ΛG and of kG correspond by reduction mod. $J(\Lambda)$. Both are usually called the ℓ-blocks of G. Their units are called the ℓ-block idempotents.

The space $\mathrm{CF}(G, K)$ of central functions is seen as the space of maps $f\colon KG \to K$ that are fixed under conjugacy by elements of G.

The ℓ-blocks induce a partition $\mathrm{Irr}(G) = \bigcup_i \mathrm{Irr}(G, B_i)$ and a corresponding orthogonal decomposition $\mathrm{CF}(G, K) = \bigoplus_i \mathrm{CF}(G, K, B_i)$ where $\mathrm{CF}(G, K, B_i) = \{f \mid \forall g \in KG, f(gb_i) = f(g)\}$. We may also write $\mathrm{Irr}(G, b_i)$ or $\mathrm{CF}(G, b_i)$.

Conversely, when $\chi \in \mathrm{Irr}(G)$, it defines a unique ℓ-block $B_G(\chi)$ (and a unique ℓ-block idempotent $b_G(\chi)$) of G not annihilated by χ.

Definition 5.1. *Let P be an ℓ-subgroup of G.*

An ℓ-subpair of G is any pair (P, b) where b is a primitive idempotent of $\mathrm{Z}(\Lambda \mathrm{C}_G(P))$.

Letting P act on kG by conjugation, denote by $(kG)^P$ the subalgebra of fixed points. The Brauer morphism $\mathrm{Br}_P : (kG)^P \to \mathrm{Z}(k\mathrm{C}_G(P))$ is defined by $\mathrm{Br}_P(\sum_{g\in G} \lambda_g g) = \sum_{g\in \mathrm{C}_G(P)} \lambda_g g$. It is a morphism of k-algebras.

Definition 5.2. *If (P, b) and (P', b') are ℓ-subpairs of G, we write $(P, b) \lhd (P', b')$ (normal inclusion) if and only if $P \lhd P'$ (therefore P' induces algebra automorphisms of $k\mathrm{C}_G(P)$), b is fixed by P' (hence $\overline{b} \in (k\mathrm{C}_G(P))^{P'}$) and $Br_{P'}(\overline{b}).\overline{b'} = \overline{b'}$.*

The inclusion relation $(P, b) \subseteq (P', b')$ between arbitrary ℓ-subpairs of G is defined from normal inclusion by transitive closure.

For the following, see [NaTs89] §5, [Thévenaz] §§18 and 41.

Theorem 5.3. *Let G, ℓ, (Λ, K, k) be as above.*

(i) If (P', b') is an ℓ-subpair of G, and $P \subseteq P'$ is a subgroup, there is a unique ℓ-subpair $(P, b) \subseteq (P', b')$.

(ii) The maximal ℓ-subpairs of G containing a given ℓ-subpair of type $(\{1\}, b)$ are G-conjugates.

(iii) An ℓ-subpair (P, b) is maximal in G if and only if $(\mathrm{Z}(P), b)$ is a maximal ℓ-subpair in $\mathrm{C}_G(P)$ and $\mathrm{N}_G(P, b)/P\mathrm{C}_G(P)$ is of order prime to ℓ.

Remark 5.4. Proving that Br_P is actually a morphism is made easier by using "relative traces" $\mathrm{Tr}_{P'}^P$. If $P' \subseteq P$ is an inclusion of ℓ-subgroups, then

$$\mathrm{Tr}_{P'}^P : kG^{P'} \to kG^P$$

is defined by $\mathrm{Tr}_{P'}^P(x) = \sum_{g\in P/P'} {}^g x$. Its image is clearly a two-sided ideal of kG^P. The kernel of Br_P is clearly $\sum_{P'\subset P} \mathrm{Tr}_{P'}^P(kG^{P'})$, a sum over $P' \neq P$. This is one of the ingredients to prove the above theorem. Another ingredient is the fact that, when P is a normal ℓ-subgroup of G, the ℓ-blocks of G, $\mathrm{C}_G(P)$ and G/P identify naturally (Brauer's "first Main Theorem", see [Ben91a] 6.2.6, [NaTs89] §5.2).

Definition 5.5. *If B_i is a block of ΛG (or the corresponding block $kG.b_i$ of kG), one calls a defect group of B_i any ℓ-subgroup $D \subseteq G$ such that there exists a maximal ℓ-subpair (D, e) containing $(\{1\}, b_i)$ (by the above, they are G-conjugates).*

Remark 5.6. (see [Ben91a] §6.3, [NaTs89] §3.6) A special case of Theorem 5.3(ii) is when $P = \{1\}$, i.e. b is a block idempotent of ΛG of trivial defect group. Then b (or the corresponding blocks of ΛG or kG) is said to be of **defect zero**. If B is an ℓ-block of G, then it is of defect zero if and only if there is some $\chi \in \mathrm{Irr}(G, B)$ vanishing on $G \setminus G_{\ell'}$. This is also equivalent to $B \otimes k = kG\overline{b}$ having a simple module which is also projective (thus implying $kG\overline{b}$ is a simple algebra). Blocks of defect zero are apparently scarce outside ℓ'-groups (see, however, a general case, "Steinberg modules," in §6.2 below).

A slightly more general case is when the defect group is *central* $P \subseteq \mathrm{Z}(G)$ (and therefore $P = \mathrm{Z}(G)_\ell$). Such blocks are in bijection with blocks of defect zero of $G/\mathrm{Z}(G)_\ell$. For each such block B, the corresponding block idempotent is written as $|G : P|^{-1}\chi(1)\sum_{g \in G_{\ell'}} \chi(g^{-1})g$ for any $\chi \in \mathrm{Irr}(G, B)$ (see [NaTs89] 3.6.22, 5.8.14). One even has $B \cong \mathrm{Mat}_d(\Lambda P)$ for some integer d, necessarily equal to $\chi(1)$ (see [Ben91a] 6.4.4, and also [Thévenaz] §49 for Puig's more general notion of "nilpotent" ℓ-blocks).

Definition 5.7. *If $x \in G_\ell$, let*

$$d^x : \mathrm{CF}(G, K) \to \mathrm{CF}(\mathrm{C}_G(x), K)$$

be defined by $d^x(f)(y) = f(xy)$ when $y \in \mathrm{C}_G(x)_{\ell'}$, $d^x(f)(y) = 0$ otherwise.

Theorem 5.8. (Brauer's "second Main Theorem") *The notation is as above. Let $B = \Lambda G.b$ be an ℓ-block of G with block idempotent b, $B_x = \Lambda \mathrm{C}_G(x) b_x$ an ℓ-block of $\mathrm{C}_G(x)$ with block idempotent b_x.*

If $d^x(\mathrm{CF}(G, K, B))$ has a non-zero projection on $\mathrm{CF}(\mathrm{C}_G(x), K, B_x)$, then there is an inclusion of ℓ-subpairs in G

$$(\{1\}, b) \subseteq (<x>, b_x).$$

For a proof, see [NaTs89] 5.4.2.

Definition 5.9. *When M is an indecomposable ΛG-module, there is a unique block $B_G(M)$ of ΛG acting by* Id *on M. This applies to indecomposable kG-modules and simple KG-modules, the latter often identified with their character $\chi \in \mathrm{Irr}(G)$. Denote by $b_G(M) \in B_G(M)$ the corresponding idempotent and $\overline{b}_G(M)$ its reduction mod. $J(\Lambda)$. One calls $B_G(1)$ (resp. $b_G(1)$) the principal block (resp. block idempotent) of G.*

Theorem 5.10. (Brauer's "third Main Theorem") *Let $(P, b) \subseteq (P', b')$ be an inclusion of ℓ-subpairs in G. If b or b' is a principal block idempotent, then both are.*

5.2. A model: blocks of symmetric groups

If X is a set, $\mathfrak{S}_X$ denotes the group of bijections $X \to X$, usually called "permutations" of X. When $X' \subseteq X$, $\mathfrak{S}_{X'}$ is considered as a subgroup of $\mathfrak{S}_X$.

Let $n \geq 1$ be an integer. One denotes $\mathfrak{S}_n := \mathfrak{S}_{\{1,\ldots,n\}}$. We assume that (Λ, K, k) is an ℓ-modular splitting system for $\mathfrak{S}_n$.

A **partition** of n is a sequence $\lambda_1 \geq \lambda_2 \geq \ldots \geq \lambda_l$ of integers ≥ 1 such that $\lambda_1 + \lambda_2 + \cdots + \lambda_l = n$. If $\lambda = \{\lambda_1 \geq \lambda_2 \geq \ldots \geq \lambda_l\}$ is a partition of n, we write $\lambda \vdash n$. The integer n is called the **size** of λ.

The set $\mathrm{Irr}(\mathfrak{S}_n)$ of irreducible characters of $\mathfrak{S}_n$ is in bijection with the set of partitions of n

$$\lambda \mapsto \chi^\lambda$$

(see [CuRe87] 75.19, [Gol93] 7, [JaKe81] 2).

In order to state properly the Murnaghan–Nakayama formula, which allows us to compute inductively the values of characters, we need to define the notion of a **hook** of a partition. In order to do that one introduces Frobenius' notion of "β-sets," i.e. finite subsets of $\mathbb{N} \setminus \{0\}$ associated with partitions. Let

$$\lambda \mapsto \beta(\lambda)$$

be the map associating the partition $\lambda = \{\lambda_1 \geq \lambda_2 \geq \ldots \geq \lambda_l\}$ with the set $\beta(\lambda) = \{\lambda_l, \lambda_{l-1} + 1, \ldots, \lambda_1 + l - 1\}$. So we have a bijection between partitions of integers greater than or equal to 1 and finite non-empty subsets of $\mathbb{N} \setminus \{0\}$. In order to treat also the case of the trivial group $\mathfrak{S}_0 := \{1\}$, we define the empty partition $\emptyset \vdash 0$ and associate with it the set $\beta(\emptyset) = \{0\}$.

Recall the notion of signature $\varepsilon \colon \mathfrak{S}_n \to \{-1, 1\}$. This extends to bijections $\sigma \colon \beta \to \beta'$ between finite subsets of $\mathbb{N}$, defined as the signature of the induced permutation of indices once β and β' have been ordered by $\geq$. Note that $\varepsilon(\sigma' \circ \sigma) = \varepsilon(\sigma')\varepsilon(\sigma)$ whenever $\sigma \colon \beta \to \beta'$ and $\sigma' \colon \beta' \to \beta''$ are bijections between finite subsets of $\mathbb{N}$.

Definition 5.11. *Let β be a finite subset of $\mathbb{N}$. Let $m \geq 1$. An m-hook of β is any subset $\gamma = \{a, a+m\} \subseteq \mathbb{N}$ such that $a + m \in \beta$ and $a \notin \beta$. One defines $\beta * \gamma$ as being $\beta \dot{+} \gamma$ (boolean sum) if $a \neq 0$ or $\beta = \{m\}$, while $\beta * \{0, m\}$ is defined as the image of $\beta \dot{+} \gamma$ under the unit translation $x \mapsto x + 1$ when $\beta \neq \{m\}$. The signature of the hook γ of β is defined as the signature of the bijection $\beta \to \beta \dot{+} \gamma$ which is the identity on $\beta \setminus \{a + m\}$. It is denoted by $\varepsilon(\beta, \gamma) = (-1)^{|[a,a+m] \cap \beta|}$.*

Definition 5.12. *Let $\lambda \vdash n$. Let $m \geq 1$. An m-hook of λ is an m-hook γ of $\beta(\lambda)$. Define $\lambda * \gamma$ by $\beta(\lambda * \gamma) = \beta(\lambda) * \gamma$ (note that $\beta(\lambda) * \gamma$ is either a subset of $\mathbb{N} \setminus \{0\}$ or equal to $\{0\}$). Then $\lambda * \gamma \vdash n - m$.*

Define $\varepsilon(\lambda, \gamma) := \varepsilon(\beta(\lambda), \gamma)$.

One says λ *is an* m*-core if and only if it has no* m*-hook.*

The following is the main tool for computing character values in symmetric groups (see, for instance, [Gol93] 12.6).

Theorem 5.13. (Murnaghan–Nakayama formula). *Let* $x \in \mathfrak{S}_n$, *let* u *be a cycle of* x, *of length* m, *so that* xu^{-1} *may be considered as an element of* $\mathfrak{S}_{n-m}$ *acting on the set of fixed points of* u. *One has*

$$\chi^\lambda(x) = \sum_\gamma \epsilon(\lambda, \gamma)\chi^{\lambda * \gamma}(xu^{-1})$$

where γ *runs on the set of* m*-hooks of* λ.

Since the integer $m \geq 1$ and the subset $\beta \subseteq \mathbb{N}$ are fixed, it is easy to see the behavior of the process $\beta \mapsto \beta * \gamma$ of removing successively all possible m-hooks. If $a \in [0, m-1]$, the subset $\beta \cap a + m\mathbb{N}$ may be replaced with $\{a, a+m, \ldots, a+m(c_a - 1)\}$ where $c_a := |\beta \cap a + m\mathbb{N}|$. One defines $\beta' = \bigcup_{a \in [0,m-1]} \{a, a+m, \ldots, a+m(c_a-1)\}$, which is clearly $\beta \dotplus \gamma_1 \dotplus \cdots \dotplus \gamma_t$ for any sequence where each γ_i is an m-hook of $\beta \dotplus \gamma_1 \dotplus \cdots \dotplus \gamma_{i-1}$, and $\beta \dotplus \gamma_1 \dotplus \cdots \dotplus \gamma_t$ has no m-hook. Note that the integer t is independent of the sequence chosen. Then the outcome of removing the m-hooks is β' if $0 \notin \beta'$ or $\beta' = \{0\}$; it is the image of β' under the unit translation $x \mapsto x+1$ if $0 \in \beta' \neq \{0\}$. It is clearly $\beta * \gamma_1 * \ldots * \gamma_t$ for any sequence of m-hooks where each γ_i is an m-hook of $\beta * \gamma_1 * \ldots * \gamma_{i-1}$, and $\beta * \gamma_1 * \ldots * \gamma_t$ has no m-hook.

Lemma 5.14. $\Pi_{i=1}^t \varepsilon(\beta * \gamma_1 * \ldots * \gamma_{i-1}, \gamma_i)$ *is independent of the sequence* $\gamma_1, \ldots, \gamma_t$.

Proof. Let $\sigma : \beta \to \beta'$ be the bijection defined by the sequence of m-hooks removals $\sigma_i : \beta \dotplus \gamma_1 \dotplus \cdots \dotplus \gamma_{i-1} \to (\beta \dotplus \gamma_1 \dotplus \cdots \dotplus \gamma_{i-1}) \dotplus \gamma_i$ (each σ_i fixes all elements but one). For each $a \in [0, m-1]$, σ restricts to the unique bijection $\beta \cap a + m\mathbb{N} \to \beta' \cap a + m\mathbb{N}$ that preserves the natural order. □

Theorem 5.15. *Let* $n \geq 0$, $m \geq 1$. *Let* $\lambda \vdash n$.

(i) If λ *is an* m*-core, it is an* mm'*-core for any* $m' \geq 1$.

(ii) For any sequence $\gamma_1, \ldots, \gamma_t$ *such that each* γ_i *is an* m*-hook of* $\lambda * \gamma_1 * \ldots * \gamma_{i-1}$ *and* $\lambda * \gamma_1 * \ldots * \gamma_t$ *has no* m*-hook, the outcome* $\lambda * \gamma_1 * \ldots * \gamma_t$ *is independent of the sequence* $\gamma_1, \ldots, \gamma_t$. *It is called the* m*-core of* λ.

(iii) (Iterated version of Murnaghan–Nakayama formula) With notation as above, let $x \in \mathfrak{S}_n$, *write* $x'c_1 \ldots c_t$ *with* $x' \in \mathfrak{S}_{n-tm}$ *and* c_i*'s being disjoint cycles of order* m *in* $\mathfrak{S}_{\{n-tm+1,\ldots,n\}}$. *Then*

$$\chi^\lambda(x'c_1 \ldots c_t) = N_{\lambda,m}\chi^{\lambda'}(x')$$

where λ' is the m-core of λ and $0 \neq N_{\lambda,m} \in \mathbb{Z}$ is a non-zero integer independent of x'.

Proof. (i) is clear from the definition of m-hooks of finite subsets of $\mathbb{N}$.

(ii) is clear from the above discussion of m-hook removal for subsets of $\mathbb{N}$.

(iii) Using the Murnaghan–Nakayama formula (Theorem 5.13), one finds $\chi^\lambda(x) = \sum_{(\gamma_i)} \Pi_{i=1}^t \varepsilon(\lambda * \gamma_1 * \ldots * \gamma_{i-1}, \gamma_i)\chi^{\lambda'}(x')$, where the sum is over all sequences $(\gamma_i)_{1\le i\le t}$ where each γ_i is an m-hook of $\lambda * \gamma_1 * \ldots * \gamma_{i-1}$ and $\lambda * \gamma_1 * \ldots * \gamma_t = \lambda'$. By Lemma 5.14, the product of signs is the same for all sequences. One finds the claimed result with N being this sign times the number of possible sequences. □

Theorem 5.16. *The ℓ-blocks of $\mathfrak{S}_n$ are in bijection*

$$\kappa \mapsto B(\kappa)$$

with ℓ-cores $\kappa \vdash s(\kappa)$ such that their sizes satisfy $s(\kappa) \le n$ and $s(\kappa) \equiv n \bmod \ell$.

The above bijection is defined by $\mathrm{Irr}(\mathfrak{S}_n, B(\kappa)) = \{\chi^\lambda \mid \kappa$ is the ℓ-core of $\lambda\}$. *The Sylow ℓ-subgroups of $\mathfrak{S}_{n-s(\kappa)}$ are defect groups of $B(\kappa)$.*

Lemma 5.17. *If κ is an ℓ-core, then χ^κ defines an ℓ-block of $\mathfrak{S}_n$ with defect zero (see Remark 5.6).*

Proof of Lemma 5.17. By Remark 5.6, we must check that $\chi^\kappa(x) = 0$ whenever $x \in \mathfrak{S}_n$ is of order a multiple of ℓ. If x is not ℓ', then it can be written $x = cx'$ where c is a cycle of order $|c|$, a multiple of ℓ, and x' has support disjoint with the one of c. Theorem 5.15(i) tells us that κ has no $|c|$-hook. The Murnaghan–Nakayama formula then implies $\chi^\kappa(cx') = 0$, thus our claim.

Another proof would consist of checking that $\chi^\kappa(1)_\ell = (n!)_\ell$ by use of the "degree formula" (see [JaKe81] 2.3.21 or [Gol93] 12.1). □

Proof of Theorem 5.16. Let $\lambda \vdash n$. Let $\kappa \vdash n - \ell w$ be its ℓ-core. Let $c_1 = (n, n-1, \ldots, n-\ell+1)$, …, $c_w = (n-\ell(w-1), \ldots, n-\ell w+1)$ be w disjoint cycles of order ℓ. Let $c = c_1 \ldots c_w$.

Lemma 5.18. *We have $\mathrm{C}_{\mathfrak{S}_n}(c) = \mathfrak{S}_{n-\ell w} \times W$ where W is a subgroup of $\mathfrak{S}_{\{n-\ell w+1,\ldots n\}}$ such that ΛW is a single block.*

Proof of Lemma 5.18. Arguing on permutations preserving the set of supports of the c_i's, it is easy to find that $\mathrm{C}_{\mathfrak{S}_n}(c) = \mathfrak{S}_{n-\ell w} \times W$ where $W = (<c_1> \times \cdots \times <c_w>) \rtimes \mathfrak{S}'_w$ and $\mathfrak{S}'_w \cong \mathfrak{S}_w$ by a map sending $(i, i+1)$ to $(n-\ell(i-1), n-\ell i)(n-\ell(i-1)+1, n-\ell i+1)\ldots(n-\ell i+1, n-\ell(i+1)+1)$ for $i = 1, \ldots, w-1$. We have $\mathrm{C}_{\mathfrak{S}'_w}(<c_1> \times \cdots \times <c_w>) = 1$.

This implies that W has a normal ℓ-subgroup containing its centralizer (in W). Then W has a single ℓ-block (see, for instance, [Ben91a] 6.2.2). □

Let $b_\kappa \in \Lambda\mathfrak{S}_{n-\ell w}$ be the block idempotent of $\mathfrak{S}_{n-\ell w}$ corresponding to χ^κ. It is of defect zero by Lemma 5.17. Since W has just one ℓ-block (Lemma 5.18), b_κ is a block idempotent of $\mathrm{C}_{\mathfrak{S}_n}(c)$. Denote $B_\kappa = \Lambda\mathrm{C}_{\mathfrak{S}_n}(c).b_\kappa$.

Let us show that $d^c(\chi^\lambda)$ has a non-zero projection on $\mathrm{CF}(\mathrm{C}_{\mathfrak{S}_n}(c), K, B_\kappa)$. Suppose the contrary. Since $\mathrm{Irr}(\mathfrak{S}_{n-\ell w}, b_\kappa) = \{\chi^\kappa\}$ (see Remark 5.6), the central function $d^c(\chi^\lambda)$ can be written as $d^c(\chi^\lambda) = \sum_{\mu\vdash n-\ell w, \mu\neq\kappa, \zeta\in\mathrm{Irr}(W)} r_{\mu,\zeta}\chi^\mu\zeta$ for scalars $r_{\mu,\zeta} \in K$. Taking restrictions to $\mathfrak{S}_{n-\ell w}$, one finds that $\mathrm{Res}_{\mathfrak{S}_{n-\ell w}}(d^c(\chi^\lambda))$ is orthogonal to χ^κ. This means $\sum_{x\in(\mathfrak{S}_{n-\ell w})_{\ell'}} \chi^\lambda(cx)\chi^\kappa(x) = 0$. Since χ^κ is in a block of defect zero, it vanishes outside ℓ'-elements (see Remark 5.6), so the above sum is $\sum_{x\in\mathfrak{S}_{n-\ell w}} \chi^\lambda(cx)\chi^\kappa(x)$. By Theorem 5.15(iii), this is $N.\langle\chi^\kappa, \chi^\kappa\rangle_{\mathfrak{S}_{n-\ell w}} = N$ where N is a non-zero integer, a contradiction.

Now Brauer's second Main Theorem implies $(\{1\}, b_{\mathfrak{S}_n}(\chi^\lambda)) \lhd (<c>, b_\kappa)$ in $\mathfrak{S}_n$.

Brauer's third Main Theorem implies the inclusion $(<c>, 1) \subseteq (S, b_0)$ in $\mathfrak{S}_{\{n-\ell w+1,\dots,n\}}$ for S a Sylow ℓ-subgroup of $\mathfrak{S}_{\{n-\ell w+1,\dots,n\}}$ containing c and b_0 the principal block idempotent of $\Lambda\mathrm{C}_{\mathfrak{S}_{\{n-\ell w+1,\dots,n\}}}(S)$. Therefore one gets easily $(<c>, b_\kappa) \subseteq (S, b_\kappa b_0)$ in $\mathfrak{S}_n$. Combining with the previous inclusion, one gets

$$\text{(M)} \qquad (\{1\}, b_{\mathfrak{S}_n}(\chi^\lambda)) \lhd (S, b_\kappa b_0).$$

Let us show that the right-hand side is a maximal ℓ-subpair. We apply Theorem 5.3(iii). We have $N_{\mathfrak{S}_n}(S) \subseteq \mathfrak{S}_{n-\ell w} \times \mathfrak{S}_{\{n-\ell w+1,\dots,n\}}$ (fixed points), and $S.\mathrm{C}_{\mathfrak{S}_n}(S) \supseteq S.\mathfrak{S}_{\{n-\ell w+1,\dots,n\}}$ has an index prime to ℓ in it. So it suffices to check that $b_\kappa b_0$ has defect $\mathrm{Z}(S)$ in $\mathrm{C}_{\mathfrak{S}_n}(S) = \mathfrak{S}_{n-\ell w} \times \mathrm{C}_{\mathfrak{S}_{\{n-\ell w+1,\dots,n\}}}(S)$. The two sides are independent. On the first, b_κ has defect zero (Lemma 5.17). On the $\mathfrak{S}_{\{n-\ell w+1,\dots,n\}}$ side, (S, b_0) is a maximal ℓ-subpair since S is a Sylow ℓ-subgroup, so that $(\mathrm{Z}(S), b_0)$ is maximal as an ℓ-subpair of $\mathrm{C}_{\mathfrak{S}_{\{n-\ell w+1,\dots,n\}}}(S)$ (Theorem 5.3(iii) again).

By Theorem 5.3(i), the above gives a map $\kappa \mapsto B(\kappa)$ where $B(\kappa)$ has the claimed defect and $\mathrm{Irr}(\mathfrak{S}_n, B(\kappa))$ contains all χ^λ where $\lambda \vdash n$ has ℓ-core κ. It remains to show that this map is injective.

Let λ, λ' be two partitions of n. Let κ, κ' be their ℓ-cores. Let us build the maximal ℓ-subpairs $(S, b_\kappa b_0)$ and $(S', b_{\kappa'} b'_0)$ as in (M) above. If χ^λ and $\chi^{\lambda'}$ are in the same ℓ-block, we have $(S, b_\kappa b_0)$ conjugate with $(S', b_{\kappa'} b'_0)$ (Theorem 5.3(ii)). Then $w = w'$ (fixed points under S and S'), and $b_\kappa = b_{\kappa'}$ hence $\chi^\kappa = \chi^{\kappa'}$ by defect zero (see Remark 5.6). Then $\kappa = \kappa'$.

5.3. Principal series and the principal block

We now give an application of the local methods described in §5.1 to ℓ-blocks of finite groups with split BN-pair of characteristic $\neq \ell$.

Theorem 5.19. *Let G be a finite group with a strongly split BN-pair ($B = U \rtimes T, N$) of characteristic p. Let (Λ, K, k) be an ℓ-modular splitting system for G. Let L be a standard Levi subgroup and $\chi \in \mathrm{Irr}(L)$ a cuspidal character with $\mathrm{Z}(L)_\ell$ in its kernel. Assume the following hypothesis*

(∗) *there exist $x_1, \ldots, x_m \in \mathrm{Z}(L)_\ell$ such that $C_i := \mathrm{C}_G(<x_1, \ldots, x_i>)$ is a standard Levi subgroup for all $i = 1, \ldots, m$, and $C_m = L$.*

Then all the irreducible characters occuring in $\mathrm{R}_L^G\chi$ are in the same block of ΛG.

Corollary 5.20. *Assume the same hypotheses as above. When $(L, \chi) = (T, 1)$, we get the following. All the indecomposable summands of $\mathrm{Ind}_B^G \Lambda$ (and all the irreducible characters occurring in $\mathrm{Ind}_B^G 1$) are in the principal block of ΛG.*

Remark 5.21. Let $G = \mathrm{GL}_n(\mathbb{F}_q)$ (see Example 2.17(i)) and assume ℓ is a prime divisor of $q-1$. Let $\omega \in \mathbb{F}_q$ be an ℓth root of unity. If $1 \leq m < n$, let d_m be the diagonal matrix with diagonal elements $(\omega, \ldots, \omega, 1, \ldots, 1)$ (m ω's and $n-m$ 1's). Then the $\mathrm{C}_G(d_i)$'s are all the maximal standard Levi subgroups of G. So, by induction, all standard Levi subgroups of G satisfy the hypothesis of Theorem 5.19.

Lemma 5.22. *H is a finite group, V a subgroup. Take $h \in \mathrm{N}_H(V)$ and assume $\mathrm{C}_H(h) \cap V = \{1\}$. Then all elements of Vh are H-conjugates.*

Proof. The map $v \mapsto h^v = v^{-1}.{}^h v.h$ sends V to Vh since h normalizes V. The map is injective since $\mathrm{C}_H(h) \cap V = \{1\}$. So its image is the whole of Vh and we get our claim. □

Recall the functors R_L^G and ${}^*\mathrm{R}_L^G$ (Notation 3.11).

Proposition 5.23. *Let $x \in G_\ell$ be such that $\mathrm{C}_G(x) = L_I$. If $J \subseteq I$ and $x \in \mathrm{Z}(L_J)$, then $d^x \circ {}^*\mathrm{R}_{L_J}^G = {}^*\mathrm{R}_{L_J}^{L_I} \circ d^x$ on $\mathrm{CF}(G, K)$.*

Proof. Assume first that x is central, i.e. $L_I = G$. Then d^x is d^1 composed with a translation by a central element. This translation clearly commutes with ${}^*\mathrm{R}_{L_J}^G$ since ${}^*\mathrm{R}_{L_J}^G$ may be seen as a multiplication by $e(U_J)$. But d^1 is self-adjoint and commutes with induction and inflation, so d^1 commutes with $\mathrm{R}_{L_J}^G$ and ${}^*\mathrm{R}_{L_J}^G$.

For a general x, we write ${}^*\mathrm{R}_{L_J}^G = {}^*\mathrm{R}_{L_J}^{L_I} \circ {}^*\mathrm{R}_{L_I}^G$ (Proposition 1.5(ii)) and it suffices to check that $d^x \circ {}^*\mathrm{R}_{L_I}^G = d^x$ on $\mathrm{CF}(G, K)$. Let f be a central function on G, let $y \in L_I = \mathrm{C}_G(x)$. If $y_\ell \neq 1$, then $d^x f(y) = d^x({}^*\mathrm{R}^G_{L_I} f)(y) = 0$. Assume $y_\ell = 1$. Then $d^x f(y) = f(xy)$, while $d^x({}^*\mathrm{R}^G_{L_I} f)(y) = ({}^*\mathrm{R}^G_{L_I} f)(xy) = |U_I|^{-1} \sum_{u \in U_I} f(uxy)$. Since f is a central function, it is enough to check that every uxy above is a G-conjugate of xy. We apply Lemma 5.22 with $V = U_I$, $h = xy$. This is possible because $xy \in \mathrm{C}_G(x) \subseteq L_I$ which normalizes U_I, and $\mathrm{C}_G(xy) \subseteq \mathrm{C}_G(x) \subseteq L_I$ has a trivial intersection with U_I. □

Proof of Theorem 5.19 and Corollary 5.20. We have clearly $L = \mathrm{C}_G(\mathrm{Z}(L)_\ell)$, so that $(\mathrm{Z}(L)_\ell, b_L(\chi))$ is an ℓ-subpair in G. We prove that, if B is a block of ΛG such that the projection of $\mathrm{R}^G_L \chi$ on B is not zero, then $(\{1\}, B) \subseteq (\mathrm{Z}(L)_\ell, b_L(\chi))$.

This proves that B is unique, as a result of Theorem 5.3(i). In the case where $\chi = 1$, $b_L(\chi)$ is the principal block and Brauer's third Main Theorem implies that B is the principal block of ΛG.

So we let ξ be an irreducible character of G occurring in $\mathrm{R}^G_L \chi$. We must prove that

$$(\{1\}, b_G(\xi)) \subseteq (\mathrm{Z}(L)_\ell, b_L(\chi)).$$

We use the following lemma, the proof of which is postponed until after this one is complete.

Lemma 5.24. *If* $\langle \xi, \mathrm{R}^G_L \chi \rangle_G \neq 0$, *then* $\langle {}^*\mathrm{R}^G_L \xi, d^1 \chi \rangle_L \neq 0$.

We may use induction on $|G : L|$, the case when $G = L$ being trivial.

Assume $L \neq G$, so that some x_i is not in $\mathrm{Z}(G)$. We may assume it is x_1. Let $C := \mathrm{C}_G(x_1) \neq G$.

The induction hypothesis in C implies that all irreducible components of $\mathrm{R}^C_L \chi$ are in a single $\mathrm{Irr}(C, b)$ for b a block idempotent of ΛC and that $(\{1\}, b) \subseteq (\mathrm{Z}(L)_\ell, b_L(\chi))$ in $C = \mathrm{C}_G(x_1)$. This inclusion is trivially equivalent to $(<x_1>, b) \subseteq (\mathrm{Z}(L)_\ell, b_L(\chi))$ in G.

However, $\langle d^{x_1} \xi, \mathrm{R}^C_L \chi \rangle_C = \langle {}^*\mathrm{R}^C_L \circ d^{x_1} \xi, \chi \rangle_L = \langle d^{x_1} \circ {}^*\mathrm{R}^G_L \xi, \chi \rangle)_L$ by Proposition 5.23. Since x_1 is in the center of L, $d^{x_1, L}$ is self-adjoint on $\mathrm{CF}(L, K)$ and, since χ has x_1 in its kernel, we have $d^{x_1} \chi = d^1 \chi$. Then $\langle d^{x_1} \xi, \mathrm{R}^C_L \chi \rangle_C = \langle {}^*\mathrm{R}^G_L \xi, d^1 \chi \rangle_L \neq 0$ by Lemma 5.24. This implies that $d^{x_1} \xi$ has a non-zero projection on $\mathrm{CF}(C, K, b)$. Then Brauer's second Main Theorem implies $(\{1\}, b_G(\xi)) \subseteq (<x_1>, b)$. Combining with the inclusion previously obtained, this gives our claim by transitivity of inclusion. □

Proof of Lemma 5.24. Let $\mathcal{I}$ be the group $\mathrm{N}_G(L)/\mathrm{N}_G(L, \chi)$. We prove first

$$(1) \qquad {}^*\mathrm{R}^G_L \xi = \langle \chi, {}^*\mathrm{R}^G_L \xi \rangle_L \sum_{\sigma \in \mathcal{I}} \chi^\sigma.$$

One has ${}^*\mathrm{R}^G_L \xi = \sum_{\chi' \in \mathrm{Irr}(L)} \langle {}^*\mathrm{R}^G_L \xi, \chi' \rangle_L \chi'$. Since $\langle {}^*\mathrm{R}^G_L \xi, \chi^\sigma \rangle_L = \langle {}^\sigma {}^*\mathrm{R}^G_L \xi, \chi \rangle_L = \langle {}^*\mathrm{R}^G_L \xi, \chi \rangle_L$ for all $\sigma \in \mathrm{N}_G(L)$, it suffices to check that $\langle {}^*\mathrm{R}^G_L \xi, \chi' \rangle_L \neq 0$

implies $\chi' = \chi^\sigma$ for some $\sigma \in \mathcal{I}$. If $\langle \mathrm{R}_L^G \chi', \xi \rangle_G \neq 0$, then transitivity of Harish-Chandra induction and Theorem 1.30 imply that χ' is cuspidal and $(P, V, \chi) \text{—} {}^g(P, V, \chi')$ (see Notation 1.10) for some $g \in G$ and a Levi decomposition $P = LV$. By Theorem 2.27(iv), there is some $g' \in \mathrm{N}_G(L)$ such that $\chi' = \chi^{g'}$.

Denote $f := {}^*\mathrm{R}_L^G \xi = \langle \chi, {}^*\mathrm{R}_L^G \xi \rangle_L \sum_{\sigma \in \mathcal{I}} \chi^\sigma$ by (1) above. Then $\langle f, d^1 \chi \rangle_L = \langle f^\sigma, d^1 \chi^\sigma \rangle_L = \langle f, d^1 \chi^\sigma \rangle_L$ for all $\sigma \in \mathcal{I}$. So $\langle f, d^1 \chi \rangle_L = |\mathcal{I}|^{-1} \langle \chi, {}^*\mathrm{R}_L^G \xi \rangle_L^{-1} \langle f, d^1 f \rangle_L = |\mathcal{I}|^{-1} \langle \chi, {}^*\mathrm{R}_L^G \xi \rangle_L^{-1} \langle d^1 f, d^1 f \rangle_L$. But f, being a character, is a central function whose values are algebraic numbers, and $f(g^{-1})$ is the complex conjugate of $f(g)$ for all $g \in L$ (see, for instance, [NaTs89] §3.2.1). Then $\langle d^1 f, d^1 f \rangle_L = |L|^{-1} \sum_{g \in L_{\ell'}} f(g) f(g^{-1})$ is a real number greater than or equal to $|L|^{-1} f(1)^2$. The latter is greater than 0 because $f(1) = {}^*\mathrm{R}_L^G \xi(1)$ is the dimension of a KL-module $\neq 0$ since $\langle f, \chi \rangle_L = \langle \xi, \mathrm{R}_L^G \chi \rangle_G \neq 0$ by hypothesis. □

5.4. Hecke algebras and decomposition matrices

We recall the notion of a decomposition matrix in order to apply it to both group algebras and Hecke algebras.

Let (Λ, K, k) be a splitting system for a Λ-algebra A, Λ-free of finite rank (see our section on Terminology). Recall that this includes the hypothesis that $A \otimes K$ is a product of matrix algebras over K.

Definition 5.25. *Let M be a finitely generated Λ-free A-module. One defines*

$$\mathrm{Dec}_A(M) = (d_{ij})$$

the matrix where i (resp. j) ranges over the isomorphism classes of simple submodules (resp. indecomposable summands) of $M \otimes K$ (resp. M) and d_{ij} denotes the multiplicity of i in $M_j \otimes K$ (M_j in the class j).

One denotes $\mathrm{Dec}(A) := \mathrm{Dec}_A({}_A A)$.

Remark 5.26. If A is the group algebra of a finite group or a block in such an algebra, $\mathrm{Dec}(A)$ has more rows than columns and in fact ${}^t\mathrm{Dec}(A)\mathrm{Dec}(A)$ is invertible (see [Ben91a] 5.3.5).

However, even among Λ-free algebras of finite rank such that $A \otimes K$ is semi-simple, this is a rather exceptional phenomenon: A being fixed, $A \otimes K$ has a finite number of simple modules but in general A has infinitely many indecomposable modules, so we may have matrices $\mathrm{Dec}_A(M)$ (M an A-module) with many more columns than rows. Those matrices are in turn of the type $\mathrm{Dec}(A)$ by Proposition 5.27 below.

Proposition 5.27. $\mathrm{Dec}_A(M) \cong \mathrm{Dec}(\mathrm{End}_A(M)^{\mathrm{opp}})$ *where the bijection between indecomposable modules is induced by* $\mathrm{Hom}_A(M, -)$*, and the bijection between simple modules is induced by* $\mathrm{Hom}_{A\otimes K}(M \otimes K, -)$.

Proof. Let us abbreviate $E := \mathrm{End}_A(M)$ and $H_M = \mathrm{Hom}_A(M, -)$. Let N be an indecomposable direct summand of M. Let S be a simple $A \otimes K$-module. The number in $\mathrm{Dec}_A(M)$ associated with the pair (N, S) is the dimension of $\mathrm{Hom}_{A\otimes K}(N \otimes K, S)$. One may use Theorem 1.25(i) for the $A \otimes K$-module $M \otimes K$ since $A \otimes K$ and therefore $\mathrm{End}_{A\otimes K}(M \otimes K) = E \otimes K$ is semi-simple (hence symmetric). One gets that $\mathrm{Hom}_{A\otimes K}(N \otimes K, S) \cong \mathrm{Hom}_{E\otimes K}(H_{M\otimes K}(N \otimes K), H_{M\otimes K}(S))$. One has clearly $H_{M\otimes K}(N \otimes K) = H_M(N) \otimes K$. Now, note that the $H_M(N)$ for N ranging over the indecomposable summands of M are the right projective indecomposable modules for E (write $N = iM$ for a primitive idempotent $i \in E$, and $H_M(iM) = iE$). The same result for the semi-simple module $M \otimes K$ gives us that $H_{M\otimes K}(S)$ ranges over the simple (= projective indecomposable) $E \otimes K$-modules, whence our claim. □

Theorem 5.28. *Let G be a finite group with split BN-pair of characteristic p, with Weyl group (W, S), $B = UT$. Let ℓ be a prime $\neq p$. Let (Λ, K, k) be an ℓ-modular splitting system for G.*

Then the decomposition matrix of $\mathcal{H}_\Lambda(G, B)$ (see Definition 3.4) embeds in that of G.

Proof. (See also Exercise 3.) By Theorem 3.3, $\mathcal{H}_\Lambda(G, B) \cong \mathrm{End}_{\Lambda G}(\mathrm{Ind}_B^G \Lambda)$ where Λ is the trivial ΛB-module.

Let $x = \sum_{u\in U} u = |U|e(U)$, $y = \sum_{t\in T} t \in \Lambda G$. Then $xy = \sum_{b\in B} b$ and therefore

$$\Lambda G e(U) \cong \Lambda G x \cong \mathrm{Ind}_U^G \Lambda$$

is projective, while $\Lambda Gxy \cong \mathrm{Ind}_B^G \Lambda$. The decomposition matrix of the module ΛGx clearly embeds in that of ΛG (it corresponds to certain columns of it). We then show that $\mathrm{Dec}_{\Lambda G}(\Lambda Gxy)$ embeds in $\mathrm{Dec}_{\Lambda G}(\Lambda Gx)$. This gives our claim by Proposition 5.27 (with $M = {}_A A$). The module ΛGxy is the image of ΛGx under the map $\mu: a \mapsto ay$. If M is an indecomposable direct summand of ΛGxy, there is an indecomposable direct summand P of ΛGx sent onto M (uniqueness of projective covers). If S is a simple component of $M \otimes K$ we have to show that it is not a component of $\mathrm{Ker}(\mu) \otimes K$. To see that, it suffices in fact to show that $\Lambda Gxy \otimes K$ and $\mathrm{Ker}(\mu) \otimes K$ have no simple component in common. In terms of characters this amounts to checking that $\mathrm{Ind}_B^G 1$ is orthogonal to $\mathrm{Ind}_U^G 1 - \mathrm{Ind}_B^G 1$. Using the Mackey formula ([Ben91a] 3.3.4), one finds $\langle \mathrm{Ind}_B^G 1, \mathrm{Ind}_U^G 1\rangle_G = |B\backslash G/U|$ and $\langle \mathrm{Ind}_B^G 1, \mathrm{Ind}_B^G 1\rangle_G = |B\backslash G/B|$. Both are equal to $|W|$ by Bruhat decomposition. This completes our proof. □

5.5. A proof of Brauer's third Main Theorem

We prove the following generalization of R. Brauer's third Main Theorem (where $H = \{1\}$, $\rho = 1_G$).

Proposition 5.29. *Let $\rho \in \mathrm{Irr}(G)$ and H be a subgroup of G such that $\mathrm{Res}_H^G \rho$ is irreducible and is in $\mathrm{Irr}(H, b_H)$, where b_H is an ℓ-block idempotent of ΛH with central defect group (in H). Let $Q \subseteq Q'$ be two ℓ-subgroups of $\mathrm{C}_G(H)$. Then $(Q, b_{\mathrm{C}_G(Q)}(\mathrm{Res}^G_{\mathrm{C}_G(Q)}\rho)) \subseteq (Q', b_{\mathrm{C}_G(Q')}(\mathrm{Res}^G_{\mathrm{C}_G(Q')}\rho))$ in G.*

Proof. For every subgroup H' such that $H \subseteq H' \subseteq G$, denote $\sigma_{H'} = \sum_{g\in H'} \rho(g^{-1})g \in \mathrm{Z}(\Lambda H')$. One has $\mathrm{Res}^G_{H'}\rho \in \mathrm{Irr}(H')$. The ℓ-block idempotent $b_{H'} \in \mathrm{Z}(\Lambda H')$ corresponding to $\mathrm{Res}^G_{H'}\rho$ satisfies $b_{H'}.e = e$, where $e = |H'|^{-1}\rho(1)\sigma_{H'}$ is the primitive idempotent of $\mathrm{Z}(KH')$ associated with $\mathrm{Res}^G_{H'}\rho \in \mathrm{Irr}(H')$ (see [NaTs89] 3.6.22). Therefore

$$\text{(E)} \qquad \overline{b_{H'}}\,\overline{\sigma_{H'}} = \overline{\sigma_{H'}}.$$

Remark 5.6 also allows us to write $b_H = \frac{\rho(1)}{|H:Z(H)_\ell|}\sum_{g\in H_{\ell'}} \rho(g^{-1})g$ and (consequently) $\frac{\rho(1)}{|H:Z(H)_\ell|}$ is a unit in Λ. Then there exists some $g \in H$ such that $\rho(g) \neq 0$. Then $\overline{\sigma_{H'}} \neq 0$ in kG, and (E) above characterizes $b_{H'}$ as a result of the orthogonality of distinct block idempotents.

Now, to check the proposition, it suffices to check the case when $Q \lhd Q'$. Then $\mathrm{Res}^G_{\mathrm{C}_G(Q)}\rho$ is fixed by Q' since ρ is a central function on G, thus $\sigma_{\mathrm{C}_G(Q)}$ and $b_{\mathrm{C}_G(Q)}$ are fixed by Q'. One has $\mathrm{Br}_{Q'}(\overline{\sigma_{\mathrm{C}_G(Q)}}) = \overline{\sigma_{\mathrm{C}_G(Q')}}$. In order to check the inclusion it suffices to check $\mathrm{Br}_{Q'}(\overline{b_{\mathrm{C}_G(Q)}})\overline{\sigma_{\mathrm{C}_G(Q')}} = \overline{\sigma_{\mathrm{C}_G(Q')}}$. Using the fact that $\mathrm{Br}_{Q'}$ induces an algebra morphism on $(kG)^{Q'}$, we have $\mathrm{Br}_{Q'}(\overline{b_{\mathrm{C}_G(Q)}})\overline{\sigma_{\mathrm{C}_G(Q')}} = \mathrm{Br}_{Q'}(\overline{b_{\mathrm{C}_G(Q)}})\mathrm{Br}_{Q'}(\overline{\sigma_{\mathrm{C}_G(Q)}}) = \mathrm{Br}_{Q'}(\overline{b_{\mathrm{C}_G(Q)}\sigma_{\mathrm{C}_G(Q)}}) = \mathrm{Br}_{Q'}(\overline{\sigma_{\mathrm{C}_G(Q)}}) = \overline{\sigma_{\mathrm{C}_G(Q')}}$ as claimed. □

Exercises

1. Let B, B' be finite subsets of $\mathbb{N}$. Fix $m \geq 1$. Assume $\sigma\colon B \to B'$ is a bijection such that, for any $b \in B$, $\sigma(b) - b \in m\mathbb{N}$. Then B may be deduced from B' by a sequence of removal of m-hooks. Prove a converse.

   Show that the map of Theorem 5.16 is actually onto.

2. If D, D' are matrices, one defines $D \subseteq D'$ by the condition that there is a permutation of rows and columns of D' producing a matrix which can be written as

$$\begin{pmatrix} D & * \\ 0 & * \end{pmatrix}.$$

If A and B are Λ-free finitely generated algebras and B is a quotient of A, show that ${}^t\mathrm{Dec}(B) \subseteq {}^t\mathrm{Dec}(A)$.

If N is a direct summand of M in $A-\mathbf{mod}$, show that $\mathrm{Dec}_A(N) \subseteq \mathrm{Dec}_A(M)$.

3. Let G be a finite group with split BN-pair of characteristic p. Let (Λ, K, k) be an ℓ-modular splitting system for G, where ℓ is a prime $\neq p$. Denote $Y := \mathrm{Ind}_U^G \Lambda$ where Λ is considered as the trivial ΛU-module. Show that Y is projective. Show that $\mathrm{End}_{\Lambda G}(Y) = \bigoplus_{n \in N} \Lambda a_n$ where the a_n's are defined in a way similar to Definition 6.7 below. Check Proposition 6.8(ii) for those a_n's. Show that $\alpha := \sum_{t \in T} a_t$ is in the center of $\mathrm{End}_{\Lambda G}(Y)$, and that $\alpha Y \cong \mathrm{Ind}_B^G \Lambda$. Deduce that $\mathrm{End}_{\Lambda G}(\mathrm{Ind}_B^G \Lambda)$ is a quotient of $\mathrm{End}_{\Lambda G}(Y)$ as a Λ-algebra. Deduce Theorem 5.28 by applying the above.

Notes

Brauer's "local" strategy for $\mathfrak{S}_n$ applies well to many finite groups G close to the simple groups.

Picking any irreducible character $\chi \in \mathrm{Irr}(G)$, one would find some non-central ℓ-element x such that $d^x \chi \neq 0$, thus starting an induction process by use of the second Main Theorem (see, for instance, [Pu87]).

There are, however, blocks where this does not work. Let G be a central non-trivial extension of $\mathfrak{S}_n$ by $Z = Z(G)$ of order 2,

$$1 \to Z \to G \to \mathfrak{S}_n \to 1.$$

For $\ell = 2$, there are many $\chi \in \mathrm{Irr}(G)$ such that $d^x \chi = 0$ for any 2-element $x \in G \setminus Z$ but the corresponding 2-block $B_G(\chi)$ has defect $\neq Z$ (see [BeOl97]).

Theorem 5.28 is due to Dipper [Dip90].

6
Simple modules in the natural characteristic

We keep G a finite group endowed with a strongly split BN-pair $B = UT, N, \ldots$ of characteristic p. In the present chapter, we give some results about representations in characteristic p ("natural" characteristic). For the moment, k is of characteristic p, U denotes the Sylow p-subgroup of B. We study the permutation module $\mathrm{Ind}_U^G k := k[G/U]$ and its endomorphism algebra $\mathcal{H}_k(G, U)$. The symmetry property of Hecke algebras mentioned in transversal characteristic (see Theorem 1.20) is now replaced by the self-injectivity of $\mathcal{H}_k(G, U)$. This allows us to relate the simple submodules of $\mathrm{Ind}_U^G k$ to the simple $\mathcal{H}_k(G, U)$-modules. The latter are one-dimensional, a feature reminiscent of the "highest weight" property of irreducible representations of complex Lie algebras. Among the direct summands of $\mathrm{Ind}_U^G k$, one finds the "Steinberg module," which is at the same time simple and projective. This leads us quite naturally to an enumeration of the blocks of kG, and a checking of J. Alperin's "weight conjecture" in this special context of BN-pairs represented in natural characteristic.

6.1. Modular Hecke algebra associated with a Sylow p-subgroup

Let G be a finite group with a strongly split BN-pair of characteristic p with subgroups $B = UT, N, S$ (see Definition 2.20). Let us recall that T is commutative.

In the remainder of the chapter, k is a field of characteristic p containing a $|G|_{p'}$th root of 1 (so that $kH/J(kH)$ is a split semi-simple algebra for any subgroup H of G; see [NaTs89] §3.6).

Proposition 6.1. *Let Q be a finite p-group. Then the unique simple kQ-module is the trivial module. One has the following consequences. The regular module kQ is indecomposable. Any kQ-module $M \neq 0$ satisfies $M^Q \neq 0$.*

Proof. See [Ben91a] 3.14.1. □

Definition 6.2. *If $\delta \in \Delta$ (see §2.1), let G_δ be the group generated by X_δ and $X_{-\delta}$. Let $T_\delta = T \cap G_\delta$.*

Proposition 6.3. *(i) There is $n_\delta \in X_\delta X_{-\delta} X_\delta \cap N$ such that its class mod. T is the reflection s_δ associated with δ.*

If each n_δ ($\delta \in \Delta$) is chosen as above, we have the following properties.

(ii) $N \cap G_\delta = T_\delta \cup n_\delta T_\delta$ and G_δ has a split BN-pair $(X_\delta T_\delta, N \cap G_\delta, \{s_\delta\})$ of characteristic p.

(iii) $n_\delta^{-1}(X_\delta \setminus \{1\})n_\delta^{-1} \subseteq X_\delta T_\delta n_\delta^{-1} X_\delta$.

(iv) $[T, n_\delta] \subseteq T_\delta$.

(v) Take $\delta_1, \ldots, \delta_l, \delta'_1, \ldots, \delta'_l \in \Delta$, and $t \in T$ such that $n_{\delta_1} \ldots n_{\delta_l} = n_{\delta'_1} \ldots n_{\delta'_l} t$ and $l(n_{\delta_1} \ldots n_{\delta_l}) = l$. Then $t \in T_{\delta_1} \ldots T_{\delta_l}$.

Proof. (i) Using Theorem 2.19(v), one has $B_{-\delta} = s_\delta B_\delta s_\delta \subseteq B_\delta \cup B_\delta s_\delta B_\delta$, while $B_\delta \cap B_{-\delta} = T \neq B_{-\delta}$. Then $B_{-\delta} \cap B_\delta s_\delta B_\delta \neq \emptyset$. So there is a representative of s_δ in $B_\delta B_{-\delta} B_\delta$. But this last expression is $X_\delta X_{-\delta} X_\delta T$, so one may take the representative in $X_\delta X_{-\delta} X_\delta$.

(ii) We have $G_\delta \subseteq L_\delta$ (see Definition 2.24) and we have seen that $L_\delta \cap N = T \cup T n_\delta$ (Proposition 2.25). So $N \cap G_\delta = T_\delta \cup T_\delta n_\delta$. It is now easy to check that $(X_\delta T_\delta, N \cap G_\delta, \{s_\delta\})$ satisfies the axioms of a split BN-pair since $n_\delta \in G_\delta$ and L_δ has a split BN-pair. One has $X_\delta \cap X_{-\delta} = \{1\}$ by Theorem 2.23(i) or Proposition 2.25.

(iii) Using the Bruhat decomposition in G_δ, one has $n_\delta^{-1}(X_\delta \setminus \{1\})n_\delta^{-1} \in X_\delta T_\delta n_\delta^{-1} X_\delta$ since $n_\delta^{-1} X_\delta n_\delta^{-1} \cap T_\delta X_\delta \subseteq T_\delta$ (use Theorem 2.23(i), for instance).

(iv) This is because $[T, n_\delta] \subseteq T$ and $[T, G_\delta] \subseteq G_\delta$.

(v) Induction on l. The case $l = 1$ is trivial. Using (i) and the exchange condition, one may assume that $(\delta'_1, \ldots, \delta'_l) = (\delta'_1, \delta_1, \ldots, \delta_{l-1})$ and $\delta'_1 = s_{\delta_1} \ldots s_{\delta_{l-1}}(\delta_l)$. Then $(G_{\delta'_1})^{n_{\delta_1} \ldots n_{\delta_{l-1}}} = G_{\delta_l}$ and $t = (n_{\delta'_1}^{-1})^{n_{\delta_1} \ldots n_{\delta_{l-1}}} n_{\delta_l} \in G_{\delta_l}$. This proves our claim. □

Lemma 6.4. *Let $\delta \in \Delta$, then $\mathrm{Ind}_{X_\delta}^{G_\delta} k$ is a direct sum of $|T_\delta| + 1$ indecomposable modules.*

Proof. Assume $G = G_\delta$. Then $\mathrm{Ind}_U^B k = \bigoplus_\lambda \lambda$ where the direct sum is over $\lambda \in \mathrm{Hom}(T, k^\times)$ and the same letter denotes the associated one-dimensional kT-, or kB-module. So it suffices to show that $\mathrm{Ind}_B^G \lambda$ is indecomposable when $\lambda \neq 1$, and is the direct sum of two indecomposable modules when $\lambda = 1$. First $\mathrm{Res}_B^G(\mathrm{Ind}_B^G \lambda) = \lambda \oplus \mathrm{Ind}_T^B \lambda^{n_\delta}$ thanks to Mackey decomposition, i.e. a sum of two indecomposable kU-modules since $\mathrm{Res}_U^B(\mathrm{Ind}_T^B \lambda^{n_\delta}) = kU$ (Mackey formula) is indecomposable by Proposition 6.1. Thus, if $\mathrm{Ind}_B^G \lambda$ is not indecomposable, it is a direct sum of two indecomposable kG-modules $M_1 \oplus M_2$ with $\mathrm{Res}^G_B M_1 = \lambda$. But then M_1 is one-dimensional. But a one-dimensional kG-module is necessarily trivial since G is generated by X_δ and $X_{-\delta}$, two p-subgroups. So, if $\lambda \neq 1$, $\mathrm{Ind}^G_B \lambda$ is indecomposable. So $\mathrm{Ind}^G_U k$ is a sum of $\leq |T_\delta| + 1$ indecomposable modules. The equality won't be used and is left to the reader. □

Definition 6.5. *Take $\delta \in \Delta$, and n_δ as in Proposition 6.3(i). Let $z_\delta \colon T_\delta \to \mathbb{N}$ be defined by $z_\delta(t) = |n_\delta X_\delta n_\delta \cap X_\delta n_\delta t X_\delta|$.*

If $n \in N$, let $U_n = U \cap U^{w_0 w}$, where w is the class of n mod. T.

Proposition 6.6. *G is a disjoint union of the double cosets UnU for $n \in N$.*

Proof. The Bruhat decomposition (Theorem 2.14) implies that the union is G and that $UnU = Un'U$ implies $n' = nt$ for a $t \in T$. But $tU \cap U^n \subseteq U$ since U is the set of elements of B of order a power of p. So $t = 1$. □

An endomorphism of $\mathrm{Ind}^G_U k = kG \otimes_{kU} k$ is defined by the image of $1 \otimes 1$, and this image must be a U-fixed element. A basis of those fixed elements is given by the sums $s_C := \sum_{x \in C/U} x \otimes 1$ for $C \in U\backslash G/U$. By the Mackey formula, each sum defines an element of $\mathrm{End}_{kG}(\mathrm{Ind}^G_U k)$, and those form a basis. Proposition 6.6 then suggests the following definition and the first point of the next proposition.

Definition 6.7. *If $n \in N$, one defines $a_n \in \mathrm{End}_{kG}(\mathrm{Ind}^G_U k)$ by $a_n(g \otimes 1) = g \sum_{u \in U_n} un^{-1} \otimes 1$.*

Proposition 6.8. *(i) One has $\mathrm{End}_{kG}(\mathrm{Ind}^G_U k) = \bigoplus_{n \in N} k a_n$.*

(ii) If $l(nn') = l(n) + l(n')$, then $a_n a_{n'} = a_{nn'}$.

(iii) Take $\delta \in \Delta$ and n_δ as in Proposition 6.3(i). Then $(a_{n_\delta})^2 = a_{n_\delta}(\sum_{t \in T_\delta} z_\delta(t) a_t)$.

Proof. (ii) It suffices to show that $a_n a_{n'}(1 \otimes 1) = a_{nn'}(1 \otimes 1)$. One has $a_n a_{n'}(1 \otimes 1) = \sum_{u \in U_n\ ,\ u' \in U_{n'}} u' n'^{-1} u n^{-1} \otimes 1$. Using Theorem 2.23(ii) and Proposition 2.3(iii), one has $U_{n'}(U_n)^{n'} = U_{nn'}$ with $U_{n'} \cap (U_n)^{n'} = \{1\}$. So $a_n a_{n'}(1 \otimes 1) = \sum_{v \in U_{nn'}} v n'^{-1} n^{-1} \otimes 1 = a_{nn'}(1 \otimes 1)$ as stated.

(iii) It suffices to show that $(a_{n_\delta})^2$ and $a_{n_\delta}(\sum_{t \in T_\delta} z_\delta(t) a_t)$ coincide on $1 \otimes 1$, which generates Y. One has $(a_{n_\delta})^2(1 \otimes 1) = \sum_{u,v \in X_\delta} u n_\delta^{-1} v n_\delta^{-1} \otimes 1$. The sum for $u \in X_\delta$ and $v = 1$ gives zero since $\sum_{u \in X_\delta} u(n_\delta)^{-2} = (n_\delta)^{-2} \sum_{u \in X_\delta} u$ and the $\sum$ acts by $|X_\delta| = 0$ on $1 \otimes 1$. If $v \in X_\delta \setminus \{1\}$, then $n_\delta^{-1} v n_\delta^{-1} \in X_\delta t(v)^{-1} n_\delta^{-1} X_\delta$ for a unique $t(v) \in T_\delta$ by Proposition 6.3(iii) and Proposition 6.6. Then $(a_{n_\delta})^2(1 \otimes 1) = \sum_{u,v \in X_\delta, v \neq 1} u t(v)^{-1} n_\delta^{-1} \otimes 1 = \sum_{1 \neq v \in X_\delta} a_{n_\delta t(v)}(1 \otimes 1)$. Therefore $(a_{n_\delta})^2 = a_{n_\delta} \sum_{1 \neq v \in X_\delta} a_{t(v)}$ by (ii). But now $z_\delta(t) = |n_\delta X_\delta n_\delta \cap X_\delta n_\delta t X_\delta| = |n_\delta^{-1} X_\delta n_\delta^{-1} \cap X_\delta t^{-1} n_\delta^{-1} X_\delta|$. So eventually $(a_{n_\delta})^2 = a_{n_\delta}(\sum_{t\in T_\delta} z_\delta(t)a_t)$ as claimed. □

Definition 6.9. *If* $\lambda \in \mathrm{Hom}(T, k^\times)$, *let* $\Delta_\lambda = \{\delta \in \Delta \mid \lambda(T_\delta) = 1\}$.

Theorem 6.10. *(i) For all* $t \in T_\delta$, $z_\delta(t).|T_\delta| \equiv -1$ *mod. p.*

(ii) $\mathcal{H}_k(G, U)$ *can be presented in terms of generators* a_n $(n \in N)$ *obeying the relations* $a_n a_{n'} = a_{nn'}$ *when* $l(nn') = l(n) + l(n')$ *and* $(a_{n_\delta})^2 = -|T_\delta|^{-1} \sum_{t\in T_\delta} a_{n_\delta t}$ *for all* $\delta \in \Delta$ *(and* n_δ *is as in Proposition 6.3(i)).*

(iii) The simple $\mathcal{H}_k(G, U)$*-modules are the one-dimensional* $\psi(\lambda, I)$ *such that* $\lambda \in \mathrm{Hom}(T, k^\times)$ *and* $I \subseteq \Delta_\lambda$ *defined as follows:*

$\psi(\lambda, I)(a_n) = (-1)^l \lambda(t)$ *if* $n = n_{\delta_1} \ldots n_{\delta_l}.t$ *with* $\delta_1 \ldots \delta_l \in I$ *and* $t \in T$,

$\psi(\lambda, I)(a_n) = 0$, *if* $n \notin N_I$.

Proof. In what follows, we consider the integers $z_\delta(t)$ mod. p, i.e. as elements of k. Note first that the intersections $n_\delta X_\delta n_\delta \cap X_\delta n_\delta t X_\delta$ $(t \in T_\delta)$ are disjoint by Proposition 6.6 and they exhaust $n_\delta(X_\delta \setminus \{1\})n_\delta$ by Proposition 6.3(iii). Then $\sum_{t\in T_\delta} z_\delta(t) = |X_\delta| - 1 = -1$, so (i) is equivalent to showing that z_δ is constant on T_δ. We show that:

(i′) $z_\delta(tt') = z_\delta(t)$ for all $t \in T_\delta$, $t' \in [n_\delta, T]$.

This is proved as follows. If $s \in T$, $a_{n_\delta} a_s = a_{{}^{n_\delta}s} a_{n_\delta}$ by Proposition 6.8(ii), so $(a_{n_\delta})^2$ commutes with a_s. Using the expression of Proposition 6.8(iii), this gives $z_\delta(t) = z_\delta(t s^{n_\delta} s^{-1})$. Thus (i′) is proved.

Let us now show the following presentation with generators $(a_n)_{n\in N}$ and relations

(ii′) $a_n a_{n'} = a_{nn'}$ when $l(nn') = l(n) + l(n')$ and

$(a_{n_\delta})^2 = \sum_{t\in T_\delta} z_\delta(t) a_{n_\delta t}$ for all $\delta \in \Delta$.

By Proposition 6.8(ii), (iii), $\mathcal{H}_k(G, U)$ is a quotient of the above k-algebra, so it suffices to show that the above has dimension less than or equal to $|N|$. For this it is enough to show that any product $a_n a_{n'}$ is a linear combination of $(a_{n''})_{n''\in N}$. When $n' \in T$, the first relation applies. Otherwise, writing $n' = n_1 n_\delta$ with $\delta \in \Delta$, and $l(n') = l(n_1) + 1$, one has $a_{n'} = a_{n_1} a_{n_\delta}$. Using induction on $l(n')$, one may therefore assume $n' = n_\delta$. If $l(n n_\delta) = l(n) + 1$, then the first relation gives $a_n a_{n'} = a_{nn'}$. Otherwise $n = n_2 n_\delta$ with $l(n) = l(n_2) + 1$. Then $a_n = a_{n_2} a_{n_\delta}$ and $a_n a_{n'} = a_{n_2}(a_{n_\delta})^2 = \sum_{t\in T_\delta} z_\delta(t) a_{n_2} a_{n_\delta t}$. But the last expression is $\sum_{t\in T_\delta} z_\delta(t) a_{n_2 n_\delta t}$, again by the case of additivity.

Using this presentation of $\mathcal{H}_k(G, U)$, one may construct the following one-dimensional representations (described as maps $\mathcal{H}_k(G, U) \to k$).

Let $\lambda \in \mathrm{Hom}(T, k^\times)$, and $I \subseteq \Delta_\lambda$. Note first that λ is fixed by N_I, by Proposition 6.3(iv). Let $\psi(\lambda, I)\colon \mathcal{H}_k(G, U) \to k$ be defined as in (iii). Let us show that this is well defined. If $n = n_{\delta_1} \ldots n_{\delta_l} t = n_{\delta'_1} \ldots n_{\delta'_l} t'$ with $l = l(n)$, then

Proposition 6.3(v) implies that $t't^{-1} \in T_{\delta_1} \dots T_{\delta_l}$. If all the δ_i are in $I \subseteq \Delta_\lambda$, then $\lambda(t't^{-1}) = 1$. Otherwise the image of the two decompositions under $\psi(\lambda, I)$ is zero since $\{\delta_1, \dots, \delta_l\} = \{\delta'_1, \dots, \delta'_l\} \not\subseteq I$.

Let us show now that $\psi(\lambda, I)$ is a morphism by using the relations of (ii′). The second relation of (ii′) is satisfied since $\sum_{t \in T_\delta} z_\delta(t) = |X_\delta| - 1 = -1$ in k. For the first, let $n = n_{\delta_1} \dots n_{\delta_l} t$, $n' = n_{\delta'_1} \dots n_{\delta'_{l'}} t'$ with $t, t' \in T$ and $l(nn') = l + l'$. Then $nn' = n_{\delta_1} \dots n_{\delta_l} n_{\delta'_1} \dots n_{\delta'_l} t^{n'} t'$. If all the δ'_i are in I, then n' fixes λ so the relation is satisfied. If one of the δ'_i is outside I, then the relation amounts to $0 = 0$.

(i) Assume that z_δ is not constant on T_δ. In the case $G = G_\delta$, we have constructed above $|T_\delta| + 1$ one-dimensional representations of $\mathcal{H}_k(G_\delta, X_\delta)$. Since z_δ is a function on $T_\delta/[T_\delta, n_\delta]$, and since $k(T_\delta/[T_\delta, n_\delta])$ is split semi-simple, there would be some n_δ-fixed linear character $\lambda_0 \in \mathrm{Hom}(T_\delta, k^\times)$ such that $b := \sum_{t \in T_\delta} z_\delta(t)\lambda_0(t) \neq 0$. Then one may define ψ_0 on $\mathcal{H}_k(G_\delta, X_\delta)$ by $\psi_0(a_t) = \lambda_0(t)$, $\psi_0(a_{tn_\delta}) = -b\lambda_0(t)$. It is easy to check that this is a well-defined (since $\lambda_0 = (\lambda_0)^{n_\delta}$) representation of $\mathcal{H}_k(G_\delta, X_\delta)$, not among the ones we defined earlier. Then $\mathcal{H}_k(G_\delta, X_\delta)$ has at least $|T_\delta| + 2$ simple modules. But the simple $\mathrm{End}_{kG}(\mathrm{Ind}_U^G k)$-modules are in bijection with the isomorphism types of indecomposable summands of $\mathrm{Ind}_U^G k$. Then Lemma 6.4 gives a contradiction. Thus (i) is proved.

(ii) is clear by combining (i) and (ii′).

(iii) The representations have already been constructed. It remains to show that they are the only ones. Let M be a simple $\mathcal{H}_k(G, U)$-module. The subalgebra generated by the $(a_t)_{t \in T}$ is isomorphic to kT, as a result of Proposition 6.8(ii). But kT is commutative, split semi-simple by hypothesis, so M is a direct sum of lines stable under the a_t's. Let L be one, let $n \in N$ be an element of maximal length such that $a_n.L \neq 0$. It suffices to show that $a_n.L$ is $\mathcal{H}_k(G, U)$-stable to obtain $M = a_n.L$ and thus of dimension 1. The a_t's stabilize $a_n.L$ since $a_t a_n.L = a_n a_{t^n}.L \subseteq a_n.L$. By Proposition 6.8(ii), it is clear that $\mathcal{H}_k(G, U)$ is generated by the a_t's and the a_{n_δ}'s. If $l(n_\delta n) = l(n) + 1$, then $a_{n_\delta} a_n.L = a_{n_\delta n}.L = 0$ by the choice of n. If $l(n_\delta n) = l(n) - 1$, then $a_{n_\delta} a_n = (a_{n_\delta})^2 a_{n_\delta^{-1} n} = \sum_{t \in T_\delta} z_\delta(t) a_n a_{n^{-1} n_\delta t n_\delta^{-1} n}$ by Proposition 6.8(ii) and (iii). Then $a_{n_\delta} a_n.L \subseteq a_n.L$ since the a_t's stabilize L.

It remains to check that any $\psi \in \mathrm{Hom}(\mathcal{H}_k(G, U), k)$ is of the form stated. Restricting to the a_t's yields a $\lambda \in \mathrm{Hom}(T, k^\times)$. Then $\psi(a_{n_\delta})$ must satisfy $\psi(a_{n_\delta}).(\psi(a_{n_\delta}) + |T_\delta|^{-1} \sum_{t \in T_\delta} \lambda(t)) = 0$. Therefore $\psi(a_{n_\delta})$ is either 0 or $-|T_\delta|^{-1} \sum_{t \in T_\delta} \lambda(t)$. Interpreting $|T_\delta|^{-1} \sum_{t \in T_\delta} \lambda(t)$ as an inner product of characters of T_δ, one sees that it is 1 or zero depending on whether $\lambda(T_\delta) = 1$ or not, i.e. $\delta \in \Delta_\lambda$ or not. □

Proposition 6.11. $\mathrm{End}_{kG}(\mathrm{Ind}_U^G k)$ *is Frobenius (in the sense of [Ben91a] 1.6.1; see also Definition 1.19).*

Proof. For any n, $n' \in N$, it is easy to show that $a_n a_{n'} \in k a_{nn'} + \sum_{n'' \in N,\ l(n'') < l(n)+l(n')} k.a_{n''}$ (use induction on $l(n) + l(n')$ and Proposition 6.8(ii) and (iii)).

Let $n_0 \in N$ be an element of maximal length. Let us show that $\mathrm{End}_{kG}(\mathrm{Ind}_U^G k)$ is Frobenius for the linear form f sending a_{n_0} to 1 and all other a_n to 0. This gives our claim ([Ben91a] 1.6.2).

If $a \in \mathcal{H}_k(G, U)$ is written as $a = \sum_{n \in N} \lambda_n a_n$ with $\lambda_n \in k$ not all zero, choose $n_1 \in N$ of maximal length such that $\lambda_{n_1} \neq 0$. Now, it is clear from the above remark and Proposition 6.8(ii) that $f(a.a_{n_1^{-1} n_0}) = \lambda_{n_1} = f(a_{n_0 n_1^{-1}} a)$. Then $f(a\mathcal{H}_k(G, U)) \neq 0$ and $f(\mathcal{H}_k(G, U)a) \neq 0$. Thus our claim is proved. □

6.2. Some modules in characteristic p

By Frobenius reciprocity and Proposition 6.1, any simple kG-module M satisfies $\mathrm{Hom}_{kG}(\mathrm{Ind}_U^G k, M) \cong M^U \neq 0$, so any simple kG-module is a quotient of $\mathrm{Ind}_U^G k$. Since $\mathrm{Ind}_U^G k$ is isomorphic with its dual, every simple kG-module injects in it. By Proposition 6.11, one may therefore apply Theorem 1.25 to kG and $\mathrm{Ind}_U^G k$. Here are some consequences on the simple kG-modules.

Theorem 6.12. *The simple kG-modules are in bijection with the one-dimensional simple $\mathcal{H}_k(G, U)$-modules.*

Let M be a simple kG-module associated with $\psi: \mathcal{H}_k(G, U) \to k$ defined as in Theorem 6.10(iii) by a linear character $\lambda: T \to k^\times$ and a subset $I \subseteq \Delta_\lambda$. One has the following results.

(i) M^U is a line and $\sum_{x \in U_n} xn^{-1}.m = \psi(a_n)m$ for all $m \in M^U$, $n \in N$.

(ii) The following three conditions are equivalent:

- *M is of dimension $|U|$,*
- *M is projective,*
- *$I = \Delta$.*

Otherwise, the dimension of M is less than $|U|$.

(iii) Let $J \subseteq \Delta$, denote $U_J^- = U^{w_0} \cap U^{w_0 w_J}$ and let $\mathrm{rad}(kU_J^-)$ be the Jacobson radical of the group algebra of U_J^- (i.e. its augmentation ideal; see, for instance, Proposition 6.1). Then

$$\mathrm{Res}^G_{L_J} M = M^{U_J} \oplus \mathrm{rad}(kU_J^-) M^{U_J}$$

and M^{U_J} is a simple kL_J-module associated with λ and $I \cap J$ (with respect to the BN-pair of L_J described in Proposition 2.25).

Proof. Theorem 1.25(ii) tells us that the functor $\mathrm{Hom}_{kG}(\mathrm{Ind}_U^G k, -)$: $kG-\mathbf{mod} \to \mathbf{mod}-\mathcal{H}_k(G, U)$ induces a bijection between the simple modules. Taking duals over k bijects the right and left modules, so the simple left $\mathcal{H}_k(G, U)$-modules are described by Theorem 6.10(iii).

Assume M is a simple kG-module such that $\mathrm{Hom}_{kG}(\mathrm{Ind}_U^G k, M)$ provides the one-dimensional $\mathcal{H}_k(G, U)$-module associated with $\lambda \in \mathrm{Hom}(T, k^\times)$ and $I \subseteq \Delta_\lambda$.

The Frobenius reciprocity allows us to identify $\mathrm{Hom}_{kG}(\mathrm{Ind}_U^G k, M)$ with the fixed points M^U. This implies that M^U has dimension 1. The explicit bijection $\mathrm{Hom}_{kG}(\mathrm{Ind}_U^G k, M) \to M^U$ is the one sending $\phi \in \mathrm{Hom}_{kG}(\mathrm{Ind}_U^G k, M)$ to $\phi(1 \otimes 1)$. According to Notation 1.24, the right action of the elements of $\mathcal{H}_k(G, U) = \mathrm{End}_{kG}(\mathrm{Ind}_U^G k)$ is by composition on the right, so the above identification $\mathrm{Hom}_{kG}(\mathrm{Ind}_U^G k, M) \to M^U$ gives $\phi(1 \otimes 1).a_n = \phi \circ a_n(1 \otimes 1) = \sum_{u \in U_n} un^{-1}\phi(1 \otimes 1)$. This gives the formula announced in (i).

Denote $X_J^- = U^{w_0} \cap U^{w_0 w_J}$. Let us show now that

(iii′) $kL_J.M^U = kX_J^-.M^U$.

One must check that $kX_J^-.M^U$ is stable under L_J. Taking the w_J-conjugate of the usual BN-pair of L_J, one sees that L_J is generated by T, X_J^- and the n_δ for $\delta \in J$. The groups T and X_J^- stabilize $kX_J^-.M^U$. Let $\delta \in J$. One has $X_J^- = (X_J^- \cap n_\delta^{-1} X_J^- n_\delta) X_{-\delta}$ (apply Theorem 2.23 to the w_J-conjugates). Then $n_\delta k X_J^-.M^U \subseteq kX_J^-.kX_\delta n_\delta.M^U$. It suffices to check $kX_\delta n_\delta.M^U \subseteq kX_{-\delta}.M^U$. If $x \in X_\delta$, $x \neq 1$, then $xn_\delta \in X_{-\delta} T X_\delta$ by Proposition 6.3(iii). Then $xn_\delta M^U \subseteq kX_{-\delta} M^U$. When $x = 1$, one has $n_\delta.m = a_{n_\delta^{-1}}(m) - \sum_{x \in X_\delta,\, x \neq 1} xn_\delta.m$ for all $m \in M^U$ by (i). The sum is in $kX_{-\delta}.M^U$ by the case $x \neq 1$ just treated. Then $n_\delta m \in kX_{-\delta} M^U$ as claimed. Thus (iii′).

(ii) Applying (iii′) with $J = \Delta$, one gets $kG.M^U = kU^{w_0}.M^U$. But $kG.M^U$ is a non-zero kG-submodule of M, so $M = kU^{w_0}.M^U$. Taking $0 \neq m \in M^U$, one has $M = kU^{w_0} m$, which means that the map

$$kU^{w_0} \to \mathrm{Res}_{U^{w_0}}^G M \quad y \mapsto y.m$$

is onto. Then $\dim_k M \leq |U|$ with equality if and only if the above map is injective. The kernel of this map is a left ideal of kU^{w_0}. This module is indecomposable so its socle is simple, hence equal to the line generated by the sum of elements of U^{w_0}. Then the map above is injective if and only if $\sum_{u \in U^{w_0}} u.m \neq 0$. By (i), this is equivalent to $\psi(a_{n_0}) \neq 0$ where n_0 is an element of N of maximal length. Then $n_0 \in N_I$, $w_0 \in W_I$, and therefore $I = \Delta$.

Since this is also the condition for the map above to be an isomorphism, we see that $I = \Delta$ implies that $\mathrm{Res}^G_{U^{w_0}} M$ is projective. This in turn is equivalent to M being projective (apply, for instance, [NaTs89] 4.2.5) since U^{w_0} is a Sylow p-subgroup of G. Conversely, if M is projective, its dimension is a multiple of $|U^{w_0}|$ (deduce this, for instance, from Proposition 6.1), thus implying that the map above is an isomorphism.

(iii) Note first that $kL_J.M^U \subseteq M^{U_J}$. Let us check now that $kL_J.M^U$ is a simple kL_J-module. If $0 \neq M' \subseteq kL_J.M^U$ is L_J-stable, M' has non-zero fixed points under the action of $U \cap U^{w_0 w_J}$ since this is a p-subgroup of L_J. But the elements of M' are all fixed by U_J, and $U_J.(U \cap U^{w_0 w_J}) = U$ (Theorem 2.23(ii)). So M' contains the line M^U and therefore $M' \supseteq kL_J.M^U$. So the latter is a simple kL_J-submodule of M^{U_J}. Its type is given by the action on M^U of the sums $\sum_{u \in U_n} xn^{-1}$ for $n \in N_J$ as a result of (i). This gives the same linear character of T and the subset $I \cap J$ in Δ.

We have seen, as a special case of (iii′), that $M = kU^{w_0}.M^U$. Now $U^{w_0} = U_J^- X_J^-$ by Theorem 2.23(ii). So $M = kX_J^-.M^U + \mathrm{rad}(kU_J^-)kX_J^-.M^U$ since $\mathrm{rad}(kQ)$ is the augmentation ideal for all p-groups Q (as annihilator of the unique simple kQ-module; see Proposition 6.1). Then $M = kX_J^-.M^U + \mathrm{rad}(kU_J^-)kX_J^-.M^U$. To complete the proof of (iii), it suffices to check that $M^{U_J} \cap \mathrm{rad}(kU_J^-)kX_J^-.M^U = 0$. Suppose it is not 0. This intersection is a kL_J-module, so it must have non-zero fixed points under $U \cap U^{w_0 w_J}$. Again, this implies that this intersection contains the line M^U. Then $kX_J^-.M^U \subseteq \mathrm{rad}(kU_J^-)kX_J^-.M^U$. By iteration, this would contradict the nilpotence of $\mathrm{rad}(kU_J^-)$. So $M^{U_J} \cap \mathrm{rad}(kU_J^-)kX_J^-.M^U = 0$. Since $M^{U_J} \supseteq kX_J^-.M^U$ and $M = kX_J^-.M^U + \mathrm{rad}(kU_J^-)kX_J^-.M^U$, this yields at once $M^{U_J} = kX_J^-.M^U$ and $M^{U_J} \oplus \mathrm{rad}(kU_J^-)kX_J^-.M^U = M$. Thus our claim is proved. □

Definition 6.13. *The Steinberg kG-module is the simple kG-module corresponding to the pair $(\lambda, I) = (1, \Delta)$.*

The Steinberg module is also projective by Theorem 6.12(ii). See also Exercises 2–4.

6.3. Alperin's weight conjecture in characteristic p

We keep G a finite group with a strongly split BN-pair of characteristic p (see Definition 2.2). We recall the existence of subgroups $B = UT$, the set Δ, and subgroups $P_I = U_I L_I$ when $I \subseteq \Delta$ (see Definition 2.24). We assume moreover the following.

Hypothesis 6.14. *If V is a subgroup of G such that it is the maximal normal p-subgroup of its normalizer $\mathrm{N}_G(V)$, then there are $g \in G$ and $I \subseteq \Delta$ such that $V = gU_Ig^{-1}$.*

Remark 6.15. Let us show how this hypothesis may be verified for groups of type $G = \mathbf{G}^F$ where $\mathbf{G}$ is a reductive linear algebraic group and F is a Frobenius endomorphism associated with the definition of the reductive group $\mathbf{G}$ over a finite field of characteristic p (see [DiMi91] §3; this context is described in more detail in A2.4 and Chapter 8, below).

For the terminology about algebraic groups, we refer to [Hum90] and [Borel]. A statement similar to Hypothesis 6.14 for reductive groups is as follows (see [Hum90] 30.3).

Let $\mathbf{B}$ be a Borel subgroup of $\mathbf{G}$. If V is a closed subgroup of $R_u(\mathbf{B})$ (unipotent radical of $\mathbf{B}$), then the ascending sequence $V_0 = V$, $V_i = V_{i-1}R_u(\mathrm{N}_{\mathbf{G}}(V_{i-1}))$ stabilizes at some group of type $R_u(\mathbf{P}(V))$ where $\mathbf{P}(V)$ is a parabolic subgroup of $\mathbf{G}$, i.e. $\mathbf{P}(V)$ contains a $\mathbf{G}$-conjugate of $\mathbf{B}$ ([Hum90] 30.3). Note that the sequence $\mathrm{N}_{\mathbf{G}}(V_i)$ is also ascending (and stops at $\mathbf{P}(V)$).

When $G = \mathbf{G}^F$ for F a Frobenius endomorphism, a BN-pair is given in G by $B := \mathbf{B}^F$ for $\mathbf{B}$ an F-stable Borel subgroup, $N = \mathrm{N}_{\mathbf{G}}(\mathbf{T})^F$ for $\mathbf{T} \subseteq \mathbf{B}$ an F-stable maximal torus. The G-conjugates of parabolic subgroups of G containing B are the subgroups $\mathbf{P}^F$ where $\mathbf{P}$ contains a $\mathbf{G}$-conjugate of $\mathbf{B}$ and is F-stable (see [DiMi91] §3).

Returning to our problem of checking Hypothesis 6.14 for G, let V be a p-subgroup of G such that it is the maximal normal p-subgroup of $\mathrm{N}_G(V)$. Since $B_p = R_u(\mathbf{B})^F$ contains a Sylow p-subgroup of G, one may assume $V \subseteq R_u(\mathbf{B})$. Then the above process may be applied (V is closed since finite). Since $F(V) = V$, we have $F(V_i) = V_i$ for all i and therefore $F(R_u(\mathbf{P}(V))) = R_u(\mathbf{P}(V))$. Taking normalizers, we find that $\mathbf{P}(V)$ is F-stable, too. Then, upon possibly replacing V with a G-conjugate, one finds that there is a parabolic subgroup $P_I \supseteq B$ such that $V \subseteq U_I$ and $\mathrm{N}_G(V) \subseteq P_I$. Now $\mathrm{N}_{U_I}(V)$ is normal in $\mathrm{N}_G(V)$ since $\mathrm{N}_G(V) \subseteq P_I$ normalizes U_I. By maximality of V, this implies $\mathrm{N}_{U_I}(V) = V$. But $V \subseteq U_I$ is an inclusion of p-groups, so we must have $V = U_I$.

J. Alperin's "weight" conjecture is as follows (see [Ben91b] §6.9).

Conjecture. Let X be a finite group, p a prime, k an algebraically closed field of characteristic p. Then the number of simple kX-modules equals the number of X-conjugacy classes of pairs (V, π) where V is a p-subgroup of X and π is a *simple projective* $k(\mathrm{N}_X(V)/V)$-module. Note that simple projective kX-modules are in bijection with blocks of kX with defect zero (see §5.2 above).

We prove the following.

Theorem 6.16. *The above conjecture is true if X is a group with a strongly split BN-pair of characteristic p satisfying Hypothesis 6.14, and k is of characteristic p.*

Proof. Let us notice first that, if there is a simple projective kX-module π, then X has no non-trivial normal p-subgroup, since such a normal subgroup should be included in all defect groups (see [Ben91a] 6.1.1).

A consequence of this remark, applied to quotients $\mathrm{N}_X(V)/V$, is that each p-subgroup V of Alperin's conjecture must be the maximal normal p-subgroup of its normalizer $\mathrm{N}_X(V)$.

Assume G is a finite group with a strongly split BN-pair of characteristic p satisfying Hypothesis 6.14 and k is an algebraically closed field of characteristic p. By Hypothesis 6.14, the pairs (V, π) are of type (U_I, π) for $I \subseteq \Delta$ and π a simple projective kL_I-module.

By Proposition 2.29, a U_I is conjugate to a U_J if and only if $I = J$. Then the number of G-conjugacy classes of pairs (V, π) of Alperin's conjecture is the number of pairs (I, π) where $I \subseteq \Delta$ and π is a simple projective kL_I-module. The group L_I has a strongly split BN-pair of characteristic p, so its simple projective modules are given by Theorem 6.12(ii). Their number is the number of $\lambda \in \mathrm{Hom}(T, k^\times)$ such that Δ_λ contains I. Taking the sum of those numbers over $I \subseteq \Delta$, we find the number of pairs (I, λ) where $\lambda \in \mathrm{Hom}(T, k^\times)$ and $I \subseteq \Delta_\lambda$. This is the number of simple kG-modules by Theorem 6.12. □

6.4. The p-blocks

Let X be a finite group and k a field of characteristic p containing a $|X|$th root of 1. We will use the results of §5.1 about the blocks of kX (*p-blocks* of X) and associated defect groups, particularly principal blocks and blocks with defect zero. One further property of defect groups is the following (see [CuRe87] 57.31 or the proof of [Ben91a] 6.1.1).

Theorem 6.17. *If D is a defect group of a p-block of X, and S is a p-group such that $D \subseteq S \subseteq X$, then there is $x \in \mathrm{C}_X(D)$ such that $D = S \cap S^x$.*

Theorem 6.18. *Let G be a finite group with a strongly split BN-pair $(B = UT, N)$ of characteristic p satisfying Hypothesis 6.14. Assume that $W = N/T$ is of irreducible type (i.e. there is no partition of Δ into two non-empty orthogonal subsets) with $\mathrm{C}_G(U) = \mathrm{Z}(U)$ (hence $\mathrm{Z}(G) = \{1\}$). Then every p-block is either the principal block or a block of defect zero.*

Remark 6.19. Theorem 2.31 tells us that, when W is irreducible of cardinality $\neq 2$, the axioms of strongly split BN-pairs imply the condition $\mathrm{C}_G(U) = \mathrm{Z}(G)\mathrm{Z}(U)$.

Proof of Theorem 6.18. Let b be a p-block of G with defect group D. We prove that the group D is either 1 or a Sylow p-subgroup of G. This is enough to establish our claim since U is a Sylow p-subgroup, $\mathrm{C}_G(U) = \mathrm{Z}(U)$ and a p-group has just one block, the principal block (use, for instance, Proposition 6.1), thus giving our claim by Brauer's third Main Theorem (see Theorem 5.10).

We may assume that $D \subseteq U$ and $\mathrm{N}_U(D)$ is a Sylow p-subgroup of $\mathrm{N}_G(D)$. As recalled above, we have $D = U \cap U^g$ with $g \in \mathrm{C}_G(D)$. Then $g \in \mathrm{N}_G(D)$ and therefore D is the maximal normal p-subgroup of $\mathrm{N}_G(D)$. Then D is G-conjugate to a U_I by Hypothesis 6.14, hence $D = U_I$ by Proposition 2.29(ii)

We have $U_I = U \cap U^g$ for some $g \in \mathrm{C}_G(U_I) \subseteq \mathrm{N}_G(U_I) = P_I$ (see Proposition 2.29(ii)). Let us write $g \in BwB$ for $w \in W_I$. Then, since B normalizes U and U_I, one has $U_I = U \cap U^w$. Theorem 2.23(i) then implies $\Phi_w = \Phi_I^+$ and therefore that w is the element of maximal length in W_I. We may now apply Lemma 2.30. This tells us that $I = \emptyset$ or Δ, i.e. $D = U$ or 1, a contradiction. □

Exercises

1. (a) Show that Theorem 6.12 can be proved under the hypothesis that k is just big enough so that kT is split. This occurs if and only if k contains roots of unity of order the exponent of T.
   (b) Find a version of Theorem 6.12 where k is any field of characteristic p (replace one-dimensional kT-modules with simple kT-modules). Deduce that kG is split if and only if kT is split.
2. Use the setting of Theorem 6.12.
   (a) For any simple kG-module M, show that $M = kB.M^{U^-}$ where $U^- = U^{w_0}$ and $M^{U^-} = w_0 M^U$ is a line.
   (b) Show that if M is the Steinberg module (see Definition 6.13), then $\mathrm{Res}^G_B M$ is a quotient of $\mathrm{Ind}^B_T k$, hence is equal to it.
   (c) Let Λ be a complete valuation ring with residue field k. Show that the Steinberg kG-module lifts to a projective ΛG-module whose restriction to B is $\mathrm{Ind}^B_T \Lambda$ (one may use properties of permutation modules; see [Thévenaz] §27, [Ben91a] §5.5).
   (d) Deduce [DiMi91] 9.2 and its corollaries.
3. Use the notation of Theorem 6.12. Let $\lambda \in \mathrm{Hom}(T, k^\times)$ and $I \subseteq \Delta_\lambda$, defining a one-dimensional representation $\psi: \mathcal{H}_k(G, U) \to k$.

(a) There is an indecomposable direct summand $Y(\lambda, I)$ of $\mathrm{Ind}_U^G k$ associated with ψ. Show that it is characterized by the property that $M(\lambda, I) := \mathrm{hd}(Y(\lambda, I))$ is a simple kG-module such that $M(\lambda, I)^U$ is a line satisfying the relations of Theorem 6.12(ii).

(b) Show that $Y(1, \emptyset) = k$, the trivial kG-module.

(c) If $I \subseteq J \subseteq \Delta$, denote by $Y_J(\lambda, I)$ the kL_J-module defined as above for $L_J, B \cap L_J, N \cap L_J, J$ instead of G, B, N, Δ. Considering $Y_J(\lambda, I)$ as a U_J-trivial kP_J-module, show that $\mathrm{Ind}_{P_J}^G \big(Y_J(\lambda, I)\big)$ is a direct summand of $\mathrm{Ind}_U^G k$.

(d) Show that $\mathrm{Ind}_{P_J}^G(Y_J(\lambda, I)) \cong \bigoplus_{I'} Y(\lambda, I')$ where the sum is over subsets $I' \subseteq \Delta_\lambda$ such that $I' \cap J = I$ (compute $\mathrm{Hom}_G(\mathrm{Ind}_{P_J}^G Y_J(\lambda, I), M(\lambda', I'))$ by use of Theorem 6.12(iii)).

(e) Show the following equation in the *representation ring* (see [Ben91a] §5.1) of kG

$$Y(\lambda, I) = \sum_{J \subseteq I} (-1)^{|J|} \mathrm{Ind}_{P_{J(\lambda)}}^G (Y_{J(\lambda)}(\lambda, \emptyset))$$

where $J(\lambda)$ denotes $J \cup (\Delta_\lambda \setminus I)$.

(f) Show that the Steinberg module (see Definition 6.13) is

$$\sum_{I \subseteq \Delta} (-1)^{|I|} \mathrm{Ind}_{P_I}^G 1$$

in the representation ring of kG (where 1 denotes the trivial module). Let Λ be a complete valuation ring with residue field k. Show that the Steinberg kG-module lifts to a projective ΛG-module defined by the same equation in the representation ring of ΛG (use [Ben91a] §5.5). Show that the irreducible character associated with the p-block of defect zero defined by the Steinberg module is also defined by the above formula.

4. We use the notation of §5.4. Let $w_0 \in W$ be the element of maximal length.

(a) Let $e = (-1)^{l(w_0)} |T|^{-1} \sum_{n \in N | nT = w_0} a_n \in \mathcal{H}_k(G, U)$. Show that $e.a_t = e$ and $e.a_{n_\delta} = -e$ for all $t \in T$ and $\delta \in \Delta$. Show that e is an idempotent and that $e.\mathrm{Ind}_U^G k$ is isomorphic with the Steinberg module $Y(1, \Delta)$.

(b) Let $\sigma \in kG$ be the sum of the elements of the double coset Bw_0B. Show that $kG\sigma$ is isomorphic with the Steinberg module.

5. Show that the invariance with regard to parabolic subgroups implied by Theorem 3.10 does not hold in characteristic p. Consider (B, U)—$(T.U^{w_0}, U^{w_0})$ and associated fixed point functors applied to a simple kG-module associated with $(\lambda, \emptyset)$ where $\lambda^{w_0} \neq \lambda$.

Notes

Representations of finite BN-pairs in natural characteristic are generally studied by use of algebraic groups, Lie algebras, and related structures (see the book [Jantzen]). Irreducible rational representations of reductive groups are described by Lusztig conjectures, see [Donk98b] for an introduction.

The elementary approach to simple modules followed here originates in [Gre78], and also borrows from [Tin80], [Sm82], and [Ca88]. See also [Tin79], and [Dip80], [Dip83]. A related subject is that of Hecke algebras where the parameter is 0; see [No79], [Donk98a] §2.2, [KrThi99] and their references.

Though it is "only" about simple modules, Alperin's weight conjecture has not been solved in general. It was introduced in [Al87]. Checkings ensued for natural characteristic ([Ca88]; see also [Ben91b] §6.9 and [LT92]), for symmetric groups and general linear groups ([AlFo90]), and many other BN-pairs in non-natural characteristic ([An93], [An98]).

Many reformulations of Alperin's weight conjecture have been given, in particular by Knörr–Robinson [KnRo89]. Dade stated stronger conjectures in terms of exterior action and character degrees ([Da92] and [Da99]), in an attempt to find a (still elusive) version that could reduce to quasi-simple groups. See also [Rob98] and the references in [An01]. For relations with Broué's abelian defect conjecture, see also [Rick01].

PART II

Deligne–Lusztig varieties, rational series, and Morita equivalences

Let G be a finite group with a split BN-pair of characteristic p. In this part, we expound the approach initiated by Deligne–Lusztig [DeLu76] which describes certain linear representations of G stemming from its action on the étale cohomology groups (see Appendix 3) of certain algebraic varieties. This approach requires us to realize G as $\mathbf{G}^{F_0}$ where $\mathbf{G}$ is a connected reductive group over an algebraic closure $\mathbf{F}$ of the field with q elements $\mathbb{F}_q$ (q a power of p) and

$$F_0 \colon \mathbf{G} \to \mathbf{G}$$

is the Frobenius endomorphism associated with a definition of $\mathbf{G}$ over $\mathbb{F}_q$ (see A2.4 and A2.5). This is analogous to $\mathrm{GL}_n(\mathbb{F}_q)$ being constructed from $\mathrm{GL}_n(\mathbf{F})$ as fixed point subgroup under the map F_0 raising matrix entries to the qth power (another example is that of the unitary group where F_0 is replaced with $F_0 \circ \sigma$, where σ is defined by transposition composed with inversion).

A basic idea in Grothendieck's algebraic geometry is to consider, together with any given algebraic variety $\mathbf{X}$, all possible ways to realize it as a reasonable quotient of another by a finite group action (thus providing a notion of fundamental group; see A3.16). The Lang map $g \mapsto \mathrm{Lan}(g) := g^{-1}F_0(g)$, for instance, realizes $\mathbf{G}^{F_0}$ as Galois group of such a covering $\mathbf{G} \to \mathbf{G}$. When $\mathbf{P} = \mathbf{V} \rtimes \mathbf{L}$ is a Levi decomposition in $\mathbf{G}$ with F_0-stable $\mathbf{L}$, one defines the Deligne–Lusztig variety $\mathbf{Y}_{\mathbf{V}}^{(\mathbf{G})}$ as $\mathrm{Lan}^{-1}(\mathbf{V}.F_0\mathbf{V})/\mathbf{V}$. The finite group $\mathbf{G}^{F_0} \times \mathbf{L}^{F_0}$ acts on it, and the associated étale cohomology groups are bimodules defining a generalization of Harish-Chandra induction.

In the case when $\mathbf{L} = \mathbf{T}$, some F_0-stable maximal torus of $\mathbf{G}$, this defines generalized characters $\mathrm{R}_{\mathbf{T}}^{\mathbf{G}}\theta$ for $\theta \colon \mathbf{T}^{F_0} \to K^\times$ any character of $\mathbf{T}^{F_0}$ (where K is a field of characteristic zero such that the group algebras KH are split for any subgroup $H \subseteq \mathbf{G}^{F_0}$). The $\mathbf{G}^{F_0}$-conjugacy classes of pairs $(\mathbf{T}, \theta)$ are in bijection with conjugacy classes of semi-simple elements in $(\mathbf{G}^*)^{F_0}$, where $\mathbf{G}^*$ is the connected reductive group dual to $\mathbf{G}$ (e.g. for $\mathbf{G} = \mathrm{SL}_n$, $\mathbf{G}^* = \mathrm{PGL}_n$, the explicit definition is given in Chapter 8 below). This implies a partition of irreducible characters over K

$$\mathrm{Irr}(\mathbf{G}^{F_0}) = \cup_s \mathcal{E}(\mathbf{G}^{F_0}, s)$$

("rational series"; see [DiMi91] 14.41) where s ranges over $(\mathbf{G}^*)^{F_0}_{\mathrm{ss}}$ mod. $(\mathbf{G}^*)^{F_0}$-conjugacy. For $s = 1$, one uses the term "unipotent characters" for the elements of $\mathcal{E}(\mathbf{G}^{F_0}, 1)$. Each of the above sets is in turn in bijection with a set of unipotent characters

$$\mathcal{E}(\mathbf{G}^{F_0}, s) \cong \mathcal{E}(\mathrm{C}_{\mathbf{G}^*}(s)^{F_0}, 1)$$

thus establishing a "Jordan decomposition" for characters (see [DiMi91] 13.23 and further results in Chapter 15).

In order to discuss modular representations of $\mathbf{G}^{F_0}$, let ℓ be a prime, and let (Λ, K, k) be an ℓ-modular splitting system for $\mathbf{G}^{F_0}$. In Chapter 9, we exhibit the relation between blocks of $\Lambda\mathbf{G}^{F_0}$ and rational series (Broué–Michel, [BrMi89]). Chapters 10 to 12 contain the proof that the above Jordan decomposition, at least when $\mathrm{C}_{\mathbf{G}^*}(s)$ is a Levi subgroup $\mathbf{L}^*$, is induced by a Morita equivalence between a block of $\Lambda\mathbf{L}^{F_0}$ and one of $\Lambda\mathbf{G}^{F_0}$ (Bonnafé–Rouquier, [BoRo03]).

Bonnafé–Rouquier's theorem essentially allows us to reduce the study of blocks of finite reductive groups to the study of blocks defined by a unipotent character ("unipotent blocks"). The remaining parts of the book focus attention on them.

The three appendices at the end of the book gather the background necessary to understand Grothendieck's theory (derived categories, algebraic geometry and étale cohomology).

7

Finite reductive groups and Deligne–Lusztig varieties

The present chapter is devoted to the basic properties of Deligne–Lusztig varieties that will be useful in the later chapters of this part. Recall from the introduction to this part that $\mathbf{G}$ is a connected reductive group over $\mathbf{F}$, an algebraic closure of $\mathbb{F}_q$. We denote by

$$F: \mathbf{G} \to \mathbf{G}$$

an endomorphism of an algebraic group such that a power of F is the Frobenius endomorphism F_0 associated with a definition of $\mathbf{G}$ over $\mathbb{F}_q$. The starting point is the surjectivity of the Lang map

$$\mathrm{Lan}: \mathbf{G} \to \mathbf{G}$$

defined by $\mathrm{Lan}(g) = g^{-1}F(g)$.

Let $\mathbf{P} = \mathbf{LV}$ be a Levi decomposition (see A2.4) with $F\mathbf{L} = \mathbf{L}$. One defines the **Deligne–Lusztig varieties**

$$\mathbf{Y}_{\mathbf{V}}^{(\mathbf{G})} := \mathrm{Lan}^{-1}(F(\mathbf{V}))/\mathbf{V} \cap F\mathbf{V}, \quad \mathbf{X}_{\mathbf{V}}^{(\mathbf{G})} := \mathrm{Lan}^{-1}(F(\mathbf{P}))/\mathbf{P} \cap F\mathbf{P} \cong \mathbf{Y}_{\mathbf{V}}^{(\mathbf{G})}/\mathbf{L}^F.$$

We show here that they are smooth of constant dimension, that of $\mathbf{V}/\mathbf{V} \cap F\mathbf{V}$ ($\mathbf{P} = \mathbf{LV}$ is a Levi decomposition with $F\mathbf{L} = \mathbf{L}$). Note that, when moreover $F\mathbf{V} = \mathbf{V}$, one finds the finite sets $\mathbf{G}^F/\mathbf{V}^F$ and $\mathbf{G}^F/\mathbf{P}^F$ relevant to Harish-Chandra induction; see Chapter 3.

We also prove a transitivity property $\mathbf{Y}_{\mathbf{VV}'}^{(\mathbf{G})} \cong \mathbf{Y}_{\mathbf{V}}^{(\mathbf{G})} \times \mathbf{Y}_{\mathbf{V}'}^{(\mathbf{L})}/\mathbf{L}^F$ (for the diagonal action of $\mathbf{L}^F$) when $\mathbf{V}'$ is the unipotent radical of a parabolic subgroup of $\mathbf{L}$ (Theorem 7.9).

An important special case is when $\mathbf{P}$ is a Borel subgroup. Then the corresponding varieties $\mathbf{X}_{\mathbf{V}}$ may be defined by an element w of the Weyl group $W = \mathrm{N}_{\mathbf{G}}(\mathbf{T}_0)/\mathbf{T}_0$, where $\mathbf{T}_0 \supseteq \mathbf{B}_0$ are a maximal torus and a Borel subgroup, both F-stable. The varieties $\mathbf{X}(w) \subseteq \mathbf{G}/\mathbf{B}_0$ are intersections of the cells $O(w)$

(see A2.6) with the graph of F on $\mathbf{G}/\mathbf{B}_0$. The closure $\overline{\mathbf{X}}(w)$ of $\mathbf{X}(w)$ in $\mathbf{G}/\mathbf{B}_0$ is smooth whenever w is a product of commuting generators in the Weyl group W, a fact that is essential for computing étale cohomology of those varieties. The closed subvariety $\overline{\mathbf{X}}(w) \setminus \mathbf{X}(w)$ is a smooth divisor with normal crossings (Proposition 7.13).

Another important property of the varieties $\mathbf{Y}_{\mathbf{V}}^{(\mathbf{G})}$ and $\mathbf{X}_{\mathbf{V}}^{(\mathbf{G})}$ is that they can be embedded as open subsets of affine varieties (Theorem 7.15, due to Haastert, [Haa86]). Here one uses the criteron of quasi-affinity in terms of invertible sheaves (see A2.10).

The reader will find the background material concerning algebraic groups and quotient varieties in Appendix 2.

7.1. Reductive groups and Lang's theorem

Recall that $\mathbf{F}$ is an algebraic closure of a finite field $\mathbb{F}_q$. Let $\mathbf{G}$ be a connected affine algebraic group over $\mathbf{F}$, and let $F_0 : \mathbf{G} \to \mathbf{G}$ be the Frobenius endomorphism associated with a definition of $\mathbf{G}$ over $\mathbb{F}_q$.

Theorem 7.1. *Let $F : \mathbf{G} \to \mathbf{G}$ be an endomorphism such that $F^m = F_0$ for some integer m. Denote* Lan: $\mathbf{G} \to \mathbf{G}$ *defined by* $\mathrm{Lan}(g) = g^{-1}F(g)$.

(i) Lan *is onto ("Lang's theorem").*

(ii) The tangent map (see A2.3) $\mathcal{T}\mathrm{Lan}_x : \mathcal{T}\mathbf{G}_x \to \mathcal{T}\mathbf{G}_{\mathrm{Lan}(x)}$ is an isomorphism for all $x \in \mathbf{G}$.

(iii) There exists a pair $\mathbf{T}_0 \subseteq \mathbf{B}_0$ consisting of a maximal torus and a Borel subgroup of $\mathbf{G}$ such that $F(\mathbf{B}_0) = \mathbf{B}_0$ and $F(\mathbf{T}_0) = \mathbf{T}_0$

(iv) F composed with any inner automorphism satisfies the same hypotheses.

Proof. Denote by $\iota : \mathbf{G} \to \mathbf{G}$ the inversion map $\iota(x) = x^{-1}$. Let $F' : \mathbf{G} \to \mathbf{G}$ be a rational group morphism such that $\mathcal{T}F'_1$ is nilpotent. Denote $\mathrm{Lan}'(x) = x^{-1}F'(x)$. We also use the notation $[\cdot x]$ (resp. $[x\cdot]$) for the map $\mathbf{G} \to \mathbf{G}$ defined by $[\cdot x](g) = gx$ (resp. $[x\cdot](g) = xg$).

Decomposing Lan' as $\mathbf{G} \to \mathbf{G} \times \mathbf{G} \to \mathbf{G}$ where the first arrow is (ι, F') and the second is multiplication, we get $\mathcal{T}\mathrm{Lan}'_x = \mathcal{T}[\cdot F'(x)]_{x^{-1}}\mathcal{T}\iota_x + \mathcal{T}[x^{-1}\cdot]_{F'(x)}\mathcal{T}F'_x$, where in addition $\mathcal{T}\iota_x = \mathcal{T}[x^{-1}\cdot]_1\mathcal{T}\iota_1\mathcal{T}[\cdot x^{-1}]_x$, $\mathcal{T}\iota_1 = -\mathrm{Id}_{\mathcal{T}\mathbf{G}}$, and $\mathcal{T}F'_x = \mathcal{T}[\cdot F'(x)]_1\mathcal{T}F'_1(\mathcal{T}[\cdot x]_1)^{-1}$ (see A2.4). Then $\mathcal{T}\mathrm{Lan}'_x = \mathcal{T}[\cdot F'(x)]_{x^{-1}}\mathcal{T}[x^{-1}\cdot]_1(-\mathrm{Id}_{\mathcal{T}\mathbf{G}} + \mathcal{T}F'_1)\mathcal{T}[\cdot x^{-1}]_x$ and all terms in this composition are bijections since $\mathcal{T}F'_1$ is nilpotent. Taking $F = F'$ is possible since $(\mathcal{T}F_1)^m = \mathcal{T}(F_0)_1 = 0$ (see A2.5). This gives (ii).

Since $\mathbf{G}$ is smooth (A2.4) and $\mathcal{T}\mathrm{Lan}'_x$ is an isomorphism, we know that Lan' is separable (see A2.6), hence dominant.

Let $a \in G$. Then one may take F' defined by $F'(x) = aF(x)a^{-1}$ since $(F')^m$ is then F^m composed with an appropriate conjugation. This tells us that the image of $g \mapsto g^{-1}aF(g)a^{-1}$ contains a non-empty open subset of $\mathbf{G}$. The same is true for the images of $g \mapsto g^{-1}aF(g)$ and $g \mapsto g^{-1}F(g)$. Arguing that $\mathbf{G}$ is irreducible because it is smooth (see A2.4) and connected, those two open subsets must have a non-empty intersection. So we get $g^{-1}aF(g) = h^{-1}F(h)$ for some $g, h \in \mathbf{G}$. One obtains $a = \mathrm{Lan}(hg^{-1})$. This gives (i).

Note that F is injective since F_0 is (see A2.5). Then $F\mathbf{G} = \mathbf{G}$ since $F\mathbf{G}$ is closed of the same dimension as $\mathbf{G}$ (see A2.4) and $\mathbf{G}$ is connected.

If $\mathbf{B}$ is any Borel subgroup, $F\mathbf{B}$ is also a connected solvable subgroup, so there exists $g \in \mathbf{G}$ such that $F\mathbf{B} \subseteq \mathbf{B}^g$. Writing $g = a^{-1}F(a)$, by (i) we get that $\mathbf{B}_0 := {}^a\mathbf{B}$ satisfies $F\mathbf{B}_0 \subseteq \mathbf{B}_0$. The equality comes from the remark above. The case of maximal tori is checked in the same fashion within $\mathbf{B}_0$. This gives (iii).

(iv) If we are looking at $x \mapsto aF(x)a^{-1}$, one decomposes $a = g^{-1}F(g)$ by (i) and gets a map sending gxg^{-1} to $gF(x)g^{-1}$. Its mth power is $gxg^{-1} \mapsto gF_0(x)g^{-1}$, i.e. a Frobenius map for a definition of $\mathbf{G}$ over $\mathbb{F}_q$ where the subalgebra A_0 (see A2.5) is now the original one conjugated by g (or better its comorphism). □

7.2. Varieties defined by the Lang map

We introduce a broad model of varieties associated with a Lang map $g \mapsto g^{-1}F(g)$ where $F\colon \mathbf{G} \to \mathbf{G}$ is an algebraic group endomorphism such that some power of F is a Frobenius endomorphism. We recall the notions of a tangent sheaf $\mathcal{T}\mathbf{X}$ of an $\mathbf{F}$-variety $\mathbf{X}$ (see A2.3) and quotient varieties (see A2.6).

Theorem 7.2. *Let $\mathbf{G}$ be a linear algebraic group defined over $\mathbb{F}_q$ with Frobenius F_0. Let $F\colon \mathbf{G} \to \mathbf{G}$ an endomorphism such that $F^m = F_0$ for some integer m.*

Let $\mathbf{V} \subseteq \mathbf{V}'$ be closed connected subgroups of $\mathbf{G}$ (not necessarily defined over $\mathbb{F}_q$) such that

(*) *$\mathbf{V} \cap F\mathbf{V}'$ is connected, and $\mathcal{T}(\mathbf{V} \cap F\mathbf{V}')_1 = \mathcal{T}\mathbf{V}_1 \cap \mathcal{T}(F\mathbf{V}')_1$ in $\mathcal{T}\mathbf{G}_1$.*

Denote

$$\mathbf{Y}_{\mathbf{V}\subseteq\mathbf{V}'} := \{g\mathbf{V} \mid g^{-1}F(g) \in \mathbf{V}.F\mathbf{V}'\} \subseteq \mathbf{G}/\mathbf{V}.$$

We abbreviate $\mathbf{Y}_{\mathbf{V}\subseteq\mathbf{V}} = \mathbf{Y}_{\mathbf{V}}$.

(i) $\mathbf{Y}_{\mathbf{V}\subseteq\mathbf{V}'}$ is a smooth locally closed subvariety of $\mathbf{G}/\mathbf{V}$ of dimension $\dim(\mathbf{V}') - \dim(\mathbf{V} \cap F\mathbf{V}')$ at every point.

(ii) $\mathbf{G}^F$ acts on $\mathbf{Y}_{\mathbf{V}\subseteq\mathbf{V}'}$ by left translation and permutes transitively its connected components.

(iii) Assume $\mathbf{V} \subseteq \mathbf{V}$ *also satisfies (*). The quotient morphism* $\mathbf{G} \to \mathbf{G}/\mathbf{V}$ *induces a surjective morphism* $\mathbf{Y}_{1\subseteq\mathbf{V}'} \to \mathbf{Y}_{\mathbf{V}\subseteq\mathbf{V}'}$ *whose differentials are also onto. The quotient morphism* $\mathbf{G}/(\mathbf{V} \cap F\mathbf{V}) \to \mathbf{G}/\mathbf{V}$ *induces an isomorphism* $\mathbf{Y}_{\mathbf{V}\cap F\mathbf{V}\subseteq\mathbf{V}} \xrightarrow{\sim} \mathbf{Y}_{\mathbf{V}\subseteq\mathbf{V}}$.

Theorem 7.3. *Let* $(\mathbf{G}, F)$ *be as in Theorem 7.2. Assume further that* $\mathbf{G}$ *is connected reductive. Let* $\mathbf{V}$ *be the unipotent radical of a parabolic subgroup with Levi decomposition* $\mathbf{P} = \mathbf{V} \rtimes \mathbf{L}$ *and assume* $\mathbf{L}$ *is* F*-stable,* $F\mathbf{L} = \mathbf{L}$*. Let* $\mathbf{V}' \subseteq \mathbf{H} \subseteq \mathbf{L}$, $\mathbf{C} \subseteq \mathbf{H}$ *be closed connected subgroups of* $\mathbf{G}$*. Assume* $\mathbf{H}$ *normalizes* $\mathbf{C}$ *and* $F\mathbf{H} = \mathbf{H}$*. Assume* $\mathcal{T}(\mathbf{C} \cap \mathbf{V}')_1 = \mathcal{T}\mathbf{C}_1 \cap \mathcal{T}\mathbf{V}'_1$*, and that the inclusions* $\mathbf{V}' \subseteq \mathbf{V}'$ *and* $\mathbf{V}' \subseteq \mathbf{C}\mathbf{V}'$ *satisfy (*).*

Denote $\mathbf{K} := \{h \in \mathbf{H} \mid h^{-1}F(h) \in F\mathbf{C}\}$*. Then the following hold.*

(i) $\mathbf{K} = \mathbf{K}^\circ.\mathbf{H}^F$.

(ii) The inclusions $\mathbf{V} \subseteq \mathbf{V}\mathbf{C}$ *and* $\mathbf{V}\mathbf{V}' \subseteq \mathbf{V}\mathbf{C}\mathbf{V}'$ *satisfy (*).*

(iii) $(\mathbf{Y}^{(\mathbf{G})}_{\mathbf{V}\subseteq\mathbf{V}\mathbf{C}} \times \mathbf{Y}^{(\mathbf{H})}_{\mathbf{V}'\subseteq\mathbf{V}'})/\mathbf{K} \cong \mathbf{Y}^{(\mathbf{G})}_{\mathbf{V}\mathbf{V}'\subseteq\mathbf{V}\mathbf{C}\mathbf{V}'}$ *by the map* $(x\mathbf{V}, x'\mathbf{V}') \mapsto xx'\mathbf{V}\mathbf{V}'$ *(the* $\mathbf{K}$*-quotient is for the diagonal action, and superscripts indicate the ambient groups used to define the varieties).*

The following will be useful.

Lemma 7.4. *Let* $\mathbf{P} = \mathbf{L}\mathbf{V}$, $\mathbf{P}' = \mathbf{L}'\mathbf{V}'$ *be two Levi decompositions. Assume* $\mathbf{L} = \mathbf{L}'$*. Then* $\mathbf{P}' \cap \mathbf{V} \subseteq \mathbf{V}'$ *and* $\mathbf{L} \cap \mathbf{V}\mathbf{V}' = \{1\}$.

Proof of Lemma 7.4. To check $\mathbf{P}' \cap \mathbf{V} \subseteq \mathbf{V}'$, one may use the ideas of Chapter 2 (the condition $\mathbf{L}' = \mathbf{L}$ implies $(\mathbf{P}, \mathbf{V})$—$(\mathbf{P}', \mathbf{V}')$) then take a limit from the finite case (write $\mathbf{G}$ as an ascending union $\bigcup_n \mathbf{G}^{F_0^{n!}}$ of finite BN-pairs). In a more classical way, one may also select a maximal torus in $\mathbf{L}$ and argue on roots (see A2.5). □

Proof of Theorem 7.2. Denote $\mathbf{Y} := \mathbf{G}/\mathbf{V}$, $\mathbf{Y}' := \mathbf{G}/F\mathbf{V}'$. We denote $\mathrm{Lan}(g) := g^{-1}F(g)$.

The group $\mathbf{G}$ acts diagonally on $\mathbf{Y} \times \mathbf{Y}'$ and we denote by Ω the orbit of $(\mathbf{V}, F\mathbf{V}')$, $\Omega := \{(g\mathbf{V}, gF\mathbf{V}') \mid g \in \mathbf{G}\}$, a locally closed subvariety of $\mathbf{G}/\mathbf{V} \times \mathbf{G}/F\mathbf{V}'$ (see A2.4). Moreover, the map $g \mapsto (g\mathbf{V}, gF\mathbf{V}')$ induces an isomorphism $\mathbf{G}/\mathbf{V} \cap F\mathbf{V}' \cong \Omega$. This is because the kernel of the tangent map at 1 of each reduction map $\mathbf{G} \to \mathbf{G}/\mathbf{V}$ and $\mathbf{G} \to \mathbf{G}/F\mathbf{V}'$ is $\mathcal{T}\mathbf{V}_1$, resp. $\mathcal{T}(F\mathbf{V}')_1$, and their intersection is $\mathcal{T}(\mathbf{V} \cap F\mathbf{V}')_1$ by condition (*), so that [Borel] 6.7(c) applies. Then Ω is smooth of dimension $\dim(\mathbf{G}) - \dim(\mathbf{V} \cap F\mathbf{V}')$, its tangent space being the image of that of $\mathbf{G}$.

The morphism F induces a morphism $F' \colon \mathbf{Y} \to \mathbf{Y}'$. We denote by Γ its graph $\Gamma := \{(g\mathbf{V}, F(g)F\mathbf{V}') \mid g \in \mathbf{G}\}$, a closed subvariety of $\mathbf{Y} \times \mathbf{Y}'$ isomorphic to

$\mathbf{G}/\mathbf{V}$ by the first projection. This isomorphism obviously sends $\Gamma \cap \Omega$ to $\mathbf{Y}_{\mathbf{V}\subseteq\mathbf{V}'}$. So $\mathbf{Y}_{\mathbf{V}\subseteq\mathbf{V}'}$ is a locally closed subvariety of $\mathbf{G}/\mathbf{V}$.

Since Γ is of dimension $\dim(\mathbf{G}) - \dim(\mathbf{V})$, Theorem 7.2(i) will follow once we check that Γ and Ω intersect transversally (see A2.3).

Let $y = (g\mathbf{V}, F(g)F\mathbf{V}') \in \Gamma \cap \Omega$. Up to an appropriate right translation by an element of $\mathbf{V}$, we may arrange $g^{-1}F(g) \in F\mathbf{V}'$ so that $y = (g\mathbf{V}, F(g)F\mathbf{V}') = (g\mathbf{V}, gF\mathbf{V}')$. Denote by $\pi: \mathbf{G} \to \mathbf{Y}$, $\pi': \mathbf{G} \to \mathbf{Y}'$ the quotient maps. By what has been said above, we have $\mathcal{T}\Omega_y = (\mathcal{T}\pi_g, \mathcal{T}\pi'_g).\mathcal{T}\mathbf{G}_g$. Similarly, we have $\mathcal{T}\Gamma_y = (\mathcal{T}\mathrm{Id}_{g\mathbf{V}}, \mathcal{T}F'_{g\mathbf{V}})\mathcal{T}\mathbf{Y}_{g\mathbf{V}} = (\mathcal{T}\pi_g, \mathcal{T}\pi'_{F(g)} \circ \mathcal{T}F_g)\mathcal{T}\mathbf{G}_g$ since $\pi' \circ F = F' \circ \pi$. Denoting by $\rho: \mathbf{G} \to \mathbf{G}$ the right translation by $F(g^{-1})g$, we have $\pi' = \pi' \circ \rho$ since $F(g^{-1})g = (g^{-1}F(g))^{-1} \in F\mathbf{V}'$. Differentiating at $F(g)$ yields $\mathcal{T}\pi'_{F(g)} = \mathcal{T}\pi'_g\mathcal{T}\rho_{F(g)}$. Then $\mathcal{T}\pi'_{F(g)} \circ \mathcal{T}F_g: \mathcal{T}\mathbf{G}_g \to \mathcal{T}\mathbf{Y}'_{gF\mathbf{V}'}$ can also be written as $\mathcal{T}\pi'_g \circ \varepsilon$, where $\varepsilon \in \mathrm{End}_{\mathbf{F}}(\mathcal{T}\mathbf{G}_g)$ is the differential at g of $\rho \circ F: t \mapsto F(t)F(g^{-1})g$. We have $(\rho \circ F)^m(t) = F^m(t)F^m(g^{-1})g$ whose differential is 0 everywhere, by A2.5. Then $\varepsilon^m = 0$.

The transversality now reduces to the following lemma in linear algebra (the proof is left as an exercise).

Lemma 7.5. *Let* $0 \to V \to G \xrightarrow{a} H \to 0$ *and* $0 \to V' \to G \xrightarrow{a'} H' \to 0$ *be exact sequences of finite-dimensional* $\mathbf{F}$*-vector spaces. Let* $\varepsilon \in \mathrm{End}(G)$ *be nilpotent with* $\varepsilon V \subseteq V'$*. Then* $(a, a')(G)$ *and* $(a, a' \circ \varepsilon)(G)$ *intersect transversally in* $H \times H'$*, their intersection being* $\{(a(x), a'(x)) \mid x \in (1 - \varepsilon)^{-1}(V')\} \cong V'/V \cap V'$.

(iii) is clear for the morphisms. For the differentials, Lemma 7.5 above gives $(\mathcal{T}\mathbf{Y}_{\mathbf{V}\subseteq\mathbf{V}'})_{g\mathbf{V}} = \mathcal{T}\pi_g((\mathrm{Id} - \varepsilon)^{-1}(\mathcal{T}gF\mathbf{V}')_g)$ with the same notation as above. Replacing $\mathbf{V}$ by $\{1\}$, we indeed get $(\mathcal{T}\mathbf{Y}_{1\subseteq\mathbf{V}'})_g = (\mathrm{Id} - \varepsilon)^{-1}(\mathcal{T}gF\mathbf{V}')_g$ and therefore $(\mathcal{T}\mathbf{Y}_{\mathbf{V}\subseteq\mathbf{V}'})_{g\mathbf{V}} = \mathcal{T}\pi_g(\mathcal{T}\mathbf{Y}_{1\subseteq\mathbf{V}'})_g$. This will also give the isomorphism we state. First the morphism is clearly a bijection. Both varieties in bijection are smooth, as a result of (i). The morphism is separable by what has been seen about differentials (see A2.6). So this is an isomorphism by the characterization of quotients given in A2.6 applied here to the action of a trivial group.

(ii) Denote $\mathbf{V}'' = F\mathbf{V}'$. We now prove that Lan is a quotient map $\mathrm{Lan}^{-1}(\mathbf{V}'') \to \mathbf{V}''$, i.e. that the variety structure on $\mathrm{Lan}^{-1}(\mathbf{V}'')/\mathbf{G}^F$ induced by the bijection (and the variety structure of $\mathbf{V}''$) is actually the quotient structure on $\mathrm{Lan}^{-1}(\mathbf{V}'')/\mathbf{G}^F$. We apply again the criterion given in A2.6. First $\mathbf{V}''$ is smooth since it is a closed subgroup (see A2.4), and $\mathrm{Lan}^{-1}(\mathbf{V}'')$ is smooth by (i). The differential criterion of separability amounts to checking that $\mathcal{T}\mathrm{Lan}_x$ is onto for every $x \in \mathrm{Lan}^{-1}(\mathbf{V}'')$. This is the case because of Theorem 7.1(iii) and the fact that $\mathcal{T}(\mathrm{Lan}^{-1}\mathbf{V}'')_x$ and $(\mathcal{T}\mathbf{V}'')_{\mathrm{Lan}(x)}$ have the same dimension (that of $\mathbf{V}''$), by (i).

Now $\mathbf{G}^F$ acts transitively on the irreducible components of $\mathrm{Lan}^{-1}(\mathbf{V}'')$ (see A2.6), i.e. its connected components, since it is smooth. This gives (ii) for $\mathrm{Lan}^{-1}(\mathbf{V}'') = \mathbf{Y}_{1\subseteq\mathbf{V}'}$. As for $\mathbf{Y}_{\mathbf{V}\subseteq\mathbf{V}'}$, (iii) tells us that it is the image of $\mathbf{Y}_{1\subseteq\mathbf{V}'}$ by the reduction map $\mathbf{G} \to \mathbf{G}/\mathbf{V}$, which is an open map (see A2.6) and a $\mathbf{G}^F$-map. Then the connected components of $\mathbf{Y}_{\mathbf{V}\subseteq\mathbf{V}'}$ are unions of images of the ones of $\mathbf{Y}_{1\subseteq\mathbf{V}'}$, thus our claim is proved. □

Proof of Theorem 7.3. (i) is Theorem 7.2(ii) in $\mathbf{H}$ since $\mathbf{K} = \mathbf{Y}^{(\mathbf{H})}_{1\subseteq\mathbf{C}}$.

(ii) First there is a maximal torus $\mathbf{T}_0$ of $\mathbf{L}$ such that $F\mathbf{T}_0 = \mathbf{T}_0$ (Theorem 7.1(iii)). Any closed connected unipotent subgroup $\mathbf{V} \subseteq \mathbf{G}$ normalized by $\mathbf{T}_0$ is of the form $\mathbf{X}_{\alpha_1} \ldots \mathbf{X}_{\alpha_r}$, where $\{\alpha_1, \ldots, \alpha_r\} \subseteq \Phi(\mathbf{G}, \mathbf{T}_0)$ is the list of roots α such that $\mathbf{X}_\alpha \subseteq \mathbf{V}$ (see A2.4 and [DiMi91] 0.34). Moreover $\mathbf{V} \cong \mathbf{X}_{\alpha_1} \times \cdots \times \mathbf{X}_{\alpha_r} \cong \mathbb{A}^r$ by the product map, and $\mathcal{T}\mathbf{V}$ (we omit the subscript 1) is the subspace of the Lie algebra $\mathcal{T}\mathbf{G} = \mathcal{T}\mathbf{T}_0 \oplus \bigoplus_\alpha \mathcal{T}\mathbf{X}_\alpha$ corresponding to $\{\alpha_1, \ldots, \alpha_r\}$ (see A2.4). Then F induces a permutation of the roots and clearly the inclusion $\mathbf{V} \subseteq \mathbf{V}$ satisfies (*). If $\mathbf{P} = \mathbf{V} \rtimes \mathbf{L}$ is a Levi decomposition with $F\mathbf{L} = \mathbf{L}$, one may take $\mathbf{T}_0 \subseteq \mathbf{L}$. As for $\mathbf{P}$, we have $F\mathbf{P} = F\mathbf{V} \rtimes \mathbf{L}$ and therefore $\mathbf{P} \cap F\mathbf{P} = (\mathbf{V} \cap F\mathbf{V}) \rtimes \mathbf{L}$. Then $\mathcal{T}(\mathbf{P} \cap F\mathbf{P}) = \mathcal{T}\mathbf{L} \oplus \mathcal{T}(\mathbf{V} \cap F\mathbf{V})$.

Now, to check that the inclusion $\mathbf{V}\mathbf{V}' \subseteq \mathbf{V}\mathbf{C}\mathbf{V}'$ satisfies (*), one first notes that Lemma 7.4 implies $\mathbf{V}\mathbf{V}' \cap F(\mathbf{V}\mathbf{C}\mathbf{V}') = (\mathbf{V} \cap F\mathbf{V}) \rtimes (\mathbf{V}' \cap F(\mathbf{C}\mathbf{V}'))$, the last expression being an algebraic semi-direct product (tangent spaces in direct sum), thanks to the above. Then we have $\mathcal{T}(\mathbf{V}\mathbf{V}' \cap F(\mathbf{V}\mathbf{C}\mathbf{V}')) = \mathcal{T}(\mathbf{V} \cap F\mathbf{V}) \oplus \mathcal{T}(\mathbf{V}' \cap F(\mathbf{C}\mathbf{V}'))$. But $\mathcal{T}(\mathbf{V}\mathbf{V}') \cap \mathcal{T}(F(\mathbf{V}\mathbf{C}\mathbf{V}'))) = (\mathcal{T}\mathbf{V} \oplus \mathcal{T}\mathbf{V}') \cap (\mathcal{T}F\mathbf{V} \oplus \mathcal{T}F(\mathbf{C}\mathbf{V}')) = (\mathcal{T}\mathbf{V} \cap \mathcal{T}F\mathbf{V}) \oplus (\mathcal{T}\mathbf{V}' \cap \mathcal{T}F(\mathbf{C}\mathbf{V}')) = (\mathcal{T}\mathbf{V} \cap \mathcal{T}F\mathbf{V}) \oplus \mathcal{T}(\mathbf{V}' \cap F(\mathbf{C}\mathbf{V}'))$, as a result of (*) for $\mathbf{V}' \subseteq \mathbf{C}\mathbf{V}'$, and the above description of $\mathcal{T}\mathbf{G}$, $\mathcal{T}\mathbf{P}$ and $\mathcal{T}F\mathbf{P}$ in terms of roots. Then (*) for $\mathbf{V}\mathbf{V}' \subseteq \mathbf{V}\mathbf{C}\mathbf{V}'$ follows from (*) for $\mathbf{V} \subseteq \mathbf{V}$. Note that the above also applies for $\mathbf{V}' = 1$.

(iii) The map is the restriction of $\mathbf{G}/\mathbf{V} \times \mathbf{H}/\mathbf{V}' \to \mathbf{G}/\mathbf{V}\mathbf{V}'$, $(x\mathbf{V}, y\mathbf{V}') \mapsto (xy\mathbf{V}\mathbf{V}')$, which in turn is induced by multiplication and the quotient morphisms $\pi: \mathbf{G} \to \mathbf{G}/\mathbf{V}, \pi': \mathbf{H} \to \mathbf{H}/\mathbf{V}', \pi'': \mathbf{G} \to \mathbf{G}/\mathbf{V}\mathbf{V}'$. This is therefore a morphism.

If $x \in \mathbf{G}$, $y \in \mathbf{H}$ and $x^{-1}F(x) \in \mathbf{V}\mathbf{C}$, $y^{-1}F(y) \in \mathbf{V}'$, then $y^{-1}x^{-1}F(xy) = (x^{-1}F(x))^y y^{-1}F(y) \in (\mathbf{V}\mathbf{C})^y\mathbf{V}' = \mathbf{V}\mathbf{C}\mathbf{V}'$ by the hypotheses. So the map is well defined and is obviously a morphism.

We check that it is onto. If $z \in \mathrm{Lan}^{-1}(\mathbf{V}\mathbf{C}\mathbf{V}')$, we have $z^{-1}F(z) = uv$ with $u \in \mathbf{V}\mathbf{C}$, $v \in \mathbf{V}'$. By Lang's theorem applied in $\mathbf{H}$, we have $v = y^{-1}F(y)$ for some $y \in \mathbf{H}$. Now, the above rearrangement shows $\mathrm{Lan}(zy^{-1}) = {}^y u \in \mathbf{V}\mathbf{C}$, so one may take $x := zy^{-1}$.

To show that the map of Theorem 7.3(iii) is a bijection up to $\mathbf{K}$-action, we take $g_1, g_2 \in \mathbf{G}$, $l_1, l_2 \in \mathbf{H}$ such that $g_i\mathbf{V} \in \mathbf{Y}_{\mathbf{V}\subseteq\mathbf{V}\mathbf{C}}$, $l_i\mathbf{V}' \in \mathbf{Y}_{\mathbf{V}'\subseteq\mathbf{V}'}$ and $g_1l_1\mathbf{V}\mathbf{C}\mathbf{V}' = g_2l_2\mathbf{V}\mathbf{C}\mathbf{V}'$. We may choose g_1 within $g_1\mathbf{V}$ so that $g_1^{-1}F(g_1) \in F(\mathbf{V}\mathbf{C})$. Since

$g_1 \in g_2\mathbf{V}\mathbf{H}$, we may also choose g_2 within $g_2\mathbf{V}$ so that $t := g_2^{-1}g_1 \in \mathbf{H}$. We then get $g_1^{-1}F(g_1) = t^{-1}g_2^{-1}F(g_2)F(t) \in t^{-1}\mathbf{V}F(\mathbf{VC})F(t) = t^{-1}F(t)\mathbf{V}F(\mathbf{VC})$ and therefore $t^{-1}F(t) \in F(\mathbf{VC})\mathbf{V}$. Now Lemma 7.4 applied to $\mathbf{P}$ and $F(\mathbf{P})$ gives $t^{-1}F(t) \in F\mathbf{C}$, i.e. $t \in \mathbf{K}$. Now $g_1l_1\mathbf{V}\mathbf{V}' = g_2l_2\mathbf{V}\mathbf{V}'$ gives $tl_1\mathbf{V}\mathbf{V}' = l_2\mathbf{V}\mathbf{V}' \in \mathbf{V} \rtimes\!\!\!\lhd \mathbf{H}$. Taking the components in $\mathbf{L}$, we get $tl_1\mathbf{V}' = l_2\mathbf{V}'$. So we have $(g_1\mathbf{V}, l_1\mathbf{V}') = (g_2\mathbf{V}, l_2\mathbf{V}').t$ (diagonal action) as claimed.

To show that $(x\mathbf{V}, y\mathbf{V}') \mapsto \mu'(x\mathbf{V}, y\mathbf{V}') = xy\mathbf{V}\mathbf{V}'$ is an isomorphism mod. diagonal action of $\mathbf{K}$, and since the varieties involved are smooth (Theorem 7.2(i)), it again suffices to check that μ' is separable (see A2.6). The differential criterion reduces this to showing that $\mathcal{T}\mu_{(x\mathbf{V},y\mathbf{V}')}$ is onto for at least one point $(x\mathbf{V}, y\mathbf{V}')$ in each connected component of $\mathbf{Y}^{(\mathbf{G})}_{\mathbf{V}\subseteq\mathbf{VC}} \times \mathbf{Y}^{(\mathbf{H})}_{\mathbf{V}'\subseteq\mathbf{V}'}$. So we may take $x \in \mathbf{G}^F$, $y \in \mathbf{H}^F$, as a result of Theorem 7.2(ii). Denote by $\mu: \mathbf{G} \times \mathbf{L} \to \mathbf{G}$ the multiplication in $\mathbf{G}$. We have $\mu' \circ (\pi, \pi') = \pi'' \circ \mu$ so, by Theorem 7.2(iii), we have to check that $\mathcal{T}\mu_{(x,y)}(\mathcal{T}(\mathbf{Y}^{(\mathbf{G})}_{1\subseteq\mathbf{VC}} \times \mathbf{Y}^{(\mathbf{H})}_{1\subseteq\mathbf{V}'})_{(x,y)}) = (\mathcal{T}\mathbf{Y}^{(\mathbf{G})}_{1\subseteq\mathbf{VCV}'})_{xy}$ in $\mathcal{T}\mathbf{G}_{xy}$.

Recall the notation $[\cdot x]$ (resp. $[x\cdot]$) for the map $\mathbf{G} \to \mathbf{G}$ defined by $[\cdot x](g) = gx$ (resp. $[x\cdot](g) = xg$).

By the classical formula, $(\mathcal{T}\mu)_{(x,y)}(\mathcal{T}\mathrm{Lan}^{-1}(\mathbf{VC})_x \times \mathcal{T}(\mathbf{H} \cap \mathrm{Lan}^{-1}(\mathbf{V}'))_y) = (\mathcal{T}[\cdot y])_x\mathcal{T}\mathrm{Lan}^{-1}(\mathbf{VC})_x + (\mathcal{T}[x\cdot])_y\mathcal{T}(\mathbf{H} \cap \mathrm{Lan}^{-1}(\mathbf{V}'))_y$. We must show that this is $\mathcal{T}\mathrm{Lan}^{-1}(\mathbf{VCV}')_{xy}$ in $\mathcal{T}\mathbf{G}_{xy}$. We apply the isomorphism $\mathcal{T}\mathrm{Lan}_{xy}$ (see Theorem 7.1(ii)). Our claim now reduces to $\mathcal{T}(\mathrm{Lan} \circ [\cdot y])_x\mathcal{T}\mathrm{Lan}^{-1}(\mathbf{VC})_x + \mathcal{T}(\mathrm{Lan} \circ [x\cdot])_y\mathcal{T}(\mathbf{H} \cap \mathrm{Lan}^{-1}(\mathbf{V}'))_y = \mathcal{T}(\mathbf{VCV}')_1$. Since $y \in \mathbf{H}^F$, $\mathrm{Lan} \circ [\cdot y](z) = [\cdot y] \circ [y^{-1}\cdot] \circ \mathrm{Lan}(z)$ for all $z \in \mathrm{Lan}^{-1}(\mathbf{VC})$, and therefore $\mathcal{T}(\mathrm{Lan} \circ [\cdot y])_x\mathcal{T}\mathrm{Lan}^{-1}(\mathbf{VC})_x = \mathcal{T}((\mathbf{VC})^y)_1 = \mathcal{T}(\mathbf{VC})_1$. Similarly $\mathrm{Lan} \circ [x\cdot] = \mathrm{Lan}$, so $\mathcal{T}(\mathrm{Lan} \circ [x\cdot])_y\mathcal{T}(\mathbf{H} \cap \mathrm{Lan}^{-1}(\mathbf{V}'))_y = \mathcal{T}\mathbf{V}'_1$. So our claim amounts to the equality $\mathcal{T}(\mathbf{VCV}') = \mathcal{T}(\mathbf{VC}) + \mathcal{T}\mathbf{V}'$. We have an inclusion, so it suffices to check dimensions. The surjection $\mathbf{VC} \times \mathbf{V}' \to \mathbf{VCV}'$ gives $\dim(\mathbf{VCV}') \leq \dim(\mathbf{VC}) + \dim(\mathbf{V}') - \dim(\mathbf{VC} \cap \mathbf{V}')$ (see A2.4). Each of the above dimensions coincides with the dimension of the tangent space, so it suffices to check that $\mathcal{T}(\mathbf{VC} \cap \mathbf{V}') = \mathcal{T}(\mathbf{VC}) \cap \mathcal{T}\mathbf{V}'$. By the Levi decomposition, we have $\mathbf{VC} \cap \mathbf{V}' = \mathbf{C} \cap \mathbf{V}'$, and the same for tangent spaces, so our claim is a consequence of the hypothesis $\mathcal{T}(\mathbf{C} \cap \mathbf{V}') = \mathcal{T}\mathbf{C} \cap \mathcal{T}\mathbf{V}'$. □

7.3. Deligne–Lusztig varieties

We keep $\mathbf{G}$ a connected reductive $\mathbf{F}$-group, $F_0: \mathbf{G} \to \mathbf{G}$ the Frobenius morphism associated with the definition of $\mathbf{G}$ over $\mathbb{F}_q$, $F: \mathbf{G} \to \mathbf{G}$ an endomorphism such that $F^m = F_0$ for some $m \geq 1$.

We prove a series of corollaries of Theorem 7.2 and Theorem 7.3.

Definition 7.6. *If* $\mathbf{P} = \mathbf{V} \rtimes \mathbf{L}$ *is a Levi decomposition in* $\mathbf{G}$ *with* $F\mathbf{L} = \mathbf{L}$, *define*

$$\mathbf{Y}_{\mathbf{V}}^{(\mathbf{G},F)} := \{g\mathbf{V} \mid g^{-1}F(g) \in \mathbf{V}.F(\mathbf{V})\},$$
$$\mathbf{X}_{\mathbf{V}}^{(\mathbf{G},F)} := \{g\mathbf{P} \mid g^{-1}F(g) \in \mathbf{P}.F(\mathbf{P})\}.$$

Clearly, $\mathbf{G}^F$ *acts on the left on* $\mathbf{Y}_{\mathbf{V}}^{(\mathbf{G},F)}$ *and* $\mathbf{X}_{\mathbf{V}}^{(\mathbf{G},F)}$. *Moreover,* $\mathbf{L}^F$ *acts on* $\mathbf{Y}_{\mathbf{V}}^{(\mathbf{G},F)}$ *on the right.*

Here is a series of corollaries of Theorem 7.3.

Theorem 7.7. *Both* $\mathbf{Y}_{\mathbf{V}}^{(\mathbf{G},F)}$ *and* $\mathbf{X}_{\mathbf{V}}^{(\mathbf{G},F)}$ *are smooth of dimension* $\dim(\mathbf{V}) - \dim(\mathbf{V} \cap F\mathbf{V})$. *Both are acted on by* $\mathbf{G}^F$ *on the left, and this action is transitive on the connected components.*

Theorem 7.8. *The finite group* $\mathbf{L}^F$ *acts freely on* $\mathbf{Y}_{\mathbf{V}}^{(\mathbf{G},F)}$ *on the right, and the map*

$$\mathbf{G}/\mathbf{V} \to \mathbf{G}/\mathbf{P}, \quad g\mathbf{V} \mapsto g\mathbf{P},$$

induces a Galois covering $\mathbf{Y}_{\mathbf{V}}^{(\mathbf{G},F)} \to \mathbf{X}_{\mathbf{V}}^{(\mathbf{G},F)}$ *of group* $\mathbf{L}^F$.

Theorem 7.9. (Transitivity) *If* $\mathbf{P} = \mathbf{V} \rtimes \mathbf{L}$ *is a Levi decomposition in* $\mathbf{G}$ *with* $F\mathbf{L} = \mathbf{L}$, *and* $\mathbf{Q} = \mathbf{V}' \rtimes \mathbf{M}$ *is a Levi decomposition in* $\mathbf{L}$ *with* $F\mathbf{M} = \mathbf{M}$, *then the multiplication induces an isomorphism*

$$\big(\mathbf{Y}_{\mathbf{V}}^{(\mathbf{G},F)} \times \mathbf{Y}_{\mathbf{V}'}^{(\mathbf{L},F)}\big)/\mathbf{L}^F \to \mathbf{Y}_{\mathbf{V}.\mathbf{V}'}^{(\mathbf{G},F)},$$

where the action of $\mathbf{L}^F$ *is the diagonal action.*

Theorem 7.10. *Let* $\mathbf{G} \subseteq \widetilde{\mathbf{G}}$ *be an inclusion of connected reductive* $\mathbf{F}$*-groups defined over* $\mathbb{F}_q$ *such that* $\widetilde{\mathbf{G}} = \mathbf{G}\mathrm{Z}(\widetilde{\mathbf{G}})$ *with Frobenius endomorphism* $F : \widetilde{\mathbf{G}} \to \widetilde{\mathbf{G}}$. *Let* $\mathbf{P} = \mathbf{V} \rtimes \mathbf{L}$ *be a Levi decomposition in* $\mathbf{G}$ *with* $F\mathbf{L} = \mathbf{L}$. *Then* $\mathrm{Z}(\widetilde{\mathbf{G}})\mathbf{P} = \mathbf{V} \rtimes (\mathrm{Z}(\widetilde{\mathbf{G}})\mathbf{L})$ *is a Levi decomposition in* $\widetilde{\mathbf{G}}$ *and* $\mathbf{Y}_{\mathbf{V}}^{(\widetilde{\mathbf{G}},F)} \cong (\widetilde{\mathbf{G}}^F \times \mathbf{Y}_{\mathbf{V}}^{(\mathbf{G},F)})/\mathbf{G}^F$ *by the obvious product map.*

Proof of Theorems 7.7–7.10. We omit superscripts $(\mathbf{G}, F)$. First note that $\mathbf{X}_{\mathbf{V}} = \mathbf{Y}_{\mathbf{P}}$ in the notation of Theorem 7.2.

Then Theorem 7.2(i) and Theorem 7.3(i) imply that $\mathbf{Y}_{\mathbf{V}}$ and $\mathbf{X}_{\mathbf{V}}$ are smooth varieties of dimensions $\dim(\mathbf{V}) - \dim(\mathbf{V} \cap F\mathbf{V})$, $\dim(\mathbf{P}) - \dim(\mathbf{P} \cap F\mathbf{P})$ respectively. But these are equal since $\mathbf{P} \cong \mathbf{L} \times \mathbf{V}$ as varieties, and $F\mathbf{L} = \mathbf{L}$. This gives Theorem 7.7, the second statement being Theorem 7.2(ii).

Theorem 7.9 is implied by Theorem 7.3(iii) with $\mathbf{H} = \mathbf{L}$ and $\mathbf{C} = 1$. Theorem 7.8 is also a consequence of Theorem 7.3(iii) with $\mathbf{V}' = \mathbf{H} = \mathbf{L}$ (for which the condition (*) is clear) and $\mathbf{C} = 1$, since then $\mathbf{Y}_{\mathbf{V}'}^{(\mathbf{L})}$ is one point fixed by $\mathbf{L}^F$, while $\mathbf{V}\mathbf{V}' = \mathbf{P}$.

For Theorem 7.10, note that Theorem 7.3(iii) for $\mathbf{V} = \mathbf{C} = \{1\}$ (and the ambient group renamed $\widetilde{\mathbf{G}}$) gives

$$((\widetilde{\mathbf{G}})^F \times \mathbf{Y}_{\mathbf{V}'}^{(\mathbf{H})})/\mathbf{H}^F \cong \mathbf{Y}_{\mathbf{V}'}^{(\widetilde{\mathbf{G}})}$$

for any $\mathbf{H}$ and $\mathbf{V}'$ satisfying certain conditions. These are clearly satisfied by $\mathbf{H} = \mathbf{G}$ and $\mathbf{V}'$ a unipotent radical normalized by an F-stable Levi subgroup (see Theorem 7.3(ii)). □

Using the usual presentation of Coxeter groups and the Word Lemma (see [Bour68] IV.1.5), we may rephrase the property mentioned at the end of A2.4 as follows. Let $\mathbf{T}$ be a maximal torus of a connected reductive $\mathbf{F}$-group $\mathbf{G}$. Recall the notation $W(\mathbf{G}, \mathbf{T}) := \mathrm{N}_{\mathbf{G}}(\mathbf{T})/\mathbf{T}$ for the associated Weyl group (see A2.4).

Theorem 7.11. *There is a map*

$$W(\mathbf{G}, \mathbf{T}) \to \mathrm{N}_{\mathbf{G}}(\mathbf{T}), \quad w \mapsto \dot{w}$$

such that $\dot{w}\mathbf{T} = w$, $\dot{1} = 1$, and $\dot{w} = \dot{w}'\dot{w}''$ whenever $w = w'w''$ with lengths adding.

Here is a proposition showing how to parametrize Deligne–Lusztig varieties a little more precisely using a pair $\mathbf{T}_0 \subseteq \mathbf{B}_0$ as in Theorem 7.1(iii).

Definition 7.12. *When $\mathbf{B}_0 \supseteq \mathbf{T}_0$ are some F-stable Borel subgroup and maximal torus, and $w \in W(\mathbf{G}, \mathbf{T}_0)$, let $\mathbf{Y}^{(\mathbf{G},F)}(w) := \{g\mathbf{U}_0 \mid g^{-1}F(g) \in \mathbf{U}_0\dot{w}\mathbf{U}_0\}$, and $\mathbf{X}^{(\mathbf{G},F)}(w) := \{g\mathbf{B}_0 \mid g^{-1}F(g) \in \mathbf{B}_0 w\mathbf{B}_0\}$, where $\mathbf{U}_0$ denotes the unipotent radical of $\mathbf{B}_0$.*

Proposition 7.13. *Let $\mathbf{T}_0 \subseteq \mathbf{B}_0$ be a pair of F-stable maximal torus and Borel subgroup, respectively (see Theorem 7.1(iii)). Recall $\Phi(\mathbf{G}, \mathbf{T}_0) \supseteq \Delta$, the associated root system and set of simple roots, respectively (see A2.4).*

Let $\mathbf{P} = \mathbf{V} \rtimes \mathbf{L}$ be a Levi decomposition with $F\mathbf{L} = \mathbf{L}$.

Let $w \in W(\mathbf{G}, \mathbf{T}_0)$.

(i) There exist $v \in W(\mathbf{G}, \mathbf{T}_0)$, $I \subseteq \Delta$, and $a \in \mathbf{G}$ such that $aF(a^{-1}) = \dot{v}$, $v^{-1}(I) \subseteq \Delta$, ${}^a\mathbf{V} = \mathbf{U}_I$, ${}^a\mathbf{L} = \mathbf{L}_I$, and $x \mapsto {}^a x$ induces an isomorphism $\mathbf{Y}_{\mathbf{V}}^{(\mathbf{G},F)} \to \mathbf{Y}_{\mathbf{U}_I}^{(\mathbf{G},\dot{v}F)}$ where $\dot{v}F$ denotes F composed with conjugation by $\dot{v}$, see Theorem 7.1(iv)).

(ii) For any $b \in \mathrm{Lan}^{-1}(\dot{w})$, the map $x \mapsto xb^{-1}$ induces isomorphisms

$$\mathbf{Y}^{(\mathbf{G},F)}(w) \to \mathbf{Y}_{{}^b\mathbf{U}_0}^{(\mathbf{G},F)} \quad \text{and} \quad \mathbf{X}^{(\mathbf{G},F)}(w) \to \mathbf{X}_{{}^b\mathbf{U}_0}^{(\mathbf{G},F)}.$$

(iii) The dimension of $\mathbf{X}(w)$ *is* $l(w)$ *(length relative to the generators* $S \subseteq W(\mathbf{G}, \mathbf{T}_0)$ *associated with* Δ*; see A2.4).*

(iv) Denote by $\leq$ *the Bruhat order on* $W(\mathbf{G}, \mathbf{T}_0)$ *relative to* S *above. Then*

$$\overline{\mathbf{X}}(w) := \bigcup_{w' \leq w} \mathbf{X}(w')$$

is the Zariski closure of $\mathbf{X}(w)$ *in* $\mathbf{G}/\mathbf{B}_0$*.*

(v) If w *is a product of pairwise commuting elements of* S*, then* $\overline{\mathbf{X}}(w)$ *is smooth and the* $\overline{\mathbf{X}}(w')$*'s for* $w' \leq w$ *and* $l(w') = l(w) - 1$ *make a smooth divisor with normal crossings equal to* $\overline{\mathbf{X}}(w) \setminus \mathbf{X}(w)$ *(see A2.3).*

Proof. (i) There are $x \in \mathbf{G}$ and $I \subseteq \Delta$ such that $\mathbf{P} = {}^x\mathbf{P}_I$, $\mathbf{L} = {}^x\mathbf{L}_I$, $\mathbf{V} = {}^x\mathbf{U}_I$. Since F induces a permutation of Δ, we have $F\mathbf{L}_I = \mathbf{L}_{I'}$ for some $I' \subseteq \Delta$. The condition $F\mathbf{L} = \mathbf{L}$ now reads ${}^{x^{-1}F(x)}\mathbf{L}_{I'} = \mathbf{L}_I$. By transitivity of $\mathbf{L}_I$ on its maximal tori (see A2.4), we have $x^{-1}F(x) \in \mathbf{L}_I\dot{v}$ where $v \in W(\mathbf{G}, \mathbf{T}_0)$ satisfies ${}^{\dot{v}}\mathbf{L}_{I'} = \mathbf{L}_I$. By transitivity of W_I on the bases of its root system (see Chapter 2), one may even assume $vI' = I$. Applying Lang's theorem to $y \mapsto {}^{\dot{v}}F(y)$ on $\mathbf{L}_I$ (see Theorem 7.1(iv)), we have $x^{-1}F(x) = y^{-1}\dot{v}F(y)$ for some $y \in \mathbf{L}_I$. Then

$$\begin{aligned} \mathbf{Y}_{\mathbf{V}} &= \{g\mathbf{V} \mid g^{-1}F(g) \in \mathbf{V}F\mathbf{V}\} \\ &= \{gx\mathbf{U}_Ix^{-1} \mid g^{-1}F(g) \in x\mathbf{U}_Ix^{-1}F(x)F\mathbf{U}_IF(x^{-1})\}. \end{aligned}$$

But $x\mathbf{U}_Ix^{-1}F(x)F\mathbf{U}_IF(x^{-1}) = xy^{-1}\mathbf{U}_I\dot{v}F\mathbf{U}_IF(yx^{-1})$ since $y \in \mathbf{L}_I$ normalizes $\mathbf{U}_I$. Then $g \mapsto {}^{yx^{-1}}g$ transforms $\mathbf{Y}_{\mathbf{V}}$ into

$$\mathbf{Y}_{\mathbf{U}_I}^{(\dot{v}F)} = \{g\mathbf{U}_I \mid g^{-1}\dot{v}F(g) \in \mathbf{U}_I\dot{v}F\mathbf{U}_I\}$$

where the exponent $\dot{v}F$ indicates the Frobenius we are taking to build this $\mathbf{Y}_{\mathbf{U}_I}$. By this change of variable, the action of $\mathbf{G}^F \times \mathbf{L}^F$ is replaced by the one of $\mathbf{G}^F \times \mathbf{L}_I^{\dot{v}F}$ since $F({}^xt) = {}^xt$ is equivalent to ${}^yt \in \mathbf{L}_I^{\dot{v}F}$ for $t \in \mathbf{L}_I$. One then takes $a = yx^{-1}$.

(ii) Easy.

(iii) By (ii) and Theorem 7.7, $\mathbf{Y}(w)$ is of dimension $\dim({}^b\mathbf{U}_0) - \dim({}^b\mathbf{U}_0 \cap F({}^b\mathbf{U}_0)) = \dim(\mathbf{U}_0) - \dim(\mathbf{U}_0 \cap {}^w\mathbf{U}_0) = l(w)$.

(iv) One may (as in the proof of Theorem 7.2) view $\mathbf{X}(w)$ as the intersection of the graph of F on $\mathbf{G}/\mathbf{B}_0$ with the $\mathbf{G}$-orbit on $\mathbf{G}/\mathbf{B}_0 \times \mathbf{G}/\mathbf{B}_0$ associated with w. That is, $O(w) = \{(g\mathbf{B}_0, gw\mathbf{B}_0) \mid g \in \mathbf{G}\}$. The graph of F is closed while the expression of $\overline{O(w)}$ in terms of the Bruhat order is well known (see [Jantzen] II.13.7, or A2.6). This shows that $\overline{\mathbf{X}}(w)$ is closed. The subvariety $\mathbf{X}(w)$ is dense in it, since $\overline{\mathbf{X}}(w) \setminus \mathbf{X}(w)$ is a union of subvarieties $\mathbf{X}(w')$ of dimensions $l(w') < l(w)$.

(v) If w is a product of commuting elements of S, then $\{w' \mid w' \leq w\}$ is a parabolic subgroup $<I>$ where $I \subseteq S$. Then $\overline{\mathbf{X}}(w) = \{g\mathbf{B}_0 \mid g^{-1}F(g) \in \mathbf{P}_I\}$.

This is isomorphic with the intersection of the graph of F on $\mathbf{G}/\mathbf{B}_0$ with $\overline{O}_I := \{(g\mathbf{B}_0, g'\mathbf{B}_0) \mid g\mathbf{P}_I = g'\mathbf{P}_I\}$. Those varieties intersect transversally since their tangent spaces are of the form $\mathrm{Im}(\mathrm{Id}_V \times \varepsilon) \subseteq V \times V$ and $E \subseteq V \times V$ (respectively) for V a vector space, $\varepsilon \in \mathrm{End}(V)$ a nilpotent endomorphism and E a subspace containing the diagonal (see Lemma 7.5). It remains to check that $\overline{O}_I$ is smooth and that the $\overline{O}_{I'}$'s for $I' \subseteq I$ with $|I| = |I'| + 1$ make a smooth divisor with normal crossings.

By [Hart] III.10.1.(d), $\overline{O}_I$, which is $\mathbf{G}/\mathbf{B}_0 \times_{\mathbf{G}/\mathbf{P}_I} \mathbf{G}/\mathbf{B}_0$ for the evident map $\mathbf{G}/\mathbf{B}_0 \to \mathbf{G}/\mathbf{P}_I$, is smooth over $\mathbf{G}/\mathbf{P}_I$ of relative dimension $2\dim(\mathbf{P}_I/\mathbf{B}_0)$ in the sense of [Hart] §III.10. Then $\overline{O}_I$ is smooth over $\mathrm{Spec}(\mathbf{F})$ of dimension $\dim(\mathbf{G}/\mathbf{B}_0) + \dim(\mathbf{G}/\mathbf{P}_I)$ by [Hart] 10.1(c).

From the smoothness of $\overline{O}_I$, it is now clear that the map $\widetilde{O}_I := \{(g, g') \in \mathbf{G} \times \mathbf{G} \mid g\mathbf{P}_I = g'\mathbf{P}_I\} \to \overline{O}_I$ is a $\mathbf{B} \times \mathbf{B}$-quotient (see A2.6). Then our statement about the $\overline{O}_{I'}$'s reduces to the same in $\widetilde{O}_I \cong \mathbf{G} \times \mathbf{P}_I$ where it is a trivial consequence of the fact that the corresponding $\mathbf{P}_{I'}$'s are of codimension 1 in $\mathbf{P}_I$ and have tangent spaces in general position (remember that the root system of $\mathbf{P}_I$ is of type $(\mathbf{A}_1)^{|I|}$ since the elements of I commute pairwise). □

The following is a slight generalization of Theorem 7.8 and will be useful in Chapter 11.

Theorem 7.14. *Let $\mathbf{P} = \mathbf{V} \rtimes \mathbf{L}$ be a Levi decomposition and $n \in \mathbf{G}$ such that $n(F\mathbf{L})n^{-1} = \mathbf{L}$. Let $\mathbf{H} \supseteq \mathbf{C}$ be closed connected subgroups of $\mathbf{L}$ such that $n(F\mathbf{H})n^{-1} = \mathbf{H}$ and $\mathbf{H}$ normalizes $\mathbf{C}$.*

Define $\mathbf{K} := \{h \in \mathbf{H} \mid h^{-1}nF(h)n^{-1} \in \mathbf{C}\}$, $\mathbf{Y} := \{g\mathbf{V} \mid g^{-1}F(g) \in \mathbf{C}.\mathbf{V}nF(\mathbf{V})\}$, and $\mathbf{X} := \{g\mathbf{H}\mathbf{V} \mid g^{-1}F(g) \in \mathbf{H}.\mathbf{V}nF(\mathbf{V})\}$. Then $\mathbf{K}$ is a closed subgroup of $\mathbf{G}$ acting on $\mathbf{Y}$ (on the right) and

(i) $\mathbf{K} = \mathbf{K}^\circ.\mathbf{H}^{nF}$ (where nF denotes the endomorphism $g \mapsto nF(g)n^{-1}$),

(ii) $\mathbf{Y}$ and $\mathbf{X}$ are smooth locally closed subvarieties of $\mathbf{G}/\mathbf{V}$, $\mathbf{G}/\mathbf{H}\mathbf{V}$ respectively, and $\mathbf{Y}/\mathbf{K} \cong \mathbf{X}$ by $g\mathbf{V} \mapsto g\mathbf{H}\mathbf{V}$.

Proof. By Lang's theorem (Theorem 7.1(i)), we may write $n = aF(a^{-1})$ for some $a \in \mathbf{G}$. Then $a\mathbf{Y} = \mathbf{Y}^{(\mathbf{G})}_{\mathbf{V}\subseteq \mathbf{V}(nF)^{-1}\mathbf{C}}$ and $a\mathbf{X} = \mathbf{Y}^{(\mathbf{G})}_{\mathbf{V}\mathbf{H}\subseteq \mathbf{V}\mathbf{H}}$ (notation of Theorem 7.2) for the endomorphism nF. This endomorphism is a Frobenius endomorphism (Theorem 7.1(iv)), so we may as well assume $n = 1$. Denoting by $\mathbf{C}'$ a closed connected normal subgroup of $\mathbf{H}$ such that $F\mathbf{C}' = \mathbf{C}$ (for instance, $(F^{-1}\mathbf{C} \cap \mathbf{H})^\circ$, left as an exercise), our claim follows from Theorem 7.3(iii) with $\mathbf{V}' = \mathbf{H}$ once we check that the inclusions $\mathbf{H} \subseteq \mathbf{H}$ and $\mathbf{H} \subseteq \mathbf{C}'\mathbf{H}$ satisfy (*) of Theorem 7.2 and that $\mathcal{T}(\mathbf{C}' \cap \mathbf{H}) = \mathcal{T}\mathbf{C}' \cap \mathcal{T}\mathbf{H}$. Those conditions are trivial. □

7.4. Deligne–Lusztig varieties are quasi-affine

We now prove another important property of Deligne–Lusztig varieties (see also Exercise 11.2).

Theorem 7.15. *Let* $(\mathbf{G}, F_0)$ *be a connected reductive* $\mathbf{F}$*-group defined over* $\mathbb{F}_q$*. Let* $F: \mathbf{G} \to \mathbf{G}$ *be an endomorphism such that* $F^m = F_0$ *for some* $m \geq 1$*. Let* $\mathbf{P} = \mathbf{V} \rtimes \mathbf{L}$ *be a Levi decomposition with* $F\mathbf{L} = \mathbf{L}$*.*

Then $\mathbf{X}_\mathbf{V}$ *and* $\mathbf{Y}_\mathbf{V}$ *(see Definition 7.6) are quasi-affine varieties.*

The proof involves the criterion of quasi-affinity of Theorem A2.11.

Write $\mathbf{P} = \mathbf{V} \rtimes \mathbf{L}$. Let $\mathbf{V}_0\mathbf{T}_0$ be an F-stable Borel subgroup of $\mathbf{L}$ (see Theorem 7.1(iii)) with $F\mathbf{T}_0 = \mathbf{T}_0$. Then $\mathbf{Y}^{(\mathbf{L})}_{\mathbf{V}_0} = \mathbf{L}^F/\mathbf{V}_0^F$. Theorem 7.9 now implies that $\mathbf{Y}_{\mathbf{V}.\mathbf{V}_0} \cong \mathbf{Y}_\mathbf{V}/\mathbf{V}_0^F$, so Corollary A2.13 implies that it suffices to check the quasi-affinity of $\mathbf{Y}_{\mathbf{V}.\mathbf{V}_0}$ to get that $\mathbf{Y}_\mathbf{V}$ is quasi-affine. This in turn implies that $\mathbf{X}_\mathbf{V}$ is quasi-affine, by Theorem 7.8 and Corollary A2.13 again. Similarly, the $\mathbf{T}_0^F$-quotient $\mathbf{Y}_{\mathbf{V}.\mathbf{V}_0} \to \mathbf{X}_{\mathbf{V}.\mathbf{V}_0}$ of Theorem 7.8 implies that it suffices to check the quasi-affinity of $\mathbf{X}_\mathbf{V}$ where $\mathbf{V}$ is the unipotent radical of a Borel subgroup of $\mathbf{G}$.

By Proposition 7.13(i),(ii) (with $I = \emptyset$), one may assume that $\mathbf{X}_\mathbf{V}$ is in the form $\mathbf{X}(w)$ for $w \in W(\mathbf{G}, \mathbf{T}_0)$, having fixed $\mathbf{B}_0 \supseteq \mathbf{T}_0$, both F-stable. So, in view of the quasi-affinity criterion of Theorem A2.11, it suffices to prove the following (see A2.8 for the notion of ample invertible sheaf on a variety).

Proposition 7.16. *If* $w \in W(\mathbf{G}, \mathbf{T}_0)$*, then the structure sheaf of* $\mathbf{X}(w)$ *is ample.*

If we have an ample invertible sheaf on $\mathbf{G}/\mathbf{B}_0$, then its restriction to $\overline{\mathbf{X}}(w)$ (i.e. inverse image by the corresponding immersion) is also ample, and its further restriction to the open subvariety $\mathbf{X}(w)$ (see Proposition 7.13(iv)) is then ample too (see A2.8). So, to prove Proposition 7.16, it suffices to arrange that this restriction is the structure sheaf of $\mathbf{X}(w)$ to have the quasi-affinity of $\mathbf{X}(w)$.

Let us show how invertible sheaves on $\mathbf{G}/\mathbf{B}_0$ can be built from linear characters of $\mathbf{T}_0$ (see A2.9). The quotient $\mathbf{G}/\mathbf{B}_0$ is locally trivial, as a result of the covering of $\mathbf{G}/\mathbf{B}_0$ by translates of the "big cell" $\mathbf{B}_0 w_0 \mathbf{B}_0/\mathbf{B}_0$ (see A2.6). If $\lambda \in X(\mathbf{T}_0)$, we denote by the same symbol the one-dimensional $\mathbf{B}_0$-module and consider $\mathcal{L}_{\mathbf{G}/\mathbf{B}_0}(\lambda)$ (see A2.9). This is an invertible sheaf over $\mathbf{G}/\mathbf{B}_0$, using the characterization in terms of tensor product with the dual (see A2.8) and since $\mathcal{L}_{\mathbf{G}/\mathbf{B}_0}(\lambda)^\vee = \mathcal{L}_{\mathbf{G}/\mathbf{B}_0}(-\lambda)$ and $\mathcal{L}_{\mathbf{G}/\mathbf{B}_0}(\lambda) \otimes \mathcal{L}_{\mathbf{G}/\mathbf{B}_0}(\lambda') = \mathcal{L}_{\mathbf{G}/\mathbf{B}_0}(\lambda + \lambda')$ (the group $X(\mathbf{T}_0)$ is denoted additively) by the general properties of the $\mathcal{L}(\lambda)$ construction.

Let us denote by $j: \mathbf{X}(w) \to \mathbf{G}/\mathbf{B}_0$ the natural immersion. We shall prove the following.

Proposition 7.17. *$j^*\mathcal{L}_{\mathbf{G}/\mathbf{B}_0}(\lambda \circ F) \cong j^*\mathcal{L}_{\mathbf{G}/\mathbf{B}_0}(w(\lambda))$, where $w(\lambda)$ is defined by $w(\lambda)(t) = \lambda(t^w)$ for all $t \in \mathbf{T}_0$.*

Let us say how this implies Proposition 7.16. From [Jantzen] II.4.4 (or even II.4.3), we know that the set of $\lambda \in X(\mathbf{T}_0)$ such that $\mathcal{L}_{\mathbf{G}/\mathbf{B}_0}(\lambda)$ is ample is non-empty. Let ω be such an element of $X(\mathbf{T}_0)$.

By Lang's theorem applied to $\mathbf{T}_0$, the map $X(\mathbf{T}_0) \to X(\mathbf{T}_0)$ defined by $\lambda \mapsto w(\lambda) - \lambda \circ F$ is injective (use Theorem 7.1(iv) and (i)), so its cokernel is finite. There exist $\lambda \in X(\mathbf{T}_0)$ and an integer $m \geq 1$ such that $w(\lambda) - \lambda \circ F = m\omega$. Then $\mathcal{L}_{\mathbf{G}/\mathbf{B}_0}(w(\lambda) - \lambda \circ F) = \mathcal{L}_{\mathbf{G}/\mathbf{B}_0}(m\omega) = \mathcal{L}_{\mathbf{G}/\mathbf{B}_0}(\omega)^{\otimes m}$ is ample since $\mathcal{L}_{\mathbf{G}/\mathbf{B}_0}(\omega)$ is ample (see A2.8). Its restriction to $\mathbf{X}(w)$, $j^*\mathcal{L}_{\mathbf{G}/\mathbf{B}_0}(w(\lambda) - \lambda \circ F)$, is the structure sheaf, because Proposition 7.17 allows us to write $j^*\mathcal{L}_{\mathbf{G}/\mathbf{B}_0}(w(\lambda) - \lambda \circ F) = j^*(\mathcal{L}_{\mathbf{G}/\mathbf{B}_0}(w(\lambda)) \otimes \mathcal{L}_{\mathbf{G}/\mathbf{B}_0}(\lambda \circ F)^\vee) = j^*\mathcal{L}_{\mathbf{G}/\mathbf{B}_0}(w(\lambda)) \otimes j^*\mathcal{L}_{\mathbf{G}/\mathbf{B}_0}(\lambda \circ F)^\vee = j^*\mathcal{L}_{\mathbf{G}/\mathbf{B}_0}(\lambda \circ F) \otimes j^*\mathcal{L}_{\mathbf{G}/\mathbf{B}_0}(\lambda \circ F)^\vee = \mathcal{O}_{\mathbf{X}(w)}$ since we are tensoring an invertible sheaf on $\mathbf{X}(w)$ with its dual (see A2.8). As said before, this completes the proof of Proposition 7.16.

Proof of Proposition 7.17. Denote by $\bar{F}: \mathbf{G}/\mathbf{B}_0 \to \mathbf{G}/\mathbf{B}_0$ the morphism induced by F. Taking $X = X' = \mathbf{G}$, $G = G' = \mathbf{B}_0$ and $\varphi = F$ in A2.9, we get

(1) $$\bar{F}^*\mathcal{L}_{\mathbf{G}/\mathbf{B}_0}(\lambda) \cong \mathcal{L}_{\mathbf{G}/\mathbf{B}_0}(\lambda \circ F).$$

As in the proof of Theorem 7.2, we consider $\Omega := \{(g\mathbf{B}_0, g\dot{w}\mathbf{B}_0) \mid g \in \mathbf{G}\} \subseteq \mathbf{G}/\mathbf{B}_0 \times \mathbf{G}/\mathbf{B}_0$ which is locally closed in $\mathbf{G}/\mathbf{B}_0 \times \mathbf{G}/\mathbf{B}_0$ (being a $\mathbf{G}$-orbit), and $\Gamma = \{(g\mathbf{B}_0, F(g\mathbf{B}_0)) \mid g \in \mathbf{G}\} \subseteq \mathbf{G}/\mathbf{B}_0 \times \mathbf{G}/\mathbf{B}_0$ (closed), so that $\mathbf{X}(w) \cong \Omega \cap \Gamma$ by the first projection $\mathbf{G}/\mathbf{B}_0 \times \mathbf{G}/\mathbf{B}_0 \to \mathbf{G}/\mathbf{B}_0$.

Now $\Omega \cong \mathbf{G}/\mathbf{B}_0 \cap {}^w\mathbf{B}_0$ is the quotient by $\mathbf{T}_0$ of $\mathbf{E}_w := \mathbf{G}/\mathbf{U}_0 \cap {}^w\mathbf{U}_0$, where $\mathbf{T}_0$ acts freely on the right (see [Borel] 6.10). This quotient is locally trivial since the "big cell" satisfies $\mathbf{B}_0 w_0 \mathbf{B}_0 \cong \mathbf{T}_0 \times \mathbf{U}_0 \times \mathbf{U}_0$ as a $\mathbf{T}_0$-variety. This allows us to define $\mathcal{L}_\Omega(\lambda)$. We have a commutative diagram, where the top map is $\mathbf{T}_0$-equivariant, vertical maps are quotient maps, and where pr_1 is the restriction of the first projection $\mathbf{G}/\mathbf{B}_0 \times \mathbf{G}/\mathbf{B}_0 \to \mathbf{G}/\mathbf{B}_0$

$$\begin{array}{ccc} \mathbf{E}_w & \longrightarrow & \mathbf{E}_1 \\ \downarrow & & \downarrow \\ \mathbf{E}_w/\mathbf{T}_0 \cong \Omega & \xrightarrow{\mathrm{pr}_1} & \mathbf{G}/\mathbf{B}_0 = \mathbf{E}_1/\mathbf{T}_0 \end{array}$$

Now A2.9(I) for $X = \mathbf{E}_w$, $X' = \mathbf{E}_1$, $G = G' = \mathbf{T}_0$, $\alpha = \mathrm{Id}$, $\varphi(x) = x\mathbf{U}_0$, gives

(2) $$\mathrm{pr}_1^*\mathcal{L}_{\mathbf{G}/\mathbf{B}_0}(\lambda) \cong \mathcal{L}_\Omega(\lambda).$$

Let $i: \Omega \cap \Gamma \to \Omega$ be the natural immersion, so that $\mathrm{pr}_1 \circ i = j \circ \pi_1$, where $\pi_1: \Omega \cap \Gamma \to \mathbf{X}(w)$ is an isomorphism and therefore (2) above gives

$$i^*\mathcal{L}_\Omega(\lambda) \cong \pi_1^* j^* \mathcal{L}_{\mathbf{G}/\mathbf{B}_0}(\lambda). \tag{3}$$

Now there is another commutative diagram

$$\begin{array}{ccc} \mathbf{E}_w & \longrightarrow & \mathbf{E}_1 \\ \downarrow & & \downarrow \\ \mathbf{E}_w/\mathbf{T}_0 \cong \Omega & \xrightarrow{\mathrm{pr}_2} & \mathbf{G}/\mathbf{B}_0 = \mathbf{E}_1/\mathbf{T}_0 \end{array}$$

where vertical maps are the same as in the first one, the top map is $\varphi = (g(\mathbf{U}_0 \cap {}^w\mathbf{U}_0) \mapsto g\dot{w}\mathbf{U}_0)$, which is compatible with the automorphism $\alpha = (t \mapsto t^w)$ of $\mathbf{T}_0$, and pr_2 is the second projection. Then A2.9(I) for $X = \mathbf{E}_w$, $X' = \mathbf{E}_1$, $G = G' = \mathbf{T}_0$, gives

$$\mathrm{pr}_2^* \mathcal{L}_{\mathbf{G}/\mathbf{B}_0}(\lambda) \cong \mathcal{L}_\Omega(w(\lambda)). \tag{4}$$

Using (3) for $\lambda \circ F$ and $w(\lambda)$, along with the fact that π_1 is an isomorphism, Proposition 7.17 reduces to

$$i^*\mathcal{L}_\Omega(\lambda \circ F) \cong i^*\mathcal{L}_\Omega(w(\lambda)).$$

Using (1) and (2) for the left-hand side, (4) for the right-hand side, we are reduced to checking

$$(\bar{F} \circ \mathrm{pr}_1 \circ i)^* \mathcal{L}_{\mathbf{G}/\mathbf{B}_0}(\lambda) \cong (\mathrm{pr}_2 \circ i)^* \mathcal{L}_{\mathbf{G}/\mathbf{B}_0}(\lambda).$$

But $\bar{F} \circ \mathrm{pr}_1 \circ i = \mathrm{pr}_2 \circ i$, both sending $(g\mathbf{B}_0, F(g)\mathbf{B}_0)$ to $F(g)\mathbf{B}_0$. This completes our proof. □

Exercises

1. When $\mathbf{G}$ is no longer connected in Theorem 7.1, express the image of $g \mapsto g^{-1}F(g)$ in terms of the same question for a finite $\mathbf{G}$.
2. Under the hypotheses of Theorem 7.2 for $(\mathbf{G}, F)$, show that, if $\mathbf{X}$ is a closed smooth subvariety of $\mathbf{G}$, then $\mathrm{Lan}^{-1}(\mathbf{X})$ is a closed smooth subvariety of $\mathbf{G}$ and Lan: $\mathrm{Lan}^{-1}(\mathbf{X}) \to \mathbf{X}$ is a $\mathbf{G}^F$-quotient map. Show that $\mathbf{Y}_{\mathbf{V} \subseteq \mathbf{V}'}$ can be considered as a $\mathbf{V}$-quotient of $\mathrm{Lan}^{-1}(\mathbf{V}.F\mathbf{V}')$.

Notes

For the more general subject of flag varieties $\mathbf{G}/\mathbf{B}$ and Schubert varieties, see the book [BiLa00] and its references.

The surjectivity of the Lang map goes back to Lang, [La56]. See also [Stein68b]. Deligne–Lusztig varieties were introduced in [DeLu76]. Most of their properties are sketched there (see also [Lu76a] 3 and [BoRo03]).

The quasi-affinity of the Deligne–Lusztig varieties is due to Haastert; see [Haa86].

8
Characters of finite reductive groups

In the present chapter, we recall some results about finite reductive groups $\mathbf{G}^F$ and their ordinary characters $\mathrm{Irr}(\mathbf{G}^F)$. The framework is close to that of Chapter 7, $\mathbf{G}$ is a connected reductive $\mathbf{F}$-group, $F\colon \mathbf{G} \to \mathbf{G}$ is the Frobenius endomorphism associated with the definition of $\mathbf{G}$ over the finite field $\mathbb{F}_q \subseteq \mathbf{F}$. The group of fixed points $\mathbf{G}^F$ is finite. We take ℓ to be a prime $\neq p$ and K to be a finite extension of $\mathbb{Q}_\ell$ assumed to be a splitting field for $\mathbf{G}^F$ and its subgroups. One considers $\mathrm{Irr}(\mathbf{G}^F)$ as a basis of the space $\mathrm{CF}(\mathbf{G}^F, K)$ of central functions $\mathbf{G}^F \to K$.

The Frobenius map F is expressed in terms of the root datum (see A2.4) associated with $\mathbf{G}$ and we recall the notion of a pair $(\mathbf{G}^*, F^*)$ **dual** to $(\mathbf{G}, F)$ around a dual pair of maximal tori $(\mathbf{T}, \mathbf{T}^*)$.

The Deligne–Lusztig induction,

$$\mathrm{R}^{\mathbf{G}}_{\mathbf{L}\subseteq\mathbf{P}}\colon \mathbb{Z}\mathrm{Irr}(\mathbf{L}^F) \to \mathbb{Z}\mathrm{Irr}(\mathbf{G}^F),$$

is defined by étale cohomology (see Appendix 3) of the varieties $\mathbf{X}_\mathbf{V}$ associated with Levi decompositions $\mathbf{P} = \mathbf{LV}$ satisfying $F\mathbf{L} = \mathbf{L}$ (see Chapter 7). We recall some basic results, such as the character formula (Theorem 8.16) and independence of the $\mathrm{R}^{\mathbf{G}}_{\mathbf{L}\subseteq\mathbf{P}}$ with respect to $\mathbf{P}$ when $\mathbf{L}$ is a torus. But the main theme of the chapter is that of **Lusztig series**. Let us begin with **geometric series**. When $(\mathbf{T}, \mathbf{T}^*)$ is a dual pair of F-stable tori, then $\mathrm{Irr}(\mathbf{T}^F)$ is isomorphic with $\mathbf{T}^{*F}$. A basic result on generalized characters $\mathrm{R}^{\mathbf{G}}_{\mathbf{T}}\theta$ (where $\theta \in \mathrm{Irr}(\mathbf{T}^F)$) is that two such generalized characters are disjoint whenever they correspond with rational elements of the corresponding tori of $\mathbf{G}^*$ that are not $\mathbf{G}^*$-conjugate. This gives a partition

$$\mathrm{Irr}(\mathbf{G}^F) = \bigcup_s \widetilde{\mathcal{E}}(\mathbf{G}^F, s)$$

indexed by classes of semi-simple elements of $(\mathbf{G}^*)^F$, two series $\widetilde{\mathcal{E}}(\mathbf{G}^F, s)$ being equal if and only if the corresponding s are $\mathbf{G}^*$-conjugate. A finer partition is given by **rational series**. One has

$$\mathrm{Irr}(\mathbf{G}^F) = \bigcup_s \mathcal{E}(\mathbf{G}^F, s)$$

indexed by semi-simple elements of $(\mathbf{G}^*)^F$, two series $\mathcal{E}(\mathbf{G}^F, s)$ being equal if and only if the corresponding s are $(\mathbf{G}^*)^F$-conjugate. The two notions differ only when the center of $\mathbf{G}$ is not connected.

In the following, we refer mainly to [Springer], [Cart85], and [DiMi91].

8.1. Reductive groups, isogenies

Recall $\mathbb{G}_\mathrm{m}$ the multiplicative group of $\mathbf{F}$, considered as an algebraic group.

Let $\mathbf{G}$ be a connected reductive $\mathbf{F}$-group (see A2.4). Let $\mathbf{T}$ be a maximal torus of $\mathbf{G}$. Then a **root datum** of $\mathbf{G}$ is defined, i.e. $(X, Y, \Phi, \Phi^\vee)$ where

- $X = X(\mathbf{T}) = \mathrm{Hom}(\mathbf{T}, \mathbb{G}_\mathrm{m})$ is the group of characters of $\mathbf{T}$,
- $Y = Y(\mathbf{T}) = \mathrm{Hom}(\mathbb{G}_\mathrm{m}, \mathbf{T})$ is the group of one parameter subgroups of $\mathbf{T}$,
- Φ is a root system in $X \otimes_\mathbb{Z} \mathbb{R}$, any $\alpha \in \Phi$ is defined by the action of $\mathbf{T}$ on some non-trivial minimal closed unipotent subgroup of $\mathbf{G}$ normalized by $\mathbf{T}$, Φ is the set of "roots of $\mathbf{G}$ relative to $\mathbf{T}$",
- $\Phi^\vee$ is a root system in $Y \otimes_\mathbb{Z} \mathbb{R}$, via the pairing between X and Y, the elements of $\Phi^\vee$, as linear forms on X are exactly the coroots $\alpha^\vee$, $\alpha \in \Phi$.

The root datum so defined characterizes $\mathbf{G}$ up to some isomorphisms and any root datum is the root datum of some reductive algebraic group. Morphisms between root data define morphisms between reductive algebraic groups (see [Springer] 9.6.2).

By the **type of G** we mean the type of the root system, a product of irreducible types among $\mathbf{A}_n$ $(n \geq 1)$, $\mathbf{B}_n$, $\mathbf{C}_n$ $(n \geq 2)$, $\mathbf{D}_n$ $(n \geq 3)$, $\mathbf{G}_2$, $\mathbf{F}_4$, $\mathbf{E}_6$, $\mathbf{E}_7$, $\mathbf{E}_8$.

One has $\mathbf{G} = \mathrm{Z}(\mathbf{G})^\circ[\mathbf{G}, \mathbf{G}]$. Let $\mathbb{Z}\Phi$ be the subgroup of $X(\mathbf{T})$ generated by the set of roots. The group is semi-simple, i.e. $\mathbf{G} = [\mathbf{G}, \mathbf{G}]$, equivalently $\mathrm{Z}(\mathbf{G})$ is finite, if and only if $\mathbb{Z}\Phi$ and $X(\mathbf{T})$ have equal ranks. Let $\Omega := \mathrm{Hom}(\mathbb{Z}\Phi^\vee, \mathbb{Z})$ be the "weight lattice;" then the cokernel of $\mathbb{Z}\Phi \to \Omega$ is the **fundamental group** of the root system, a finite abelian group. When $\mathbf{G}$ is semi-simple the restriction map $X \to \Omega$ is injective and $\Omega/X \cong (\Omega/\mathbb{Z}\Phi)/(X/\mathbb{Z}\Phi)$ is in duality with $Y/\mathbb{Z}\Phi^\vee$. The group $\mathbf{G}$ is said to be **adjoint**, denoted $\mathbf{G}_\mathrm{ad}$, if $X = \mathbb{Z}\Phi$ and then $\mathrm{Z}(\mathbf{G}) = \{1\}$. The group $\mathbf{G}$ is simply connected, denoted $\mathbf{G}_\mathrm{sc}$, if $X = \Omega$, and then $\mathrm{Z}(\mathbf{G}) \cong \mathrm{Hom}(\Omega/\mathbb{Z}\Phi, \mathbb{G}_\mathrm{m})$. For any $\mathbf{G}$, corresponding to $\mathbb{Z}\Phi \to X([\mathbf{G}, \mathbf{G}]) \to \Omega$ and to $\mathbb{Z}\Phi \to X(\mathbf{T})$, one has surjective morphisms

$$\mathbf{G}_\mathrm{sc} \to [\mathbf{G}, \mathbf{G}] \to \mathbf{G}_\mathrm{ad}, \quad \mathbf{G} \to \mathbf{G}_\mathrm{ad}$$

and the latter has kernel the center of $\mathbf{G}$. That does not imply that $\mathbf{G}_{\mathrm{ad}}^F$ is a quotient of $\mathbf{G}^F$. In any case, the groups $\mathbf{G}_{\mathrm{sc}}^F$, $[\mathbf{G},\mathbf{G}]^F$ and $\mathbf{G}_{\mathrm{ad}}^F$ have equal orders ([Cart85] §2.9). Frequent use will be made of the following elementary result.

Proposition 8.1. *Let f be an endomorphism of a group G. Denote by $[G, f]$ the set of $g.f(g^{-1})$ for $g \in G$.*

(i) Let Z be a central f-stable subgroup of G, so that f acts on G/Z. One has a natural exact sequence

$$1 \to G^f/Z^f \longrightarrow (G/Z)^f \longrightarrow ([G, f] \cap Z)/[Z, f] \to 1.$$

(ii) Assume G is finite commutative and H is an f-stable subgroup. Then $|H| = |H^f|.|[H, f]|$, in particular $|H : [H, f]|$ divides $|G^f|$.

References to Proposition 8.1 are numerous and generally kept vague. Often (i) is used in the case of central product (G being a direct product). Also, Lang's theorem allows us to simplify the commutator groups $[Z, f]$ when Z is a connected $\mathbf{F}$-group and f is a Frobenius endomorphism.

Any $\alpha \in \Phi$ defines a reflection s_α: $X(\mathbf{T}) \otimes \mathbb{R} \to X(\mathbf{T}) \otimes \mathbb{R}$, and Φ is stable under s_α. The group generated by the s_α ($\alpha \in \Phi$) is the Weyl group of the root system; denote it by $W(\Phi)$. Then $W(\Phi)$ acts by restriction on $X(\mathbf{T})$, a left action, hence on $Y(\mathbf{T})$, by transposition, a right action. Then $W(\Phi)$ is canonically isomorphic to (and frequently identified with) $\mathrm{N}_{\mathbf{G}}(\mathbf{T})/\mathbf{T}$; denote it by $W(\mathbf{G},\mathbf{T})$, or W when there is no ambiguity. The actions of $W(\mathbf{T})$ on $\mathbf{T}$, by conjugacy from $\mathrm{N}_{\mathbf{G}}(\mathbf{T})$, on $X(\mathbf{T})$ and on $Y(\mathbf{T})$ are linked as follows, where $(\chi, t, \eta, a) \in X(\mathbf{T}) \times \mathbf{T} \times Y(\mathbf{T}) \times \mathbf{F}$,

$$(w\chi)(t) = \chi(w^{-1}tw),\ \ (\eta w)(a) = w^{-1}\eta(a)w,\ \ \langle \eta, w\chi\rangle = \langle \eta w, \chi\rangle.$$

A Borel subgroup $\mathbf{B}$ of $\mathbf{G}$ containing $\mathbf{T}$ corresponds to a basis of Φ, the sets of simple roots with respect to $(\mathbf{T},\mathbf{B})$. Let S be the set of reflections defined by the simple roots with respect to such a pair and $(\mathbf{T},\mathbf{B})$, and let $N = \mathrm{N}_{\mathbf{G}}(\mathbf{T})$. We may consider S as a subset of $W = N/\mathbf{T}$. Then $(\mathbf{G}, \mathbf{B}, N, S)$ is a split BN-pair of characteristic p, with a finite Coxeter group W, in the sense of Definitions 2.12 and 2.20 (see [DiMi91] 0.12).

In the non-twisted case $\mathbf{G}$ is defined over $\mathbb{F}_q$ by a Frobenius map F (see [DiMi91] 3 and references) and we are interested in the group of points of $\mathbf{G}$ over $\mathbb{F}_q$, or fixed points of F. More generally we shall have to consider an endomorphism $F: \mathbf{G} \to \mathbf{G}$ such that some power F^δ of F is the Frobenius endomorphism F_1 of $\mathbf{G}$ defining a rational structure on $\mathbb{F}_{q_1}$. Note that the positive real number $(q_1)^{1/\delta}$ is well defined.

The endomorphism F gives rise to — or may be defined by — an endomorphism of the root datum of $\mathbf{G}$ defined around an F-stable maximal torus $\mathbf{T}$. First F acts on the set of one-dimensional connected unipotent subgroups of $\mathbf{G}$ that are normalized by $\mathbf{T}$; that action defines a permutation f of Φ, $F(X_\alpha) = X_{f\alpha}$ ($\alpha \in \Phi$). Then F acts on $X(\mathbf{T})$ by $(\chi \mapsto \chi \circ F)$ and the transpose of F, denoted $F^\vee$, acts on $Y(\mathbf{T})$. One has

$$\textbf{(8.2)} \qquad F: X \to X, \quad F^\vee: Y \to Y, \quad q: \Phi \to \{p^n\}_{n \in \mathbb{N}}, \quad f: \Phi \to \Phi,$$
$$F(f\alpha)) = q(\alpha)\alpha, \quad F^\vee(\alpha^\vee) = q(\alpha)(f\alpha)^\vee, \quad \alpha \in \Phi \subset X(\mathbf{T}),$$

where $q(\alpha)$ is a power of p — recall that X is a contravariant functor. Furthermore, F acts as an automorphism f' of $W(\mathbf{T})$ and the group $W.\langle f' \rangle$ acts on $X(\mathbf{T})$ as $W.\langle F \rangle$, hence frequently we write F instead of f'. If F is a Frobenius morphism with respect to $\mathbb{F}_q$, then $q(\alpha) = q$ for all $\alpha \in \Phi$ and $F(\alpha) = qf(\alpha)$. Such an automorphism of Φ composed with multiplication by a power of p is sometimes called a p**-morphism**.

Conversely, let $(\mathbf{G}, F)$ be a connected reductive $\mathbf{F}$-group defined over $\mathbb{F}_q$. There exist $\mathbf{T} \subset \mathbf{B}$ a maximal torus and Borel subgroup, both F-stable (see Theorem 7.1(iii)). The Weyl group $N_{\mathbf{G}}(\mathbf{T})/\mathbf{T}$ and its set of generators associated with $\mathbf{B}$ are F-stable. That is how F induces a permutation of the set of roots and of the set simple roots.

The isogeny theorem ([Springer] 9.6.5) says that if $\mathbf{G}$ is defined by the root datum $(X, Y, \Phi, \Phi^\vee)$, any $(F, F^\vee, f, q)$ that satisfies (8.2) may be realized by an isogeny $F: \mathbf{G} \to \mathbf{G}$. The isogeny is defined modulo interior automorphisms defined by elements of the torus $\mathbf{T}$.

To every orbit ω of F on the set of connected components of the Dynkin diagram of $\mathbf{G}$ there corresponds a well-defined F-stable subgroup $\mathbf{G}'_\omega$ of $[\mathbf{G}, \mathbf{G}]$ and a component $\mathbf{G}_\omega = Z^\circ(\mathbf{G})\mathbf{G}'_\omega$ of $\mathbf{G}$. Recall that the only non-trivial automorphisms f of irreducible root systems have order 2 for types $\mathbf{A}_n$ ($n \geq 2$), $\mathbf{D}_n$ ($n \geq 3$), $\mathbf{E}_6$, or order 3 for type $\mathbf{D}_4$. The finite group $(\mathbf{G}(\omega)/Z(\mathbf{G}_\omega))^F$ is characterized by its simple type $\mathbf{X}_\omega \in \{\mathbf{A}_n, {}^2\mathbf{A}_n, \mathbf{B}_n, \mathbf{C}_n, \mathbf{D}_n, {}^2\mathbf{D}_n, \mathbf{G}_2, {}^3\mathbf{D}_4, \mathbf{F}_4, \mathbf{E}_6, {}^2\mathbf{E}_6, \mathbf{E}_7, \mathbf{E}_8\}$ and an extension field $\mathbb{F}_{q^{m(\omega)}}$ of $\mathbb{F}_q$ of degree $m(\omega)$ equal to the length of the orbit ω. Call $(\mathbf{X}_\omega, q^{m(\omega)})$ the irreducible **rational type** of $\mathbf{G}_\omega$, and $\times_\omega(\mathbf{X}_\omega, q^{m(\omega)})$ the rational type of $(\mathbf{G}, F)$.

The split BN-pair of $\mathbf{G}^F$ is obtained by taking fixed points under F in a maximal torus and a Borel subgroup $\mathbf{T} \subset \mathbf{B}$, both F-stable. The type of the Coxeter group of $\mathbf{G}^F$ is then the same as that describing the rational type, except for twisted types that obey the following rules ${}^2\mathbf{A}_n \mapsto \mathrm{BC}_{[n+1/2]}$, ${}^2\mathbf{D}_n \mapsto \mathrm{BC}_{n-1}$ (see the notation of Example 2.1), ${}^2\mathbf{E}_6 \mapsto \mathrm{F}_4$ (standard notation), while ${}^3\mathbf{D}_4$ gives a dihedral Coxeter group of order 12.

8.2. Some exact sequences and groups in duality

Let $(\mathbf{G}, F)$ be defined around a maximal torus $\mathbf{T}$ by a root datum and p-morphism. The endomorphism F gives rise to four short exact sequences. They are well defined after a coherent choice of primitive roots of unity via monomorphisms of multiplicative groups

(8.3) $$\overline{\mathbb{Q}}_\ell^{\,\times} \xleftarrow{\iota'} (\mathbb{Q}/\mathbb{Z})_{p'} \xrightarrow{\iota} \mathbf{F}^\times, \quad \kappa = \iota' \circ \iota^{-1} \colon \mathbf{F}^\times \longrightarrow \overline{\mathbb{Q}}_\ell^{\,\times}.$$

Let $D = (\mathbb{Q}/\mathbb{Z})_{p'}$. Using ι, one has an isomorphism

(8.4) $$\begin{array}{ccc} Y(\mathbf{T}) \otimes_{\mathbb{Z}} D & \longrightarrow & \mathbf{T}, \\ \eta \otimes a & \mapsto & \eta(\iota(a)) \end{array}$$

and natural isomorphisms

(8.5) $$\begin{array}{c} X(\mathbf{T}) \longleftrightarrow \mathrm{Hom}(Y(\mathbf{T}) \otimes D, D), \\ Y(\mathbf{T}) \longleftrightarrow \mathrm{Hom}(X(\mathbf{T}) \otimes D, D) \end{array}$$

with F-action. The first short exact sequence describes $\mathbf{T}^F$ as the kernel of the endomorphism $(F-1)$ of $\mathbf{T}$, $\mathbf{T}$ viewed as in (8.4):

(8.6) $$1 \longrightarrow \mathbf{T}^F \longrightarrow Y(\mathbf{T}) \otimes_{\mathbb{Z}} D \xrightarrow{F-1} Y(\mathbf{T}) \otimes_{\mathbb{Z}} D \longrightarrow 0.$$

By the functor $\mathrm{Hom}(-, D)$ (or $\mathrm{Hom}(-, \mathbb{G}_m)$) one gets from (8.5) and (8.6) the second sequence

(8.7) $$0 \longrightarrow X(\mathbf{T}) \xrightarrow{F-1} X(\mathbf{T}) \xrightarrow{\mathrm{Res}} \mathrm{Irr}(\mathbf{T}^F) \longrightarrow 1$$

where Res is just the "restriction from $\mathbf{T}$ to $\mathbf{T}^F$" through the morphism κ.

Applying the snake lemma, from (8.6) one gets $\mathbf{T}^F$ as a cokernel of $(F-1)$ on $Y(\mathbf{T})$. Assume that F^d is a split Frobenius endomorphism with respect to $\mathbb{F}_{q'}$ for some q' (i.e. is multiplication by q' on $Y(\mathbf{T})$), put $\zeta' = \iota(1/(q'-1))$, then the explicit morphism $N_1 \colon Y(\mathbf{T}) \to \mathbf{T}^F$ one gets is

(8.8) $$0 \longrightarrow Y(\mathbf{T}) \xrightarrow{F-1} Y(\mathbf{T}) \xrightarrow{N_1} \mathbf{T}^F \longrightarrow 1,$$

(8.9) $$N_1(\eta) = N_{F^d/F}(\eta(\zeta')) \quad (\eta \in Y(\mathbf{T})).$$

N_1 depends on ι, ι', but not on the choice of d.

Similarly, using the snake lemma and (8.7), or applying $\mathrm{Hom}(-, D)$ to (8.8), one gets

(8.10) $$1 \longrightarrow \mathrm{Irr}(\mathbf{T}^F) \longrightarrow X(\mathbf{T}) \otimes_{\mathbb{Z}} D \xrightarrow{F-1} X(\mathbf{T}) \otimes_{\mathbb{Z}} D \longrightarrow 0.$$

Note that the pairing between X and Y reduces modulo $(F-1)$ to the natural pairing between $\mathrm{Irr}(\mathbf{T})^F$ and $\mathbf{T}^F$.

Recall the classification under $\mathbf{G}^F$ of F-stable maximal tori of $\mathbf{G}$ (see [DiMi91] §3, [Cart85] §3.3, [Srinivasan] II). Let $\mathbf{T}'$ be such a torus in $\mathbf{G}$. Let $g \in \mathbf{G}$ be such that $\mathbf{T}' = g\mathbf{T}g^{-1}$. Then $g^{-1}F(g) \in \mathrm{N}_{\mathbf{G}}(\mathbf{T})$, so let $w = g^{-1}F(g)\mathbf{T} \in W$. With this notation g^{-1} sends $(\mathbf{T}', F)$ to $(\mathbf{T}, wF)$. The $\mathbf{G}^F$-conjugacy class of $\mathbf{T}'$ corresponds to the $W.{<}F{>}$-conjugacy class of wF. One says that $\mathbf{G}^F$-conjugacy classes of F-stable maximal tori of $\mathbf{G}$ are parametrized by F-conjugacy classes of W and that $\mathbf{T}'$ is of type w with respect to $\mathbf{T}$, denoted by $\mathbf{T}_w$ ([DiMi91] 3.23).

The preceding constructions from $(\mathbf{T}, F)$ apply to the endomorphism wF of $\mathbf{T}$ for any $w \in \mathrm{N}_{\mathbf{G}}(\mathbf{T})/\mathbf{T}$. Let d be some natural integer such that F^d is split $t \mapsto t^{q^d}$ on any F-stable maximal torus of $\mathbf{G}$. Let $\zeta = \iota(1/(q^d - 1))$. The sequences (8.7) and (8.8) become

(8.11) $$0 \longrightarrow Y(\mathbf{T}) \xrightarrow{wF-1} Y(\mathbf{T}) \xrightarrow{N_w} \mathbf{T}^{wF} \longrightarrow 1,$$

(8.12) $$N_w^{(F)}(\eta) = N_{F^d/wF}(\eta(\zeta)) \quad (\eta \in Y(\mathbf{T})).$$

For convenience we fix ζ, but N_w is defined independently of the choice of d. The superscript F is often omitted.

If $(X, Y, \Phi, \Phi^\vee)$ is a root datum, then $(Y, X, \Phi^\vee, \Phi)$ is a root datum. The algebraic groups they define are said to be in duality. More generally we say that $\mathbf{G}^*$ is a **dual** of $\mathbf{G}$, or that $\mathbf{G}$ and $\mathbf{G}^*$ are **in duality** (around $\mathbf{T}, \mathbf{T}^*$), when $\mathbf{T}, \mathbf{T}^*$ are maximal tori of $\mathbf{G}$ and $\mathbf{G}^*$ respectively, with a given isomorphism of root data

(8.13) $$(X(\mathbf{T}), Y(\mathbf{T}), \Phi(\mathbf{G}, \mathbf{T}), \Phi(\mathbf{G}, \mathbf{T})^\vee) \longleftrightarrow (Y(\mathbf{T}^*), X(\mathbf{T}^*), \Phi(\mathbf{G}^*, \mathbf{T}^*)^\vee, \Phi(\mathbf{G}^*, \mathbf{T}^*)).$$

The isomorphism has to preserve the pairing between the groups X, Y and to exchange the maps $(\alpha \mapsto \alpha^\vee)$ and $(\beta^\vee \mapsto \beta)$ $(\alpha \in \Phi(\mathbf{G}, \mathbf{T}), \beta \in \Phi(\mathbf{G}^*, \mathbf{T}^*))$.

If $\mathbf{B}$ is a Borel subgroup of $\mathbf{G}$ containing $\mathbf{T}$, then the bijection $\Phi(\mathbf{G}, \mathbf{T}) \to \Phi(\mathbf{G}^*, \mathbf{T}^*)^\vee$ (resp. $\Phi(\mathbf{G}, \mathbf{T})^\vee \to \Phi(\mathbf{G}^*, \mathbf{T}^*)$) carries the set Δ of simple roots (resp. the set $\Delta^\vee$ of simple coroots) defined by $\mathbf{B}$ to a basis $(\Delta^*)^\vee$ of $\Phi(\mathbf{G}^*, \mathbf{T}^*)^\vee$ (resp. a basis Δ^* of $\Phi(\mathbf{G}^*, \mathbf{T}^*)$) that defines a Borel subgroup $\mathbf{B}^*$ of $\mathbf{G}^*$ containing $\mathbf{T}^*$. The bijection between the sets of simple roots gives rise to a bijection between the sets of parabolic subgroups containing the corresponding Borel subgroups (see Definition 2.15 and Theorem 2.16). We frequently identify Δ and Δ^*, considering Δ as a unique set of indices for sets of simple roots (or coroots) or simple reflections. Thus for any subset I of Δ are defined corresponding parabolic subgroups of $\mathbf{G}$ containing $\mathbf{B}$, and of $\mathbf{G}^*$ containing $\mathbf{B}^*$, and standard Levi subgroups $\mathbf{L}_I$ and $\mathbf{L}_I^*$. One sees easily that the given duality (8.13) restricts to a duality between $\mathbf{L}_I$ and $\mathbf{L}_I^*$ around $\mathbf{T}$ and $\mathbf{T}^*$.

In this situation the tori $\mathbf{T}$ and $\mathbf{T}^*$ are in duality (we may consider that they are defined by the functors X, Y and an empty root system). The given isomorphism induces an anti-isomorphism $(w \mapsto w^*)$ between Weyl groups $W(\mathbf{G}, \mathbf{T})$ (left-acting on $X(\mathbf{T})$) and $W(\mathbf{G}^*, \mathbf{T}^*)$ (right-acting on $Y(\mathbf{T}^*)$). The map $(w \mapsto w^*)$ takes the set of reflections onto the set of reflections, simple ones onto simple ones, in the case of corresponding Borel subgroups.

If furthermore $(F, F^\vee, f, q)$ satisfying (8.2) is given by an isogeny F of $\mathbf{G}$, then $(F^\vee, F, f^{-1}, q \circ f^{-1})$ defines a quadruple $(F^*, (F^*)^\vee, f^*, q^*)$ that is realized by an isogeny $F^*: \mathbf{G}^* \to \mathbf{G}^*$. Then we say that $(\mathbf{G}, F)$ and $(\mathbf{G}^*, F^*)$ are **in duality over** $\mathbb{F}_q$. If $\eta \in Y(\mathbf{T}^*)$ corresponds to $\chi \in X(\mathbf{T})$, then $F^* \circ \eta$ corresponds to $\chi \circ F$ (i.e. F corresponds to $(F^*)^\vee$). One has $F^*((F(w))^*) = w^*$ for any $w \in W$. In other words the Frobenius maps operate in inverse ways on the (anti)-isomorphic Weyl groups.

When $\mathbf{T}$ and $\mathbf{T}^*$ are in duality with endomorphisms F and F^*, the isomorphism $(X(\mathbf{T}), Y(\mathbf{T}), F, F^\vee) \to (Y(\mathbf{T}^*), X(\mathbf{T}^*), (F^*)^\vee, F^*)$ sends the short exact sequence (8.7) (resp. (8.6)) for $\mathbf{T}$ to (8.8) (resp. (8.10)) for $\mathbf{T}^*$ and vice versa.

Let $\mathbf{B}$ be a Borel subgroup of $\mathbf{G}$ such that $\mathbf{T} \subseteq \mathbf{B}$ and $F(\mathbf{B}) = \mathbf{B}$. Then F stabilizes the corresponding basis of $\Phi(\mathbf{G}, \mathbf{T})$ and the dual Borel so defined is F^*-stable. More generally, if $I \subseteq \Delta$ and $F(I) = I$, then the duality between $\mathbf{L}_I$ and $\mathbf{L}_I^*$ extends to $(\mathbf{L}_I, F)$ and $(\mathbf{L}_I^*, F^*)$.

Therefore there is a well-defined isomorphism

(8.14)
$$\begin{array}{ccc} (\mathbf{T}^*)^{F^*} & \longleftrightarrow & \mathrm{Irr}(\mathbf{T})^F \\ s & \mapsto & \theta = \hat{s} \end{array}$$

such that $\theta = \hat{s} \in \mathrm{Irr}(\mathbf{T}^F)$ and $s \in (\mathbf{T}^*)^{F^*}$ correspond as follows, after some identifications, for any $\eta \in Y(\mathbf{T}) = X(\mathbf{T}^*)$ and any $\lambda \in X(\mathbf{T}) = Y(\mathbf{T}^*)$

(8.15)
$$\theta(N_{F^d/F}(\eta(\zeta))) = \kappa(\zeta^{\langle \lambda, N_{F^d/F}(\eta)\rangle}) = \kappa(\eta(s))$$

In (8.15) ζ is a fixed root of unity as in (8.12), because the duality may be extended to other pairs of tori. Indeed, once the duality with p-morphisms between $(\mathbf{G}, F)$ and $(\mathbf{G}^*, F^*)$ is defined around maximal tori $\mathbf{T}$ and $\mathbf{T}^*$ by an isomorphism (8.13) between root data with p-morphisms, for every maximal F-stable torus $\mathbf{T}'$ of $\mathbf{G}$ there exists a maximal F^*-stable torus $\mathbf{S}$ of $\mathbf{G}^*$ such that the duality between $\mathbf{G}$ and $\mathbf{G}^*$ may be defined around $\mathbf{T}''$ and $\mathbf{S}$. Precisely, if $\mathbf{T}' = g\mathbf{T}g^{-1}$ is of type w with respect to $\mathbf{T}$, and $\mathbf{S} = h\mathbf{T}^*h^{-1}$ is of type $F^*(w^*)$ with respect to $\mathbf{T}^*$, where $g \in \mathbf{G}, h \in \mathbf{G}^*$ and $(w \mapsto w^*)$ are as described above, then the given isomorphism (8.13) between root data is send by conjugacy, using (g, h), to another one between root data around $\mathbf{T}'$ and $\mathbf{S}$, with p-morphisms induced by the action of F and F^* respectively ([Cart85] 4.3.4).

Recall that $\mathbf{G}^F$-conjugacy classes of F-stable Levi subgroups $\mathbf{L}$ of not-necessarily F-stable parabolic subgroups of $\mathbf{G}$ are classified by F-conjugacy classes of cosets $W_I w$, where $I \subseteq \Delta$ and w satisfies $wF(W_I)w^{-1} = W_I$ ([DiMi91] 4.3). Here $\mathbf{L}$ is conjugate to $\mathbf{L}_I$ by some $g \in \mathbf{G}$ such that $\mathbf{T}_w = g^{-1}\mathbf{T}g$ is a maximal torus of $\mathbf{L}$, of type w with respect to $\mathbf{T}$. We shall say that $W_I w$ is a type of $\mathbf{L}$. To such a class of F-stable Levi subgroups in $\mathbf{G}$ there corresponds a class of F^*-stable Levi subgroups $\mathbf{L}^*$ in $\mathbf{G}^*$, whose parameter is the coset $W_I^* F^*(w^*)$. In this context the outer automorphism groups $\mathrm{N}_{\mathbf{G}}(\mathbf{L})/\mathbf{L} \cong \mathrm{N}_{W(\mathbf{G})}(W(\mathbf{L}))/W(\mathbf{L})$ and $\mathrm{N}_{\mathbf{G}^*}(\mathbf{L}^*)/\mathbf{L}^*$ are isomorphic, with F- and F^*-actions, via $(w \mapsto w^*)$ around $(\mathbf{T}_w, \mathbf{T}^*{}_{F^*(w^*)})$.

8.3. Twisted induction

Let $\mathbf{P} = \mathbf{LV}$ be a Levi decomposition in $\mathbf{G}$ with $F\mathbf{L} = \mathbf{L}$. The methods of étale cohomology allow us to define a "twisted induction"

$$\mathrm{R}^{\mathbf{G}}_{\mathbf{L}\subseteq\mathbf{P}} \colon \mathbb{Z}\mathrm{Irr}(\mathbf{L}^F) \to \mathbb{Z}\mathrm{Irr}(\mathbf{G}^F)$$

generalizing Harish-Chandra induction. Its adjoint for the usual scalar product is denoted by ${}^*\mathrm{R}^{\mathbf{G}}_{\mathbf{L}\subseteq\mathbf{P}}$.

The construction is as follows. Recall the variety $\mathbf{Y}_{\mathbf{V}} := \{g\mathbf{V} \mid g^{-1}F(g) \in \mathbf{V}.F(\mathbf{V})\}$ of Definition 7.6. It is acted on by $\mathbf{G}^F$ on the left and $\mathbf{L}^F$ on the right; each action is free. We recall briefly how étale cohomology associates with such a situation an element of $\mathbb{Z}\mathrm{Irr}(\mathbf{G}^F \times (\mathbf{L}^F)^{\mathrm{opp}})$, which, in turn, by tensor product provides the above $\mathrm{R}^{\mathbf{G}}_{\mathbf{L}\subseteq\mathbf{P}}$.

Let (Λ, K, k) be an ℓ-modular splitting system for $\mathbf{G}^F \times \mathbf{L}^F$. Let $n \geq 1$, and denote $\Lambda^{(n)} = \Lambda/J(\Lambda)^n$. The constant sheaf $\Lambda^{(n)}$ for the étale topology on $\mathbf{Y}_{\mathbf{V}}$ (see A3.1 and A3.2) defines an object $R_n := \mathrm{R}_{\mathrm{c}}\Gamma(\mathbf{Y}_{\mathbf{V}}, \Lambda^{(n)})$ of the derived category $D^b(\Lambda^{(n)}[\mathbf{G}^F \times (\mathbf{L}^F)^{\mathrm{opp}}])$ (see A3.7 and A3.14) such that $R_n = \mathrm{R}_{n+1} \bigotimes_{\Lambda^{(n+1)}} \Lambda^{(n)}$. The limit over n of each cohomology group $\mathrm{H}^i(R_n)$ gives a $\Lambda\mathbf{G}^F$-$\Lambda\mathbf{L}^F$-bimodule, which, once tensored with K, is denoted by $\mathrm{H}^i(\mathbf{Y}_{\mathbf{V}}, K)$, or simply $\mathrm{H}^i(\mathbf{Y}_{\mathbf{V}}, \overline{\mathbb{Q}}_\ell)$ if one tensors with $\overline{\mathbb{Q}}_\ell$. The element of $\mathbb{Z}\mathrm{Irr}(\mathbf{G}^F \times (\mathbf{L}^F)^{\mathrm{opp}})$ is then $\sum_i (-1)^i \mathrm{H}^i(\mathbf{Y}_{\mathbf{V}}, K)$, i.e.

$$\mathrm{R}^{\mathbf{G}}_{\mathbf{L}\subseteq\mathbf{P}}(-) = \sum_i (-1)^i \mathrm{H}^i(\mathbf{Y}_{\mathbf{V}}, K) \otimes_{\mathbf{L}^F} -$$

The subscript c in $\mathrm{R}_{\mathrm{c}}\Gamma$ above indicates that direct images with compact support are considered. This has in general more interesting properties, for instance with regard to base changes. But, here, it coincides with ordinary cohomology by Poincaré–Verdier duality (A3.12) since the variety $\mathbf{Y}_{\mathbf{V}}$ is smooth (see

Theorem 7.7). Similarly, the above definition of $R^{\mathbf{G}}_{\mathbf{L}\subseteq\mathbf{P}}$ coincides with that of [DiMi91] 11.1 since the variety $\mathbf{Y}_{1\subseteq\mathbf{V}}$ (see Theorem 7.2) used there is such that $\mathbf{Y}_{\mathbf{V}} = \mathbf{Y}_{1\subseteq\mathbf{V}}/\mathbf{V}\cap F\mathbf{V}$, a locally trivial quotient (see Lemma 12.15 below).

The following is to be found in [DiMi91] 12.4, 12.17, and [Cart85] 7.2.8. Let $\epsilon_{\mathbf{G}} = (-1)^{\sigma(\mathbf{G})}$ where $\sigma(\mathbf{G})$ is the $\mathbb{F}_q$-rank of $\mathbf{G}$, see [DiMi91] 8.3–8.6.

Theorem 8.16. *Let $\mathbf{P} = \mathbf{L}\mathbf{V}$ be a Levi decomposition in $\mathbf{G}$ with $F\mathbf{L} = \mathbf{L}$.*

(i) If $f \in \mathrm{CF}(\mathbf{L}^F, K)$, and s is the semi-simple component of an element g of $\mathbf{G}^F$, then

$$R^{\mathbf{G}}_{\mathbf{L}\subseteq\mathbf{P}} f\,(g)$$
$$= |C^\circ_{\mathbf{G}}(s)^F|^{-1}|\mathbf{L}^F|^{-1} \sum_{\{h\in\mathbf{G}^F \mid s\in {}^h\mathbf{L}\}} |C^\circ_{{}^h\mathbf{L}}(s)^F|\, R^{C^\circ_{\mathbf{G}}(s)}_{C^\circ_{{}^h\mathbf{L}}(s)\subseteq C^\circ_{{}^h\mathbf{P}}(s)}({}^h f)(g).$$

(ii) $R^{\mathbf{G}}_{\mathbf{L}\subseteq\mathbf{P}} f(1) = \varepsilon_{\mathbf{G}}\varepsilon_{\mathbf{L}}|\mathbf{G}^F : \mathbf{L}^F|_{p'}\, f(1)$.

For the following see [DiMi91] 11.15 and 12.20.

Theorem 8.17. *(i) Let $\mathbf{T}$ be an F-stable maximal torus of $\mathbf{G}$. Let θ be a linear representation of $\mathbf{T}^F$ with values in $\overline{\mathbb{Q}}_\ell$. The generalized character $R^{\mathbf{G}}_{\mathbf{T}\subseteq\mathbf{B}}\,\theta$ is called a* **Deligne–Lusztig character** *and is independent of the choice of the Borel subgroup $\mathbf{B}$. Hence $R^{\mathbf{G}}_{\mathbf{T}\subseteq\mathbf{B}}$ is simplified as $R^{\mathbf{G}}_{\mathbf{T}}$.*

(ii) Let s be a semi-simple element of $\mathbf{G}^F$. For any subgroup H of $\mathbf{G}^F$ containing s, let π^H_s be the central function on H with value $|C_H(s)|$ on the H-conjugacy class of s and 0 elsewhere. One has

$$\varepsilon_{C^\circ_{\mathbf{G}}(s)}|C_{\mathbf{G}^F}(s)|_p.\pi^{\mathbf{G}^F}_s = \sum_{\mathbf{T}} \varepsilon_{\mathbf{T}} R^{\mathbf{G}}_{\mathbf{T}}(\pi^{\mathbf{T}^F}_s)$$

where the sum is over all F-stable maximal tori $\mathbf{T}$ of $\mathbf{G}$ with $s \in \mathbf{T}$.

Remark 8.18. (i) When $\mathbf{P} = \mathbf{L}\mathbf{V}$ is a Levi decomposition in $\mathbf{G}$ with $F\mathbf{L} = \mathbf{L}$ and $F\mathbf{P} = \mathbf{P}$, the outcome of the above construction is Harish-Chandra induction denoted by $R^{\mathbf{G}^F}_{\mathbf{L}^F}$ in Notation 3.11. This comes from the fact that $\mathbf{Y}_{\mathbf{V}} = \mathbf{G}^F$ and étale topology is trivial in dimension 0.

(ii) The space of central functions $\sum_{(\mathbf{T},\theta)} K.R^{\mathbf{G}}_{\mathbf{T}}\,\theta \subseteq \mathrm{CF}(\mathbf{G}^F, K)$ is usually called the space of **uniform functions** on $\mathbf{G}^F$. By Theorem 8.17 (ii), the characteristic function of a semi-simple conjugacy class of $\mathbf{G}^F$ is a uniform function. The regular character of $\mathbf{G}^F$ is a uniform function.

Let $\tau\colon \mathbf{G}_{\mathrm{sc}} \to [\mathbf{G},\mathbf{G}]$ be a simply connected covering. One has $\mathbf{G} = \mathbf{T}\tau(\mathbf{G}_{\mathrm{sc}})$, hence $\mathbf{G}^F = \mathbf{T}^F\tau(\mathbf{G}_{\mathrm{sc}}{}^F)$ by Proposition 8.1(i) and Lang's theorem since $\mathbf{T}\cap\tau(\mathbf{G}_{\mathrm{sc}})$ is connected. Thus $\mathrm{Irr}(\mathbf{G}^F/\tau(\mathbf{G}_{\mathrm{sc}}{}^F))$ may be identified with a subgroup of $\mathrm{Irr}(\mathbf{T}^F)$. As $X(\mathbf{T} \cap [\mathbf{G}, \mathbf{G}])$ may be identified with $\mathbb{Z}\Phi$, the isomorphism (8.14) $(\mathbf{T}^*)^{F^*} \to \mathrm{Irr}(\mathbf{T}^F)$ defines by restrictions an isomorphism

(8.19)
$$\begin{array}{ccc} \mathrm{Z}(\mathbf{G}^*)^{F^*} & \longrightarrow & \mathrm{Irr}(\mathbf{G}^F/\tau({\mathbf{G}_{\mathrm{sc}}}^F)) \\ z & \mapsto & \hat{z} \end{array}$$

As a consequence of Theorem 8.16(i) and with this notation one has, for any pair $(\mathbf{S}, \theta)$ defining a Deligne–Lusztig character,

(8.20)
$$\hat{z} \otimes \mathrm{R}_{\mathbf{S}}^{\mathbf{G}}\,\theta = \mathrm{R}_{\mathbf{S}}^{\mathbf{G}}\big(\mathrm{Res}_{\mathbf{S}^F}^{\mathbf{G}^F}\,\hat{z} \otimes \theta\big)$$

8.4. Lusztig's series

We now introduce several partitions of $\mathrm{Irr}(\mathbf{G}^F)$ induced by the Deligne–Lusztig characters $\mathrm{R}_{\mathbf{T}}^{\mathbf{G}}\theta$.

Proposition 8.21. *Let* $\mathbf{T}$ *and* $\mathbf{S}$ *be two maximal* F*-stable tori of* $\mathbf{G}$, $\theta \in \mathrm{Irr}(\mathbf{T}^F)$, $\xi \in \mathrm{Irr}(\mathbf{S}^F)$. *Let* $\mathbf{T}^*$ *and* $\mathbf{S}^*$ *be maximal* F^**-stable tori in* $\mathbf{G}^*$, *in dual classes of* $\mathbf{T}$ *and* $\mathbf{S}$ *respectively, and* $t \in \mathbf{T}^*$ *(resp.* $s \in \mathbf{S}^*$*) corresponding by duality, i.e. by formula (8.15), to* θ *(resp.* ξ*). The pairs* $(\mathbf{T}, \theta)$ *and* $(\mathbf{S}, \xi)$ *are said to be geometrically conjugate if and only if* s *and* t *are* $\mathbf{G}^*$*-conjugate.*

Thus the geometric conjugacy classes of pairs $(\mathbf{T}, \theta)$ *are in one-to-one correspondence with* F^**-stable conjugacy classes of semi-simple elements of* $\mathbf{G}^*$. *Similarly, the* $\mathbf{G}^F$*-conjugacy classes of pairs* $(\mathbf{T}, \theta)$ *are in one-to-one correspondence with the* ${\mathbf{G}^*}^{F^*}$*-conjugacy classes of pairs* $(\mathbf{T}^*, t)$, *where* $t \in {\mathbf{T}^*}^{F^*}$.

Remark 8.22. (i) The last assertion of Proposition 8.21 allows us to write $\mathrm{R}_{\mathbf{T}^*}^{\mathbf{G}}\, s$ for $\mathrm{R}_{\mathbf{T}}^{\mathbf{G}}\,\theta$ when $(\mathbf{T}^*, s)$ corresponds with $(\mathbf{T}, \theta)$.

(ii) Let π be a set of prime numbers. If $(\mathbf{T}, \theta)$ corresponds with s, then $(\mathbf{T}, \theta_\pi)$ corresponds with s_π.

Definition 8.23. *Let* s *be some semi-simple element of* ${\mathbf{G}^*}^{F^*}$.

The geometric Lusztig series associated to the $\mathbf{G}^*$*-conjugacy class of* s *is the set of irreducible characters of* $\mathbf{G}^F$ *occurring in some* $\mathrm{R}_{\mathbf{T}}^{\mathbf{G}}\,\theta$, *where* $(\mathbf{T}, \theta)$ *is in the geometric conjugacy class defined by* s. *It is denoted by* $\widetilde{\mathcal{E}}(\mathbf{G}^F, s)$.

The rational Lusztig series associated to the ${\mathbf{G}^*}^{F^*}$*-conjugacy class* $[s]$ *of* s *is the set of irreducible characters of* $\mathbf{G}^F$ *occurring in some* $\mathrm{R}_{\mathbf{T}}^{\mathbf{G}}\,\theta$, *where* $(\mathbf{T}, \theta)$ *corresponds by duality to a pair* $(\mathbf{T}^*, t)$, *where* $t \in {\mathbf{T}^*}^{F^*} \cap [s]$. *It is denoted by* $\mathcal{E}(\mathbf{G}^F, s)$.

The elements of $\widetilde{\mathcal{E}}(\mathbf{G}^F, 1) = \mathcal{E}(\mathbf{G}^F, 1)$ *are called* **unipotent irreducible characters**.

Indeed the $\mathbf{G}^F$-conjugacy classes of pairs $(\mathbf{S}, \xi)$ are in one-to-one correspondence with the $(\mathbf{G}^*)^{F^*}$-conjugacy classes of pairs $(\mathbf{S}^*, s)$ where $\mathbf{S}^*$ is an F^*-stable maximal torus of $\mathbf{G}^*$ and $s \in (\mathbf{S}^*)^{F^*}$. The correspondence $(\mathbf{S}, \xi) \mapsto (\mathbf{S}^*, s)$ is such that $\mathbf{S}^*$ is in duality with $\mathbf{S}$ as described in the preceding section and ξ maps to s by the isomorphism $\mathrm{Irr}(\mathbf{S}^F) \mapsto (\mathbf{S}^*)^{F^*}$ of formula (8.14) (see [DiMi91] 11.15, 13.13).

Theorem 8.24. *(i) The set of geometric Lusztig series $\widetilde{\mathcal{E}}(\mathbf{G}^F, s)$ is a partition of* $\mathrm{Irr}(\mathbf{G}^F)$. *One has $\widetilde{\mathcal{E}}(\mathbf{G}^F, s) = \widetilde{\mathcal{E}}(\mathbf{G}^F, s')$ if and only if s and s' are conjugate in $\mathbf{G}^*$.*

(ii) Let s be a semi-simple element of ${\mathbf{G}^*}^{F^*}$. The geometric Lusztig series $\widetilde{\mathcal{E}}(\mathbf{G}^F, s)$ is the disjoint union of the rational Lusztig series $\mathcal{E}(\mathbf{G}^F, t)$ such that t is $\mathbf{G}^*$-conjugate to s. One has $\widetilde{\mathcal{E}}(\mathbf{G}^F, t) = \widetilde{\mathcal{E}}(\mathbf{G}^F, t')$ if and only if t and t' are conjugate in ${\mathbf{G}^*}^{F^*}$.

(iii) If the center of $\mathbf{G}$ is connected, then any geometric series is a rational series.

Proof. Assertion (i) is proved by Deligne–Lusztig in a fundamental paper [DeLu76], as a consequence of a stronger property. Let $\mathbf{B} = \mathbf{U}.\mathbf{T}$ and $\mathbf{B}' = \mathbf{U}'.\mathbf{T}'$ be Levi decompositions of Borel subgroups of $\mathbf{G}$, with F-stable maximal tori, assume that $(\mathbf{T}, \theta)$ and $(\mathbf{T}', \theta')$ correspond to s and s' respectively. If $\mathrm{H}^i(\mathbf{Y}_\mathbf{U}, \overline{\mathbb{Q}}_\ell)) \otimes \theta$ and $\mathrm{H}^i(\mathbf{Y}'_\mathbf{U}, \overline{\mathbb{Q}}_\ell)) \otimes \theta'$ have a common irreducible constituent, then s and s' are $\mathbf{G}^*$-conjugate; see also [DiMi91] 13.3, and §12.4 below.

A connection between geometric conjugacy and rational conjugacy is described in §15.1.

Note that any F^*-stable conjugacy class of $\mathbf{G}^*$ contains an element of ${\mathbf{G}^*}^{F^*}$ by Lang's theorem (Theorem 7.1(i)) and ${\mathbf{G}^*}^{F^*}$-conjugacy classes of rational elements that are geometrically conjugate to s are parametrized by the F^*-conjugacy classes of $C_\mathbf{G}(s)/C^\circ_\mathbf{G}(s)$ ([DiMi91] 3.12, 3.21). By a theorem of Steinberg (see [Cart85] 3.5.6), if the derived group of $\mathbf{G}$ is simply connected, then the centralizer of any semi-simple element of $\mathbf{G}$ is connected. If the center of $\mathbf{G}$ is connected, then the simply connected covering of the derived group of $\mathbf{G}^*$ is bijective. An equivalent condition on the root datum is that the quotient $X(\mathbf{T})/\mathbb{Z}\Phi$ has no p'-torsion (see [Cart85] 4.5.1). Hence when the center of $\mathbf{G}$ is connected, an F^*-stable conjugacy class of semi-simple elements of $\mathbf{G}^*$ contains exactly one $(\mathbf{G}^*)^{F^*}$-conjugacy class, hence (iii). □

Proposition 8.25. *Let $\mathbf{P} = \mathbf{V} \rtimes \mathbf{L}$ be a Levi decomposition where $\mathbf{L}$ is F-stable. Let s (resp. t) be a semi-simple element of ${\mathbf{G}^*}^{F^*}$ (resp. ${\mathbf{L}^*}^{F^*}$, $\mathbf{L}^*$ a Levi subgroup of $\mathbf{G}^*$ in duality with $\mathbf{L}$), and let $\eta \in \widetilde{\mathcal{E}}(\mathbf{L}^F, t)$. One has*

$$\mathrm{R}^\mathbf{G}_{\mathbf{L}\subseteq\mathbf{P}}\, \eta \in \mathbb{Z}\widetilde{\mathcal{E}}(\mathbf{G}^F, t)$$

If η *occurs in* ${}^*\mathrm{R}^{\mathbf{G}}_{\mathbf{L}\subseteq\mathbf{P}}\,\chi$ *and* $\chi \in \widetilde{\mathcal{E}}(\mathbf{G}^F, s)$*, then* t *is conjugate to* s *in* $\mathbf{G}^*$.

Proof. The second assertion follows from the first by adjunction.

Let $\mathbf{B}_1 = \mathbf{U}_1\mathbf{T} \subset \mathbf{B} = \mathbf{U}\mathbf{T} \subset \mathbf{P}$ be Levi decompositions of some Borel subgroups of $\mathbf{L}$ and $\mathbf{G}$ such that the $\mathbf{L}$-geometric class of $(\mathbf{T}, \theta)$ corresponds to that of t (see Proposition 8.21). By Definition 8.23, Theorem 7.9 and the Künneth formula (A3.11) any irreducible constituent of $\mathrm{H}^i(\mathbf{Y}_{\mathbf{V}}^{(\mathbf{G},F)}, \overline{\mathbb{Q}}_\ell) \otimes \eta$ appears in some $\mathrm{H}^j(\mathbf{Y}_{\mathbf{V}}^{(\mathbf{G},F)}, \overline{\mathbb{Q}}_\ell) \otimes_{\mathbf{L}^F} \mathrm{H}^k(\mathbf{Y}_{\mathbf{U}_1}^{(\mathbf{L},F)}, \overline{\mathbb{Q}}_\ell) \otimes_{\mathbf{T}^F} \theta$ hence belongs to $\widetilde{\mathcal{E}}(\mathbf{G}^F, t)$ (see the proof of Theorem 8.24 (i)). This proves $\mathrm{R}^{\mathbf{G}}_{\mathbf{L}\subseteq\mathbf{P}}\,\eta \in \mathbb{Z}\widetilde{\mathcal{E}}(\mathbf{G}^F, t)$. □

From formula (8.20) one deduces the following (see [DiMi91] 13.30 and its proof).

Proposition 8.26. *Let* $z \in \mathrm{Z}(\mathbf{G}^*)^{F^*}$*, let* $\hat{z}$ *be the corresponding linear character of* $\mathbf{G}^F$*, (8.19). For any semi-simple element* s *in* $(\mathbf{G}^*)^{F^*}$ *multiplication by* $\hat{z}$ *defines a bijection* $\mathcal{E}(\mathbf{G}^F, s) \to \mathcal{E}(\mathbf{G}^F, (sz))$.

Theorem 8.27. *We keep the hypotheses of Proposition 8.25. Assume* $\mathrm{C}^\circ_{\mathbf{G}^*}(s).\mathrm{C}_{\mathbf{G}^*}(s)^F \subseteq \mathbf{L}^*$*, then the map* $\varepsilon_{\mathbf{G}}\varepsilon_{\mathbf{L}}\mathrm{R}^{\mathbf{G}}_{\mathbf{L}\subseteq\mathbf{P}}$ *induces a bijection* $\mathcal{E}(\mathbf{L}^F, s) \to \mathcal{E}(\mathbf{G}^F, s)$.

Proof. See [DiMi91] 13.25 and its proof. See also Exercise 2 for a translation of the hypothesis on s.

Exercises

1. Prove Proposition 8.1. As a corollary, show that if $\mathbf{G}$ is a semi-simple reductive group with a Frobenius F then $|\mathbf{G}^F| = |\mathbf{G}^F_{\mathrm{ad}}|$.
2. Let $\mathbf{L}^*$ be an F^*-stable Levi subgroup in $\mathbf{G}^*$, let $\mathbf{S}^*$ be a maximal F^*-stable torus in $\mathbf{L}^*$. Assume that $(\mathbf{T}_w, \mathbf{L})$ and $(\mathbf{S}^*, \mathbf{L}^*)$ are in duality by restriction from the duality between $\mathbf{G}$ and $\mathbf{G}^*$. Here $w = g^{-1}F(g)\mathbf{T} \in W$ is a type of $\mathbf{T}_w = g\mathbf{T}g^{-1}$ with respect to an F-stable maximal torus $\mathbf{T}$ of an F-stable Borel subgroup of $\mathbf{G}$. Let $W_I w$ be a type of $\mathbf{L}$. Let $(\mathbf{T}, \xi)$ and $(\mathbf{S}^*, s)$ be corresponding pairs ($\xi \in \mathrm{Irr}(\mathbf{T}^F_w)$, $s \in (\mathbf{S}^*)^{F^*}$) and let $\theta = \xi \circ \mathrm{ad}\, g^{-1} \in \mathrm{Irr}(\mathbf{T}^{wF})$. Show that
   (a) $\mathrm{C}^\circ_{\mathbf{G}^*}(s) \subseteq \mathbf{L}^*$ if and only if for all $\alpha \in \Phi$ such that $\theta(N_w(\alpha^\vee)) = 1$ one has $\alpha \in \Phi_I$,
   (b) $\mathrm{C}_{\mathbf{G}^*}(s) \subseteq \mathbf{L}^*$ if and only if for all $v \in W$ such that $(\theta \circ N_w)^v = \theta \circ N_w$, one has $v \in W_I$.

Hint. By [Borel] II, 4.1, $\mathrm{C}^\circ_{\mathbf{G}^*}(s)$ is generated by a torus $\mathbf{T}$ containing s and the set of root subgroups of $\mathbf{G}$ for roots vanishing on s. Use (8.12), (8.15).

Notes

Reductive groups in duality over $\mathbf{F}$ were first used by Deligne–Lusztig in [DeLu76], thus extending to arbitrary fields a construction over $\mathbb{C}$ due to Langlands. We have also borrowed from [Cart85] §4 and [DiMi91] §13.

As said before, the methods of étale cohomology in finite reductive groups, and most of the theorems in this chapter, are due to Deligne–Lusztig ([DeLu76]; see also [Lu76a] for Theorem 8.27).

Two natural problems were to be solved after Deligne–Lusztig's paper. The first is to describe fully, at least when the center of $\mathbf{G}$ is connected (and therefore all centralizers of semi-simple elements in $\mathbf{G}^*$ are connected), the set $\mathrm{Irr}(\mathbf{G}^F)$ and the decomposition of the generalized charcters $\mathrm{R}_\mathbf{T}^\mathbf{G}(\theta)$ in this basis. This was done by Lusztig in his book [Lu84], the parametrization being by pairs (s, λ), where s ranges over $(\mathbf{G}^*)^F_{\mathrm{ss}}$ mod. $\mathbf{G}^{*F}$-conjugacy and λ ranges over $\mathcal{E}(\mathrm{C}_{\mathbf{G}^*}(s)^F, 1)$. This is called "Jordan decomposition" of characters (see our Chapter 15 below). This goes with a combinatorial description of unipotent characters and Harish-Chandra series; see [Lu77] for the classical types. The fairly unified treatment in [Lu84] involves a broad array of methods, mainly intersection cohomology (see [Rick98] for an introduction), and Kazhdan-Lusztig's bases in Hecke algebras. The case when $\mathrm{Z}(\mathbf{G})$ is no longer connected was treated in [Lu88] (see also Chapter 16 below).

A second problem is to describe the integers $\langle \mathrm{R}^\mathbf{G}_{\mathbf{L}\subseteq\mathbf{P}}\zeta, \chi\rangle_{\mathbf{G}^F}$ for $\zeta \in \mathrm{Irr}(\mathbf{L}^F)$ and $\chi \in \mathrm{Irr}(\mathbf{G}^F)$. This was done essentially by Asai and Shoji, see [As84a], [As84b], [Sho85], and [Sho87]. The proofs involve a delicate analysis of the combinatorics of Fourier transforms (see [Lu84] §12), Hecke algebras and Shintani descent. Remaining problems, such as the case of special linear groups, the Mackey formula, or independence with regard to $\mathbf{P}$, were solved only recently (see [Bo00] and its references).

9

Blocks of finite reductive groups and rational series

We now come to ℓ-blocks and ℓ-modular aspects of ordinary characters for a prime ℓ. Let (Λ, K, k) be an ℓ-modular splitting system for the finite group G. The decomposition of the group algebra

$$\Lambda G = B_1 \times \cdots \times B_\nu$$

as a product of blocks ("ℓ-blocks of G") induces a corresponding partition of irreducible characters

$$\mathrm{Irr}(G) = \bigcup_i \mathrm{Irr}(G, B_i) \quad \text{(see §5.1).}$$

Take now $(\mathbf{G}, F)$ a connected reductive $\mathbf{F}$-group defined over $\mathbb{F}_q$ (see A2.4 and A2.5). For $G = \mathbf{G}^F$, we recall the decomposition

$$\mathrm{Irr}(\mathbf{G}^F) = \bigcup_s \mathcal{E}(\mathbf{G}^F, s)$$

into rational series (see §8.4) where s ranges over semi-simple elements of $\mathbf{G}^{*F}$, and where $(\mathbf{G}^*, F)$ is in duality with $(\mathbf{G}, F)$. If s is a semi-simple ℓ'-element of $(\mathbf{G}^*)^F$, one defines $\mathcal{E}_\ell(\mathbf{G}^F, s) := \bigcup_t \mathcal{E}(\mathbf{G}^F, st)$ where t ranges over the ℓ-elements of $\mathrm{C}_{\mathbf{G}^*}(s)^F$.

A first theorem, due to Broué–Michel, on blocks of finite reductive groups tells us that $\mathcal{E}_\ell(\mathbf{G}^F, s)$ is a union of sets $\mathrm{Irr}(\mathbf{G}^F, B_i)$; see [BrMi89]. The proof uses several elementary properties of the duality for irreducible characters (see Chapter 4 or [DiMi91] §8) along with some classical corollaries of the character formula, in order to check that the central function $\sum_{\chi \in \mathcal{E}_\ell(\mathbf{G}^F, s)} \chi(1)\chi$ sends $\mathbf{G}^F$ into $|\mathbf{G}^F|\Lambda$.

We then turn to the "isometric case" where the semi-simple element $s \in (\mathbf{G}^*)^F$ defining the series $\mathcal{E}(\mathbf{G}^F, s)$ satisfies $\mathrm{C}_{\mathbf{G}^*}(s) \subseteq \mathbf{L}^*$ for some F-stable Levi subgroup of $\mathbf{G}^*$. We show that the isometry of Theorem 8.27 induces an isometry on the unions of rational series defined above, and that Morita equivalence between the corresponding product of ℓ-blocks holds as long as a certain bi-projectivity property is checked ([Bro90b]). This prepares the way for the "first reduction" of Chapter 10.

9.1. Blocks and characters

We briefly recall some notation about ℓ-blocks and characters (see Chapter 5 and the classical textbooks [Ben91a], [CuRe87], [NaTs89]). Let G be a finite group and ℓ be a prime. Let (Λ, K, k) be an ℓ-modular splitting system for G.

We denote by $\mathrm{CF}(G, K)$ the space of G-invariant linear maps $KG \to K$, a K-basis being given by the set $\mathrm{Irr}(G)$ of irreducible characters. It is orthonormal for the scalar product on $\mathrm{CF}(G, K)$ defined by $\langle f, f' \rangle_G = |G|^{-1} \sum_{g \in G} f(g) f'(g^{-1})$. Another way of stating that is to define the following idempotents.

Definition 9.1. *If $\chi \in \mathrm{Irr}(G)$, let $e_\chi = |G|^{-1}\chi(1) \sum_{g \in G} \chi(g^{-1})g \in KG$ be the primitive idempotent of $\mathrm{Z}(KG)$ acting by* Id *on the representation space of χ (see [Thévenaz] 42.4).*

Recall the partition of $\mathrm{Irr}(G)$ induced by blocks of ΛG

Definition 9.2. $\mathrm{Irr}(G) = \bigcup_i \mathrm{Irr}(G, B_i)$ *where* $\mathrm{Irr}(G, B_i) = \mathrm{Irr}(G, b_i) = \{\chi \in \mathrm{Irr}(G) \mid \chi(b_i) = \chi(1)\}$ *whenever $B_i = \Lambda G.b_i$ for the primitive central idempotent b_i (see §5.1).*

Proposition 9.3. *Let E be a subset of* $\mathrm{Irr}(G)$ *and* $\mathrm{pr}_E \colon \mathrm{CF}(G, K) \to \mathrm{CF}(G, K)$ *be the associated orthogonal projection of image KE. The following are equivalent.*

(i) E is a union of sets $\mathrm{Irr}(G, B_i)$ *where B_i's are ℓ-blocks of G,*

(ii) $\mathrm{pr}_E(\mathrm{reg}_G)(g) \in |G|\Lambda = |G|_\ell \Lambda$ *for any $g \in G$,*

(iii) $\sum_{\chi \in E} e_\chi \in \Lambda G$.

Proof. (ii) and (iii) are clearly equivalent since $\mathrm{reg}_G = \sum_{\chi \in \mathrm{Irr}(G)} \chi(1)\chi$.

Denote each block of ΛG by $B_i = \Lambda G.b_i$ where b_i is the corresponding central idempotent. One has $b_i e_\chi = e_\chi$ or 0 according to whether $\chi \in \mathrm{Irr}(G, B_i)$ or not. Then $\sum_{\chi \in \mathrm{Irr}(G, B_i)} e_\chi = b_i$. So (i) implies (iii)

Assume (iii). Then $\sum_{\chi \in E} e_\chi = |G|^{-1} \sum_{g \in G} \mathrm{pr}_E(\mathrm{reg}_G)(g) g^{-1}$ is a central idempotent $b \in \Lambda G$, so it is a sum of block idempotents. This gives (i). □

9.2. Blocks and rational series

For the remainder of the chapter, we fix $(\mathbf{G}, F)$ a connected reductive $\mathbf{F}$-group defined over $\mathbb{F}_q$ (see A2.4 and A2.5). Let $(\mathbf{G}^*, F)$ be in duality with $(\mathbf{G}, F)$ (see (8.3)).

Let ℓ be a prime not dividing q and let (Λ, K, k) be an ℓ-modular splitting system for $\mathbf{G}^F$.

Definition 9.4. *Define $\mathcal{E}(\mathbf{G}^F, \ell')$ as the union of rational series $\mathcal{E}(\mathbf{G}^F, s)$ (see Definition 8.23) such that s ranges over the semi-simple elements in $(\mathbf{G}^*)^F$ whose order is prime to ℓ. If s is any such element, define $\mathcal{E}_\ell(\mathbf{G}^F, s)$ as the union of rational series $\mathcal{E}(\mathbf{G}^F, t)$ such that $s = t_{\ell'}$.*

Definition 9.5. *A uniform function is any K-linear combination of the $\mathrm{R}_\mathbf{T}^\mathbf{G}\theta$'s for $\mathbf{T}$ an F-stable maximal torus and $\theta \colon \mathbf{T}^F \to K^\times$ a linear character (see Remark 8.18(ii)).*

A p-constant function is any $f \in \mathrm{CF}(\mathbf{G}^F, K)$ such that $f(us) = f(s)$ for any Jordan decomposition $us = su$ with unipotent u and semi-simple s in $\mathbf{G}^F$.

We recall some corollaries of the character formula (Theorem 8.16). See §8.3 for twisted induction and its adjoint.

Proposition 9.6. *Let $f \in \mathrm{CF}(\mathbf{G}^F, K)$ be p-constant.*

(i) f is uniform.

(ii) ${}^*\mathrm{R}_{\mathbf{L}\subseteq\mathbf{P}}^\mathbf{G} f = \mathrm{Res}_{\mathbf{L}^F}^{\mathbf{G}^F} f$.

(iii) If $\zeta \colon \mathbf{L}^F \to K$ is a central function, then $\mathrm{R}_{\mathbf{L}\subseteq\mathbf{P}}^\mathbf{G}(\zeta).f = \mathrm{R}_{\mathbf{L}\subseteq\mathbf{P}}^\mathbf{G}(\zeta.\mathrm{Res}_{\mathbf{L}^F}^{\mathbf{G}^F} f)$.

References for proof. (i) [DiMi91] 12.21. (ii) Combine [DiMi91] 12.6(ii) and 12.7. (iii) [DiMi91] 12.6(i).

Definition 9.7. *Let $D_G = \sum_I (-1)^{|I|} \mathrm{R}_{L_I}^G \circ {}^*\mathrm{R}_{L_I}^G$ defined on $\mathbb{Z}\mathrm{Irr}(G)$ for a finite split BN-pair (G, B, N, S), the sum being over subsets of the set S.*

We list below the properties of D_G, where $G = \mathbf{G}^F$ that will be useful to us.

Proposition 9.8. *(i) $D_G^2 = \mathrm{Id}$, D_G permutes the characters up to signs.*

(ii) If f is a p-constant map on $\mathbf{G}^F$, then $D_G(f.\chi) = f.D_G(\chi)$.

(iii) $|\mathbf{G}^F|_{p'}^{-1}\mathrm{reg}_{\mathbf{G}^F}$ is the image by D_G of the characteristic function of unipotent elements of $\mathbf{G}^F$.

(iv) Let $\mathbf{T}$ be an F-stable maximal torus of $\mathbf{G}$. Then $D_G \circ \mathrm{R}_\mathbf{T}^\mathbf{G} = \varepsilon_\mathbf{G}\varepsilon_\mathbf{T}\mathrm{R}_\mathbf{T}^\mathbf{G}$ and therefore D_G preserves rational series.

Proof. (i) is [DiMi91] 8.14 and 8.15. One may also use Corollary 4.19 for the split semi-simple algebra KG. (iii) is [DiMi91] 9.4 (see also Exercise 6.2). For (ii) apply [DiMi91] 12.6 (or see the proof of [DiMi91] 9.4). (iv) is [DiMi91] 12.8.

Definition 9.9. *Let $e_{\ell'}^{\mathbf{G}^F} \in K\mathbf{G}^F$ be the sum of central idempotents associated to the characters in $\mathcal{E}(\mathbf{G}^F, \ell')$ (see Definition 9.4). If M is a $\Lambda\mathbf{G}^F$-module, one may define $e_{\ell'}^{\mathbf{G}^F}.M \subseteq M \otimes K$. If M is Λ-free of finite rank, then $e_{\ell'}^{\mathbf{G}^F}.M$ is a Λ-free $\Lambda\mathbf{G}^F$-submodule of $M \otimes_\Lambda K$.*

If $s \in (\mathbf{G}^*)^F$ is a semi-simple ℓ'-element, let $b_\ell(\mathbf{G}^F, s) = \sum_\chi e_\chi \in K\mathbf{G}^F$ where χ ranges over $\mathcal{E}_\ell(\mathbf{G}^F, s)$.

Theorem 9.10. *If P is a projective $\Lambda\mathbf{G}^F$-module, then the map*

$$P \to e_{\ell'}^{\mathbf{G}^F}.P, \quad x \mapsto e_{\ell'}^{\mathbf{G}^F}.x$$

defines a projective cover whose kernel is stable under $\mathrm{End}_{\Lambda\mathbf{G}^F}(P)$.

When (Λ, K, k) is an ℓ-modular splitting system for a finite group G, recall the map $d^1\colon \mathrm{CF}(G, K) \to \mathrm{CF}(G, K)$ consisting of restriction of central functions to $G_{\ell'}$ and extension by 0 elsewhere (see Definition 5.7).

Lemma 9.11. *If $\mathbf{T}$ is an F-stable maximal torus and $\theta \in \mathrm{Irr}(\mathbf{T}^F)$, then $d^1\mathrm{R}_\mathbf{T}^\mathbf{G}\theta = d^1\mathrm{R}_\mathbf{T}^\mathbf{G}\theta_{\ell'} = |\mathbf{T}^F|_\ell^{-1}\sum_{\theta'}\mathrm{R}_\mathbf{T}^\mathbf{G}(\theta\theta')$ where θ' ranges over the set of irreducible characters of $\mathbf{T}^F$ with $\mathbf{T}_{\ell'}^F$ in their kernel.*

Proof of Lemma 9.11. Viewing d^1 as multiplication by the p-constant function $d^1(1)$, Proposition 9.6(iii) gives $d^1\mathrm{R}_\mathbf{T}^\mathbf{G}\theta = \mathrm{R}_\mathbf{T}^\mathbf{G}(d^1\theta)$. One has $d^1\theta = d^1\theta_{\ell'}$, whence our first equality. The second comes from $d^1(1) = |\mathbf{T}^F|_\ell^{-1}\sum_{\theta'}\theta'$, where θ' ranges over the linear characters with $\mathbf{T}_{\ell'}^F$ in their kernel (regular character of $\mathbf{T}^F/\mathbf{T}_{\ell'}^F$). □

Proof of Theorem 9.10. The map is clearly onto. Its kernel Ω satisfies $\Omega \otimes K = (1 - e_{\ell'}^{\mathbf{G}^F}).(P \otimes K)$ and has no irreducible component in common with $e_{\ell'}^{\mathbf{G}^F}(P \otimes K)$. So $\Omega \otimes K$ is stable under $\mathrm{End}_{K\mathbf{G}^F}(P \otimes K)$. Then Ω is stable under $\mathrm{End}_{\Lambda\mathbf{G}^F}(P)$.

Now, in order to show that the map is a projective cover, it suffices to check the case of an indecomposable P. Then it suffices to check that $e_{\ell'}^{\mathbf{G}^F}.P \neq \{0\}$. Let us denote by ψ the character of P (the trace map on the elements of $\mathbf{G}^F$). If $e_{\ell'}^{\mathbf{G}^F}.P = \{0\}$, then $e_\chi P = \{0\}$ for any $\chi \in \mathcal{E}(\mathbf{G}^F, \ell')$ and therefore $\langle\psi, \chi\rangle_{\mathbf{G}^F} = 0$ for any $\chi \in \mathcal{E}(\mathbf{G}^F, \ell')$. This in turn implies $\langle\psi, \mathrm{R}_\mathbf{T}^\mathbf{G}\theta'\rangle_{\mathbf{G}^F} = 0$ for any $(\mathbf{T}, \theta')$ such that θ' is of order prime to ℓ. But $\mathrm{reg}_{\mathbf{G}^F}$ is a uniform function (Theorem 8.17(ii)), so, if $\psi(1) \neq 0$, i.e. $P \neq \{0\}$, there is some $(\mathbf{T}, \theta)$ such that $\langle\psi, \mathrm{R}_\mathbf{T}^\mathbf{G}\theta\rangle_{\mathbf{G}^F} \neq 0$.

However, using Lemma 9.11 and the fact that ψ is zero outside ℓ'-elements (see [NaTs89] 3.6.9(ii)), we get $\langle\psi, \mathrm{R}_{\mathbf{T}}^{\mathbf{G}}\theta\rangle_{\mathbf{G}^F} = \langle d^1\psi, \mathrm{R}_{\mathbf{T}}^{\mathbf{G}}\theta\rangle_{\mathbf{G}^F} = \langle\psi, d^1\mathrm{R}_{\mathbf{T}}^{\mathbf{G}}\theta\rangle_{\mathbf{G}^F} = \langle\psi, d^1\mathrm{R}_{\mathbf{T}}^{\mathbf{G}}\theta_{\ell'}\rangle_{\mathbf{G}^F} = \langle d^1\psi, \mathrm{R}_{\mathbf{T}}^{\mathbf{G}}\theta_{\ell'}\rangle_{\mathbf{G}^F} = \langle\psi, \mathrm{R}_{\mathbf{T}}^{\mathbf{G}}\theta_{\ell'}\rangle_{\mathbf{G}^F} = 0$ by our hypothesis. A contradiction. □

Theorem 9.12. *Let s be a semi-simple ℓ'-element of $\mathbf{G}^{*F}$.*

(i) $\mathcal{E}_\ell(\mathbf{G}^F, s)$ is a union of ℓ-blocks $\mathrm{Irr}(\mathbf{G}^F, B_i)$, i.e. $b_\ell(\mathbf{G}^F, s) \in \Lambda\mathbf{G}^F$.

(ii) For each ℓ-block B such that $\mathrm{Irr}(\mathbf{G}^F, B) \subseteq \mathcal{E}_\ell(\mathbf{G}^F, s)$, one has $\mathrm{Irr}(\mathbf{G}^F, B) \cap \mathcal{E}(\mathbf{G}^F, s) \neq \emptyset$.

Definition 9.13. *A unipotent ℓ-block of $\mathbf{G}^F$ is any ℓ-block B of $\mathbf{G}^F$ such that $\mathrm{Irr}(\mathbf{G}^F, B) \cap \mathcal{E}(\mathbf{G}^F, 1) \neq \emptyset$. By the above, this condition is equivalent to $\mathrm{Irr}(\mathbf{G}^F, B)$ being included in $\bigcup_{t\in(\mathbf{G}^*)^F_\ell}\mathcal{E}(\mathbf{G}^F, t)$.*

Proof of Theorem 9.12. For (ii) we apply Theorem 9.10 with $P = \Lambda\mathbf{G}^F.b \neq \{0\}$ where b is the unit of B. One has $e_{\ell'}^{\mathbf{G}^F}.P \neq \{0\}$, therefore $e_{\ell'}^{\mathbf{G}^F}.b \neq \{0\}$. That is, $\mathrm{Irr}(\mathbf{G}^F, B) \cap \mathcal{E}(\mathbf{G}^F, \ell') \neq \emptyset$. Once (i) is proved, this gives (ii).

Let us denote by pr the orthogonal projection

$$\mathrm{CF}(\mathbf{G}^F, K) = K(\mathrm{Irr}(\mathbf{G}^F)) \to K(\mathcal{E}_\ell(\mathbf{G}^F, s)).$$

Lemma 9.14. *If f is a uniform function on $\mathbf{G}^F$, then $\mathrm{pr}(d^1.f) = d^1\mathrm{pr}(f)$.*

Proof of Lemma 9.14. One may assume that $f = \mathrm{R}_{\mathbf{T}}^{\mathbf{G}}\theta$ for some pair $(\mathbf{T}, \theta)$ (Definition 9.5). One must show that, if $(\mathbf{T}, \theta)$ corresponds with some semi-simple $(\mathbf{T}^*, s')$ where $\mathbf{T}^*$ is an F-stable maximal torus in $\mathbf{G}^*$, $s' \in \mathbf{T}^{*F}$ (see §8.2), and such that $s'_{\ell'} = s$, then $\mathrm{pr}(d^1 f) = d^1 f$ – and that $\mathrm{pr}(d^1 f) = 0$ otherwise. Since the various $\mathcal{E}_\ell(\mathbf{G}^F, s)$ make a partition of $\mathrm{Irr}(\mathbf{G}^F)$, it suffices to check that $d^1\mathrm{R}_{\mathbf{T}}^{\mathbf{G}}\theta \in K(\mathcal{E}_\ell(\mathbf{G}^F, s))$.

By Lemma 9.11, one has $d^1\mathrm{R}_{\mathbf{T}}^{\mathbf{G}}\theta = \sum_{\theta'}\mathrm{R}_{\mathbf{T}}^{\mathbf{G}}\theta\theta'$ with $(\theta\theta')_{\ell'} = \theta_{\ell'}$ for each θ' in the sum. Then each $\mathrm{R}_{\mathbf{T}}^{\mathbf{G}}(\theta\theta') \in K(\mathcal{E}_\ell(\mathbf{G}^F, s))$, whence our claim. □

Let us now prove Theorem 9.12(i). If π is a set of primes, denote by $\delta_\pi \in \mathrm{CF}(\mathbf{G}^F, K)$ the function defined by $\delta_\pi(g) = 1$ if $g \in \mathbf{G}^F_\pi$, $\delta_\pi(g) = 0$ otherwise. Note that it is p-constant (see Definition 9.5) as long as $p \in \pi$. Note also that $d^1(1) = \delta_{\ell'}$.

The central function $\delta_{\{p,\ell\}}$ is uniform (Proposition 9.6(i)), so we may apply Lemma 9.14 with $f = \delta_{\{p,\ell\}}$. This gives

$$\mathrm{pr}(\delta_p) = \delta_{\ell'}.\mathrm{pr}(\delta_{\{p,\ell\}}). \tag{1}$$

Let us now apply the duality functor D_G to (1). By Proposition 9.8(iii) and (iv), the left-hand side gives $D_G \circ \mathrm{pr}(\delta_p) = \mathrm{pr} \circ D_G(\delta_p) = |\mathbf{G}^F|_{p'}^{-1}\mathrm{pr}(\mathrm{reg}_{\mathbf{G}^F})$.

By Proposition 9.8(ii), the right-hand side gives $\delta_{\ell'}.\mathrm{pr}\circ D_G(\delta_{\{p,\ell\}})$. Then

$$\mathrm{pr}(\mathrm{reg}_{\mathbf{G}^F}) = |\mathbf{G}^F|_{p'}.\delta_{\ell'}.\mathrm{pr}\circ D_G(\delta_{\{p,\ell\}}). \tag{2}$$

We have $\delta_{\{p,\ell\}}\in\Lambda(\mathrm{Irr}(\mathbf{G}^F))$ by a classical result on "ℓ-constant" functions (see [NaTs89] 3.6.15(iii)). Then also $\mathrm{pr}\circ D_G(\delta_{\{p,\ell\}})\in\Lambda(\mathrm{Irr}(\mathbf{G}^F))$ by Proposition 9.8(i). Now, (2) implies that $\mathrm{pr}(\mathrm{reg}_{\mathbf{G}^F})$ takes values in $|\mathbf{G}^F|_{p'}\Lambda = |\mathbf{G}^F|\Lambda$. Then Proposition 9.3(ii) gives our claim (i). □

Let $\mathbf{P}=\mathbf{LV}$ be a Levi decomposition with $F\mathbf{P}=\mathbf{P}$ and $F\mathbf{L}=\mathbf{L}$. Recall (Remark 8.18(i)) that, in this case, $\mathrm{R}^{\mathbf{G}}_{\mathbf{L}\subseteq\mathbf{P}}$ coincides on characters with the classical Harish-Chandra functor denoted by $\mathrm{R}^{\mathbf{G}^F}_{\mathbf{L}^F}$ in Chapter 3 (see Notation 3.11). In the following we use the notation $\mathrm{R}^{\mathbf{G}}_{\mathbf{L}\subseteq\mathbf{P}}$ (resp. ${}^*\mathrm{R}^{\mathbf{G}}_{\mathbf{L}\subseteq\mathbf{P}}$) to denote Harish-Chandra induction $\mathrm{R}^{\mathbf{G}^F}_{\mathbf{L}^F}$ (resp. restriction ${}^*\mathrm{R}^{\mathbf{G}^F}_{\mathbf{L}^F}$) applied to modules.

Proposition 9.15. *Let M be a Λ-free $\Lambda\mathbf{L}^F$-module. Then $e^{\mathbf{G}^F}_{\ell'}.\mathrm{R}^{\mathbf{G}}_{\mathbf{L}\subseteq\mathbf{P}}M\cong \mathrm{R}^{\mathbf{G}}_{\mathbf{L}\subseteq\mathbf{P}}(e^{\mathbf{L}^F}_{\ell'}.M)$. Similarly $e^{\mathbf{L}^F}_{\ell'}.{}^*\mathrm{R}^{\mathbf{G}}_{\mathbf{L}\subseteq\mathbf{P}}N\cong {}^*\mathrm{R}^{\mathbf{G}}_{\mathbf{L}\subseteq\mathbf{P}}(e^{\mathbf{G}^F}_{\ell'}.N)$ for any Λ-free $\Lambda\mathbf{L}^F$-module N.*

Proof. By Proposition 8.25, $(1-e^{\mathbf{G}^F}_{\ell'})\mathrm{R}^{\mathbf{G}}_{\mathbf{L}\subseteq\mathbf{P}}(e^{\mathbf{L}^F}_{\ell'}M)=e^{\mathbf{G}^F}_{\ell'}\mathrm{R}^{\mathbf{G}}_{\mathbf{L}\subseteq\mathbf{P}}((1-e^{\mathbf{L}^F}_{\ell'})M)=\{0\}$ since this is the case for $M\otimes K$. Now, regarding $\mathrm{R}^{\mathbf{G}}_{\mathbf{L}\subseteq\mathbf{P}}M=\Lambda\mathbf{G}^Fe(R_u(\mathbf{P})^F)\otimes_{\mathbf{P}^F}M$ as a subgroup of $\mathrm{R}^{\mathbf{G}}_{\mathbf{L}\subseteq\mathbf{P}}(M\otimes K)=K\mathbf{G}^Fe(R_u(\mathbf{P})^F)\otimes_{\mathbf{P}^F}M\otimes K$, one has the equality $e^{\mathbf{G}^F}_{\ell'}\mathrm{R}^{\mathbf{G}}_{\mathbf{L}\subseteq\mathbf{P}}M=\mathrm{R}^{\mathbf{G}}_{\mathbf{L}\subseteq\mathbf{P}}(e^{\mathbf{L}^F}_{\ell'}M)$ in $\mathrm{R}^{\mathbf{G}}_{\mathbf{L}\subseteq\mathbf{P}}M\otimes K$. This is because, if $m\in M$ and $x\in\Lambda\mathbf{G}^Fe(R_u(\mathbf{P})^F)$, then $e^{\mathbf{G}^F}_{\ell'}x\otimes m=e^{\mathbf{G}^F}_{\ell'}x\otimes e^{\mathbf{L}^F}_{\ell'}m=e^{\mathbf{G}^F}_{\ell'}x\otimes m$ by what is recalled in the beginning of this proof.

The statement concerning ${}^*\mathrm{R}^{\mathbf{G}}_{\mathbf{L}\subseteq\mathbf{P}}$ is proved in the same fashion, replacing the bimodule $\Lambda\mathbf{G}^Fe(R_u(\mathbf{P})^F)$ with $e(R_u(\mathbf{P})^F)\Lambda\mathbf{G}^F$. □

9.3. Morita equivalence and ordinary characters

We keep $(\mathbf{G}, F)$ a connected reductive $\mathbf{F}$-group defined over $\mathbb{F}_q$, and $(\mathbf{G}^*, F)$ in duality with $(\mathbf{G}, F)$.

Let $\mathbf{L}$ be an F-stable Levi subgroup of $\mathbf{G}$ in duality with $\mathbf{L}^*$ in $\mathbf{G}^*$ (see §8.2). In the following, we show that the isometry of Theorem 8.27 extends to the sets $\mathcal{E}_\ell(\mathbf{G}^F, s)$ (see also Exercise 5).

Theorem 9.16. *Let $s\in(\mathbf{L}^*)^F$ be a semi-simple ℓ'-element. Assume $\mathrm{C}^\circ_{\mathbf{G}^*}(s).\mathrm{C}_{\mathbf{G}^*}(s)^F\subseteq\mathbf{L}^*$. Then, for any Levi decomposition $\mathbf{P}=\mathbf{VL}$, the map $\varepsilon_{\mathbf{G}}\varepsilon_{\mathbf{L}}\mathrm{R}^{\mathbf{G}}_{\mathbf{L}\subseteq\mathbf{P}}$ induces a bijection between $\mathcal{E}_\ell(\mathbf{G}^F, s)$ and $\mathcal{E}_\ell(\mathbf{L}^F, s)$.*

In particular, when $\mathrm{C}_{\mathbf{G}^*}(s)=\mathbf{C}^*$ is a Levi subgroup, $\varepsilon_{\mathbf{G}}\varepsilon_{\mathbf{C}}\mathrm{R}^{\mathbf{G}}_{\mathbf{C}\subseteq\mathbf{P}}\hat{s}$ induces a bijection between $\mathcal{E}_\ell(\mathbf{G}^F, s)$ and $\mathcal{E}_\ell(\mathbf{C}^F, 1)$ (see (8.14) for the notation $\hat{s}$).

Proof. Let $st \in \mathbf{G}^{*F}$ be a semi-simple element such that $(st)_\ell = t$. Then $t \in \mathrm{C}_{\mathbf{G}^*}(s)^F \subseteq \mathbf{L}^{*F}$. This implies that $\mathcal{E}_\ell(\mathbf{G}^F, s) = \bigcup_t \mathcal{E}(\mathbf{G}^F, st)$ and $\mathcal{E}_\ell(\mathbf{L}^F, s) = \bigcup_t \mathcal{E}(\mathbf{L}^F, st)$ are both indexed by $\mathrm{C}_{\mathbf{L}^*}(s)^F_\ell$. Moreover two sets $\mathcal{E}(\mathbf{G}^F, st)$ and $\mathcal{E}(\mathbf{G}^F, st')$ are equal if and only if st and st' are $\mathbf{G}^{*F}$-conjugate. But a rational element bringing st to st' must centralize $s = (st)_{\ell'}$, so it belongs to $\mathrm{C}_{\mathbf{G}^*}(s)^F$. This is included in $\mathbf{L}^{*F}$ by our hypothesis, so st and st' are $\mathbf{L}^{*F}$-conjugate. This shows that the disjoint unions $\mathcal{E}_\ell(\mathbf{G}^F, s) = \bigcup_t \mathcal{E}(\mathbf{G}^F, st)$ and $\mathcal{E}_\ell(\mathbf{L}^F, s) = \bigcup_t \mathcal{E}(\mathbf{L}^F, st)$ have the same number of distinct terms. Moreover $\mathrm{C}^\circ_{\mathbf{G}^*}(st)\mathrm{C}_{\mathbf{G}^*}(st)^F \subseteq \mathrm{C}^\circ_{\mathbf{G}^*}(s)\mathrm{C}_{\mathbf{G}^*}(s)^F \subseteq \mathbf{L}^*$, so Theorem 8.27 implies that $\varepsilon_\mathbf{G}\varepsilon_\mathbf{L}\mathrm{R}^\mathbf{G}_{\mathbf{L}\subseteq\mathbf{P}}$ induces a bijection $\mathcal{E}_\ell(\mathbf{L}^F, s) \to \mathcal{E}_\ell(\mathbf{G}^F, s)$. □

The next theorem sets the framework in which we will prove a Morita equivalence in subsequent chapters. For Morita equivalences, we refer to [Thévenaz] §1.9.

The following lemma is trivial.

Lemma 9.17. *Let R be a ring, $\phi\colon L' \to L''$ a map in* **mod** $-$ *R. Assume L is a left R-module isomorphic with ${}_RR$. Then ϕ is an isomorphism if and only if $\phi \otimes_R L\colon L' \otimes_R L \to L'' \otimes_R L$ is an isomorphism.*

Theorem 9.18. *Let G, H be two finite groups, let ℓ be a prime, and let (Λ, K, k) be an ℓ-modular splitting system for $G \times H$. Let $e \in \mathrm{Z}(\Lambda G)$, $f \in \mathrm{Z}(\Lambda H)$ be central idempotents. Denote by $A = \Lambda Ge$, $B = \Lambda Hf$ the corresponding products of blocks. Let M be a ΛG-ΛH-bimodule, projective on each side. Denote $A_K := A \otimes_\Lambda K$, etc. Assume that $I \mapsto M_K \otimes_{B_K} I$ sends* $\mathrm{Irr}(H, B)$ *bijectively into* $\mathrm{Irr}(G, A)$. *Then $Mf \otimes_B -$ induces a Morita equivalence between $B-$***mod** *and $A-$***mod**.

Proof. Note first that $(1 - e)Mf = 0$ since this holds once tensored with K by the hypothesis on M_K and the fact that A_K and B_K are semi-simple. So we may replace M with $Mf = eMf$ and consider it as an A-B-bimodule.

Denote $M^\vee = \mathrm{Hom}_\Lambda(M, \Lambda)$, considered as a B-A-bimodule.

Then $M^\vee$ is bi-projective, and $M^\vee \otimes_A -$ is left and right adjoint to $M \otimes_B -$ (see [KLRZ98] 9.2.4). The same is true for $(M^\vee)_K = (M_K)^\vee$ with regard to A_K and B_K over K.

It suffices to show that $M \otimes_B M^\vee \cong {}_AA_A$ and $M^\vee \otimes_A M \cong {}_BB_B$ (see [Thévenaz] 1.9.1 and 1.9.2).

Denote $N := M^\vee$. The algebras A_K and B_K are split semi-simple, so the hypothesis that $M_K \otimes_{B_K} -$ bijects simple modules translates into the isomorphisms

$$M_K \otimes_{B_K} N_K \cong A_K \quad \text{and} \quad N_K \otimes_{A_K} M_K \cong B_K$$

as bimodules. The hypothesis on $M_K \otimes_{B_K} -$ implies that $M_K \cong \bigoplus_{i=1}^{\nu} S_i \otimes_K T_i^\vee$ where $i \mapsto S_i$ and $i \mapsto T_i$ are indexations of simple modules for A_K and B_K, their (common) number being ν. Then $N_K = M_K^\vee = \bigoplus_i T_i \otimes_K S_i^\vee$ and, for instance, $M_K \otimes_{B_K} N_K \cong \bigoplus_i S_i \otimes_K S_i^\vee \cong A_K$ as a bimodule since A_K is the corresponding product of matrix algebras.

Let us consider $M \otimes_B N$ as a left A-module. It is projective since M and N are bi-projective (see Exercise 4.9). We have seen that $(M \otimes_B N) \otimes_\Lambda K = M_K \otimes_{B_K} N_K$, as a left A_K-module is isomorphic with A_K. By invertibility of the Cartan matrix for group algebras (see [Ben91a] 5.3.6), this implies that ${}_A(M \otimes_B N) \cong {}_A A$. Similarly, $(M \otimes_B N)_A \cong A_A$, ${}_B(N \otimes_A N) \cong {}_B B$, and $(N \otimes_A N)_B \cong B_B$.

Now, take $\epsilon \colon B \to N \otimes_A M$ and $\eta \colon M \otimes_B N \to A$ the unit and co-unit associated with the (right and left) adjunctions between $M \otimes_B -$ and $N \otimes_A -$. The composition of the following maps

$$N \xrightarrow{\epsilon \otimes_B N} N \otimes_A M \otimes_B N \xrightarrow{N \otimes_A \eta} N$$

is the identity by the usual properties of adjunctions (see [McLane97] IV.1). As right A-module, the middle term is N since $(N \otimes_A M)_B \cong B_B$. Then all three terms are isomorphic in $\mathbf{mod} - A$, and therefore the two maps are inverse isomorphisms (another proof would consist in tensoring by K and using $N_K \otimes_{A_K} M_K \otimes_{B_K} N_K = N_K$ to show that $\epsilon \otimes_B N$, and therefore $N \otimes_A \eta$, are isomorphisms).

Since the first morphism above is an isomorphism, upon tensoring with M on the right, one gets that

$$N \otimes_A M \xrightarrow{\epsilon \otimes_B N \otimes_A M} N \otimes_A M \otimes_B N \otimes_A M$$

is an isomorphism. The above Lemma 9.17 for $R = B$, $L = L'' = N \otimes_A M$ and $L' = B$ tells us that ϵ was an isomorphism in the first place. The same can be done for η. □

Corollary 9.19. *Assume $C^\circ_{\mathbf{G}^*}(s).C_{\mathbf{G}^*}(s)^F \subseteq \mathbf{L}^*$, and that $\mathbf{L}^*$ is in duality with a Levi subgroup $\mathbf{L}$ of an F-***stable** *parabolic subgroup $\mathbf{P} = \mathbf{VL}$. Then the sums of blocks $\Lambda\mathbf{G}^F.b_\ell(\mathbf{G}^F, s)$ and $\Lambda\mathbf{L}^F.b_\ell(\mathbf{L}^F, s)$ are Morita equivalent, i.e.*

$$\Lambda\mathbf{G}^F.b_\ell(\mathbf{G}^F, s)\text{–}\mathbf{mod} \cong \Lambda\mathbf{L}^F.b_\ell(\mathbf{L}^F, s)\text{–}\mathbf{mod}.$$

Proof. Let $M = \Lambda\mathbf{G}^F\varepsilon$ where $\varepsilon = |\mathbf{V}^F|^{-1}\sum_{v \in \mathbf{V}^F} v \in \Lambda\mathbf{G}^F$. Since ε is an idempotent fixed by $\mathbf{L}^F$-conjugacy, M is a bi-projective $\Lambda\mathbf{G}^F$-$\Lambda\mathbf{L}^F$-bimodule. When I is any $\Lambda\mathbf{L}^F$-module, $M \otimes_{\Lambda\mathbf{L}^F} I$ is the $\mathbf{G}^F$-module obtained by inflation from $\mathbf{L}^F$ to $\mathbf{P}^F$, then induction from $\mathbf{P}^F$ to $\mathbf{G}^F$. On characters, this is $R_{\mathbf{L}}^{\mathbf{G}}$ (see Remark 8.18(i)), so Theorem 8.27 implies that $M \otimes_\Lambda K$ induces a bijection $\mathrm{Irr}(\mathbf{L}^F, b_\ell(\mathbf{L}^F, s)) \to \mathrm{Irr}(\mathbf{G}^F, b_\ell(\mathbf{G}^F, s))$. Now Theorem 9.18 tells us that $\Lambda\mathbf{G}^F b_\ell(\mathbf{G}^F, s)$ and $\Lambda\mathbf{L}^F b_\ell(\mathbf{L}^F, s)$ are Morita equivalent. The latter algebra is isomorphic with $\Lambda\mathbf{L}^F b_\ell(\mathbf{L}^F, 1)$ by the map $x \mapsto \lambda(x)x$ (for $x \in \mathbf{L}^F$) corresponding with the linear character $\lambda = \hat{s}\colon \mathbf{L}^F \to \Lambda^\times$ (see Proposition 8.26). □

Remark 9.20. One has $\Lambda\mathbf{G}^F.b_\ell(\mathbf{G}^F, s) \cong \mathrm{Mat}_{|\mathbf{G}^F:\mathbf{P}^F|}(\Lambda\mathbf{L}^F.b_\ell(\mathbf{L}^F, s))$; see Exercise 6.

Exercises

1. As a consequence of Theorem 9.12(i), show Lemma 9.14 for any central function f.
2. Show that $\delta_{\{p,\ell\}} \in K\mathcal{E}(\mathbf{G}^F, \ell')$. Deduce that formula (2) in the proof of Theorem 9.12(i) can be refined with pr replaced by the projection on $K(\mathcal{E}(\mathbf{G}^F, s))$ in the right-hand side (only):

   (3) $$\mathrm{pr}(\mathrm{reg}_{\mathbf{G}^F}) = |\mathbf{G}^F|_{p'}\delta_{\ell'}.\mathrm{pr}_s \circ D(\delta_{\{\ell,p\}}).$$

   Deduce Theorem 9.12(ii) from this new formula.
3. Prove directly equation (3) above by showing that the scalar products with all the $R_{\mathbf{T}}^{\mathbf{G}}\theta$'s are equal.
4. Show Proposition 9.15 by showing first that $e_{\ell'}^{\mathbf{G}^F}\Lambda\mathbf{G}^F e(R_u(P)^F) = \Lambda\mathbf{G}^F e(R_u(P)^F)e_{\ell'}^{\mathbf{L}^F} = e_{\ell'}^{\mathbf{G}^F}\Lambda\mathbf{G}^F e(R_u(P)^F)e_{\ell'}^{\mathbf{L}^F}$ as Λ-submodules of $K\mathbf{G}^F$. Give a generalization with two finite groups G and L, sets of characters E_G and E_L and a G-L-bimodule B over Λ such that $B \otimes K$ has adequate properties with regard to E_G and E_L.

   Show that $e_{\ell'}^G M = M/N$ where N is a unique Λ-pure submodule such that $N \otimes K$ has irreducible components only outside $\mathcal{E}(G, \ell')$.
5. Let G, H be two finite groups. Let ℓ be a prime, and (Λ, K, k) be an ℓ-modular splitting system for $G \times H$. Let b (resp. c) be a block idempotent of ΛG (resp. ΛH).

   Let $\mu = \sum_{\chi,\psi} m_{\chi,\psi}\chi \otimes \psi$ be a linear combination where χ ranges over $\mathrm{Irr}(G, b)$, ψ over $\mathrm{Irr}(H, c)$ and $m_{\chi,\psi} \in \mathbb{Z}$, i.e. an element of $\mathbb{Z}\mathrm{Irr}(G \times H, b \otimes c)$.

   We say that μ is **perfect** if and only if it satisfies the following for any $g \in G, h \in H$:
   (i) $\mu(g, h) \in |C_G(g)|\Lambda \cap |C_H(h)|\Lambda$,
   (ii) if $\mu(g, h) \neq 0$, then $g_\ell = 1$ if and only if $h_\ell = 1$.

   The conditions above clearly define a subgroup of $\mathbb{Z}\mathrm{Irr}(G \times H, b \otimes c)$.

(a) Let M be a ΛGb-ΛHc-bimodule. Show that, if M is bi-projective, then its character is perfect (for (i) reduce to $g \in Z(G)$, $H = <h>$ and use Higman's criterion (see [Ben91a] 3.6.4, [NaTs89] 4.2.2); for (ii) one may also assume $G = <g>$).

(b) Let $\mathbf{P} = \mathbf{LV}$ be a Levi decomposition in a connected reductive group $(\mathbf{G}, F)$ defined over $\mathbb{F}_q$. Assume $F(\mathbf{L}) = \mathbf{L}$ and let us consider $R^{\mathbf{G}}_{\mathbf{L}\subseteq\mathbf{P}}: \mathbb{Z}\mathrm{Irr}(\mathbf{L}^F) \to \mathbb{Z}\mathrm{Irr}(\mathbf{G}^F)$. Show that the associated generalized character of $\mathbf{G}^F \times \mathbf{L}^F$ is perfect (use A3.15 and the above).

(c) Let $\mu \in \mathbb{Z}\mathrm{Irr}(G \times H)$ (not necessarily perfect). Considering it as a linear combination of KG-KH-bimodules, it induces a map $\mathrm{Irr}(H) \to \mathbb{Z}\mathrm{Irr}(G)$ and therefore also a (K-linear) map $I_\mu: Z(KH) \to Z(KG)$, since $Z(KH) = \bigoplus_{\psi\in\mathrm{Irr}(H)} K.e_\psi$ (see Definition 9.1). Show that $I_\mu(\sum_h \lambda_h h) = \sum_{g\in G}(|H|^{-1}\sum_{h\in H}\mu(g,h)\lambda_h)g$ for $\sum_h \lambda_h h \in Z(KH)$ with $\lambda_h \in K$ (reduce to $\mu \in \mathrm{Irr}(G \times H)$). Show that, if μ satisfies (i), then $I_\mu(Z(\Lambda H)) \subseteq Z(\Lambda G)$.

(d) If $\mu \in \mathbb{Z}\mathrm{Irr}(G \times H)$ is perfect and induces an isometry $\mathbb{Z}\mathrm{Irr}(H, c) \to \mathbb{Z}\mathrm{Irr}(G, b)$ (this is sometimes called a **perfect isometry**), show that it preserves the partition induced by ℓ-blocks.

(e) Deduce from the above that the bijection of Theorem 9.16 preserves the partition induced by ℓ-blocks.

(f) Show that a derived equivalence (see Chapter 4) between $\Lambda G.b$ and $\Lambda H.c$ induces a perfect isometry.

6. Assume the hypotheses of Theorem 9.18. Assume moreover that there is an integer $n \geq 1$ such that $\dim(M_K \otimes_{B_K} S) = n.\dim(S)$ for any simple B_K-module S. Show that $A \cong \mathrm{Mat}_n(B)$ (assuming $M = Mf$, prove $M_B \cong (B_B)^n$ as a projective right B-module).
7. Show Theorem 9.18 for arbitrary Λ-free algebras A, B of finite rank; see [Bro90b].

Notes

Theorem 9.12(i), and the proof we give, are due to Broué–Michel; see [BrMi89]. Theorem 9.12(ii) is due to Hiss; see [Hi90]. Theorem 9.18 is due to Broué [Bro90b].

Perfect isometries (see Exercise 5 and [KLRZ98] §6.3) were introduced by Broué; see [Bro90a].

10

Jordan decomposition as a Morita equivalence: the main reductions

We recall the notation of Chapter 9. Let $(\mathbf{G}, F)$ be a connected reductive $\mathbf{F}$-group defined over $\mathbb{F}_q$ (see A2.4 and A2.5). Let $(\mathbf{G}^*, F)$ be in duality with $(\mathbf{G}, F)$ (see Chapter 8). Let ℓ be a prime not dividing q, let (Λ, K, k) be an ℓ-modular splitting system for $\mathbf{G}^F$. Let s be a semi-simple ℓ'-element of $(\mathbf{G}^*)^F$. We have seen that the irreducible characters in rational series $\mathcal{E}(\mathbf{G}^F, t)$ with $t_{\ell'} = s$ are the irreducible representations of a sum of blocks $\Lambda\mathbf{G}^F.b_\ell(\mathbf{G}^F, s)$ in $\Lambda\mathbf{G}^F$. Assume $\mathrm{C}_{\mathbf{G}^*}(s) \subseteq \mathbf{L}^*$, the latter an F-stable Levi subgroup of $\mathbf{G}^*$. Then $\mathrm{R}_\mathbf{L}^\mathbf{G}$ induces a bijection

$$\mathrm{Irr}(\mathbf{L}^F, b_\ell(\mathbf{L}^F, s)) \to \mathrm{Irr}(\mathbf{G}^F, b_\ell(\mathbf{G}^F, s))$$

(Theorem 9.16). The aim of this chapter is to establish the main reductions towards the following.

Theorem 10.1. (Bonnafé–Rouquier) *$\Lambda\mathbf{G}^F.b_\ell(\mathbf{G}^F, s)$ and $\Lambda\mathbf{L}^F.b_\ell(\mathbf{L}^F, s)$ are Morita equivalent.*

In view of Theorem 9.18, one has essentially to build a $\Lambda\mathbf{G}^F$-$\Lambda\mathbf{L}^F$-bimodule, projective on each side and such that the induced tensor product functor provides over K the above bijection of characters.

The $\mathrm{R}_\mathbf{L}^\mathbf{G}$ functor is obtained from an object of $D^b(\Lambda(\mathbf{G}^F \times \mathbf{L}^F)\text{−}\mathbf{mod})$ provided by the étale cohomology of the variety $\mathbf{Y}_\mathbf{V} := \mathbf{Y}_\mathbf{V}^{(\mathbf{G},F)}$ ($\mathbf{LV}$ being a Levi decomposition; see Chapter 7). It can be represented by a complex Ω of $\Lambda(\mathbf{G}^F \times \mathbf{L}^F)$-modules, projective on each side (see A3.15).

It is easily shown that our claim is now equivalent to checking that Ω can be taken to have only a single non-zero term. To obtain this, one reduces the claim to showing that the sheaf $\mathcal{F}_s$ on $\mathbf{X}_\mathbf{V} := \mathbf{Y}_\mathbf{V}/\mathbf{L}^F$ naturally associated with the constant sheaf on $\mathbf{Y}_\mathbf{V}$, and the representation $\Lambda\mathbf{L}^F.b_\ell(\mathbf{L}^F, s)$ of $\mathbf{L}^F$, satisfy a certain condition relative to a compactification $\mathbf{X}_\mathbf{V} \xrightarrow{j} \overline{\mathbf{X}}$, namely that its extension by 0 coincides with its direct image (see A3.3)

$$j_!\mathcal{F}_s = j_*\mathcal{F}_s$$

and that its higher direct images vanish:

$$\mathrm{R}^i j_*\mathcal{F}_s = 0 \quad \text{for } i \geq 1.$$

This kind of problem is known as a problem of ramification, related to the possibility of extending $\mathcal{F}_s$ into a locally constant sheaf on intermediate subvarieties of $\overline{\mathbf{X}}_\mathbf{V}$ (see A3.17). But here $\mathcal{F}_s$ is associated with a representation of $\mathbf{L}^F$ which is not of order prime to p (“wild” ramification). One further reduction, due to Bonnafé–Rouquier, is then checked (§10.5), showing that the above question on direct images is implied by a theorem of ramification and generation (Theorem 10.17(a) and (b) in §10.4 below) pertaining only to Deligne–Lusztig varieties $\mathbf{X}(w)$ and their Galois coverings $\mathbf{Y}(w) \to \mathbf{X}(w)$ of group $\mathbf{T}^{wF}$ (see Definition 7.12), clearly of order prime to p. Theorem 10.17 is proved in the next two chapters.

Recall from Chapter 9 that we have fixed $p \neq \ell$ two primes, q a power of p, $\mathbf{F}$ an algebraic closure of $\mathbb{F}_q$, K a finite extension of $\mathbb{Q}_\ell$, Λ its subring of integers over $\mathbb{Z}_\ell$, $k = \Lambda/J(\Lambda)$, such that (Λ, K, k) is an ℓ-modular splitting system for all finite groups encountered.

10.1. The condition $i^*\,\mathrm{R}\,j_*\,\mathcal{F} = 0$

In this section, we establish a preparatory result that rules out the bi-projectivity and $D^b(\Lambda(\mathbf{G}^F \times \mathbf{L}^F)\text{–}\mathbf{mod})$ vs $\Lambda(\mathbf{G}^F \times \mathbf{L}^F)\text{–}\mathbf{mod}$ questions, thus leading to a purely sheaf-theoretic formulation. We use the notation of Appendix 3.

Condition 10.2. *Let $\mathbf{X}$ be an $\mathbf{F}$-variety and $j\colon \mathbf{X} \to \overline{\mathbf{X}}$ be an open immersion with $\overline{\mathbf{X}}$ a complete variety. Let $i\colon \overline{\mathbf{X}} \setminus \mathbf{X} \to \overline{\mathbf{X}}$ be the associated closed immersion. Let $\mathcal{F}$ be a locally constant sheaf in $Sh_k(\mathbf{X}_{\text{ét}})$. Then the following two conditions are equivalent.*

(a) The natural map $j_!\mathcal{F} \to j_*\mathcal{F}$ (see A3.3) induces an isomorphism $j_!\mathcal{F} \cong \mathrm{R}j_*\mathcal{F}$ in $D^b_k(\overline{\mathbf{X}})$.

(b) $i^*\mathrm{R}j_*\mathcal{F} = 0$ in $D^b_k(\overline{\mathbf{X}} \setminus \mathbf{X})$.

Both imply

(c) the natural map $\mathrm{R_c}\Gamma(\mathbf{X}, \mathcal{F}) \to \mathrm{R}\Gamma(\mathbf{X}, \mathcal{F})$ in $D^b(k\text{–}\mathbf{mod})$ is an isomorphism.

Proof. By A3.9, we have the exact sequence

$$0 \to j_! \to j_* \to i_*i^*j_* \to 0.$$

All those functors are left-exact. Taking the derived functors (see A1.10) yields a distinguished triangle in $D_k^b(\overline{\mathbf{X}})$

$$j_!\mathcal{F} \to \mathrm{R}j_*\mathcal{F} \to i_*i^*\mathrm{R}j_*\mathcal{F} \to j_!\mathcal{F}[1]$$

where we have used the fact that $j_!$, i^* and i_* are exact (see A3.3).

We get at once the equivalence between (a) and (b) since $i_*i^*\mathrm{R}j_*\mathcal{F} = 0$ in $D_k^b(\overline{\mathbf{X}})$ if and only if $i^*\mathrm{R}j_*\mathcal{F} = 0$ in $D_k^b(\overline{\mathbf{X}} \setminus \mathbf{X})$ by applying i^* (see A3.9).

We now check (c). Denote by $\sigma : \mathbf{X} \to \mathrm{Spec}(\mathbf{F})$, $\overline{\sigma} : \overline{\mathbf{X}} \to \mathrm{Spec}(\mathbf{F})$ the structure morphisms. By (A3.4) and the definition of direct images with compact support (A3.6), the natural transformation

$$(\mathrm{c}') \qquad \mathrm{R_c}\sigma_* \to \mathrm{R}\sigma_*,$$

is the image by $\mathrm{R}\overline{\sigma}_*$ of the natural transformation

$$(\mathrm{a}') \qquad j_! \to \mathrm{R}j_*.$$

Applying this to $\mathcal{F}$ satisfying (a), we get that (a′) and therefore (c′) are isomorphisms. □

Proposition 10.3. *Let* $\mathbf{X}$ *be a smooth quasi-affine* $\mathbf{F}$*-variety. Let* A *be* $\Lambda/J(\Lambda)^n$ *for some integer* $n \geq 1$*. Let* $\mathcal{F}$ *be a locally constant sheaf on* $\mathbf{X}_{\text{ét}}$ *(see A3.8) with* A*-free stalks at closed points of* $\mathbf{X}$*.*

Assume that we have a compactification $\mathbf{X} \to \overline{\mathbf{X}}$ *of* $\mathbf{X}$ *satisfying (c) of Condition 10.2 for* $\mathcal{F} \otimes k$ *and its dual* $(\mathcal{F} \otimes k)^\vee$ *(see A3.12).*

Then $\mathrm{H}_c^{\dim(\mathbf{X})}(\mathbf{X}, \mathcal{F})$ *is* A*-free of finite rank, and* $\mathrm{H}_c^i(\mathbf{X}, \mathcal{F}) = 0$ *when* $i \neq \dim(\mathbf{X})$*.*

Proof. Let $\mathbf{X} \xrightarrow{j'} \mathbf{X}'$ be an open immersion with $\mathbf{X}'$ an affine $\mathbf{F}$-variety of dimension d. Note that d is also the dimension of $\mathbf{X}$. Let $\mathbf{X}' \xrightarrow{\overline{j}} \overline{\mathbf{X}}'$ be an open immersion with $\overline{\mathbf{X}}'$ a complete $\mathbf{F}$-variety (for instance, the projective variety associated with $\mathbf{X}'$; see A2.2). Then $\overline{j} \circ j'$ is an open immersion of $\mathbf{X}$ into a complete $\mathbf{F}$-variety and this may be used to define $\mathrm{R_c}$ (see A3.6) on $Sh(\mathbf{X}_{\text{ét}})$. The natural transformation $\overline{j}_!j'_! \to \overline{j}_*j'_*$ is the composition of the natural transformations $\overline{j}_!j'_! \to \overline{j}_*j'_! \to \overline{j}_*j'_*$. Then the natural transformation of corresponding derived functors $\overline{j}_!j'_! = (\overline{j}j')_! \to \mathrm{R}(\overline{j}j')_*$ is the composition

$$\overline{j}_!j'_! \to (\mathrm{R}\overline{j}_*)j'_! \to \mathrm{R}(\overline{j}_*j'_*).$$

Composing with $R\sigma_*$, where $\sigma:\overline{\mathbf{X}}' \to \mathrm{Spec}(\mathbf{F})$, and taking homology at $\mathcal{F}$, we get (see A3.4)

$$\mathrm{H}_c(\mathbf{X}, \mathcal{F}) \to \mathrm{H}(\mathbf{X}', j'_!\mathcal{F}) \to \mathrm{H}(\mathbf{X}, \mathcal{F})$$

where the composition is the natural map $\mathrm{H}_c(\mathbf{X}, \mathcal{F}) \to \mathrm{H}(\mathbf{X}, \mathcal{F})$. By (c) of Condition 10.2, we know that it is an isomorphism of commutative groups, so the second map

$$\mathrm{H}(\mathbf{X}', j'_!\mathcal{F}) \to \mathrm{H}(\mathbf{X}, \mathcal{F})$$

above is **onto**.

Since $\mathbf{X}'$ is affine of dimension d, the finiteness theorem (A3.7) implies that $\mathrm{H}^i(\mathbf{X}', \mathcal{F}') = 0$ as long as $i > d$ and $\mathcal{F}'$ is a constructible sheaf on $\mathbf{X}'$. We may then take $\mathcal{F}' = j'_!\mathcal{F}$ (see A3.2 and A3.3). Applying the surjection mentioned above, we get the same property for the homology of $\Gamma(\mathbf{X}, \mathcal{F})$. Since $\mathcal{F}$ has A-free stalks, $R\Gamma(\mathbf{X}, \mathcal{F})$ is therefore represented by a complex of A-free (i.e. projective) modules (see A3.15). Since it has zero homology in degrees $i > d$, it may be represented by a complex of A-free modules in degrees $\in [0, d]$, zero outside $[0, d]$ (use Exercise A1.2). The same applies to $\mathcal{F}^\vee$.

Since $\mathbf{X}$ is smooth, the Poincaré–Verdier duality (A3.12) implies that $R_c\Gamma(\mathbf{X}, \mathcal{F}) \cong R\Gamma(\mathbf{X}, \mathcal{F}^\vee)^\vee[-2d]$. By the above, this implies that $R_c\Gamma(\mathbf{X}, \mathcal{F})$ is represented by a complex C of free A-modules, zero outside $[d, 2d]$. Its homology is in the corresponding degrees but since $R_c\Gamma(\mathbf{X}, \mathcal{F} \otimes k) \cong R\Gamma(\mathbf{X}, \mathcal{F} \otimes k)$, the universal coefficient formula (see A3.8) implies that this homology is also in $[0, d]$. So eventually $\mathrm{H}(C) = \mathrm{H}^d(C)[-d]$. But since C has null terms in degree $< d$, C is homotopically equivalent to a perfect complex in only one degree d (use Exercise A1.2 again). Then $\mathrm{H}(C) = \mathrm{H}^d(C)[-d]$ and is A-free. Thus our Proposition. □

Remark 10.4. When $\mathbf{X}$ is affine, the first part of the above proof may be skipped, and the hypothesis that $\mathcal{F}^\vee$ satisfies Condition 10.2(c) can be avoided.

10.2. A first reduction

We now fix $(\mathbf{G}, F)$, $\mathbf{G}^*$, and (Λ, K, k) as in the introduction to the chapter. We assume that $s \in (\mathbf{G}^*)^F$ is a semi-simple ℓ'-element such that $\mathrm{C}_{\mathbf{G}^*}(s) \subseteq \mathbf{L}^*$ where $\mathbf{L}^*$ is a Levi subgroup in duality with $\mathbf{L}$ (see §8.2) and such that $F\mathbf{L} = \mathbf{L}$. Let $\mathbf{P} = \mathbf{VL}$ be a Levi decomposition.

Recall (Chapter 7) the Deligne–Lusztig varieties $\mathbf{Y}_\mathbf{V}^{(\mathbf{G},F)}$ and $\mathbf{X}_\mathbf{V}^{(\mathbf{G},F)}$ (often abbreviated by omitting the superscript $(\mathbf{G}, F)$) and the locally trivial $\mathbf{L}^F$-quotient map $\pi:\mathbf{Y}_\mathbf{V}^{(\mathbf{G},F)} \to \mathbf{X}_\mathbf{V}^{(\mathbf{G},F)}$ (Theorem 7.8).

Definition 10.5. *Let* $\mathcal{F} = \pi_*^{\mathbf{L}^F}\Lambda_{\mathbf{Y_V}}$, *a sheaf of* $\Lambda\mathbf{L}^F$*-modules on* $\mathbf{X_V}$ *(see A3.14). Let* $\mathcal{F}_s = \mathcal{F}.b_\ell(\mathbf{L}^F, s)$ *(see Definition 9.9). Denote* $\mathcal{F}_s^k = \mathcal{F}_s \otimes_\Lambda k$.

The following is then clear from the definition of twisted induction (see §8.3).

Lemma 10.6. *Let* $n \geq 1$; *then* $\mathrm{R_c\Gamma}(\mathbf{X_V}, \mathcal{F}_s \otimes_\Lambda \Lambda/J(\Lambda)^n) = \mathrm{R_c\Gamma}(\mathbf{Y_V}, \Lambda/J(\Lambda)^n).b_\ell(\mathbf{L}^F, s)$ *and may be considered as an object of* $D^b(\Lambda/J(\Lambda)^n\mathbf{G}^F-\mathbf{mod}-\Lambda/J(\Lambda)^n\mathbf{L}^F)$. *The limit (over n) of its homology induces the* $\mathrm{R}^{\mathbf{G}}_{\mathbf{L}\subseteq\mathbf{P}}$ *functor.*

Here is our first main reduction.

Theorem 10.7. *Let* $\overline{\mathbf{X}}_\mathbf{V}$ *be the Zariski closure of* $\mathbf{X_V}$ *in* $\mathbf{G/P}$, *and* $j_\mathbf{V}: \mathbf{X_V} \to \overline{\mathbf{X}}_\mathbf{V}$ *be the associated open immersion. Assume that* $(\mathbf{X_V}, j_\mathbf{V}, \mathcal{F}_s^k)$ *and* $(\mathbf{X_V}, j_\mathbf{V}, \mathcal{F}_{s^{-1}}^k)$ *satisfy Condition 10.2.*

Then there is a Morita equivalence

$$\Lambda\mathbf{L}^F.b_\ell(\mathbf{L}^F, s)-\mathbf{mod} \xrightarrow{\sim} \Lambda\mathbf{G}^F.b_\ell(\mathbf{G}^F, s)-\mathbf{mod}.$$

Proof. The Λ-duality functor permutes the blocks of $\Lambda\mathbf{G}^F$ and sends $b_\ell(\mathbf{G}^F, s)$ to $b_\ell(\mathbf{G}^F, s^{-1})$ since the conjugate of the generalized character $\mathrm{R}^{\mathbf{G}}_{\mathbf{T}}\theta$ is $\mathrm{R}^{\mathbf{G}}_{\mathbf{T}}\theta^{-1}$ (exercise: use the character formula or the original definition), and $(\mathbf{T}, \theta^{-1})$ corresponds to s^{-1} when $(\mathbf{T}, \theta)$ corresponds to s (see §8.2). So $(\mathcal{F}_s)^\vee \cong \mathcal{F}_{s^{-1}}$ and $(\mathcal{F}_s^k)^\vee \cong \mathcal{F}_{s^{-1}}^k$.

Denote by $i_\mathbf{V}: \overline{\mathbf{X}}_\mathbf{V} \setminus \mathbf{X_V} \to \mathbf{X_V}$ the closed immersion associated with $j_\mathbf{V}$.

Denote $\Lambda^{(n)} := \Lambda/J(\Lambda)^n$ and $\mathcal{F}_s^{(n)} := \mathcal{F}_s \otimes_\Lambda \Lambda^{(n)}$.

The variety $\mathbf{X_V}$ is smooth (Theorem 7.7) and quasi-affine (Theorem 7.15). Moreover $\overline{\mathbf{X}}_\mathbf{V}$ is complete, being closed in the complete variety $\mathbf{G/P}$ (see A2.6). We may apply Proposition 10.3. This ensures that $\mathrm{R_c\Gamma}(\mathbf{X_V}, \mathcal{F}_s^{(n)}) \in D^b(\Lambda^{(n)}-\mathbf{mod})$ is represented by a complex in the single degree $d := \dim(\mathbf{X_V}) = \dim(\mathbf{Y_V})$ which is moreover $\Lambda^{(n)}$-free. Both $\mathbf{G}^F$ and $\mathbf{L}^F$ act on it by A3.14. Let us show that it is projective as a $\Lambda^{(n)}\mathbf{G}^F$-module (and as a $\Lambda^{(n)}\mathbf{L}^F$-module).

The stabilizers of closed points of $\mathbf{Y_V}$ in $\mathbf{G}^F$ are intersections with $\mathbf{G}^F$ of conjugates of $\mathbf{V}$, since $\mathbf{Y_V} \subseteq \mathbf{G/V}$. So they are finite p-groups. By A3.15, this implies that $\mathrm{R_c\Gamma}(\mathbf{Y_V}, \Lambda^{(n)}) \in D^b(\Lambda^{(n)}\mathbf{G}^F-\mathbf{mod})$ can be represented by a complex of projective $\Lambda^{(n)}\mathbf{G}^F$-modules. So $\mathrm{H}_c^d(\mathbf{X_V}, \mathcal{F}_s^{(n)}) = \mathrm{H}_c^d(\mathbf{Y_V}, \Lambda^{(n)}).b_\ell(\mathbf{G}^F, s)$ is a projective $\Lambda^{(n)}\mathbf{G}^F$-module (apply Exercise A1.4(c) with $m = m'$). Similarly for the (free) right action of $\mathbf{L}^F$. So $\mathrm{H}_c^d(\mathbf{X_V}, \mathcal{F}_s^{(n)})$ is a projective right $\Lambda^{(n)}\mathbf{L}^F$-module.

The projective limit $\lim_n \mathrm{H}_c^d(\mathbf{X_V}, \mathcal{F}_s^{(n)})$ is both a projective $\Lambda\mathbf{G}^F$-module and a projective $\Lambda\mathbf{L}^F$-module, its rank being the same for all n. To complete our proof, we must show that Theorem 9.18 may be applied, taking

$A = \Lambda\mathbf{G}^F.b_\ell(\mathbf{G}^F, s)$, $B = \Lambda\mathbf{L}^F.b_\ell(\mathbf{L}^F, s)$ and $M = \lim_n \mathrm{H}_c^d(\mathbf{X_V}, \mathcal{F}_s^{(n)})$. This is a matter of looking at $M_K \otimes_{\mathbf{L}^F} -$ on simple $K\mathbf{L}^F$-modules (see also the proof of Corollary 9.19). On characters of $\mathbf{L}^F$, $M_K \otimes_{\mathbf{L}^F} -$ is by definition the projection on $K\mathbf{L}^F.b_\ell(\mathbf{L}^F, s)$ followed by the twisted induction $\mathrm{R}^{\mathbf{G}}_{\mathbf{L}\subseteq\mathbf{P}}$ times the sign $(-1)^d$ (see Lemma 10.6 above). Then Theorem 9.16 tells us at the same time that $M_K \otimes_{\mathbf{L}^F} -$ sends the simple $K\mathbf{L}^F.b_\ell(\mathbf{L}^F, s)$-modules into simple $K\mathbf{G}^F.b_\ell(\mathbf{G}^F, s)$-modules, and that it bijects them. Thus our theorem. □

10.3. More notation: smooth compactifications

We keep $(\mathbf{G}, F)$ a connected reductive $\mathbf{F}$-group defined over $\mathbb{F}_q$. Let us fix $\mathbf{B}_0 \supseteq \mathbf{T}_0$ a Borel subgroup and maximal torus, both F-stable (see Theorem 7.1(iii)). Let $\mathbf{U}_0$ be the unipotent radical of $\mathbf{B}_0$. Denote by S the set of generating reflections of $W(\mathbf{G}, \mathbf{T}_0)$ associated with $\mathbf{B}_0$.

Notation 10.8. Let $\Sigma(S)$ be the set of finite (possibly empty) sequences of elements of $S \cup \{1\}$. We often abbreviate $\Sigma(S) = \Sigma$. We denote by Σ_{red} the subset consisting of reduced decompositions.

Concatenation is denoted by $w \cup w'$ for $w, w' \in \Sigma$. This monoid acts on $\mathbf{T}_0$ through the evident map $\Sigma \to W(\mathbf{G}, \mathbf{T}_0)$. Recalling the map $w \mapsto \dot{w}$ from W to $\mathrm{N}_{\mathbf{G}}(\mathbf{T}_0)$ (see Theorem 7.11), we define $\mathbf{T}_0^{wF} \subseteq \mathbf{T}_0$ as $\{t \in \mathbf{T}_0 \mid \dot{s}_1 \ldots \dot{s}_r F(t) \dot{s}_r^{-1} \ldots \dot{s}_1^{-1} = t\}$ for $w = (s_1, \ldots, s_r)$. (If, moreover, $s_1 \ldots s_r$ is a reduced expression, the product $\dot{s}_1 \ldots \dot{s}_r$ depends only on $s_1 \ldots s_r$, thus allowing us to define wF as an automorphism of any F-stable subgroup containing $\mathbf{T}_0$.)

We denote by $l(w)$ the number of indices i such that $s_i \neq 1$. If $w' = (s'_1, \ldots, s'_{r'}) \in \Sigma$, denote $w' \leq w$ if and only if $r = r'$ and, for all i, $s'_i \in \{1, s_i\}$.

If $X, Y \in \mathbf{G}/\mathbf{U}_0$ (resp. $\mathbf{G}/\mathbf{B}_0$) and $w \in W$, denote $X \xrightarrow{w} Y$ if and only if $X^{-1}Y = \mathbf{U}_0 \dot{w} \mathbf{U}_0$ (resp. $X^{-1}Y = \mathbf{B}_0 w \mathbf{B}_0$).

When $w = (s_1, \ldots, s_r) \in \Sigma$, denote by $\mathbf{Y}(w)$ (resp. $\mathbf{X}(w)$) the set of r-tuples $(Y_1, \ldots, Y_r) \in (\mathbf{G}/\mathbf{U}_0)^r$ (resp. $(\mathbf{G}/\mathbf{B}_0)^r$) such that

$$Y_1 \xrightarrow{s_1} Y_2 \xrightarrow{s_2} \ldots \xrightarrow{s_{r-1}} Y_r \xrightarrow{s_r} F(Y_1).$$

Let $\overline{\mathbf{X}}(w) := \bigcup_{w' \leq w} \mathbf{X}(w')$. When $w' \leq w$ in Σ, denote $j_{w'}^{\overline{w}} \colon \mathbf{X}(w') \to \overline{\mathbf{X}}(w)$.

Let us show how to interpret the above varieties associated with $w \in \Sigma$ as examples of the varieties defined in Chapter 7.

Consider now, for some integer r, the group $\mathbf{G}^{(r)} := \mathbf{G} \times \cdots \times \mathbf{G}$ (r times) with the following endomorphism

$$F_r \colon \mathbf{G}^{(r)} \longrightarrow \mathbf{G}^{(r)}$$
$$(g_1, g_2, \ldots, g_r) \mapsto (g_2, \ldots, g_r, F(g_1)).$$

Since $\mathbf{G}$ is defined over $\mathbb{F}_q$ by F, $\mathbf{G}^{(r)}$ is defined over $\mathbb{F}_q$ by $(F_r)^r$. Furthermore $\mathbf{G}^F$ is isomorphic to $(\mathbf{G}^{(r)})^{F_r}$ by the diagonal morphism (but $(\mathbf{G}^{(r)})^{(F_r)^r} \cong (\mathbf{G}^F)^r$). We identify the Weyl group of $\mathbf{G}^{(r)}$ with respect to $\mathbf{T}_0^{(r)}$ with $W(\mathbf{G},\mathbf{T}_0)^r$, as well as the variety $\mathbf{G}^{(r)}/\mathbf{B}^{(r)}$ with $(\mathbf{G}/\mathbf{B})^r$, and so on.

If $w \in (S \cup \{1\})^r$, one may consider it as an element of $W(\mathbf{G}^{(r)},\mathbf{T}_0^{(r)})$ and form the varieties of Definition 7.12. Note that w is a product of pairwise commuting generators of the Weyl group of $\mathbf{G}^{(r)}$ with respect to $\mathbf{T}_0^{(r)}$. Using the above group $\mathbf{G}^{(r)}$ and morphism $F_r\colon \mathbf{G}^{(r)} \to \mathbf{G}^{(r)}$, then $\mathbf{Y}^{(\mathbf{G}^{(r)},F_r)}(w)$ and $\mathbf{X}^{(\mathbf{G}^{(r)},F_r)}(w)$ are the varieties defined above and denoted by $\mathbf{Y}(w)$, $\mathbf{X}(w)$ respectively. Note that the commuting actions of the finite groups $(\mathbf{G}^{(r)})^{F_r}$ and $(\mathbf{T}_0^{(r)})^{wF_r}$ on $\mathbf{Y}^{(\mathbf{G}^{(r)},F_r)}(w)$ may be identified with actions of $\mathbf{G}^F$ and $\mathbf{T}_0^{wF}$ (see Notation 10.8) on $\mathbf{Y}(w)$. For the action of $\mathbf{G}^F$ this is the isomorphism $(\mathbf{G}^{(r)})^{F_r} \cong \mathbf{G}^F$ mentioned above. For the action of $\mathbf{T}_0^{wF}$, we have clearly

Lemma 10.9. *Let $w = (w_j)_{1\le j\le r} \in W(\mathbf{G},\mathbf{T}_0)^r$, denote $\mathbf{T}_0^{wF} = \mathbf{T}_0^{w_1\ldots w_rF}$ and define $\iota_w\colon \mathbf{T}_0 \to \mathbf{T}_0^r$ by*

$$\iota_w(t) = (t, {w_1}^{-1}tw_1, \ldots, (w_1\ldots w_{r-1})^{-1}tw_1\ldots w_{r-1}) \quad (t \in \mathbf{T}_0).$$

One has $\iota_w(\mathbf{T}_0^{wF}) = (\mathbf{T}_0^r)^{wF_r}$. Hence ι_w defines an isomorphism $\mathbf{T}_0^{wF} \to (\mathbf{T}_0^r)^{wF_r}$.

Theorem 7.8 and Proposition 7.13 now give

Proposition 10.10. *Let $w \in \Sigma$. Then*

(i) $\mathbf{Y}(w)$ (and $\mathbf{X}(w)$) are smooth, quasi-affine, of dimension $l(w)$,

(ii) $\mathbf{Y}(w) \to \mathbf{X}(w)$ is a Galois covering of group $\mathbf{T}_0^{wF}$,

(iii) $\overline{\mathbf{X}}(w)$ is closed in $(\mathbf{G}/\mathbf{B}_0)^r$,

(iv) the $\overline{\mathbf{X}}(w')$ for $w' \le w$ and $l(w') = l(w) - 1$ form a smooth divisor with normal crossings making $\overline{\mathbf{X}}(w) \setminus \mathbf{X}(w)$.

The next proposition needs a little more work. We show how the transitivity theorem on varieties $\mathbf{Y}_\mathbf{V}$ (see Theorem 7.9) applies to varieties $\mathbf{Y}(w)$.

Definition 10.11. *If $I \subseteq S$ and $v \in W(\mathbf{G},\mathbf{T}_0)$ is such that $vF(I)v^{-1} = I$, let $\mathbf{Y}_{I,v} = \{g\mathbf{U}_I \mid g^{-1}F(g) \in \mathbf{U}_I\dot{v}F(\mathbf{U}_I)\} \subseteq \mathbf{G}/\mathbf{U}_I$, a variety with a $\mathbf{G}^F \times (\mathbf{L}_I)^{\dot{v}F}$-action, and $\mathbf{X}_{I,v} = \{g\mathbf{P}_I \mid g^{-1}F(g) \in \mathbf{P}_I\dot{v}F(\mathbf{P}_I)\} \subseteq \mathbf{G}/\mathbf{P}_I$.*

Note that, if $a \in \mathbf{G}$ is such that $a^{-1}F(a) = \dot{v}$, then $a.\mathbf{Y}_{I,v} = \mathbf{Y}_{\mathbf{U}_I}^{(\mathbf{G},\dot{v}F)}$ (see Definition 7.6). This left translation by a, $\mathbf{Y}_{I,v} \to \mathbf{Y}_{\mathbf{U}_I}^{(\mathbf{G},\dot{v}F)}$, transforms $\mathbf{G}^F \times {\mathbf{L}_I}^{\dot{v}F}$-action into $\mathbf{G}^{\dot{v}F} \times {\mathbf{L}_I}^{\dot{v}F}$-action.

Proposition 10.12. *Let I, v, $\mathbf{Y}_{I,v}$ be as above. Let $d_v \in \Sigma$ be a minimal decomposition of v. Let $w \in (I \cup \{1\})^r$. Then we have an isomorphism*

$$(\mathbf{Y}_{I,v} \times \mathbf{Y}^{\mathbf{L}_I,\dot{v}F}(w))/\mathbf{L}_I^{\dot{v}F} \to \mathbf{Y}(w \cup d_v)$$

uniquely defined on $\mathbf{Y}_{I,v} \times \mathbf{Y}^{\mathbf{L}_I, \dot{v}F}(w)$ *by*

$$(\mathbf{x}, (\mathbf{V}_1, \mathbf{V}_2, \ldots; \mathbf{V}_r)) \mapsto (\mathbf{V}'_1, \ldots, \mathbf{V}'_{r+l(v)}),$$

with $\mathbf{V}'_i = \mathbf{x}\mathbf{V}_i$ *for* $i \leq r$.

As a consequence, $\mathbf{Y}^{(\mathbf{G},F)}(w) \cong \mathbf{Y}^{(\mathbf{G},F)}(d_w)$ *for any* $w \in W(\mathbf{G}, \mathbf{T}_0)$ *and* $d_w \in S^{l(w)}$ *a reduced decomposition of* w.

Proof. Note that, if $I = \emptyset$, then $\mathbf{U}_I = \mathbf{U}_0$, $\mathbf{Y}_{I,v} = \mathbf{Y}^{(\mathbf{G},F)}(v)$, $\mathbf{L}_I = \mathbf{T}_0$, $w = 1$ and $\mathbf{Y}^{(\mathbf{L}_I,\dot{v}F)}(w)$ is a $\mathbf{T}_0^{\dot{v}F}$-orbit. The isomorphism indeed reduces to $\mathbf{Y}^{(\mathbf{G},F)}(v) \cong \mathbf{Y}^{(\mathbf{G}^{(l)},F_l)}(d_v) = \mathbf{Y}(d_v)$ where l is the length of v, whence the last assertion.

Let us now return to the general case with $w = (s_1, \ldots, s_r) \in (I \cup \{1\})^r$. Note that $l(s_r v) = l(s_r) + l(v)$ since v^{-1} sends the simple roots corresponding to I to positive roots. Define

$$\mathbf{Y}' = \{(g_1\mathbf{U}_0, \ldots, g_r\mathbf{U}_0) \mid g_1\mathbf{U}_0 \xrightarrow{s_1} g_2\mathbf{U}_0 \xrightarrow{s_2} \ldots g_r\mathbf{U}_0 \xrightarrow{s_r v} F(g_1)\mathbf{U}_0\}$$
$$\subseteq (\mathbf{G}/\mathbf{U}_0)^r.$$

By the arguments used above for the varieties $\mathbf{Y}(w)$ ($w \in \Sigma$), it is a locally closed subvariety in a variety of the type defined in Chapter 7 for the group $\mathbf{G}^{(r)}$ (with $w \mapsto \dot{w}$ defined on the whole of $W(\mathbf{G}^{(r)}, \mathbf{T}_0^{(r)}) = W(\mathbf{G}, \mathbf{T}_0)^r$ instead of just $(S \cup \{1\})^r$), hence smooth of dimension $l(w) + l(v)$ (see Proposition 7.13).

The existence of a natural bijective map $\mathbf{Y}(w \cup d_v) \to \mathbf{Y}'$ is given by the following lemma.

Lemma 10.13. *Assume* $l(w_1w_2) = l(w_1) + l(w_2)$ *in* $W(\mathbf{G}, \mathbf{T}_0)$. *Then* $x \xrightarrow{w_1w_2} y$ *in* $\mathbf{G}/\mathbf{U}_0$, *if and only if there is a* $z \in \mathbf{G}/\mathbf{U}_0$ *such that* $x \xrightarrow{w_1} z \xrightarrow{w_2} y$. *This* z *is unique.*

Proof. Denote $w_3 := w_1w_2$. It is enough to check that $\mathbf{U}_0\dot{w}_1\mathbf{U}_0\dot{w}_2\mathbf{U}_0 = \mathbf{U}_0\dot{w}_3\mathbf{U}_0$. Since $\dot{w}_3 = \dot{w}_1\dot{w}_2$ (see Theorem 7.11), it suffices to check $\mathbf{U}_0\dot{w}_1\mathbf{U}_0\dot{w}_2\mathbf{U}_0 = \mathbf{U}_0\dot{w}_1\dot{w}_2\mathbf{U}_0$. This is a consequence of $\mathbf{U}_0 = (\mathbf{U}_0 \cap \mathbf{U}_0^{\dot{w}_1})(\mathbf{U}_0 \cap {}^{\dot{w}_2}\mathbf{U}_0)$ which in turn can be deduced from the corresponding partition of positive roots (see Proposition 2.3(iii)) as in the finite case — or even as a limit of the finite case. □

As a result of Lemma 10.13, the map $(g_1\mathbf{U}_0, \ldots, g_{r+l(v)}\mathbf{U}_0) \mapsto (g_1\mathbf{U}_0, \ldots, g_r\mathbf{U}_0)$ is a bijective map $\mathbf{Y}(w \cup d_v) \to \mathbf{Y}'$. As a bijective morphism, clearly separable, between smooth varieties, one obtains an isomorphism (see A2.6).

It remains to check that multiplication induces an isomorphism $(\mathbf{Y}_{I,v} \times \mathbf{Y}^{\mathbf{L}_I,\dot{v}F}(w))/\mathbf{L}_I^{\dot{v}F} \to \mathbf{Y}'$.

Consider $\dot{v}' = (1, 1, \ldots, 1, \dot{v}) \in W(\mathbf{G}, \mathbf{T}_0)^r$. One has $\dot{v}'F_r = (\dot{v}F)_r$ as endomorphisms of $\mathbf{G}^{(r)}$. The diagonal morphism $g\mathbf{U}_I \mapsto (g\mathbf{U}_I, \ldots, g\mathbf{U}_I) \in \mathbf{G}^{(r)}/{\mathbf{U}_I}^r$ restricts to an isomorphism of varieties $\mathbf{Y}_{I,\dot{v}} \to \mathbf{Y}_{I^r,\dot{v}'}$, which preserves $\mathbf{G}^F \times {\mathbf{L}_I}^{\dot{v}F} \cong (\mathbf{G}^{(r)})^{F_r} \times (\mathbf{L}_{I^r})^{(\dot{v}F)_r}$-actions.

Let $a \in \mathbf{G}^{(r)}$ be such that $aF_r(a)^{-1} = \dot{v}'$. We have seen that $a\mathbf{Y}_{I^r,\dot{v}'} = \mathbf{Y}_{\mathbf{U}_{I^r}}^{\mathbf{G}^r,(\dot{v}F)_r}$. For $b \in {\mathbf{L}_I}^r$ such that $b(\dot{v}F)_r(b)^{-1} = \dot{w}$, one has $\mathbf{Y}^{({\mathbf{L}_I}^r,(\dot{v}F)_r)}(w)b = \mathbf{Y}_{\mathbf{V}'}^{(\mathbf{L}_I^r,(\dot{v}F)_r)}$ where $\mathbf{V}' = b^{-1}(\mathbf{L}_{I^r} \cap \mathbf{U}_0^r)b$ (Proposition 7.13(ii)). Then $\mathbf{U}_I\mathbf{V}' = b^{-1}\mathbf{U}_0^r b$. Theorem 7.9 applies, there is an isomorphism, given by multiplication,

$$\mathbf{Y}_{\mathbf{U}_{I^r}}^{\mathbf{G}^r,(\dot{v}F)_r} \times \mathbf{Y}_{\mathbf{V}'}^{(\mathbf{L}_I^r,(\dot{v}F)_r)}/({\mathbf{L}_I}^r)^{(\dot{v}F)_r} \cong \mathbf{Y}_{b^{-1}\mathbf{U}_0^r b}^{(\mathbf{G}^r,(\dot{v}F)_r)}.$$

The initial product is then isomorphic to $a^{-1}\mathbf{Y}_{b^{-1}\mathbf{U}_0^r b}^{(\mathbf{G}^r,(\dot{v}F)_r)}b^{-1}$. But $baF_r(ba)^{-1} = \dot{w}\dot{v}'$. So $(ba)^{-1}b\mathbf{Y}_{b^{-1}\mathbf{U}_0^r b}^{(\mathbf{G}^r,(\dot{v}F)_r)}b^{-1} = (ba)^{-1}\mathbf{Y}_{\mathbf{U}_0}^{(\mathbf{G}^r,\dot{w}\dot{v}'F_r)} = \mathbf{Y}'$ by Proposition 7.13(ii). □

10.4. Ramification and generation

We now give some notation in order to state Theorem 10.17. We keep $(\mathbf{G}, F)$ a connected reductive $\mathbf{F}$-group defined over $\mathbb{F}_q$. We keep $(\mathbf{G}^*, F)$ in duality with $(\mathbf{G}, F)$ (see Chapter 8). We recall that $\mathbf{T}_0 \subseteq \mathbf{B}_0$ are a maximal torus and a Borel subgroup, both F-stable. We introduce the sheaves $\mathcal{F}_w(M)$ on $\mathbf{X}(w)$, for M a $k\mathbf{T}_0^{wF}$-module, and the related complexes of $k\mathbf{G}^F$-modules $\mathcal{S}_{(w,\theta)}$.

Definition 10.14. *Let $\Theta(\mathbf{G}, F)$ denote the set of pairs (w, θ) where $w \in \Sigma$ (see Notation 10.8) and θ is a linear character $\mathbf{T}_0^{wF} \to k^\times$.*

Since k is big enough so that we get as θ all possible ℓ'-characters of all finite groups $\mathbf{T}_0^{wF}$ ($w \in W$), the partition into rational series (see §8.4) implies a partition $\Theta(\mathbf{G}, F) = \cup_s \Theta(\mathbf{G}, F, s)$ where s ranges over $(\mathbf{G}^*)^F$-conjugacy classes of semi-simple ℓ'-elements of $(\mathbf{G}^*)^F$.

Definition 10.15. *If $w = (s_1, \ldots, s_r) \in \Sigma$, let, for each i such that $s_i \neq 1$, α_i be the image by $s_1 \ldots s_{i-1}$ of the positive root corresponding to s_i. Define $w_\theta := (s_1', \ldots, s_r') \in \Sigma$ by $s_i' = 1$ if $s_i \neq 1$ and $\theta \circ N_{s_1\ldots s_r}(\alpha_i^\vee) = 1$, let $s_i' = s_i$ otherwise (see (8.11) and (8.12) for the definition of $N_v\colon Y(\mathbf{T}_0) \to \mathbf{T}_0^{vF}$ for $v \in W(\mathbf{G}, \mathbf{T}_0)$).*

Definition 10.16. *Let $(w, \theta) \in \Theta(\mathbf{G}, F)$. Define k_θ as the one-dimensional $k\mathbf{T}_0^{wF}$-module corresponding to θ. Let $b_\theta := |\mathbf{T}_0^{wF}|_{\ell'}^{-1} \sum_{t\in(\mathbf{T}_0^{wF})_{\ell'}} \theta(t)t^{-1} \in k\mathbf{T}_0^{wF}$, i.e. the primitive idempotent of $k\mathbf{T}_0^{wF}$ acting non-trivially on k_θ. Since $\mathbf{Y}(w)$ is a variety with right $\mathbf{T}_0^{wF}$-action, with associated quotient $\pi\colon \mathbf{Y}(w) \to$*

$\mathbf{X}(w)$, the direct image $\pi_* k_{\mathbf{Y}(w)}$ may be considered as a sheaf of (right) $k\mathbf{T}_0^{wF}$-modules on $\mathbf{X}(w)$ (a locally constant sheaf; see A3.16). One defines

$$\mathcal{F}_w : k\mathbf{T}_0^{wF}\mathbf{-mod} \to Sh_{k\mathbf{G}^F}(\mathbf{X}(w)_{\text{ét}})$$

by $\mathcal{F}_w(M) = \pi_*^{\mathbf{T}_0^{wF}} k_{\mathbf{Y}(w)} \otimes_{k\mathbf{T}_0^{wF}} M_{\mathbf{X}(w)} = \pi_*^{\mathbf{T}_0^{wF}} k_{\mathbf{Y}(w)} \overset{\mathrm{L}}{\otimes}_{k\mathbf{T}_0^{wF}} M_{\mathbf{X}(w)}$ *(notation of A3.14).*

We abbreviate $\mathcal{F}_w(k_\theta) = \mathcal{F}_w(\theta)$ *and* $\mathcal{F}_{(w,\theta)} = \mathcal{F}_w(k\mathbf{T}_0^{wF} b_\theta) = (\pi_* k_{\mathbf{Y}(w)}) b_\theta \in Sh_k(\mathbf{X}(w))$.

Since the quasi-projective variety $\mathbf{Y}(w)$ is acted on by $\mathbf{G}^F \times (\mathbf{T}_0^{wF})^{\mathrm{opp}}$, A3.14 tells us that $\mathrm{R}\Gamma(\mathbf{Y}(w), k_{\mathbf{Y}(w)})$ is represented by a complex of $k\mathbf{G}^F$-$k\mathbf{T}_0^{wF}$-bimodules. So we may define $\mathcal{S}_{(w,\theta)} := \mathrm{R}\Gamma(\mathbf{Y}(w), k_{\mathbf{Y}(w)}).b_\theta$ as an object of $D^b(k\mathbf{G}^F\mathbf{-mod})$.

Keep the hypotheses and notation of the above definitions. Note that the following theorem does not mention Levi subgroups or the condition $\mathrm{C}_{\mathbf{G}^*}(s) \subseteq \mathbf{L}^*$. For the notion of a subcategory of a derived category generated by a set of objects, see A1.7.

Theorem 10.17. *(a) (Ramification) Let $w' \leq w$ in Σ and $\theta : \mathbf{T}_0^{wF} \to k^\times$. Then $(j_{w'}^{\overline{w}})^* \mathrm{R}(j_w^{\overline{w}})_* \mathcal{F}_w(\theta) = 0$ unless $w_\theta \leq w'$.*

(b) (Generation) Let s be a semi-simple ℓ'-element of $(\mathbf{G}^*)^F$. The subcategory of $D^b(k\mathbf{G}^F)$ generated by the $\mathcal{S}_{(w,\theta)}$'s for $(w,\theta) \in \Theta(\mathbf{G}, F, s)$ contains $k\mathbf{G}^F.b_\ell(\mathbf{G}^F, s)$ (see Definition 9.9).

10.5. A second reduction

We now show that Theorem 10.1 reduces to the above Theorem 10.17, that is a question on the varieties $\mathbf{X}(w)$, i.e. varieties $\mathbf{X}_\mathbf{V}$ where $\mathbf{V}$ is a unipotent radical of a Borel subgroup.

Theorem 10.18. *Theorem 10.17 implies Theorem 10.1.*

Proof. Let $\mathbf{P} = \mathbf{V} \rtimes \mathbf{L}$ with $F\mathbf{L} = \mathbf{L}$, and $\mathbf{L}$ in duality with $\mathbf{L}^*$ such that $\mathrm{C}_{\mathbf{G}^*}(s) \subseteq \mathbf{L}^*$ as in Theorem 10.1. In view of Theorem 10.7, it suffices to check that $\mathbf{X}_\mathbf{V}$, the immersions $i_\mathbf{V} : \overline{\mathbf{X}}_\mathbf{V} \setminus \mathbf{X}_\mathbf{V} \to \overline{\mathbf{X}}_\mathbf{V}$, $j_\mathbf{V} : \mathbf{X}_\mathbf{V} \to \overline{\mathbf{X}}_\mathbf{V}$, and $\mathcal{F}_s^k$ satisfy Condition 10.2, i.e.

$$i_\mathbf{V}^* \mathrm{R}(j_\mathbf{V})_* \mathcal{F}_s^k = 0.$$

We give the proof of the following at the end of the chapter.

Proposition 10.19. *Assume $\mathbf{P} = \mathbf{LV}$ is a Levi decomposition with $F\mathbf{L} = \mathbf{L}$. Let $v \in W(\mathbf{G}, \mathbf{T}_0)$, $I \subseteq \Delta$, $a \in \mathrm{Lan}^{-1}(\dot{v})$ such that $v^{-1}(I) \subseteq \Delta$, ${}^a\mathbf{V} = \mathbf{U}_I$,*

${}^a\mathbf{L} = \mathbf{L}_I$, and $x \mapsto {}^a x$ induces an isomorphism $\mathbf{Y}_{\mathbf{V}}^{(\mathbf{G},F)} \to \mathbf{Y}_{\mathbf{U}_I}^{(\mathbf{G},\dot{v}F)}$ (see Proposition 7.13(i)). Let $s \in (\mathbf{L}^*)^F$ be a semi-simple element in a Levi subgroup of $\mathbf{G}^*$ in duality with $\mathbf{L}$ and such that $\mathrm{C}_{\mathbf{G}^*}(s) \subseteq \mathbf{L}^*$. Then, for all $w \in \Sigma(I)$ and $\theta\colon \mathbf{T}_0^{wvF} \to k^\times$ such that $(w, \theta) \in \Theta(\mathbf{L}_I, d_v F, s)$ (see Definition 10.14), we have $(w \cup d_v)_\theta = w_\theta \cup d_v$, where $(w \cup d_v)_\theta$ is computed in $\mathbf{G}$ relative to F, and w_θ is computed in $\mathbf{L}_I$ relative to $d_v F$.

In view of the above, it suffices to show the theorem with $(\mathbf{G}, F)$ replaced by $(\mathbf{G}, \dot{v}F)$ and $(\mathbf{P}, \mathbf{L})$ by $(\mathbf{P}_I, \mathbf{L}_I)$. We use the same notation for $\mathcal{F}_s$ but omit the exponent.

We abbreviate $L := \mathbf{L}_I^{\dot{v}F}$. We denote by i, j the closed and open immersions associated with $\mathbf{X}_{I,v} \subseteq \overline{\mathbf{X}}_{I,v}$ where $\overline{\mathbf{X}}_{I,v}$ denotes the Zariski closure of $\mathbf{X}_{I,v}$ in $\mathbf{G}/\mathbf{P}_I$ (see Definition 10.11). We must check

$$i^*\mathrm{R}j_*\mathcal{F}_s = 0. \tag{1}$$

Let $\pi^L\colon \mathbf{Y}_{I,v} \to \mathbf{X}_{I,v}$ be the map defined by $\pi^L(g\mathbf{U}_I) = g\mathbf{P}_I$, an L-quotient map by Theorem 7.8 and Definition 10.11. We have $\mathcal{F}_s = (\pi^L_* k_{\mathbf{Y}_{I,v}}).b_\ell(\mathbf{L}^F, s)$ where $\pi^L_* k_{\mathbf{Y}_{I,v}} \in D^b_{kL}(\mathbf{X}_{I,v})$. So our claim will result from checking that

$$i^*\mathrm{R}j_*\big((\pi^L_* k_{\mathbf{Y}_{I,v}}) \overset{\mathrm{L}}{\otimes}_{kL} M\big) = 0 \tag{1_M}$$

for the $kL.b_\ell(L, s)$-module $M = kL.b_\ell(L, s)$. The above, seen as a functor $D^b(kL\mathbf{-mod}) \to D^b_k(\overline{\mathbf{X}}_{I,v} \setminus \mathbf{X}_{I,v})$, preserves distinguished triangles as a composition of derived functors. So Theorem 10.17(b) applied to $(\mathbf{L}_I, \dot{v}F)$ implies that it suffices to check

$$i^*\mathrm{R}j_*\big((\pi^L_* k_{\mathbf{Y}_{I,v}}) \overset{\mathrm{L}}{\otimes}_{kL} \mathcal{S}^{(\mathbf{L}_I,\dot{v}F)}_{(w,\theta)}\big) = 0 \tag{2}$$

for any $(w, \theta) \in \Theta(\mathbf{L}_I, \dot{v}F, s)$ (see A1.7 and A1.8).

For the remainder of the proof of the theorem, we fix such a $(w, \theta) \in \Theta(\mathbf{L}_I, \dot{v}F, s)$.

Lemma 10.20. *Assume $w \in (I \cup \{1\})^r$. Let $\tau\colon \mathbf{X}^{(\mathbf{G},F)}(w \cup d_v) \to \mathbf{X}_{I,v}$ be the map defined by $\tau(g_1\mathbf{B}_0, \ldots, g_{r+l(v)}\mathbf{B}_0) = g_1\mathbf{P}_I$. Then $\mathrm{R}\tau_*(\mathcal{F}_{(w\cup d_v,\theta)}) \cong (\pi^L_* k_{\mathbf{Y}_{I,v}}) \otimes_{kL} \mathcal{S}^{(\mathbf{L}_I,\dot{v}F)}_{(w,\theta)}$.*

Proof of Lemma 10.20. Let

$$\mathbf{Y}' := \mathbf{Y}_{I,v} \times \mathbf{Y}^{(\mathbf{L}_I,\dot{v}F)}(w) \xrightarrow{\pi'} \mathbf{Y}(w \cup d_v) \xrightarrow{\pi''} \mathbf{X}(w \cup d_v) \xrightarrow{\tau} \mathbf{X}_{I,v}$$

where π' and π'' are defined by Proposition 10.12 and Proposition 10.10(ii) respectively. Then, π' being a L-quotient, we have $\mathrm{R}\tau_* k_{\mathbf{X}(w\cup d_v)} = \mathrm{R}(\tau\pi'')_*(\pi'_* k_{\mathbf{Y}'} \otimes_{kL} k)$ (see equation (1) in A3.15).

The above composite also decomposes as

$$\mathbf{Y}_{I,v} \times \mathbf{Y}^{(\mathbf{L}_I,\dot{v}F)}(w) \xrightarrow{\pi^L \times \mathrm{Id}} \mathbf{X}_{I,v} \times \mathbf{Y}^{(\mathbf{L}_I,\dot{v}F)}(w) \xrightarrow{\sigma} \mathbf{X}_{I,v}$$

where σ is the first projection. Then $\mathrm{R}(\tau\pi''\pi')_* k_{\mathbf{Y}'} = \mathrm{R}(\sigma(\pi^L \times \mathrm{Id}))_* k_{\mathbf{Y}'} = (\mathrm{R}\sigma_*)\big((\pi^L \times \mathrm{Id})_* k_{\mathbf{Y}'}\big)$ by the usual properties of direct images (see A3.4). We have $(\pi^L \times \mathrm{Id})_* k_{\mathbf{Y}'} = \pi^L_* k_{\mathbf{Y}_{I,v}} \times k_{\mathbf{Y}^{(\mathbf{L}_I,\dot{v}F)}(w)}$ on $\mathbf{X}_{I,v} \times \mathbf{Y}^{(\mathbf{L}_I,\dot{v}F)}(w)$, and in turn its image under $\mathrm{R}\sigma_*$ is $\pi^L_* k_{\mathbf{Y}_{I,v}} \otimes_k \mathrm{R}\Gamma(\mathbf{Y}^{(\mathbf{L}_I,\dot{v}F)}(w), k)$.

So $\mathrm{R}\tau_* k_{\mathbf{X}(w\cup d_v)} = (\pi^L_* k_{\mathbf{Y}_{I,v}} \otimes \mathrm{R}\Gamma(\mathbf{Y}^{(\mathbf{L}_I,\dot{v}F)}(w), k)) \otimes_{kL} k$. Since the action of L on the tensor product inside the parentheses is diagonal, we have $\mathrm{R}\tau_* k_{\mathbf{X}(w\cup d_v)} = \pi^L_* k_{\mathbf{Y}_{I,v}} \otimes_{kL} \mathrm{R}\Gamma(\mathbf{Y}^{(\mathbf{L}_I,\dot{v}F)}(w), k) \cong \pi^L_* k_{\mathbf{Y}_{I,v}} \overset{\mathrm{L}}{\otimes}_{kL} \mathrm{R}\Gamma(\mathbf{Y}^{(\mathbf{L}_I,\dot{v}F)}(w), k)$ (see A3.15). The action of $\mathbf{T}_0^{wvF}$ is the one on the right side induced by the right-sided action on $\mathbf{Y}^{(\mathbf{L}_I,\dot{v}F)}(w)$. So, applying $- \overset{\mathrm{L}}{\otimes}_{k\mathbf{T}_0^{wvF}} k\mathbf{T}_0^{wvF} b_\theta$, we get $\mathrm{R}\tau_*(\mathcal{F}_{(w\cup d_v,\theta)}) \cong \pi^L_* k_{\mathbf{Y}_{I,v}} \otimes_{kL} \mathcal{S}^{\mathbf{L}_I,\dot{v}F}_{(w,\theta)} \cong \pi^L_* k_{\mathbf{Y}_{I,v}} \overset{\mathrm{L}}{\otimes}_{kL} \mathcal{S}^{\mathbf{L}_I,\dot{v}F}_{(w,\theta)}$ by our definitions and A3.15. This is our initial claim. □

In view of Lemma 10.20, (2) will be implied by

(3) $$i^*\mathrm{R}j_*\mathrm{R}\tau_*\mathcal{F}_{(w\cup d_v,\theta)} = 0.$$

Consider now the diagram

$$\begin{array}{ccccc}
\mathbf{X}(w\cup d_v) & \xrightarrow{j^{\overline{w\cup d_v}}_{w\cup d_v}} & \overline{\mathbf{X}}(w\cup d_v) & \xleftarrow{i_{\mathbf{Z}}} & \mathbf{Z} \\
\tau\downarrow & & \downarrow\overline{\tau} & & \downarrow\tau' \\
\mathbf{X}_{I,v} & \xrightarrow{j} & \overline{\mathbf{X}}_{I,v} & \xleftarrow{i} & \overline{\mathbf{X}}_{I,v}\setminus\mathbf{X}_{I,v}
\end{array}$$

where $\overline{\tau}$ is defined by $\overline{\tau}(g_1\mathbf{B}_0, \ldots) = g_1\mathbf{P}_I$, and the right square is a fibered product, thus implying $\mathbf{Z} = \overline{\mathbf{X}}(w\cup d_v) \setminus \overline{\tau}^{-1}(\mathbf{X}_{I,v})$. The left square commutes, so (3) can be rewritten as

(4) $$i^*\mathrm{R}\overline{\tau}_*\mathrm{R}\Big(j^{\overline{w\cup d_v}}_{w\cup d_v}\Big)_*\mathcal{F}_{(w\cup d_v,\theta)} = 0.$$

The varieties $\overline{\mathbf{X}}(w\cup d_v)$ and $\overline{\mathbf{X}}_{I,v}$ are closed subvarieties (Proposition 7.13(iv)) of the complete varieties $(\mathbf{G}/\mathbf{B}_0)^{l(w)+l(v)}$ and $\mathbf{G}/\mathbf{P}_I$ respectively (see A2.6). Then $\overline{\tau}$ is a proper morphism (see A2.7). The base change theorem for proper morphisms (A3.5) then yields $i^*\mathrm{R}\overline{\tau}_* \cong (\mathrm{R}\tau'_*)i^*_{\mathbf{Z}}$ so that (4) will be implied by

(5) $$i_{\mathbf{Z}}^{*}\mathrm{R}\left(j_{w\cup d_v}^{\overline{w\cup d_v}}\right)_{*}\mathcal{F}_{(w\cup d_v,\theta)} = 0.$$

The above can be written as $i_{\mathbf{Z}}^{*}\mathrm{R}(j_{w\cup d_v}^{\overline{w\cup d_v}})_{*}\mathcal{F}_{w\cup d_v}$, which is a composition of derived functors applied to the projective module $k\mathbf{T}_0^{w(vF)}b_\theta$. But $D^b(k\mathbf{T}_0^{w(vF)}b_\theta-\mathbf{mod})$ is generated by the only simple $k\mathbf{T}_0^{w(vF)}b_\theta$-module k_θ (see A1.12 or Exercise A1.3), so it suffices to check the following

(6) $$i_{\mathbf{Z}}^{*}\mathrm{R}\left(j_{w\cup d_v}^{\overline{w\cup d_v}}\right)_{*}\mathcal{F}_{w\cup d_v}(\theta) = 0.$$

If $Z\xrightarrow{i_Z}X$ is a closed immersion, $Z=\bigcup_t Z_t$ is a covering by locally closed subvarieties with associated immersions $Z_t\xrightarrow{i_t}X$, and if $\mathcal{C}\in D_k^b(X)$, then the criterion of exactness in terms of stalks (see A3.2) easily implies that $i_Z^*\mathcal{C}=0$ in $D_k^b(Z)$ if and only if $i_t^*\mathcal{C}=0$ in $D_k^b(Z_t)$ for all t.

So our claim (6) is now a consequence of Theorem 10.17(a) and the following.

Lemma 10.21. $\overline{\mathbf{X}}(w\cup d_v)\setminus\overline{\tau}^{-1}(\mathbf{X}_{I,v}) = \bigcup_{w',v'}\mathbf{X}(w'\cup v')$ *where the union ranges over* $w'\leq w$ *and* $v'<d_v$ *in* Σ*. In particular,* $(w\cup d_v)_\theta\not\leq w'\cup v'$*.*

Proof of Lemma 10.21. Concerning the first equality, the inclusion $\overline{\mathbf{X}}(w\cup d_v)\setminus\overline{\tau}^{-1}(\mathbf{X}_{I,v})\subseteq\bigcup_{w',v'}\mathbf{X}(w'\cup v')$ is enough for our purpose (the converse is left as an exercise). By definition of the Bruhat order, we clearly have $\overline{\mathbf{X}}(w\cup d_v)=\bigcup_{w',v'}\mathbf{X}(w'\cup v')$ where the union is over $w'\leq w$ and $v'\leq d_v$. We must check that $\overline{\tau}(\mathbf{X}(w'\cup d_v))\subseteq\mathbf{X}_{I,v}$ for all $w'\leq w$. Let $(g_1\mathbf{B}_0,\ldots,g_{r+l(v)}\mathbf{B}_0)\in\mathbf{X}(w'\cup d_v)$. Since w' is a sequence of elements of $I\cup\{1\}$, we have $g_1^{-1}g_{r+1}\in\mathbf{P}_I$ while $g_{r+1}\mathbf{B}_0\xrightarrow{v}F(g_1)\mathbf{B}_0$ by Lemma 10.13, i.e. $g_{r+1}^{-1}F(g_1)\in\mathbf{B}_0\dot{v}\mathbf{B}_0$. Then $g_1^{-1}F(g_1)\in\mathbf{P}_I\dot{v}\mathbf{B}_0$ and therefore $g_1\mathbf{P}_I\in\mathbf{X}_{I,v}$.

The last assertion comes from Proposition 10.19. □

Proof of Proposition 10.19. Let $w=(s_1,\ldots,s_r)$, $w\cup d_v=(s_1,\ldots,s_r,s_{r+1},\ldots s_{r+l})$. For $i\in[1,r+l]$ such that $s_i\neq 1$, denote by δ_i the simple root corresponding with s_i and let $\alpha_i=s_1\ldots s_{i-1}(\delta_i)$. The claimed equality amounts to showing the following two statements.

(1) If $i\leq r$ and $s_i\neq 1$, then $\theta\circ N_{w\cup d_v}^{(F)}(\alpha_i^\vee)=1$ if and only if $\theta\circ N_w^{(vF)}(\alpha_i^\vee)=1$ (where the exponent in the norm map indicates which Frobenius endomorphism is considered).

(2) If $i>r$ (and therefore $s_i\neq 1$) then $\theta\circ N_{w\cup d_v}^{(F)}(\alpha_i^\vee)\neq 1$.

The first statement is clear from the definition of the norm maps (8.12) implying $N_b^F=N_1^{bF}$ for any $b\in\Sigma$.

Let us check the second (see §2.1 for the elementary properties of roots used below). Let $\beta_i := s_{r+1} \dots s_{i-1}(\delta)$. The latter is positive since d_v is a reduced decomposition. But $v^{-1}(\beta_i) = s_{r+l}s_{r+l-1} \dots s_i(\delta_i) = -s_{r+l}s_{r+l-1} \dots s_{i+1}(\delta_i)$ is negative since $s_{r+l}s_{r+l-1} \dots s_i$ is a reduced decomposition. This implies that $\beta_i \notin \Phi_I$ since $v^{-1}(I) \subseteq \Delta$. Then any element of W_I sends it to an element of $\Phi \setminus \Phi_I$, so $\alpha_i \notin \Phi_I$. By Exercise 8.2, this implies $\theta(N_{w \cup d_v}(\alpha_i^\vee)) \neq 1$. □

Exercises

1. Assume that **L** is a torus in Theorem 10.1. Show that Theorem 10.17(a) (ramification) is enough to get Theorem 10.1 (note that, in the notation of §10.5, $(d_v)_\theta = d_v$, $\mathcal{F}_s = \mathcal{F}_{d_v}(b_\theta)$, and therefore $(j_{v'}^{\overline{d}_v})^* \mathrm{R}(j_{d_v}^{\overline{d}_v})_* \mathcal{F}_s = (j_{v'}^{\overline{d}_v})^* \mathrm{R}(j_{d_v}^{\overline{d}_v})_* \mathcal{F}_{s^{-1}} = 0$ for all $v' < d_v$).
2. Show that the condition in Theorem 10.17(a) is equivalent to the existence of some $\theta' \colon \mathbf{T}_0^{w'F} \to k^\times$ such that (w, θ) and (w', θ') are in the same rational series.

Notes

Proposition 10.3 mixes a classical argument ([SGA.4$\frac{1}{2}$] p. 180; see also [Bro90b] 3.5) with some adaptations in the quasi-affine case (see [Haa86]). The first reduction (Theorem 10.7), assuming affinity of Deligne–Lusztig varieties, is due to Broué, thus covering the case of a torus ([Bro90b], see Exercise 1). The second reduction (§10.5) is taken from [BoRo03].

11

Jordan decomposition as a Morita equivalence: sheaves

This chapter is devoted to determining where the locally constant sheaves $\mathcal{F}_w(\theta)$ on $\mathbf{X}(w)_{\text{ét}}$ (see §10.5) ramify. This includes the proof of Theorem 10.17(a), due to Deligne–Lusztig.

Recall that $(\mathbf{G}, F)$ is a connected reductive $\mathbf{F}$-group defined over $\mathbb{F}_q$. We have fixed $\mathbf{T} \subseteq \mathbf{B}$ a maximal torus and a Borel subgroup of $\mathbf{G}$, both F-stable. This allows us to define the Weyl group $W(\mathbf{G}, \mathbf{T})$ and its subset S of simple reflections relative to $\mathbf{B}$. Recall the notation Σ for the set of finite sequences of elements of $S \cup \{1\}$ and the partial ordering $\leq$ on Σ. When $w' \leq w$ in Σ, recall the varieties and the immersion

$$j_{w'}^{\overline{w}}: \mathbf{X}(w') \to \overline{\mathbf{X}}(w)$$

(see Notation 10.8). Recall that, with $\theta: \mathbf{T}^{wF} \to k^\times$ considered as a one-dimensional representation of $\mathbf{T}^{wF}$, $\mathcal{F}_w(\theta)$ is the locally constant sheaf on $\mathbf{X}(w)_{\text{ét}}$ associated with θ and the $\mathbf{T}^{wF}$-torsor $\mathbf{Y}(w) \to \mathbf{X}(w)$.

We prove the following theorem. The first statement is the "ramification" part of Theorem 10.17. The second statement, due to Bonnafé–Rouquier, will contribute to the proof of the "generation" part of Theorem 10.17, completed in the next chapter.

Theorem 11.1. *Let $w \in \Sigma$, $\theta \in \mathrm{Hom}(\mathbf{T}^{wF}, k^\times)$.*

(a) If $w' \leq w$, then $(j_{w'}^{\overline{w}})^* \mathrm{R}(j_w^{\overline{w}})_* \mathcal{F}_w(\theta) = 0$ unless $w_\theta \leq w'$ (see Definition 10.16).

(b) The mapping cone of $(j_w^{\overline{w}})_! \mathcal{F}_w(\theta) \to \mathrm{R}(j_w^{\overline{w}})_* \mathcal{F}_w(\theta)$ is in the subcategory of $D_k^b(\overline{\mathbf{X}}(w))$ generated by the $(j_{\overline{w}'}^{\overline{w}})_* \circ (j_{w'}^{\overline{w}'})_! \mathcal{F}_{w'}(\theta')$'s for $w_\theta \leq w' < w$ and $\theta' \in \mathrm{Hom}(\mathbf{T}^{w'F}, k^\times)$.

The proof of (a) is §11.1 below. It involves the techniques of A3.16 and A3.17 about ramification of locally constant sheaves along divisors with normal crossings. The underlying idea is to reduce $\mathbf{X}(w) \to \overline{\mathbf{X}}(w)$ to the "one-dimensional" case of $\mathbb{G}_m \hookrightarrow \mathbb{G}_a$, or even the more drastic reduction, suited to étale cohomology, to the embedding of the generic point $\{\eta\} \hookrightarrow \mathrm{Spec}(\mathbf{F}[z]^{\mathrm{sh}})$.

The proof of (b) occupies the next three sections. This time, one constructs a $\mathbf{T}^{wF}/\mathrm{Ker}(\theta)$-torsor for $\mathbf{X}[w_\theta, w]$, the open subvariety of $\overline{\mathbf{X}}(w)$ corresponding to the $\mathbf{X}(w')$'s with $w_\theta \leq w' \leq w$.

11.1. Ramification in Deligne–Lusztig varieties

Let $(\mathbf{G}, F, \mathbf{B}, \mathbf{T})$ be as in §10.5. Recall $S \subseteq W(\mathbf{G}, \mathbf{T})^F$ and the corresponding simple roots $\Delta(\mathbf{G}, \mathbf{T})$ in the root system $\Phi(\mathbf{G}, \mathbf{T})$. Let $r \geq 1$, $w = (s_i)_{1\leq i\leq r} \in (S \cup \{1\})^r$. Recall $\mathbf{X}(w)$ (resp. $\overline{\mathbf{X}}(w)$) the locally closed (resp. closed) subvariety of $(\mathbf{G}/\mathbf{B})^r$ (see Notation 10.8).

Lemma 11.2. *Let $a \leq w$ in $(S \cup \{1\})^r$ with $l(w) > 1$. Then $[1, w] \setminus [a, w] = \bigcup_v [1, v]$ where the union is over $v \in [1, w] \setminus [a, w]$ such that $l(v) = l(w) - 1$. The corresponding union $\bigcup_v \overline{\mathbf{X}}(v)$ is a smooth divisor with normal crossings in $\overline{\mathbf{X}}(w)$.*

Proof. The first equality is an easy property of the product ordering in $(S \cup \{1\})^r$. The consequence on $\overline{\mathbf{X}}$ comes from Proposition 10.10(iv). □

To prove Theorem 11.1(a), as a result of the above lemma (with $a = w_\theta$) and since $j^{\overline{w}}_{w'} = j^{\overline{w}}_{\overline{v}} \circ j^{\overline{v}}_{w'}$ for $w' \leq v \leq w$, it suffices to show that $(j^{\overline{w}}_{\overline{v}})^* \mathrm{R}(j^{\overline{w}}_{w})_* \mathcal{F}_w(\theta) = 0$ when $l(v) = l(w) - 1$, $v \leq w$, $w_\theta \not\leq v$. By the realization of $\overline{\mathbf{X}}(w) \setminus \mathbf{X}(w)$ as a smooth divisor with normal crossings, and Theorem A3.19, it is equivalent to showing that $\mathcal{F}_w(\theta)$ (see Definition 10.16) ramifies along those $\mathbf{X}(v)$'s. Inputting also the definition of w_θ, it is enough to prove the following.

Theorem 11.1(a′). *Let $\nu \in [1, r[$ such that $s_\nu \neq 1$. Define w'_ν from $w = (s_i)_{1\leq i\leq r}$ by replacing the νth component s_ν by 1, and let*

$$D_\nu = \overline{\mathbf{X}}(w'_\nu).$$

Let δ_ν be the simple root corresponding to the reflection s_ν and let $\beta_\nu = (s_1 \dots s_{\nu-1})(\delta_\nu)$. Let $N_w \colon Y(\mathbf{T}) \to \mathbf{T}^{wF}$ be as defined in (8.12). Then $\mathcal{F}_w(\theta)$ ramifies along $D_\nu \cap \overline{\mathbf{X}}(w)$ when $\theta(N_w(\beta_\nu^\vee)) \neq 1$ (see (8.11) for the definition of $N_w \colon Y(\mathbf{T}) \to \mathbf{T}^{wF}$).

Let $w' = (s_i)_{1\leq i\leq r, s_i\neq 1}$. Recall the varieties $\mathbf{Y}(w)$ of Notation 10.8. One has clearly a canonical isomorphism between $(\mathbf{Y}(w') \to \mathbf{X}(w') \to \overline{\mathbf{X}}(w'))$ and $(\mathbf{Y}(w) \to \mathbf{X}(w) \to \overline{\mathbf{X}}(w))$. Thanks to the following lemma we shall replace

$(\mathbf{T}, wF)$ by $(\mathbf{T}^{(r)}, wF_r)$, θ by $\theta \circ \iota_w{}^{-1}$, where $\iota_w \colon \mathbf{T}^{wF} \to (\mathbf{T}^{(r)})^{wF_r}$ is the isomorphism defined in Lemma 10.9, and replace $(\mathbf{G}, F)$ by$(\mathbf{G}^{(r)}, F_r)$.

Lemma 11.3. *Let* $\alpha \in \Delta(\mathbf{G}, \mathbf{T})$*, let* $\beta = s_1 \dots s_{\nu-1}(\alpha)$ *and let* $\alpha_\nu = (\eta_i)_{1\le i\le r} \in Y(\mathbf{T}^{(r)}) = Y(\mathbf{T})^r$ *be the coroot of* $\mathbf{G}^{(r)}$ *defined by*

$$\eta_i = \begin{cases} 0 & \text{if } i \neq \nu, \\ \alpha^\vee & \text{if } i = \nu. \end{cases}$$

One has $\iota_w(N_w(\beta^\vee)) = N_w^{(r)}(\alpha^\vee)$.

Proof. Here $N_w^{(r)}$ is the canonical morphism $Y(\mathbf{T}^{(r)}) \to (\mathbf{T}^{(r)})^{wF_r}$.

Let d be such that $(wF)^d = F^d$ is a split Frobenius map over $\mathbb{F}_{q^d}$, i.e. amounts to $t \mapsto t^{q^d}$ on $\mathbf{T}$. Then $(wF_r)^{dr} = (F_r)^{dr}$ induces multiplication by q^d on $Y(\mathbf{T}^{(r)})$. Let ζ be a selected primitive $(q^d - 1)$th root of unity in $\overline{\mathbb{Q}}_\ell$. By (8.12), $N_w(\beta^\vee) = N_{F^d/wF}(\beta^\vee(\zeta))$ and $N_w^{(r)}(\alpha_\nu^\vee) = N_{F_r^{dr}/wF_r}(\alpha_\nu^\vee(\zeta))$. We compute the last expression. One has $\alpha_\nu^\vee(\zeta) = (1, \dots, 1, t_0, 1, \dots, 1)$ where $t_0 = \alpha^\vee(\zeta)$ is in component number ν. Hence $(wF_r)^h(\alpha_\nu^\vee(\zeta))$ has all components equal to 1 apart from the component of index $\nu - h$ modulo r, which may be written $w_h(t_0)$. Here w_h is the word of length h on the r "letters" $s_1, s_2, \dots, (s_r F)$ which is obtained by left truncation of the long word $s_1 s_2 \dots (s_r F) s_1 s_2 \dots (s_r F) s_1 s_2 \dots s_{\nu-1}$ ($(\nu - 1 + Cr)$ letters, C large enough). The equality $N_w^{(r)}(\alpha_\nu^\vee) = \iota_w(N_w(\beta^\vee))$ follows. □

Replacing now $\mathbf{G}^r$ with $\mathbf{G}$ (or, better, denoting $\mathbf{G}^r$ by $\mathbf{G}$), one is led to prove Theorem 11.1(a′) above with $w \in W(\mathbf{G}, \mathbf{T})$ a product $w = s_1 s_2 \dots s_l$ of commuting simple reflections, $\mathbf{X}(w) \to \overline{\mathbf{X}}(w)$ being therefore smooth subvarieties of $\mathbf{G}/\mathbf{B}$ and $\mathbf{Y}(w)$ a $\mathbf{T}^{wF}$-torsor over $\mathbf{X}(w)$ defined as in Chapter 7. Note that now $\beta_\nu = \alpha_\nu$.

We abbreviate by writing just $\mathbf{Y} \to \mathbf{X} \to \overline{\mathbf{X}}$ instead of $\mathbf{Y}(w) \to \mathbf{X}(w) \to \overline{\mathbf{X}}(w)$.

As $\mathbf{T}^{wF}$ is of order prime to p, the ramification of $\mathbf{Y} \to \mathbf{X}$ relative to D is tame.

Recall k_θ, the $k\mathbf{T}^{wF}$-module with support k defined by θ, and $\mathcal{F}_w(\theta)$ the locally constant sheaf on $\mathbf{X}_{\text{ét}}$ defined by the $\mathbf{T}^{wF}$-torsor $\mathbf{Y}_\theta := \mathbf{Y} \times k_\theta/\mathbf{T}^{wF} \to \mathbf{X}$.

Let $\overline{\mathbf{X}}_\nu = \overline{\mathbf{X}} \setminus \bigcup_{i\neq\nu} D_i$ (see Theorem 11.1(a′)), $D'_\nu = D_\nu \cap \overline{\mathbf{X}}_\nu$. Let d_ν be the generic point of the irreducible divisor D'_ν and let $\overline{d}_\nu \colon \mathrm{Spec}(\mathbf{F}(z_1, \dots, z_{r-1})) \to \overline{\mathbf{X}}_\nu$ be a geometric point of $\overline{\mathbf{X}}_\nu$ of image d_ν. Following the general procedure of A3.17, the ramification to compute is the ramification of the map $\mathbb{A}_{\mathbf{F}}^{\text{sh}} \times_{\overline{\mathbf{X}}_\nu} \mathbf{Y}_\theta \to \mathbb{A}_{\mathbf{F}}^{\text{sh}}$ where $x \colon \mathbb{A}_{\mathbf{F}}^{\text{sh}} \to \overline{\mathbf{X}}_\nu$ is such that the closed point (z) of $\mathbb{A}_{\mathbf{F}}^{\text{sh}} = \mathrm{Spec}(\mathbf{F}[z]^{\text{sh}})$ is mapped onto d_ν and $\overline{d}_\nu = x \circ \overline{d}'_\nu$, where $\overline{d}'_\nu$ is a geometric point of $\mathbb{A}^{\text{sh}}$.

We further prove that Theorem 11.1(a′) will be deduced from the following proposition.

Proposition 11.4. *Let* $\mathbb{A}^{\mathrm{sh}} \to \overline{\mathbf{X}}_\nu$ *as above. Let* $\pi' \colon \mathbf{Y}' := \mathbb{G}_{\mathrm{m}} \times_{\mathbf{T}} \mathbf{T} \to \mathbb{G}_{\mathrm{m}}$ *be the pull-back of the Lang covering* $\mathbf{T} \to \mathbf{T}$, $(t \mapsto t^{-1} wF(t))$, *(with group* $\mathbf{T}^{wF}$*), under the morphism* $\alpha_\nu^\vee \colon \mathbb{G}_{\mathrm{m}} \to \mathbf{T}$. *Consider* $\mathbb{A}^{\mathrm{sh}}$ *as* $\mathrm{Spec}(\mathcal{O}_{\mathbb{G}_a,\overline{0}})$. *The* $\mathbb{A}^{\mathrm{sh}}$*-schemes* $\mathbb{A}^{\mathrm{sh}} \times_{\overline{\mathbf{X}}_\nu} \mathbf{Y}$ *and* $\mathbb{A}^{\mathrm{sh}} \times_{\mathbb{G}_a} \mathbf{Y}'$ *have isomorphic fibers over the generic point of* $\mathbb{A}^{\mathrm{sh}}$.

Let us say how this implies Theorem 11.1(a′). By Proposition 11.4, the covering $\mathbf{Y} \to \mathbf{X}$ ramifies along $D_\nu \cap \overline{\mathbf{X}}$ in the same way as $\mathbf{Y}' \to \mathbb{G}_{\mathrm{m}}$ ramifies at 0 (considered as divisor of $\mathbb{G}_a$ with $\mathbb{G}_a = \mathbb{G}_{\mathrm{m}} \cup \{0\}$ as $\mathbf{F}$-varieties). Let $\mathcal{F}'_w(\theta)$ be the locally constant sheaf over $\mathbb{G}_{\mathrm{m}}$ defined by the $\mathbf{T}^{wF}$-torsor $\mathbf{Y}' \to \mathbb{G}_{\mathrm{m}}$ and θ. To prove Theorem 11.1 we consider $\mathcal{F}'_w(\theta)$ instead of $\mathcal{F}_w(\theta)$.

By functoriality of fundamental groups (see A3.16), $y \in Y(\mathbf{T}) = \mathrm{Hom}(\mathbb{G}_{\mathrm{m}}, \mathbf{T})$ defines a map $\hat{y} \colon \pi_1^{\mathrm{t}}(\mathbb{G}_{\mathrm{m}}) \to \pi_1^{\mathrm{t}}(\mathbf{T})$ between tame fundamental groups. The Lang covering $\mathbf{T} \to \mathbf{T}$ defined above is a $\mathbf{T}^{wF}$-torsor, so it defines a quotient $\rho \colon \pi_1^{\mathrm{t}}(\mathbf{T}) \to \mathbf{T}^{wF}$. The ramification of $\mathcal{F}'_w(\theta)$ at 0 is that of $\mathbf{Y}'/\mathrm{Ker}\,\theta \to \mathbb{G}_{\mathrm{m}}$ along the divisor $\{0\} \subseteq \mathbb{G}_a$. So it is given by the composed map $\theta \circ \rho \circ \widehat{(\alpha_\nu^\vee)}$.

We show that, for any $y \in Y(\mathbf{T})$, $N_w(y)$ is a generator of the image of $\rho \circ \hat{y}$. Remember that $\pi_1^{\mathrm{t}}(\mathbb{G}_{\mathrm{m}})$ is the closure of $(\mathbb{Q}/\mathbb{Z})_{p'} \cong \mathbf{F}^\times$ (see (8.3)) with regard to finite quotients (see A3.16). If $\mathbf{T} = \mathbb{G}_{\mathrm{m}}$ and $wF = F$, ρ is just the identification $\mu_{q-1} = \mathbb{F}_q^\times \cong \mathbf{T}^F$. Then $y \in Y(\mathbf{T})$ is defined by an exponent $h \in \mathbb{Z}$. Let a be the generator of $\mathbb{F}_q^\times$ defined by $\iota(a) = 1/(q-1)$. Then $N(y) = a^h$. However, $\hat{y}$ is $(\zeta \mapsto \zeta^h)$ on any nth root of unity. The split case ($F(t) = t^q$ for all $t \in \mathbf{T}$) follows. The non-split case reduces to the split one. Assume that $(wF)^d = F_0$ is a split Frobenius. Let $\rho_0 \colon \pi_1^{\mathrm{t}}(\mathbf{T}) \to \mathbf{T}^{F_0}$ be defined by the Lang covering of $\mathbf{T}$ relative to F_0. There are surjective norm maps $N_T \colon \mathbf{T}^{F_0} \to T^{wF}$, $N_Y \colon Y(\mathbf{T}) \to Y(\mathbf{T})$ such that, with $N_0 \colon Y(\mathbf{T}) \to \mathbf{T}^{F_0}$ defined by (8.8), one has $N_w \circ N_Y = N_T \circ N_0$. Furthermore N_Y defines $\hat{N}_Y \colon \pi_1^{\mathrm{t}}(\mathbf{T}) \to \pi_1^{\mathrm{t}}(\mathbf{T})$ such that $\rho \circ \hat{N}_Y = N_T \circ \rho_0$. Assume $y = N_Y(y_0) \in Y(\mathbf{T})$, then $\hat{y} = \hat{N}_Y \circ \hat{y}_0$. As $N_0(y_0)$ is a generator of the image of $\rho_0 \circ \hat{y}_0$, $N_T(N_0(y_0))$ is a generator of the image of $N_T \circ \rho_0 \circ \hat{y}_0$. But $N_T(N_0(y_0)) = N_w(y)$ and $N_T \circ \rho_0 \circ \hat{y}_0 = \rho \circ \hat{N}_Y \circ \hat{y}_0 = \rho \circ \hat{y}$.

To prove Proposition 11.4, we view $\mathbf{X}(w)$ as the intersection of the $\mathbf{G}$-orbit of type w, $O(w) \subseteq \mathbf{G}/\mathbf{B} \times \mathbf{G}/\mathbf{B}$, with the graph Γ of the map $F \colon \mathbf{G}/\mathbf{B} \to \mathbf{G}/\mathbf{B}$ (see the proof of Theorem 7.2 and Proposition 7.13(iv)). So we write $\mathbf{Y} = \{g\mathbf{U} \mid F(g\mathbf{U}) = g\dot{w}\mathbf{U}\}$ (see §10.3) and the map $\mathbf{Y} \to \mathbf{X}$ is $g\mathbf{U} \mapsto (g\mathbf{B}, F(g)\mathbf{B})$. Then $\mathbf{Y}$ appears to be the subvariety of $\mathbf{G}/\mathbf{U}$ defined by the equation $F(u) = \psi(u)$, where $\psi = \psi^{\dot{w}}$ is well defined. We easily get the following.

Proposition 11.5. *Let* pr_i^w $(i = 1, 2)$ *be the projection on* $\mathbf{G}/\mathbf{B}$ *of* $O(w) \subset (\mathbf{G}/\mathbf{B})^2$. *Let* $\mathcal{Y}_i \to O(w)$ $(i = 1, 2)$ *be the pull-back under* pr_i^w *of the* $\mathbf{T}$*-torsor* $\mathbf{G}/\mathbf{U} \to \mathbf{G}/\mathbf{B}$. *The right multiplication by* $\dot{w}$ *induces a map between* $\mathbf{T}$*-torsors over* $O(w)$, $\psi^{\dot{w}}: \mathcal{Y}_1 \to \mathcal{Y}_2$, *and* $\psi^{\dot{w}}$ *is compatible with the automorphism* $(t \mapsto t^w)$ *of* $\mathbf{T}$.

Proof of Proposition 11.5. The scheme and $\mathbf{T}$-torsor $\mathcal{Y}_1$ (resp. $\mathcal{Y}_2$) is defined by a subvariety of $\mathbf{G}/\mathbf{U} \times \mathbf{G}/\mathbf{B}$: $\{(g\mathbf{U}, g\dot{w}\mathbf{B}) \mid g \in \mathbf{G}\}$ with the morphism $(g\mathbf{U}, g\dot{w}\mathbf{B}) \mapsto (g\mathbf{B}, g\dot{w}\mathbf{B}) \in O(w)$ (resp. $\{(g\dot{w}\mathbf{U}, g\mathbf{B}) \mid g \in \mathbf{G}\}$ with $(g\dot{w}\mathbf{U}, g\mathbf{B}) \mapsto (g\mathbf{B}, g\dot{w}\mathbf{B})$). The map $\psi^{\dot{w}}$ sends $(g\mathbf{U}, g\dot{w}\mathbf{B})$ onto $(g\dot{w}\mathbf{U}, g\mathbf{B})$. □

If $\dot{w}$ is fixed then we may write ψ instead of $\psi^{\dot{w}}$.

As the Frobenius map extends all over $\mathbf{G}/\mathbf{B}$, the ramification of $\mathbf{Y}$ relative to D_ν depends on the local behavior of ψ near d_ν. The image in $\mathbf{T}^{wF}$ of the tame fundamental group $\pi_1^D(\mathbf{X})$ (see A3.17) is defined by its various quotients in $\mathbf{T}^{wF}/\mathrm{Ker}(\theta)$ for θ a linear character (see [GroMur71] 1.5.6). For such a θ there is some $\lambda \in X(\mathbf{T})$ such that $\mathbf{T}^{wF} \cap \mathrm{Ker}(\lambda) = \mathrm{Ker}(\theta)$.

Proposition 11.6. *For* $\lambda \in X(\mathbf{T})$ *let* $E_\lambda \to \mathbf{G}/\mathbf{B}$ *be the line bundle (A2.9) over* $\mathbf{G}/\mathbf{B}$ *defined by* λ *and the* $\mathbf{T}$*-torsor* $\mathbf{G}/\mathbf{U} \to \mathbf{G}/\mathbf{B}$. *Let* $E_{\lambda,i} \to \mathbf{X}$ $(i = 1, 2)$ *be their pull-back under* pr_i^w. *The map* ψ *restricts to an isomorphism of line bundles* $\psi_\lambda: E_{\lambda,1} \to E_{\lambda \circ \mathrm{ad}w,2}$.

Proof of Proposition 11.6. Recall the notation $w^{-1}(\lambda) = \lambda \circ \mathrm{ad}w$. There are natural morphisms of schemes over $\mathbf{G}/\mathbf{B}$: $\mathbf{G}/\mathbf{U} \to E_\lambda$ and $\mathbf{G}/\mathbf{U} \to E_{w^{-1}(\lambda)}$, hence $\mathcal{Y}_1 \to E_{\lambda,1}$. Using the description of $\mathcal{Y}_i$ we gave in the proof of Proposition 11.5, we see that the following diagram is commutative

$$\begin{array}{ccc} \mathcal{Y}_1 & \xrightarrow{\psi} & \mathcal{Y}_2 \\ \downarrow & & \downarrow \\ E_{\lambda,1} & \xrightarrow{\psi_\lambda} & E_{w^{-1}(\lambda),2} \end{array}$$ □

The map ψ_λ may be viewed as a section of $E_{\lambda^{-1},1} \otimes E_{w^{-1}(\lambda),2}$. It extends to $\overline{\mathbf{X}}_\nu$ if and only if its local order along D_ν is zero (see Exercise 3).

That order is given by the following.

Proposition 11.7. *Let* α_ν *be as in Theorem 11.1(a′), let* ψ_λ *be as in Proposition 11.6. The order of* ψ_λ *along the divisor* D_ν *is* $\langle \lambda, \alpha_\nu^\vee \rangle$.

Proof of Proposition 11.7. The fiber over any $b \in \mathbf{G}/\mathbf{B}$ is a principal homogeneous space for $\mathbf{T}$; denote it $U(b)$. If $\dot{w} = \dot{w}_1\dot{w}_2$ and $l(w) = l(w_1) + l(w_2)$, then

$O(w)$ may be seen as the product over $\mathbf{G}/\mathbf{B}$ of $O(w_1)$ with its second projection, and $O(w_2)$ with its first projection. One has clearly, for any $u \in U(b)$,

$$\psi^{\dot{w}}(u) = (\psi^{\dot{w}_2} \circ \psi^{\dot{w}_1})(u)$$

here $\psi^{\dot{w}_1}(u) \in U(b')$ and $U(b')$ is a fiber of $\mathcal{Y}_1^{\dot{w}_2}$ identified with a fiber of $\mathcal{Y}_2^{\dot{w}_1}$ (evident notations).

Let $\dot{w}_{\nu-1} = \dot{s}_1 \ldots \dot{s}_{\nu-1}$, let $\lambda_\nu = w_{\nu-1}^{-1}(\lambda)$, and define w' by $w = w_{\nu-1} s_\nu w'$, so that $\dot{w} = \dot{w}_{\nu-1} \dot{s}_\nu \dot{w}'$. That decomposition allows us to write $O(w)$ as a fiber product $O(w_\nu) \times O(s_\nu) \times O(w')$ and view $O(w)$ as a subvariety of $(\mathbf{G}/\mathbf{B})^4$. The four projections give rise to four $\mathbf{T}$-torsors over $\overline{\mathbf{X}}_\nu$, and the composition formula $\psi^{\dot{w}} = \psi^{\dot{w}'} \psi^{\dot{s}_\nu} \psi^{\dot{w}_{\nu-1}}$ makes sense: for any $(b, b') \in O(w)$ and any $u \in U(b)$ one has $\psi(u) = (\psi^{\dot{w}'} \psi^{\dot{s}_\nu} \psi^{\dot{w}_{\nu-1}})(u) \in U(b')$ where $\psi^{\dot{w}'} : \mathcal{Y}_1^{w'} \to \mathcal{Y}_2^{w'}$, $\psi^{\dot{w}_{\nu-1}} : \mathcal{Y}_1^{w_{\nu-1}} \to \mathcal{Y}_2^{w_{\nu-1}}$ and $\psi^{\dot{s}_\nu}$ are defined by Proposition 11.5. The maps $\psi^{\dot{w}_{\nu-1}}$, $\psi^{\dot{w}'}$ and $\psi^{\dot{s}_\nu}$ define isomorphisms of line bundles $E_\lambda \to E_{\lambda_\nu}$, $E_{s_\nu(\lambda_\nu)} \to E_{w^{-1}(\lambda)}$, $E_{\lambda_\nu} \to E_{s_\nu(\lambda_\nu)}$. As $\psi^{\dot{w}_{\nu-1}}$ and $\psi^{\dot{w}'}$ have null order near a general point of D_ν the order to compute is the order of $\psi_\nu^{\dot{s}_\nu}$ along D_ν.

Then the restriction of $\psi^{\dot{s}_\nu}$ to a fiber $U(b)$ is defined inside a minimal parabolic subgroup $\mathbf{P}$ containing the Borel subgroup corresponding to b and a reflection in the conjugacy class of s_ν. Clearly $\psi^{\dot{s}_\nu}$ may be described in the quotient $\mathbf{P}/R_u(\mathbf{P})$, or in its derived subgroup, or in its universal covering $\mathrm{SL}_2(\mathbf{F})$. So we study that *minimal case*.

We take $\mathrm{SL}_2(\mathbf{F})$ acting on $\mathbf{F}^2$, let e be the first element of the canonical basis of $\mathbf{F}^2$. Then $\mathbf{B}$ is the subgroup of unimodular upper triangular matrices and is the stabilizer of $\mathbf{F}e$, $\mathbf{U}$ is the stabilizer of e, $\mathbf{T}$ is the subgroup of unimodular diagonal matrices and acts on $\mathbf{F}e$. One may identify $\mathbf{G}/\mathbf{B}$ with the projective line, or the variety of subspaces of dimension 1 of $\mathbf{F}^2$. For any $g \in \mathrm{SL}_2(\mathbf{F})$, $\mathbf{T}$ acts on $g\mathbf{B}/g\mathbf{U}$ as $g\mathbf{T}g^{-1}$ acts on $\mathbf{F}ge$. Hence we identify $\mathbf{G}/\mathbf{U}$ with $\mathbf{F}^2 \setminus \{0\}$.

Take $\dot{s} = \begin{pmatrix} 0 & 1 \\ -1 & 0 \end{pmatrix}$. The relation $b \overset{s}{-} b'$ between two projective points is equivalent to $b \neq b'$. The divisor is the diagonal set in $(\mathbf{G}/\mathbf{B})^2$. Let ρ be defined by $te = \rho(t)e$, any $t \in \mathbf{T}$, where ρ is a character of $\mathbf{T}$. Let α be the simple root corresponding to s; one has $\langle \rho, \alpha^\vee \rangle = 1$. With the preceding identifications and $\psi = \psi^{\dot{s}}$, when $u, v \in \mathbf{F}^2$ and $u \notin \mathbf{F}v$, $\psi(u, \mathbf{F}v) = (\mathbf{F}u, cv)$ where $c \in \mathbf{F}$ is such that $ge = u$ and $g(\dot{s}e) = cv$ for some $g \in \mathrm{SL}_2(\mathbf{F})$ (see the proof of Proposition 11.3). Thus $-c\det(u, v) = 1$ and ψ_ρ is induced by the map $(u, \mathbf{F}v, a) \mapsto (\mathbf{F}u, v, -\det(u, v)a)$ where $a \in \mathbf{F}$. We see that ψ_ρ vanishes with order 1, equal to $\langle \rho, \alpha^\vee \rangle$, for $\mathbf{F}u \to \mathbf{F}v$. More generally $\psi_{c\rho}$ is of order $c = \langle \alpha^\vee, c\rho \rangle$.

Coming back to the general situation we see that the order of ψ_λ along D_ν is $\langle \rho, \alpha_\nu^\vee \rangle = \langle \lambda, w_{\nu-1}(\alpha_\nu)^\vee \rangle = \langle \lambda, \alpha_\nu^\vee \rangle$ (recall that we assume that the s_i's commute). □

Proof of Proposition 11.4. Begin with $h'\colon \mathbb{A}^{\mathrm{sh}} \times_{\mathbb{G}_a} \mathbf{Y}' \to \mathbb{A}^{\mathrm{sh}}$, the easiest to compute. As a variety $\mathbf{Y}'$ is $\mathbb{G}_{\mathrm{m}} \times_{\mathbf{T}} \mathbf{T} = \{(z,t) \in \mathbb{G}_{\mathrm{m}} \times \mathbf{T} \mid t^{-1}(\dot{w}F)(t) = \alpha_\nu^\vee(z)\}$. The equation of the fiber of h' above the generic point of $\mathbb{A}^{\mathrm{sh}}$ is

$$t^{-1}(\dot{w}F)(t) = \alpha_\nu^\vee(z) \tag{1}$$

Consider now the fibred product

$$\begin{array}{ccc} \mathbb{A}^{\mathrm{sh}} \times_{\overline{\mathbf{X}}_\nu} \mathbf{Y} & \longrightarrow & \mathbf{Y} \\ \big\downarrow{\scriptstyle h} & & \big\downarrow \\ \mathbb{A}^{\mathrm{sh}} & \xrightarrow{\;x\;} & \overline{\mathbf{X}}_\nu \end{array}$$

By base change, ψ defines a morphism Ψ between $\mathbf{T}$-torsors above the generic point η, i.e. over $\mathbf{F}[z]^{\mathrm{sh}}[z^{-1}]$, and the fiber of h over η is the $\mathbf{T}^{wF}$-torsor $\{\eta\} \times_X \mathbf{Y}$, kernel of (Ψ, F). As x is transverse, in the composition formula $\psi^{\dot{w}}(u) = (\psi^{\dot{w}'}\psi^{\dot{s}_\nu}\psi^{\dot{w}_{\nu-1}})(u)$ the maps $\psi^{\dot{w}'}$ and $\psi^{\dot{w}_{\nu-1}}$ extend over $\mathbb{A}^{\mathrm{sh}}$ (may be defined over $\mathbf{F}[z]^{\mathrm{sh}}$) and we have to consider $\psi^{\dot{s}_\nu}$. So, as in the proof of Proposition 11.7 and with the same notation, we go down to the minimal case. The image of the generic point of $\mathbb{A}^{\mathrm{sh}}$ is the generic point of an affine curve in $(\mathbf{G}/\mathbf{B})^2$. Without loss of generality we may assume that x is defined by $x(z) = (b(z), b'(z))$ $(z \in \mathbf{F})$ with $b'(0) = b(z) = \mathbf{F}e$ and $b'(z) = \mathbf{F}u(z)$ where $u(z) = \begin{pmatrix} 1 \\ z \end{pmatrix}$. Now, as is shown in the proof of Proposition 11.7, Ψ is given by the equation $\Psi(ae, b'(z)) = (\mathbf{F}e, -a^{-1}z^{-1}u(z))$. As $ae = \alpha^\vee(a)$ for $a \in \mathbf{F}$, one defines an isomorphism Ψ_0 of order zero at $z = 0$ by the formula $\Psi_0(u) = \Psi(u\alpha^\vee(z)^{-1})$ for $u \in U(b(z))$, $z \neq 0$. By construction Ψ_0 extends over $\mathbb{A}^{\mathrm{sh}}$.

Coming back to the initial $\dot{w}$, we define $\Phi\colon E_0 \to E_r$ over η by

$$\Phi(u) = \Psi\big(u\alpha_\nu^\vee(z)^{-1}\big)$$

where Φ is the composition of the pull-back of $\psi^{w_{\nu-1}}$, Ψ_0 and the pull-back of $\psi^{w'}$, and Φ extends over $\mathbb{A}^{\mathrm{sh}}$. By pull-back under x the fiber to compute is defined by the equation

$$F(u) = \Phi\big(u\alpha_\nu^\vee(z)^{-1}\big). \tag{2}$$

One has $\Phi(ut) = \Phi(u)\dot{w}^{-1}t\dot{w}$, $F(ut) = F(u)F(t)$ and $F(u) \in \Phi(u)\mathbf{T}$ for any $u \in U(b(z))$, $z \neq 0$, $t \in \mathbf{T}$. The equation $F(u) = \Phi(u)$ has a solution by the Lang theorem applied to the endomorphism wF of $\mathbf{T}$. As F and Ψ_0 extend over $\mathbb{A}^{\mathrm{sh}}$ and $\mathbb{A}^{\mathrm{sh}}$ is a strictly henselian ring, there exists u_0 with the equality $F(u_0) = \Phi(u_0)$ over $\mathbb{A}^{\mathrm{sh}}$. For any $u \in E_0$, there is $t \in \mathbf{T}$ such that $ut = u_0$, hence $F(u) = F(u_0)F(t)^{-1}$. The equation (2) becomes $\Phi(u_0) = \Phi(u_0 t^{-1}\alpha_\nu^\vee(z)^{-1})F(t)$, but $\Phi(u_0 t^{-1}\alpha_\nu^\vee(z)^{-1})F(t) = (u_0 t^{-1}.\dot{w})F(t) = u_0 t^{-1}(\dot{w}F(t)\dot{w}^{-1}).\dot{w} = \Phi(u_0 t^{-1}(\dot{w}F)(t)\alpha_\nu^\vee(z)^{-1})$. Hence the equation is equivalent to

$$u_0 = u_0 t^{-1}(\dot{w}F)(t)\alpha_\nu^\vee(z)^{-1},$$

equivalent to (1).

The fibers of h and h' over η are isomorphic $\mathbf{T}^{wF}$-torsors. □

11.2. Coroot lattices associated with intervals

We now begin the proof of Theorem 11.1(b). We return to the general setting where $(\mathbf{G}, F, \mathbf{B}, \mathbf{T})$, $S \subseteq W(\mathbf{G}, \mathbf{T})^F$ are as in §10.5. Recall the simple roots $\Delta(\mathbf{G}, \mathbf{T})$ in the root system $\Phi(\mathbf{G}, \mathbf{T})$. Let $r \geq 1$, $w = (s_i)_{1 \leq i \leq r} \in (S \cup \{1\})^r$. Here we consider various subgroups and quotients of $\mathbf{T}^{wF}$. The notation is that of Chapter 10.

Recall that we write $wF \colon \mathbf{T} \to \mathbf{T}$ for the endomorphism defined by $wF(t) = {}^{w_1 \ldots w_r}F(t)$. Similarly, the norm map $Y(\mathbf{T}) \to \mathbf{T}^{wF}$ in the short exact sequence

(11.8) $$0 \longrightarrow Y(\mathbf{T}) \xrightarrow{wF-1} Y(\mathbf{T}) \xrightarrow{N_w} \mathbf{T}^{wF} \longrightarrow 1$$

will be denoted by $N_w := N_{w_1 \ldots w_r}$ (see (8.12)).

Notation and Definition 11.9. *Let $v = (v_j)_j$, $w = (w_j)_j$ in $(S \cup \{1\})^r$ be such that $v \leq w$.*

(a) Denote

$$I(v, w) := \{j \mid 1 \leq j \leq r,\ v_j = 1 \neq w_j\}.$$

(b) For $j \in I(1, w)$ let $\delta(w_j)$ be the simple root that defines the reflection w_j and denote

$$\eta_{w,j} = w_1 \ldots w_{j-1}(\delta(w_j)^\vee).$$

(c) Denote

$$Y_{[v,w]} = \sum_{j \in I(v,w)} \mathbb{Z}\,\eta_{w,j}.$$

If $v \leq x \leq y \leq w$, then $I(x, y) \subseteq I(v, w)$ and $Y_{[x,w]} \subseteq Y_{[v,w]}$.

Proposition 11.10. *Let $v \leq x \leq w$ in $(S \cup \{1\})^r$.*

(i) The quotient groups $\mathbf{T}^{wF}/N_w(Y_{[v,w]})$ and $\mathbf{T}^{xF}/N_x(Y_{[v,w]})$ are naturally isomorphic to the same quotient of $Y(\mathbf{T})$.

(ii) One has $Y_{[v,x]} + Y_{[x,w]} = Y_{[v,w]}$.

Proof. We prove first the inclusion

$$(\star) \qquad (w-v)Y(\mathbf{T}) \subseteq Y_{[v,w]}$$

We use induction on the number of elements of $I(v,w)$. $I(v,w)$ is empty if and only if $v=w$ and then the inclusion is evident. If $I(v,w)$ is not empty, let m be its smallest element. Let $x=(x_j)_j \in (S\cup\{1\})^r$ such that $v\leq x\leq w$ and $I(v,x)=\{m\}$, so that $Y_{[x,w]}\subseteq Y_{[v,w]}$, $I(x,w)=I(v,w)\setminus\{m\}$. One has $(w-v)Y(\mathbf{T})\subseteq (w-x)Y(\mathbf{T})+(x-v)Y(\mathbf{T})$. By the induction hypothesis, $(w-x)Y(\mathbf{T})\subseteq Y_{[x,w]}$. As $v\leq x$, $Y_{[x,w]}\subseteq Y_{[v,w]}$. One has $\delta(x_m)=\delta(w_m)$ and $\eta_{x,m}=\eta_{w,m}$ because x and w coincide on the first m components. By definition of x, $(x-v)Y(\mathbf{T})\subseteq w_1\ldots w_{m-1}(w_m-1)Y(\mathbf{T})$, hence $(x-v)Y(\mathbf{T})\subseteq \mathbb{Z}\eta_{w,m}$ and $\eta_{w,m}\in Y_{[v,w]}$ by Definition 11.9.

(i) is equivalent to the equality $(xF-1)Y(\mathbf{T})+Y_{[v,w]}=(wF-1)Y(\mathbf{T})+Y_{[v,w]}$ by (11.8) above. Using $(\star)$ and the inclusion $Y_{[x,w]}\subseteq Y_{[v,w]}$, one has

$$\begin{aligned}(xF-1)Y(\mathbf{T}) &\subseteq (wF-1)Y(\mathbf{T})+(x-w)FY(\mathbf{T})\\ &\subseteq (wF-1)Y(\mathbf{T})+Y_{[v,w]},\\ (wF-1)Y(\mathbf{T}) &\subseteq (xF-1)Y(\mathbf{T})+(w-x)FY(\mathbf{T})\\ &\subseteq (xF-1)Y(\mathbf{T})+Y_{[v,w]}.\end{aligned}$$

To prove (ii) we prove

$$(\star\star) \qquad Y_{[v,w]}=\sum_{j\in I(v,w)}\mathbb{Z}x_1\ldots x_{j-1}(\delta(w_j)^\vee)$$

Let $\eta_{w,j}=w_1\ldots w_{j-1}(\delta(w_j)^\vee)$ be some generator of $Y_{[v,w]}$ ($j\in I(v,w)$). One has

$$(w_1\ldots w_{j-1}-x_1\ldots x_{j-1})(\delta(w_j)^\vee)\subseteq \sum_{i\in I(x,w)\cap[1,j-1]}\mathbb{Z}\eta_{w,i}$$

by $(\star)$ applied *with* $r=j$ to the sequences of the first j components of x and w. Furthermore $I(x,w)\subseteq I(v,w)$. Then the family $(x_1\ldots x_{j-1}(\delta(w_j)^\vee))_{j\in I(v,w)}$ is expressed as a linear combination of generators of $Y_{[v,w]}$ by means of a unipotent triangular matrix with coefficients in $\mathbb{Z}$. This implies $(\star\star)$.

Clearly $I(v,w)$ is the disjoint union of $I(v,x)$ and $I(x,w)$, and $x_j=w_j$ when $j\in I(v,x)$. Applying $(\star\star)$ twice, to the triples (v,x,w) and (v,v,w), one obtains (ii). □

In the following we go beyond (i) of Proposition 11.10 to define some diagonalizable subgroups of $\mathbf{T}^{(r)}$. Recall (see §10.3) that the endomorphism F_r is defined on $\mathbf{G}^{(r)}=\mathbf{G}^r$, hence on $\mathbf{T}^{(r)}$ by $F_r(g_1,g_2,\ldots,g_r)=(g_2,\ldots,g_r,F(g_1))$.

Then $(S \cup \{1\})^r$ may be identified with a subset of $W(\mathbf{G}^{(r)}, \mathbf{T}^{(r)})$.

Recall the following from Lemma 10.9.

For $x = (x_j)_{1 \leq j \leq r}$ an element of $(S \cup \{1\})^r$, define $\iota_x : \mathbf{T} \to \mathbf{T}^{(r)}$ by

$$\iota_x(t) = (t, x_1{}^{-1} t x_1, \ldots, (x_1 \ldots x_{r-1})^{-1} t x_1 \ldots x_{r-1}) \quad (t \in \mathbf{T}).$$

For $v \leq w$ in $(S \cup \{1\})^r$, define $\mathbf{T}[v, w] \subseteq \mathbf{T}^{(r)}$ as the image of $\prod_{j \in I(v,w)} \delta(w_j)^\vee$ where $\delta(w_j)^\vee$ takes values in the jth component of $\mathbf{T}^{(r)}$. Finally denote

$$\mathbf{S}[v, w] = \{t \in \mathbf{T}^{(r)} \mid t^{-1}({}^{wF_r}t) \in \mathbf{T}[v, w]\}.$$

Proposition 11.11. *Let $v \leq x = (x_j)_j \leq w$ in $(S \cup \{1\})^r$. One has*

(i) $\iota_x(\mathbf{T}^{xF}) = (\mathbf{T}^{(r)})^{xF_r}$ and $\mathbf{S}[v, w] = \iota_x(\mathbf{T}^{xF})\mathbf{S}[v, w]^\circ$,

(ii) $\iota_x(N_x(Y_{[v,w]})) \subseteq \mathbf{S}[v, w]^\circ$ and the product of homomorphisms

$$Y(\mathbf{T}) \xrightarrow{N_x} \mathbf{T}^{xF} \xrightarrow{\iota_x} \mathbf{S}[v, w] \longrightarrow \mathbf{S}[v, w]/\mathbf{S}[v, w]^\circ$$

is independent of $x \in [v, w]$,

(iii) if the coroots of $\mathbf{G}$ are injections $\mathbb{G}_m \to \mathbf{T}$, then $\iota_x(N_x(Y_{[v,w]})) = \mathbf{S}[v, w]^\circ$.

Proof. The verification of the equality $\iota_x(\mathbf{T}^{xF}) = (\mathbf{T}^{(r)})^{xF_r}$ is immediate (and is valid for any $(x_j)_j \in N_{\mathbf{G}}(\mathbf{T})^r$). One has $\mathbf{T}[x, w] \subseteq \mathbf{T}[v, w]$ because $I(x, w) \subseteq I(v, w)$ and ${}^w t \in \mathbf{T}[x, w]^x t$ for all $t \in \mathbf{T}^{(r)}$. Hence

$$\text{(1)} \qquad \begin{aligned} \mathbf{S}[v, w] &= \{t \in \mathbf{T}^{(r)} \mid t^{-1}({}^{xF_r}t) \in \mathbf{T}[v, w]\}, \\ (w - x)Y(\mathbf{T}^{(r)}) &\subseteq Y(\mathbf{T}[v, w]). \end{aligned}$$

Hence (i) by Theorem 7.14(i) with $\mathbf{C} = \mathbf{T}[v, w]$ and $\mathbf{P} = \mathbf{L} = \mathbf{G}^{(r)}$, $\mathbf{H} = \mathbf{T}^{(r)}$, $n = (\dot{x}_1, \ldots, \dot{x}_r)$.

(ii) Let $d \in \mathbb{N}$ be such that $(wF)^d$ is split, i.e. $(wF)^d(t) = t^{q^d}$, for all $t \in \mathbf{T}$ and $w \in W$. Then $(wF_r)^{rd}$ is split. This allows us to define $N'_x : Y(\mathbf{T}^{(r)}) \to (\mathbf{T}^{(r)})^{wF_r}$, as N_w is defined in (11.8), by

$$N'_x(\eta) = N_{F_r^{rd}/xF_r}(\eta(\omega)) \quad (\eta \in Y(\mathbf{T}^{(r)}))$$

where ω is the selected element of order $q^d - 1$ in $\mathbf{F}^\times$. As ${}^{xF_r}N_{F_r^{rd}/xF_r}(t) = N_{F_r^{rd}/xF_r}(t)t^{q^d-1}$, $N_{F_r^{rd}/xF_r}(\mathbf{T}[v, w]) \subseteq \mathbf{S}[v, w]$. But $N_{F_r^{rd}/xF_r}$ preserves connectedness and has a finite kernel, contained in $(\mathbf{T}^{(r)})^{F_r^{rd}}$. Hence $N_{F_r^{rd}/xF_r}(\mathbf{T}[v, w]) \subseteq \mathbf{S}[v, w]^\circ$, and the tori $\mathbf{T}[v, w]$ and $N_{F_r^{rd}/xF_r}(\mathbf{T}[v, w])$ have equal dimension. Via Lang's map, $\mathbf{T}[v, w]$ and $\mathbf{S}[v, w]$ have equal dimension. Hence

$$\text{(2)} \qquad N_{F_r^{rd}/xF_r}(\mathbf{T}[v, w]) = \mathbf{S}[v, w]^\circ.$$

As a consequence $N'_x(Y(\mathbf{T}[v, w])) \subseteq \mathbf{S}[v, w]^\circ$.

Now we compare $N_{F_r^{rd}/xF_r}$ and $N_{F_r^{rd}/wF_r}$ on $Y(\mathbf{T}^{(r)})$. These maps may be extended to the vector space $V = \mathbb{Q} \otimes_{\mathbb{Z}} Y(\mathbf{T}^{(r)})$. Let $\eta \in Y(\mathbf{T}^{(r)})$ and let $\xi \in V$ be such that $\eta = (wF_r - 1)\xi$. One has $N_{F_r^{rd}/wF_r}((wF_r - 1)\xi) = (q^{d/m_0} - 1)\xi = N_{F_r^{rd}/xF_r}((xF_r - 1)\xi)$ and $\eta = (xF_r - 1)\xi + (w - x)F_r(\xi)$. Hence $N_{F_r^{rd}/wF_r}(\eta) - N_{F_r^{rd}/xF_r}(\eta) = -N_{F_r^{rd}/xF_r}((w - x)F_r(\xi))$. From the inclusion in (1) above, we see that $(w - x)F_r(\xi) \in \mathbb{Q} \otimes_{\mathbb{Z}} Y(\mathbf{T}[v, w])$, then by (2) above, $N_{F_r^{rd}/wF_r}(\eta) - N_{F_r^{rd}/xF_r}(\eta) \in \mathbb{Q} \otimes_{\mathbb{Z}} Y(\mathbf{S}[v, w]^\circ) \cap Y(\mathbf{T}^{(r)}) = Y(\mathbf{S}[v, w]^\circ)$.

The last relation implies that the composition of morphisms

$$\begin{aligned} Y\left(\mathbf{T}^{(r)}\right) &\xrightarrow{N'_x} \left(\mathbf{T}^{(r)}\right)^{xF_r} \longrightarrow \left(\mathbf{T}^{(r)}\right)^{xF'} / \left(\mathbf{T}^{(r)}\right)^{xF_r} \cap \mathbf{S}[v, w]^\circ \\ &\cong \mathbf{S}[v, w]/\mathbf{S}[v, w]^\circ \end{aligned}$$

is independent of x.

To obtain (ii) from the last result, factor N'_x via $\iota_x \circ N_x$. Let $\eta \in Y(\mathbf{T})$, let $\xi(\eta, j) = (0, \ldots, 0, \eta, 0, \ldots, 0) \in Y(\mathbf{T}^{(r)})$ with η on component j. The first component of $N'_x(\xi)$ is precisely $N_x(x_1 \ldots x_{j-1}(\eta))$. Hence the map $p_x \colon Y(\mathbf{T}^{(r)}) \to Y(\mathbf{T})$, such that $p_x(\xi(\eta, j)) = x_1 \ldots x_{j-1}(\eta)$ for all (η, j), satisfies $\pi_1 \circ N'_x = N_x \circ p_x$ where π_1 is the first projection. But N'_x have values in $(\mathbf{T}^{(r)})^{xF_r} = \iota_x(\mathbf{T}^{xF})$ and $\pi_1 \circ \iota_x$ is the identity, hence

$$N'_x = \iota_x \circ N_x \circ p_x. \tag{3}$$

Clearly $p_x(Y(\mathbf{T}[v, w])) = Y_{[v,w]}$, hence $\iota_x(N_x(Y_{[v,w]})) \subseteq \mathbf{S}[v, w]^\circ$ and p_x is surjective.

(iii) For a coroot $\alpha^\vee \in Y(\mathbf{T})$ (α a root) one has $\alpha^\vee(\mathbb{G}_m)^{F^d} = \alpha^\vee(\mathbb{F}^\times_{q^{d/m_0}})$. Hence $\mathbf{T}[v, w]^{F_r^{rd}} = \{\eta(\omega)\}_{\eta \in Y(\mathbf{T}[v,w])}$. Then by definition of N'_x one has $N'_x(Y(\mathbf{T}[v, w])) = N_{F_r^{rd}/xF_r}(\mathbf{T}[v, w]^{F_r^{rd}})$. As $(xF_r \circ N_{F_r^{rd}/xF_r})(t) = (N_{F_r^{rd}/xF_r} \circ xF_r)(t) = N_{F_r^{rd}/xF_r}(t)t^{q^d - 1}$, one has $N_{F_r^{rd}/xF_r}(\mathbf{T}[v, w]^{F_r^{rd}}) = N_{F_r^{rd}/xF_r}(\mathbf{T}[v, w]) \cap (\mathbf{T}^{(r)})^{xF_r}$. Thus one gets

$$N_{F_r^{rd}/xF_r}(\mathbf{T}[v, w]) \cap \left(\mathbf{T}^{(r)}\right)^{xF_r} = N'_x(Y(\mathbf{T}[v, w])).$$

With (2) and (3) above, this implies (iii). □

11.3. Deligne–Lusztig varieties associated with intervals

We keep the notation of the preceding section. We write $\mathbf{B} = \mathbf{U}\mathbf{T}$ where $\mathbf{U}$ is the unipotent radical of $\mathbf{B}$.

For $r \geq 1$ and $w \in (S \cup \{1\})^r$, recall $F_r \colon \mathbf{G}^{(r)} \to \mathbf{G}^{(r)}$ and, for $w = (w_1, \ldots w_r) \in (S \cup \{1\})^r$, $\mathbf{Y}(w) = \{g\mathbf{U}^{(r)} \mid g \in \mathbf{G}^{(r)},\ g^{-1}F_r(g) \in \mathbf{U}^{(r)}(\dot{w}_1, \ldots, \dot{w}_r)\mathbf{U}^{(r)}\}$, $\mathbf{X}(w) = \{g\mathbf{B}^{(r)} \mid g \in \mathbf{G}^{(r)},\ g^{-1}F_r(g) \in \mathbf{B}^{(r)}w\mathbf{B}^{(r)}\}$.

Definition 11.12. *Let $v \leq w$ in $(S \cup \{1\})^r$.*

(i) Let

$$\mathbf{X}[v, w] := \bigcup_{v \leq v' \leq w} \mathbf{X}(v'),$$

a locally closed subvariety of $(\mathbf{G}/\mathbf{B})^r$.

(ii) For any simple root δ let $\mathbf{G}_\delta$ be the subgroup of semi-simple rank 1 of $\mathbf{G}$ generated by $\mathbf{X}_\delta$ and $\mathbf{X}_{-\delta}$ (a central quotient of $\mathrm{SL}_2(\mathbf{F})$; see A2.4 or [Springer] §7.2). Recall the definition of $\delta(w_j)$ (see Definition 11.9). Let $\mathcal{U} = \mathcal{U}_1 \times \cdots \times \mathcal{U}_r \subseteq \mathbf{G}^{(r)}$ be defined by (v, w) and the components

$$\mathcal{U}_j = \begin{cases} \mathbf{G}_{\delta(w_j)}\mathbf{U} & \text{if } j \in I(v, w), \\ \mathbf{U}\dot{w}_j\mathbf{U} & \text{if } j \notin I(v, w). \end{cases}$$

Define $\mathbf{Y}[v, w]$ *by*

$$\mathbf{Y}[v, w] = \left\{ g\mathbf{U}^{(r)} \,\middle|\, g \in \mathbf{G}^{(r)},\ g^{-1}F_r(g) \in \mathcal{U} \right\}.$$

Proposition 11.13. *Keep $v \leq w$ in $(S \cup \{1\})^r$.*

(i) $\mathbf{Y}[v, w]$ is a locally closed smooth subvariety of $\mathbf{G}^{(r)}/\mathbf{U}^{(r)}$ and the morphism $\mathbf{Y}(w) \to \mathbf{Y}[v, w]$ is an open dominant immersion. The group $\mathbf{G}^F \cong (\mathbf{G}^{(r)})^{F_r}$ acts on $\mathbf{Y}[v, w]$ on the left.

(ii) The group $\mathbf{S}[v, w]$ (see Proposition 11.11) stabilizes $\mathbf{Y}[v, w]$, and the projection $\mathbf{G}/\mathbf{B} \to \mathbf{G}/\mathbf{U}$ defines a map $\mathbf{Y}[v, w] \to \mathbf{X}[v, w]$ that factors through the quotient $\mathbf{Y}[v, w] \to \mathbf{Y}[v, w]/\mathbf{S}[v, w]$. The induced map $\mathbf{Y}[v, w]/\mathbf{S}[v, w] \to \mathbf{X}[v, w]$ is an isomorphism.

Proof. If $j \in I(v, w)$, denote $\mathbf{T}_j = \mathbf{T} \cap \mathbf{G}_{\delta(w_j)}$. Then $\mathcal{U}_j = \mathbf{U}\dot{w}_j\mathbf{U}\mathbf{T}_j \cup \mathbf{U}\mathbf{T}_j$ by the BN-pair structure in the parabolic group $\mathbf{P}_{\delta(w_j)}$ and since $\dot{w}_j \in \mathbf{G}_{\delta(w_j)}$. Then (i) is obtained by applying Theorem 7.14 to $\mathbf{V} = \mathbf{U}^{(r)}$, $\mathbf{L} = \mathbf{T}^{(r)}$, $n = (\dot{w}_j)_j$, $\mathbf{C} = \prod_{j \in I(v,w)} \mathbf{G}_{\delta(w_j)}$ embedded in $\mathbf{G}^{(r)}$ by completing with 1's when $j \notin I(v, w)$.

(ii) The projection $(\mathbf{G}/\mathbf{U})^r \to (\mathbf{G}/\mathbf{B})^r$ sends $\mathbf{Y}[v, w]$ in $\mathbf{X}[v, w]$. From the definition of $\mathbf{Y}[v, w]$ it follows that, for any $y \in \mathbf{Y}[v, w]$ and $t \in \mathbf{T}^{(r)}$, $y.t$ is in $\mathbf{Y}[v, w]$ if and only if $t \in \mathbf{S}[v, w]$. Hence the morphism $\mathbf{Y}[v, w]/\mathbf{S}[v, w] \to \mathbf{X}[v, w]$ is defined and is bijective. It remains to check that it is separable (see A2.6). It suffices to check that it is an isomorphism over some open dense subvariety of $\mathbf{X}[v, w]$. We take $\mathbf{X}(w)$. Then our claim is a consequence of Theorem 7.14 with $\mathbf{P} = \mathbf{B}^{(r)}$, $\mathbf{L} = \mathbf{H} = \mathbf{T}^{(r)}$, $\mathbf{C} = \mathbf{T}[v, w]$ since then $\mathbf{K} = \mathbf{S}[v, w]$ (see the proof of Proposition 11.11(i)). □

It is classical to embed a given $(\mathbf{G}, F)$ in a connected reductive $(\widetilde{\mathbf{G}}, F)$ with connected center $\mathrm{Z}(\widetilde{\mathbf{G}})$ and $\widetilde{\mathbf{G}} = \mathrm{Z}(\widetilde{\mathbf{G}}).\mathbf{G}$ (see [DiMi91] p. 140 or §15.1 below). Taking the dual of this construction, one finds $\hat{\mathbf{G}} \to \mathbf{G}$, a surjection of

$\mathbf{F}$-groups defined over $\mathbb{F}_q$ with kernel a subtorus of $\mathrm{Z}^\circ(\hat{\mathbf{G}})$ and all coroots of $\hat{\mathbf{G}}$ being injections.

Theorem 11.14. *Let $v \le w$ in $(S \cup \{1\})^r$. Let $\hat{\mathbf{G}} \to \mathbf{G}$ be a covering as above, with standard identifications and kernel $\mathbf{C}$. Let $\hat{\mathbf{Y}}[v, w] \to \hat{\mathbf{X}}[v, w]$ (see Definition 11.12) be defined from $\hat{\mathbf{G}}$, identify $\mathbf{X}[v, w]$ and $\hat{\mathbf{X}}[v, w]$. Let $\mathbf{Y}_0[v, w] := \hat{\mathbf{Y}}[v, w]/\mathbf{C}^F\hat{\mathbf{S}}[v, w]^\circ$. Let*

$$p_{v,w}\colon \mathbf{Y}_0[v, w] \longrightarrow \mathbf{X}[v, w]$$

be induced by the natural map (Proposition 11.13). The above is a $\mathbf{T}^{wF}/N_w(Y_{[v,w]})$-torsor.

Let $x \in [v, w]$. The natural morphism $\mathbf{Y}(x) \to \mathbf{Y}_0[v, w]$ has values in $p_{v,w}^{-1}(\mathbf{X}(x))$. Via the isomorphism of groups $\mathbf{T}^{wF}/N_w(Y_{[v,w]}) \cong \mathbf{T}^{xF}/N_x(Y_{[v,w]})$ obtained in Proposition 11.10(i), the $\mathbf{T}^{xF}/N_x(Y_{[v,w]})$-torsor

$$\mathbf{Y}(x)/N_x\big(Y_{[v,w]}\big) \longrightarrow \mathbf{X}(x)$$

is isomorphic to the $\mathbf{T}^{wF}/N_w(Y_{[v,w]})$-torsor defined by restriction of $p_{v,w}$

$$p_{v,w}\colon p_{v,w}^{-1}(\mathbf{X}(x)) \longrightarrow \mathbf{X}(x).$$

Proof. From Proposition 11.11 one sees that the injection $\mathbf{Y}(x) \to \mathbf{Y}[v, w]$ is a morphism of $\mathbf{G}^F$-varieties-$\mathbf{T}^{xF}$ via $\mathbf{G}^F \cong (\mathbf{G}^{(r)})^{F_r}$ and $\iota_x\colon \mathbf{T}^{xF} \to \mathbf{S}[v, w]$. Using Proposition 11.11(i) and definition of $p_{v,w}$, one has isomorphisms of coverings of $\mathbf{X}(x)$

$$\begin{aligned}\mathbf{Y}(x)/(\iota_x(\mathbf{T}^{xF}) \cap \mathbf{S}[v, w]^\circ) &\cong ((\mathbf{Y}(x) \times \mathbf{S}[v, w])/\mathbf{T}^{xF})/\mathbf{S}[v, w]^\circ\\ &\cong p_{v,w}^{-1}(\mathbf{X}(x)).\end{aligned}$$

Assume first that $\hat{\mathbf{G}} = \mathbf{G}$. By Proposition 11.11(ii) and (iii), $\mathbf{Y}(x)/N_x(Y_{[v,w]})$ is isomorphic to $p_{v,w}^{-1}(\mathbf{X}(x))$ as $\mathbf{S}[v, w]/\mathbf{S}[v, w]^\circ$-torsor, independently of x.

In the general case, with $\rho\colon \hat{\mathbf{G}} \to \mathbf{G}$, $\mathbf{C} = \mathrm{Ker}(\rho)$ and defining $\rho(\dot{s}) = \rho(\dot{s})$ for all $s \in S$ (s and $\rho(s)$ may be identified) one has a natural isomorphism $\hat{\mathbf{Y}}(x)/\mathbf{C}^F \to \mathbf{Y}(x)$ by Theorem 7.14. Furthermore, for any x between v and w, by functoriality of the functor Y and relations between root data of $\hat{\mathbf{G}}$ and of $\mathbf{G}$, $\hat{\mathbf{T}}^{xF}/\hat{N}_x(\hat{Y}_{[v,w]})$ and $\mathbf{T}^{xF}/N_x(Y_{[v,w]})$ are isomorphic via ρ. □

Corollary 11.15. *Let $v \le x \le y \le w$ in $(S \cup \{1\})^r$. The canonical immersion $\mathbf{Y}_0[x, y] \hookrightarrow \mathbf{Y}_0[v, w]$ defines a $\mathbf{G}^F$-equivariant isomorphism of $\mathbf{T}^{wF}/N_w(Y_{[v,w]})$-torsors over $\mathbf{X}[x, y]$ via the isomorphism $\mathbf{T}^{yF}/N_y(Y_{[v,w]}) \cong \mathbf{T}^{wF}/N_w(Y_{[v,w]})$ (Proposition 11.10(i))*

$$\mathbf{Y}_0[x, y]/\big(N_y\big(Y_{[v,w]}\big)/N_y\big(Y_{[x,y]}\big)\big) \longrightarrow p_{v,w}^{-1}(\mathbf{X}[x, y]).$$

Proof. One has $I(x, y) \subseteq I(v, w)$ hence $Y_{[x,y]} \subseteq Y_{[v,w]}$ and an immersion $\mathbf{Y}_0[x, y] \hookrightarrow \mathbf{Y}_0[v, w]$ of image $p_{v,w}^{-1}(\mathbf{X}[x, y])$, over $\mathbf{X}[x, y] \hookrightarrow \mathbf{X}[v, w]$ by Theorem 11.14. The map in Corollary 11.15 is well defined. It is an isomorphism over the open subvariety $\mathbf{X}(y)$ by Theorem 11.14 again. □

11.4. Application: some mapping cones

We keep the notation of the preceding section (see Definition 11.12). When $A \subseteq B$ are subsets of $(S \cup \{1\})^r$, we denote by j_A^B the immersion $\bigcup_{w\in A} \mathbf{X}(w) \to \bigcup_{w\in B} \mathbf{X}(w)$. We abbreviate $[1^r, w]$ as $\overline{w}$.

We now fix $w \in (S \cup \{1\})^r$ and $\theta\colon \mathbf{T}^{wF} \to k^\times$ a morphism. Recall $w_\theta \in (S \cup \{1\})^r$ (see Definition 10.15). In the notation of Definition 11.9, a defining property of w_θ is as follows

(11.16) $$[w_\theta, w] = \big\{y \in (S \cup \{1\})^r \bigm| y \leq w,\ \theta \circ N_w(Y_{[y,w]}) = 1\big\}.$$

By Proposition 11.10(i), any $y \in [w_\theta, w]$ defines a morphism $\theta_y\colon \mathbf{T}^{yF} \to k^\times$ with $N_y(Y_{[w_\theta,w]})$ in its kernel.

Lemma 11.17. $y_{\theta_y} = w_\theta$.

Proof. If $x \leq y$, then $\theta \circ N_w(Y_{[y,w]}) = \theta_y \circ N_y(Y_{[x,y]})$ by Proposition 11.10(i) and (11.16) above. This implies $[y_{\theta_y}, y] = [w_\theta, y]$, again by (11.16). □

Proposition 11.18. *For $[x, y] \subseteq [w_\theta, w]$, the locally constant sheaf $\mathcal{F}_{[x,y]}(\theta)$ on $\mathbf{X}[x, y]_{\text{ét}}$ associated with $\theta\colon \mathbf{T}^{xF}/N_x(Y_{[w_\theta,w]}) \to k^\times$ and the $\mathbf{T}^{xF}/N_x(Y_{[w_\theta,w]})$-torsor on $\mathbf{X}[x, y]$ of Corollary 11.15 satisfy*

$$\mathcal{F}_{[x',y']}(\theta) = \big(j_{[x',y']}^{[x,y]}\big)^* \mathcal{F}_{[x,y]}(\theta)$$

for all $[x', y'] \subseteq [x, y]$.

Proof. The definition of $\mathcal{F}_{[x,y]}(\theta)$ is consistent by Proposition 11.11(ii) and Lemma 11.17. Using stalks, the claimed equality is clear from the definition of the locally constant sheaf. □

If $v \leq x \leq w$ in $(S \cup \{1\})^r$, we define $(j_{[v,x]}^{\overline{w}})_!\colon D_k^b(\mathbf{X}[v, x]) \to D_k^b(\overline{\mathbf{X}}(w))$ by $(j_{[v,x]}^{\overline{w}})_! = (j_{\overline{x}}^{\overline{w}})_* \circ (j_{[v,x]}^{\overline{x}})_!$. Note that $(j_{[v,x]}^{\overline{w}})_! = \mathrm{R_c}(j_{[v,x]}^{\overline{w}})_*$ since $j_{\overline{x}}^{\overline{w}}$ is a proper morphism.

Lemma 11.19. *If $x \in [w_\theta, w]$, the natural map $(j_{[w_\theta,x]}^{\overline{w}})_! \mathcal{F}_{[w_\theta,x]}(\theta) \to \mathrm{R}(j_{[w_\theta,x]}^{\overline{w}})_* \mathcal{F}_{[w_\theta,x]}(\theta)$ is an isomorphism.*

Proof. We have $(j^{\overline{w}}_{[w_\theta,x]})_! = (j^{\overline{w}}_{\overline{x}})_* \circ (j^{\overline{x}}_{[w_\theta,x]})_!$ and $(j^{\overline{w}}_{[w_\theta,x]})_* = (j^{\overline{w}}_{\overline{x}})_* \circ (j^{\overline{x}}_{[w_\theta,x]})_*$, where $j^{\overline{w}}_{\overline{x}}$ is a closed immersion, so that $(j^{\overline{w}}_{\overline{x}})_*$ is exact. The natural map $(j^{\overline{w}}_{[w_\theta,x]})_!\mathcal{F}_{[w_\theta,x]}(\theta) \to \mathrm{R}(j^{\overline{w}}_{[w_\theta,x]})_*\mathcal{F}_{[w_\theta,x]}(\theta)$ is the image by $(j^{\overline{w}}_{\overline{x}})_*$ of the natural map $(j^{\overline{x}}_{[w_\theta,x]})_!\mathcal{F}_{[w_\theta,x]}(\theta) \to \mathrm{R}(j^{\overline{x}}_{[w_\theta,x]})_*\mathcal{F}_{[w_\theta,x]}(\theta)$. Since $x_\theta = w_\theta$ (see Lemma 11.17), our claim now reduces to the case $w = x$, i.e., we must show

$$(1)\qquad \left(j^{\overline{w}}_{[w_\theta,w]}\right)_!\mathcal{F}_{[w_\theta,w]}(\theta) \xrightarrow{\sim} \mathrm{R}\left(j^{\overline{w}}_{[w_\theta,w]}\right)_*\mathcal{F}_{[w_\theta,w]}(\theta).$$

By the criterion of isomorphism in terms of stalks (see A3.2) and the expression of $\overline{\mathbf{X}}(w) \setminus \mathbf{X}[w_\theta, w]$ as $\bigcup_{w'} \overline{\mathbf{X}}(w')$ where w' ranges over elements of $[1^r, w] \setminus [w_\theta, w]$ such that $l(w') = l(w) - 1$ (see Lemma 11.2), it suffices to check that

$$(2)\qquad \left(j^{\overline{w}}_{\overline{w'}}\right)^*\mathrm{R}\left(j^{\overline{w}}_{[w_\theta,w]}\right)_*\mathcal{F}_{[w_\theta,w]}(\theta) = 0,$$

for any $w' \in [1, w] \setminus [w_\theta, w]$ such that $l(w') = l(w) - 1$. For such a w', Theorem 11.1(a′) tells us that $\mathcal{F}_w(\theta)$ ramifies along $\overline{\mathbf{X}}(w')$, and therefore (see Theorem A3.19) $(j^{\overline{w}}_{\overline{w'}})^*(j^{\overline{w}}_{w})_*\mathcal{F}_w(\theta) = 0$. Since $\mathcal{F}_w(\theta) = (j^{[w_\theta,w]}_{w})^*\mathcal{F}_{[w_\theta,w]}(\theta)$ (see Proposition 11.18), this can be rewritten as

$$\left(j^{\overline{w}}_{\overline{w'}}\right)^*\left(j^{\overline{w}}_{w}\right)_*\left(j^{[w_\theta,w]}_{w}\right)^*\mathcal{F}_{[w_\theta,w]}(\theta) = 0.$$

By cohomological purity (A3.13), we have $(j^{[w_\theta,w]}_{w})_*(j^{[w_\theta,w]}_{w})^*\mathcal{F}_{[w_\theta,w]}(\theta) \cong \mathcal{F}_{[w_\theta,w]}(\theta)$, so the above gives

$$\left(j^{\overline{w}}_{\overline{w'}}\right)^*\left(j^{\overline{w}}_{[w_\theta,w]}\right)_*\mathcal{F}_{[w_\theta,w]}(\theta) = 0.$$

By Theorem A3.19, this implies that $\mathcal{F}_{[w_\theta,w]}(\theta)$ ramifies along the divisor $\overline{\mathbf{X}}(w')$, thus implying (2) by Theorem A3.19 again. □

Let us define several subcategories of $D^b_k(\overline{\mathbf{X}}(w))$.

Definition 11.20. *Let $\mathcal{D}_1$ be the subcategory generated by the $\mathrm{R}(j^{\overline{w}}_{[x',x]})_*\mathcal{F}_{[x',x]}(\theta)$'s for $x' \leq x$ in $[w_\theta, w]$.*

Let $\mathcal{D}_2$ be generated by the $\mathrm{R}(j^{\overline{w}}_{[w_\theta,x]})_*\mathcal{F}_{[w_\theta,x]}(\theta)$'s for $x \in [w_\theta, w]$.

Let $\mathcal{D}'$ be generated by the $(j^{\overline{w}}_{x})_!\mathcal{F}_x(\theta)$'s for $x \in [w_\theta, w]$.

Let $\mathcal{D}'_1$ be generated by the $(j^{\overline{w}}_{[x',x]})_!\mathcal{F}_{[x',x]}(\theta)$'s for $x' \leq x$ in $[w_\theta, w]$.

Lemma 11.19 implies at once that $\mathcal{D}_2$ equals the category generated by the $(j^{\overline{w}}_{[w_\theta,x]})_!\mathcal{F}_{[w_\theta,x]}(\theta)$'s for $x \in [w_\theta, w]$, and therefore $\mathcal{D}_2 \subseteq \mathcal{D}'_1$.

Let us show that $\mathcal{D}'_1 \subseteq \mathcal{D}'$. Suppose we have $w_\theta \leq v \leq v' \leq x \leq w$ in $(S \cup \{1\})^r$ with $l(v') = l(v) + 1$. Then $\mathbf{X}[v, x] = \mathbf{X}[v', x] \sqcup \mathbf{X}[v, x']$ where $x' \in (S \cup \{1\})^r$ (see Lemma 11.2). By Proposition 11.18, the associated open-closed exact sequence (A3.9) can then be written as

$$0 \to \left(j^{[v,x]}_{[v',x]}\right)_!\mathcal{F}_{[v',x]}(\theta) \to \mathcal{F}_{[v,x]}(\theta) \to \left(j^{[v,x]}_{[v,x']}\right)_*\mathcal{F}_{[v,x']}(\theta) \to 0.$$

Applying the exact functor $(j^{\overline{w}}_{[v,x]})_! = (j^{\overline{w}}_{\overline{x}})_*(j^{\overline{x}}_{[v,x]})_!$ we get the exact sequence

$$0 \to \big(j^{\overline{w}}_{[v',x]}\big)_!\mathcal{F}_{[v',x]}(\theta) \to \big(j^{\overline{w}}_{[v,x]}\big)_!\mathcal{F}_{[v,x]}(\theta) \to \big(j^{\overline{w}}_{[v,x']}\big)_!\mathcal{F}_{[v,x']}(\theta) \to 0$$

since, regarding the fourth term, we have

$$\big(j^{\overline{x}}_{[v,x]}\big)_!\big(j^{[v,x]}_{[v,x']}\big)_* = \mathrm{R_c}\big(j^{\overline{x}}_{[v,x]}\big)_*\mathrm{R_c}\big(j^{[v,x]}_{[v,x']}\big)_* = \mathrm{R_c}\big(j^{\overline{x}}_{[v,x']}\big)_* = \big(j^{\overline{x}}_{\overline{x}'}\big)_*\big(j^{\overline{x}'}_{[v,x']}\big)_!$$

(see A3.6 and use the fact that $j^{[v,x]}_{[v,x']}$ is proper, being a closed immersion). This exact sequence implies that the middle term is in the bounded derived category generated by the others (see Exercise A1.3(c)). Then $(j^{\overline{w}}_{[v,x]})_!\mathcal{F}_{[v,x]}(\theta)$ is in the category generated by the $(j^{\overline{w}}_{[a,b]})_!\mathcal{F}_{[a,b]}(\theta)$'s where $[a,b] \subset [v,x]$ (a strict inclusion). Using induction, this implies that $(j^{\overline{w}}_{[v,x]})_!\mathcal{F}_{[v,x]}(\theta)$ is in $\mathcal{D}'$. Thus our claim is proved.

Let us show that $\mathcal{D}_1 \subseteq \mathcal{D}_2$. Taking again $w_\theta \leq v \leq v' \leq x \leq w$ in $(S \cup \{1\})^r$ with $l(v') = l(v) + 1$ and the associated decomposition $\mathbf{X}[v,x] = \mathbf{X}[v',x] \sqcup \mathbf{X}[v,x']$ (Lemma 11.2), the fact that $\mathcal{F}_{[v',x]}(\theta) = (j^{[v,x]}_{[v',x]})^*\mathcal{F}_{[v,x]}(\theta)$ (Proposition 11.18) does not ramify along $\mathbf{X}[v,x']$ implies that $\mathrm{R}(j^{[v,x]}_{[v',x]})_*\mathcal{F}_{[v',x]}(\theta)$ is represented by a complex

$$0 \to \mathcal{F}_{[v,x]} \to \big(j^{[v,x]}_{[v,x']}\big)_*\mathcal{F}_{[v,x']}(\theta) \to 0$$

(see Theorem A3.19). So $\mathrm{R}(j^{[v,x]}_{[v',x]})_*\mathcal{F}_{[v',x]}(\theta)$ is in the category generated by $\mathcal{F}_{[v,x]}$ and $(j^{[v,x]}_{[v,x']})_*\mathcal{F}_{[v,x']}(\theta)$ (being the mapping cone of the map of complexes concentrated in degree 0 associated with the above). Applying the derived functor $\mathrm{R}(j^{\overline{w}}_{[v,x]})_*$, one finds that $\mathrm{R}(j^{\overline{w}}_{[v',x]})_*\mathcal{F}_{[v',x]}(\theta)$ is in the category generated by $\mathrm{R}(j^{\overline{w}}_{[v,x]})_*\mathcal{F}_{[v,x]}$ and $\mathrm{R}(j^{\overline{w}}_{[v,x']})_*\mathcal{F}_{[v,x']}(\theta)$. Thus, for any $[v',x] \subseteq\,]w_\theta, w]$, this implies that $\mathrm{R}(j^{\overline{w}}_{[v',x]})_*\mathcal{F}_{[v',x]}(\theta)$ is in the subcategory generated by the $\mathrm{R}(j^{\overline{w}}_{[a,b]})_*\mathcal{F}_{[a,b]}(\theta)$'s for $[a,b] \subseteq [w_\theta, x]$ with $a < v'$. By induction this tells us that $\mathrm{R}(j^{\overline{w}}_{[v',x]})_*\mathcal{F}_{[v',x]}(\theta)$ is in $\mathcal{D}_2$. Thus our claim is proved.

Summing up the inclusions we have proved, we get $\mathcal{D}_1 \subseteq \mathcal{D}'$.

We now finish the proof of Theorem 11.1(b). Let C be the mapping cone of the map

$$\big(j^{\overline{w}}_w\big)_!\mathcal{F}_w(\theta) \to \mathrm{R}\big(j^{\overline{w}}_w\big)_*\mathcal{F}_w(\theta).$$

We now know that, given that this map is in $\mathcal{D}'$, C is in $\mathcal{D}'$, i.e. in the category generated by the $(j^{\overline{w}}_x)_!\mathcal{F}_x(\theta)$'s for $x \in [w_\theta, w]$. But this mapping cone is annihilated by the exact functor $(j^{\overline{w}}_w)^*$ since the image of the natural transformation $(j^{\overline{w}}_w)_! \to (j^{\overline{w}}_w)_*$ is the identity transformation $\mathrm{Id} \to \mathrm{Id}$. Also clearly $(j^{\overline{w}}_w)^*(j^{\overline{w}}_x)_!\mathcal{F}_x(\theta) = 0$ when $x \in [w_\theta, w[$ (use stalks). So, denoting by i the closed immersion $\overline{\mathbf{X}}(w) \setminus \mathbf{X}(w) \to \overline{\mathbf{X}}(w)$, we have $i_*i^*C = C$ and $i_*i^*(j^{\overline{w}}_x)_!\mathcal{F}_x(\theta) = (j^{\overline{w}}_x)_!\mathcal{F}_x(\theta)$ when $x \in [w_\theta, w[$, as a result of the open-closed exact sequence. But i_*i^* annihilates $(j_w^{\overline{w}})_!\mathcal{F}_w(\theta)$ since $i^*(j_w^{\overline{w}})_! = 0$. So C is in the image of $\mathcal{D}'$ by the exact functor i_*i^*, which is actually the category generated by the $(j_x^{\overline{w}})_!\mathcal{F}_x(\theta)$'s for $x \in [w_\theta, w[$. This completes our proof of Theorem 11.1. □

Exercises

1. Let $\mathbf{G} = \mathrm{GL}_2(\mathbf{F})$, and $\mathbf{B}$, resp. $\mathbf{T}$, be its upper triangular, resp. diagonal, subgroup. Let F be the Frobenius endomorphism raising matrix entries to the qth power. Let $s := \begin{pmatrix} 0 & 1 \\ 1 & 0 \end{pmatrix}$, so that $\mathbf{T}^{sF}$ is isomorphic with μ_{q+1} the group of $(q+1)$th roots of 1 in $\mathbf{F}^\times$. One considers the $\mathbf{T}^{sF}$-torsor $\mathbf{Y}(s)\xrightarrow{\pi}\mathbf{X}(s)$ and $\overline{\mathbf{X}}(s)$ as in Definition 7.12 and Proposition 7.13(iv).
   (a) Show that one may identify $\mathbf{Y}(s)$, resp. $\mathbf{X}(s)$, with the subvariety of $\mathbb{A}^2_{\mathbf{F}}$, resp. $\mathbb{P}^1_{\mathbf{F}}$, defined by $xy^q - yx^q = 1$, resp. $xy^q - yx^q \neq 0$, and the above $\mathbf{T}^{sF}$-torsor with the μ_{q+1}-action on $\mathbf{Y}(s)$ defined by $a.(x, y) = (ax, ay)$. Then $\overline{\mathbf{X}}(s)$ identifies with $\mathbb{P}^1_{\mathbf{F}}$.
   (b) Let $\mathbf{Y}'(s)$ be the closed subvariety of $\mathbb{P}^2_{\mathbf{F}}$ defined by $xy^q - yx^q = z^{q+1}$. Show that there is an open embedding $\mathbf{Y}(s)\xrightarrow{\rho}\mathbf{Y}'(s)$, which μ_{q+1}-action extends, and that $\mathbf{Y}'(s)$ is smooth. Show that we have a commutative square
$$\begin{array}{ccc} \mathbf{Y}(s) & \xrightarrow{\rho} & \mathbf{Y}'(s) \\ {\scriptstyle\pi}\downarrow & & \downarrow{\scriptstyle\overline{\pi}} \\ \mathbf{X}(s) & \xrightarrow{j} & \overline{\mathbf{X}}(s) \end{array}$$
   where $\overline{\pi}$ is induced by $(x, y, z) \mapsto (x, y)$ and is a μ_{q+1}-quotient (but not a torsor).
   (c) Let ω be a closed point of $\overline{\mathbf{X}}(s) \setminus \mathbf{X}(s)$, so that $|\overline{\pi}^{-1}(\omega)| = 1$. Use the arguments in the proof of Theorem A3.19 to check that $(\mathrm{R}j_*\mathcal{F}_s(\theta))_\omega = 0$ when $\theta\colon \mathbf{T}^{sF} \to k^\times$ is $\neq 1$.
   (d) Noting that the positive coroot of $Y(\mathbf{T})$ is sent to a generator of $\mathbf{T}^{sF}$ by the norm N_s, deduce that the above gives Theorem 11.1(a) in that case.
2. We use the notation of §11.1. Let $\lambda \in X(\mathbf{T})$.
   (a) Consider ψ_λ as a section of an invertible sheaf on $\mathbf{G}/\mathbf{B}$. Show that if $\langle\lambda, \beta_\nu^\vee\rangle > 0$ for any $\nu \in [1, r]$, then ψ_λ vanishes outside $O(w)$.
   (b) Use Proposition 11.7 to show that ψ induces a section of the invertible sheaf $\mathrm{pr}_1^*(\mathcal{L}_{\mathbf{G}/\mathbf{B}}(w^{-1}(\lambda) \circ F - \lambda))$ on $\overline{\mathbf{X}}(w)$ which is zero outside $\mathbf{X}(w)$.

   (c) Use the criterion of affinity ([EGA] I.5.5.7) to prove that $\mathbf{X}(w)$ is affine whenever $\mathcal{L}_{\mathbf{G}/\mathbf{B}}(w^{-1}(\lambda) \circ F - \lambda)$ is ample and $\langle \lambda, \beta_\nu^\vee \rangle > 0$ for any $\nu \in [1, r]$.
   (d) Show that the above two conditions can be achieved as long as there is some $\mu \in X(\mathbf{T})$ such that $\langle \mu, \delta^\vee \rangle > 0$ and $\langle \mu - \mu \circ F, \delta^\vee \rangle > 0$ for any simple root δ (use [Jantzen] II.4.4). Show that this is in turn possible as long as q is bigger than the sum of coordinates in $\Delta(\mathbf{G}, \mathbf{T})$ of any root (*Coxeter number* of $W(\mathbf{G}, \mathbf{T})$, see [Bour68] VI.1.11, [Hum90] 3.20).
3. Let X be a smooth $\mathbf{F}$-variety such that $X = U \sqcup D$ where $D = \overline{\{d\}}$ is of codimension 1. Let $Y \to U$ be a $\mathbb{G}_{\mathrm{m}}$-torsor. Let ψ be an automorphism of Y over U.
   (a) Show that the possibility of extending ψ over D reduces to the study of some morphism $\mathbb{G}_{\mathrm{a}} \to X$ with $0 \mapsto d$ and its pull-back to Y (see A3.17).
   (b) With $X = \mathbb{G}_{\mathrm{a}}$ and $U = \mathbb{G}_{\mathrm{m}}$, show that ψ is defined in a neighborhood of 0 by an endomorphism of $\mathbf{F}[u, u^{-1}] \otimes_{\mathbf{F}} \mathbf{F}[z, z^{-1}]$ such that $z \mapsto z$, $u \mapsto az^m u$ for some $a \in \mathbf{F}^\times$, $m \in \mathbb{Z}$. Call m the *local order* of ψ near 0. Then ψ extends to $\mathbf{F}[z]$ if and only if $m = 0$.
4. (a) Using Proposition 11.18, show the converse of Theorem 11.1(a′).
   (b) Compute $(j_{w'}^{\overline{w}})^* \mathrm{R}^m (j_w^{\overline{w}})_* \mathcal{F}_w(\theta)$ for $w_\theta \leq w' < w$, $l(w') = l(w) - 1$, $m = 0, 1$ (use A3.11 and the case of the constant sheaf; see [SGA.4$\frac{1}{2}$] p. 255).

Notes

Section 11.1 draws on [DeLu76] §9.5 and §9.6. See [BoRo04] for an alternative approach. The rest of the chapter is taken from [BoRo03]. Exercise 1 was communicated to us by Cédric Bonnafé.

12

Jordan decomposition as a Morita equivalence: modules

This chapter is essentially about modules for $k\mathbf{G}^F$ and $k\mathbf{T}_0^{wF}$, the corresponding derived categories and the functors defined by Deligne–Lusztig varieties between them. We expound Bonnafé–Rouquier's proof of Theorem 10.17(b), which is the last step to check Theorem 10.1 establishing a Morita equivalence between $\Lambda\mathbf{G}^F.b_\ell(\mathbf{G}^F, s)$ and $\Lambda\mathbf{L}^F.b_\ell(\mathbf{L}^F, s)$ when $s \in \mathbf{G}^*$ satisfies $\mathrm{C}_{\mathbf{G}^*}(s) \subseteq \mathbf{L}^*$.

Theorem 10.17(b) requires us to show that $k\mathbf{G}^F.b_\ell(\mathbf{G}^F, s)$ is generated by the elements $\mathcal{S}_{(w,\theta)}$ of $D^b(k\mathbf{G}^F)$ defined by étale cohomology of varieties $\mathbf{X}(w)$ and pairs (w, θ) (see §10.4) with $\theta\colon \mathbf{T}^{wF} \to k^\times$ in a rational series associated with s.

One first proves a general criterion of generation of $D^b(A\text{-}\mathbf{proj})$ (A a finite-dimensional algebra over a field) by complexes satisfying a kind of triangularity with respect to simple A-modules. The remainder of the chapter consists in showing that the complexes $\mathcal{S}_{(w,\theta)}$ satisfy the three main hypotheses of this general criterion.

The first hypothesis is checked in §12.2 (Proposition 12.3).

The second hypothesis is checked in §12.3 (Proposition 12.9). A key fact is that, given P a projective indecomposable $k\mathbf{G}^F$-module, then the smallest w such that $\mathcal{S}_{(w,\theta)}$ displays P has it essentially in degree $l(w)$. The main argument uses quasi-affinity through Proposition 10.3. This is also where Theorem 11.1(b), established in the preceding chapter, is used.

The third hypothesis of the generation criterion of Proposition 12.1 is checked in §12.4. The important fact is the disjunction of the complexes $\mathcal{S}_{(w,\theta)}$ according to rational series. The proof is modeled on the case of characters (disjunction of $\mathrm{R}_\mathbf{T}^\mathbf{G}\theta$'s). Several arguments involve characters over k or Λ (i.e. the Grothendieck group of $K\mathbf{G}^F$ or $k\mathbf{G}^F$) and especially the independence of twisted induction $\mathrm{R}_{\mathbf{T}\subseteq\mathbf{B}}^\mathbf{G}$ with regard to $\mathbf{B}$ (see Theorem 8.17(i)).

12.1. Generating perfect complexes

In the following, A is a finite-dimensional algebra over a field k. Note that the objects of $D^b(k-\mathbf{mod})$ may be identified with their homology (see for instance Exercise 4.12).

Proposition 12.1. *Let $\mathcal{C}$ be a finite set of objects of $C^b(A$-$\mathbf{proj})$, endowed with a map $l:\mathcal{C} \to \mathbb{N}$ and an equivalence relation $\sim$ such that*

(1) if S is a simple A-module, there is at least some $C \in \mathcal{C}$ such that $\mathrm{RHom}_A(C, S) \neq 0$,

(2) if C is of minimal $l(C)$ satisfying the above for a given S, then $\mathrm{RHom}_A(C, S)$ *has non-zero homology in degree 0 and only there,*

(3) if $C \not\sim C'$ in $\mathcal{C}$, then $\mathrm{RHom}_A(C, C') = 0$.

Then $A = \prod_{\mathbf{c}\in\mathcal{C}/\sim} A_{\mathbf{c}}$, a direct product where the simple $A_{\mathbf{c}}$-modules are the simple A-modules S such that $\mathrm{RHom}_A(C, S) \neq 0$ *for some $C \in \mathbf{c}$. If $\mathbf{c} \in \mathcal{C}/\sim$, denote by <$\mathbf{c}$> the smallest full subcategory of $C^b(A-\mathbf{mod})$ containing $\mathbf{c}$, and stable under direct sums, direct summands, shifts and mapping cones (see A1.7). Then <$\mathbf{c}$> contains the regular module ${}_{A_{\mathbf{c}}}A_{\mathbf{c}}[0]$.*

Proof of Proposition 12.1. Let us first forget about $\sim$, i.e. we assume that $\mathcal{C}/\sim = \{\mathcal{C}\}$.

In view of the hypotheses and the conclusion, one may assume that the objects in $\mathcal{C}$ have no direct summand null homotopic. We prove the following.

Lemma 12.2. *If C is a perfect complex with no summand null homotopic and S is a simple A-module, then* $\mathrm{H}^{-i}(\mathrm{RHom}_A(C, S)) \neq 0$ *for both the smallest and the biggest i such that* $\mathrm{Hom}_A(C^i, S) \neq 0$.

Note that the above implies that, in addition to (1) and (3), we have, for any simple A-module S,

(2′) if $C \in \mathcal{C}$ is of minimal $l(C)$ satisfying $\mathrm{RHom}_A(C, S) \neq 0$, then
$\mathrm{Hom}_A(C^i, S) \neq 0$ if and only if $i = 0$.

Proof of Lemma 12.2. Let i_0 be the smallest integer such that $\mathrm{Hom}_A(C^{i_0}, S) \neq 0$ and let us check that $\mathrm{H}^{-i_0}(\mathrm{RHom}_A(C, S)) \neq 0$. Assume that on the contrary $\mathrm{Hom}_A(C^{i_0+1}, S) \to \mathrm{Hom}_A(C^{i_0}, S) \to 0$ is exact (see A1.11). Choose a surjection $C^{i_0} \to S$. Then the above exact sequence allows us to extend our $C^{i_0} \to S$ to a morphism $C \to S_{[i_0,i_0+1]}$ (see the notation $M_{[i,i+1]}$ for A-modules M in Exercise A1.2) onto in each degree (this is where we use minimality of i_0). Denote by P a projective cover of S. Then the projectivity of $P_{[i_0,i_0+1]}$ in $C^b(A-\mathbf{mod})$ (see Exercise 4.11) implies that we have morphisms

$$P_{[i_0,i_0+1]} \xrightarrow{t} C \to S_{[i_0,i_0+1]}$$

whose composition is onto. By projectivity of C^{i_0+1}, there is a retraction $t' : C^{i_0+1} \longrightarrow P$ of t^{i_0+1}, and $t' \circ \partial^{i_0} : C^{i_0} \longrightarrow P$ is also a retraction of t^{i_0}. This yields a retraction $C \to P_{[i_0,i_0+1]}$ of the above t in $C^b(A-\mathbf{mod})$, and therefore C has a direct summand $\cong P_{[i_0,i_0+1]}$, hence null homotopic: a contradiction. Similarly, if i_0 is now the biggest integer such that P is a direct summand of C^{i_0} and if $\mathrm{H}^{-i_0}(\mathrm{RHom}_A(C, S)) = 0$, this implies that $0 \to \mathrm{Hom}_A(C^{i_0}, S) \to \mathrm{Hom}_A(C^{i_0-1}, S)$ is exact. Then a surjection $C^{i_0} \to S$ will yield a morphism $C \to S_{[i_0-1,i_0]}$ onto in each degree, and the same arguments as above would imply that C has a direct summand $\cong P_{[i_0-1,i_0]}$, a contradiction.

We now replace (2) with (2′).

We must check that ${}_AA \in <\mathcal{C}>$, or equivalently $<\mathcal{C}>$ contains any projective indecomposable module. Assume this is not the case. By (1), this means that some C^i for some $C \in \mathcal{C}$ is not in $<\mathcal{C}>$. Let $\mathcal{E} \subseteq \mathcal{C}$ be defined by $C \in \mathcal{E}$ if and only if some C^i is not in $<\mathcal{C}>$. Let $C \in \mathcal{E}$ be such that $l(C)$ is minimal. Now (2′) implies that any C^i with $i \neq 0$ is in $<\mathcal{C}>$, and that $C^0 \notin <\mathcal{C}>$.

Assume there is $i_0 < 0$ with $C^{i_0} \neq 0$. We may assume that $C^i = 0$ for $i < i_0$. Let $f : C \to C^{i_0}[-i_0]$ be defined by Id at degree i_0. Then $\mathrm{Cone}(f) \cong (C^{i_0})_{[i_0-1,i_0]} \oplus C^{>i_0}[1]$, where $C^{>i_0}$ is the complex obtained from C by replacing C^{i_0} with 0 and leaving other terms unchanged (see Exercise A1.3(a)). By iteration, we get that the complex $C' := \ldots \to 0 \to 0 \to C^0 \to C^{-1} \to \ldots$ is in $<\mathcal{C}>$. Taking now i_0 to be the highest degree such that $C^i \neq 0$, we may consider $g : C^{i_0}[-i_0] \to C'$ defined by Id at degree i_0 (same as above on k-duals) and easily find that the mapping cone of g is isomorphic with $(C^{i_0})_{[i_0-1,i_0]} \oplus (C')^{<i_0}$. Iterating this as long as $i_0 > 0$ (and therefore $C^{i_0}[-i_0] \in <\mathcal{C}>$), we get finally that $C^0 \in <\mathcal{C}>$: a contradiction.

Assume now that $\sim$ is non-trivial. In view of our claim, we may assume that $\mathcal{C} = \mathcal{C}' \cup \mathcal{C}''$ with $\mathrm{RHom}_A(C', C'') = 0$ for all $C' \in \mathcal{C}'$, $C'' \in \mathcal{C}''$. This can be written as

$$\mathrm{Hom}_{D^b(A)}(C', C''[i]) = 0$$

for all $i \in \mathbb{Z}$ (see A1.11). The above is true for any $C' \in <\mathcal{C}'>$, $C'' \in <\mathcal{C}''>$ by the basic properties of derived functors. Then $<\mathcal{C}' \cup \mathcal{C}''> = <\mathcal{C}'> \times <\mathcal{C}''>$.

We know that ${}_AA \in <\mathcal{C}>$. So we may write ${}_AA = C_0' \oplus C_0''$ with $C_0' \in <\mathcal{C}'>$ and $C_0'' \in <\mathcal{C}''>$. Taking endomorphism algebras, we get $A \cong B' \times B''$ as k-algebras with $B'C'' = B''C' = 0$ in $D^b(A)$ for all $C' \in \mathcal{C}'$, $C'' \in \mathcal{C}''$. Then the simple B'-modules are exactly the simple A-modules S such that $\mathrm{RHom}_A(C', S) \neq 0$ for some $C' \in \mathcal{C}'$ by Lemma 12.2. □

12.2. The case of modules induced by Deligne–Lusztig varieties

Recall that $(\mathbf{G}, F)$ is a connected reductive $\mathbf{F}$-group defined over $\mathbb{F}_q$, with $\mathbf{T} \subseteq \mathbf{B}$ a maximal torus and Borel subgroup, both F-stable and with respect to which the Weyl group $W := \mathrm{N}_{\mathbf{G}}(\mathbf{T})/\mathbf{T}$ with set S of generators is defined (see §10.3)

We plan to apply Proposition 12.1 in the following framework.

For $A = k\mathbf{G}^F$, we take $\mathcal{C}$ and l as follows. Recall the set $\Sigma(S)_{\text{red}}$ of reduced decompositions of elements of W. One takes $\mathcal{C}$ as a set of representatives of the $\mathcal{S}_{(w,\theta)}[l(w)]$'s for $w \in \Sigma(S)_{\text{red}}$, $\theta \in \mathrm{Hom}(\mathbf{T}^{wF}, k^\times)$ (see Notation 10.8 and Definition 10.16). One defines $l(\mathcal{S}_{(w,\theta)}) = l(w)$, and $\sim$ to be the relation on the (w, θ)'s defined by rational series (see Definition 10.14).

Proposition 12.3. *Theorem 10.17(b) reduces to showing that the $\mathcal{S}_{(w,\theta)}$'s above satisfy conditions (2) and (3) of Proposition 12.1.*

Proof of Proposition 12.3. It suffices to show that condition (1) is satisfied and that $A_{\mathbf{c}} \supseteq k\mathbf{G}^F.b_\ell(\mathbf{G}^F, s)$ (see Definition 9.9) whenever $\mathbf{c}$ corresponds to $s \in \mathbf{G}^{*F}$. Note that the latter inclusion will imply equality $A_{\mathbf{c}} = k\mathbf{G}^F.b_\ell(\mathbf{G}^F, s)$ by Proposition 12.1 and the fact that the $b_\ell(\mathbf{G}^F, s)$'s are all the block idempotents of $\Lambda\mathbf{G}^F$.

Lemma 12.4. *If $(w, \theta) \in \Theta(\mathbf{G}, F)$, the Brauer character (see [Ben91a] §5.3, [NaTs89] §3.6) of the element of the Grothendieck group of $k\mathbf{G}^F$ associated with $\mathcal{S}_{(w,\theta)}$ is the generalized character of $\varprojlim_n \mathrm{R}\Gamma(\mathbf{Y}(w), (\Lambda/J(\Lambda)^n)_{\mathbf{Y}(w)}).b_\theta^\Lambda$ (where b_θ^Λ denotes the primitive idempotent of $\Lambda\mathbf{T}^{wF}$ lifting $b_\theta \in k\mathbf{T}^{wF}$).*

Proof of Lemma 12.4. This is a straightforward consequence of the fact that $\mathrm{R}\Gamma(\mathbf{Y}(w), \Lambda_{\mathbf{Y}(w)}).b_\theta^\Lambda$ may be represented by a complex Γ of projective $\Lambda\mathbf{G}^F$-modules (using the fact that the stabilizers of closed points of $\mathbf{Y}(w)$ in $\mathbf{G}^F$ are ℓ'-groups; see A3.15). Then the Brauer character of $\mathcal{S}_{(w,\theta)} = \mathrm{R}\Gamma(\mathbf{Y}(w), k_{\mathbf{Y}(w)}).b_\theta$ is that of $\Gamma \otimes k$, therefore it is $\sum_i (-1)^i [\Gamma^i]$. The same reasoning gives that this is also the character of $\mathrm{R}\Gamma(\mathbf{Y}(w), \Lambda_{\mathbf{Y}(w)}).b_\theta^\Lambda$. □

Let M be a simple $k\mathbf{G}^F$-module; denote by ρ its Brauer character (a central function $\mathbf{G}^F \to K$ which vanishes outside ℓ'-elements; see [Ben91a] §5.3). By Theorem 8.17(ii) for the regular character, there is some $w \in W(\mathbf{G}, \mathbf{T})$ and some $\theta \in \mathrm{Irr}(\mathbf{T}^{wF})$ such that $\langle \mathrm{R}_w^{\mathbf{G}}\theta, \rho\rangle_{\mathbf{G}^F} \neq 0$, where $\mathrm{R}_w^{\mathbf{G}}\theta = \mathrm{R}_{\mathbf{T}_w}^{\mathbf{G}}\theta$ for $\mathbf{T}_w$ of type w with regard to $\mathbf{T}$ (see §8.2). Since $\rho = \rho\delta_{\ell'}$ for $\delta_{\ell'}$ the characteristic function of ℓ'-elements and since $\delta_{\ell'}\mathrm{R}_w^{\mathbf{G}}\theta = \mathrm{R}_w^{\mathbf{G}}(\delta_{\ell'}\theta)$ (see Proposition 9.6(iii)), we may replace θ by its ℓ'-part. So we assume that θ is ℓ'. We use the same letter to denote the associated linear characters $\mathbf{T}^{wF} \to \Lambda^*$ or k^*. If we define the

$(\mathbf{G}^*)^F$-conjugacy class of $s \in (\mathbf{G}^*)^F_{\mathrm{ss}}$ as corresponding to (w, θ), then $b_\ell(\mathbf{G}^F, s).\mathrm{R}_w^{\mathbf{G}}\theta = \mathrm{R}_w^{\mathbf{G}}\theta$ and M is a $k\mathbf{G}^F.b_\ell(\mathbf{G}^F, s)$-module by the above about ρ.

Let d_w be a reduced decomposition of w. Then $(d_w, \theta) \in \Theta_k(\mathbf{G}, F)$. Let us show that $\mathrm{RHom}_{k\mathbf{G}^F}(\mathcal{S}_{(d_w,\theta)}, M)) \neq 0$.

By its definition, $\mathrm{R}_w^{\mathbf{G}}\theta$ is the Lefschetz character (see A1.12) of $[\mathrm{H}(\mathrm{R}\Gamma(\mathbf{X}(w), \pi_*^{\mathbf{T}^{wF}} \Lambda_{\mathbf{Y}(w)})), \otimes_\Lambda K].e_\theta$. Therefore, since $b_\theta = \sum_\tau e_{\theta\tau}$ (the sum being over ℓ-characters of $\mathbf{T}^{wF}$), we have that $\sum_\tau \mathrm{R}_w^{\mathbf{G}}(\theta\tau)$ is the Lefschetz character of $\mathrm{H}(\mathrm{R}\Gamma(\mathbf{X}(w), \pi_*^{\mathbf{T}^{wF}} \Lambda_{\mathbf{Y}(w)}.b_\theta)) \otimes_\Lambda K$. Denoting by Γ a representative of $\mathrm{R}\Gamma(\mathbf{X}(w), \pi_*^{\mathbf{T}^{wF}} \Lambda_{\mathbf{Y}(w)}.b_\theta)$ in $C^b(\Lambda\mathbf{G}^F\text{–}\mathbf{proj})$ (see A3.15) without non-zero summand null homotopic, we have that $\Gamma \otimes_\Lambda k$ represents $\mathcal{S}_{(d_w,\theta)}$ (use Proposition 10.12 with $I = \emptyset$, then $\mathbf{Y}(w) \cong \mathbf{Y}(d_w)$). Using again that $\rho = \rho\delta_{\ell'}$ and that multiplication with $\delta_{\ell'}$ commutes with the twisted induction $\mathrm{R}_\mathbf{L}^\mathbf{G}$ (see Proposition 9.6(iii)), we get $\langle \sum_\tau \mathrm{R}_w^{\mathbf{G}}(\theta\tau), \rho\rangle_{\mathbf{G}^F} = |\mathbf{T}^{wF}|_\ell \langle \mathrm{R}_w^{\mathbf{G}}\theta, \rho\rangle_{\mathbf{G}^F} \neq 0$. By the above, this implies that there is some i such that $\langle \mathrm{H}^i(\Gamma) \otimes K, \rho\rangle_{\mathbf{G}^F} \neq 0$ (we identify $K\mathbf{G}^F$-modules and characters). Then $\langle \Gamma^i \otimes K, \rho\rangle_{\mathbf{G}^F} \neq 0$. But Γ^i being a projective $\Lambda\mathbf{G}^F$-module, and ρ the Brauer character of M, this means that the projective cover of M is a direct summand of $\Gamma^i \otimes k$ (see for instance [NaTs89] 3.6.10(i), [Thévenaz] 42.9). By Lemma 12.2, this gives condition (1) of Proposition 12.1 for our situation. □

12.3. Varieties of minimal dimension inducing a simple module

In view of Proposition 12.3, we now show that condition (2) of Proposition 12.1 is satisfied by the class $\mathcal{C}$ defined in §12.2.

Let us first introduce some more notation.

Definition 12.5. *Let $w \in \Sigma(S)$. This defines $\mathbf{Y}(w)$ (see Notation 10.8), a variety with action of $\mathbf{G}^F$ on the left and of $\mathbf{T}^{wF}$ on the right. One denotes by $\mathcal{S}_w$, $\mathcal{R}_w$: $D^b(k\mathbf{T}^{wF}) \to D^b(k\mathbf{G}^F)$ the associated "inductions." That is,*

$$\mathcal{S}_w M = \mathrm{R}\Gamma(\mathbf{Y}(w), k) \overset{\mathrm{L}}{\otimes}_{k\mathbf{T}^{wF}} M, \quad \mathcal{R}_w M = \mathrm{R_c}\Gamma(\mathbf{Y}(w), k) \overset{\mathrm{L}}{\otimes}_{k\mathbf{T}^{wF}} M$$

for M in $D^b(k\mathbf{T}^{wF})$. Whenever M is a projective $k\mathbf{T}^{wF}$-module, we choose representatives of $\mathcal{S}_w M$ and $\mathcal{R}_w M$ in $C^b(k\mathbf{G}^F\text{–}\mathbf{proj})$ (see A3.7 and A3.15) without non-zero summand null homotopic. Recall (§10.4) the notation $\mathcal{S}_{(w,\theta)} = \mathcal{S}_w(k\mathbf{T}^{wF} b_\theta)$ and $\mathcal{R}_{(w,\theta)} = \mathcal{R}_w(k\mathbf{T}^{wF} b_\theta)$ where b_θ is the primitive idempotent of $k\mathbf{T}^{wF}$ associated with $\theta\colon \mathbf{T}^{wF} \to k^\times$.

The following is what will be used from Theorem 11.1(b).

Theorem 12.6. *The mapping cone of the morphism $\mathcal{R}_w(\theta) \to \mathcal{S}_w(\theta)$ is in the subcategory of $D^b(k\mathbf{G}^F)$ generated by the $\mathcal{S}_{w'}(\theta')$'s for $w' < w$ and $\theta' \in \mathrm{Hom}(\mathbf{T}^{w'F}, k^\times)$.*

Proof. If $w' \leq w$ in S, denote by $j_w^{\overline{w}}: \mathbf{X}(w) \to \overline{\mathbf{X}}(w)$ and by $j_{\overline{w}'}^{\overline{w}}: \overline{\mathbf{X}}(w') \to \overline{\mathbf{X}}(w)$ the open and closed immersions, respectively. By Theorem 11.1(b), the morphism

$$\left(j_w^{\overline{w}}\right)_! \mathcal{F}_w(\theta) \to \mathrm{R}\left(j_w^{\overline{w}}\right)_* \mathcal{F}_w(\theta)$$

in $D_k^b(\overline{\mathbf{X}}(w))$ has a mapping cone which is in the subcategory generated by the sheaves $\mathcal{F}'_{w'}(\theta') := (j_{\overline{w}'}^{\overline{w}})_* (j_{w'}^{\overline{w}'})_! \mathcal{F}_{w'}(\theta')$ for $w' < w$ and $\theta': \mathbf{T}^{w'F} \to k^\times$. The morphism $\mathcal{R}_w(\theta) \to \mathcal{S}_w(\theta)$ is the image $\mathrm{R}\Gamma(\overline{\mathbf{X}}(w), -)$ of the above morphism (see A3.4 and A3.6), so its mapping cone is in the subcategory of $D^b(k\mathbf{G}^F)$ generated by the $\mathrm{R}\Gamma(\overline{\mathbf{X}}(w), \mathcal{F}'_{w'}(\theta'))$. We have $\mathrm{R}\Gamma(\overline{\mathbf{X}}(w), \mathcal{F}'_{w'}(\theta')) \cong \mathrm{R}\Gamma(\overline{\mathbf{X}}(w'), (j_{w'}^{\overline{w}'})_! \mathcal{F}_{w'}(\theta')) = \mathrm{R_c}\Gamma(\mathbf{X}(w'), \mathcal{F}_{w'}(\theta')) = \mathcal{R}_{w'}(\theta')$ by A3.6 again. So the mapping cone of the morphism $\mathcal{R}_w(\theta) \to \mathcal{S}_w(\theta)$ is in the subcategory of $D^b(k\mathbf{G}^F)$ generated by the $\mathcal{S}_{w'}(\theta')$ for $w' < w$ and $\theta' \in \mathrm{Hom}(\mathbf{T}^{w'F}, k^\times)$. But $\mathcal{R}_{w'}(\theta') \cong \mathcal{S}_{w'}(\theta^{-1})[-2l(w')]$ by Poincaré–Verdier duality (A3.12) and smoothness of $\mathbf{X}(w')$. Thus our claim is proved. □

Lemma 12.7. *Let $\mathbf{X}$ be a quasi-affine $\mathbf{F}$-variety of dimension d. Let $\mathcal{F}$ be a constructible sheaf of k-spaces on $\mathbf{X}_{\text{ét}}$. Assume that $\mathcal{D}(\mathcal{F})$ is quasi-isomorphic to $\mathcal{G}[-2d]$ for $\mathcal{G}$ a sheaf (see the notation $\mathcal{D}(\mathcal{F})$ in A3.12). Assume that both $\mathcal{F}$ and $\mathcal{G}$ satisfy Condition 10.2(c). Then $\mathrm{H}^i(\mathbf{X}, \mathcal{F}) = 0$ for $i \neq d$.*

Proof. The proof is almost identical to the one for Proposition 10.3. We do not have that $\mathbf{X}$ is smooth or that $\mathcal{D}(\mathcal{F})$ is the naive k-dual $\mathcal{H}om(\mathcal{F}, k_{\mathbf{X}}[-2d])$ but we do have $\mathrm{H}_c^i(\mathbf{X}, \mathcal{G}) \cong \mathrm{H}^{2d-i}(\mathbf{X}, \mathcal{F})^*$ by Poincaré–Verdier duality (in the form of A3.12) and since $\mathcal{D}(\mathcal{F})$ is concentrated in degree $-2d$. This allows us to proceed with the same arguments on i's as in Proposition 10.3. □

Here is a case where $\mathcal{D}(\mathcal{F})$ is easily computed without the variety being supposed smooth, but a finite quotient of a smooth variety.

Lemma 12.8. *Let G be a finite group acting on a variety $\mathbf{X}$ such that the stabilizers of closed points are of order invertible in k. Assume $\mathbf{X}$ is quasi-projective, smooth, with all connected components of same dimension d. Let $\pi: \mathbf{X} \to \mathbf{X}/G$ be the finite quotient. If M is a kG-module, denote $\mathcal{F}_M := \pi_*^G k_{\mathbf{X}} \otimes_{kG} M_{\mathbf{X}/G} = \pi_*^G k_{\mathbf{X}} \overset{\mathrm{L}}{\otimes}_{kG} M_{\mathbf{X}/G}$ (see A3.15(1)) on $(\mathbf{X}/G)_{\text{ét}}$. Then*

(i) $\mathrm{R}\Gamma(\mathbf{X}/G, \mathcal{F}_M) = \mathrm{RHom}_{kG}(M^*, \mathrm{R}\Gamma(\mathbf{X}, k))$, *and* $\mathrm{R_c}\Gamma(\mathbf{X}/G, \mathcal{F}_M) = \mathrm{RHom}_{kG}(M^*, \mathrm{R_c}\Gamma(\mathbf{X}, k))$,

(ii) $\mathcal{D}(\mathcal{F}_M) \cong \mathcal{F}_{M^*}[2d]$.

Proof. (i) See A3.15(1).

(ii) We have

$$\begin{aligned}\mathcal{D}(\mathcal{F}_M) &= \mathrm{R}\mathcal{H}om\big(\pi_* k_{\mathbf{X}} \overset{\mathrm{L}}{\otimes}_{kG} M_{\mathbf{X}/G}, \sigma^!_{\mathbf{X}/G} k\big) \\ &\cong \mathrm{R}\mathcal{H}om\big(M_{\mathbf{X}/G}, \mathrm{R}\mathcal{H}om(\pi_* k_{\mathbf{X}}, \sigma^!_{\mathbf{X}/G} k)\big)\end{aligned}$$

by the right–left adjunction between the functors $\mathcal{H}om(\pi_* k_{\mathbf{X}}, -)$ and $\pi_* k_{\mathbf{X}} \otimes_{kG}$ – inherited from the corresponding assertion for modules. Since π is finite, π_* is exact (see A3.3) and $\pi_* = \mathrm{R_c}\pi_*$. So the adjunction formula between $\mathrm{R_c}\pi_*$ and $\pi^!$ (see A3.12) gives $\mathrm{R}\mathcal{H}om(\pi_* k_{\mathbf{X}}, \sigma^!_{\mathbf{X}/G} k) \cong \pi_*(\mathrm{R}\mathcal{H}om(k_{\mathbf{X}}, \pi^! \sigma^!_{\mathbf{X}/G} k)) = \pi_*(\mathrm{R}\mathcal{H}om(k_{\mathbf{X}}, \sigma^!_{\mathbf{X}} k)) = \pi_* \mathcal{D}(k_{\mathbf{X}})$. Since the variety $\mathbf{X}$ is smooth, one has $\mathcal{D}(k_{\mathbf{X}}) = k_{\mathbf{X}}[2d]$ (see A3.12).

This gives $\mathcal{D}(\mathcal{F}_M) \cong \mathrm{R}\mathcal{H}om(M_{\mathbf{X}/G}, \pi_* k_{\mathbf{X}}[2d])$ and our claim. □

Proposition 12.9. *The $\mathcal{S}_{(w,\theta)}$'s for $w \in \Sigma(S)_{\mathrm{red}}$, $\theta \in \mathrm{Hom}(\mathbf{T}_0^{wF}, k^\times)$ satisfy (2) of Proposition 12.1.*

The following is reminiscent (and a consequence) of independence of the $\mathrm{R}^{\mathbf{G}}_{\mathbf{T}\subseteq\mathbf{B}}$ functor with regard to $\mathbf{B}$ (see Theorem 8.17(i)).

Proposition 12.10. *Let $(w, \theta) \in \Theta_k(\mathbf{G}, F)$. Let $w' \in \Sigma(\overline{S})$ be such that the associated product in $W(\mathbf{G}, \mathbf{T}_0)$ is the same as the one for w (thus $(w', \theta) \in \Theta_k(\mathbf{G}, F)$). Then $\mathcal{S}_{(w,\theta)}$ and $\mathcal{S}_{(w',\theta)}$ have the same image in the Grothendieck group of $k\mathbf{G}^F$.*

Proof of Proposition 12.10. Using Lemma 12.4, it suffices to check that $\mathrm{R}\Gamma(\mathbf{Y}(w), \Lambda_{\mathbf{Y}(w)}).b_\theta^\wedge$ and $\mathrm{R}\Gamma(\mathbf{Y}(w'), \Lambda_{\mathbf{Y}(w')}).b_\theta^\wedge$ have the same character. Let us write $w = (s_1, \dots, s_r)$, $w' = (s'_1, \dots, s'_{r'})$. Up to completing one of those sequences with 1's, we may assume $r = r'$. From Proposition 7.13(ii), we know that $\mathbf{Y}(w)$ is of type $\mathbf{Y}^{(\mathbf{G}^r, F_r)}_{a(U_0^r)}$ where $F_r \colon \mathbf{G}^r \to \mathbf{G}^r$ is defined by $(g_1, \dots, g_r) \mapsto (g_2, \dots, g_r, F(g_1))$ and $a^{-1} F_r(a) = \dot{w} := (\dot{s}_1, \dots, \dot{s}_r)$ in $\mathbf{G}^r$ (see also §10.3). This is a situation where $\mathbf{G}^r$ is a reductive group over $\mathbf{F}$ and $(F_r)^r = F \times \dots \times F$ has in turn a power which is a Frobenius map for a definition of $\mathbf{G}^r$ over a finite subfield of $\mathbf{F}$. Let $a_1 \in \mathbf{G}$ be such that $a_1^{-1} F(a_1) = \dot{s}_1 \dots \dot{s}_r$. Then $a := (a_1, a_1\dot{s}_1, \dots, a_1\dot{s}_1 \dots \dot{s}_{r-1})$ satisfies $a^{-1} F_r(a) = \dot{w}$.

Denote $\nu := \dot{s}_1 \dots \dot{s}_r \in \mathrm{N}_{\mathbf{G}}(\mathbf{T}_0)$. By Theorem 7.11, we have $\dot{s}'_1 \dots \dot{s}'_r = t_0 \nu$ for some $t_0 \in \mathbf{T}_0$. By Lang's theorem (Theorem 7.1(i)) applied to $\mathbf{T}_0$ and νF, there is some $t \in \mathbf{T}_0$ such that $t^{-1} \nu F(t) = t_0 \nu$. Then $a'_1 := a_1 t$ satisfies $a_1'^{-1} F(a'_1) = \dot{s}'_1 \dots \dot{s}'_r$, and therefore $a' := (a'_1, a'_1\dot{s}'_1, \dots, a'_1\dot{s}'_1 \dots \dot{s}'_{r-1})$ satisfies $a'^{-1} F_r(a') = \dot{w}'$.

The characters we are looking at are of type $\mathrm{R}^{\mathbf{G}^r}_{{}^a(\mathbf{T}_0^r)\subseteq {}^a(\mathbf{B}_0^r)}(\sum_\tau \tau\theta)$ and $\mathrm{R}^{\mathbf{G}^r}_{{}^{a'}(\mathbf{T}_0^r)\subseteq {}^{a'}(\mathbf{B}_0^r)}(\sum_\tau \tau\theta)$ and this does not depend on the Borel subgroups, by Theorem 8.17(i). (The τ's range over ℓ-characters of $\mathbf{T}_0^{wF}$.)

We clearly have ${}^a(\mathbf{T}_0^r) = {}^{a'}(\mathbf{T}_0^r) = ({}^{a_1}\mathbf{T}_0)^r$. So the above characters coincide. □

Remark. The context of Theorem 8.17(i) is slightly different from the present one, since there, as in [DiMi91] §11, $F\colon \mathbf{G} \to \mathbf{G}$ is the Frobenius map associated with a definition of $\mathbf{G}$ over $\mathbb{F}_q$. Here we need to apply this theorem in a case where $\mathbf{G}$ is replaced with $\mathbf{G}^r$ and F with F_r. But the arguments of [DiMi91] can still be used in this situation (see Exercise 7).

Proof of Proposition 12.9. Let M be a simple $k\mathbf{G}^F$-module. Let $w \in \Sigma(S \cup \{1\})$ with minimal $l(w)$ such that $\mathrm{RHom}_{k\mathbf{G}^F}(\mathcal{S}_w k\mathbf{T}^{wF}, M) \neq 0$. We are going to prove that this complex is concentrated in degree $-l(w)$, and that $w \in \Sigma_{\mathrm{red}}$. This will clearly give our claim.

Lemma 12.11. *For any $w' \in \Sigma(S \cup \{1\})$ of length $< l(w)$, and for any $k\mathbf{T}^{w'F}$-module M',* $\mathrm{RHom}_{k\mathbf{G}^F}(\mathcal{S}_{w'} M', M) = 0$.

Proof of Lemma 12.11. (See also Exercise 2.) By the definition of w, we have the lemma for $M' = k\mathbf{T}^{w'F}$. This can be expressed as the fact that $(\mathcal{S}_{w'} k\mathbf{T}^{w'F})^* \overset{\mathrm{L}}{\otimes}_{\mathbf{G}^F} M$ is acyclic as a complex of k-spaces. By Poincaré duality (A3.12), $\mathrm{R_c}\Gamma(\mathbf{Y}(w')^{\mathrm{opp}}, k) \overset{\mathrm{L}}{\otimes}_{\mathbf{G}^F} M = 0$ in $D^b(k\mathbf{T}^{w'F})$, where $\mathbf{Y}(w')$ is considered as a $\mathbf{G}^F \times (\mathbf{T}^{w'F})^{\mathrm{opp}}$-variety. Let $\theta'\colon \mathbf{T}^{w'F} \to k^\times$. Recall the notation $k_{\theta'}$ to denote the one-dimensional $k\mathbf{T}^{w'F}$-module associated with θ'. Applying the derived functor $k_{\theta'} \overset{\mathrm{L}}{\otimes}_{\mathbf{T}^{w'F}} -$ to the above, one finds that $k_{\theta'} \overset{\mathrm{L}}{\otimes}_{\mathbf{T}^{w'F}} \mathrm{R_c}\Gamma(\mathbf{Y}(w')^{\mathrm{opp}}, k) \overset{\mathrm{L}}{\otimes}_{\mathbf{G}^F} M$ is acyclic. The first $\overset{\mathrm{L}}{\otimes}$ can be written as $\mathcal{R}_{w'}(k_{\theta'})^{\mathrm{opp}}$ by equation (1) of A3.15. So we get our claim for $k_{\theta'}$ by Poincaré duality. Our claim for arbitrary M' follows by generation in $D^b(k\mathbf{T}^{w'F})$ (see A1.12). □

By Lemma 12.2, the condition $\mathrm{RHom}_{k\mathbf{G}^F}(\mathcal{S}_w k\mathbf{T}^{wF}, M) \neq 0$ is equivalent to the projective cover of M being a summand of $\mathcal{S}_w k\mathbf{T}^{wF}$. The same lemma applied to k-duals shows that this is now also equivalent to $\mathrm{RHom}_{k\mathbf{G}^F}(M, \mathcal{S}_w k\mathbf{T}^{wF}) \neq 0$. By Theorem 12.6, we have a morphism $\mathcal{R}_w k\mathbf{T}^{wF} \to \mathcal{S}_w k\mathbf{T}^{wF}$ whose mapping cone is in the subcategory of $D^b(k\mathbf{G}^F)$ generated by the $\mathcal{S}_{w'} k_{\theta'}$ with $w' < w$. For each, we have $\mathrm{RHom}_{k\mathbf{G}^F}(M, \mathcal{S}_{w'} k_{\theta'}) = 0$ by Lemma 12.11.

Now, using preservation of mapping cones by derived functors (see A1.8), the following canonical map is an isomorphism

$$\text{(I)} \qquad \mathrm{RHom}_{k\mathbf{G}^F}(M, \mathcal{R}_w k\mathbf{T}^{wF}) \to \mathrm{RHom}_{k\mathbf{G}^F}(M, \mathcal{S}_w k\mathbf{T}^{wF}).$$

By Lemma 12.8(i), this means that $\mathcal{F}_{M^*} := \pi_* k \overset{\mathrm{L}}{\otimes}_{k\mathbf{T}^{wF}} M^*_{\mathbf{X}(w)}$ satisfies Condition 10.2(c).

Let us show that w is also of minimal length such that $\mathrm{RHom}_{k\mathbf{G}^F}(\mathcal{S}_w k\mathbf{T}^{wF}, M^*) \neq 0$. It suffices to check that $\mathrm{RHom}_{k\mathbf{G}^F}(\mathcal{S}_w k\mathbf{T}^{wF}, M^*) \neq 0$, because then, by symmetry of hypothesis, a w' of smaller length satisfying that would give $\mathrm{RHom}_{k\mathbf{G}^F}(\mathcal{S}_{w'} k\mathbf{T}^{w'F}, M) \neq 0$, a contradiction.

What we have to check is that the projective cover of M^* is a direct summand in some component of the perfect complex $\mathcal{S}_w k\mathbf{T}^{wF}$. Since k-duality permutes simple and projective indecomposable modules in the same way within $k\mathbf{G}^F$**–mod**, this is equivalent to the projective cover of M being a summand in $(\mathcal{S}_w k\mathbf{T}^{wF})^*$. But Lemma 12.8(i) and Poincaré–Verdier duality tell us that $(\mathcal{S}_w k\mathbf{T}^{wF})^* \cong \mathcal{R}_w k\mathbf{T}^{wF}[2l(w)]$. So we are reduced to checking that $\mathrm{RHom}_{k\mathbf{G}^F}(S, \mathcal{R}_w k\mathbf{T}^{wF}) \neq 0$. This is now clear from (I) above and our hypothesis on w.

This gives $\mathrm{RHom}_{k\mathbf{G}^F}(M^*, \mathcal{S}_w k\mathbf{T}^{wF}) \neq 0$ as claimed.

Applying now to M^* what we had for M, we know that $\mathcal{F}_M$ also satisfies Condition 10.2(c). Recall that $\mathbf{X}(w)$ is quasi-affine by Proposition 10.10(i). Then Lemma 12.7 and Lemma 12.8(ii) give that the homology of $\mathcal{F}_{M^*}$ is concentrated in degree $\dim(\mathbf{Y}(w)) = l(w)$. Lemma 12.8(i) tells us that $\mathrm{RHom}_{k\mathbf{G}^F}(M, \mathcal{S}_{(w,\theta)})$ is concentrated in degree $l(w)$, or equivalently (Lemma 12.2) that the projective cover of M is present at degree $l(w)$ and only there. This gives that $\mathrm{RHom}_{k\mathbf{G}^F}(\mathcal{S}_{(w,\theta)}[l(w)], M)$ has non-zero homology at degree 0 and only there.

It remains to show that w is reduced. Let $v \in \Sigma_{\mathrm{red}}$ represent the product $s_1 s_2 \ldots s_r \in W(\mathbf{G}, \mathbf{T}_0)$, where $w = (s_1, s_2, \ldots, s_r)$. Then $(v, \theta) \in \Theta_k(\mathbf{G}, F)$, and Proposition 12.10 implies that $\mathcal{S}_{(v,\theta)}$ has the same Brauer character as $\mathcal{S}_{(w,\theta)}$. Since $\mathrm{RHom}_{k\mathbf{G}^F}(M, \mathcal{S}_{(w,\theta)})$ is in a single degree, this means that the projective cover of M occurs in only one degree of $\mathcal{S}_{(w,\theta)}$ (Lemma 12.2 again). Then, by linear independence of Brauer characters of projective indecomposable modules (see [Ben91a] 5.3.6), the Brauer character of M occurs in that of $\mathcal{S}_{(w,\theta)}$, hence in that of $\mathcal{S}_{(v,\theta)}$. This in turn clearly forces the projective cover of M to be present in at least one $\mathcal{S}^i_{(v,\theta)}$. Then $\mathrm{RHom}_{k\mathbf{G}^F}(\mathcal{S}_{(v,\theta)}, M) \neq 0$. But the minimality of $l(w)$ now implies $v = w$. □

12.4. Disjunction of series

In the present section, we show that the $\mathcal{S}_{(w,\theta)}$'s satisfy condition (3) of Proposition 12.1.

We have to check that $\mathrm{RHom}_{k\mathbf{G}^F}(\mathcal{S}_{(w,\theta)}, \mathcal{S}_{(w',\theta')}) = 0$ whenever $w, w' \in \Sigma(S)_{\mathrm{red}}$, $\theta \colon \mathbf{T}_0^{wF} \to k^\times$, $\theta' \colon \mathbf{T}_0^{w'F} \to k^\times$ and (w, θ), (w', θ') correspond to semi-simple elements of $(\mathbf{G}^*)^F$, not $(\mathbf{G}^*)^F$-conjugate. Let us first check the following.

Proposition 12.12. *Let* $\mathbf{VS}$, $\mathbf{V}'\mathbf{S}'$ *be Levi decompositions of Borel subgroups of* $\mathbf{G}$ *with* $\mathbf{S}$, $\mathbf{S}'$ *some* F*-stable maximal tori. Let* $\theta\colon \mathbf{S}^F \to k^\times$ *and* $\theta'\colon \mathbf{S}'^F \to k^\times$. *Denote by* b_θ *and* $b_{\theta'}$ *the associated primitive idempotents of the group algebra* $k\mathbf{S}^F$ *(resp.* $k\mathbf{S}'^F$*); see Definition 10.16.*

Recall $\mathbf{Y}_{1\subseteq\mathbf{V}} = \{z \in \mathbf{G} \mid z^{-1}F(z) \in F\mathbf{V}\}$, *a closed subvariety of* $\mathbf{G}$, $\mathbf{G}^F \times \mathbf{S}^F$*-stable (see §7.1). Denote by* $\pi\colon \mathbf{Y}_{1\subseteq\mathbf{V}} \to \mathbf{Y}_{1\subseteq\mathbf{V}}/\mathbf{S}^F$, $\pi'\colon \mathbf{Y}_{1\subseteq\mathbf{V}'} \to \mathbf{Y}_{1\subseteq\mathbf{V}'}/\mathbf{S}'^F$ *the associated quotients. Let* $\mathcal{R} := \mathrm{R_c}\Gamma(\mathbf{Y}_{1\subseteq\mathbf{V}}/\mathbf{S}^F, \pi_* k_{\mathbf{Y}_{1\subseteq\mathbf{V}}}.b_\theta)$, $\mathcal{R}' := \mathrm{R_c}\Gamma(\mathbf{Y}_{1\subseteq\mathbf{V}'}/\mathbf{S}'^F, \pi'_* k_{\mathbf{Y}_{1\subseteq\mathbf{V}'}}.b_{\theta'})$ *in* $D^b(k\mathbf{G}^F\text{-}\mathbf{mod})$.

If $(\mathbf{S}, \theta)$ *and* $(\mathbf{S}', \theta')$ *are not in the same geometric series, then* $\mathrm{RHom}_{k\mathbf{G}^F}(\mathcal{R}, \mathcal{R}') = 0$.

Proof of Proposition 12.12. The proof is very similar to the classic one over K (see [Srinivasan] 6.12, [DiMi91] §11 and §13). We restate the main steps for the convenience of the reader and take the opportunity to make more precise a couple of arguments.

We have

$$\begin{aligned}\mathrm{RHom}_{k\mathbf{G}^F}(\mathcal{R}, \mathcal{R}') &\cong \mathcal{R}^\vee \overset{\mathrm{L}}{\otimes}_{k\mathbf{G}^F} \mathcal{R}' \\ &= ((\mathrm{R_c}\Gamma(\mathbf{Y}_{1\subseteq\mathbf{V}}, k_{\mathbf{Y}_{1\subseteq\mathbf{V}}}).b_\theta)^\vee \otimes_k \mathrm{R_c}\Gamma(\mathbf{Y}_{1\subseteq\mathbf{V}'}, k_{\mathbf{Y}_{1\subseteq\mathbf{V}'}}).b_{\theta'}))_{\mathbf{G}^F}\end{aligned}$$

(co-invariants) since $\mathrm{R_c}\Gamma(\mathbf{Y}_{1\subseteq\mathbf{V}}, k_{\mathbf{Y}_{1\subseteq\mathbf{V}}}).b_\theta$ is represented by a complex of projective $k\mathbf{G}^F$-modules (see A3.15). Using also the Künneth formula (A3.11), we get $\mathrm{RHom}_{k\mathbf{G}^F}(\mathcal{R}, \mathcal{R}') \cong \mathrm{R_c}\Gamma(\mathbf{Y}_{1\subseteq\mathbf{V}} \times \mathbf{Y}_{1\subseteq\mathbf{V}'}/\mathbf{G}^F, k).b_{\theta^{-1}} \otimes b_{\theta'}$ where $\mathbf{Y}_{1\subseteq\mathbf{V}} \times \mathbf{Y}_{1\subseteq\mathbf{V}'}/\mathbf{G}^F$ is the quotient for the diagonal action and the quotient is still endowed with the action of $\mathbf{S}^F \times \mathbf{S}'^F$ (on the right) making $\mathrm{R_c}\Gamma(\mathbf{Y}_{1\subseteq\mathbf{V}} \times \mathbf{Y}_{1\subseteq\mathbf{V}'}/\mathbf{G}^F, k)$ into a class of bounded complexes of (right) $k(\mathbf{S}^F \times \mathbf{S}'^F)$-modules. We are going to study $\mathbf{Y}_{1\subseteq\mathbf{V}} \times \mathbf{Y}_{1\subseteq\mathbf{V}'}/\mathbf{G}^F$.

Let $\mathbf{Z} := \{(u, u', g) \in F\mathbf{V} \times F\mathbf{V}' \times \mathbf{G} \mid uF(g) = gu'\}$. By differentiating the defining equation, it is easily proved that it is a closed smooth subvariety of $F\mathbf{V} \times F\mathbf{V}' \times \mathbf{G}$. It is stable under the action of $\mathbf{S}^F \times \mathbf{S}'^F$ given by $(u, u', g) \mapsto (u^s, u'^{s'}, s^{-1}gs')$ for $s \in \mathbf{S}^F$, $s' \in \mathbf{S}'^F$. The morphism $\mathbf{Y}_{1\subseteq\mathbf{V}} \times \mathbf{Y}_{1\subseteq\mathbf{V}'} \to \mathbf{Z}$ defined by $(x, x') \mapsto (x^{-1}F(x), x'^{-1}F(x'), x^{-1}x')$ is clearly bijective between $\mathbf{Y}_{1\subseteq\mathbf{V}} \times \mathbf{Y}_{1\subseteq\mathbf{V}'}/\mathbf{G}^F$ and $\mathbf{Z}$ (see also [DiMi91] 11.7). By differentiating the morphism above, we easily find that it is a separable map. So we have an isomorphism $\mathbf{Y}_{1\subseteq\mathbf{V}} \times \mathbf{Y}_{1\subseteq\mathbf{V}'}/\mathbf{G}^F \xrightarrow{\sim} \mathbf{Z}$ (see A2.6). This is an isomorphism of $\mathbf{S}^F \times \mathbf{S}'^F$-sets.

If $x \in \mathbf{G}$ satisfies $\mathbf{S}^x = \mathbf{S}'$, one defines $\mathbf{Z}_x := \{(u, u', g) \in \mathbf{Z} \mid g \in \mathbf{SV}x\mathbf{S}'\mathbf{V}'\}$. We have clearly $\mathbf{Z} = \bigcup_x \mathbf{Z}_x$, a disjoint union when x ranges over a (finite) representative system of double cosets $x \in \mathbf{S}\backslash\mathbf{G}/\mathbf{S}'$ such that $\mathbf{S}^x = \mathbf{S}'$. Those x are of the form nx_0 where $n \in \mathrm{N}_{\mathbf{G}}(\mathbf{S})$ and x_0 is some fixed element of $\mathbf{G}$ such that $\mathbf{S}^{x_0} = \mathbf{S}'$ and $\mathbf{V}^{x_0} = \mathbf{V}'$. The Bruhat order on $\mathrm{N}_{\mathbf{G}}(\mathbf{S})/\mathbf{S}$ with respect to $\mathbf{VS}$ allows us to list the distinct $\mathbf{Z}_x$ so as to get $\mathbf{Z} = \mathbf{Z}_1 \cup \mathbf{Z}_2 \cup \ldots$ a (finite) disjoint union where each union $\mathbf{Z}_1 \cup \mathbf{Z}_2 \cup \ldots \cup \mathbf{Z}_i$ is closed since the corresponding union $\bigcup_{i'=1}^{i} \mathbf{VS}n_{i'}x_0\mathbf{V}'\mathbf{S}' = (\bigcup_{i'=1}^{i}\mathbf{VS}n_{i'}\mathbf{VS})x_0$ is closed. These $\mathbf{Z}_i$ are $\mathbf{S}^F \times \mathbf{S}'^F$-stable, so the open-closed exact sequence (see A3.9) implies that there is some x such that $\mathbf{S}^x = \mathbf{S}'$ and $\mathrm{R}_c\Gamma(\mathbf{Z}_x, k).b_{\theta^{-1}} \otimes b_{\theta'} \neq 0$.

Let $\mathbf{V}'_-$ be the unipotent radical of the Borel subgroup opposite $\mathbf{V}'\mathbf{S}'$ with regard to $\mathbf{S}'$ (i.e. $\mathbf{V}'_- = \mathbf{V}^{n_0x_0}$ where $n_0 \in \mathrm{N}_{\mathbf{G}}(\mathbf{S})$ is such that $\mathbf{V} \cap \mathbf{V}^{n_0} = 1$). One defines

$$\mathbf{S}^F \times \mathbf{S}'^F \subseteq \mathbf{H}_x := \{(s, s') \mid sF(s)^{-1} = F(x)s'F(s')^{-1}F(x)^{-1}\} \subseteq \mathbf{S} \times \mathbf{S}'.$$

Then $\mathbf{H}_x$ is obviously a group which is made to act on $\mathbf{Z}_x$ as follows. Write $\mathbf{VS}x\mathbf{V}'\mathbf{S}' = (\mathbf{V} \cap x\mathbf{V}'_-x^{-1})\mathbf{S}x\mathbf{V}' \cong \mathbf{V} \cap x\mathbf{V}'_-x^{-1} \times \mathbf{S} \times \mathbf{V}'$ by the unique decomposition $g = b_gxv'_g = v_gs_gxv'_g$ where $v_g \in \mathbf{V} \cap x\mathbf{V}'_-x^{-1}$, $s_g \in \mathbf{S}$, $v'_g \in \mathbf{V}'$, and $b_g = v_gs_g$. If $s \in \mathbf{S}$, $s' \in \mathbf{S}'$, define $(u, u', g).(s, s') = ((uF(v_g))^sF(v_g^s)^{-1}, (u'F(v'_g)^{-1})^{s'}F(v'^{s'}_g), s^{-1}gs')$. It is easily checked that this defines an action of $\mathbf{S} \times \mathbf{S}'$ on $F\mathbf{V} \times F\mathbf{V}' \times \mathbf{VS}x\mathbf{V}'\mathbf{S}'$ (note that $(v_g)^s = v_{s^{-1}gs'}$), and (less easily) that $\mathbf{Z}_x$ is preserved by $\mathbf{H}_x$. This last action extends the action of $\mathbf{S}^F \times \mathbf{S}'^F$ on $\mathbf{Z}_x$. See also [DiMi91] 11.8 where a $\mathbf{V} \cap \mathbf{V}'$-torsor makes the above formulae more natural.

Let m be an integer such that $F^m(x) = x$ (see A2.5). We show that $\theta \circ N_{F^m/F} = \theta' \circ N_{F^m/F} \circ \mathrm{ad}(F(x)^{-1})$.

Let $H_{x,m} := \{(N_{F^m/F}(s), N_{F^m/F}(s^{F(x)})) \mid s \in \mathbf{S}^{F^m}\} \subseteq \mathbf{S}^F \times \mathbf{S}'^F$ (one has $(\mathbf{S}^{F^m})^{F(x)} \subseteq \mathbf{S}'^{F^m}$ since $\mathbf{S}^{F(x)} = \mathbf{S}'$). Let $\mathbf{H}_{x,m} := \{(N_{F^m/F}(s), N_{F^m/F}(s^{F(x)})) \mid s \in \mathbf{S}\}$ a subtorus of $\mathbf{S} \times \mathbf{S}'$. One has $\mathbf{H}_{x,m} \subseteq \mathbf{H}_x$ since, denoting by Lan_F the Lang map $g \mapsto g^{-1}F(g)$, it is easily checked that $\mathrm{Lan}_F \circ N_{F^m/F} = \mathrm{Lan}_{F^m}$ on $\mathbf{S}$ and therefore $F(x)\mathrm{Lan}_F(N_{F^m/F}((s^{-1})^{F(x)}))F(x)^{-1} = F(x)\mathrm{Lan}_{F^m}((s^{-1})^{F(x)})F(x)^{-1} = F(x)s^{F(x)}F^m(s^{-1})^{F(x)}F(x)^{-1} = \mathrm{Lan}_{F^m}(s^{-1})$ for any $s \in \mathbf{S}$.

The proof of the following lemma will be given later.

Lemma 12.13. *Let $\mathbf{X}$ be a quasi-projective $\mathbf{F}$-variety and $\mathbf{H}$ be a torus acting on it. Let k be a finite commutative ring whose characteristic is invertible in $\mathbf{F}$. Let H' be a finite subgroup of $\mathbf{H}$. Then H' acts trivially on $\mathrm{H}^i_c(\mathbf{X}/H', \pi_*^{H'}k_{\mathbf{X}})$.*

Applying the above for $\mathbf{H}_{x,m}$ and $H' = H_{x,m}$ acting on $\mathbf{Z}_x$, we find that $H_{x,m}$ acts trivially on any $\mathrm{H}^i(\mathbf{Z}_x, k)$. But $\mathrm{H}^i(\mathbf{Z}_x, k).b_{\theta^{-1}\otimes\theta'}$ only involves simple $k[\mathbf{S}^F \times \mathbf{S}'^F]$-modules isomorphic to the one associated with the linear character $\theta^{-1} \otimes \theta'$, so $\mathrm{Res}^{\mathbf{S}^F \times \mathbf{S}'^F}_{H_{x,m}} \mathrm{H}^i(\mathbf{Z}_x, k).b_{\theta^{-1}\otimes\theta'} \neq 0$ only if $(\theta^{-1} \otimes \theta')(H_{x,m}) = 1$. This means $\theta \circ N_{F^m/F}(s) = \theta' \circ N_{F^m/F}(s^{F(x)})$ for all $s \in \mathbf{S}^{F^m}$. This contradicts the fact that $(\mathbf{S}, \theta)$ and $(\mathbf{S}', \theta')$ are not geometrically conjugate (see Proposition 8.21 and (8.15)).

Then $\mathrm{R_c}\Gamma(\mathbf{Z}, k).b_{\theta^{-1}} \otimes b_{\theta'}$ is acyclic, i.e. equal to 0 in $D^b(k)$. □

Proof of Lemma 12.13. If $H' \subseteq H''$ are finite subgroups of $\mathbf{H}$, then $\mathrm{Res}^{H''}_{H'} \mathrm{H}^i_c(\mathbf{X}/H'', \pi_*^{H''} k_{\mathbf{X}}) \cong \mathrm{H}^i_c(\mathbf{X}/H', \pi_*^{H'} k_{\mathbf{X}})$ (see A3.8, A3.14). So our claim will follow from the cyclic case. Assume $H' = <h'>$ for some $h' \in \mathbf{H}$. The sum $\bigoplus_i \mathrm{H}^i_c(\mathbf{X}, k_{\mathbf{X}})$ is a finite group by A3.7. Let N be the order of its automorphism group. Since $\mathbf{H}$ is a torus, hence a divisible group, there is $h \in \mathbf{H}$ such that $h^N = h'$. But h induces an automorphism of $\mathrm{H}^i_c(\mathbf{X}/<h>, \pi_*^{<h>} k_{\mathbf{X}})$ whose Nth power is trivial. This implies that $h^N = h'$ acts trivially. □

Lemma 12.14. *Let* $\mathbf{V}$ *be a connected unipotent group acting on an* $\mathbf{F}$*-variety* $\mathbf{X}$ *such that there exists a locally trivial* $\mathbf{V}$*-quotient* $\pi : \mathbf{X} \to \mathbf{X}'$ *with* $\mathbf{X}'$ *smooth. Then* $\mathrm{R_c}\pi_*$ *induces an isomorphism* $\mathrm{R_c}\Gamma(\mathbf{X}, k) \xrightarrow{\sim} \mathrm{R_c}\Gamma(\mathbf{X}', k)[-2d]$ *where* d *denotes the dimension of* $\mathbf{V}$.

Proof. By A3.6, it suffices to show that $\mathrm{R_c}\pi_* k \cong k[-2d]$. By [Milne80] VI.11.18, we have a "trace map" $\eta : \mathrm{R_c}\pi_* k \to k[-2d]$ in $D^b(\mathbf{X}')$ which is an isomorphism at degree $2d$ cohomology (or use A3.12). To get our claim, it suffices to show that $\mathrm{R_c}\pi_* k$ and $k[-2d]$ have isomorphic stalks at closed points of $\mathbf{X}$ (see A3.2). Let $\mathrm{Spec}(\mathbf{F}) \xrightarrow{x} \mathbf{X}$ be a closed point and let us form the fiber of π, i.e. the fibered product

$$\begin{array}{ccc} \mathbf{X}_x & \longrightarrow & \mathbf{X} \\ \big\downarrow{\scriptstyle \pi_x} & & \big\downarrow{\scriptstyle \pi} \\ \mathrm{Spec}(\mathbf{F}) & \xrightarrow{x} & \mathbf{X}' \end{array}$$

The base change for direct images with proper support (A3.6) gives $(\mathrm{R_c}\pi_* k)_x \cong \mathrm{R_c}(\pi_x)_* k$. But $\mathbf{X}_x \cong \mathbf{V} \cong \mathbb{A}^d_{\mathbf{F}}$ since locally everything is trivial. Then $\mathrm{R_c}(\pi_x)_* k = k[-2d]$ by the classical result on cohomology with compact support for affine spaces (see A3.13, giving $\mathrm{R}\Gamma(\mathbb{A}^1_{\mathbf{F}}, k) \cong k[0]$, then use the Künneth formula to get $\mathrm{R}\Gamma(\mathbb{A}^d_{\mathbf{F}}, k) \cong k[0]$, and apply Poincaré duality). □

Let us embed $\mathbf{G} \subseteq \widetilde{\mathbf{G}} = \mathbf{G}.\mathrm{Z}(\widetilde{\mathbf{G}})$ where $\mathrm{Z}(\widetilde{\mathbf{G}})$ is connected (use a central quotient of $\mathbf{G} \times \mathbf{T}$ where $\mathbf{T}$ is an F-stable maximal torus; see §15.1 below).

Lemma 12.15. *Let* $\mathbf{VS}$ *be a Levi decomposition of a Borel subgroup of* $\mathbf{G}$ *with* $F\mathbf{S} = \mathbf{S}$*, and let* b *be an idempotent of the group algebra* $k\mathbf{S}^F$*. Then*

(i) $\mathrm{R}\Gamma(\mathbf{Y}^{(\mathbf{G})}_{1\subseteq\mathbf{V}}, k) \cong \mathrm{R}\Gamma(\mathbf{Y}^{(\mathbf{G})}_{\mathbf{V}\subseteq\mathbf{V}}, k)[-2\dim\mathbf{V}]$,
(ii) $\mathrm{R_c}\Gamma(\mathbf{Y}^{(\mathbf{G})}_{1\subseteq\mathbf{V}}, k) \cong \mathrm{R_c}\Gamma(\mathbf{Y}^{(\mathbf{G})}_{\mathbf{V}\subseteq\mathbf{V}}, k)[-2\dim\mathbf{V}]$,
(iii) $\mathrm{R_c}\Gamma(\mathbf{Y}^{(\widetilde{\mathbf{G}})}_{1\subseteq\mathbf{V}}, k).b \cong \mathrm{Ind}_{\mathbf{G}^F}^{\widetilde{\mathbf{G}}^F}\mathrm{R_c}\Gamma(\mathbf{Y}^{(\mathbf{G})}_{1\subseteq\mathbf{V}}, k).b$.

Proof. (i) is a consequence of (ii) using Poincaré duality (A3.12) since $\mathbf{Y}_{1\subseteq\mathbf{V}}$ is smooth (see Theorem 7.2(i)).

(ii) is a consequence of Lemma 12.14, knowing that the map $\mathbf{Y}_{1\subseteq\mathbf{V}} \to \mathbf{Y}_{\mathbf{V}\subseteq\mathbf{V}}$ induced by $\mathbf{G} \to \mathbf{G}/\mathbf{V}$ is locally trivial (see A2.6).

(iii) We have $\mathbf{Y}^{(\mathbf{G})}_{\mathbf{V}\subseteq\mathbf{V}} \xrightarrow{\sim} \widetilde{\mathbf{G}}^F \times \mathbf{Y}^{(\mathbf{G})}_{1\subseteq\mathbf{V}}/\mathbf{G}^F$ as a $\widetilde{\mathbf{G}}^F \times \mathbf{S}^F$-variety, by Theorem 7.10. So, using equation (1) of A3.15 and the Künneth formula (A3.11), we get

$$\mathrm{R_c}\Gamma\big(\mathbf{Y}^{(\widetilde{\mathbf{G}})}_{\mathbf{V}\subseteq\mathbf{V}}, k\big) \xrightarrow{\sim} \mathrm{R_c}\Gamma(\widetilde{\mathbf{G}}^F, k)\overset{\mathrm{L}}{\otimes}_{k\mathbf{G}^F}\mathrm{R_c}\Gamma\big(\mathbf{Y}^{(\mathbf{G})}_{\mathbf{V}\subseteq\mathbf{V}}, k\big)$$

in $D^b(k\widetilde{\mathbf{G}}^F \otimes_k k\mathbf{S}^F)$. We have $\mathrm{R_c}\Gamma(\widetilde{\mathbf{G}}^F, k) = k\widetilde{\mathbf{G}}^F$ trivially (use finite quotient), so $\mathrm{R_c}\Gamma(\mathbf{Y}^{(\widetilde{\mathbf{G}})}_{\mathbf{V}\subseteq\mathbf{V}}, k) \xrightarrow{\sim} k\widetilde{\mathbf{G}}^F \otimes_{k\mathbf{G}^F} \mathrm{R_c}\Gamma(\mathbf{Y}^{(\mathbf{G})}_{\mathbf{V}\subseteq\mathbf{V}}, k)$ and therefore $\mathrm{R_c}\Gamma(\mathbf{Y}^{(\widetilde{\mathbf{G}})}_{\mathbf{V}\subseteq\mathbf{V}}, k)b \xrightarrow{\sim} k\widetilde{\mathbf{G}}^F \otimes_{k\mathbf{G}^F} \mathrm{R_c}\Gamma(\mathbf{Y}^{(\mathbf{G})}_{\mathbf{V}\subseteq\mathbf{V}}, k)b$ by further tensoring $-\otimes_{\mathbf{S}^F} k\mathbf{S}^F b$. We may replace the index $\mathbf{V} \subseteq \mathbf{V}$ with $1 \subseteq \mathbf{V}$ by (ii). This completes our proof. □

Let us now check condition (3) of Proposition 12.1, thus completing the proof of Theorem 10.17(b) (see Proposition 12.3). Let $(w, \theta), (w', \theta') \in \Theta_k(\mathbf{G}, F)$ be in distinct rational series with $w, w' \in \Sigma(S)_{\mathrm{red}}$ (see Notation 10.8).

Denote $w = (s_1, \ldots, s_m)$. Taking $a \in \mathbf{G}$ such that $a^{-1}F(a) = \dot{s}_1 \ldots \dot{s}_m$ and denoting $\mathbf{S} := \mathbf{T}_0^a$, $\mathbf{V} := \mathbf{U}_0^a$, we have $\mathbf{Y}(w) \cong \mathbf{Y}_\mathbf{V}$ as a $\mathbf{G}^F \times \mathbf{T}_0^{wF}$-variety (see Proposition 10.12 for $I = \emptyset$). We do the same for (w', θ'). This gives $(\mathbf{S}', \theta')$ which is in a rational series different from the one of $(\mathbf{S}, \theta)$. Take an embedding $\mathbf{G} \subseteq \widetilde{\mathbf{G}}$ as above. Let $\widetilde{\mathbf{S}} := \mathrm{Z}(\widetilde{\mathbf{G}}).\mathbf{S}$, $\widetilde{\mathbf{S}}' := \mathrm{Z}(\widetilde{\mathbf{G}}).\mathbf{S}'$, and let $\widetilde{\theta}$ (resp. $\widetilde{\theta}'$) be linear characters extending θ (resp. θ') on $\mathbf{S}^F$ (resp. $\mathbf{S}'^F$). Then $(\widetilde{\mathbf{S}}, \widetilde{\theta})$ and $(\widetilde{\mathbf{S}}', \widetilde{\theta}')$ are in distinct rational (=geometric) series (see Theorem 8.24(iii)). Proposition 12.12 applied in $\widetilde{\mathbf{G}}$ gives that $\mathrm{RHom}_{k\widetilde{\mathbf{G}}^F}(\mathrm{R_c}\Gamma(\mathbf{Y}^{(\widetilde{\mathbf{G}})}_{1\subseteq\mathbf{V}}, k).b_{\widetilde{\theta}}, \mathrm{R_c}\Gamma(\mathbf{Y}^{(\widetilde{\mathbf{G}})}_{1\subseteq\mathbf{V}'}, k).b_{\widetilde{\theta}'}) = 0$. Summing over all extensions $\widetilde{\theta}, \widetilde{\theta}'$ of θ, θ', we find $\mathrm{RHom}_{k\widetilde{\mathbf{G}}^F}(\mathrm{R_c}\Gamma(\mathbf{Y}^{(\widetilde{\mathbf{G}})}_{1\subseteq\mathbf{V}}, k).b_\theta, \mathrm{R_c}\Gamma(\mathbf{Y}^{(\widetilde{\mathbf{G}})}_{1\subseteq\mathbf{V}'}, k).b_{\theta'}) = 0$ since $b_\theta = \sum_{\widetilde{\theta}} b_{\widetilde{\theta}}$. Using Lemma 12.15(iii) and the Mackey formula giving $\mathrm{Res}_{\mathbf{G}^F}^{\widetilde{\mathbf{G}}^F} \circ \mathrm{Ind}_{\mathbf{G}^F}^{\widetilde{\mathbf{G}}^F} = \sum_{g\in\widetilde{\mathbf{G}}^F/\mathbf{G}^F} \mathrm{ad}(g)$, we have

$$\bigoplus_{g\in\widetilde{\mathbf{G}}^F/\mathbf{G}^F} \mathrm{RHom}_{k\mathbf{G}^F}\big({}^g\big(\mathrm{R_c}\Gamma(\mathbf{Y}^{(\mathbf{G})}_{1\subseteq\mathbf{V}}, k).b_\theta\big), \mathrm{R_c}\Gamma\big(\mathbf{Y}^{(\mathbf{G})}_{1\subseteq\mathbf{V}'}, k\big).b_{\theta'}\big) = 0.$$

Then $\mathrm{RHom}_{k\mathbf{G}^F}(\mathrm{R_c}\Gamma(\mathbf{Y}^{(\mathbf{G})}_{1\subseteq\mathbf{V}}, k).b_\theta, \mathrm{R_c}\Gamma(\mathbf{Y}^{(\mathbf{G})}_{1\subseteq\mathbf{V}'}, k).b_{\theta'}) = 0$. We may replace $\mathrm{R_c}\Gamma$ with $\mathrm{R}\Gamma$ by Poincaré duality, and $\mathbf{Y}_{1\subseteq\mathbf{V}}$ with $\mathbf{Y}_\mathbf{V}$ by Lemma 12.15(i). This is now our claim. □

Exercises

1. Prove Lemma 12.2 for all degrees where $\mathrm{Hom}_A(C^i, S) \neq 0$.
2. Show Lemma 12.11 by using free resolutions of $k\mathbf{T}^{w'F}$-modules and the fact that $\mathcal{S}_{w'}$ is a functor represented by a bounded complex of bimodules. Find a connection with Lemma 1.15.
3. Let $\mathbf{G}$ be any connected $\mathbf{F}$-group acting on an $\mathbf{F}$-variety $\mathbf{X}$. Show that $\mathbf{G}$ acts trivially on $\mathrm{H}^i(\mathbf{X}, \mathbb{Z}/n)$'s.
4. Show that $\mathbf{H}_x^\circ = \mathbf{H}_{x,m}$ (notation of the proof of Proposition 12.12).
5. Show that the hypothesis that $\mathcal{G}$ satisfies Condition 10.2(c) is implied by the other hypotheses of Lemma 12.7 (use Exercise A3.1).
6. Prove Theorem 9.12(i) as a consequence of Proposition 12.1 being satisfied by the $S_{(w,\theta)}$'s.
7. We use the notation of Proposition 12.12 and its proof. We identify θ with a morphism $\mathbf{S}^F \to \Lambda^\times$ and denote by K_θ the one-dimensional $K\mathbf{S}^F$ it defines. Let $\Lambda^{(n)} = \Lambda/J(\Lambda)^n$ for $n \geq 1$. Assume $\mathbf{S} = \mathbf{S}'$. Show that $[\lim_{\leftarrow n} \mathrm{R_c}\Gamma(\mathbf{Y}_{1\subseteq\mathbf{V}} \times \mathbf{Y}_{1\subseteq\mathbf{V}'}/\mathbf{G}^F, \Lambda^{(n)})] \otimes_{\mathbf{S}^F\times\mathbf{S}^F} (K_\theta \otimes K_{\theta^{-1}})$ is of dimension $|\mathbf{N}_{\mathbf{G}^F}(\mathbf{S}, \theta) : \mathbf{S}^F|$ (use the arguments of [DiMi91] §11).

   Deduce that Theorem 8.17(i) holds in the context of the proof of Proposition 12.10.

Notes

The whole chapter is taken from [BoRo03]. §12.4 is based on the proof in [DeLu76] of the corresponding statement for representations in characteristic 0 (see also [DiMi91] §11).

PART III

Unipotent characters and unipotent blocks

Let $(\mathbf{G}, F)$ be a connected reductive $\mathbf{F}$-group defined over $\mathbb{F}_q$.

We have seen in the preceding part that unipotent blocks constitute a model for blocks of finite reductive groups. We now focus a bit more on characters, and show some properties of unipotent characters among ordinary representations of a unipotent block.

Let us recall the partition

$$\mathrm{Irr}(\mathbf{G}^F) = \bigcup_s \mathcal{E}(\mathbf{G}^F, s)$$

of irreducible characters of $\mathbf{G}^F$ into rational series, where s ranges over conjugacy classes of semi-simple elements in $(\mathbf{G}^*)^F$ (see §8.4). We denote by ℓ a prime not dividing q, and by $\mathcal{E}(\mathbf{G}^F, \ell')$ the subset of the above union corresponding to semi-simple elements of order prime to ℓ. Let $(\mathcal{O}, K, k)$ be an ℓ-modular splitting system for $\mathbf{G}^F$.

When ℓ does not divide the order of $(\mathrm{Z}(\mathbf{G})/\mathrm{Z}^\circ(\mathbf{G}))^F$, the elements of $\mathcal{E}(\mathbf{G}^F, \ell')$ are approximations of the (Brauer) characters of simple $k\mathbf{G}^F$-modules (see Theorem 14.4). These results prepare the ground for the determination of decomposition numbers (see Part IV). The proof of Theorem 14.4 involves a comparison of centralizers of ℓ-elements in $\mathbf{G}^F$ and $(\mathbf{G}^*)^F$. These are mainly Levi subgroups, so we need to relate types of maximal F-stable tori to the possibility of containing rational ℓ-elements, hence the necessity of considering cyclotomic polynomials as divisors of the polynomial order of tori. This analysis of tori will also be needed in Part V.

An important fact about unipotent characters is that many of their properties only "depend" on the type of the group, not on the size of q or $\mathrm{Z}(\mathbf{G})$. We prove some "standard" isomorphisms for unipotent blocks showing that they do not depend on $\mathrm{Z}^\circ(\mathbf{G})$. The proof needs, however, a thorough discussion of non-unipotent characters and how they restrict from $\mathbf{G}^F$ to $[\mathbf{G}, \mathbf{G}]^F$. Embedding $\mathbf{G}$ in a group with connected center and same $[\mathbf{G}, \mathbf{G}]$ may also be necessary in order to deal with non-connected centralizers of semi-simple elements $C_{\mathbf{G}^*}(s)$ in $\mathbf{G}^*$. This leads to establishing a fundamental property of Jordan decomposition of characters (see Chapter 8) with regard to Deligne–Lusztig generalized characters $R_{\mathbf{T}}^{\mathbf{G}}\theta$ (see §15.2). A theorem of Lusztig shows that restrictions to $\mathbf{G}^F$ of irreducible characters of $[\mathbf{G}, \mathbf{G}]^F$ are sums *without multiplicities* of irreducible characters. The combinatorial proof of this result in spin groups is in Chapter 16.

13

Levi subgroups and polynomial orders

Let $(\mathbf{G}, F)$ be a connected reductive $\mathbf{F}$-group defined over $\mathbb{F}_q$. As is well-known, the cardinality of $\mathbf{G}^F$ is a polynomial expression of q, with coefficients in $\mathbb{Z}$. We call it the polynomial order of $(\mathbf{G}, F)$. In this polynomial $P_{(\mathbf{G},F)}(x)$, the prime divisors $\neq x$ are cyclotomic polynomials.

If $\mathbf{S}$ is an F-stable subgroup whose polynomial order is a power of the dth cyclotomic polynomial ϕ_d, then $\mathbf{S}$ is a torus. It is natural to study those "ϕ_d-tori" like ℓ-elements of a finite group (ℓ a prime). This leads to an analogue of Sylow's theorem, due to Broué–Malle; see Theorem 13.18 below. If $\mathbf{S}$ is a ϕ_d-torus, then $C_{\mathbf{G}}(\mathbf{S})$ is an F-stable Levi subgroup, called a "d-split" Levi subgroup of $\mathbf{G}$. We show how this class relates to centralizers of actual ℓ-elements of $\mathbf{G}^F$.

A related type of problem is to compute more generally centralizers of semi-simple elements.

This is to be done in view of the local methods of Part V, but also in connection with the Jordan decomposition of characters (see Chapter 15).

13.1. Polynomial orders of F-stable tori

We need to give a formal definition of the fact that the order of $\mathbf{G}^F$ is a "polynomial in q."

Theorem 13.1. *For each connected reductive group $(\mathbf{G}, F)$ defined over $\mathbb{F}_q$, there exists a unique "polynomial order" $P_{(\mathbf{G},F)}(x) \in \mathbb{Z}[x]$ satisfying the following:*

- *there is $a \geq 1$ such that $|\mathbf{G}^{F^m}| = P_{(\mathbf{G},F)}(q^m)$ for all $m \geq 1$ such that $m \equiv 1$ (mod. a).*

One has $P_{(\mathbf{G},F)} = P_{([\mathbf{G},\mathbf{G}],F)}P_{(\mathbf{Z}^\circ(\mathbf{G}),F)}$ and $P_{([\mathbf{G},\mathbf{G}],F)}$ has the form $x^N \prod_{d\geq 1} \phi_d^{v_d}$ for v_d, $N \in \mathbb{N}$, where ϕ_d denotes the dth cyclotomic polynomial.

Proposition 13.2. *(i) Polynomial orders are unchanged by isogenies.*

(ii) If $\mathbf{H}$ *is a connected* F*-stable reductive subgroup of* $\mathbf{G}$*, then* $P_{(\mathbf{H},F)}$ *divides* $P_{(\mathbf{G},F)}$*, with equality if and only if* $\mathbf{H} = \mathbf{G}$.

Proofs of Theorem 13.1 and Proposition 13.2. If $(\mathbf{T}, F)$ is a torus defined over $\mathbb{F}_q$, then $P_{(\mathbf{T},F)}$ is the characteristic polynomial of $q^{-1}F$ acting on $Y(\mathbf{T}) \otimes \mathbb{R}$ (see [Cart85] (3.3.5)).

For more general connected reductive $\mathbf{F}$-groups $(\mathbf{G}, F)$ defined over $\mathbb{F}_q$, let $\mathbf{T}$ be an F-stable maximal torus of an F-stable Borel subgroup. Then ([Cart85] 2.9), one has $|\mathbf{G}^F| = |\mathrm{Z}^\circ(\mathbf{G})^F| q^N \prod_{1\leq i\leq l}(q^{d_i} - \epsilon_i)$ where $2N = |\Phi(\mathbf{G}, \mathbf{T})|$, the exponents d_i are so-called exponents of $W(\mathbf{G}, \mathbf{T})$ acting on the symmetric algebra of $Y(\mathbf{T}) \otimes \mathrm{R}$ ([Bour68] §V.6), and the roots of unity ϵ_i are defined from the action of $q^{-1}F$ on the subalgebra of $W(\mathbf{G}, \mathbf{T})$-invariants. Let a be the order of $q^{-1}F$, then $q^{-1}F$ and $q^{-m}F^m$ are equal symmetries of $Y(\mathbf{T}) \otimes \mathrm{R}$ whenever $m \equiv 1$ (mod. a), hence the (d_i, ϵ_i) are the same for F and F^m. Clearly $P_{(\mathbf{G},F)}$ is uniquely defined by an infinity of values.

Note that the degree $N + \sum_i d_i = 2N + l$ of $P_{([\mathbf{G},\mathbf{G}],F)}$ is the dimension of $[\mathbf{G}, \mathbf{G}]$.

One has $P_{\mathbf{G},F} = P_{[\mathbf{G},\mathbf{G}],F} P_{\mathrm{Z}^\circ(\mathbf{G}),F}$ by Proposition 8.1 (with $G = [\mathbf{G}, \mathbf{G}] \times \mathrm{Z}^\circ(\mathbf{G})$, $Z = H = [\mathbf{G}, \mathbf{G}] \cap \mathrm{Z}^\circ(\mathbf{G})$) and Lang's theorem (see also Exercise 8.1).

Proposition 13.2(i) follows from the above.

Let $\mathbf{H} \subset \mathbf{G}$ be as in Proposition 13.2(ii); $P_{\mathbf{H},F}$ and $P_{\mathbf{G},F}$ are elements of $\mathbb{Z}[x]$ with leading coefficient 1. There are infinitely many $q' = q^m$ such that $P_{\mathbf{H},F}(q')$ is non-zero and divides $P_{\mathbf{G},F}(q')$. Now, using euclidean division $P_{\mathbf{G},F} = P_{\mathbf{H},F}.Q + R$ in $\mathbb{Z}[x]$ one sees that $R = 0$. When equality holds between degrees of $P_{\mathbf{H},F}$ and $P_{\mathbf{G},F}$, the groups have equal dimension. □

Definition 13.3. *Let* $\emptyset \neq E \subseteq \{1, 2, 3, \ldots\}$*. One calls a* ϕ_E*-subgroup of* $\mathbf{G}$ *any* F*-stable torus* $\mathbf{S}$ *such that* $P_{(\mathbf{S},F)}$ *can be written as* $\Pi_{d\in E}(\phi_d)^{n_d}$ *for some integers* n_d.

An E*-split Levi subgroup of* $\mathbf{G}$ *is the centralizer in* $\mathbf{G}$ *of some* ϕ_E*-subgroup of* $\mathbf{G}$.

When E is a singleton $\{e\}$, we may write e-split instead of $\{e\}$-split. A 1-subgroup is just a split torus, i.e. F acts as multiplication by q on its character group. The group $\mathbf{G}$ itself is E-split for any E. A 1-split Levi subgroup is an F-stable Levi subgroup of an F-stable parabolic subgroup of $\mathbf{G}$ ([BoTi] 4.15).

Example 13.4. (i) Let $\mathbf{T}$ be a reference torus. If $\mathbf{T}' = g\mathbf{T}g^{-1}$ is obtained from $\mathbf{T}$ by twisting with $w \in W$, i.e. $g^{-1}F(g)\mathbf{T} = w$, then $P_{(\mathbf{T}',F)}$ is the characteristic polynomial of $(q^{-1}F)w$ acting on $Y(\mathbf{T}) \otimes \mathbb{R}$.

(ii) Let $\mathbf{G} = \mathrm{GL}_n(\mathbf{F})$ and F be the usual map $(x_{ij}) \mapsto (x_{ij}^q)$ raising all matrix entries to the qth power. Let $\mathbf{T}$ be the diagonal torus of $\mathbf{G}$ and let us use it to parametrize $\mathbf{G}^F$-conjugacy classes of F-stable maximal tori by conjugacy classes of $N_{\mathbf{G}}(\mathbf{T})/\mathbf{T} \cong \mathfrak{S}_n$ (see §8.2). If the conjugacy class of $w \in \mathfrak{S}_n$ corresponds with the partition $\lambda_1, \ldots, \lambda_l$ of n (i.e. w is a product of t disjoint cycles of orders $\lambda_1, \ldots, \lambda_l$), the polynomial order of the tori of type w is $(x^{\lambda_1} - 1) \ldots (x^{\lambda_l} - 1)$. Note that the polynomial order determines the $\mathbf{G}^F$-conjugacy class of the torus. The Coxeter torus $\mathbf{T}_{(n)}$ is associated with the class of cycles of order n. This maximal torus $\mathbf{T}_{(n)}$ is the only n-split proper Levi subgroup of $\mathbf{G}$. To see this, note that its polynomial order $x^n - 1$ is the only polynomial order of an F-stable maximal torus divisible by ϕ_n.

Let $e \geq 1$, $m \geq 0$ be such that $me \leq n$. Let $\mathbf{S}_{(e)}$ be a Coxeter torus of $\mathrm{GL}_e(\mathbf{F})$. Then let $\mathbf{L}^{(m)}$ be $\mathrm{GL}_{n-me}(\mathbf{F}) \times (\mathbf{S}_{(e)})^m$ embedded in $\mathbf{G} = \mathrm{GL}_n(\mathbf{F})$ via any isomorphism $\mathbf{F}^n \cong \mathbf{F}^{n-me} \times (\mathbf{F}^e)^m$. Then $\mathbf{L}^{(m)}$ is e-split as a result of the above. A maximal e-split proper Levi $\mathbf{L}$ subgroup of $\mathbf{G}$ is isomorphic to $(\mathrm{GL}_m)^e \times \mathrm{GL}_{n-me}$ with $\mathbf{L}^F \cong \mathrm{GL}_m(q^e) \times \mathrm{GL}_{n-me}(q)$.

Now let F' be the Frobenius endomorphism $(x_{ij}) \mapsto {}^t(x_{ij}^q)^{-1}$. Then $\mathbf{G}^{F'}$ is the general unitary group on $\mathbb{F}_{q^2}$, often denoted by $\mathrm{GL}_n(-q)$, and with rational type $({}^2\mathbf{A}_{n-1}, -q)$. The diagonal torus has polynomial order $(X+1)^n$ and is not 1-split, but it is a maximal 2-split torus and its polynomial order determines its $\mathbf{G}^{F'}$-conjugacy class.

(iii) When the type of $\mathbf{G}$ is a product of types $\mathbf{A}_n$, we define a diagonal torus of $\mathbf{G}$ as a product of diagonal tori of components. Then all diagonal tori are $\mathbf{G}^F$-conjugate, and if an F-stable Levi subgroup $\mathbf{L}$ contains a diagonal torus $\mathbf{T}$ of $\mathbf{G}$, then $\mathbf{T}$ is a diagonal torus of $\mathbf{L}$.

(iv) Assume $(\mathbf{G}, F)$ is semi-simple and irreducible and has type $\mathbf{X} \in \{\mathbf{A}_n, \mathbf{D}_{2n+1}, \mathbf{E}_6\}$. Let $\mathbf{T}_0$ be a reference maximal F-stable torus in $\mathbf{G}$. An element w_0 in the Weyl group which is of greatest length with respect to some basis of Φ acts on $Y(\mathbf{T}) \otimes \mathrm{R}$ as $-\sigma$, where σ is a symmetry that restricts to Φ as an automorphism of order 2 of the root system and Dynkin diagram [Bour68]. If $(\mathbf{G}, F)$ has rational type $(\mathbf{X}, q)$, $(\mathbf{G}, \sigma F)$ is defined over $\mathbb{F}_q$ with rational type $({}^2\mathbf{X}, q)$. One sees that if $\mathbf{T}$ (resp. $\mathbf{S}$) is obtained in $(\mathbf{G}, F)$ (resp. $(\mathbf{G}, \sigma F)$) from $\mathbf{T}_0$ by twisting by w (resp. ww_0), then $P_{\mathbf{T},F}(X) = P_{\mathbf{S},\sigma F}(-X)$. The polynomial orders of $(\mathbf{G}, F)$ and of $(\mathbf{G}, \sigma F)$ are related in the same way.

Proposition 13.5. *If $\mathbf{T}$ is a torus defined over $\mathbb{F}_q$ with Frobenius F, if E is a non-empty set of integers, then there is a unique maximal ϕ_E-subgroup in $\mathbf{T}$. One denotes it by $\mathbf{T}_{\phi_E}$.*

Proof. By Proposition 13.2(ii), the polynomial order of the subtorus $\mathbf{T}_{\phi_E}$ has to be the product P_E of biggest powers of cyclotomic polynomials ϕ_e, $e \in E$, dividing $P_{\mathbf{T},F}$. That property defines $Y(\mathbf{T}_{\phi_E})$ as a pure subgroup of $Y(\mathbf{T})$: there is a unique subspace V_E of $Y(\mathbf{T} \otimes \mathrm{R})$ such that the restriction of $(q^{-1}F)^{-1}$ to V_E has characteristic polynomial P_E and $Y(\mathbf{T}_{\phi_E}) = Y(\mathbf{T}) \cap V_E$. □

In the following proposition, we show that an F-orbit of length m on the set of irreducible components of $\mathbf{G}$ induces the substitution $x \mapsto x^m$ in polynomial orders.

Proposition 13.6. *Let $(\mathbf{G}, F)$ be a connected reductive group defined over $\mathbb{F}_q$. Assume $\mathbf{G} = \mathbf{G}_1.\mathbf{G}_2 \ldots \mathbf{G}_m$ is a central product of connected reductive subgroups $\mathbf{G}_i$ such that $F(\mathbf{G}_i) = \mathbf{G}_{i+1}$ (i is taken mod. m). If the product is direct or if $\mathbf{G}$ is semi-simple, then $P_{(\mathbf{G},F)}(x) = P_{(\mathbf{G}_1,F^m)}(x^m)$. The F-stable Levi subgroups of $(\mathbf{G}, F)$ are the $\mathbf{L}_1.F(\mathbf{L}_1)\ldots F^{m-1}(\mathbf{L}_1)$'s, where $\mathbf{L}_1$ is any F^m-stable Levi subgroup of $(\mathbf{G}_1, F^m)$.*

Proof. By Theorem 13.1 and Proposition 13.2, it is enough to prove this for direct products. It is clear that the projection $\pi_1 \colon \mathbf{G} \to \mathbf{G}_1$ bijects $\mathbf{G}^{F^k}$ and $\mathbf{G}_1^{F^{mk}}$ for any $k \geq 1$, so $P_{(\mathbf{G},F)}(q^k) = P_{(\mathbf{G}_1,F^m)}(q^{mk})$ for infinitely many k's, whence the claimed equality.

Consider $\mathbf{L} = \mathrm{C}_{\mathbf{G}}(\mathbf{S})$ where $\mathbf{S}$ is a torus of $\mathbf{G}$. One has $\mathbf{L} = \mathrm{C}_{\mathbf{G}_1}(\pi_1(\mathbf{S}))$ $.F(\mathrm{C}_{\mathbf{G}_1}(\pi_1(\mathbf{S}))\ldots F^{m-1}(\mathrm{C}_{\mathbf{G}_1}(\pi_1(\mathbf{S}))$; this gives the second statement with $\mathbf{L}_1 = \mathrm{C}_{\mathbf{G}_1}(\pi_1(\mathbf{S}))$. □

Proposition 13.7. *Let E be any non-empty set of integers. A Levi subgroup of $\mathbf{G}$ is E-split if and only if its image in $\mathbf{G}_{\mathrm{ad}}$ is so, or if and only if $\mathbf{L} \cap [\mathbf{G}, \mathbf{G}]$ is E-split.*

Proof. Let $\pi \colon \mathbf{G} \to \mathbf{G}_{\mathrm{ad}}$ be the natural map. If $\mathbf{S}$ is an F-stable torus in $\mathbf{G}$, then $\pi(\mathbf{S})$ is an F-stable torus in $\pi(\mathbf{G})$ and one has $P_{\mathbf{S}\mathrm{Z}^\circ(\mathbf{G}),F} = P_{\pi(\mathbf{S}),F} P_{\mathrm{Z}^\circ(\mathbf{G}),F}$. Furthermore $\mathrm{C}_{\mathbf{G}}(\mathbf{S}) = \mathrm{C}_{\mathbf{G}}(\mathbf{S}\mathrm{Z}^\circ(\mathbf{G}))$ and $\pi(\mathrm{C}_{\mathbf{G}}(\mathbf{S})) = \mathrm{C}_{\pi(\mathbf{G})}(\pi(\mathbf{S}))$. The first assertion follows. The second equivalence follows from $\mathrm{C}_{\mathbf{G}}(\mathbf{S}) = \mathrm{C}_{\mathbf{G}}(\mathbf{S} \cap [\mathbf{G}, \mathbf{G}])$. □

Proposition 13.8. *Groups in duality over $\mathbb{F}_q$ (and their connected centers) have equal polynomial orders.*

Proof. See §8.2 and [Cart85] 4.4.

Proposition 13.9. *Let $(\mathbf{G}, F)$ be a connected reductive group defined over $\mathbb{F}_q$, let $(\mathbf{G}^*, F)$ be in duality with $(\mathbf{G}, F)$. For each non-empty set of integers E the bijection $\mathbf{L} \mapsto \mathbf{L}^*$ between $\mathbf{G}^F$-conjugacy classes of F-stable Levi subgroups of $\mathbf{G}$ and $(\mathbf{G}^*)^F$-conjugacy classes of F-stable dual Levi subgroups of $\mathbf{G}^*$ (see §8.2) restricts to a bijection between respective classes of E-split Levi subgroups.*

Proof. By symmetry, it suffices to check that, if $\mathbf{L}$ is E-split, then $\mathbf{L}^*$ is E-split. Let $\mathbf{S}^* := \mathrm{Z}^\circ(\mathbf{L}^*)_{\phi_E}$, then $\mathbf{M}^* := \mathrm{C}_{\mathbf{G}^*}(\mathbf{S}^*) \supseteq \mathbf{L}^*$ is in duality with $\mathbf{M}$ such that $\mathbf{G} \supseteq \mathbf{M} \supseteq \mathbf{L}$. One has $\mathrm{Z}^\circ(\mathbf{M}^*)_{\phi_E} = \mathrm{Z}^\circ(\mathbf{L}^*)_{\phi_E}$, while $P_{\mathrm{Z}^\circ(\mathbf{L}^*),F} = P_{\mathrm{Z}^\circ(\mathbf{L}),F}$ by Proposition 13.8 and $P_{\mathrm{Z}^\circ(\mathbf{M}^*),F} = P_{\mathrm{Z}^\circ(\mathbf{M}),F}$, so $\mathrm{Z}^\circ(\mathbf{M})_{\phi_E} = \mathrm{Z}^\circ(\mathbf{L})_{\phi_E}$ and $\mathbf{M} \subseteq \mathrm{C}_{\mathbf{G}}(\mathrm{Z}^\circ(\mathbf{L})_{\phi_E}) = \mathbf{L}$ since $\mathbf{L}$ is E-split. So $\mathbf{L} = \mathbf{M}$ and $\mathbf{L}^* = \mathbf{M}^*$, which is E-split. □

13.2. Good primes

The notion of "good primes" for root systems was defined by Springer–Steinberg in order to study certain unipotent classes. Here we are mainly interested in semi-simple elements (see in particular Proposition 13.16(ii) below).

Definition 13.10. *A prime ℓ is said to be* **good** *for a root system Φ if and only if $(\mathbb{Z}\Phi/\mathbb{Z}A)_\ell = \{0\}$ for every subset $A \subseteq \Phi$. If $\mathbf{G}$ is a connected reductive $\mathbf{F}$-group, a prime is said to be good for $\mathbf{G}$ if and only if it is good for its root system.*

Note that the notion does not depend on the rational structure.

In the following table, $(\mathbf{G}, F)$ is a connected reductive $\mathbf{F}$-group defined over $\mathbb{F}_q$, it has irreducible rational type $(\mathbf{X}, r)$ with r a power of q (see §8.1). We recall the list of good primes for $\mathbf{G}$ (see [Cart85] 1.14), along with $|\mathrm{Z}(\mathbf{G}_{\mathrm{sc}})^F|$, the number of rational points of order prime to p in the fundamental group (see §8.1). The group is of rational type $(\mathbf{X}, r)$.

Table 13.11

type $\mathbf{X}$	$\mathbf{A}_n$	${}^2\mathbf{A}_n$	$\mathbf{B}_n$, $\mathbf{C}_n$	$\mathbf{D}_n$, ${}^2\mathbf{D}_n$	${}^3\mathbf{D}_4$
good ℓ's	all	all	$\neq 2$	$\neq 2$	$\neq 2$
$\lvert\mathrm{Z}(\mathbf{G}_{\mathrm{sc}})^F\rvert$	$(n+1, r-1)$	$(n+1, r+1)$	$(2, r-1)$	$(4, r^2-1)$	1

type	${}^2\mathbf{E}_6$, $\mathbf{F}_4$, $\mathbf{G}_2$	$\mathbf{E}_6$	$\mathbf{E}_7$	$\mathbf{E}_8$
good ℓ's	$\neq 2, 3$	$\neq 2, 3$	$\neq 2, 3$	$\neq 2, 3, 5$
$\lvert\mathrm{Z}(\mathbf{G}_{\mathrm{sc}})^F\rvert$	1	$3_{p'}$	$2_{p'}$	1

Proposition 13.12. *Let $(\mathbf{G}, F)$ be a connected reductive group defined over $\mathbb{F}_q$. Let ℓ be a prime good for $\mathbf{G}$.*

(i) If $\mathbf{G}$ has no component of type $\mathbf{A}$, then ℓ divides neither $|\mathrm{Z}(\mathbf{G})/\mathrm{Z}^\circ(\mathbf{G})|$ nor $|\mathrm{Z}(\mathbf{G}^*)/\mathrm{Z}^\circ(\mathbf{G}^*)|$.

Let $\mathbf{H}$ *be an* F*-stable reductive subgroup of* $\mathbf{G}$ *which contains an* F*-stable maximal torus of* $\mathbf{G}$.

(ii) If $|(\mathrm{Z}(\mathbf{G})/\mathrm{Z}^\circ(\mathbf{G}))^F|$ *is prime to* ℓ*, then* $|(\mathrm{Z}(\mathbf{H})/\mathrm{Z}^\circ(\mathbf{H}))^F|$ *is prime to* ℓ.

(iii) If $|(\mathrm{Z}(\mathbf{G}^*)/\mathrm{Z}^\circ(\mathbf{G}^*))^F|$ *is prime to* ℓ*, then* $|(\mathrm{Z}(\mathbf{H}^*)/\mathrm{Z}^\circ(\mathbf{H}^*))^F|$ *is prime to* ℓ.

(iv) If $\mathbf{H}$ *is a Levi subgroup, then (ii) and (iii) above hold for any prime* ℓ*, without the assumption that* ℓ *is good for* $\mathbf{G}$.

Proof. (i) As $\mathbf{G} = \mathrm{Z}^\circ(\mathbf{G})[\mathbf{G}, \mathbf{G}]$, $\mathrm{Z}(\mathbf{G})/\mathrm{Z}^\circ(\mathbf{G})$ is a quotient of $\mathrm{Z}([\mathbf{G}, \mathbf{G}])$, itself a quotient of $\mathrm{Z}(\mathbf{G}_{\mathrm{sc}})$. Hence (i) follows from Table 13.11.

(ii) and (iii) Let $\mathbf{T}$ be a maximal F-stable torus of $\mathbf{G}$ contained in $\mathbf{H}$. Let $\Phi \subseteq X(\mathbf{T})$ (resp. $\Phi^\vee \subseteq Y(\mathbf{T})$) be the set of roots (resp. coroots) of $\mathbf{G}$ relative to $\mathbf{T}$. We may assume that $(\mathbf{H}^*, F)$ is in duality with $(\mathbf{H}, F)$ and is defined by the root datum $(Y(\mathbf{T}), X(\mathbf{T}), {\Phi_{\mathbf{H}}}^\vee, \Phi_{\mathbf{H}})$, where $\Phi_{\mathbf{H}}$ is a subsystem of Φ and ${\Phi_{\mathbf{H}}}^\vee$ is the set of roots of $\mathbf{H}^*$ relative to $\mathbf{T}^*$. But if ℓ is good for Φ, then it is good for $\Phi^\vee$ and for any subsystem. So, if it is good for $\mathbf{G}$, then it is good for $\mathbf{G}^*$, $\mathbf{H}$ and $\mathbf{H}^*$.

The groups of characters of the finite abelian groups $\mathrm{Z}(\mathbf{G})/\mathrm{Z}^\circ(\mathbf{G})$, $\mathrm{Z}(\mathbf{H})/\mathrm{Z}^\circ(\mathbf{H})$, $\mathrm{Z}(\mathbf{G}^*)/\mathrm{Z}^\circ(\mathbf{G}^*)$ and $\mathrm{Z}(\mathbf{H}^*)/\mathrm{Z}^\circ(\mathbf{H}^*)$ are isomorphic, with F-action, to the p'-torsion-groups of $X(\mathbf{T})/\mathbb{Z}\Phi$, $X(\mathbf{T})/\mathbb{Z}\Phi_{\mathbf{H}}$, $Y(\mathbf{T})/\mathbb{Z}\Phi^\vee$ and $Y(\mathbf{T})/\mathbb{Z}{\Phi_{\mathbf{H}}}^\vee$ respectively ([Cart85] 4.5.8). Under hypotheses (ii) $(X(\mathbf{T})/\mathbb{Z}\Phi^F)_\ell = \{0\}$. But ℓ is good for Φ, hence $(\mathbb{Z}\Phi/\mathbb{Z}\Phi_{\mathbf{H}})_\ell^F = \{0\}$. Then $(X(\mathbf{T})/\mathbb{Z}\Phi_{\mathbf{H}})_\ell^F = \{0\}$, so $(\mathrm{Z}(\mathbf{H})/\mathrm{Z}^\circ(\mathbf{H}))_\ell^F = \{1\}$. So (ii) is proved, and similarly (iii) because ℓ is good for $\Phi^\vee$.

(iv) If $\mathbf{H}$ is a Levi subgroup of $\mathbf{G}$ then $\mathbb{Z}\Phi/\mathbb{Z}\Phi_{\mathbf{H}}$ and $\mathbb{Z}\Phi^\vee/\mathbb{Z}{\Phi_{\mathbf{H}}}^\vee$ have no torsion, hence the torsion groups of $Y(\mathbf{T})/\mathbb{Z}\Phi^\vee$ and $Y(\mathbf{T})/\mathbb{Z}{\Phi_{\mathbf{H}}}^\vee$ — and of $X(\mathbf{T})/\mathbb{Z}\Phi$ and $X(\mathbf{T})/\mathbb{Z}\Phi_{\mathbf{H}}$ — are isomorphic. □

13.3. Centralizers of ℓ-subgroups and some Levi subgroups

For references we gather some classical results and give some corollaries.

Proposition 13.13. *Let* $\mathbf{T}$ *be a maximal torus of* $\mathbf{G}$*. Let* $\pi : \mathbf{G} \to \mathbf{G}_{\mathrm{ad}}$*. Let* S *be a subset of* $\mathbf{T}$*. Then the following hold.*

(i) $\mathrm{C}^\circ_{\mathbf{G}}(S) = \langle \mathbf{T}, \mathbf{X}_\alpha \,;\, \alpha \in \Phi(\mathbf{G}, \mathbf{T}),\ \alpha(S) = 1 \rangle$ *and*
$\mathrm{C}_{\mathbf{G}}(S) = \mathrm{C}^\circ_{\mathbf{G}}(S). \langle w \in W(\mathbf{G}, \mathbf{T});\ w(S) = 1 \rangle$*. Both groups are reductive.*

(ii) If $S' \subseteq \mathbf{T}$*, then* $\mathrm{C}^\circ_{\mathrm{C}^\circ_{\mathbf{G}}(S')}(S) = \mathrm{C}^\circ_{\mathbf{G}}(S'S)$.

(iii) $\pi(\mathrm{C}^\circ_{\mathbf{G}}(S)) = \mathrm{C}^\circ_{\pi(\mathbf{G})}(\pi(S))$.

(iv) If S *is a torus, then* $\mathrm{C}_{\mathbf{G}}(S)$ *is a Levi subgroup of* $\mathbf{G}$*, hence is connected.*

Proof. (i) See [Cart85] Theorems 3.5.3, 3.5.4 and the proof of Proposition 3.6.1.
(ii) Clear since $\mathbf{T} \subset \mathrm{C}^\circ_{\mathbf{G}}(S)$.
(iii) follows from (i).
(iv) [DiMi91] Proposition 1.22. □

Here is Steinberg's theorem on centralizers (see [Cart85] 3.5.6).

Theorem 13.14. *If the derived group of* $\mathbf{G}$ *is simply connected, then the centralizer in* $\mathbf{G}$ *of any semi-simple element is connected.*

Example 13.15. Let s be a semi-simple element in $\mathrm{GL}_n(\mathbf{F})$. One has an isomorphism $\mathrm{C}_{\mathbf{G}}(s) \cong \prod_{\alpha \in \mathcal{S}} \mathrm{GL}(V_\alpha)$, where $\mathcal{S}$ is the set of eigenvalues of s and V_α is an eigenspace of s acting on $\mathbf{F}^n$. The centralizer of s is a Levi subgroup of $\mathbf{G}$ and its center $\mathbf{S}$ is connected.

Let F be a Frobenius endomorphism as in Example 13.4(ii) so that $\mathrm{GL}_n(q) = \mathrm{GL}_n(\mathbf{F})^F$ and assume that $F(s) = s$. Then $\mathrm{C}_{\mathbf{G}}(s)^F \cong \prod_{\omega \in \mathcal{S}/<F>} \mathrm{GL}_{m(\omega)}(q^{|\omega|})$ where ω is an orbit under F in $\mathcal{S}$ and $m(\omega)$ is the dimension of V_α for any $\alpha \in \omega$. Hence $\mathbf{S}$ is an F-stable torus with polynomial order $\prod_\omega (X^{|\omega|} - 1)$.

Let F' be as in Example 13.4(ii) so that $\mathrm{GL}_n(\mathbf{F})^{F'} = \mathrm{GL}_n(-q)$, and assume now that $F'(s) = s$. For $\omega \in \mathcal{S}$ let $\bar{\omega} = \{\alpha^{-1} \mid \alpha \in \omega\}$ and let $q_\omega = q^{|\omega|}$ if $\bar{\omega} \neq \omega$ and $q_\omega = -q^{|\omega|/2}$ if $\bar{\omega} = \omega$. Then $\mathrm{C}_{\mathbf{G}}(s)^{F'} \cong \prod_{\{\omega, \bar{\omega}\}} \mathrm{GL}_{m(\omega)}(q_\omega)$ and $\mathbf{S}$ has polynomial order $\prod_{\omega \neq \bar{\omega}} (X^{|\omega|} - 1) \prod_{\omega = \bar{\omega}} (X^{|\omega|/2} + 1)$.

Let e be the order of q modulo ℓ. If s has order ℓ and $\omega \neq \{1\}$, then $e = |\omega|$, hence $\mathrm{C}_{\mathbf{G}}(s) = \mathrm{C}_{\mathbf{G}}(\mathbf{S}) = \mathrm{C}_{\mathbf{G}}(\mathbf{S}_e)$ is an e-split Levi subgroup of $\mathbf{G}$ in both cases.

Proposition 13.16. *Let E be a set of primes not dividing q. Let Y be an E-subgroup of* $\mathbf{G}^F$ *included in a torus.*

(i) $\mathrm{C}_{\mathbf{G}}(Y)/\mathrm{C}^\circ_{\mathbf{G}}(Y)$ *is a finite E-group and it is F-isomorphic with a section of* $\mathrm{Z}(\mathbf{G}^*)/\mathrm{Z}^\circ(\mathbf{G}^*)$.

(ii) If any $\ell \in E$ is good for $\mathbf{G}$, *then* $\mathrm{C}^\circ_{\mathbf{G}}(Y)$ *is a Levi subgroup of* $\mathbf{G}$.

Proof. (i) Follows from Theorem 13.14 and Proposition 8.1; see [DiMi91] 13.14(iii) and 13.15(i).

(ii) By Proposition 13.13(ii), it suffices to check the case of a cyclic ℓ-group $Y = <y>$. Proposition 13.13(iii) also allows us to switch from $\mathbf{G}$ to any group of the same type. So we may reduce the proof to the case of simple types. If the type is $\mathbf{A}$, we may take $\mathbf{G}$ to be a general linear group. Then the statement comes from Example 13.15.

Assume now that the type of $\mathbf{G}$ does not include type $\mathbf{A}$. By Proposition 13.12(i), ℓ does not divide the order of $(\mathrm{Z}(\mathbf{G}^*)/\mathrm{Z}^\circ(\mathbf{G}^*))^F$. Then (i) above implies that $\mathbf{H} := \mathrm{C}_{\mathbf{G}}(y)$ is connected. One has $y \in \mathrm{Z}(\mathbf{H})^F$. But

Proposition 13.12(ii) implies that $(\mathrm{Z}(\mathbf{H})/\mathrm{Z}^\circ(\mathbf{H}))^F$ is of order prime to ℓ, so $y \in \mathrm{Z}^\circ(\mathbf{H})$. Then clearly $\mathbf{H} = \mathrm{C}_\mathbf{G}(\mathrm{Z}^\circ(\mathbf{H}))$. This is a characterization of Levi subgroups. □

From now on, assuming that the prime ℓ does not divide q, let $E_{q,\ell} := \{d \mid \ell | \phi_d(q)\}$. This is $\{e, e\ell, e\ell^2, \ldots, e\ell^m, \ldots\}$ where e is the order of q mod. ℓ.

Lemma 13.17. *Assume that ℓ does not divide $|(\mathrm{Z}(\mathbf{G})/\mathrm{Z}^\circ(\mathbf{G}))^F|$. Let $\pi : \mathbf{G} \to \mathbf{G}_{\mathrm{ad}}$ be the canonical morphism.*

(i) If X is an F-stable subgroup of $\mathbf{G}$, then $\pi(X)^F_\ell = \pi(X^F_\ell)$.

(ii) Assume ℓ is good for $\mathbf{G}$. Let $\mathbf{S}$ be a $\phi_{E_{q,\ell}}$-subgroup of $\mathbf{G}$. Then $\mathrm{C}^\circ_\mathbf{G}(\mathbf{S}) = \mathrm{C}^\circ_\mathbf{G}(\mathbf{S}^F_\ell)$.

Proof. (i) By Proposition 8.1, $\pi(X)^F = (X\mathrm{Z}(\mathbf{G})/\mathrm{Z}(\mathbf{G}))^F$ is an extension of $([X\mathrm{Z}(\mathbf{G}), F] \cap \mathrm{Z}(\mathbf{G}))/[\mathrm{Z}(\mathbf{G}), F]$ by $(X\mathrm{Z}(\mathbf{G}))^F/\mathrm{Z}(\mathbf{G})^F$. By Lang's theorem, $\mathrm{Z}^\circ(\mathbf{G}) \subseteq [\mathrm{Z}(\mathbf{G}), F]$. Therefore the hypothesis on ℓ implies that $([X\mathrm{Z}(\mathbf{G}), F] \cap \mathrm{Z}(\mathbf{G}))/[\mathrm{Z}(\mathbf{G}), F]$ is ℓ' ; thus $\pi(X)^F_\ell \subseteq \pi(X^F)$. Moreover, if s is of finite order, then $\pi(s_\ell) = \pi(s)_\ell$. This implies $\pi(X)^F_\ell = \pi(X^F_\ell)$.

(ii) Let us show first that $\mathbf{S}^F_\ell \subseteq \mathrm{Z}(\mathbf{G})$ if and only if $\mathbf{S} \subseteq \mathrm{Z}(\mathbf{G})$.

The "if" part is clear, so assume $\pi(\mathbf{S}) \neq \{1\}$. In $\mathbf{G}_{\mathrm{ad}}$, $\pi(\mathbf{S})$ is a non-trivial $\phi_{E_{q,\ell}}$-subgroup (Proposition 13.7), so $|\pi(\mathbf{S})^F|$, which is the value at q of some non-trivial product of cyclotomic polynomials ϕ_d with $d \in E_{q,\ell}$, is a multiple of ℓ. Then, by (i), $\pi(\mathbf{S})^F_\ell = \pi(\mathbf{S}^F_\ell) \neq \{1\}$. Hence $\mathbf{S}^F_\ell \not\subseteq \mathrm{Z}(\mathbf{G})$.

Let us now prove (ii) by induction on dim $\mathbf{G}$. If $\mathbf{S}$ is central in $\mathbf{G}$, then everything is clear. Otherwise, we now know that $\mathbf{S}^F_\ell \not\subseteq \mathrm{Z}(\mathbf{G})$. Let $\mathbf{L} = \mathrm{C}^\circ_\mathbf{G}(\mathbf{S}^F_\ell)$, this is a Levi subgroup by Proposition 13.16 (ii). By definition of good primes ℓ is good for $\mathbf{L}$. Then one may apply the induction hypothesis to $(\mathbf{L}, F)$ by Proposition 13.12(iv). It provides the equality sought. □

Theorem 13.18. *Let $(\mathbf{G}, F)$ be a connected reductive $\mathbf{F}$-group defined over $\mathbb{F}_q$. Given an integer $d \geq 1$, there is a unique $\mathbf{G}^F$-conjugacy class of F-stable tori $\mathbf{S}$ such that $P_{\mathbf{S},F}$ is the biggest power of ϕ_d dividing $P_{\mathbf{G},F}$. They coincide with the maximal ϕ_d-subgroups of $\mathbf{G}$.*

Proof. In view of Proposition 13.5 the theorem reduces to a statement about F-stable maximal tori, and can be therefore expressed in terms of the root datum of $\mathbf{G}$ around an F-stable maximal torus $\mathbf{T}$ and the p-morphism induced by F (see §8.1).

We prove the proposition by induction on the dimension of $\mathbf{G}$ in connection with preceding results. Given d, we may choose q and ℓ such that

(a) ℓ is a prime good for $\mathbf{G}$,

(b) $\ell \equiv 1$ mod. d and ℓ is bigger than the order of the Weyl group of $\mathbf{G}$,

(c) q is of order d mod. ℓ.

Condition (a) excludes only a finite number of primes, hence (b) is possible by Dirichlet's theorem on arithmetic progressions (see for instance [Serre77b] §VI). The same argument and the fact that $(\mathbb{Z}/\ell\mathbb{Z})^\times$ is cyclic allows (c).

Then the order formula (proof of Proposition 13.2) and (b) imply that $|\mathbf{G}^F|_\ell = (\phi_d(q)^a)_\ell$ where a is the biggest power of ϕ_d in $P_{\mathbf{G},F}$. Moreover, ℓ does not divide the order of $\mathrm{Z}(\mathbf{G})/\mathrm{Z}^\circ(\mathbf{G})$ (see Table 13.11).

Assume that $\mathbf{S}$ and $\mathbf{S}'$ are ϕ_d-subgroups such that $P_{\mathbf{S},F} = {\phi_d}^a$. We show by induction that $\mathbf{S}$ contains a $\mathbf{G}^F$-conjugate of $\mathbf{S}'$. Once the existence of $\mathbf{S}$ is established (see below), this will give all the claims of our theorem.

Applying Propositions 13.2 and 13.5 to $\mathbf{S}.\mathrm{Z}^\circ(\mathbf{G})_{\phi_d}$, we see first that $\mathbf{S} \supseteq \mathrm{Z}^\circ(\mathbf{G})_{\phi_d}$. However, $\mathbf{S}_\ell^F$ is a Sylow ℓ-subgroup of $\mathbf{G}^F$. So we may assume that $\mathbf{S}_\ell^F \supseteq {\mathbf{S}'}_\ell^F$. By Lemma 13.17(ii), $\mathrm{C}^\circ_{\mathbf{G}}({\mathbf{S}'}_\ell^F)$ is a Levi subgroup of $\mathbf{G}$. If $\mathbf{L} = \mathbf{G}$, then $\mathbf{S}' \subseteq \mathrm{Z}^\circ(\mathbf{G})_{\phi_d} \subseteq \mathbf{S}$. Otherwise, $\mathrm{C}^\circ_{\mathbf{G}}({\mathbf{S}'}_\ell^F)$ is a proper Levi subgroup containing both $\mathbf{S}$ and $\mathbf{S}'$. The induction hypothesis then gives our claim.

For the existence, one may assume that $\mathbf{G} = \mathbf{G}_{\mathrm{ad}}$ by Proposition 13.7. If $a \neq 0$, then $|\mathbf{G}^F|$ is divisible by ℓ. Let S be a Sylow ℓ-subgroup of $\mathbf{G}^F$ and let $x \in \mathrm{Z}(S)$, $x \neq 1$. Then $\mathbf{L} := \mathrm{C}^\circ_{\mathbf{G}}(x)$ is a proper F-stable Levi subgroup of $\mathbf{G}$ with $\mathbf{L}^F = \mathrm{C}_{\mathbf{G}^F}(x)$ (apply Proposition 13.16 knowing that ℓ is good and bigger than $|\mathrm{Z}(\mathbf{G}_{\mathrm{sc}})|$; see Table 13.11). Then $P_{\mathbf{L},F}$ divides $P_{\mathbf{G},F}$ (Proposition 13.2(ii)) and is divisible by ${\phi_d}^a$ since $|\mathbf{G}^F : \mathbf{L}^F| = P_{\mathbf{G},F}(q)/P_{\mathbf{L},F}(q)$ is prime to ℓ. The induction hypothesis then implies our claim. □

Proposition 13.19. *If an F-stable Levi subgroup $\mathbf{L}$ of $\mathbf{G}$ satisfies $\mathbf{L} = \mathrm{C}^\circ_{\mathbf{G}}(\mathrm{Z}(\mathbf{L})_\ell^F)$, then it is $E_{q,\ell}$-split. If, moreover, ℓ is good for $\mathbf{G}$ and $(\mathrm{Z}(\mathbf{G})/\mathrm{Z}^\circ(\mathbf{G}))^F$ is of order prime to ℓ, then the converse is true.*

Proof. We use the abbreviated notation $E = E_{q,\ell}$. By definition of $E_{q,\ell}$, one has $\mathbf{T}_\ell^F = (\mathbf{T}_{\phi_E})_\ell^F$ for any F-stable torus $\mathbf{T}$.

Assume first that $(\mathrm{Z}(\mathbf{G})/\mathrm{Z}^\circ(\mathbf{G}))^F$ is of order prime to ℓ. By Proposition 13.13 (iv), one has $\mathrm{Z}(\mathbf{L})_\ell^F = \mathrm{Z}^\circ(\mathbf{L})_\ell^F$, hence $\mathbf{L} = \mathrm{C}^\circ_{\mathbf{G}}(\mathrm{Z}(\mathbf{L})_\ell^F)$ implies that $\mathbf{L} = \mathrm{C}^\circ_{\mathbf{G}}(\mathrm{Z}^\circ(\mathbf{L})_{\phi_E})$ so that $\mathbf{L}$ is E-split.

Conversely let $\mathbf{S}$ be a ϕ_E-subgroup of $\mathbf{G}$ and let $\mathbf{L} = \mathrm{C}_{\mathbf{G}}(\mathbf{S})$. If ℓ is good then $\mathbf{L} = \mathrm{C}^\circ_{\mathbf{G}}(\mathbf{S}_\ell^F)$ by Lemma 13.17(ii).

Now assume only that $\mathbf{L} = \mathrm{C}^\circ_{\mathbf{G}}(\mathrm{Z}(\mathbf{L})_\ell^F)$. There is an embedding $\mathbf{G} \to \widetilde{\mathbf{G}}$, defined on $\mathbb{F}_q$, with isomorphic derived groups and such that $\mathrm{Z}(\widetilde{\mathbf{G}}) = \mathrm{Z}^\circ(\widetilde{\mathbf{G}})$ (take $\widetilde{\mathbf{G}} = \mathbf{G} \times \mathbf{T}/\mathrm{Z}(\mathbf{G})$ where $\mathbf{T}$ is an F-stable maximal torus of $\mathbf{G}$; see §15.1 below). Then $\widetilde{\mathbf{L}} := \mathrm{Z}^\circ(\widetilde{\mathbf{G}})\mathbf{L}$ is a Levi subgroup of $\widetilde{\mathbf{G}}$. One has $\widetilde{\mathbf{L}} = \mathrm{C}^\circ_{\widetilde{\mathbf{G}}}(\mathrm{Z}(\mathbf{L})_\ell^F)$, hence $\widetilde{\mathbf{L}} = \mathrm{C}^\circ_{\widetilde{\mathbf{G}}}(\mathrm{Z}(\widetilde{\mathbf{L}})_\ell^F)$. By the first part of the proof, $\widetilde{\mathbf{L}}$ is E-split. By Proposition 13.7, $\mathbf{L}$ is E-split. □

Exercises

1. Give an example of a centralizer of a semi-simple element which is not connected (take $\mathbf{G} = \mathrm{PGL}$). Deduce an example of a finite commutative (non-cyclic) subgroup which is made of semi-simple elements but is not included in a torus. Show that, if $q \neq 2$, then $\mathrm{PGL}_2(\mathbb{F}_q)$ has Klein subgroups but none is in a torus.
2. Assume ℓ is good and does not divide $q|(\mathrm{Z}(\mathbf{G}^*)/\mathrm{Z}^\circ(\mathbf{G}^*))^F|$. Let Y be a commutative ℓ-subgroup of $\mathbf{G}^F$. Show that Y is included in a torus.
3. Let W be a Weyl group. Let $w \in W$. Let d be the order of one of its eigenvalues in the reflection representation. Show that d divides at least one "exponent" (see [Bour68] §V.6) of W. *Hint*: apply Proposition 13.2(ii) and the expresssion for $P_{\mathbf{G},F}$ in terms of exponents.

Notes

The results of this chapter are mostly a formalization of easy computations. The vocabulary of polynomial orders was introduced by Broué–Malle (see [BrMa92]), with the underlying idea that many theorems about unipotent characters should be expressed and checked at the level of the root datum, forgetting about q, and ultimately be given analogues in the case of complex reflection groups instead of just Weyl groups of reductive groups (see [BrMaRo98], [Bro00], and their references).

Theorem 13.18 is related to the existence of certain "regular" elements in Weyl groups, a notion introduced by Springer. See [Sp74], where the proof uses arguments of elementary algebraic geometry (over $\mathbb{C}$).

14

Unipotent characters as a basic set

We now come back to irreducible characters of groups $\mathbf{G}^F$. Recall from Chapter 8 that $\mathrm{Irr}(\mathbf{G}^F)$ decomposes as $\mathrm{Irr}(\mathbf{G}^F) = \bigcup_s \mathcal{E}(\mathbf{G}^F, s)$ where s ranges over $(\mathbf{G}^*)^F$-conjugacy classes of semi-simple elements, for $\mathbf{G}^*$ in duality with $\mathbf{G}$.

Let ℓ be as usual a prime different from the characteristic of the field over which $\mathbf{G}$ is defined. Denote by $\mathcal{E}(\mathbf{G}^F, \ell')$ the union of the $\mathcal{E}(\mathbf{G}^F, s)$ above where s ranges over semi-simple ℓ'-element.

The main property of $\mathcal{E}(\mathbf{G}^F, \ell')$ proved in this chapter is that, when ℓ does not divide the order of $(\mathrm{Z}(\mathbf{G})/\mathrm{Z}^\circ(\mathbf{G}))^F$, the restriction of the elements of $\mathcal{E}(\mathbf{G}^F, \ell')$ to ℓ'-elements of $\mathbf{G}^F$ form a K-basis of central functions $\mathbf{G}^F_{\ell'} \to K$ (Theorem 14.4, due to Geck–Hiss). We prove a stronger statement, namely, that $\mathcal{E}(\mathbf{G}^F, \ell')$ is a *basic set* of characters (see Definition 14.3). This is a key property that will allow us to consider the elements of $\mathcal{E}(\mathbf{G}^F, \ell')$ as approximations of simple $k\mathbf{G}^F$-modules (compare Theorem 9.12(ii)). More in this direction will be obtained in Part IV, when we study decomposition matrices. The proof of Theorem 14.4 involves a comparison between centralizers of ℓ-elements in $\mathbf{G}^F$ and $(\mathbf{G}^*)^F$.

14.1. Dual conjugacy classes for ℓ-elements

Let $(\mathbf{G}, F)$ be a connected reductive $\mathbf{F}$-group defined over $\mathbb{F}_q$. Assume $(\mathbf{G}, F)$ and $(\mathbf{G}^*, F)$ are in duality. We write W^* and F for actions on $X(\mathbf{T}) \cong Y(\mathbf{T}^*)$ (see §8.2).

We show the following result.

Proposition 14.1. *Let ℓ be a prime not dividing q, or $|(\mathrm{Z}(\mathbf{G})/\mathrm{Z}^\circ(\mathbf{G}))^F| \times |(\mathrm{Z}(\mathbf{G}^*)/\mathrm{Z}^\circ(\mathbf{G}^*))^F|$, and good for $\mathbf{G}$ (see Definition 13.10).*

There exists a one-to-one map from the set of $\mathbf{G}^F$-conjugacy classes of ℓ-elements of $\mathbf{G}^F$ onto the set of $(\mathbf{G}^*)^F$-conjugacy classes of ℓ-elements of $(\mathbf{G}^*)^F$

such that, if the class of $x \in \mathbf{G}_\ell^F$ *maps onto the class of* y, *then* $\mathrm{C}^\circ_{\mathbf{G}}(x)$ *and* $\mathrm{C}^\circ_{\mathbf{G}^*}(y)$ *are Levi subgroups in dual classes.*

Proof of Proposition 14.1. Let $E = E_{q,\ell}$ be the set of integers $d \geq 1$ such that ℓ divides $\phi_d(q)$ (see Lemma 13.17). Then $(x \in \mathbf{G}_\ell^F \mapsto \mathrm{C}^\circ_{\mathbf{G}}(x))$ defines a map from the set of $\mathbf{G}^F$-conjugacy classes of ℓ-elements of $\mathbf{G}^F$ to the set of $\mathbf{G}^F$-conjugacy classes of E-split Levi subgroups of $\mathbf{G}$ (Proposition 13.19). Now $\mathbf{G}^F$-conjugacy of elements x, x' such that $\mathbf{L} = \mathrm{C}^\circ_{\mathbf{G}}(x) = \mathrm{C}^\circ_{\mathbf{G}}(x')$ is induced by conjugacy under $\mathrm{N}_{\mathbf{G}}(\mathbf{L})^F$. It is sufficient to show that, for any E-split Levi subgroups $\mathbf{L}$ and $\mathbf{L}^*$ in dual classes, the number of orbits of $\mathrm{N}_{\mathbf{G}}(\mathbf{L})^F$ acting on $X := \{x \in \mathbf{G}_\ell^F \mid \mathbf{L} = \mathrm{C}^\circ_{\mathbf{G}}(x)\}$ is equal to the number of orbits of $\mathrm{N}_{\mathbf{G}^*}(\mathbf{L}^*)^F$ acting on $Y := \{y \in (\mathbf{G}^*)_\ell^F \mid \mathbf{L}^* = \mathrm{C}^\circ_{\mathbf{G}^*}(y)\}$.

Let $\mathbf{T}$ be a quasi-split maximal torus (see [DiMi91] 8.3) in $\mathbf{L}$ and put $A = \mathrm{Z}(\mathbf{L})_\ell^F$. The intersection $(\mathrm{N}_{\mathbf{G}}(\mathbf{T}) \cap \mathrm{N}_{\mathbf{G}}(\mathbf{L}))/\mathbf{T}$ is a semi-direct product $W(\mathbf{L}, \mathbf{T}).V$ of F-stable subgroups of $W(\mathbf{G}, \mathbf{T})$. This can be seen by looking at the action of $\mathrm{N}_{\mathbf{G}}(\mathbf{L})$ on the Borel subgroups of $\mathbf{L}$, or by proving, in the notation of Chapter 2, that $\mathrm{N}_W(W_I) = W_I \rtimes W^I$ (see Definition 2.26 and Theorem 2.27(iv)). Now the action of $\mathrm{N}_{\mathbf{G}}(\mathbf{L})^F$ on the center of $\mathbf{L}$ reduces to the action of V^F. Recall that the hypothesis on $\mathrm{Z}(\mathbf{G}^*)$ implies that $\mathrm{C}_{\mathbf{G}}(Z)^F = \mathrm{C}^\circ_{\mathbf{G}}(Z)^F$ for any ℓ-subgroup Z of $\mathbf{G}^F$ included in a torus (Proposition 13.16(i)). Hence, if $x \in A$ is fixed under some $v \in V^F \setminus \{1\}$, then $v \in \mathrm{C}^\circ_{\mathbf{G}}(x)$ and $x \notin X$. Thus the number of orbits in X under the action of $\mathrm{N}_{\mathbf{G}}(\mathbf{L})^F$ is $|X|/|V^F|$. For Levi subgroups $\mathbf{L}$ and $\mathbf{L}^*$ in dual classes, the quotients $\mathrm{N}_{\mathbf{G}}(\mathbf{L})/\mathbf{L}$ and $\mathrm{N}_{\mathbf{G}^*}(\mathbf{L}^*)/\mathbf{L}^*$ are in natural correspondence, i.e. $\mathrm{N}_{\mathbf{G}^*}(\mathbf{L}^*)/\mathbf{L}^*$ is isomorphic to $V^* := \{v^* \mid v \in V\}$ (see the end of §8.2). We have to show that $|X| = |Y|$.

If $x \in A \setminus X$, then $\mathrm{C}^\circ_{\mathbf{G}}(x)$ is an E-split Levi subgroup of $\mathbf{G}$ that strictly contains $\mathbf{L}$, hence there exists $v \in W(\mathrm{C}^\circ_{\mathbf{G}}(x), \mathbf{T})^F \setminus W(\mathbf{L}, \mathbf{T})^F$ with $v(x) = x$. So X is defined by the equality

$$X = A \setminus \bigcup_{v \in W^F, v \notin W(\mathbf{L},\mathbf{T})} A^v$$

where $A^v = A \cap \mathrm{C}_{\mathbf{T}}(v)$. Of course one has, with $B = \mathrm{Z}(\mathbf{L}^*)_\ell^F$,

$$Y = B \setminus \bigcup_{v^* \in (W^*)^F, v^* \notin W(\mathbf{L}^*,\mathbf{T}^*)} B^{v^*}.$$

As $W(\mathbf{L}^*, \mathbf{T}^*) = W(\mathbf{L}, \mathbf{T})^*$, by the inclusion-exclusion formula, if one has $|A \cap \mathrm{C}_{\mathbf{G}}(U)| = |B \cap \mathrm{C}_{\mathbf{G}}(U^*)|$ for any subset of subgroup U of W^F, then $|X| = |Y|$. We prove the following (after the proof of Proposition 14.1).

Lemma 14.2. *Let* $\mathbf{L}$ *be any* E*-split Levi subgroup of* $\mathbf{G}$, *let* $\mathbf{T}$ *be an* F*-stable maximal torus in* $\mathbf{L}$, *and let* V' *be a subset of* $\mathrm{N}_{\mathbf{G}}(\mathbf{T})^F$. *Then* $\mathrm{C}^\circ_{\mathbf{G}}(\mathrm{Z}(\mathbf{L})_\ell^F \cap \mathrm{C}_{\mathbf{G}}(V'))$ *is the smallest of the* E*-split Levi subgroups of* $\mathbf{G}$ *that contains* $\mathbf{L}$ *and* V'.

Let $\mathbf{S}$ be the maximal ϕ_E-subgroup of the center of $\mathbf{L}$. We write A^U (resp. $\mathbf{S}^U$) for $A \cap \mathrm{C}_{\mathbf{G}}(U)$ (resp. $\mathbf{S} \cap \mathrm{C}_{\mathbf{G}}(U)$), and $\mathbf{L}_U$ for $\mathrm{C}^\circ_{\mathbf{G}}(A^U)$. Lemma 14.2 implies that $\mathbf{L}_U = \mathrm{C}_{\mathbf{G}}((\mathbf{S}^U)^\circ)$. In a duality between $\mathbf{L}$ and $\mathbf{L}^*$ around rational maximal tori, the ϕ_E-subgroups $\mathbf{S}$ and $\mathrm{Z}(\mathbf{L}^*)_{\phi_E}$ correspond as well as $(\mathbf{S}^U)^\circ$ and $(\mathrm{Z}(\mathbf{L}^*)_{\phi_E}^{U^*})^\circ$, so that $\mathbf{L}_U$ and $\mathrm{C}^\circ_{\mathbf{G}^*}(B^{U^*}) = \mathrm{C}_{\mathbf{G}^*}((\mathrm{Z}(\mathbf{L}^*)^U_{\phi_E})^\circ)$ are in dual classes. As $U \subseteq \mathrm{C}^\circ_{\mathbf{G}}(A^U)$ because A^U is an ℓ-group, $A^U = \mathrm{Z}^\circ(\mathrm{C}^\circ_{\mathbf{G}}(A^U))^F_\ell$ and similarly $B^{U^*} = \mathrm{Z}^\circ(\mathrm{C}^\circ_{\mathbf{G}^*}(B^{U^*}))^F_\ell$ by Proposition 13.9 and Proposition 13.12(ii). Now $|A^U| = |B^{U^*}|$ follows from Proposition 13.8. □

Proof of Lemma 14.2. Let $A' = A \cap \mathrm{C}_{\mathbf{G}}(V')$. We know that $\mathbf{L} = \mathrm{C}^\circ_{\mathbf{G}}(A)$ and that $\mathrm{C}^\circ_{\mathbf{G}}(A')$ is an $E_{q,\ell}$-split Levi subgroup of $\mathbf{G}$ that contains $\mathbf{L}$ by Proposition 13.19. As A' is an ℓ-subgroup of $\mathbf{G}^F$ and $V' \subseteq \mathrm{C}_{\mathbf{G}}(A')^F$, Proposition 13.16(i) implies $V' \subseteq \mathrm{C}^\circ_{\mathbf{G}}(A')$. Conversely, let $\mathbf{M}$ be an E-split Levi subgroup of $\mathbf{G}$ that contains $\mathbf{L}$ and V'. One has $\mathbf{M} = \mathrm{C}^\circ_{\mathbf{G}}(\mathrm{Z}(\mathbf{M})^F_\ell)$ and clearly $\mathrm{Z}_{\mathbf{G}}(\mathbf{M})^F_\ell \subseteq A'$, hence $\mathrm{C}^\circ_{\mathbf{G}}(A') \subseteq \mathbf{M}$. □

14.2. Basic sets in the case of connected center

Let G a finite group. Let ℓ be a prime. Let $(\mathcal{O}, K, k)$ be an ℓ-modular splitting system for G. Recall the space of central functions $\mathrm{CF}(G, K)$ (see §5.1) and the decomposition map $d^1\colon \mathrm{CF}(G, K) \to \mathrm{CF}(G, K)$ (see Definition 5.7).

Definition 14.3. *Let b be a central idempotent of $\mathcal{O}G$, so that $\mathcal{O}G.b$ is a product of ℓ-blocks of G (see §5.1). Any $\mathbb{Z}$-basis of the lattice $\mathbb{Z}d^1(\mathrm{Irr}(G, b))$ is called a* **basic set** *of b. A subset $\mathcal{B} \subseteq \mathrm{Irr}(b)$ is called a basic set of characters if and only if the $(d^1\chi)_{\chi\in\mathcal{B}}$ are distinct and a basis of the lattice in $\mathrm{CF}(G, K, b)$ generated by $d^1\chi$'s for $\chi \in \mathrm{Irr}(G, b)$.*

Note that it is enough to check that $(d^1\chi)_{\chi\in\mathcal{B}}$ generates the same lattice as $(d^1\chi)_{\chi\in\mathrm{Irr}(G,b)}$ and that $|\mathcal{B}|$ has a cardinality greater than or equal to the expected dimension, i.e. the number of simple $kG.b$-modules (see [NaTs89] 3.6.15).

Note that a set of characters $\mathcal{B} \subseteq \mathrm{Irr}(G, b)$ as above is a basic set for b if and only if it is one for each central idempotent $b' \in \mathrm{Z}(\mathcal{O}G.b)$.

Theorem 14.4. *Let ℓ be a prime good for $\mathbf{G}$ and not dividing the order of $(\mathrm{Z}(\mathbf{G})/\mathrm{Z}^\circ(\mathbf{G}))^F$. Let $s \in (\mathbf{G}^*)^F$ be a semi-simple ℓ'-element. Then $\{d^1(\chi)\}_{\chi\in\mathcal{E}(\mathbf{G}^F,s)}$ is a basic set for $\mathcal{O}\mathbf{G}^F.b_\ell(\mathbf{G}^F, s)$ (see Definition 9.4 and Theorem 9.12(i)).*

The "generating" half of Theorem 14.4 is an easy consequence of commutation of the decomposition map d^1 and the $\mathrm{R}^{\mathbf{G}}_{\mathbf{L}}$ induction.

Proposition 14.5. *Let ℓ be a prime good for* $\mathbf{G}$. *Assume the order of* $(\mathrm{Z}(\mathbf{G})/\mathrm{Z}^\circ(\mathbf{G}))^F$ *is prime to* ℓ. *Let* $s \in (\mathbf{G}^*)^F$ *be a semi-simple* ℓ'*-element. Let* $\chi \in \mathcal{E}_\ell(\mathbf{G}^F, s) = \mathrm{Irr}(\mathbf{G}^F, b_\ell(\mathbf{G}^F, s))$. *Then* $d^1\chi$ *is in the group generated by the* $d^1\chi'$*'s for* $\chi' \in \mathcal{E}(\mathbf{G}^F, s)$.

Proof of Proposition 14.5. By definition, there exists $t \in \mathrm{C}_{G^*}(s)^F_\ell$ such that $\chi \in \mathcal{E}(G^F, st)$. Let $\mathbf{L}$ be a Levi subgroup of $\mathbf{G}$ such that $\mathbf{L}$ and $\mathrm{C}^\circ_{\mathbf{G}^*}(t)$ are in dual classes (see Proposition 13.16(ii)). One has $\mathrm{C}_{\mathbf{G}^*}(t)^F \subseteq \mathrm{C}^\circ_{\mathbf{G}^*}(t)$ by Proposition 13.16(i) and the hypothesis on ℓ, hence $\mathrm{C}^\circ_{\mathbf{G}^*}(st)\mathrm{C}_{\mathbf{G}^*}(st)^F \subseteq \mathrm{C}^\circ_{\mathbf{G}^*}(t)$. By Theorem 8.27, for any choice of a parabolic subgroup having $\mathbf{L}$ as Levi subgroup, $\varepsilon_\mathbf{L}\varepsilon_\mathbf{G}\mathrm{R}^\mathbf{G}_\mathbf{L}$ induces a one-to-one map from $\mathcal{E}(\mathbf{L}^F, st)$ onto $\mathcal{E}(\mathbf{G}^F, st)$. Since t is central and rational in the dual of $\mathbf{L}$ there exists a linear character θ of $\mathbf{L}^F$, in duality with t, such that $\theta.\mathcal{E}(\mathbf{L}^F, s) = \mathcal{E}(\mathbf{L}^F, st)$ (Proposition 8.26). Thus one has $\chi = \varepsilon_\mathbf{L}\varepsilon_\mathbf{G}\mathrm{R}^\mathbf{G}_\mathbf{L}(\theta \otimes \xi)$ for some $\xi \in \mathcal{E}(\mathbf{L}^F, s)$. The order of θ is the order of t. We get $d^1\chi = \varepsilon_\mathbf{L}\varepsilon_\mathbf{G}\mathrm{R}^\mathbf{G}_\mathbf{L}(d^1(\theta \otimes \xi)) = \varepsilon_\mathbf{L}\varepsilon_\mathbf{G}\mathrm{R}^\mathbf{G}_\mathbf{L}(d^1\xi) = d^1(\varepsilon_\mathbf{L}\varepsilon_\mathbf{G}\mathrm{R}^\mathbf{G}_\mathbf{L}\xi)$ by Proposition 9.6(iii). By Proposition 8.25, $\mathrm{R}^\mathbf{G}_\mathbf{L}\xi$ decomposes on elements of $\widetilde{\mathcal{E}}(\mathbf{G}^F, s)$, so that $d^1\chi \in d^1(\mathbb{Z}\mathcal{E}(\mathbf{G}^F, \ell'))$ (see Definition 9.4). But if $\chi' \in \mathcal{E}(\mathbf{G}^F, s')$ for some semi-simple ℓ'-element of $(\mathbf{G}^*)^F$, then $d^1\chi' \in \mathrm{CF}(\mathbf{G}^F, b_\ell(\mathbf{G}^F, s'))$ by Brauer's second Main Theorem (Theorem 5.8). So $d^1\chi \in \mathbb{Z}d^1(\mathcal{E}(\mathbf{G}^F, s))$ (see also Proposition 15.7 below). □

Proof of Theorem 14.4. Let us introduce the following notation. If G is a finite group and π is a set of primes, let $[G]_\pi$ denote any representative system of the G-conjugacy classes of π-elements of G. Recall that π' denotes the complementary set of primes.

In view of Proposition 14.5 above, it remains to check that $|\mathcal{E}(\mathbf{G}^F, \ell')|$ equals the dimension of $d^1(\mathrm{CF}(\mathbf{G}^F, K))$, i.e. $|[\mathbf{G}^F]_{\ell'}|$. In other words, we must prove the equality

(BS(1)) $$|[\mathbf{G}^F]_{\ell'}| = \sum_{s \in [(\mathbf{G}^*)^F]_{\{\ell,p\}'}} |\mathcal{E}(\mathbf{G}^F, s)|.$$

Let $t \in (\mathbf{G}^*)^F_\ell$, and define $\mathbf{G}(t)$ in duality with $\mathrm{C}^\circ_{\mathbf{G}^*}(t)$ as in the proof of Proposition 14.5. Then Proposition 14.5 applies to $\mathbf{G}(t)$, by Proposition 13.12(ii). So the set of $d^1(\xi)$, where $\xi \in \mathcal{E}(\mathbf{G}(t)^F, s)$, is a generating set for $b_\ell(\mathbf{G}(t)^F, s)$ whenever s is an ℓ'-element in $\mathrm{C}^\circ_{\mathbf{G}^*}(t)^F$. With evident notation this yields

(GS(t)) $$|[\mathbf{G}(t)^F]_{\ell'}| \leq \sum_{s \in [\mathrm{C}^\circ_{\mathbf{G}^*}(t)^F]_{\{\ell,p\}'}} |\mathcal{E}(\mathbf{G}(t)^F, s)|.$$

Recall that $\mathrm{C}_{\mathbf{G}^*}(t)^F = \mathrm{C}^\circ_{\mathbf{G}^*}(t)^F$ if $t \in (\mathbf{G}^*)^F_\ell$; see Proposition 13.16(i). So we may assume that the union of the $[\mathrm{C}^\circ_{\mathbf{G}^*}(t)^F]_{\{\ell,p\}'}.t$ when t runs over $[(\mathbf{G}^*)^F]_\ell$ is a set of representatives of $(\mathbf{G}^*)^F$-conjugacy classes of semi-simple elements of $(\mathbf{G}^*)^F$, i.e. equals $[(\mathbf{G}^*)^F]_{p'}$. From $\mathrm{Irr}(\mathbf{G}^F) = \bigcup_{(s,t)} \mathcal{E}(\mathbf{G}^F, st)$ and $|\mathcal{E}(\mathbf{G}^F, st)| = |\mathcal{E}(\mathbf{G}(t)^F, s)|$ when $t = (st)_\ell$ (see §8.4), we get

$$\text{(GS)} \qquad \sum_{t \in [(\mathbf{G}^*)^F]_\ell} |[\mathbf{G}(t)^F]_{\ell'}| \leq |\mathrm{Irr}(\mathbf{G}^F)|.$$

Let us assume that $(\mathrm{Z}(\mathbf{G}^*)/\mathrm{Z}^\circ(\mathbf{G}^*))^F$ is prime to ℓ, for instance $\mathrm{Z}(\mathbf{G}^*)$ is connected. Then we may apply Proposition 14.1. So there is a one-to-one map $(t \mapsto t')$, from $[(\mathbf{G}^*)^F]_\ell$ onto $[\mathbf{G}^F]_\ell$, with $\mathbf{G}(t) = \mathrm{C}^\circ_{\mathbf{G}}(t')$. Since the number of irreducible characters equals the number of conjugacy classes, which in turn can be computed using the decomposition of each element into its ℓ and ℓ'-parts, we get

$$|\mathrm{Irr}(\mathbf{G}^F)| = \sum_{t' \in [\mathbf{G}^F]_\ell} |[\mathrm{C}_{\mathbf{G}^F}(t')]_{\ell'}|$$

with $\mathrm{C}_{\mathbf{G}^F}(t') = \mathbf{G}(t)^F$. Therefore there is equality in (GS), and this implies that there is equality in (GS(t)) for all $t \in [(\mathbf{G}^*)^F]_\ell$. With $t = 1$, we get (BS(1)).

We no longer assume that $\mathrm{Z}(\mathbf{G}^*)$ is connected. Let $\mathbf{G}^* \to \mathbf{H}^*$ be a closed embedding of reductive **F**-groups defined over $\mathbb{F}_q$, such that $\mathbf{H}^* = \mathrm{Z}(\mathbf{H}^*).\mathbf{G}^*$ and $\mathrm{Z}(\mathbf{H}^*)$ is connected (take for instance $\mathbf{H}^* := \mathbf{G}^* \times \mathbf{S}/\mathrm{Z}(\mathbf{G}^*)$ where $\mathbf{S}$ is an F-stable torus of $\mathbf{G}^*$ containing $\mathrm{Z}(\mathbf{G}^*)$, the action of $\mathrm{Z}(\mathbf{G}^*)$ on $\mathbf{G}^* \times \mathbf{S}$ being diagonal). By duality, $\mathbf{G}$ is a quotient of a dual $\mathbf{H}$ of $\mathbf{H}^*$, with kernel a central torus $\mathbf{K}$ defined on $\mathbb{F}_q$, so that $\mathbf{G}^F \cong \mathbf{H}^F/\mathbf{K}^F$ (see §15.1). If $s \in \mathbf{G}^{*F}$ is a semi-simple element, then $s \in \mathbf{H}^{*F}$ and the elements of $\mathcal{E}(\mathbf{H}^F, s)$ have $\mathbf{K}^F$ in their kernel, and they provide bijectively all of $\mathcal{E}(\mathbf{G}^F, s)$. This is easily seen from the corresponding property of the Deligne–Lusztig characters (see §8.4). Concerning the centers, we have $\mathrm{Z}(\mathbf{G}) = \mathrm{Z}(\mathbf{H})/\mathbf{K}$ and $\mathrm{Z}^\circ(\mathbf{G}) = \mathrm{Z}^\circ(\mathbf{H})/\mathbf{K}$. Then $(\mathrm{Z}(\mathbf{H})/\mathrm{Z}^\circ(\mathbf{H}))^F$ is of order prime to ℓ. So the preceding proof applies to $\mathbf{H}$. This tells us that the $d^1\chi$'s for $\chi \in \mathcal{E}(\mathbf{H}^F, s)$ are distinct and linearly independent. This gives the same for $\mathcal{E}(\mathbf{G}^F, s)$ since central functions on $\mathbf{G}^F_{\ell'}$ are in bijection with $\mathbf{K}^F$-constant central functions on $\mathbf{H}^F_{\ell'}.\mathbf{K}^F$. □

As a conclusion, let us state a generalization without condition on $\mathrm{Z}(\mathbf{G})/\mathrm{Z}^\circ(\mathbf{G})$, which is not used in the remainder of the book.

Theorem 14.6. *Let $(\mathbf{G}, F)$ be a connected reductive **F**-group defined over $\mathbb{F}_q$. Let ℓ be a prime good for $\mathbf{G}$ and not dividing q. Let $\mathbf{G} \subseteq \widetilde{\mathbf{G}}$ be an embedding of reductive groups defined over $\mathbb{F}_q$ such that $\widetilde{\mathbf{G}} = \mathrm{Z}(\widetilde{\mathbf{G}})$ and $\mathrm{Z}(\widetilde{\mathbf{G}})$ is connected. Let $\mathbf{G}^F \subseteq H \subseteq \widetilde{\mathbf{G}}^F$ such that $H/\mathbf{G}^F$ is a Sylow ℓ-subgroup of $\widetilde{\mathbf{G}}^F/\mathbf{G}^F$. Then the restrictions of the elements of $\mathcal{E}(\mathbf{G}^F, \ell')$ to $\mathbf{G}^F_{\ell'}$ are distinct, linearly independent, and generate the space of H-stable maps $\mathbf{G}^F_{\ell'} \to K$.*

For the fact that characters in $\mathcal{E}(\mathbf{G}^F, \ell')$ are fixed by H, see Exercise 17.1. See also Exercise 5.

Exercises

1. Let $\mathbf{G} \hookrightarrow \mathbf{H}$ be a closed group embedding defined on $\mathbb{F}_q$ between groups with connected centers such that $\mathbf{H} = \mathrm{Z}(\mathbf{H}).\mathbf{G}$. Let $\mathbf{H}^* \to \mathbf{G}^*$ be a dual map. Let $t \in \mathbf{H}^*$ be a semi-simple rational ℓ'-element, with image s in $\mathbf{G}^*$. Show that the linear independence of the $d^1(\chi)$ for $\chi \in \mathcal{E}(\mathbf{G}^F, s)$ is equivalent to linear independence of the $d^1(\xi)$ for $\xi \in \mathcal{E}(\mathbf{H}^F, t)$.
2. Assume that $\mathbf{G}$ is a group with connected center such that each simple factor of the derived group of $\mathbf{G}$ is of type $\mathbf{A}$. Show that $\{d^1(\chi)\}_{\chi \in \mathcal{E}(\mathbf{G}^F, 1)}$ is a basic set for $b_\ell(\mathbf{G}^F, 1)$ (*Hint*: use the linear independence of unipotent Deligne–Lusztig characters; see [DiMi91] 15.8). Then, using the isometry of Theorem 9.16, prove Theorem 14.4 for $\mathbf{G}$.
3. Assume that $\mathbf{G}$ is simple and that ℓ is an odd prime, good for $\mathbf{G}$ and not dividing the determinant of the Cartan matrix of the root system of $\mathbf{G}$. Let $(X, R, Y, R^\vee, F)$ be a root datum with F-action for $(\mathbf{G}, F)$. Let Δ be a basis of R.
   (a) Identifying $\mathbf{T}$ with $Y \otimes \mathbf{F}^\times$ and $\mathbf{T}^*$ with $X \otimes \mathbf{F}^\times$, show that any element of $\mathbf{T}^*_\ell$ has a unique expression as $\Sigma_{\alpha \in \Delta} \alpha \otimes \mu(\alpha)$, with $\mu(\alpha) \in \mathbf{F}^\times_\ell$.
   (b) Prove that the correspondence $(\Sigma_{\alpha \in \Delta} \alpha \otimes \mu(\alpha) \mapsto \Sigma_{\alpha \in \Delta} \alpha^\vee \otimes \mu(\alpha)^{\langle \alpha, \alpha \rangle})$ induces a W-equivariant one-to-one map from $\mathbf{T}_\ell$ to $\mathbf{T}^*_\ell$ such that the connected centralizers of corresponding elements are in dual classes. Then prove Proposition 14.1 for $\mathbf{G}$.
4. Assuming that the center of $\mathbf{G}$ is connected, deduce the conclusion of Proposition 14.1 from Exercises 1 to 3.
5. Find a prime ℓ not dividing q, $n \geq 2$ and two distinct (unipotent) classes of $\mathrm{SL}_n(q)$ that are fused by some $g \in \mathrm{GL}_n(q)$ with $\det(g)$ an ℓ-element of $\mathbb{F}_q^\times$.

Notes

Considering the elements of $\mathcal{E}(\mathbf{G}^F, \ell')$ as a basic set is one of the main arguments of Fong–Srinivasan in [FoSr82]. It was generalized by Geck–Hiss (see [GeHi91] and [Geck93a]). Their proof is slightly different from the one given here (see Exercise 3). For Theorem 14.6, see [CaEn99a] 1.7.

15

Jordan decomposition of characters

The aim of this chapter is to give one further property of the partition into rational series

$$\operatorname{Irr}(\mathbf{G}^F) = \bigcup_s \mathcal{E}(\mathbf{G}^F, s)$$

(see Chapter 8).

Each rational series is in bijection

$$\mathcal{E}(\mathbf{G}^F, s) \longleftrightarrow \mathcal{E}(\mathrm{C}_{\mathbf{G}^*}(s)^F, 1).$$

This is the "Jordan decomposition of characters," thus reducing the study of series $\mathcal{E}(\mathbf{G}^F, s)$ to series $\mathcal{E}(\mathbf{H}^F, 1)$. One must, however, pay attention to the fact that the centralizers of semi-simple elements, such as $\mathrm{C}_{\mathbf{G}^*}(s)$ when $\mathrm{Z}(\mathbf{G})$ is not connected, are reductive but generally not connected, in contrast with $\mathbf{G}$. The bijection above is therefore a complex statement which requires some preparation, even to make sense.

Let us embed our connected reductive $\mathbf{F}$-group $\mathbf{G}$ defined over $\mathbb{F}_q$ into a group $\widetilde{\mathbf{G}}$ with the same properties, such that $\widetilde{\mathbf{G}} = \mathrm{Z}(\widetilde{\mathbf{G}})\mathbf{G}$, and $\mathrm{Z}(\widetilde{\mathbf{G}}) = \mathrm{Z}^\circ(\widetilde{\mathbf{G}})$. Then each rational series $\mathcal{E}(\mathbf{G}^F, s)$ is obtained by restriction of a single series for $\widetilde{\mathbf{G}}^F$.

A crucial intermediate theorem to get the above "Jordan decomposition" states that $\operatorname{Res}^{\widetilde{\mathbf{G}}^F}_{\mathbf{G}^F}$ sends the elements of $\operatorname{Irr}(\widetilde{\mathbf{G}}^F)$ to sums of irreducible characters *without multiplicities*. This is a general fact in all cases where the quotient $\widetilde{\mathbf{G}}^F/\mathbf{G}^F$ may be taken to be cyclic, so it remains essentially the case where $\mathbf{G}$ is a spin group in dimension $4n$ (and $\widetilde{\mathbf{G}}$ is some associated conformal group). The checking in this case will be done in the next chapter.

Jordan decomposition of irreducible characters of finite reductive groups was proved by Lusztig (see [Lu84] — connected center — and [Lu88] — general case).

Note that the Jordan decomposition of characters is basically a bijection whose canonicity is not fully established in the form we prove. It will be used in the rest of the book only through the "multiplicity one" statement given by Theorem 15.11.

15.1. From non-connected center to connected center and dual morphism

Hypothesis 15.1. *Let*

$$\sigma : \mathbf{G} \longrightarrow \widetilde{\mathbf{G}}$$

be a morphism between reductive **F**-*groups defined over* $\mathbb{F}_q$ *such that*

(15.1(σ))

$$\sigma \text{ is a closed immersion,} \quad \sigma([\mathbf{G},\mathbf{G}]) = [\widetilde{\mathbf{G}},\widetilde{\mathbf{G}}],\, \mathrm{Z}(\widetilde{\mathbf{G}}) = \mathrm{Z}^\circ(\widetilde{\mathbf{G}}).$$

Given $\mathbf{G}$, one may define $\widetilde{\mathbf{G}}$ as follows. Choose a torus $\mathbf{S} \subseteq \mathbf{G}$ containing $\mathrm{Z}(\mathbf{G})$ (for instance a maximal torus) and let $\widetilde{\mathbf{G}}$ be the quotient $\mathbf{S} \times \mathbf{G}/\mathrm{Z}(\mathbf{G})$ where $\mathrm{Z}(\mathbf{G})$ acts diagonally. If $\mathbf{G}$ is defined over $\mathbb{F}_q$ by a Frobenius endomorphism $F : \mathbf{G} \to \mathbf{G}$, one takes an F-stable $\mathbf{S}$, so that the embedding $\mathbf{G} \to \widetilde{\mathbf{G}}$ is defined over $\mathbb{F}_q$.

If $\mathbf{T} \subseteq \mathbf{B}$ are F-stable maximal torus and Borel subgroup, respectively, in $\mathbf{G}$, the root datum of $(\widetilde{\mathbf{G}}, F)$ around $\widetilde{\mathbf{T}} := \sigma(\mathbf{T}).\mathrm{Z}(\widetilde{\mathbf{G}}) = \mathbf{T} \times \mathbf{S}/\mathrm{Z}(\mathbf{G})$ is easily defined from the root datum of $(\mathbf{G}, F)$ around $\mathbf{T}$. (A stronger condition on $\widetilde{\mathbf{G}}$ would be that the quotient $X(\mathbf{T})/\mathbb{Z}\Phi(\mathbf{G},\mathbf{T})$ (see §8.1) has no torsion at all, but this is not what we ask for in general.)

On the dual side there exist dual groups $\mathbf{G}^*$ and $\widetilde{\mathbf{G}}^*$ and a surjective morphism of algebraic groups

$$\sigma^* : \widetilde{\mathbf{G}}^* \longrightarrow \mathbf{G}^*$$

such that

(15.1*(σ^*))

$$\sigma^*(\widetilde{\mathbf{G}}^*) = \mathbf{G}^*, \quad \mathrm{Ker}(\sigma^*) = \mathrm{Ker}(\sigma^*)^\circ \subset \mathrm{Z}(\widetilde{\mathbf{G}}^*), (\widetilde{\mathbf{G}}^*)_{\mathrm{sc}} \longleftrightarrow [\widetilde{\mathbf{G}}^*, \widetilde{\mathbf{G}}^*].$$

One might first define σ^* from a simply connected covering of $[\mathbf{G}^*, \mathbf{G}^*]$, and then σ by duality (see also Exercise 1).

More precisely consider tori $\widetilde{\mathbf{T}}^*$ and $\mathbf{T}^*$ in duality, with $\widetilde{\mathbf{T}}$ and $\mathbf{T}$ respectively, with F-action, and $\sigma^*(\widetilde{\mathbf{T}}^*) = \mathbf{T}^*$. That means that one has dual sequences of torsion-free groups

$$0 \longrightarrow X(\widetilde{\mathbf{T}}/\mathbf{T}) \longrightarrow X(\widetilde{\mathbf{T}}) \xrightarrow{X(\sigma)} X(\mathbf{T}) \longrightarrow 0$$

and

$$0 \longrightarrow X(\mathbf{T}^*) \xrightarrow{X(\sigma^*)} X(\widetilde{\mathbf{T}}^*) \longrightarrow \mathrm{Hom}(X(\widetilde{\mathbf{T}}/\mathbf{T}), \mathbb{Z}) \longrightarrow 0.$$

Hence $\mathrm{Ker}(\sigma^*)$ is a central torus in $\widetilde{\mathbf{G}}^*$ in duality with $\widetilde{\mathbf{T}}/\mathbf{T}$, a group isomorphic with $\widetilde{\mathbf{G}}/\mathbf{G}$ because $\widetilde{\mathbf{G}} = \sigma(\mathbf{G})\widetilde{\mathbf{T}}$, and this duality is compatible with the Frobenius endomorphisms (both denoted by F). It follows that σ^* induces a surjective morphism $(\widetilde{\mathbf{G}}^*)^F \to (\mathbf{G}^*)^F$ and isomorphisms (see (8.19))

$$\textbf{(15.2)}\qquad \begin{array}{ccccc} (\mathrm{Ker}(\sigma^*))^F & \to & \mathrm{Irr}(\widetilde{\mathbf{T}}^F/\mathbf{T}^F) & \longleftrightarrow & \mathrm{Irr}(\widetilde{\mathbf{G}}^F/\mathbf{G}^F) \\ z & \mapsto & \hat{z} & \mapsto & \hat{z} \end{array}$$

As $Y(\widetilde{\mathbf{T}}^*)/\mathbb{Z}\Phi^* \cong X(\widetilde{\mathbf{T}})/\mathbb{Z}\Phi$ has no p'-torsion, the simply connected covering $(\widetilde{\mathbf{G}}^*)_{\mathrm{sc}} \to [\widetilde{\mathbf{G}}^*, \widetilde{\mathbf{G}}^*]$ (see §8.1) is a bijection. As $\mathrm{Z}(\widetilde{\mathbf{G}})^F$ is in the kernel of $\hat{z}$ if and only if $z \in [\widetilde{\mathbf{G}}^*, \widetilde{\mathbf{G}}^*]$, the isomorphism (15.2) restricts to $(\mathrm{Ker}(\sigma^*))^F \cap [\widetilde{\mathbf{G}}^*, \widetilde{\mathbf{G}}^*] \to \mathrm{Irr}(\widetilde{\mathbf{G}}^F/\mathbf{G}^F\mathrm{Z}(\widetilde{\mathbf{G}})^F)$. By Proposition 8.1, $\widetilde{\mathbf{G}}^F/\mathbf{G}^F\mathrm{Z}(\widetilde{\mathbf{G}})^F$ is naturally isomorphic to the group of F-co-invariant points on $\mathbf{G} \cap \mathrm{Z}(\widetilde{\mathbf{G}}) = \mathrm{Z}(\mathbf{G})$, i.e. the maximal F-trivial quotient of $\mathrm{Z}(\mathbf{G})/\mathrm{Z}^\circ(\mathbf{G})$. These groups are finite groups, hence some power of F acts trivially on them and one obtains a well-defined isomorphism

$$\textbf{(15.3)}\qquad \mathrm{Ker}(\sigma^*) \cap [\widetilde{\mathbf{G}}^*, \widetilde{\mathbf{G}}^*] \longleftrightarrow \mathrm{Irr}(\mathrm{Z}(\mathbf{G})/\mathrm{Z}^\circ(\mathbf{G})).$$

Using the adjoint group $\mathbf{G}_{\mathrm{ad}}$, one may recover the quotient $\widetilde{\mathbf{G}}^F/\mathbf{G}^F\mathrm{Z}(\widetilde{\mathbf{G}})^F$. Let $\pi_0 \colon \widetilde{\mathbf{G}} \to \mathbf{G}_{\mathrm{ad}}$ be the quotient morphism. Let $\pi \colon \mathbf{G} \to \widetilde{\mathbf{G}} \to \mathbf{G}_{\mathrm{ad}}$. By Proposition 8.1(i) and Lang's theorem, one has $\pi_0(\widetilde{\mathbf{G}}^F) = \mathbf{G}_{\mathrm{ad}}^F$, and $\mathbf{G}_{\mathrm{ad}}^F/\pi(\mathbf{G}^F)$ is isomorphic to the group of F-coinvariants of $\mathrm{Z}(\mathbf{G})/\mathrm{Z}^\circ(\mathbf{G})$, giving natural isomorphisms

$$\textbf{(15.4)}\qquad \widetilde{\mathbf{G}}^F/\mathbf{G}^F\mathrm{Z}(\widetilde{\mathbf{G}})^F \longleftrightarrow \mathbf{G}_{\mathrm{ad}}^F/\pi(\mathbf{G}^F) \longleftrightarrow (\mathrm{Z}(\mathbf{G})/\mathrm{Z}^\circ(\mathbf{G}))_F.$$

Let s be a semi-simple element of $(\widetilde{\mathbf{G}}^*)^F$ and $t = \sigma^*(s)$. To the $\widetilde{\mathbf{G}}^*$-conjugacy class of s is associated a geometric class of pairs $(\widetilde{\mathbf{S}}, \xi)$, where $\mathbf{S}$ is an F-stable maximal torus in $\widetilde{\mathbf{G}}$ and $\xi \in \mathrm{Irr}(\mathbf{S}^F)$. Let $\mathbf{T} = \sigma^{-1}(\mathbf{S})$. Then the geometric class associated to t in $\mathbf{G}$ is the class of the pair $(\mathbf{T}, \mathrm{Res}^{\mathbf{S}}_{\mathbf{T}}\,\xi)$. Indeed one has the equality $\mathrm{Res}^{\widetilde{\mathbf{G}}^F}_{\mathbf{G}^F}\,\mathrm{R}^{\widetilde{\mathbf{G}}}_{\widetilde{\mathbf{S}}} = \mathrm{R}^{\mathbf{G}}_{\mathbf{T}}\,\mathrm{Res}^{\mathbf{S}^F}_{\mathbf{T}^F}$ (restrictions via σ, see [DiMi91] 13.22). More generally, let $\widetilde{\mathbf{P}} = \widetilde{\mathbf{V}}.\widetilde{\mathbf{L}}$ be a Levi decomposition in $\widetilde{\mathbf{G}}$, with $F(\widetilde{\mathbf{L}}) = \widetilde{\mathbf{L}}$, then $\sigma^{-1}(\widetilde{\mathbf{V}}.\widetilde{\mathbf{L}}) = \mathbf{V}.\mathbf{L}$ is a Levi decomposition in $\mathbf{G}$ and one has

$$\mathrm{Res}^{\widetilde{\mathbf{G}}^F}_{\mathbf{G}^F}\,\mathrm{R}^{\widetilde{\mathbf{G}}}_{\widetilde{\mathbf{L}}\subset\widetilde{\mathbf{P}}} = \mathrm{R}^{\mathbf{G}}_{\mathbf{L}\subset\mathbf{P}}\,\mathrm{Res}^{\widetilde{\mathbf{L}}^F}_{\mathbf{L}^F}. \tag{15.5}$$

Using the notation of §8.3, formula (15.5) follows essentially from the isomorphism $\mathrm{H}^i_c(\mathbf{Y}^{\widetilde{\mathbf{G}}}_{\widetilde{\mathbf{V}}}, K) \simeq K\widetilde{\mathbf{G}}^F \otimes_{K\mathbf{G}^F} \mathrm{H}^i_c(\mathbf{Y}^{(\mathbf{G},F)}_{\mathbf{V}}, K)$, itself a consequence of the decomposition $\mathbf{Y}^{(\widetilde{\mathbf{G}})}_{\mathbf{V}} = \sum_{g\in\widetilde{\mathbf{G}}^F/\mathbf{G}^F} g.\mathbf{Y}^{(\mathbf{G})}_{\mathbf{V}}$, Theorem 7.10 (see the proof of Lemma 12.15(iii)).

When s runs over a $(\widetilde{\mathbf{G}}^*)^F$-conjugacy class, $t = \sigma^*(s)$ runs over a $(\mathbf{G}^*)^F$-conjugacy class — a consequence of Lang's theorem in the kernel of σ^*, which is connected — and the pair $(\sigma^{-1}(\mathbf{S}), \mathrm{Res}^{\mathbf{S}^F}_{\sigma^{-1}(\mathbf{S})^F}\,\xi)$ runs over a rational conjugacy class. When s runs over the set of all $(\widetilde{\mathbf{G}}^*)^F$-conjugates of elements of $(\sigma^*)^{-1}(t) \cap (\widetilde{\mathbf{G}}^*)^F$, the pair $(\sigma^{-1}(\mathbf{S}), \mathrm{Res}^{\mathbf{S}^F}_{\sigma^{-1}(\mathbf{S})^F}\,\xi)$ runs over a geometric conjugacy class. As for Lusztig's series, one obtains the following.

Proposition 15.6. *Let $\sigma\colon \mathbf{G} \to \widetilde{\mathbf{G}}$ satisfying (15.1(σ)), and let $\sigma^*\colon \widetilde{\mathbf{G}}^* \to \mathbf{G}^*$ be a dual morphism. Let $\tilde{s}$ be a semi-simple element of $(\widetilde{\mathbf{G}}^*)^F$, and let $s = \sigma^*(\tilde{s})$.*

(i) The rational Lusztig series $\mathcal{E}(\mathbf{G}^F, s)$ is the set of irreducible components of the restrictions to $\mathbf{G}^F$ of elements of $\mathcal{E}(\widetilde{\mathbf{G}}^F, \tilde{s})$.

(ii) The geometric Lusztig series $\widetilde{\mathcal{E}}(\mathbf{G}^F, s)$ is the set of irreducible components of the restrictions to $\mathbf{G}^F$ of elements of $\bigcup_{\tilde{t}} \mathcal{E}(\widetilde{\mathbf{G}}^F, \tilde{t})$, where $\tilde{t}$ runs over the set of rational elements of $(\sigma^*)^{-1}(s)$.

Note that the commutative group $\mathrm{Irr}(\widetilde{\mathbf{G}}^F/\mathbf{G}^F)$ acts on $\mathrm{Irr}(\widetilde{\mathbf{G}}^F)$ by tensor product. The isomorphic group $(\mathrm{Ker}(\sigma^*))^F$ acts on conjugacy classes of $(\widetilde{\mathbf{G}}^*)^F$ by translation; a connection is given by formulae (8.20) and (15.5).

The following strengthens Proposition 8.25.

Proposition 15.7. *Let $\mathbf{P} = \mathbf{V} \rtimes \mathbf{L}$ be a Levi decomposition where $\mathbf{L}$ is F-stable. Let s (resp. t) be a semi-simple element of $\mathbf{G}^{*F}$ (resp. of $\mathbf{L}^{*F}$, $\mathbf{L}^*$ a Levi subgroup of $\mathbf{G}^*$ in duality with $\mathbf{L}$), let $\zeta \in \mathcal{E}(\mathbf{L}^F, t)$. One has*

$$\mathrm{R}^{\mathbf{G}}_{\mathbf{L}\subseteq\mathbf{P}}\,\zeta \in \mathbb{Z}\mathcal{E}(\mathbf{G}^F, t).$$

If ζ occurs in ${}^*\mathrm{R}^{\mathbf{G}}_{\mathbf{L}\subseteq\mathbf{P}}\,\chi$ and $\chi \in \mathcal{E}(\mathbf{G}^F, s)$, then t is conjugate to s in $\mathbf{G}^{*F}$.

Proof. The second assertion follows from the first by adjunction. If the center of $\mathbf{G}$ is connected, Theorem 8.24 and Proposition 8.25 give our claim. When the center of $\mathbf{G}$ is not connected, use an embedding $\mathbf{G} \to \widetilde{\mathbf{G}}$, as in Hypothesis 15.1, and a coherent choice of $\widetilde{\mathbf{T}} \subset \widetilde{\mathbf{L}}$, $\tilde{\theta} \in \mathrm{Irr}(\widetilde{\mathbf{T}}^F)$, $\widetilde{\mathbf{B}} = \widetilde{\mathbf{U}}.\widetilde{\mathbf{T}} \subset \widetilde{\mathbf{P}} = \widetilde{\mathbf{V}}.\widetilde{\mathbf{L}}$ with $(\widetilde{\mathbf{T}}, \tilde{\theta})$ corresponding to $\sigma^*(t)$, and $(\mathbf{T}, \theta)$ to t. Now ζ occurs in some $\mathrm{H}^j(\mathbf{Y}_{\mathbf{U}}^{(\mathbf{L})}, K) \otimes_{\mathbf{T}^F} \theta$. By Theorems 7.9 and 7.10 χ occurs in the restriction from $\widetilde{\mathbf{G}}^F$ to $\mathbf{G}^F$ of the similarly defined tensor product (from $\tilde{\theta}, \widetilde{\mathbf{T}}, \widetilde{\mathbf{L}}, \ldots$) and Proposition 15.6 implies that any irreducible constituent of $\mathrm{H}^j(\mathbf{Y}_{\mathbf{U}}^{(\mathbf{L})}, K) \otimes_{\mathbf{T}^F} \theta$ is in the series $\mathcal{E}(\mathbf{L}^F, t)$. Now one may mimic the proof of Proposition 8.25. □

15.2. Jordan decomposition of characters

The following fundamental theorem is an immediate corollary of the classification of $\mathrm{Irr}(\mathbf{G}^F) = \bigcup_s \mathcal{E}(\mathbf{G}^F, s)$ in the book [Lu84]. One of the main theorems, [Lu84] 4.23, describes series and projections of each irreducible character on the space on uniform functions (linear combinations of Deligne–Lusztig characters $\mathrm{R}_{\mathbf{T}}^{\mathbf{G}}\theta$). Applying this theorem in $(\mathbf{G}, F)$ and in $(\mathrm{C}_{\mathbf{G}^*}(s), F)$ one has the following.

Theorem 15.8. Jordan decomposition of irreducible representations. *Let $\mathbf{G}$ be a connected reductive $\mathbf{F}$-group defined over $\mathbb{F}_q$ with a Frobenius F. Let $(\mathbf{G}^*, F)$ be in duality with $(\mathbf{G}, F)$. Assume that the center of $\mathbf{G}$ is connected. Let s be a semi-simple element of $(\mathbf{G}^*)^F$ and let $\mathbf{G}_s = \mathrm{C}_{\mathbf{G}^*}(s)$. There exists a bijection*

$$\psi_s : \mathcal{E}(\mathbf{G}^F, s) \to \mathcal{E}({\mathbf{G}_s}^F, 1)$$

such that, for any F-stable maximal torus $\mathbf{S}$ of $\mathbf{G}^*$ containing s,

$$\epsilon_{\mathbf{G}}\langle \chi, \mathrm{R}_{\mathbf{S}}^{\mathbf{G}} s\rangle_{\mathbf{G}^F} = \epsilon_{\mathbf{G}_s}\langle \psi_s(\chi), \mathrm{R}_{\mathbf{S}}^{\mathbf{G}_s} 1\rangle_{\mathbf{G}_s^F} .$$

If all components of $\mathbf{G}_{\mathrm{ad}}$ are of classical type, any $\chi \in \mathcal{E}(\mathbf{G}^F, s)$ is uniquely defined by the scalar products $\langle \chi, \mathrm{R}_{\mathbf{S}}^{\mathbf{G}} s\rangle_{\mathbf{G}^F}$.

For the notation $\mathrm{R}_{\mathbf{S}}^{\mathbf{G}} s$ see Remark 8.22(i). Such a bijection ψ_s is said to be a **Jordan decomposition** of elements of $\mathcal{E}(\mathbf{G}^F, s)$.

On unipotent characters, the fundamental result is the following, which reduces their classification to the study of adjoint groups.

Proposition 15.9. *Let $f : \mathbf{G} \to \mathbf{G}_0$ be a morphism of algebraic groups between two reductive $\mathbf{F}$-groups such that $\mathbf{G}$ and $\mathbf{G}_0$ are defined over $\mathbb{F}_q$ by Frobenius F and F_0, f is defined over $\mathbb{F}_q$, the kernel of f is central, and $[\mathbf{G}_0, \mathbf{G}_0] \subseteq f(\mathbf{G})$.*

Then the restriction map $\chi \mapsto \chi \circ f$ for $\chi \in \mathcal{E}(\mathbf{G}_0^{F_0}, 1)$ is a bijection $\mathcal{E}(\mathbf{G}_0^{F_0}, 1) \to \mathcal{E}(\mathbf{G}^F, 1)$. It commutes with twisted induction and its adjoint.

Proof. The first statement is well-known (apply [DiMi91] 13.20). For the commutation with R and *R, apply [DiMi91] 13.22 to the inclusion $[\mathbf{G},\mathbf{G}]\to\mathbf{G}$, to an embedding $\mathbf{G}\to\widetilde{\mathbf{G}}$ as described in Hypothesis 15.1 and to the quotient $\widetilde{\mathbf{G}}\to\widetilde{\mathbf{G}}_{\mathrm{ad}}=\mathbf{G}_{\mathrm{ad}}$. □

Theorem 15.8 and Proposition 15.9 show that when $(\mathbf{G},F)$ and $(\mathbf{G}^*,F)$ are dual groups over $\mathbb{F}_q$ there is a bijection between series of unipotent irreducible characters, with a commutativity property with respect to orthogonal projections on spaces of uniform functions.

In many cases the centralizer of s, or its connected component containing 1, is a Levi subgroup of $\mathbf{G}^*$ (see Proposition 13.16(ii)). Then the Jordan decomposition reduces the computation of twisted induction on $\mathcal{E}(\mathbf{G}^F,s)$ to that on unipotent characters.

Proposition 15.10. *Let s be some semi-simple element of $\mathbf{G}^{*F}$. Assume that $\mathrm{C}^\circ_{\mathbf{G}^*}(s)$ is a Levi subgroup of $\mathbf{G}^*$ and let $\mathbf{G}(s)$ be a Levi subgroup of $\mathbf{G}$ in duality with it. Let $\mathbf{P}$ be a parabolic subgroup for which $\mathbf{G}(s)$ is the Levi complement. Then let $\hat{s}\in\mathrm{Irr}(\mathbf{G}(s)^F)$ be defined by $s\in\mathrm{Z}(\mathrm{C}^\circ_{\mathbf{G}^*}(s))^F$ thanks to (8.19).*

(i) For any $\lambda\in\mathcal{E}(\mathbf{G}(s)^F,1)$, $\epsilon_{\mathbf{G}}\epsilon_{\mathbf{G}(s)}\mathrm{R}^{\mathbf{G}}_{\mathbf{G}(s)\subset\mathbf{P}}(\hat{s}\lambda)$ is a sum of distinct elements of $\mathcal{E}(\mathbf{G}^F,s)$.

(ii) If the center of $\mathbf{G}$ is connected, then $\chi_{s,\lambda}:=\epsilon_{\mathbf{G}}\epsilon_{\mathbf{G}(s)\subset\mathbf{P}}\mathrm{R}^{\mathbf{G}}_{\mathbf{G}(s)\subset\mathbf{P}}(\hat{s}\lambda)\in\mathcal{E}(\mathbf{G}^F,s)$. Moreover, $\lambda\mapsto\chi_{s,\lambda}$ is a bijection $\mathcal{E}(\mathbf{G}(s)^F,1)\to\mathcal{E}(\mathbf{G}^F,s)$. It commutes with the orthogonal projection on the spaces of uniform functions.

Proof. Assume first that $\mathrm{C}_{\mathbf{G}^*}(s)$ is a Levi subgroup of $\mathbf{G}^*$. Then combining Theorem 8.27 and Proposition 8.26 (with $\big(s,1,\mathbf{G}(s)\big)$ instead of $(z,s,\mathbf{G})$) one obtains a bijection $\mathcal{E}(\mathbf{G}(s)^F,1)\to\mathcal{E}(\mathbf{G}^F,s)$ such that λ goes to $\varepsilon_{\mathbf{G}}\varepsilon_{\mathbf{G}(s)}\mathrm{R}^{\mathbf{G}}_{\mathbf{G}(s)\subseteq\mathbf{P}}(\hat{s}\lambda)$. If the center of $\mathbf{G}$ is connected, then $\mathrm{C}_{\mathbf{G}^*}(s)$ is connected. Then, for any pair $(\mathbf{T},\theta)$, $\mathbf{T}$ a maximal F-stable torus and $\theta\in\mathrm{Irr}(\mathbf{T}^F)$, one has

$$\big\langle\mathrm{R}^{\mathbf{G}}_{\mathbf{T}}\theta,\mathrm{R}^{\mathbf{G}}_{\mathbf{G}(s)}(\hat{s}\lambda)\big\rangle_{\mathbf{G}^F}=\begin{cases}\langle\lambda,\mathrm{R}^{\mathbf{G}(s)}_{\mathbf{T}}1\rangle_{\mathbf{G}(s)^F} & \text{if } \mathbf{T}\subset\mathbf{G}(s),\\ 0 & \text{if not}\end{cases}$$

as a consequence of the Mackey formula and the fact that $\mathbf{T}^g\subset\mathbf{G}(s)$ and $\theta^g=\mathrm{Res}^{\mathbf{G}(s)^F}_{(\mathbf{T}^g)^F}\hat{s}$ is equivalent to $g\in\mathbf{G}(s)$ ([DiMi91] 11.13).

When $\mathrm{Z}(\mathbf{G})\neq\mathrm{Z}^\circ(\mathbf{G})$, let $\sigma:\mathbf{G}\to\widetilde{\mathbf{G}}$ and $\sigma^*:\widetilde{\mathbf{G}}^*\to\mathbf{G}^*$ be the usual dual morphisms satisfying (15.1(σ)), (15.1*(σ^*)), let $t\in\widetilde{\mathbf{G}}^{*F}_{\mathrm{ss}}$ be such that $\sigma^*(t)=s$. One has $\sigma^*(\mathrm{C}_{\widetilde{\mathbf{G}}^*}(t))=\mathrm{C}^\circ_{\mathbf{G}^*}(s)$ and there exists a Levi subgroup $\widetilde{\mathbf{G}}(t)$ of $\widetilde{\mathbf{G}}$ in duality with $\mathrm{C}_{\widetilde{\mathbf{G}}^*}(t)$ and such that $\widetilde{\mathbf{G}}(t)\cap\mathbf{G}=\mathbf{G}(s)$. Then $\mathrm{Res}^{\widetilde{\mathbf{G}}(t)^F}_{\mathbf{G}(s)^F}$ restricts to bijections $(\tilde{\lambda}\mapsto\lambda)$ (resp. $(\hat{t}\lambda\mapsto\hat{s}\lambda)$), from $\mathcal{E}(\widetilde{\mathbf{G}}(t)^F,t)$ to $\mathcal{E}(\mathbf{G}(s)^F,s)$ (resp. from $\mathcal{E}(\widetilde{\mathbf{G}}(t)^F,1)$ to $\mathcal{E}(\mathbf{G}(s)^F,1)$) and $\mathrm{R}^{\mathbf{G}}_{\mathbf{G}(s)\subset\mathbf{P}}(\hat{s}\lambda)=\mathrm{Res}^{\widetilde{\mathbf{G}}(t)^F}_{\mathbf{G}(s)^F}(\mathrm{R}^{\widetilde{\mathbf{G}}}_{\widetilde{\mathbf{G}}(t)}(\hat{t}\tilde{\lambda}))$. The claim follows from the case of $\widetilde{\mathbf{G}}$, Proposition 15.6(ii) and Theorem 15.11 to come. □

The following is a crucial step to go down from connected center to non-connected center. The proof will be continued in Chapter 16.

Theorem 15.11. *For any $\chi \in \mathrm{Irr}(\mathbf{G}^F)$, the restriction of χ to $[\mathbf{G},\mathbf{G}]^F$ is a sum of distinct elements of* $\mathrm{Irr}([\mathbf{G},\mathbf{G}]^F)$.

If the conclusion of Theorem 15.11 is true then we say that "$\mathrm{Res}^{\mathbf{G}^F}_{[\mathbf{G},\mathbf{G}]^F}$ is multiplicity free".

Proof of Theorem 15.11: Reduction to a single case.

(a) Reduction to an arbitrary embedding of $[\mathbf{G},\mathbf{G}]$ in a group with connected center. Theorem 15.11 is equivalent to the following.

Theorem 15.11′. *If a morphism $\mathbf{G}\to\widetilde{\mathbf{G}}$ is defined over $\mathbb{F}_q$ and induces an isomorphism between derived subgroups, then the restriction from $\widetilde{\mathbf{G}}^F$ to $\mathbf{G}^F$ of any $\chi\in\mathrm{Irr}(\widetilde{\mathbf{G}}^F)$ is a sum of distinct irreducible characters.*

Indeed the multiplicity one property of $\mathrm{Res}^{\widetilde{\mathbf{G}}^F}_{[\widetilde{\mathbf{G}},\widetilde{\mathbf{G}}]^F}$ implies the same property for $\mathrm{Res}^{\widetilde{\mathbf{G}}^F}_{\mathbf{G}^F}$, and of $\mathrm{Res}^{\mathbf{G}^F}_{[\mathbf{G},\mathbf{G}]^F}$. We keep the notation of Theorem 15.11′.

Consider first a monomorphism $\sigma:\mathbf{G}\to\widetilde{\mathbf{G}}$ between two groups with connected centers and which satisfies (15.1(σ)) as in §15.1. Let $\sigma^*:\mathbf{G}^*\to\widetilde{\mathbf{G}}^*$ be a dual morphism. Thanks to the isomorphism in (15.2), any linear character of $\widetilde{\mathbf{G}}^F/\mathbf{G}^F$ can be written as $\hat{z}$ for some $z\in(\mathrm{Ker}(\sigma^*))^F$. Let $\tilde{s}$ be a semi-simple element of $(\widetilde{\mathbf{G}}^*)^{F^*}$. As the center of $\widetilde{\mathbf{G}}$ is connected, $C_{\widetilde{\mathbf{G}}^*}(\tilde{s})$ is connected, hence, by the isomorphism given in Theorem 15.13, $\mathrm{Ker}\,\sigma^*\cap[\tilde{s},(\widetilde{\mathbf{G}}^*)^{F^*}]=\{1\}$. If $z\neq 1$, then $\tilde{s}$ is not conjugate to $\tilde{s}z$, so $\mathcal{E}(\widetilde{\mathbf{G}}^F,\tilde{s})\neq\mathcal{E}(\widetilde{\mathbf{G}}^F,\tilde{s}z)$ (Theorem 8.24(ii)). By (8.20) one has $\chi\otimes\hat{z}\neq\chi$ for any $\chi\in\mathcal{E}(\widetilde{\mathbf{G}}^F,\tilde{s})$ and any non-trivial $\hat{z}$ in $\mathrm{Irr}(\widetilde{\mathbf{G}}^F/\mathbf{G}^F)$. By Clifford theory, between $\mathbf{G}^F$ and $\widetilde{\mathbf{G}}^F$ the restriction of χ to $\mathbf{G}^F$ is irreducible, hence the multiplicity one property holds. Now Exercise 15.2 shows that, if the property is true for at least one embedding satisfying condition (15.1(σ)), then it is true for all others.

By Theorem 15.11′, it is sufficient to consider the map $[\mathbf{G},\mathbf{G}]\to\widetilde{\mathbf{G}}$, i.e. to prove the conclusion of Theorem 15.11 for only one embedding of the given semi-simple group $[\mathbf{G},\mathbf{G}]$ in a group $\widetilde{\mathbf{G}}$ with connected center.

(b) Reduction to a simply connected group $\mathbf{G}=[\mathbf{G},\mathbf{G}]$.

When $\mathbf{G}=[\mathbf{G},\mathbf{G}]$ is not simply connected let $\eta:\mathbf{G}_0\to\widetilde{\mathbf{G}}$ be a surjective morphism defined over $\mathbb{F}_q$, satisfying (15.1*(η)). A dual morphism $\eta^*:\widetilde{\mathbf{G}}^*\to\mathbf{G}_0^*$ satisfies (15.1(η^*)). Thus the kernel of η is connected, hence $\eta(\mathbf{G}_0^F)=\widetilde{\mathbf{G}}^F$,

$[\mathbf{G}_0, \mathbf{G}_0]$ is simply connected and $\eta([\mathbf{G}_0, \mathbf{G}_0]) = [\widetilde{\mathbf{G}}, \widetilde{\mathbf{G}}] \cong \mathbf{G}$. Furthermore $[\mathbf{G}_0^*, \mathbf{G}_0^*] \cong [\widetilde{\mathbf{G}}^*, \widetilde{\mathbf{G}}^*]$ is simply connected, hence the center of $\mathbf{G}_0$ is connected. Any $\tilde{\chi}$ in $\mathrm{Irr}(\widetilde{\mathbf{G}}^F)$ is the restriction of some χ_0 in $\mathrm{Irr}(\mathbf{G}_0^F)$. Clearly the natural embedding of $[\mathbf{G}_0, \mathbf{G}_0]$ in $\mathbf{G}_0$ satisfies Hypothesis 15.1. If some multiplicity occurs in the restriction of $\tilde{\chi}$ to $\mathbf{G}^F$, it occurs in the restriction of χ_0 to $[\mathbf{G}_0, \mathbf{G}_0]^F$.

So assume now that $\mathbf{G}$ is simply connected but not necessarily equal to $[\mathbf{G}, \mathbf{G}]$. $\mathbf{G}$ is a direct product and the map F acts on the set of components. An F-orbit $(\mathbf{H}_j)_{1\leq j\leq d}$ of length d defines a component of $\mathbf{G}^F$, isomorphic to $\mathbf{H}_1^{F^d}$. Let $\mathbf{G} \to \widetilde{\mathbf{G}}$ be a direct product of embeddings $\mathbf{H}_i \to \widetilde{\mathbf{H}}_i$ compatible with F-action and for which multiplicity 1 holds. Then multiplicity one holds from $\widetilde{\mathbf{G}}^F$ to $\mathbf{G}^F$.

(c) Assume $\mathbf{G} = [\mathbf{G}, \mathbf{G}]$, simply connected.

From $\mathrm{Z}(\widetilde{\mathbf{G}})^F \mathbf{G}^F$ to $\mathbf{G}^F$ the restriction is multiplicity free since the question clearly reduces to the inclusion of commutative groups $\mathrm{Z}(\widetilde{\mathbf{G}})^F \supseteq \mathbf{G}^F \cap \mathrm{Z}(\widetilde{\mathbf{G}})^F$. If $\widetilde{\mathbf{G}}^F/\mathrm{Z}(\widetilde{\mathbf{G}})^F\mathbf{G}^F$ is cyclic, the restriction from $\widetilde{\mathbf{G}}^F$ to $\mathrm{Z}(\widetilde{\mathbf{G}})^F\mathbf{G}^F$ is multiplicity free by a general theorem (see for instance [NaTs89] Problem 3.11(i) or Lemma 18.35 below). The above occurs when $\mathrm{Z}(\mathbf{G})^F$ is cyclic and the embedding is defined so that the center of $\widetilde{\mathbf{G}}$ is of minimal rank. Indeed if the center Z of $\mathbf{G}$ is cyclic, then it is contained in an F-stable torus $\mathbf{S}$ of rank 1 of $\mathbf{G}$ and one may consider $\widetilde{\mathbf{G}} = \mathbf{S} \times \mathbf{G}/Z$. If the center of $\mathbf{G}$ is not cyclic, then $\mathbf{G}$ is a spin group, of type $\mathbf{D}$ and even rank in odd characteristic. Then $\mathrm{Z}(\mathbf{G})$ is contained in a torus $\mathbf{S}$ of rank 2, but in the rational type ${}^2\mathbf{D}$ (where the quadratic form on $\mathbb{F}_q$ has no maximal Witt index) $\mathrm{Z}(\mathbf{G})^F$ is of order 2, $\mathbf{S}$ is non-split so that $\widetilde{\mathbf{G}}^F/\mathrm{Z}(\widetilde{\mathbf{G}})^F\mathbf{G}^F$ is of order 2. Thus the only case to consider to prove Theorem 15.11 is the following: *q is odd, the center of $\mathbf{G}$ is connected, $([\mathbf{G}, \mathbf{G}], F)$ is a split spin group of even rank defined over $\mathbb{F}_q$, and the quadratic form on $\mathbb{F}_q$ has maximal Witt index.*

Clearly the restriction from $[\mathbf{G}, \mathbf{G}]^F\mathrm{Z}(\mathbf{G})^F$ to $[\mathbf{G}, \mathbf{G}]^F$ is multiplicity free. Applying Proposition 8.1 with $(G, f, G/Z) = (\mathrm{Z}(\mathbf{G}) \times [\mathbf{G}, \mathbf{G}], F, \mathbf{G})$ one gets that the quotient $\mathbf{G}^F/[\mathbf{G}, \mathbf{G}]^F\mathrm{Z}(\mathbf{G})^F$ is isomorphic to $\mathrm{Z}(\mathbf{G}) \cap [\mathbf{G}, \mathbf{G}]$, hence of order 4 and exponent 2.

To show the multiplicity one property in this particular hypothesis and that it goes through the Jordan decomposition we need the following consequences of standard "Clifford theory" (see [Ben91a] §3.13, [NaTs89] §3.3).

Proposition 15.12. *Let H be a finite group, let H' be a normal subgroup of H with a commutative quotient $A = H/H'$. Then A acts naturally on* $\mathrm{Irr}(H)'$ *and $A^\vee$ (i.e.* $\mathrm{Irr}(A)$ *with tensor product) acts on* $\mathrm{Irr}(H)$ *by tensor product. Let $\mathcal{E}$ be an A-stable subset of* $\mathrm{Irr}(H')$*, let $\mathcal{F}$ be the $A^\vee$-stable set of elements of* $\mathrm{Irr}(H)$ *whose restriction to H' contains some element of $\mathcal{E}$.*

(i) Assume that H/H' is of order 4. When $j \in \{1, 2, 4\}$, let y_j be the number of elements of $\mathcal{F}$ whose stabilizer in $A^\vee$ is of order j. The restriction from H to H' of any element of $\mathcal{F}$ is multiplicity free if and only if

$$4|\mathcal{F}| = y_1 + 4y_2 + 16y_4.$$

(ii) Assume that any χ in some subset $\mathcal{F}$ of $\operatorname{Irr}(H)$ restricts to a multiplicity free representation of H'. Then there is a unique bijection between sets of orbits

$$\mathcal{E}/A \longleftrightarrow \mathcal{F}/A^\vee$$

such that the A-orbit of χ corresponds to the set of constituents of $\operatorname{Ind}_{H'}^H \chi$ for any $\chi \in \mathcal{E}$. Moreover, the stabilizers of elements in corresponding orbits are orthogonal to each other.

Proof. (i) The action of $\lambda \in A^\vee$ on $\operatorname{Irr}(H)$ is $(\chi \mapsto \lambda \otimes \chi)$. The dual group A acts on $\operatorname{Irr}(H')$ by H-conjugacy. By Clifford's theory, the condition $\langle \operatorname{Res}_{H'}^H \chi, \chi' \rangle_{H'} \neq 0$ for $(\chi, \chi') \in \operatorname{Irr}(H) \times \operatorname{Irr}(H')$ defines a bijection from the set of $A^\vee$-orbits on $\operatorname{Irr}(H)$ to the set of A-orbits on $\operatorname{Irr}(H')$. Assuming $m := \langle \operatorname{Res}_{H'}^H \chi, \chi' \rangle_{H'} \neq 0$, let $I_{\chi'}$ be the normalizer of χ' in H, and let A_χ be the stabilizer of χ in $A^\vee$. Using the Frobenius reciprocity theorem, one sees that the following four cases may occur.

- $I_{\chi'} = H$, $A_\chi = \{1_{H/H'}\}$, $m = 1$, $\operatorname{Res}_{H'}^H \chi = \chi'$, and $\operatorname{Ind}_{H'}^H \chi'$ is a sum of four distinct elements of $\operatorname{Irr}(H')$.
- $I_{\chi'} = H$, $A_\chi = A^\vee$, $m = 2$, $\operatorname{Res}_{H'}^H \chi = 2\chi'$, and $\operatorname{Ind}_{H'}^H \chi' = 2\chi$.
- $I_{\chi'}/H'$ is of order 2, A_χ is the dual of $I_{\chi'}/H'$, $m = 1$, $\operatorname{Ind}_{H'}^H \chi'$ and $\operatorname{Res}_{H'}^H \chi$ are sums of two distincts irreducible characters.
- $I_{\chi'} = H'$, $m = 1$, $\operatorname{Res}_{H'}^H \chi$ is a sum of four distinct elements of $\operatorname{Irr}(H')$, $\chi = \operatorname{Ind}_{H'}^H \chi'$, $A_\chi = \operatorname{Irr}(H/H')$.

Let y be the number of elements χ of $\mathcal{E}$ for which the second case occurs. One has

$$|\mathcal{F}| = y_1/4 + y + y_2 + 4(y_4 - y).$$

Thus $y = 0$ is equivalent to the equality of the proposition.

The proof of (ii) is left to the reader. □

A second key result establishes a quasi-uniqueness of Jordan decomposition of irreducible representations:

Theorem 15.13. *Let $\sigma : \mathbf{G} \to \widetilde{\mathbf{G}}$ be an embedding over $\mathbb{F}_q$ that satisfies (15.1(σ)), let $\sigma^* : \widetilde{\mathbf{G}}^* \to \mathbf{G}^*$ be the dual of σ. Let s be a semi-simple element of*

$(\widetilde{\mathbf{G}}^*)^F$ *and* $t = \sigma^*(s)$. *Let* $\widetilde{\mathbf{G}}_s = \mathrm{C}_{\widetilde{\mathbf{G}}^*}(s)$,

$$A(t) = (\mathrm{C}_{\mathbf{G}^*}(t)/\mathrm{C}^\circ_{\mathbf{G}^*}(t))^F, \quad B(s) = \mathrm{Ker}(\sigma^*) \cap [s, \widetilde{\mathbf{G}}^{*F}].$$

There is a natural isomorphism from $A(t)$ *to* $B(s)$, *actions of* $A(t)$ *on* $\mathcal{E}(\widetilde{\mathbf{G}}_s^F, 1)$, *of* $B(s)$ *on* $\mathcal{E}(\widetilde{\mathbf{G}}^F, s)$, *and a Jordan decomposition* $\mathcal{E}(\widetilde{\mathbf{G}}^F, s) \to \mathcal{E}(\widetilde{\mathbf{G}}_s^F, 1)$ *that is compatible with those isomorphism and actions.*

On the proof of Theorem 15.13. The isomorphism is defined as follows. Let $\tilde{g} \in \widetilde{\mathbf{G}^*}^F$ be such that $\sigma^*(g) \in \mathrm{C}_{\mathbf{G}^*}(t)$, then $\sigma^*(g)\mathrm{C}^\circ_{\mathbf{G}^*}(t)$ maps to $[s, g]$ (details of the proof are left to the reader). Then g normalizes $\mathrm{C}^\circ_{\mathbf{G}^*}(t)$ and commutes with the action of F, hence g acts on $\mathcal{E}(\mathrm{C}^\circ_{\mathbf{G}^*}(t)^F, 1)$, so on $\mathcal{E}(\widetilde{\mathbf{G}}_s^F, 1)$ by Proposition 15.9. It is clearly an action of $A(t)$.

Any $z \in \mathrm{Ker}(\sigma^*)^F$ defines a linear character $\hat{z}$ of $\widetilde{\mathbf{G}}^F/\mathbf{G}^F$ by (15.2) hence acts on $\mathrm{Irr}(\widetilde{\mathbf{G}^*}^F)$ by $(\chi \mapsto \hat{z} \otimes \chi)$. Thus $B(s)$ is precisely the stabilizer of $\mathcal{E}(\widetilde{\mathbf{G}}^F, s)$ (Proposition 8.26). Let $\bar{g} := \sigma^*(g)\mathrm{C}^\circ_{\mathbf{G}^*}(t)^F \in A(t)$, Theorem 15.13 says that there is a bijection ψ_s that satisfies $\psi_s(\hat{z} \otimes \chi) = \bar{g}.\psi_s(\chi)$ for any $\chi \in \mathcal{E}(\widetilde{\mathbf{G}}^F, s)$, where $gsg^{-1} = zs$.

By definition of the action of $A(t)$ one has

$$\langle \bar{g}.\eta, \mathrm{R}^{\widetilde{\mathbf{G}}_s}_{g\mathbf{T}g^{-1}} 1\rangle_{\widetilde{\mathbf{G}}_s^F} = \langle \eta, \mathrm{R}^{\widetilde{\mathbf{G}}_s}_{\mathbf{T}} 1\rangle_{\widetilde{\mathbf{G}}_s^F}$$

for any $\eta \in \mathcal{E}(\widetilde{\mathbf{G}}_s^F, 1)$ and maximal F-stable torus $\mathbf{T}$ in $\widetilde{\mathbf{G}}_s$. From equality (8.20), for any $\chi \in \mathcal{E}(\widetilde{\mathbf{G}}^F, s)$ and central z in $(\widetilde{\mathbf{G}}^*)^F$ one has $\langle \hat{z} \otimes \chi, \mathrm{R}^{\widetilde{\mathbf{G}}}_{\mathbf{T}} s\rangle_{\widetilde{\mathbf{G}}^F} = \langle \chi, \mathrm{R}^{\widetilde{\mathbf{G}}}_{\mathbf{T}}(z^{-1}s)\rangle_{\widetilde{\mathbf{G}}^F}$. As $g(\mathbf{T}, z^{-1}s)g^{-1} = (g\mathbf{T}g^{-1}, s)$ and $g \in (\widetilde{\mathbf{G}}^*)^F$ one has $\mathrm{R}^{\widetilde{\mathbf{G}}}_{\mathbf{T}}(z^{-1}s) = \mathrm{R}^{\widetilde{\mathbf{G}}}_{g\mathbf{T}g^{-1}}(s)$. Now Theorem 15.8 gives

$$\langle \psi_s(\hat{z} \otimes \chi), \mathrm{R}^{\widetilde{\mathbf{G}}_s}_{\mathbf{T}} 1\rangle_{\widetilde{\mathbf{G}}_s^F} = \langle \psi_s(\chi), \mathrm{R}^{\widetilde{\mathbf{G}}_s}_{g\mathbf{T}g^{-1}} 1\rangle_{\widetilde{\mathbf{G}}_s^F}.$$

Hence $\langle \psi_s(\hat{z} \otimes \chi) - \bar{g}.\psi_s(\chi), \mathrm{R}^{\widetilde{\mathbf{G}}_s}_{\mathbf{T}} 1\rangle_{\widetilde{\mathbf{G}}_s^F} = 0$. In case ψ_s is uniquely determined by the scalar product with Deligne–Lusztig characters, ψ_s has to exchange the actions of $B(s)$ and $A(t)$. That is the case for simply connected groups $\mathbf{G} = [\mathbf{G}, \mathbf{G}]$ of classical types. There is nothing to prove when $A(t)$ is $\{1\}$. There remains one exceptional case in type E_7; see the book [Lu84] and Lusztig's article [Lu88].

Then Theorem 15.13 is proved by a reduction process to the case of a simply connected $\mathbf{G} = [\mathbf{G}, \mathbf{G}]$, as in the paragraphs (a) and (b) of the proof of Theorem 15.11 above.

The last assertion is given by isomorphisms (15.3), (15.4). □

Corollary 15.14. *Let* $(\mathbf{G}, F)$ *and* $(\mathbf{G}^*, F)$ *be in duality, let* t *be a semi-simple element of* $\mathbf{G}^{*F}$, *let* $A(t) = (\mathrm{C}_{\mathbf{G}^*}(t)/\mathrm{C}^\circ_{\mathbf{G}^*}(t))^F$. *Let* $\pi : \mathbf{G} \to \mathbf{G}_{\mathrm{ad}}$ *be the canonical* *morphism. There is a one-to-one correspondence between sets of orbits*

$$\mathcal{E}(\mathbf{G}^F, t)/(\mathrm{Z}(\mathbf{G})/\mathrm{Z}^\circ(\mathbf{G}))_F \xrightarrow{\sim} \mathcal{E}(\mathrm{C}_{\mathbf{G}^*}(t)^F, 1)/A(t)$$

such that

(i) the group $(\mathrm{Z}(\mathbf{G})/\mathrm{Z}(\mathbf{G})^\circ))_F$ *acts on* $\mathcal{E}(\mathbf{G}^F, t)$ *via the isomorphism (15.4) as* $\mathbf{G}_{\mathrm{ad}}^F/\pi(\mathbf{G}^F)$,

(ii) if $\Omega \mapsto \omega$, *the number of elements in the orbit* Ω *is the order of the stabilizer in* $A(t)$ *of any* $\lambda \in \omega$,

(iii) for any F-stable maximal torus $\mathbf{T}$ *of* $\mathrm{C}^\circ_{\mathbf{G}^*}(t)$ *and any* $\chi \in \mathcal{E}(\mathbf{G}^F, t)$ *whose* $(\mathrm{Z}(\mathbf{G})/\mathrm{Z}(\mathbf{G}))_F$*-orbit corresponds to the* $A(t)$*-orbit* ω, *one has*

$$\epsilon_{\mathbf{G}}\langle \chi, \mathrm{R}_{\mathbf{T}}^{\mathbf{G}}\, t\rangle_{\mathbf{G}^F} = \epsilon_{\mathrm{C}^\circ_{\mathbf{G}^*}(t)} \sum_{\lambda\in\omega} \langle \lambda, \mathrm{R}_{\mathbf{T}}^{\mathrm{C}^\circ_{\mathbf{G}^*}(t)}\, 1\rangle_{\mathrm{C}^\circ_{\mathbf{G}^*}(t)^F}$$

Proof. Once more let $\sigma\colon \mathbf{G} \to \widetilde{\mathbf{G}}$ be an embedding in a group with connected center (see §15.1). Clearly $\widetilde{\mathbf{G}}^F$ acts on $\mathbf{G}^F$ and leaves fixed any Deligne–Lusztig character $\mathrm{R}_{\mathbf{T}}^{\mathbf{G}}\, t$ ($t \in \mathbf{T}^F$, $\mathbf{T} \subset \mathbf{G}^*$) by Proposition 15.6. It is an action of $(\widetilde{\mathbf{G}}/\mathrm{Z}(\widetilde{\mathbf{G}}))^F \cong \mathbf{G}_{\mathrm{ad}}^F$ and $\pi(\mathbf{G}^F)$ acts by interior automorphisms of $\mathbf{G}^F$, hence one has an action of $\mathbf{G}_{\mathrm{ad}}^F/\pi(\mathbf{G}^F)$ on $\mathcal{E}(\mathbf{G}^F, t)$.

Proposition 15.12 applies to $(H', H) = (\mathbf{G}^F, \widetilde{\mathbf{G}}^F)$ with $\mathcal{F} = \mathcal{E}(\mathbf{G}^F, t)$ by Theorem 15.11 and then $\mathcal{E} = \cup_{\sigma^*(s)=t}\mathcal{E}(\widetilde{\mathbf{G}}^F, s)$ by Proposition 15.6(ii). Thus one has a bijection between $\mathcal{E}(\mathbf{G}^F, t)/\mathbf{G}_{\mathrm{ad}}^F$ and $\cup_{\sigma^*(s)=t}\mathcal{E}(\widetilde{\mathbf{G}}^F, s)/[\widetilde{\mathbf{G}}^*, \widetilde{\mathbf{G}}^*] \cap (\mathrm{Ker}(\sigma^*))^F$. By intersection with one series $\mathcal{E}(\widetilde{\mathbf{G}}^F, s_0)$ on the right-hand side the orbits are those of $B(s_0)$, as defined in Theorem 15.13, and the stabilizers of elements in an orbit are unchanged. One obtains a bijection

$$\mathcal{E}(\mathbf{G}^F, t)/\mathbf{G}_{\mathrm{ad}}^F \longleftrightarrow \mathcal{E}(\widetilde{\mathbf{G}}^F, s_0)/B(s_0) \tag{15.15}$$

such that the length of an orbit on the left-hand side is the order of a stabilizer of an element of the corresponding orbit on the right-hand side (Proposition 15.12(ii)). One may identify $\mathcal{E}(\mathbf{G}_{s_0}^F, 1)$ and $\mathcal{E}(\mathrm{C}^\circ_{\mathbf{G}^*}(t)^F, 1)$, with $A(t)$-action (Proposition 15.9). Thus by Theorem 15.13 one has a bijection

$$\mathcal{E}(\widetilde{\mathbf{G}}^F, s_0)/B(s_0) \longleftrightarrow \mathcal{E}(\mathrm{C}_{\mathbf{G}^*}(t)^F, 1)/A(t) \tag{15.16}$$

By composition of (15.15) and (15.16) and the isomorphism (15.4) one has the bijection of the proposition satisfying (i) and (ii).

Let $\Omega \subseteq \mathcal{E}(\mathbf{G}^F, t)$, $\Omega_0 \subseteq \mathcal{E}(\widetilde{\mathbf{G}}^F, s_0)$, $\omega_0 \subseteq \mathcal{E}(\mathrm{C}^\circ_{\mathbf{G}^*}(s)^F, 1)$, and $\omega \subseteq \mathcal{E}(\mathrm{C}_{\mathbf{G}^*}(s)^F, 1)$ be corresponding orbits by the bijections (15.15), (15.16) and

Proposition 15.9, and let $\chi \in \Omega$, $\eta \in \omega$. Let $\widetilde{\mathbf{T}} = \sigma^{*-1}(\mathbf{T})$; one has

$$\begin{aligned}\langle \chi, \mathrm{R}_{\mathbf{T}}^{\mathbf{G}} t \rangle_{\mathbf{G}^F} &= \langle \chi, \mathrm{Res}_{\mathbf{G}^F}^{\widetilde{\mathbf{G}}^F}(\mathrm{R}_{\widetilde{\mathbf{T}}}^{\widetilde{\mathbf{G}}} s_0) \rangle_{\mathbf{G}^F} = \langle \mathrm{Ind}_{\mathbf{G}^F}^{\widetilde{\mathbf{G}}^F} \chi, \mathrm{R}_{\widetilde{\mathbf{T}}}^{\widetilde{\mathbf{G}}} s_0 \rangle_{\mathbf{G}^F} \\ &= \sum_{\chi' \in \Omega_0} \langle \chi', \mathrm{R}_{\widetilde{\mathbf{T}}}^{\widetilde{\mathbf{G}}} s_0 \rangle_{\mathbf{G}^F} = \epsilon_{\widetilde{\mathbf{G}}} \epsilon_{\widetilde{\mathbf{G}}_{s_0}} \sum_{\chi' \in \Omega_0} \langle \psi_{s_0}(\chi'), \mathrm{R}_{\widetilde{\mathbf{T}}}^{\widetilde{\mathbf{G}}_{s_0}} 1 \rangle_{\widetilde{\mathbf{G}}_{s_0}^F} \\ &= \epsilon_{\mathbf{G}} \epsilon_{\mathrm{C}^\circ_{\mathbf{G}^*}(t)} \sum_{\eta \in \omega} \langle \eta, \mathrm{R}_{\mathbf{T}}^{\mathrm{C}^\circ_{\mathbf{G}^*}(s)} 1 \rangle_{\mathrm{C}^\circ_{\mathbf{G}^*}(s)^F}\end{aligned}$$

where we have applied successively formula (15.5), the Frobenius reciprocity theorem, Theorem 15.11 and the definition of $\Omega \mapsto \Omega_0$, Theorem 15.8 with its notation and $\omega_0 = \psi_{s_0}(\Omega_0)$, Proposition 15.9 giving $\omega_0 \mapsto \omega$. □

The following description of the action of non-special transformations on the unipotent series of some special orthogonal groups will be used in the next chapter. Recall briefly the parametrization of $\mathcal{E}(\mathbf{G}^F, 1)$ when $(\mathbf{G}, F)$ is of rational type $(\mathbf{D}_n, q)$ ([Lu84] §4, [Cart85] 16.8). To each $\chi \in \mathcal{E}(\mathbf{G}^F, 1)$ there corresponds a class of symbols, defined by an integer c and a set of two partitions α, β of $a \in \mathbb{N}$ and $b \in \mathbb{N}$ respectively such that $a + b = n - 4c^2$. Fixing $c > 0$ one obtains the constituents of a Harish-Chandra series defined by the unique unipotent cuspidal irreducible representation of a Levi subgroup of type $(\mathbf{D}_{4c^2}, q)$, the corresponding Hecke algebra being of type BC_{n-4c^2} (see [Lu84] §8). Each symbol determines only one element of $\mathcal{E}(\mathbf{G}^F, 1)$, except that there are two unipotent characters corresponding to a symbol when $\alpha = \beta$ and $c = 0$. Those are called degenerate symbols and the corresponding unipotent characters twin characters. The unipotent characters that correspond to symbols for which $c = 0$ are constituents of the principal series, whose Hecke algebra is of type D_n, hence are in one-to-one correspondence $\zeta \mapsto \phi_\zeta$ with elements of $\mathrm{Irr}(W(\mathrm{D}_n))$. The degenerate symbols correspond to irreducible representations of $W(\mathrm{BC}_n)$ whose restriction to $W(\mathrm{D}_n)$ is not irreducible.

Proposition 15.17. *Let $(\mathbf{G}, F)$ be a connected conformal group with respect to a quadratic space V of dimension $2n$ and defined over $\mathbb{F}_q$, with maximal Witt index on $\mathbb{F}_q$: $(\mathbf{G}, F)$ has rational type $(\mathbf{D}_n, q)$, a connected center and $[\mathbf{G}, \mathbf{G}] = \mathrm{SO}(V)$. Let $(\mathbf{G}^*, F)$ be a dual group. Let $\sigma \in \mathrm{O}(V)^F \setminus \mathrm{SO}(V)^F$. Then σ acts on $\mathcal{E}(\mathbf{G}^F, 1)$ in the following way: if $\zeta \in \mathcal{E}(\mathbf{G}^F, 1)$ corresponds to a degenerate symbol, then $\sigma(\zeta)$ and ζ are twin characters and, if $\zeta \in \mathcal{E}(\mathbf{G}^F, 1)$ corresponds to a non-degenerate symbol, then $\sigma(\zeta) = \zeta$.*

Sketch of a proof of Proposition 15.17. One may assume that σ centralizes a maximal F-stable torus $\mathbf{T}_0$ of $\mathbf{G}$. Thus σ acts on $W = W(\mathbf{T}_0, \mathbf{G})$ and $<W, \sigma>$ is a group of type B_n. By transposition, σ^* acts on a dual torus $\mathbf{T}_0^*$ in $\mathbf{G}^*$. The dual group is a Clifford group on a quadratic space V^* and we may assume that σ^* is a non-special element of the rational Clifford group, acting on $\mathbf{G}^*$.

For any pair $(\mathbf{T}^*, s)$ one may define ${}^\sigma(\mathrm{R}_\mathbf{T}^\mathbf{G} s)$ by

$$ {}^\sigma(\mathrm{R}_\mathbf{T}^\mathbf{G} s)(g) = \mathrm{R}_\mathbf{T}^\mathbf{G} s\,(\sigma^{-1} g \sigma). $$

The correspondence between classes of maximal F-stable tori in $\mathbf{G}$ and $\mathbf{G}^*$, classified by conjugacy classes of the Weyl group $W = W(\mathbf{T}_0, \mathbf{G})$ (see §8.2), is such that (σ, σ^*) preserves duality and one has

$$ {}^\sigma(\mathrm{R}_\mathbf{T}^\mathbf{G} s) = \mathrm{R}^\mathbf{G}_{\sigma^*\mathbf{T}\sigma^{*-1}} \sigma^* s \sigma^{*-1}. \tag{15.18} $$

Let R_w denote $\mathrm{R}_\mathbf{T}^\mathbf{G}\,1$ for $\mathbf{T}$ of type $w \in W$ with respect to $\mathbf{T}_0$. The scalar products $\langle \zeta, \mathrm{R}_w \rangle_{\mathbf{G}^F}$ are given by Fourier transforms as follows. When $\phi \in \mathrm{Irr}(W)$ let $R_\phi = |W|^{-1} \sum_w \phi(w) R_w$. There is a partition of $\mathrm{Irr}(W)$ in families $\mathrm{Irr}(W) = \coprod_\mathcal{F} \mathcal{F}$ such that for any $\zeta \in \mathcal{E}(\mathbf{G}^F, 1)$, there is a family $\mathcal{F}(\zeta)$ with

$$ \langle \zeta, \mathrm{R}_w \rangle_{\mathbf{G}^F} = \sum_{\phi \in \mathcal{F}(\zeta)} \phi(w) \langle \zeta, \mathrm{R}_\phi \rangle_{\mathbf{G}^F} \tag{15.19} $$

because ζ is orthogonal to R_ϕ when $\phi \in (\mathrm{Irr}(W) \setminus \mathcal{F}(\zeta))$ (see [Lu84] §4). Now (15.18) implies ${}^\sigma R_\phi = R_{{}^\sigma\phi}$ and from (15.19) one has clearly

$$ \langle {}^\sigma\zeta, \mathrm{R}_w \rangle_{\mathbf{G}^F} = \sum_{\phi \in \mathcal{F}(\Lambda)} {}^\sigma\phi(w) \langle \zeta, \mathrm{R}_\phi \rangle_{\mathbf{G}^F} \tag{15.20} $$

Examination of the action of σ on $\mathrm{Irr}(W)$ shows that if ζ has no twin then σ fixes any $\phi \in \mathcal{F}(\zeta)$, hence σ fixes ζ. But if ζ has a twin ζ', then $\mathcal{F}(\zeta)$ reduces to $\{\phi_\zeta\}$ and $\sigma(\phi_\zeta) = \phi_{\zeta'}$. Furthermore $\zeta = R_{\phi_\zeta}$ and $\zeta' = R_{\phi_{\zeta'}}$. Formula (15.20) reduces to $\langle {}^\sigma\zeta, R_w \rangle_{\mathbf{G}^F} = {}^\sigma\phi_\zeta(w)$ hence ζ and ζ' have equal projection on the space of uniform functions, hence are equal. □

Exercises

1. Translate Hypothesis 15.1 into the language of root data. Deduce the fact that (15.1($\to$)) and (15.1*($\to^*$)) are equivalent when $\to$ and $\to^*$ are dual.
2. Let $\mathbf{G} \to \widetilde{\mathbf{G}}$ and $\mathbf{G} \to \mathbf{G}_0$ be two embeddings between connected reductive groups defined over $\mathbb{F}_q$ satisfying Hypothesis 15.1. Prove that there exist a connected reductive group $\widetilde{\mathbf{G}}_0$ defined over $\mathbb{F}_q$ and embeddings $\widetilde{\mathbf{G}} \to \widetilde{\mathbf{G}}_0$, $\mathbf{G}_0 \to \widetilde{\mathbf{G}}_0$ such that the square diagram so defined is commutative and Hypothesis 15.1 is satisfied for all four morphisms.
3. We use the hypotheses and notation of Corollary 15.14. Let $\mathcal{E}(\mathrm{C}_{\mathbf{G}^*}(t)^F, 1)$ be the set of $\chi \in \mathrm{Irr}(\mathrm{C}_{\mathbf{G}^*}(t)^F)$ such that some $\chi_0 \in \mathcal{E}(\mathrm{C}^\circ_{\mathbf{G}^*}(t)^F, 1)$ occurs in the

restriction of χ to $\mathrm{C}^\circ_{\mathbf{G}^*}(t)^F$. Assume that the restriction of any χ to $\mathrm{C}^\circ_{\mathbf{G}^*}(s)^F$ is a sum of distinct irreducible characters. Show the existence of a bijective map

$$\psi_t\colon \mathcal{E}(\mathbf{G}^F, t) \to \mathcal{E}(\mathrm{C}_{\mathbf{G}^*}(t)^F, 1)$$

such that, for any F-stable maximal torus $\mathbf{T}$ of $\mathrm{C}^\circ_{\mathbf{G}^*}(t)$ and any $\chi \in \mathcal{E}(\mathbf{G}^F, t)$, one has

$$\epsilon_{\mathbf{G}}\langle \chi, \mathrm{R}_{\mathbf{T}}^{\mathbf{G}}\, t\rangle_{\mathbf{G}^F} = \epsilon_{\mathrm{C}^\circ_{\mathbf{G}^*}(t)}\langle \psi_t(\chi), \mathrm{Ind}_{\mathrm{C}^\circ_{\mathbf{G}^*}(t)^F}^{\mathrm{C}_{\mathbf{G}^*}(t)^F} \mathrm{R}_{\mathbf{T}}^{\mathrm{C}^\circ_{\mathbf{G}^*}(t)}\, 1\rangle_{\mathrm{C}_{\mathbf{G}^*}(t)^F}.$$

4. We use the notation and hypotheses of Theorem 15.8 for $\mathbf{G}$, $\mathbf{G}^*$, s, ψ_s. Assume that any $\chi \in \mathrm{Irr}(\mathbf{G}^F)$ is uniquely defined by its orthogonal projection on the space of uniform functions. Let $z \in \mathrm{Z}(\mathbf{G}^*)^F$, let $\lambda_z \in \mathrm{Irr}(\mathbf{G}^F)$ be defined by (8.19), let $\chi \in \mathcal{E}(\mathbf{G}^F, s)$. Let $g \in (\mathbf{G}^*)^F$ be such that $gsg^{-1} = zs$. Show that there is then a natural one-to-one map $\mathcal{E}(\mathrm{C}_{\mathbf{G}^*}(s)^F, 1) \to \mathcal{E}(\mathrm{C}_{\mathbf{G}^*}(zs)^F, 1)$, $(\zeta \mapsto g.\zeta)$ such that $\psi_{sz}(\lambda_z \otimes \chi) = g.\psi_s(\chi)$.

Notes

For general notes about the classification of $\mathrm{Irr}(\mathbf{G}^F)$, see Chapter 8.

For a more detailed study of canonicality and uniqueness of the Jordan decomposition of characters, see [DiMi90]. Most of the present chapter is due to Lusztig, see [Lu88]. The results were annouced in 1983. Proposition 15.17 may be found in [FoSr89].

According to the conclusion of [Lu88], it is expected that the theory of character sheaves might help settle the question of canonicality of Jordan decomposition, and simplify the proof of Theorem 15.11. For character sheaves, see [Lu90] and its references. Most constructions on character sheaves are defined on $\mathbf{G}$ itself. But it is often necessary to assume that q is large to derive results about representations of $\mathbf{G}^F$; see [Sho97] and its references, and see also [Bo00].

16

On conjugacy classes in type D

This chapter is devoted to the second part of the proof of Theorem 15.11. By the first part of the proof and Proposition 15.12(i), it is sufficient to prove the following.

Theorem 16.1. *Let* $(\mathbf{G}, F)$ *be a connected reductive group defined over* $\mathbb{F}_q$. *Assume* q *is odd, the center of* $\mathbf{G}$ *is connected,* $([\mathbf{G}, \mathbf{G}], F)$ *is a split spin group of even rank defined over* $\mathbb{F}_q$ *with respect to a quadratic form which has maximal Witt index on* $\mathbb{F}_q$. *Let* $A := \mathbf{G}^F/\mathrm{Z}(\mathbf{G})^F[\mathbf{G}, \mathbf{G}]^F$, *a group of order 4 and exponent 2. When* $j \in \{1, 2, 4\}$ *let* y_j *be the number of elements of* $\mathrm{Irr}(\mathbf{G}^F)$ *whose stabilizer in* $\mathrm{Irr}(A)$ *is of order* j. *One has*

$$4|\mathrm{Irr}(\mathrm{Z}(\mathbf{G})^F[\mathbf{G}, \mathbf{G}]^F)| = y_1 + 4y_2 + 16y_4.$$

As the equality has to be proved in any even rank, one checks an equality between generating functions that are power series in q and an indeterminate t whose degree denotes the dimension of the orthogonal space.

The left-hand side of the equality is obtained by enumeration of the number of conjugacy classes of the spin group. To do this one uses the standard parametrization of elements of the orthogonal group, going by elementary arguments from the orthogonal group $\mathrm{O}_{2n}(q)$ to the spin group through the special orthogonal group $\mathrm{SO}_{2n}(q)$ and its derived group $\Omega_{2n}(q)$, of which the spin group is an extension. Doing that classification of orthogonal transformations, one cannot avoid considering simultaneously all types of quadratic forms and all dimensions. In the process we give formulae for the number of conjugacy classes of orthogonal, special orthogonal, conformal and Clifford groups in odd characteristic (see §16.2 to §16.4). The reader may easily obtain similar formulae for the symplectic groups (see Exercise 3). Nevertheless, in view of the length of the proof, we focus on the critical group $\mathrm{Spin}_{2n,\mathbf{0}}(q)$. In this notation the symbol $\mathbf{0}$ in index denotes a Witt symbol on $\mathbb{F}_q$ (see the beginning of §16.2).

To compute the numbers y_j one uses Jordan decomposition of irreducible characters of $\mathbf{G}^F$ (Theorem 15.8). Thus one has to classify conjugacy classes of rational semi-simple elements of a dual group $(\mathbf{G}^*, F)$ defined over $\mathbb{F}_q$ and use the classification of unipotent characters of classical groups (see [Lu84]). As $[\mathbf{G}, \mathbf{G}]$ is simply connected and the center of $\mathbf{G}$ is connected, one may assume that $\mathbf{G}^*$ is isomorphic to $\mathbf{G}$.

16.1. Notation; some power series

A fundamental function is the series of partitions. Let $p(n)$ be the number of partitions of the natural number n, put $p(0) = 0$. Define

$$\mathcal{P}(t) := \sum_{n\in\mathbb{N}} p(n)t^n = \prod_{j\geq 1} \frac{1}{1-t^j}, \quad \mathcal{P}_n := \mathcal{P}(t^n), \quad \mathcal{P}_n^- := \mathcal{P}(-t^n),$$
$$\mathcal{G}(t) := \mathcal{P}_2(\mathcal{P}_1^-)^{-2}.$$

Recall a classical relation

$$\mathcal{P}_1\mathcal{P}_1^-\mathcal{P}_4 = {\mathcal{P}_2}^3 \tag{16.2}$$

and the Gauss identity (see [And98] Corollary 2.10) $\mathcal{G}(t) = \sum_{j\in\mathbb{Z}} t^{j^2}$. Hence

$$\mathcal{G}(t)\mathcal{G}(-t) = \mathcal{G}(-t^2)^2, \quad \mathcal{G}(t) + \mathcal{G}(-t) = 2\mathcal{G}(t^4). \tag{16.3}$$

To classify representations of semi-simple elements, let $\mathcal{F}$ (resp. $\mathcal{F}_0$) be the set of irreducible monic elements f of $\mathbb{F}_q[t]$ with all roots non-zero (resp. with no root in $\{-1, 0, 1\}$). We may identify f with its set of roots in the algebraic closure $\mathbf{F}$, an orbit under the Frobenius map ($x \mapsto x^q$) on $\mathbf{F}$. The degree of f is denoted by $|f|$. We define two involutive maps on $\mathcal{F}$

$$f \mapsto \tilde{f}, \quad f \mapsto \bar{f} \quad (f \in \mathbb{F}_q[t])$$

that are induced by

$$x \mapsto x^{-1}, \quad x \mapsto -x \quad (x \in \mathbf{F}, x \neq 0)$$

respectively. The roots of $\tilde{f}$ (resp. $\bar{f}$) are the inverses (resp. the opposites) of the roots of f.

Notation 16.4. *Let $d \in \mathbb{N}^*$. Let $N_d(q) = N_d$ be the number of $f \in \mathcal{F}_0$ of degree d such that $f = \tilde{f}$. Let $M_d(q) = M_d$ be the number of pairs $(f, \tilde{f}) \in {\mathcal{F}_0}^2$ of degree d such that $f \neq \tilde{f}$. Let*

$$F_0^{(\Sigma)}(t) = \prod_{d\geq 1} {\mathcal{P}_{2d}}^{M_d+N_{2d}}.$$

The elementary proofs of the following lemmas are left to the reader.

Lemma 16.5. *(i) One has $N_d = 0$ if d is odd and $2dN_{2d} = \sum_{C \subset D}(-1)^{|C|}(q^{d/\prod_{p \in C} p} - 1)$ where D is the set of odd prime divisors of d. Hence $N_{2d}(q^2) = 2N_{4d}(q)$.*

(ii) One has

$$M_d = \begin{cases} (q-3)/2 = N_2 - 1 & \text{if } d = 1, \\ N_{2d} & \text{if } d \text{ is odd and } d > 1, \\ N_{2d} - N_d & \text{if } d \text{ is even.} \end{cases}$$

Lemma 16.6. *One has*

$$\mathcal{P}_2 F_0^{(\Sigma)}(t^2) = \prod_{\omega} \mathcal{P}_{|\omega||f|}$$

where ω runs over the set of orbits under the four-group $<(f \mapsto \tilde{f}), (f \mapsto \bar{f})>$ acting on $\mathcal{F}_0$ and $f \in \omega$.

Lemma 16.7. *Let $\mathcal{F}_0' \subset \mathbb{F}_{q^2}[t]$ be defined in a similar way to $\mathcal{F}_0 \subset \mathbb{F}_q[t]$. Let A (resp. B) be the group $<(f' \mapsto \tilde{f}'), (f' \mapsto \bar{f}'), (f' \mapsto F(f'))>$ (resp. $<(f' \mapsto \tilde{f}'), (f' \mapsto F(\bar{f}'))>$) acting on $\mathcal{F}_0'$.*

(i) If $F(\bar{f}') = f'$, then $|f'|$ is even. If $F(f') = f'$ then $|f'|$ is odd. If the order of a root α of f' does not divide 4, then the stabilizer of f' in A has at most two elements.

(ii) One has

$$\mathcal{P}_4{}^{-1}\mathcal{P}_2{}^2 F_0^{(\Sigma)} = \prod_{\omega'} \mathcal{P}_{|\omega'||f'|}$$

where ω' runs over the set of orbits under B in $\mathcal{F}_0'$ and $f' \in \omega'$.

(iii) One has

$$\mathcal{P}_2 F_0^{(\Sigma)}(t^2) = \prod_{\omega'} \mathcal{P}_{|\omega'||f'|}$$

where ω' runs over the set of orbits under A in $\mathcal{F}_0'$ and $f' \in \omega'$.

16.2. Orthogonal groups

If $V(\mathbb{F}_q)$ is a non-degenerate orthogonal space of dimension $n \in \{2m, 2m+1\}$ over $\mathbb{F}_q$ (q odd), then the symmetric bilinear form is equivalent to one of the forms

(0) $$x_1x_2 + x_3x_4 + \cdots + x_{2m-1}x_{2m}$$

(w) $$x_1x_2 + x_3x_4 + \cdots + x_{2m-3}x_{2m-2} + x_{2m-1}^2 - \delta x_{2m}^2$$

(1) $$x_1x_2 + x_3x_4 + \cdots + x_{2m-1}x_{2m} + x_{2m+1}^2$$

(d) $$x_1x_2 + x_3x_4 + \cdots + x_{2m-1}x_{2m} + \delta x_{2m+1}^2$$

where δ has no square root in $\mathbb{F}_q$. We call the corresponding Witt symbol written above on the left the **Witt type** of the form, an element of $\mathbf{V} = \{\mathbf{0}, \mathbf{w}, \mathbf{1}, \mathbf{d}\}$. Clearly the form has maximal Witt index if and only if its Witt type $\mathbf{v}$ is in $\{\mathbf{0}, \mathbf{1}\}$. The orthogonal sum of orthogonal spaces defines a structure of a commutative group on $\mathbf{V}$ for which $\mathbf{0}$ is the null element and $\mathbf{w} + \mathbf{d} = \mathbf{1}$. It is convenient to assign the Witt type $\mathbf{0}$ to the null space. If 4 divides $(q - 1)$, then $\mathbf{V}$ has exponent 2 and the discriminant has square roots in $\mathbb{F}_q^\times$ if and only if $\mathbf{v} \in \{\mathbf{0}, \mathbf{1}\}$. If 4 divides $(q + 1)$, then $\mathbf{V}$ is cyclic with generators $\mathbf{1}$ and $\mathbf{d}$ and the discriminant has square roots in $\mathbb{F}_q^\times$ if and only if ($\mathbf{v} \in \{\mathbf{0}, \mathbf{1}\}$ and m is even) or ($\mathbf{v} \in \{\mathbf{w}, \mathbf{d}\}$ and m is odd). Nevertheless a unique notation will be used for all q.

We may assume that the space $V(\mathbb{F}_q)$ is the space of rational points of an orthogonal $\mathbf{F}$-space $V(\mathbf{F})$.

The corresponding orthogonal groups will be denoted by $\mathrm{O}_{n,\mathbf{v}}(q)$, sometimes O or $\mathrm{O}(q)$ or O_n when other parameters among $\mathbf{v}$, n, q are well defined. The groups $\mathrm{O}_{2m+1,\mathbf{1}}(q)$ and $\mathrm{O}_{2m+1,\mathbf{d}}(q)$ are isomorphic.

The special orthogonal groups $\mathrm{SO}_{n,\mathbf{v}}(q)$ ($\mathbf{v} = \mathbf{0}, \mathbf{w}, \mathbf{1}, \mathbf{d}$) may be obtained as finite reductive groups of respective rational types $(\mathbf{D}_m, q)$, $({}^2\mathbf{D}_m, q)$, $(\mathbf{C}_m, q)$ and $(\mathbf{C}_m, q)$.

We now give parameters for conjugacy classes of $\mathrm{O}_{n,\mathbf{v}}(q)$.

Consider first a semi-simple element s. The space of representation V decomposes into an orthogonal sum $V = V_1 \perp V_{-1} \perp V^0$, where V_a is the eigenspace of s for the eigenvalue a. As $\mathbb{F}_q{<}s{>}$-module, V^0 has a semi-simple decomposition. A simple representation of ${<}s{>}$ on $\mathbb{F}_q$ is defined by some $f \in \mathcal{F}$. Let $\mu(f) = \mu(\tilde{f})$ be the multiplicity in V of the representation defined by f. Let ψ_+, ψ_- be the Witt types of the restriction of the quadratic form to the spaces V_1 and V_{-1} respectively. We write $\mu(1)$, $\mu(-1)$ for $\mu(t-1)$ and $\mu(t+1)$ respectively. When $\mu(1) = 0$ (resp. $\mu(-1) = 0$), put $\psi_+ = \mathbf{0}$ (resp. $\psi_- = \mathbf{0}$). The centralizer of s in $G(q)$ is isomorphic, via the restriction to the decomposition of V as $\mathbb{F}_q{<}s{>}$-module, to a direct product $\mathrm{O}_{\mu(1),\psi_+}(q)\,\mathrm{O}_{\mu(-1),\psi_-}(q) \prod_{\{f,\tilde{f}\}\subset\mathcal{F}_0} \mathrm{GL}(\mu(f), \kappa(f))$, where $\kappa(f) = q^{|f|}$ if $f \neq \tilde{f}$ and $\kappa(f) = -q^{|f|/2}$ if $f = \tilde{f}$, and by convention the component relative to f is $\{1\}$ if $\mu(f) = 0$.

Proposition 16.8. *A conjugacy class of semi-simple elements of an orthogonal group $O_{n,\mathbf{v}}(q)$ is defined by a triple (μ, ψ_+, ψ_-), where*

$$\mu\colon \mathcal{F} \to \mathbb{N}, \quad \psi_+, \psi_- \in \mathbf{V}$$

such that

$$\psi_\epsilon \begin{cases} = \mathbf{0} & \text{if } \mu(\epsilon 1) = 0, \\ \in \{\mathbf{0}, \mathbf{w}\} & \text{if } \mu(\epsilon 1) \in 2\mathbb{N}, \\ \in \{\mathbf{1}, \mathbf{d}\} & \text{if } \mu(\epsilon 1) \notin 2\mathbb{N} \end{cases}$$

for $\epsilon \in \{+, -\}$ *and*

$$\forall f \in \mathcal{F},\ \mu(f) = \mu(\tilde{f}),\quad \sum_{f\in\mathcal{F}} \mu(f)|f| = n,$$
$$\psi_+ + \psi_- + \sum_{f\in\mathcal{F}_0} \mu(f)\mathbf{w} = \mathbf{v}.$$

The last equality holds in the Witt group. As $\mathbf{w}$ is of order 2, the last sum may be restricted to the set of f with $f = \tilde{f}$.

Proposition 16.9. *The conjugacy class of a unipotent element in an orthogonal group $O_{n,\mathbf{v}}(q)$ is uniquely defined by a couple (m_1, Ψ) where*

(i) m_1 is the function of multiplicity of Jordan blocks of a given size, $m_1 : \mathbb{N}^* \to \mathbb{N}$ with the following conditions

$$\forall j \in \mathbb{N}^*,\ m_1(2j) \in 2\mathbb{N},\quad \sum_{j>0} j m_1(j) = n$$

(ii) the function $\Psi : \mathbb{N} \to \mathbf{V}$ gives the Witt type of a bilinear symmetric form in dimension $m_1(2j+1)$ for any $j \in \mathbb{N}$ with the condition

$$\sum_{j\in\mathbb{N}} \Psi(j) = \mathbf{v}.$$

The centralizer of a unipotent with parameter (m_1, Ψ) is a direct product on the set of k such that $m_1(k) \neq 0$. If $k = 2j+1$ is odd, the reductive quotient of the component is an orthogonal group in dimension $m_1(2j+1)$, and $\Psi(j)$ gives the Witt type of the form on the multiplicity space. If k is even, then the reductive quotient of the component is a symplectic group on the multiplicity space, in even dimension $m_1(k)$.

The conjugacy class of an element of any of the considered groups is defined by the conjugacy class of the semi-simple component s and the conjugacy class of the unipotent component in the centralizer of s. We obtain the following.

Proposition 16.10. *A conjugacy class in $O_{n,\mathbf{v}}(q)$ is uniquely defined by three applications m, Ψ_+, Ψ_- such that*

$$m : \mathcal{F} \times \mathbb{N}^* \to \mathbb{N},\quad \forall (f, j)\ \ m(\tilde{f}, j) = m(f, j),\quad \sum_{(f,j)\in\mathcal{F}\times\mathbb{N}^*} j m(f, j)|f| = n,$$
$$\forall j \geq 1,\quad m(1, 2j) \in 2\mathbb{N},\ m(-1, 2j) \in 2\mathbb{N},$$
$$\Psi_+, \Psi_- : \mathbb{N} \to \mathbf{V},\quad \sum_{(f,j)\in\mathcal{F}_0\times\mathbb{N}^*} j m(f, j)\mathbf{w} + \sum_{j\in\mathbb{N}} (\Psi_+(j) + \Psi_-(j)) = \mathbf{v}.$$

Here $\Psi_+(j)$ and $\Psi_-(j)$ are Witt types of forms in respective dimensions $m(1, 2j+1)$, $m(-1, 2j+1)$.

Remarks. (a) The parameter (μ, ψ_+, ψ_-) of the conjugacy class of the semi-simple component is given by

(16.11) $$\mu(f) = \sum_{j \in \mathbb{N}} j m(f, j), \quad \psi_+ = \sum_{j \in \mathbb{N}} \Psi_+(j), \quad \psi_- = \sum_{j \in \mathbb{N}} \Psi_-(j).$$

The parameter (m_1, Ψ) of the conjugacy class of the unipotent component is given by

(16.12) $$\begin{aligned} m_1(j) &= \sum_f m(f, j)|f|, \\ \Psi(j) &= \sum_{f \in \mathcal{F}_0} m(f, 2j+1)\mathbf{w} + \Psi_+(j) + \Psi_-(j). \end{aligned}$$

Let u be a unipotent element of $\mathrm{O}_{n,\mathbf{v}}(q)$. Let (m_1, Ψ) be the parameter of the class of u. The centralizer of u contains a semi-simple element in the conjugacy class with parameter (μ, ψ_+, ψ_-) if and only if there exists (m, Ψ_+, Ψ_-) such that (16.11) and (16.12) hold.

(b) A parameter defines one and only one conjugacy class of one and only one type $(\mathbf{v}, n, q)$ of orthogonal group. Let g be an element of $\mathrm{O}_{n,\mathbf{v}}(q)$ such that $(g^2 - 1)$ is non-singular. The conjugacy class of g in the orthogonal group is the intersection with the conjugacy class of g in the full linear group.

Let $k_{n,\mathbf{v}}$ be the number of conjugacy classes of the group $\mathrm{O}_{n,\mathbf{v}}(q)$. Our first goal is to compute $1 + \Sigma_{n>0} k_{n,\mathbf{v}} t^n$. In odd dimension $(2n+1)$, $k_{2n+1,\mathbf{v}}$ is independent of $\mathbf{v} \in \{\mathbf{1}, \mathbf{d}\}$; we write k_{2n+1}.

By Proposition 16.9 the number of unipotent conjugacy classes in orthogonal groups is given by the generating function

$$\left(1 + 2\left(\sum_m t^m\right)\right)\left(1 + \sum_m t^{4m}\right)\left(1 + 2\left(\sum_m t^{3m}\right)\right)\cdots$$
$$= \prod_{j \geq 1} \frac{(1 + t^{2j-1})^2}{(1 - t^{2(2j-1)})(1 - t^{4j})}.$$

Hence, using the function $\mathcal{G}$ (§16.1), let

(16.13) $$F_1^{(\Sigma)}(t) = \mathcal{P}_2{}^2\mathcal{G}.$$

$F_1^{(\Sigma)}$ is the generating function for the number of parameters of unipotent classes, the superscript (Σ) recalls that it is the sum over all Witt types, and here we don't take into account the last equality of Proposition 16.9.

The number of parameters of conjugacy classes of any orthogonal group, for all Witt types, as described in Proposition 16.10, is given by a product of generating functions. The number of classes of elements without eigenvalues 1 or -1 is given by $\sum_m \sum_{f,j} t^{|f|\Sigma_{f,j} jm(f,j)}$ with some conditions on the parameter $m\colon \mathcal{F} \times \mathbb{N}^* \to \mathbb{N}$, i.e.

$$\prod_{(f,\tilde{f}), f\neq\tilde{f}} \left(\sum_{m\in\mathbb{N}} t^{2|f|\Sigma_j jm(f,j)}\right) \prod_{f=\tilde{f}} \left(\sum_{m\in\mathbb{N}} t^{|f|\Sigma_j jm(f,j)}\right)$$

exactly $F_0^{(\Sigma)}(t)$ in its definition (Notation 16.4, §16.1). Hence $F_0^{(\Sigma)}$ is the generating function for parameters m such that $m(1, j) = m(-1, j) = 0$ for all $j > 0$. By decomposition of any orthogonal transformation as a product, one has

(16.14)
$$\begin{aligned} 1+\sum_{n\geq 1}(2k_{2n-1}t^{2n-1} + (k_{2n,\mathbf{0}} + k_{2n,\mathbf{w}})t^{2n}) \\ = \left(F_1^{(\Sigma)}(t)\right)^2 F_0^{(\Sigma)}(t). \end{aligned}$$

One may compute $F_0^{(\Sigma)}$ using Remark (b) following Proposition 16.10. Let $g_1, g_2, \ldots$ be the invariant factors of an element x such that $x^2 - 1$ is non-singular. Here g_{i+1} is a divisor of g_i for each i and $g_i(0)g_i(1)g_i(-1) \neq 0$. Let $f_i = g_i/g_{i+1}$. The condition $m(f, k) = m(\tilde{f}, k)$ for all $(f, k) \in \mathcal{F}_0 \times \mathbb{N}^*$ becomes $g_i = \tilde{g}_i$ for all i. Let $N(d)$ be the number of monic polynomials g of degree d such that $g = \tilde{g}$ and $g(0)g(1)g(-1) \neq 0$. The coefficient in degree n of $F_0^{(\Sigma)}(t)$ is $\sum_{d_1+2d_2+\ldots=n} N(d_1)N(d_2)\ldots$. Hence

$$F_0^{(\Sigma)}(t) = \prod_{j>0}\left(1+\sum_{d>0} N(d)t^{jd}\right).$$

Let $N'(d)$ be the number of monic polynomials g of degree d such that $g = \tilde{g}$ and $g(0) \neq 0$. Out of these $N'(d)$ polynomials of degree d, $N'(d-1)$ are divisible by $(t-1)$, $N'(d-1)$ by $t+1$ and $N'(d-2)$ by t^2-1. Hence

$$N(d) = N'(d) - 2N'(d-1) + N'(d-2)$$

and so

$$1+\sum_{d>0} N(d)t^d = (1-t)^2\left(1+\sum_{d>0} N'(d)t^d\right).$$

It is easily seen that $N'(d) = q^{d/2} - q^{(d-2)/2}$ if d is even, $N'(d) = 2q^{(d-1)/2}$ if d is odd, so that $1 + \sum_d N'(d)t^d = (1+t)^2(1-qt^2)^{-1}$. A new expression for

$F_0^{(\Sigma)}$ is therefore

(16.15)
$$F_0^{(\Sigma)}(t) = \prod_{j\geq 1} \frac{(1-t^{2j})^2}{1-qt^{2j}}.$$

A similar argument in the full linear group, omitting the condition $\tilde{g} = g$, gives

(16.16)
$$\Phi(t) := \prod_{d\geq 1} \mathcal{P}_d{}^{2M_d+N_d} = \prod_{j\geq 1} \frac{(1-t^j)^3}{1-qt^j}$$

hence

(16.17)
$$F_0^{(\Sigma)}(t) = \Phi(t^2)\mathcal{P}_2 = \mathcal{P}_2 \prod_{d\geq 1} \mathcal{P}_{2d}{}^{2M_d+N_d}.$$

From (16.17) and the definition of $F_0^{(\Sigma)}$ we get $\prod_d \mathcal{P}_d{}^{N_d} = \mathcal{P}_2 \prod_d (\mathcal{P}_{2d}{}^{M_d}\mathcal{P}_{2d}{}^{N_d})$, hence

(16.18)
$$H_0(t) := \prod_{d\geq 1} \mathcal{P}_{2d}{}^{N_{2d}}, \quad \mathcal{P}_2 H_0(t^2) F_0^{(\Sigma)}(t) = H_0(t)^2.$$

Formula (16.18) may be deduced from Lemma 16.5.

To restrict to even dimensions, consider the even part of $F_1^{(\Sigma)}(t)^2$, i.e. $(F_1^{(\Sigma)}(t)^2 + F_1^{(\Sigma)}(-t)^2)/2$. From (16.14), (16.3) and (16.14) it follows that

(16.19)
$$1 + \sum_{n\geq 1} (k_{2n,\mathbf{0}} + k_{2n,\mathbf{w}})t^{2n} = \left(2\mathcal{G}(t^4)^2 - \mathcal{G}(-t^2)^2\right)\mathcal{P}_2{}^4 F_0^{(\Sigma)}.$$

To compute $k_{2n,\mathbf{0}} - k_{2n,\mathbf{w}}$, consider first the classes of elements without the eigenvalue 1 or -1. One has to substract $2t^{|f|jm(f,j)}$ when j is odd and $f = \tilde{f}$. The contribution of f becomes $\sum_{m:\mathbb{N}^*\to\mathbb{N}} (-t^{|f|})^{\Sigma_j jm(j)}$. Hence let $F_0^{(\Delta)}(t) = \prod_{d\in\mathbb{N}^*} \mathcal{P}(-t^d)^{N_d}\mathcal{P}_{2d}{}^{M_d}$ (the superscript (Δ) stands for "difference"). Now (16.2) and (16.17) imply $F_0^{(\Delta)}(t)F_0^{(\Sigma)}(t)\Phi(t^4) = F_0^{(\Sigma)}(t^2)^2\Phi(t^2)$ and we have

(16.20)
$$\mathcal{P}_4 F_0^{(\Sigma)}(t^2) = \mathcal{P}_2 F_0^{(\Delta)}(t),$$

which defines $F_0^{(\Delta)}$. By (16.15), (16.20) becomes

(16.21)
$$F_0^{(\Delta)}(t) = \prod_{j\geq 1} \frac{(1-t^{4j})(1-t^{2j})}{1-qt^{4j}}.$$

As for parameters such that $m(1, 2j+1) \neq 0$ or $m(-1, 2j+1) \neq 0$ for some $j \in \mathbb{N}$, they are in equal number in each type of group (see

Propositions 16.9 and 16.10). Hence

$$1 + \sum_{n\geq 1}(k_{2n,\mathbf{0}} - k_{2n,\mathbf{w}})t^{2n} = \mathcal{P}_4{}^2 F_0^{(\Delta)}. \tag{16.22}$$

The numbers $k_{2n,\mathbf{w}}$ and $k_{2n,\mathbf{0}}$ are given by (16.19) and (16.22).

16.3. Special orthogonal groups and their derived subgroup; Clifford groups

Proposition 16.23. *The semi-simple conjugacy class of the orthogonal group of parameter* (μ, ψ_+, ψ_-) *(Proposition 16.8) is contained in the special group if and only if* $\mu(-1) \in 2\mathbb{N}$. *It splits into two conjugacy classes of the special group O if and only if* $\mu(1) = \mu(-1) = 0$.

The generating function for the number of parameters of O-conjugacy classes of elements with unique eigenvalue -1 *and determinant 1 is* $\mathcal{P}_2{}^2\mathcal{G}(t^4)$.

Proof. A conjugacy class splits if and only if the centralizer of an element of the class is contained in the special group. The only components of the centralizer of a semi-simple element that are not contained in the special group SO are those corresponding to $f \in \mathcal{F} \setminus \mathcal{F}_0$, if non-trivial.

From (16.14) and (16.3), we deduce that the even part of $F_1^{(\Sigma)}(t)$ is $\mathcal{P}_2{}^2\mathcal{G}(t^4)$. □

Any unipotent element is special. By (16.11) the centralizer of a unipotent element is contained in the special group if and only if the parameter (m_1, Ψ) of its conjugacy class satisfies $m_1(2j+1) = 0$ for all $j \in \mathbb{N}$. Now if s is semi-simple, with centralizer $C(s)$ in the orthogonal group, and u a unipotent in $C(s)$, then $\mathrm{C_O}(su) = \mathrm{C}_{C(s)}(u)$; hence we have the following.

Proposition 16.24. *A conjugacy class of* $\mathrm{O}_{n,\mathbf{v}}(q)$ *with parameter* (m, Ψ_+, Ψ_-) *(Proposition 16.10) is contained in the special group if and only if*

$$\sum_{j\in\mathbb{N}} m(-1, 2j+1) \in 2\mathbb{N}$$

and it splits into two classes of $\mathrm{SO}_{n,\mathbf{v}}(q)$ *if and only if*

$$\forall j \in \mathbb{N}, \quad m(1, 2j+1) = m(-1, 2j+1) = 0.$$

Note that the splitting condition is independent of q.

By Propositions 16.23 and 16.24, the sum for $\mathbf{v} = \mathbf{0}, \mathbf{w}$ of the numbers of conjugacy classes of the special groups $\mathrm{SO}_{2n,\mathbf{v}}(q)$ is given by the generating function $\mathcal{P}_2{}^4\mathcal{G}(t^4)^2 F_0^{(\Sigma)}$.

Let $\mathbf{H}$ be the special Clifford group of an orthogonal space V of dimension $2m$ ($m \neq 0$) on $\mathbf{F}$ with a form defined on $\mathbb{F}_q$, as described in §16.2. The Clifford group on V, denoted by $\mathrm{CL}(V)$, is the normalizer of V in the subgroup of units of the Clifford algebra on V (V is considered as a subspace of its Clifford algebra); see [Bour59] Chapitre 9.

Denote by e the neutral element in $\mathrm{CL}(V)$, to distinguish it from $1 = 1_V \in \mathrm{SO}(V)$. A non-isotropic vector v of V defines a unit in the Clifford algebra that acts on V as the opposite of the reflection defined by v. So an exact sequence of groups

$$1 \longrightarrow \mathbf{F}^\times e \longrightarrow \mathrm{CL}(V) \xrightarrow{\pi} \mathrm{O}(V) \longrightarrow 1$$

is obtained, whose restriction to $\mathbf{H}$ is

$$1 \longrightarrow \mathbf{F}^\times e \longrightarrow \mathbf{H} \xrightarrow{\pi} \mathrm{SO}(V) \longrightarrow 1.$$

The center of $\mathbf{H}$ is $\mathbf{F}^\times e$. The derived group of $\mathbf{H}$ is the spinor group $\mathrm{Spin}(V)$ and the restriction of π to $\mathrm{Spin}(V)$ gives the exact sequence

$$1 \longrightarrow <-e> \longrightarrow [\mathbf{H}, \mathbf{H}] = \mathrm{Spin}(V) \xrightarrow{\pi} \Omega(V) \longrightarrow 1$$

where $-e := (-1).e$ and $\Omega(V)$ is the commutator subgroup of $\mathrm{SO}(V)$. As the orthogonal space is defined over $\mathbb{F}_q$, a Frobenius F acts on $\mathbf{H}$ and V, $\mathbf{H}^F$ is the special Clifford group of $V(\mathbb{F}_q)$, the first and second sequences restrict to

$$1 \longrightarrow \mathbb{F}_q^\times e \longrightarrow \mathrm{CL}(V(\mathbb{F}_q)) \xrightarrow{\pi} \mathrm{O}_{2n,\mathbf{v}}(q) \longrightarrow 1,$$
$$1 \longrightarrow \mathbb{F}_q^\times e \longrightarrow \mathbf{H}^F \xrightarrow{\pi} \mathrm{SO}_{2n,\mathbf{v}}(q) \longrightarrow 1,$$

and the last sequence restricts to

$$1 \longrightarrow <-e> \longrightarrow \mathrm{Spin}_{2n,\mathbf{v}}(q) \xrightarrow{\pi} \Omega_{2n,\mathbf{v}}(q) \longrightarrow 1$$

where $\Omega_{2n,\mathbf{v}}(q)$ is the derived subgroup of $\mathrm{SO}_{2n,\mathbf{v}}(q)$. One has $F(-e) = -e$.

Proposition 16.25. *Let s be a semi-simple element of the rational special Clifford group $\mathbf{H}^F = \mathrm{CL}(V(\mathbb{F}_q))$, let C be the conjugacy class of $\pi(s)$ in $SO_{2n,\mathbf{v}}(q)$, and let $C' = \pi^{-1}(C)$.*

(i) s and $-es$ are conjugate in $\mathbf{H}^F$ if and only if 1 and -1 are eigenvalues of $\pi(s)$.

(ii) If 1 and -1 are eigenvalues of $\pi(s)$, then C' is the union of $\frac{1}{2}(q-1)$ conjugacy classes of $\mathbf{H}^F$. If 1 or -1 is not an eigenvalue of $\pi(s)$, then C' is the union of $(q-1)$ conjugacy classes of $\mathbf{H}^F$.

Proof. For any semi-simple element s of $\mathbf{H}$, let

$$\mathrm{C}^*(s) = \{g \in \mathrm{Spin}(V) \mid g^{-1}sg \in \{s, -es\}\}.$$

By Lang's theorem in an F-stable maximal torus $\mathbf{S}$ containing s, and Proposition 8.1, any $s \in \mathbf{H}^F$ can be written as $s = zt$, with $z \in \mathbf{S}$ and $t \in [\mathbf{H}, \mathbf{H}]^F$. Then the first assertion of (i) is true for s in $\mathbf{H}^F$ if and only if it is true for t in $\mathrm{Spin}_{2n,\mathrm{v}}(q) = [\mathbf{H}, \mathbf{H}]^F$.

By definition $\mathrm{C}_{\mathrm{Spin}(V)}(t) \subset \mathrm{C}^*(t), \pi(\mathrm{C}^*(t)) = \mathrm{C}_{\mathrm{SO}(V)}(\pi(t))$ and t is conjugate to $-et$ if and only if $|\mathrm{C}^*(t) : \mathrm{C}_{\mathrm{Spin}(V)}(t)| = 2$ — if not, the index is 1.

Now $\mathrm{C}_{\mathrm{Spin}(V)}(t)$ is connected because the spinor group is simply connected (Theorem 13.14) and $\pi(\mathrm{C}_{\mathrm{Spin}(V)}(t))$ is also connected since π is continuous. Finally the conjugacy between t and $-et$ is equivalent to $|\mathrm{C}_{\mathrm{SO}(V)}(\pi(t))/\mathrm{C}^\circ_{\mathrm{SO}(V)}(\pi(t))| = 2$ — and if not, the order is 1.

By examination of the structure of $\mathrm{C}_{\mathrm{SO}(V)}(\pi(t))$, we see that the direct components acting on the sum of eigenspaces of $\pi(t)$, $V_a + V_{a^{-1}}$, isomorphic to general linear groups, are connected, apart from $a = a^{-1}$. The non-connectedness appears in $\mathrm{SO}(V_1 + V_{-1}) \cap (\mathrm{O}(V_1) \times \mathrm{O}(V_{-1}))$ when V_1 and V_{-1} are non-zero spaces, that is to say when 1 and -1 are eigenvalues of $\pi(t)$. Furthermore the $g \in \mathrm{Spin}(V)$ such that $gtg^{-1} = -et$ are precisely those g such that $\pi(g)$ centralizes $\pi(t)$ and the two components of $\pi(g)$ acting respectively on the eigenspaces V_1 and V_{-1} have determinant -1.

Assuming now that $t \in \mathrm{Spin}(V(\mathbb{F}_q))$, if such a g exists in $\mathrm{SO}(V_1 + V_{-1})$, there is one in $\mathrm{SO}(V_1 + V_{-1})^F$. This proves (i).

If two elements as and bs in $\mathbb{F}_q^\times s$ are conjugate by some $g \in \mathbf{H}^F$, then $\pi(g)$ centralizes $\pi(as) = \pi(s)$ in $\mathrm{SO}(V(\mathbb{F}_q))$. One has $\pi^{-1}(\mathrm{C}_{\mathrm{SO}(V)}(\pi(s))) = \mathbb{F}_q^\times e.\mathrm{C}^*(s)$ and $\pi(\mathbb{F}_q^\times e.\mathrm{C}_{\mathrm{Spin}(V)}(t)) = \pi(\mathrm{C}_{\mathbf{H}}(s)) = \mathrm{C}^\circ_{\mathrm{SO}(V)}(\pi(s))$ by connectedness, so that $g \in \mathbb{F}_q^\times e.\mathrm{C}^*(s)$. Thus (ii) follows from (i) and its proof. □

From Propositions 16.25, 16.24 and formulae (16.14), (16.19) and (16.22), one deduces easily the generating functions for the number of conjugacy classes of the finite Clifford groups in odd characteristic. In even dimensions with all Witt types:

(16.26)
$$\frac{q-1}{2}\left(\mathcal{P}_2{}^4\mathcal{G}(t^4)^2 + 2\mathcal{P}_2{}^2\mathcal{G}(t^4) - 1\right)F_0^{(\Sigma)}.$$

Note that, from now on, the term of null degree in the series has no significance.

The group $\Omega_{n,\mathrm{v}}(q)$ — we may write simply Ω — is the derived group of the orthogonal group, and of index 2 in the special group (recall that q is odd). It may be obtained as the kernel of the restriction to the special group of the spinor norm

$$\theta : \mathrm{O}_{n,\mathrm{v}}(q) \to \mathbb{F}_q^\times/(\mathbb{F}_q^\times)^2.$$

Clearly an orthogonal transformation belongs to Ω if and only if its semisimple component belongs to Ω. The spinor norm appears in the theory of Clifford algebras ([Artin], Chapter V). If $g \in \mathrm{O}(V(\mathbb{F}_q))$ is written as a product of symmetries with respect to (non-isotropic) vectors v_j $(1 \leq j \leq r)$, then $\theta(g) = \prod_j \langle v_j, v_j\rangle(\mathbb{F}_q^\times)^2$. The spinor norm is multiplicative with respect to the orthogonal sum. An elegant and general characterization of the spinor norm has been given by H. Zassenhaus; see [Za62].

Proposition 16.27. *Let $g \in \mathrm{O}(V(\mathbb{F}_q))$, of semi-simple component s. Let V_{-1} be the eigenspace $\{x \in V(\mathbb{F}_q) \mid s(x) = -x\}$, and let $V_2 = V_{-1}{}^\perp$. Let Δ be the discriminant of the restriction of the form to V_{-1}, and let D be the determinant of the restriction of $(1+s)/2$ to V_2. The value on g of the spinor norm is $\Delta D(\mathbb{F}_q^\times)^2$.*

As a corollary, assuming $n \in \{2m, 2m+1\}$, we obtain that $-1_{V(\mathbb{F}_q)}$ belongs to the kernel of the spinor norm if and only if $m(q-1) \equiv 0$ (mod. 4) for $\mathbf{v} \in \{\mathbf{0}, \mathbf{1}\}$ or $m(q-1) \equiv 2$ (mod. 4) for $\mathbf{v} \in \{\mathbf{w}, \mathbf{d}\}$. For n even the condition is equivalent to $-1_{V(\mathbb{F}_q)} \in \Omega_{2m,\mathbf{v}}(q)$.

By computing the determinant in the isotypic case, we have the following.

Proposition 16.28. *Let s be an element of $\mathrm{O}(V(\mathbb{F}_q))$. Assume that, for some $f \in \mathcal{F}_0$, s has minimal polynomial f if $f = \tilde{f}$, or $f\tilde{f}$ if $f \neq \tilde{f}$. Let $\alpha \in f$ (α a root of f). One has*

$$\theta(s) = (\mathbb{F}_q^\times)^2 \text{ if and only if } \begin{cases} \alpha \in (\mathbb{F}_{q^{|f|}}^\times)^2 & \text{if } f \neq \tilde{f}, \\ \alpha^{(q^{|f|/2}+1)/2} = 1 & \text{if } f = \tilde{f}. \end{cases}$$

The formula in Proposition 16.28 justifies the definition of a new application $\sigma\colon \mathcal{F}_0 \to \{-1, +1\}$. The proof of the following proposition is left to the reader.

Proposition 16.29. *Let $\sigma\colon \mathcal{F}_0 \to \{-1, 1\}$ be defined by*

$$\sigma(f) = 1 \text{ if and only if } \begin{cases} f \subset (\mathbb{F}_{q^{|f|}}^\times)^2 & \text{if } f \neq \tilde{f}, \\ \alpha^{(q^{|f|/2}+1)/2} = 1 & \text{if } f = \tilde{f} \text{ and } \alpha \in f. \end{cases}$$

For $\epsilon \in \{+, -\}$, let $2M_{d,\epsilon}$ (resp. $N_{d,\epsilon}$) be the number of $f \in \mathcal{F}_0$ of degree d and such that $\sigma(f) = \epsilon 1$, $f \neq \tilde{f}$ (resp. $f = \tilde{f}$). One has $M_{d,+} + M_{d,-} = M_d$, $N_{d,+} + N_{d,-} = N_d$,

$$M_{1,+} \in \{(q-5)/4, (q-3)/4\}, \quad N_{2,+} \in \{(q-1)/4, (q-3)/4\},$$
$$M_{1,-} = N_{2,+}$$

and, when $d > 1$,

$$N_{2d,-} = N_{2d,+} = M_{d,-}$$

Proposition 16.30. *Let (μ, ψ_+, ψ_-) be the parameter of a semi-simple conjugacy class C of $\mathrm{SO}_{n,\mathrm{v}}(q)$ $(\mu(-1) \in 2\mathbb{N})$. Define $\upsilon_- \in \mathbb{N}$ by*

$$\upsilon_- = \begin{cases} \frac{(q-1)\mu(-1)}{4} & \text{if } \psi_- = \mathbf{0}, \\ 1 + \frac{(q-1)\mu(-1)}{4} & \text{if } \psi_- = \mathbf{w}. \end{cases}$$

Then C is contained in $\Omega_{2n,\mathrm{v}}(q)$ if and only if

$$\upsilon_- + \sum_{\{(f,\tilde{f})|\sigma(f)\neq 1\}} \mu(f) \in 2\mathbb{N}\,.$$

The parameter (μ, ψ_+, ψ_-) defines one or two classes of the derived group, and it defines two classes if and only if $\mu(1) = \mu(-1) = 0$.

Proof. The integer υ_- is defined so that $\upsilon_- \in 2\mathbb{N}$ if and only if the discriminant of the restriction of the form to V_{-1} is a square in $\mathbb{F}_q$ (see the beginning of §16.2 and note that with our conventions $\upsilon_- = 0$ if $\mu(-1) = 0$). By Proposition 16.28 and the definition of σ, the sum of multiplicities $\mu(f)$ on $\{f, \tilde{f}\}$ such that $\sigma(f) \neq 1$ is even if and only if the determinant D, as defined in Proposition 16.27, is a square. Hence the first assertion follows from Proposition 16.27.

The centralizer of a semi-simple element of the special group $\mathrm{SO}_{2n,\mathrm{v}}(q)$ is never contained in the kernel of the spinor norm, hence the SO-conjugacy class doesn't split in Ω and Proposition 16.23 applies. □

A conjugacy class of the orthogonal group with parameter (m, Ψ_+, Ψ_-) is contained in $\Omega_{2n,\mathrm{v}}(q)$ if and only if the function (μ, ψ_+, ψ_-) deduced from (m, Ψ_+, Ψ_-) by (16.11) satisfies the condition of Proposition 16.30. The centralizer of an element g is in the kernel of the spinor norm if and only if it is the case in any "$\{f, \tilde{f}\}$-component" ($f \in \mathcal{F}$) of the centralizer of the semi-simple part of g. For $f \in \mathcal{F}_0$, use the following.

Proposition 16.31. *Let u be a unipotent element of some general linear or unitary group $G = \mathrm{GL}(\mu, \pm q)$. Let H be the unique subgroup of G of index 2. The centralizer of u in G is contained in H if and only if it has no Jordan block of odd size.*

On unipotents in orthogonal groups, one has the following result, by conditions (16.11) and (16.12).

Proposition 16.32. *Let u be a unipotent element of $\mathrm{O}_{n,\mathrm{v}}(q)$, with parameter (m_1, Ψ) (see Proposition 16.9).*

(i) The centralizer of u in the orthogonal group is contained in the special group if and only if $m_1(2j+1) = 0$ for all $j \in \mathbb{N}$.

(ii) The centralizer of u in the orthogonal group is contained in the kernel of the spinor norm if and only if, for all $j \in \mathbb{N}$,

$$m_1(2j+1) = 0 \text{ or}$$
$$(m_1(2j+1) = 1 \text{ and } (\Psi(j) = \mathbf{1} \text{ if and only if } (-1)^{j(q-1)/2} = 1)).$$

(iii) The centralizer of u in the orthogonal group is contained in the kernel of the product of the spinor norm by the determinant if and only if, for all $j \in \mathbb{N}$,

$$m_1(2j+1) = 0 \text{ or}$$
$$(m_1(2j+1) = 1 \text{ and } (\Psi(j) = \mathbf{1} \text{ if and only if } (-1)^{j(q-1)/2} = -1)).$$

As a consequence, if the centralizer of u is contained in the special group, then it is contained in the derived subgroup.

Recall that if $q \equiv 1$ (mod. 4), then $-\mathbf{v} = \mathbf{v}$ for all Witt symbols $\mathbf{v}$, and if $q \equiv 3$ (mod. 4), then $-\mathbf{1} = \mathbf{d}$. So Proposition 16.32 introduces a new condition on (m, Ψ_+, Ψ_-) that we formulate as follows: let $(\epsilon, \mathbf{v}) \in \{-, +\} \times \{\mathbf{1}, \mathbf{d}\}$

$(R(\epsilon, \mathbf{v}))$

$\forall j \in \mathbb{N}, \ \big(m(\epsilon 1, 2j+1) = 1 \text{ and } \Psi_\epsilon(j) = (-1)^j \mathbf{v}\big) \text{ or } m(\epsilon 1, 2j+1) = 0.$

Hence $R(+, \mathbf{1})$ is the condition in Proposition 16.32 (ii) and $R(+, \mathbf{d})$ is the condition in Proposition 16.32 (iii). Using Propositions 16.30 and 16.10 the following can be proved.

Proposition 16.33. *Let $(\epsilon, \mathbf{v}) \in \{-, +\} \times \{\mathbf{1}, \mathbf{d}\}$, let $\mathbf{v}' \in \{\mathbf{0}, \mathbf{w}\}$ and $x \in \mathrm{SO}_{2n,\mathbf{v}'}(q)$ be such that $(\epsilon 1)x$ is unipotent. Assume the parameter of the conjugacy class of s satisfies $R(\epsilon, \mathbf{v})$. Then x is in the kernel of the spinor norm and one has $\mathbf{v}' = \mathbf{w}$ if and only if $(q-1)n \notin 4\mathbb{N}$.*

When $g \in \Omega_{n,\mathbf{v}}(q)$, the number of conjugacy classes of $\Omega_{n,\mathbf{v}}(q)$ contained in the conjugacy class of g in $\mathrm{O}_{n,\mathbf{v}}(q)$ is $|\mathrm{O} : \mathrm{C_O}(g)\Omega|$. Using the structure of the centralizer of a semi-simple element and Propositions 16.31 and 16.32, one may compute $\mathrm{O}/\mathrm{C_O}(g)\Omega$ and obtain the following.

Proposition 16.34. *Let (m, Ψ_+, Ψ_-) be the parameter of a conjugacy class C of $\mathrm{O}_{n,\mathbf{v}}(q)$ and let (μ, ψ_+, ψ_-) be deduced from (m, Ψ_+, Ψ_-) by (16.11). Assume that (μ, ψ_+, ψ_-) satisfies the condition of Proposition 16.30, hence $C \subset \Omega_{n,\mathbf{v}}(q)$. The class C is the union of*

- *four conjugacy classes of $\Omega_{n,\mathbf{v}}(q)$ if and only if $m(f, 2j+1) = 0$ for all $f \in \mathcal{F}$ and all $j \in \mathbb{N}$,*
- *two or four conjugacy classes of $\Omega_{2n,\mathbf{v}}(q)$ if and only if at least one of the following two conditions hold*

(i) for all $(f, j) \in \mathcal{F}_0 \times \mathbb{N}$, $m(f, (2j+1)) = 0$ *and there exists* $\mathbf{v} \in \{\mathbf{1}, \mathbf{d}\}$ *such that* $R(+, \mathbf{v})$ *and* $R(-, \mathbf{v})$ *hold*
(ii) $m(1, 2j+1) = m(-1, 2j+1) = 0$ *for all* $j \in \mathbb{N}$,

- *one conjugacy class of* $\Omega_{2n,\mathbf{v}}(q)$ *in other cases.*

By Proposition 16.30 and formula (16.11), the parameter (m, Ψ_+, Ψ_-) of a conjugacy class of $\mathrm{O}_{2n,\mathbf{v}}(q)$ contained in $\Omega_{2n,\mathbf{v}}(q)$ has to satisfy the relation

$$\upsilon_- + \sum_{\{(f,\tilde{f}) | \sigma(f) \neq 1\}} \sum_j jm(f, j) \in 2\mathbb{N},$$

where υ_- is deduced from $\psi_- = \Sigma_j \Psi(-1, j)$ and $\mu(-1) = \Sigma_j m(-1, j) j$, as in Proposition 16.30.

By Proposition 16.29 and definition of H_0 (16.18), one has $H_0 = \prod_{d \geq 1} \mathcal{P}_d{}^{N_{d,-}} \mathcal{P}_{2d}{}^{M_{d,-}}$. Let

(16.35)
$$H_1(t) = \prod_{d \geq 1} (\mathcal{P}_{2d}^-)^{N_{2d}}.$$

The number of parameters with $\mu(1) = \mu(-1) = 0$ of conjugacy classes contained in a group Ω is given by the generating function $F_0^{(\Sigma)}[\Omega] := \frac{1}{2}(F_0^{(\Sigma)} + F_0^{(\Sigma)} H_1 H_0^{-1})$.

Furthermore we have by (16.18), (16.35), (16.2), (16.18) again and (16.20):

$$\begin{aligned}\left(\mathcal{P}_2 F_0^{(\Sigma)} H_1 H_0^{-1}\right)(t) &= H_1(t) H_0(t) H_0(t^2)^{-1} = \Pi_d (\mathcal{P}_{2d} \mathcal{P}_{2d}^- (\mathcal{P}_{4d})^{-1})^{N_{2d}} \\ &= \Pi_d \left(\mathcal{P}_{8d}{}^{-1} \mathcal{P}_{4d}{}^2\right)^{N_{2d}} = H_0(t^2)^2 H_0(t^4)^{-1} = \mathcal{P}_4 F_0^{(\Sigma)}(t^2) = \mathcal{P}_2 F_0^{(\Delta)}(t).\end{aligned}$$

Hence

(16.36)
$$F_0^{(\Sigma)}[\Omega] = \tfrac{1}{2}\left(F_0^{(\Sigma)} + F_0^{(\Delta)}\right).$$

We now want to make a distinction between the two Witt types of forms, at least for classes of elements without the eigenvalue 1 or -1. By Proposition 16.8, (16.11) and Proposition 16.30, a parameter with $\mu(1) = \mu(-1) = 0$ defines a conjugacy class contained in $\Omega_{2n,\mathbf{0}}(q)$ if and only if

$$\begin{aligned}\sum_{\sigma(f)=1, f=\tilde{f}} m(f, 2j+1) &\equiv \sum_{\sigma(f)=-1, f=\tilde{f}} m(f, 2j+1) \\ &\equiv \sum_{\sigma(f)=-1, f\neq\tilde{f}} m(f, 2j+1) \pmod{2}.\end{aligned}$$

and a conjugacy class in $\Omega_{2n,\mathbf{w}}(q)$ if and only if

$$\sum_{\sigma(f)=1, f=\tilde{f}} m(f,2j+1) \not\equiv \sum_{\sigma(f)=-1, f=\tilde{f}} m(f,2j+1)$$
$$\equiv \sum_{\sigma(f)=-1, f\neq\tilde{f}} m(f,2j+1) \pmod{2}.$$

Let

$$H_{0,-}(t) = \prod_{d\geq 1} {\mathcal{P}_{2d}}^{N_{2d,-}}, \quad H_{0,+}(t) = \prod_{d\geq 1} {\mathcal{P}_{2d}}^{N_{2d,+}} = \prod_{d\geq 1} {\mathcal{P}_{2d}}^{M_{d,-}}$$

hence $H_{0,-}H_{0,+} = H_0$ and ${\mathcal{P}_2}^{\eta} H_{0,+}(t) = H_{0,-}(t)$ with $2\eta = 1 + (-1)^{(q+1)/2}$ (see (16.18), Proposition 16.29 and Lemma 16.5). Define $H_{1,\pm}$ from $H_{0,\pm}$ as H_1 from H_0 (16.35): $H_{1,\epsilon}(t) = \prod_{d\geq 1}(\mathcal{P}_d^-)^{N_{d,\epsilon}}$. Let $L(t) = \Pi_{d\geq 1}{\mathcal{P}_{2d}}^{M_{d,+}}$, so that $F_0^{(\Sigma)}(t) = H_0(t)H_{0,+}(t)L(t)$. By substitution in (16.18), one has $H_{0,-} = \mathcal{P}_2 H_0(t^2)L$.

The generating series for the number of parameters m that satisfy the first two congruences is

$$((H_{0,+} + H_{1,+})^2(H_{0,-} + H_{1,-}) + (H_{0,+} - H_{1,+})^2(H_{0,-} - H_{1,-}))L/8,$$

i.e.

$$\big((H_{0,+}^2 + H_{1,+}^2)H_{0,-} + 2H_{0,+}H_{1,+}H_{1,-}\big)L/4.$$

Put $h_0 = H_{0,+}$, $h_1 = H_{1,+}$.

Assume $q \equiv 1 \pmod{4}$.

One has $h_0 = H_{0,-}$, hence $h_0^2 = H_0$, and $h_1 = H_{1,-}$. The last sum becomes $\frac{1}{4}(h_0^3 + 3h_0h_1^2)L$. But one has $F_0^{(\Sigma)} = h_0^3 L$ by definition of $H_{0,\pm}$ and $F_0^{(\Sigma)} = {\mathcal{P}_2}^{-1} h_0^4 h_0(t^2)^{-2}$ by (16.18). Hence $\mathcal{P}_2 h_0 h_1^2 L = h_0^2 h_1^2 h_0(t^2)^{-2} = h_0(t^2)^4 h_0(t^4)^{-2} = \mathcal{P}_4 F_0^{(\Sigma)}(t^2) = \mathcal{P}_2 F_0^{(\Delta)}(t)$, using (16.2) and (16.20). So the number of parameters with $\mu(1) = \mu(-1) = 0$ of O-conjugacy classes contained in $\Omega_{2n,\mathbf{0}}(q)$ is given by

(16.37) $$F_0[\Omega_{\mathbf{0}}] = \tfrac{1}{4}\big(F_0^{(\Sigma)} + 3F_0^{(\Delta)}\big) \quad (q \equiv 1 \pmod{4}).$$

The generating function for the number of parameters m that satisfy the last two congruences is

$$((H_{0,+} - H_{1,+})(H_{0,-} + H_{1,-})(H_{0,+} + H_{1,+})$$
$$+ (H_{0,+} + H_{1,+})(H_{0,-} - H_{1,-})(H_{0,+} - H_{1,+}))L/8,$$

i.e.

$$\tfrac{1}{4}H_{0,-}\big(H_{0,+}^2 - H_{1,+}^2\big)L.$$

So, when $q \equiv 1$ (mod. 4), the last sum can be written as $\frac{1}{4}(h_0^3 - h_0 h_1^2)L$, hence the number of parameters with $\mu(1) = \mu(-1) = 0$ of O-conjugacy classes contained in $\Omega_{2n,\mathrm{w}}(q)$ is given by

(16.38) $$F_0[\Omega_{\mathrm{w}}] = \tfrac{1}{4}\big(F_0^{(\Sigma)} - F_0^{(\Delta)}\big) \quad (q \equiv 1 \text{ (mod. 4)}).$$

(compare with (16.37) and (16.36)!).

Assume now $q \equiv 3$ (mod. 4). A similar computation, with

$$\mathcal{R} := \mathcal{P}_2{}^2\mathcal{P}_4{}^{-3}\mathcal{P}_8$$

so that $\mathcal{R}(t) = \mathcal{P}_2(\mathcal{P}_2^-)^{-1}$ by (16.2), gives

(16.39) $$F_0[\Omega_0] = \tfrac{1}{4}\big(F_0^{(\Sigma)} + (2+\mathcal{R})F_0^{(\Delta)}\big) \quad (q \equiv 3 \text{ (mod. 4)})$$

and

(16.40) $$F_0[\Omega_{\mathrm{w}}] = \tfrac{1}{4}\big(F_0^{(\Sigma)} - \mathcal{R}F_0^{(\Delta)}\big) \quad (q \equiv 3 \text{ (mod. 4)})$$

16.4. $\mathrm{Spin}_{2n}(\mathbf{F})$

The algebraic simply connected groups of types $\mathbf{D}_m$ or $\mathbf{C}_m$ are spinor groups $\mathrm{Spin}(2m)$ or $\mathrm{Spin}(2m+1)$, central extensions of the special orthogonal groups (see the beginning of §16.3). If the dimension n of V is even, hence SO of type $\mathbf{D}_{n/2}$, then the center of $\mathrm{Spin}(V)$ has order 4, and has exponent 4 when $n/2$ is odd, exponent 2 when $n/2$ is even.

Proposition 16.41. *Let (μ, ψ_+, ψ_-) be the parameter of a conjugacy class of semi-simple elements of $\mathrm{O}_{2n,\mathrm{v}}(q)$ contained in the derived subgroup $\Omega_{2n,\mathrm{v}}(q)$. The parameter defines exactly*

- *one conjugacy class of $\mathrm{Spin}_{2n,\mathrm{v}}(q)$ if $\mu(1)\mu(-1) \neq 0$,*
- *four conjugacy classes if $\mu(1) = \mu(-1) = 0$,*
- *two classes in the other cases.*

Proof. Let t be a semi-simple element of the algebraic Spin group. We have seen in the proof of Proposition 16.25 that t is conjugate to $(-e)t$ if and only if $C_{\mathrm{SO}}(\pi(t))$ is not connected. Assume now that $t \in \mathrm{Spin}_{2n,\mathrm{v}}(q)$. Examination of the centralizer of $\pi(t)$ in SO and in $\Omega_{2n,\mathrm{v}}(q)$ shows that non-connexity is equivalent to the existence of eigenvalues 1 and -1, and conjugacy of t and $(-e)t$ in the algebraic group is equivalent to conjugacy in the group over $\mathbb{F}_q$. Now Proposition 16.34 gives our claim. □

Proposition 16.42. *Let (m, Ψ_+, Ψ_-) be the parameter of a conjugacy class C of the orthogonal group contained in $\Omega_{2n,\mathrm{v}}(q)$. The number of conjugacy* *classes of* $\mathrm{Spin}_{2n,\mathbf{v}}(q)$ *contained in* $\pi^{-1}(C)$ *is*

- *8 if and only if* $m(f, 2j+1) = 0$ *for all* $f \in \mathcal{F}$ *and all* $j \in \mathbb{N}$,
- *4 or 8 if and only if*
  (i) $m(1, 2j+1) = m(-1, 2j+1) = 0$ *for all* $j \in \mathbb{N}$ *or*
  (ii) $m(f, 2j+1) = m(-\epsilon 1, 2j+1) = 0$ *and* $R(\epsilon, \mathbf{v})$ *holds for all* $(f, j) \in \mathcal{F}_0 \times \mathbb{N}$ *and some* $(\epsilon, \mathbf{v}) \in \{-, +\} \times \{\mathbf{d}, \mathbf{1}\}$,
- *2 or 4 or 8 if and only if*
  (iii) $m(\epsilon, 2j+1) = 0$ *for all* $j \in \mathbb{N}$ *and some* $\epsilon \in \{-1, 1\}$ *or*
  (iv) $m(f, 2j+1) = 0$ *for all* $(f, j) \in \mathcal{F}_0 \times \mathbb{N}$ *and there exist* $\mathbf{v}_+, \mathbf{v}_- \in \{\mathbf{d}, \mathbf{1}\}$ *such that* $R(+, \mathbf{v}_+)$ *and* $R(-, \mathbf{v}_-)$ *hold,*
- *1 in other cases.*

Proof. Let $g \in \mathrm{Spin}_{2n,\mathbf{v}}(q)$, in the $\mathrm{O}_{2n,\mathbf{v}}(q)$-conjugacy class C with parameter (m, Ψ_+, Ψ_-), and let $g = tv$ be its decomposition into semi-simple component t and unipotent component v. $\pi^{-1}(C \cap \Omega_{2n,\mathbf{v}}(q))$ is one conjugacy class of the spin group if and only if g is conjugate to $(-e)g$, or, equivalently, t is conjugate to $(-e)t$ in the centralizer of v. By the proof of Proposition 16.41, one knows that $yty^{-1} = (-e)t$ is equivalent to $\pi(y) \in (\mathrm{C}(\pi(t)) \setminus \mathrm{C}^\circ(\pi(t))$ — centralizers in the algebraic special group. Any $z \in \mathrm{C}(\pi(t)) \cap \mathrm{O}_{2n,\mathbf{v}}(q)$ decomposes in a product $z = z_+ z_- z_0$, where z_+, z_- and z_0 act respectively on eigenspaces $V_1(\pi(t))$, $V_{-1}(\pi(t))$ and on $(V_1(\pi(t)) + V_{-1}(\pi(t)))^\perp$. One sees that $z_0 \in \mathrm{SO}$ and the conjugacy condition is equivalent to: $z_+ \notin \mathrm{SO}$ and $z_- \notin \mathrm{SO}$. It is satisfied by some $z \in \Omega_{2n,\mathbf{v}}(q)$ if and only if there exists some triple (j_+, j_-, j_0) of integers and $f \in \mathcal{F}_0$ such that $m(1, 2j_+ + 1)m(-1, 2j_- + 1)m(f, 2j_0 + 1) \neq 0$ — and then by Proposition 16.31 eventually z_0 is not in the kernel of the spinor norm — or only $m(1, 2j_+ + 1)m(-1, 2j_- + 1) \neq 0$ but without the two conditions $R(+, \mathbf{v}_+)$ and $R(-, \mathbf{v}_-)$ where $\{\mathbf{v}_+, \mathbf{v}_-\} = \{\mathbf{1}, \mathbf{d}\}$ (see Proposition 16.33).

For other parameters that define classes of $\Omega_{2n,\mathbf{v}}(q)$, each class is the image of two conjugacy classes of the spin group. Now Proposition 16.42 follows from Proposition 16.34. □

We now compute the various generating functions for parameters in view of Proposition 16.42.

Assume $q \equiv 1$ (mod. 4).

The generating function $S_8(t)$ that gives the number of parameters that satisfy the first condition of Proposition 16.42 is $\mathcal{P}_4{}^2 F_0^{(\Sigma)}(t^2)$

$$S_8 = \mathcal{P}_2 \mathcal{P}_4 F_0^{(\Delta)}$$

(see (16.20)). All these classes are contained in $\mathrm{Spin}_{\mathbf{0}}$ (Proposition 16.10).

The generating function that gives the number of parameters of conjugacy classes in the groups $\mathrm{Spin}_{2n,\mathbf{0}}(q)$ that satisfy condition (i) is

$$S_4 = \mathcal{P}_4{}^2 F_0[\Omega_{\mathbf{0}}]$$

(on x_1x_{-1} the spinor norm is trivial and $\psi_+ = \psi_- = \mathbf{0}$).

To compute the number of parameters that satisfy condition (ii) we use the equality $\Pi_{j\in\mathbb{N}}(1+t^{2j+1}) = \mathcal{P}_1\mathcal{P}_2{}^{-2}\mathcal{P}_4$. The number of parameters that satisfy (ii) but not (i) is given by the series $4(\mathcal{P}_1\mathcal{P}_2{}^{-2}\mathcal{P}_4{}^3 - \mathcal{P}_4{}^2)F_0^{(\Sigma)}(t^2)$. The even part of that series is by (16.20)

$$S_4'(t) = \left(2(\mathcal{P}_1 + \mathcal{P}_1^-)\mathcal{P}_2{}^{-1}\mathcal{P}_4{}^2 - 4\mathcal{P}_2\mathcal{P}_4\right) F_0^{(\Delta)}.$$

The generating function that gives the number of parameters for conjugacy classes in the groups $\mathrm{Spin}_{2n,\mathbf{0}}(q)$ and such that $m(-1,2j+1)=0$ for all $j\in\mathbb{N}$ is a sum of $\mathcal{P}_4(\mathcal{P}_2{}^2\mathcal{G}(t^4)+\mathcal{P}_4)F_0[\Omega_{\mathbf{0}}]/2$ (parameters such that $\psi_+ = \mathbf{0}$; recall the condition $\mu(\pm 1)\in 2\mathbb{N}$ from Proposition 16.23) and $\mathcal{P}_4(\mathcal{P}_2{}^2\mathcal{G}(t^4)-\mathcal{P}_4)F_0[\Omega_{\mathbf{w}}]/2$ (parameters such that $\psi_+ = \mathbf{w}$). The sum is $\mathcal{P}_2{}^2\mathcal{P}_4\mathcal{G}(t^4)(F_0^{(\Sigma)}+F_0^{(\Delta)})/4 + \mathcal{P}_4{}^2(F_0[\Omega_{\mathbf{0}}]-F_0[\Omega_{\mathbf{w}}])/2$. By (16.37) and (16.38) we obtain $\mathcal{P}_2{}^2\mathcal{P}_4\mathcal{G}(t^4)(F_0^{(\Sigma)}+F_0^{(\Delta)})/4 + \mathcal{P}_4{}^2F_0^{(\Delta)}/4$.

Consider now the condition $m(1,2j+1)=0$ for all $j\in\mathbb{N}$ but $m(-1,2j+1)\neq 0$ at least for one j. By Proposition 16.8 and §16.3, especially Proposition 16.30, the number of such parameters for conjugacy classes in $\Omega_{2n,\mathbf{0}}(q)$ is given by a sum of $\frac{1}{2}\mathcal{P}_4(\mathcal{P}_2{}^2\mathcal{G}(t^4)-\mathcal{P}_4)F_0[\Omega_{\mathbf{0}}]$ (parameters such that $\psi_- = \mathbf{0}$) and $\frac{1}{2}\mathcal{P}_4(\mathcal{P}_2{}^2\mathcal{G}(t^4)-\mathcal{P}_4)((F_0^{(\Sigma)}-F_0^{(\Delta)})/2 - F_0[\Omega_{\mathbf{w}}])$ (parameters such that $\psi_- = \mathbf{w}$). Hence from (16.37) and (16.38) we obtain $(\mathcal{P}_2{}^2\mathcal{P}_4\mathcal{G}(t^4)-\mathcal{P}_4{}^2)(F_0^{(\Sigma)}+F_0^{(\Delta)})/4$.

The generating function that gives the number of parameters satisfying condition (iii), to define two or four or eight conjugacy classes in the groups $\mathrm{Spin}_{2n,\mathbf{0}}(q)$, is therefore

$$S_2(t) = \mathcal{P}_2{}^2\mathcal{P}_4\mathcal{G}(t^4)\left(F_0^{(\Sigma)}+F_0^{(\Delta)}\right)/2 - \mathcal{P}_4{}^2\left(F_0^{(\Sigma)}-F_0^{(\Delta)}\right)/4.$$

Condition (iv) in Proposition 16.42, excluding the preceding one, selects the series $\left((\mathcal{P}_1+\mathcal{P}_1^-)\mathcal{P}_2{}^{-2}\mathcal{P}_4{}^2 - 2\mathcal{P}_4\right)^2 F_0^{(\Sigma)}(t^2)$. Note that, from (16.2), the definition of $\mathcal{G}$ and (16.3), $\mathcal{P}_1{}^2 + (\mathcal{P}_1^-)^2 = ((\mathcal{P}_1^-)^{-2}+\mathcal{P}_1{}^{-2})\mathcal{P}_2{}^6\mathcal{P}_4{}^{-2} = (\mathcal{G}(t)+\mathcal{G}(-t))\mathcal{P}_2{}^5\mathcal{P}_4{}^{-2} = 2\mathcal{G}(t^4)\mathcal{P}_2{}^5\mathcal{P}_4{}^{-2}$. With (16.20) we get

$$S_2'(t) = \left(2\mathcal{P}_2{}^2\mathcal{P}_4\mathcal{G}(t^4) + 2\mathcal{P}_4{}^2 - 4(\mathcal{P}_1+\mathcal{P}_1^-)\mathcal{P}_2{}^{-1}\mathcal{P}_4{}^2 + 4\mathcal{P}_2\mathcal{P}_4\right) F_0^{(\Delta)}.$$

The total number of parameters is given by a sum:

$$(\mathcal{P}_4+\mathcal{P}_2{}^2\mathcal{G}(t^4))^2F_0[\Omega_0]/4, \text{ for } (\psi_+,\psi_-)=(\mathbf{0},\mathbf{0}),$$
$$(\mathcal{P}_2{}^4\mathcal{G}(t^4)^2-\mathcal{P}_4{}^2)F_0[\Omega_{\mathbf{w}}]/4, \text{ for } (\psi_+,\psi_-)=(\mathbf{w},\mathbf{0}),$$
$$(\mathcal{P}_2{}^4\mathcal{G}(t^4)^2-\mathcal{P}_4{}^2)((F_0^{(\Sigma)}-F_0^{(\Delta)})/2-F_0[\Omega_{\mathbf{w}}])/4, \text{ for } (\psi_+,\psi_-)=(\mathbf{0},\mathbf{w}),$$
$$(\mathcal{P}_2{}^2\mathcal{G}(t^4)-\mathcal{P}_4)^2((F_0^{(\Sigma)}+F_0^{(\Delta)})/2-F_0[\Omega_0])/4, \text{ for } (\psi_+,\psi_-)=(\mathbf{w},\mathbf{w}).$$

The sum is formally $\mathcal{P}_2{}^4\mathcal{G}(t^4)^2F_0^{(\Sigma)}/4+\mathcal{P}_2{}^2\mathcal{P}_4\mathcal{G}(t^4)(4F_0[\Omega_0]-F_0^{(\Sigma)}-F_0^{(\Delta)})/4+\mathcal{P}_4{}^2F_0^{(\Delta)}/4$. From (16.37) and (16.38) we get

$$S_1(t)=\tfrac{1}{4}\mathcal{P}_2{}^4\mathcal{G}(t^4)^2F_0^{(\Sigma)}+\tfrac{1}{4}\mathcal{P}_4\big(2\mathcal{P}_2{}^2\mathcal{G}(t^4)+\mathcal{P}_4\big)F_0^{(\Delta)}.$$

Finally, the generating function that gives the number of conjugacy classes in the groups $\mathrm{Spin}_{\mathbf{0}}(q)$ is $S_1+S_2+S_2'+2S_4+2S_4'+4S_8$:

(16.43)
$$S(t)=\tfrac{1}{4}(\mathcal{G}(-t^2)+\mathcal{G}(t^4))^2\mathcal{P}_2{}^4F_0^{(\Sigma)}$$
$$+(4\mathcal{G}(-t^2)^2+3\mathcal{G}(-t^2)\mathcal{G}(t^4))\mathcal{P}_2{}^4F_0^{(\Delta)}.$$

Assume now that $q\equiv 3$ (mod. 4). By Proposition 16.30 and formulae (16.39), (16.40), the formulae are different short of S_8.

$$S_4=\frac{\mathcal{P}_4{}^2}{4}\big(F_0^{(\Sigma)}+(2+\mathcal{R})F_0^{(\Delta)}\big).$$

Let $E(t)$ be the series sum of terms with degree in $4\mathbb{N}$ in $(\mathcal{P}_1+\mathcal{P}_1^-)\mathcal{P}_2{}^{-2}/2$. Condition (ii) without (i) gives

$$S_4'=4\big(E\mathcal{P}_4{}^2-\mathcal{P}_4\big)\mathcal{P}_2F_0^{(\Delta)}.$$

Condition (iii) in Proposition 16.42, with $m(1,2j+1)=0$ for all $j\in\mathbb{N}$ but $m(-1,2j+1)\neq 0$ at least for one j, gives

$$S_2(t)=\mathcal{P}_2{}^2\mathcal{P}_4\mathcal{G}(t^4)F_0^{(\Sigma)}/2+\big(\mathcal{P}_2{}^2\mathcal{P}_4\mathcal{G}(t^4)+\mathcal{P}_4{}^4(\mathcal{P}_4^-)^{-2}\big)F_0^{(\Delta)}/4$$
$$-\mathcal{P}_4{}^2\big(F_0^{(\Sigma)}-F_0^{(\Delta)}\big)/4.$$

The sum for S_1 is modified and from (16.39) and (16.40) we get

$$4S_1(t)=\mathcal{P}_2{}^4\mathcal{G}(t^4)^2F_0^{(\Sigma)}+\big(\mathcal{P}_2{}^2\mathcal{P}_4\mathcal{G}(t^4)+\mathcal{P}_4{}^4(\mathcal{P}_4^-)^{-2}+\mathcal{P}_4{}^2\big)F_0^{(\Delta)}.$$

As final sum, one obtains

(16.44)
$$S(t)=\tfrac{1}{4}(\mathcal{G}(-t^2)+\mathcal{G}(t^4))^2\mathcal{P}_2{}^4F_0^{(\Sigma)}$$
$$+(4\mathcal{G}(-t^2)+3\mathcal{G}(t^4))\mathcal{P}_2{}^4\mathcal{G}(-t^2)F_0^{(\Delta)}$$
$$-((\mathcal{P}_2-\mathcal{P}_2^-)\mathcal{P}_4\mathcal{G}(t^4)+(\mathcal{P}_2{}^{-1}-(\mathcal{P}_2^-)^{-1})\mathcal{P}_4{}^2\mathcal{P}_2F_0^{(\Delta)}.$$

Proposition 16.45. *The number of conjugacy classes of the group* $\mathrm{Spin}_{2n,\mathbf{0}}(q)$ *is the coefficient of* t^{2n} *in the series given by formula (16.43) if* $q\equiv 1$ (mod. 4) *and by formula (16.44) if* $q \equiv 3$ (mod. 4). *These numbers are polynomials in* q *whose coefficients are independent of* q *modulo* 4 *when* n *is even.*

16.5. Non-semi-simple groups, conformal groups

In this section we assume the following.

Hypothesis and notation 16.46. *Let* $\mathbf{H}$ *be an algebraic connected group defined over* $\mathbb{F}_q$, *with Frobenius* F *and a connected center. Let* $Z := \mathrm{Z}([\mathbf{H}, \mathbf{H}]) = \mathrm{Z}(\mathbf{H}) \cap [\mathbf{H}, \mathbf{H}]$. *Elements of* $\mathbf{H}^F$ *are products* zg, *with* $z \in \mathrm{Z}(\mathbf{H})$, $g \in [\mathbf{H}, \mathbf{H}]$ *and* $z^{-1}F(z) = gF(g)^{-1} \in Z$. *Let*

$$G := \{g \in [\mathbf{H}, \mathbf{H}] \mid gF(g)^{-1} \in Z\}, \quad \rho\colon G \to Z, \quad \rho(g) = gF(g)^{-1}.$$

Clearly G *is a subgroup of* $[\mathbf{H}, \mathbf{H}]$, ρ *is a morphism with kernel* $[\mathbf{H}, \mathbf{H}]^F$ *and, by Lang's theorem in the connected group* $[\mathbf{H}, \mathbf{H}]$, *one has*

$$G/[\mathbf{H}, \mathbf{H}]^F \cong Z.$$

Proposition 16.47. $\mathbf{H}^F$ *is the disjoint union of* $\mathbf{H}^F$*-invariant sets* $\mathrm{Z}(\mathbf{H})^F zC$ *where* C *is a* G*-conjugacy class and* $z \in \mathrm{Z}(\mathbf{H})$. *The number of* $\mathbf{H}^F$*-conjugacy classes contained in each such set is* $|\mathrm{Z}(\mathbf{H})^F/Z^F|$, *the number of* G*-conjugacy classes contained in* $Z^F C$.

Proof. One has an isomorphism (Proposition 8.1)

$$\mathbf{H}^F \cong \{(z, g) \in \mathrm{Z}(\mathbf{H}) \times G \mid \rho(z^{-1}) = \rho(g)\}/Z$$

and a morphism $\rho\colon \mathbf{H}^F \to Z/[Z, F]$ defined by the map $(z, g)Z \mapsto \rho(g)[Z, F]$, whose kernel is $\mathrm{Z}(\mathbf{H})^F[\mathbf{H}, \mathbf{H}]^F \cong (\mathrm{Z}(\mathbf{H})^F \times [\mathbf{H}, \mathbf{H}]^F)Z/Z$ and whose image is $Z/[Z, F]$ by Lang's theorem.

Clearly $\rho(g)$ is an invariant of the conjugacy class of zg. Furthermore, as $\mathbf{H}^F \subset \mathrm{Z}(\mathbf{H})G \subset \mathrm{Z}(\mathbf{H})\mathbf{H}^F$, the conjugacy class of zg in $\mathbf{H}^F$ is the set zC, where C is the conjugacy class of g in G. Let $D(g)$ be the set of $y \in \mathrm{Z}(\mathbf{H})$ such that g is conjugate to yg in G. If $y \in D(g)$, then $y \in [\mathbf{H}, \mathbf{H}]$, hence $y \in Z$. Furthermore, $zyg \in \mathbf{H}^F$ because $\mathbf{H}^F$ is normal in $\mathrm{Z}(\mathbf{H})G$, hence $D(g)$ is a subgroup of Z^F. One has $\mathrm{Z}(\mathbf{H})zC \cap \mathbf{H}^F = \mathrm{Z}(\mathbf{H})^F zC$, and $\mathrm{Z}(\mathbf{H})^F zC$ is the union of exactly $|\mathrm{Z}(\mathbf{H})^F/R(g)|$ conjugacy classes of $\mathbf{H}^F$. Finally note that $|Z^F/D(g)|$ is the number of conjugacy classes of G contained in $Z^F C$ and that for two G-conjugacy classes C and C', z being defined modulo Z^F, the equalities $\mathrm{Z}(\mathbf{H})^F zC = \mathrm{Z}(\mathbf{H})^F zC'$ and $Z^F C = Z^F C'$ are equivalent. □

We have to study conjugacy classes of G. We get the following result immediately.

Proposition 16.48. *Let $g \in G$ and let K be any F-stable subgroup of $\mathbf{H}$ containing g. Let $x \in K$.*

(i) Let $L := \mathrm{Lan}_K^{-1}(\mathrm{C}_K(g))$, then $g^L = g^K \cap \mathbf{H}^F g$. Assume $\mathbf{H}^F \subseteq K$. Then

$$([K, F] \cap \mathrm{C}_K(g)) \colon [\mathrm{C}_K(g), F]).|g^K \cap \mathbf{H}^F g| = |\mathbf{H}^F \colon \mathrm{C}_{\mathbf{H}^F}(g)|.$$

(ii) There exists $t \in \mathbf{H}^F$ such that $xgx^{-1} = tgt^{-1}$ if and only if $x^{-1}F(x) \in [\mathrm{C}_{\mathbf{H}}(g), F]$.

Hence if $g \in \mathbf{H}^F$ and $\rho(g') = z \in Z$, then $\mathrm{C}_G(g) \cap \mathbf{H}^F g'$ is not empty if and only if $z^{-1} \in [\mathrm{C}_{\mathbf{H}}(g), F]$.

Proof. Let $x \in K$. One verifies that $xgx^{-1}F(xgx^{-1})^{-1} = gF(g)^{-1}$ if and only if $x^{-1}F(x) \in \mathrm{C}_K(g)$ and (i) follows. The relation $xgx^{-1} = tgt^{-1}$ with $t \in K^F$ is equivalent to $x^{-1}t \in \mathrm{C}_K(g)$ with $x^{-1}F(x) = x^{-1}tF(t^{-1})F(x)$, hence the equivalence in (ii). □

We apply the preceding analysis to conformal and Clifford groups.

Let $\mathrm{SO}(V)$ be a special orthogonal group with respect to one of the forms described in §16.2. A linear endomorphism g of V is said to be conformal if there exists a scalar λ_g such that $\langle gv, gv' \rangle = \lambda_g \langle v, v' \rangle$ for all v, v' in V. Then $\mathrm{CSO}(V)$ is the group of conformal g with $\det g = (\lambda_g)^m$. Its center is connected.

When $[\mathbf{H}, \mathbf{H}]$ is a special orthogonal group (resp. a spin group) we index ρ, G with 1 (resp. 0). Hence, with $\mathbf{H} = \mathrm{CSO}_{2n}(\mathbf{F})$, the form and the group being defined on $\mathbb{F}_q$, with a Frobenius endomorphism F,

$$G_1 := \{x \in \mathrm{SO}_{2n}(\mathbf{F}) \mid F(x) \in \{-x, x\}\}, \quad \rho_1 \colon G_1 \to \, < -1 >$$

and ρ_1 has kernel $\mathrm{SO}_{2n,\mathbf{v}}(q) = \mathrm{SO}_{2n}(\mathbf{F})^F$. Put $\mathrm{CSO}_{2n,\mathbf{v}}(q) = \mathrm{CSO}_{2n}(\mathbf{F})^F$.

Write SO for $\mathrm{SO}_{2n}(\mathbf{F})$, CSO for $\mathrm{CSO}_{2n}(\mathbf{F})$, and $\mathrm{SO}(q)$ for $\mathrm{SO}^F = \mathrm{SO}_{2n,\mathbf{v}}(q)$, etc. Clearly $\mathrm{SO}(q) \subset G_1 \subset \mathrm{SO}(q^2)$. Note that $\mathrm{SO}(q^2)$ here is defined with respect to a split form on $\mathbb{F}_{q^2}$.

Proposition 16.49. *Let $\mathcal{F}'$, $\mathcal{F}'_0$ be defined from $(\mathbf{F}, F^2)$, as $\mathcal{F}$ and $\mathcal{F}_0$ are defined from $(\mathbf{F}, F)$.*

Let $t = zs$ be a semi-simple element of $\mathrm{CSO}_{2n,\mathbf{v}}(q)$, where $(z, s) \in \mathrm{Z(CSO)}_{\mathbf{v}} \times G_1$ (see Proposition 16.47). For $s \in \mathrm{SO}_{2n,\mathbf{v}}(q)$, let (μ, ψ_+, ψ_-) be its parameter in $\mathrm{O}_{2n,\mathbf{v}}(q)$ (see Proposition 16.10). For $F(s) = -s$, let (μ', ψ'_+, ψ'_-) be its parameter in $\mathrm{O}_{2n,\mathbf{0}}(q^2)$. We shall say that the parameter and the G_1-conjugacy class of t are associated.

When $s \in \mathrm{SO}_{2n,\mathbf{v}}(q)$, (μ, ψ_+, ψ_-) is associated to exactly one semi-simple conjugacy class of G_1, hence $(q-1)/2$ semi-simple conjugacy classes of

$\mathrm{CSO}_{2n,\mathbf{v}}(q)$, *if and only if* $\mu(1) \neq 0$ *or* $\mu(-1) \neq 0$, *and to exactly two semi-simple conjugacy classes of* G_1, *hence* $(q-1)$ *semi-simple conjugacy classes of* $\mathrm{CSO}_{2n,\mathbf{v}}(q)$, *if and only if* $\mu(1) = \mu(-1) = 0$.

A semi-simple conjugacy class of $\mathrm{SO}_{2n,\mathbf{0}}(q^2)$ *contains an element of* $(G_1 \setminus \mathrm{SO}_{2n,\mathbf{v}}(q))$ *if and only if its parameter* (μ', ψ'_+, ψ'_-) *verifies*

$$\mu'(f') = \mu'(F(\bar{f}')) = \mu'(\tilde{f}') = \mu'(F(\bar{\tilde{f}}')), \quad \psi'_+ = \psi'_- \in \{\mathbf{0}, \mathbf{w}\}$$

for all $f' \in \mathcal{F}'$. *and, in case* $q \equiv 1 \pmod 4$,

$$\psi'_+ + \left(\sum_{\tilde{f}' \in \{f', F(\bar{f}')\}} \mu'(f') \right) \mathbf{w} = \mathbf{v}$$

or, in case $q \equiv 3 \pmod 4$,

$$\psi'_+ + \left(\sum_{\tilde{f}' = f'} \mu'(f') \right) \mathbf{w} = \mathbf{v}$$

where the sums run over a set of orbits under the group $< (f' \mapsto \tilde{f}'), (f' \mapsto F(\bar{f}') >$ *acting on* $\mathcal{F}'_0$.

Then (μ', ψ'_+, ψ'_-) *is associated to exactly two* G_1*-conjugacy classes, hence* $(q-1)$ *conjugacy classes of* $\mathrm{CSO}_{2n,\mathbf{v}}(q)$.

Proof. Let C be the G_1-conjugacy class of s.

If $F(s) = s$, then any direct component C_f $(f \in \mathcal{F})$ of $\mathrm{C}_{\mathbf{H}}(s)$ is F-stable, so that $-1 \in [F, \mathrm{C}_{\mathbf{H}}(s)]$, which implies $\mathrm{C}(s) \cap (G_1 \setminus \mathrm{SO}(q)) \neq \emptyset$ by Proposition 16.48 (ii). Hence the conjugacy class of s in $\mathrm{SO}(q)$ is a conjugacy class of G_1 and the parameter of s is well defined by C. By Propositions 16.23 and 16.47, $\mathrm{Z}(\mathbf{H})^F C$ is the union of $(q-1)$ (resp. $(q-1)/2$) conjugacy classes of $\mathbf{H}^F$ if and only if $\mu(1) = \mu(-1) = 0$ (resp. $\mu(1) \neq 0$ or $\mu(-1) \neq 0$).

Assume now $F(s) = -s$. One has $\mathrm{C}_{\mathbf{H}}(s) \subset [F, \mathbf{H}]$ and $[F, \mathrm{C}_{\mathbf{H}}(s)] = \mathrm{C}^\circ_{\mathbf{H}}(s)$. By Proposition 16.48(i) (with $K = \mathrm{SO}$) we get that the SO-conjugacy class of g intersects $(G_1 \setminus \mathrm{SO}_{2n,\mathbf{v}}(q))$ in one or two G_1-conjugacy classes, and in exactly two conjugacy classes of G_1 if and only if $\mathrm{C}_{\mathrm{SO}}(s) \neq \mathrm{C}^\circ_{\mathrm{SO}}(s)$. The last condition is equivalent to "1 and -1 are eigenvalues of s." Note that s and $-s$ are conjugate in $\mathrm{O}_\mathbf{0}(q^2)$, hence $\mu'(1) = \mu'(-1)$.

If $\mu'(1) = 0$, then the class of s in SO contains a single class of $\mathrm{SO}_\mathbf{0}(q^2)$ and the function μ', giving multiplicities of eigenvalues, is decided by $\mathbf{v}$. The parameter of s in $\mathrm{O}_\mathbf{0}(q^2)$ gives two classes in a group G_1 for only one value of $\mathbf{v}$.

If $\mu'(1) \neq 0$, then the class of s in SO contains two classes of $\mathrm{SO}_\mathbf{0}(q^2)$, with parameters (μ', ψ'_+, ψ'_-) that differ on ψ'_+. We shall see that each one intersects

G_1 in one class because $\mathbf{v} \in \{\mathbf{0}, \mathbf{w}\}$ is fixed. Hence the conjugacy class of g in $\mathrm{SO}(q^2)$ intersects G_1 in one class.

Applying Proposition 16.47, one gets the first part of the proposition.

Let $(f' \mapsto \bar{f}')$ be induced on $\mathcal{F}'$ by $(\alpha \mapsto -\alpha)$ in $\mathbf{F}$. Let (μ', ψ'_+, ψ'_-) be the parameter of some semi-simple s in $\mathrm{O}_\mathbf{0}(q^2)$. Let f' be an irreducible polynomial on $\mathbb{F}_{q^2}$ such that $\mu'(f') \neq 0$. Let $V_\alpha = V_\alpha(s)$ be the eigenspace of s for some $\alpha \in f'$; one has $F(V_\alpha) = V_{F(\alpha)}(F(t))$. The relation $F(s) = -s$ is equivalent to $F(V_\alpha(s_1)) = V_{F(-\alpha)}(s_1)$ for any α, any such f', and then $\mu'(F(\bar{f}')) = \mu'(f')$. When $V_1 \neq \{0\}$, then $F(V_1) = V_{-1}$, hence $\psi'_- = F(\psi'_+) = \psi'_+$. Under the preceding conditions on (μ, ψ'_+, ψ'_-) any decomposition of the representation space in an orthogonal sum $\bigoplus_\alpha V_\alpha$, where $\dim(V_\alpha) = \mu'(f')$ for $\alpha \in f'$, $F(V_\alpha) = V_{F(-\alpha)}$ for all α, and $V_\alpha \oplus V_{\alpha^{-1}}$ is an hyperbolic sum of two isotropic spaces if $\alpha \neq \alpha^{-1}$, is the decomposition into eigenspaces of some semi-simple element s of $\mathrm{SO}(q^2)$, in a class of parameter $(\mu', \psi'_\pm)$ in $\mathrm{O}(q^2)$, and such that $F(s)$ is conjugate to $-s$ in $\mathrm{O}_\mathbf{0}(q^2)$. The conjugacy class of s in $\mathrm{O}_\mathbf{0}(q^2)$ is one class in $\mathrm{SO}_\mathbf{0}(q^2)$ when $\mu'(1) \neq 0$. Then by Lang's theorem in the connected group SO, the class of s in SO intersects G_1 in some s_1. Clearly the parameters of conjugacy classes of s and s_1 in the group $\mathrm{O}_\mathbf{1}(q^2)$ can differ only on $\psi'_+ = \psi'_-$. But G_1 has a semi-simple element of parameter $(\mu', \mathbf{0}, \mathbf{0})$ if and only if it has an element of parameter $(\mu', \mathbf{w}, \mathbf{w})$. These elements, with different parameters in $\mathrm{O}_\mathbf{0}(q^2)$, are conjugate under the algebraic special group. Hence the two G_1-conjugacy classes contained in the SO-conjugacy class are different $\mathrm{SO}_\mathbf{0}(q^2)$-conjugacy classes with different parameters in $\mathrm{O}_\mathbf{0}(q^2)$.

Let (m', Ψ'_+, Ψ'_-) be a parameter of a conjugacy class of $\mathrm{SO}_{2n,\mathbf{0}}(q^2)$ contained in G_1. Assuming that the bilinear form is defined on $\mathbb{F}_q$, we have to compute its Witt type on $\mathbb{F}_q$, or equivalently, as the dimension is known, its discriminant.

Assume first that $\mu'(1) = \mu'(-1) = 0$, where μ' is part of the parameter of the semi-simple component conjugacy class. As the possible parameters (m', Ψ'_+, Ψ'_-) such that (16.11) holds depend only on μ', we may consider only semi-simple classes. Thus let $\mu' : \mathcal{F}'_0 \to \mathbb{N}$ with $\mu'(f') = \mu'(F(\bar{f}')) = \mu'(\tilde{f}') = \mu'(F(\bar{\tilde{f}}'))$. Let x be in the class and, for $\mu'(f') \neq 0$, let $V_{f'} = \Sigma_{\alpha \in f'} V_\alpha$ be the sum of eigenspaces V_α for $\alpha \in f'$ of x.

Consider the restriction of the form to the subspace $V' := V_{f'} + V_{F(\bar{f}')} + V_{\tilde{f}'} + V_{F(\bar{\tilde{f}}')}$. V' is defined on $\mathbb{F}_q$ because $F(x) = -x$ implies $F(V_{f'}) = V_{F(\bar{f}')}$ and $F(V_{\tilde{f}'}) = V_{F(\bar{\tilde{f}}')}$. We are interested in the Witt type of $V'(\mathbb{F}_q)$.

(1) Assume $\tilde{f}' \notin \{f', F(\bar{f}')\}$.

The subspaces $V_{f'} + V_{F(\bar{f}')}$ and $V_{\tilde{f}'} + V_{F(\bar{\tilde{f}}')}$ are defined on $\mathbb{F}_q$, are totally isotropic and in duality by the form. As the form is defined on $\mathbb{F}_q$, the dual of a basis of $\mathbb{F}_q$ of one of these spaces (i.e. where each element of the basis is fixed by F) is defined on $\mathbb{F}_q$ too. Hence $V'(\mathbb{F}_q)$ is of Witt type $\mathbf{0}$.

(2) Assume now $\tilde{f}' = f'$ or $\tilde{f}' = F(\bar{f}')$ (the two equalities cannot be satisfied simultaneously because $f' \in \mathcal{F}_0'$). One has $V' = V_{f'} + F(V_{f'})$. Fix a basis of $V_{f'}(\mathbb{F}_{q^2})$ and take its image by F to define a basis $(e_i)_i$ of $V'(\mathbb{F}_{q^2})$. Let D be the matrix of scalar products of the e_i, the Gramian matrix. The matrix with respect to $(e_i)_i$ of a basis of $V'(\mathbb{F}_q)$ may be written with four square blocks $M = \begin{pmatrix} A & F(B) \\ F(A) & B \end{pmatrix}$ where $F^2(A) = A$ and $F^2(B) = B$. Its Gramian matrix is $MD\,{}^tM$ and has determinant $\mathrm{Det}(M)^2\mathrm{Det}(D)$. One has $F(\mathrm{Det}(M)) = \mathrm{Det}(F(M)) = (-1)^m\mathrm{Det}(M)$ where m is the size of A. Hence $\mathrm{Det}(M) \in \mathbb{F}_q$ if and only if m, the dimension of $V_{f'}$, is even. One has $m = \mu'(f')|f'|$.

(a) $\tilde{f}' = f'$.

Then $|f'|$ and m are even, $\mathrm{Det}(M)^2$ is a square in $\mathbb{F}_q$, and $V' = V_{f'} + F(V_{f'})$ is an orthogonal sum. As the dimension of V' is a multiple of 4, the Witt type of $V'(\mathbb{F}_q)$ is $\mathbf{0}$ if and only if $\mathrm{Det}(D)$ is a square in $\mathbb{F}_q$. As F exchanges the two orthogonal subspaces, D is a diagonal of two square blocks D_0 and $F(D_0)$, hence $\mathrm{Det}(D) \in \mathbb{F}_q$. $\mathrm{Det}(D)$ is a square in $\mathbb{F}_q$, if and only if $\mathrm{Det}(D_0)$ is a square in $\mathbb{F}_{q^2}$, and $\mathrm{Det}(D_0)$ is a square if and only if the Witt type of the $\mathbb{F}_{q^2}$-form on $V_{f'}$ is $\mathbf{0}$ (one has $q^2 \equiv 1 \pmod{4}$), hence if and only if $\mu'(f')$ is even (see Proposition 16.8).

So the Witt type of $V'(\mathbb{F}_q)$ is $\mathbf{0}$ if and only if $\mu'(f')$ is even, whatever q modulo 4 is.

(b) $\tilde{f}' = F(\bar{f}')$.

Then $|f'|$ is odd and $F(V_{f'}) = V_{\bar{f}'}$, $V'(\mathbb{F}_{q^2})$ is an hyperbolic sum of Witt type $\mathbf{0}$. One has $F^{|f'|}(V_a) = V_{a^{-1}}$. On a suitable basis of eigenvectors in $V_{f'}$ and its transform by $F^{|f'|}$, the matrix D has determinant $(-1)^m$.

If $\mu'(f')$ is odd, then m is odd, so that $\mathrm{Det}(M) \notin \mathbb{F}_q$, hence $\mathrm{Det}(M)^2$ is not a square in $\mathbb{F}_q$. As the size of M is even but not divisible by 4, the Witt type is $\mathbf{w}$ if $q \equiv 1 \pmod{4}$ and $\mathbf{0}$ if $q \equiv 3 \pmod{4}$.

If $\mu'(f')$ is even, then m is even, $\mathrm{Det}(M^2G)$ is a square in $\mathbb{F}_q$. As the size of M is divisible by 4, the Witt type is $\mathbf{0}$ whatever q modulo 4 is.

Assume now that $\mu'(1) = \mu'(-1) \neq 0$, and $\psi'_+ = \psi'_-$.

We have $F(V_1) = V_{-1}$, $F(V_{-1}) = V_1$, hence we consider the restriction of the form to $V' := V_1 + V_{-1}$, where $V'(\mathbb{F}_{q^2})$ is of Witt type $\mathbf{0}$. As $\mu'(-1)$ is even, the discriminant is a square in $\mathbb{F}_q$ if and only if the discriminant of ψ'_+ is a square in $\mathbb{F}_{q^2}$. As 4 divides $q^2 - 1$ and the dimension of V', we see that the space $V'(\mathbb{F}_q)$ is of Witt type ψ'_+. □

Proposition 16.50. *(i) Let (m, Ψ_+, Ψ_-) be a parameter of the conjugacy class of some $x \in \mathrm{SO}_{2n,\mathbf{v}}(q)$. Then* *(a)* (m, Ψ_+, Ψ_-) *is the unique parameter associated to two conjugacy classes of* G_1 *if and only if* $m(1, 2j+1) = m(-1, 2j+1) = 0$ *for all* $j \in \mathbb{N}$,

(b) (m, Ψ_+, Ψ_-) *is the unique parameter associated to one or two conjugacy classes of* G_1 *if and only if* $m(1, j) \in 2\mathbb{N}$ *and* $m(-1, j) \in 2\mathbb{N}$ *for all* $j > 0$,

(c) if there exists $j \in \mathbb{N}$ *such that* $m(1, 2j+1) \notin 2\mathbb{N}$ *or* $m(-1, 2j+1) \notin 2\mathbb{N}$, *then the class of* x *in* G_1 *is the union of two classes of* $\mathrm{SO}_{2n,\mathbf{v}}(q)$ *with different parameters.*

(ii) A conjugacy class of $\mathrm{SO}_{2n,\mathbf{0}}(q^2)$ *contains an element of* $(G_1 \setminus \mathrm{SO}_{2n,\mathbf{v}}(q))$ *if and only if its parameter* (m', Ψ'_+, Ψ'_-) *in* $\mathrm{O}_{2n,\mathbf{0}}(q^2)$ *satisfies*

$$m'(f', j) = m'(F(\bar{f}'), j), \quad \Psi'_+(j) = \Psi'_-(j)$$

for all $f' \in \mathcal{F}'_0$.

Then the parameter (m', Ψ'_+, Ψ'_-) *is associated to, and is the unique one associated to, exactly two conjugacy classes of* G_1.

Proof. When $x \in \mathrm{SO}_{2n,\mathbf{v}}(q)$ has parameter (m, Ψ_+, Ψ_-), by Proposition 16.48 the condition $\mathrm{C}_{G_1}(x) \not\subset \mathrm{SO}_{2n,\mathbf{v}}(q)$ is equivalent to $-1_V \in [F, \mathrm{C}_{\mathrm{SO}}(x)]$, i.e. $m(1, j), m(-1, j) \in 2\mathbb{N}$ for all $j \in \mathbb{N}^*$. Then the conjugacy class of x in $\mathrm{SO}(q)$ is a conjugacy class of G_1.

When $F(x) = -x$, and u is the unipotent component of x, then $F(u) = u$. Clearly two unipotent elements of $\mathrm{O}_{\mathbf{v}}(q)$ with equal parameters in $\mathrm{O}_{\mathbf{0}}(q^2)$ have equal parameters in $\mathrm{O}_{\mathbf{0}}(q)$. It follows that if x and x' are elements of G_1 with equal semi-simple components and equal parameters in $\mathrm{O}_{2n,\mathbf{0}}(q^2)$, then x and x' are conjugate in G_1.

Now apply Propositions 16.48 and 16.23. □

For our purpose we need the number of parameters of conjugacy classes in G_0, hence in G_1. In finding this, we also obtain the number of conjugacy classes of the conformal orthogonal group.

Proposition 16.51. *The sum for the two Witt types* **0** *and* **w** *of the number of conjugacy classes of the conformal special orthogonal groups in even dimension on* $\mathbb{F}_q$ *is given by the following generating function*

$$\frac{q-1}{8}(\mathcal{G}(t^2) + 3\mathcal{G}(-t^2))^2 \mathcal{P}_2{}^4 F_0^{(\Sigma)}.$$

Proof. The number of parameters that satisfy condition (a) of Proposition 16.50 is given by the generating function $\mathcal{P}_4^2 F_0^{(\Sigma)}$. The condition $m_1(j) \in 2\mathbb{N}$ selects the function $\mathcal{P}_4 \mathcal{P}_8^{-1} \mathcal{P}_4^3 \mathcal{G}(t^2) = \mathcal{P}_2^2 \mathcal{G}(-t^4)$ by (16.13), (16.2), the definition of $\mathcal{G}$

and (16.3). Hence the number of parameters that satisfy condition (b) is given by the generating function $(\mathcal{P}_2{}^4\mathcal{G}(-t^4)^2 + \mathcal{P}_4{}^2)F_0^{(\Sigma)}$. Other SO-conjugacy classes (condition (c)) are fused in G_1 and their number is given by $\mathcal{P}_2{}^4(\mathcal{G}(t^4)^2 - \mathcal{G}(-t^4)^2)F_0^{(\Sigma)}$.

The total number of G_1-conjugacy classes contained in some $\mathrm{SO}_{2n,\mathbf{v}}(q)$ for some $\mathbf{v}$ is given by

$$\tfrac{1}{2}\mathcal{P}_2{}^4(\mathcal{G}(t^4)^2 + \mathcal{G}(t^{-4})^2 + 2\mathcal{G}(t^{-2})^2)F_0^{(\Sigma)}.$$

We now have to enumerate parameters (m', Ψ'_+, Ψ'_-) of classes with a non-empty intersection with $(G_1 \setminus \mathrm{SO}_{2n,\mathbf{v}}(q))$. The conditions are $m'(f', j) = m'(F(\bar{f}'), j) = m'(\tilde{f}', j)$ for all $f' \in \mathcal{F}'$ and $j \in \mathbb{N}^*$, and, $\Psi_+(j) = \Psi_-(j)$ for all $j > 0$.

The number of parameters (m', Ψ'_+, Ψ'_-) such that $m(1, j) = 0$ for all j is given by the function $\prod_{\omega'} \mathcal{P}(t^{|\omega'||f'|})$, where ω' runs over the set of orbits $\{f', F(\bar{f}'), \tilde{f}', F(\bar{\tilde{f}}')\}$ in $\mathcal{F}'_0$ and $f' \in \omega'$. As $f' = \tilde{f}'$ implies $f' \neq F'(\bar{f}')$, the corresponding class is the split group. By Lemma 16.7 (ii), the generating function is $\mathcal{P}_4{}^{-1}\mathcal{P}_2{}^2F_0^{(\Sigma)}$. The number of parameters of classes with only eigenvalues 1 and -1 is given by $\mathcal{P}_4{}^2\mathcal{G}(t^8)$. For all classes we obtain $\mathcal{G}(t^8)\mathcal{P}_4\mathcal{P}_2{}^2F_0^{(\Sigma)} = \mathcal{P}_2{}^4\mathcal{G}(-t^2)\mathcal{G}(t^8)F_0^{(\Sigma)}$. Each parameter corresponds to two G_1-conjugacy classes in exactly one of the two Witt types of forms.

Now we have $2(\mathcal{G}(t^4)^2 + \mathcal{G}(-t^4)^2) = (\mathcal{G}(t^2) + \mathcal{G}(-t^2))^2$ and $2\mathcal{G}(t^8) = \mathcal{G}(t^2) + \mathcal{G}(-t^2)$, hence $\frac{1}{2}(\mathcal{G}(t^4)^2 + \mathcal{G}(-t^4)^2) + \mathcal{G}(-t^2)^2 + 2\mathcal{G}(-t^2)\mathcal{G}(t^8) = \frac{1}{4}(\mathcal{G}(t^2) + 3\mathcal{G}(-t^2))^2$.

The generating function for the sum of numbers of conjugacy classes in the two groups G_1 is therefore

$$\left(\frac{\mathcal{G}(t^2)}{2} + \frac{3\mathcal{G}(-t^2)}{2}\right)^2 \mathcal{P}_2{}^4F_0^{(\Sigma)}(t).$$

In the conformal group $\mathbf{H} := \mathrm{CSO}_{2n,\mathbf{v}}(q)$, one has $\mathrm{Z}(\mathbf{H})^F = q - 1$. An elementary argument (see Proposition 16.47) gives the claim of the proposition. □

16.6. Group with connected center and derived group $\mathrm{Spin}_{2n}(\mathrm{F})$; conjugacy classes

In this section we assume Hypothesis 16.46 where $[\mathbf{H}, \mathbf{H}]$ is the Spin group as described at the beginning of §16.4, with $\pi\colon \mathrm{Spin}(V) \to \mathrm{SO}(V)$.

First we enumerate conjugacy classes of the group $\mathbf{H}^F$. Let

$$z_0 \in \pi^{-1}(-1_V), \quad Z_0 = \mathrm{Z}([\mathbf{H}, \mathbf{H}]) = \{e, -e, z_0, (-e)z_0\}.$$

If the rank n of $[\mathbf{H},\mathbf{H}]$ is odd, then z_0 is of order 4 and $z_0^2 = -e$. If the rank of $[\mathbf{H},\mathbf{H}]$ is even, then z_0 is of order 2. Recall that $-1_V \in \Omega_{2n,\mathbf{0}}(q)$ — equivalently $z_0 \in \mathrm{Spin}_{2n,\mathbf{0}}(q)$ — (resp. $-1_V \in \Omega_{2n,\mathbf{w}}(q)$ and $z_0 \in \mathrm{Spin}_{2n,\mathbf{w}}(q)$) if and only if 4 divides $(q^n - 1)$ (resp. 4 divides $(q^n + 1)$). When $z_0 \notin \mathrm{Spin}_{2n,\mathbf{0}}(q)$, $F(z_0) = (-e)z_0$. Define

$$G_0 := \{x \in \mathrm{Spin(V)} \mid F(x) \in xZ_0\},\ \ \rho_0\colon G_0 \to Z_0,\ \ \rho_0(g) = gF(g)^{-1}.$$

One has $\pi(G_0) = G_1$, G_1 defined as earlier. From Proposition 16.48 we have the following.

Proposition 16.52. *Two semi-simple elements g and g' of G_0 are conjugate in G_0 if and only if $\rho_0(g) = \rho_0(g')$ and g and g' are conjugate in* $\mathbf{H}$.

Proof. The centralizer $\mathrm{C}_{\mathbf{H}}(g)$ is connected because $[\mathbf{H},\mathbf{H}]$ is simply connected (Theorem 13.14). Thus $\mathrm{C}_{\mathbf{H}}(g) = [\mathrm{C}_{\mathbf{H}}(g), F]$. Hence Proposition 16.52 follows from Proposition 16.48. □

Proposition 16.53. *Let $|\mathrm{Z}(\mathbf{H})^F| = |Z^F|N$. Let (μ, ψ_+, ψ_-) or (μ', ψ'_+, ψ'_-) be the parameter of the class of a semi-simple element of G_1, as described in Proposition 16.49. It defines exactly*

- *one conjugacy class of G_0 and N conjugacy classes of $\mathbf{H}^F$ if $\mu(1)\mu(-1) \neq 0$,*
- *four conjugacy classes of G_0 and $4N$ conjugacy classes of $\mathbf{H}^F$ if $\mu(1) = \mu(-1) = 0$ or $\mu'(1) = 0$,*
- *two conjugacy classes of G_0 and $2N$ conjugacy classes of $\mathbf{H}^F$ otherwise.*

Proof. Let g be semi-simple in G_0. Let $D(g)$ be defined as in the proof of Proposition 16.47. Recall that $(-e) \in D(g)$ if and only if 1 and -1 are eigenvalues of $\pi(g)$ (proof of Proposition 16.41). Clearly $D(g) \cap \{z_0, (-e)z_0\} \neq \emptyset$ if and only if $\pi(g)$ and $-\pi(g)$ are conjugate in SO.

The equality $\rho_0(g) = (-e)$ is equivalent to $F(\pi(g)) = \pi(g)$ with $F(g) \neq g$, i.e. $\pi(g) \in \mathrm{SO}(q)$ and $\pi(g) \notin \Omega(q)$ (clearly $\mathrm{SO}(q) \subset \Omega(q^2)$). Hence $\rho_0(g) \in \,<-e>$ is equivalent to $\pi(g) \in \mathrm{SO}_{2n,\mathbf{v}}(q)$.

The relation $\rho_0(g) \in \{z_0, (-e)z_0\}$ is equivalent to $\pi(g)) \in (G_1 \setminus \mathrm{SO}(q))$. The distinction between z_0 and $(-e)z_0$ corresponds to the two classes modulo $\Omega(q)$ contained in $(G_1 \setminus \mathrm{SO}(q))$.

By Proposition 16.52, a semi-simple conjugacy class of $[\mathbf{H},\mathbf{H}]$ (or of $\mathbf{H}$) intersects $\mathrm{Spin}_{2n,\mathbf{v}}(q)$ in at most one conjugacy class of $\mathrm{Spin}_{2n,\mathbf{v}}(q)$. So Propositions 16.49 and 16.52 give our claim. □

Proposition 16.54. *Let $x \in G_0$, and let (m, Ψ_+, Ψ_-) or (m', Ψ'_+, Ψ'_-) be a parameter associated to the conjugacy class of $\pi(x) \in G_1$ (see Proposition 16.50). Then*

(1) (m, Ψ_+, Ψ_-) *is the (unique) parameter associated to exactly four conjugacy classes of* G_0 *if and only if* $m(1, 2j+1) = m(-1, 2j+1) = 0$ *for all* $j \in \mathbb{N}$*;*

(2) (m, Ψ_+, Ψ_-) *is the (unique) parameter associated to two or four conjugacy classes of* G_0 *if and only if* $m(1, 2j+1) = 0$ *and* $m(-1, 2j+1) \in 2\mathbb{N}$ *for all* $j \in \mathbb{N}$*, or* $m(-1, 2j+1) = 0$ *and* $m(1, 2j+1) \in 2\mathbb{N}$ *for all* $j \in \mathbb{N}$*;*

(3) the conjugacy class of x *in* G_0 *is associated to two parameters if and only if* $m(-1, 2j_- + 1) \notin 2\mathbb{N}$ *for some* $j_- \in \mathbb{N}$ *and* $m(1, 2j_+ + 1) \neq 0$ *for some* $j_+ \in \mathbb{N}$ *or* $m(1, 2j_+ + 1) \notin 2\mathbb{N}$ *for some* $j_+ \in \mathbb{N}$ *and* $m(-1, 2j_- + 1) \neq 0$ *for some* $j_- \in \mathbb{N}$*;*

(4) in all other cases (m, Ψ_+, Ψ_-) *is the unique parameter associated to exactly one conjugacy class of* G_0*;*

(5) (m', Ψ'_+, Ψ'_-) *is the (unique) parameter associated to exactly four conjugacy classes of* G_0 *if and only if* $m'(1, 2j+1) = 0$ *for all* $j \in \mathbb{N}$*;*

(6) in all other cases (m', Ψ'_+, Ψ'_-) *is the (unique) parameter associated to exactly two conjugacy classes of* G_0.

Proof. Let s be the semi-simple component of x. Unipotent elements and conjugacy classes of unipotent elements of $C_{G_0}(s)$ and of $\pi(C_{G_0}(s))$ are in one-to-one correspondence. One has $\pi(C_{G_0}(s)) \neq C_{G_1}(\pi(s))$ if and only if 1 and -1 are eigenvalues of s. Let u be a unipotent element of $C_{G_0}(s)$. Then su is conjugate to $(-e)su$ in G_0 if and only if $C_{G_1}(\pi(u)) \cap C_{G_1}(\pi(s)) \not\subset \pi(C_{G_0}(s)) = C^\circ_{\mathbf{H}}(\pi(s)) \cap G_1$. By Proposition 16.32, this fails if and only if the parameter (m, Ψ_+, Ψ_-) or (m', Ψ'_+, Ψ'_-) associated to the class of $\pi(su)$ satisfies $m(1, 2j+1) = 0$ for all $j \in \mathbb{N}$ or $m(-1, 2j+1) = 0$ for all $j \in \mathbb{N}$ (or $m'(1, 2j+1) = 0$ for all $j \in \mathbb{N}$). Thus Proposition 16.54 follows from Proposition 16.50. □

From now on we consider only the split case $\mathbf{v} = \mathbf{0}$.

Proposition 16.55. *Let* $\mathbf{H}$ *be an algebraic group of type* $\mathbf{D}_n$*, defined over* $\mathbb{F}_q$*, with a connected center and a simply connected derived group, so that* $[\mathbf{H}, \mathbf{H}]^F$ *is the split rational spin group* $\mathrm{Spin}_{2n,\mathbf{0}}(q)$*. Let* $N = |\mathrm{Z}(\mathbf{H})^F|$*. The number of conjugacy classes of* $\mathbf{H}^F$ *is the coefficient of* t^{2n} *in the series*

$$\frac{N}{4}\left[\mathcal{P}_2{}^4\left(\frac{(\mathcal{G}(t^2) + 3\mathcal{G}(-t^2))^2}{8} + \frac{\mathcal{G}(-t^2)(\mathcal{G}(t^4) + 3\mathcal{G}(-t^4))}{2}\right)F_0^{(\Sigma)} + 4\mathcal{P}_4{}^2 F_0^{(\Delta)}\right].$$

Proof. Using Proposition 16.47, we consider conjugacy classes of G_0 under the conditions of Proposition 16.54. By Propositions 16.10 and 16.23, the total number of parameters (m, Ψ_+, Ψ_-) of conjugacy classes of $\mathrm{SO}_{2n,\mathbf{0}}(q)$ is given by the function

$$S(t) = \tfrac{1}{2}\big(\mathcal{P}_2{}^4\mathcal{G}(t^4)^2 F_0^{(\Sigma)} + \mathcal{P}_4{}^2 F_0^{(\Delta)}\big).$$

The parameters of conjugacy classes of G_0 with two different parameters, given by (3) in Proposition 16.54, are equally distributed between the two Witt types of groups. The number of such parameters is given by $S_{1/2}(t) := F_0^{(\Sigma)}\mathcal{P}_2{}^4((\mathcal{G}(t^4) - \mathcal{G}(-t^4))(\mathcal{G}(t^4) - \mathcal{G}(-t^2)) - (\mathcal{G}(t^4) - \mathcal{G}(-t^4))^2/2$, i.e.

$$2S_{1/2}(t) = \mathcal{P}_2{}^4(\mathcal{G}(t^4) - \mathcal{G}(-t^4))(\mathcal{G}(t^4) + \mathcal{G}(-t^4) - 2\mathcal{G}(-t^2))F_0^{(\Sigma)}.$$

The number of parameters that give two or four conjugacy classes, condition (2), in the split group is given by $S_2(t) := (\mathcal{P}_2{}^2\mathcal{G}(-t^4)\mathcal{P}_4 - \mathcal{P}_4{}^2)F_0^{(\Sigma)} + \mathcal{P}_4{}^2(F_0^{(\Sigma)} + F_0^{(\Delta)})/2$, i.e.

$$2S_2(t) = \mathcal{P}_2{}^4\mathcal{G}(-t^2)(2\mathcal{G}(-t^4) - \mathcal{G}(-t^2))F_0^{(\Sigma)} + \mathcal{P}_4{}^2 F_0^{(\Delta)}.$$

Among them the number of parameters that give four conjugacy classes, condition (1), is given by the function

$$S_4(t) = \mathcal{P}_4{}^2\big(F_0^{(\Sigma)} + F_0^{(\Delta)}\big)/2.$$

The condition $m'(1, 2j+1) = 0$ for all $j \in \mathbb{N}$, along with condition (b) in Proposition 16.50, selects parameters (m', Ψ'_+, Ψ'_-) whose number is given by the function

$$2S'_4(t) = \mathcal{P}_8\mathcal{P}_2{}^2\mathcal{P}_4{}^{-1} F_0^{(\Sigma)} + \mathcal{P}_4{}^2 F_0^{(\Delta)}.$$

The total number of parameters (m', Ψ'_+, Ψ'_-) for conjugacy classes of G_1 or G_0 is given by the function

$$2S'(t) = \mathcal{P}_2{}^2\mathcal{P}_4\mathcal{G}(t^8)F_0^{(\Sigma)} + \mathcal{P}_4{}^2 F_0^{(\Delta)}.$$

The function that gives the number of conjugacy classes of G_0 is the linear combination $S - \frac{1}{2}S_{1/2} + S_2 + 2S_4 + 2S' + 2S'_4$. After simplifications we obtain, up to the factor $N/4$, the series given in Proposition 16.55. □

16.7. Group with connected center and derived group $\mathrm{Spin}_{2n}(\mathbf{F})$; Jordan decomposition of characters

To compute the numbers y_j (Theorem 16.1), we use Lusztig's Jordan decomposition of irreducible characters. Let $\mathbf{H}^*$ be in duality with $\mathbf{H}$ over $\mathbb{F}_q$. One may assume that $\mathbf{H}$ and $\mathbf{H}^*$ are isomorphic over $\mathbb{F}_q$. Let $K = \mathrm{Z}(\mathbf{H})^F\mathbf{G}^F$ and $A = \mathbf{H}^F/K$. By Proposition 8.1, A is isomorphic to the group of F-cofixed points of $\mathrm{Z}(\mathbf{H}) \cap \mathbf{G}$, hence to $\mathrm{Z}(\mathrm{Spin})^F = \mathrm{Z}(\mathrm{Spin}_{2n,\mathbf{v}}(q))$. Thus A is of order 4 if and only if $-1_V \in \Omega_{2n,\mathbf{v}}(q)$, i.e. $\mathbf{v} = \mathbf{0}$ or ($q \equiv 1$ (mod. 4) and n is odd). A is non-cyclic if and only if n is even and $\mathbf{v} = \mathbf{0}$, and this is the case we consider. We may use the duality between $\mathcal{A} := \mathrm{Irr}(A)$ and the center $\mathcal{A}^*$ of $[\mathbf{H}^*, \mathbf{H}^*]^F \cong \mathrm{Spin}_{2n,\mathbf{v}}(q)$; denote this correspondence $\mathcal{A}^* \to \mathcal{A}$ by $z \mapsto \lambda_z$ (8.19). Here $z \in \langle -e, z_0 \rangle$ where $\pi(-e) = 1, \pi(z_0) = -1$, notation of §16.4.

One has the decomposition $\mathrm{Irr}(\mathbf{H}^F) = \dot{\bigcup}_{(s)}\mathcal{E}(\mathbf{H}^F, s)$ in Lusztig series, where (s) runs over the set of F-stable semi-simple conjugacy classes of $\mathbf{H}^*$. As $[\mathbf{H}^*, \mathbf{H}^*]$ is simply connected any F-stable conjugacy class of $\mathbf{H}^*$ (resp. $[\mathbf{H}^*, \mathbf{H}^*]$) contains exactly one conjugacy class of the finite group $(\mathbf{H}^*)^F$ (resp. $[\mathbf{H}^*, \mathbf{H}^*]^F$) (Theorem 13.14). Let $\psi_s : \mathcal{E}(\mathbf{H}^F, s) \to \mathcal{E}(\mathrm{C}_{\mathbf{H}^*}(s)^F, 1)$ be a Jordan decomposition of $\mathcal{E}(\mathbf{H}^F, s)$ (Theorem 15.8). If $\eta \in \mathcal{E}(\mathbf{H}^F, s)$ and $\lambda_z \in I_\eta$, then there exists $g \in (\mathbf{H}^*)^F$ with $gsg^{-1} = zs$. Then g acts on $\mathcal{E}(\mathrm{C}_{\mathbf{H}^*}(s)^F, 1)$ (see Proposition 15.9) and one has

(16.56) $$\psi_{sz}(\lambda_z \otimes \chi) = g.\psi_s(\chi) \quad (\mu \in \mathcal{E}(\mathrm{C}_{\mathbf{H}^*}(s)^F, 1),\ gsg^{-1} = zs)$$

by Theorem 15.8 (see Exercise 15.4).

We need generating series for the number of irreducible unipotent characters. The elements of $\mathcal{E}(\mathbf{G}^F, 1)$ are parametrized by "symbols" (see [Cart85] §16.8) and by the last assertion of Theorem 15.8 the map $\mu \mapsto g.\mu$ in (16.56) is the identity on symbols.

Denote by $\mathbf{\Phi}(n, \mathbf{v})$ the number of unipotent irreducible characters of an adjoint group of type $\mathbf{D}_n$ if $\mathbf{v} = \mathbf{0}$, and ${}^2\mathbf{D}_n$ if $\mathbf{v} = \mathbf{w}$. Define the application p_2 by

$$\mathcal{P}(t)^2 = \sum_{n\in\mathbb{N}} p_2(n)t^n.$$

Lusztig proved ([Lu77] 3.3)

$$\mathbf{\Phi}(n, \mathbf{0}) = \begin{cases} \sum_{d>0} p_2(n-4d^2) + \frac{1}{2}p_2(n) + \frac{3}{2}p(n/2) & \text{if } n \text{ is even,} \\ \sum_{d>0} p_2(n-4d^2) + \frac{1}{2}p_2(n) & \text{if } n \text{ is odd.} \end{cases}$$

$$\mathbf{\Phi}(n, \mathbf{w}) = \sum_{d>0} p_2(n-(2d-1)^2),$$

where $p(a) = 0$ when $a \notin \mathbb{N}$.

Let $\tilde{\mathbf{\Phi}}(n, \mathbf{0}) = \mathbf{\Phi}(n, \mathbf{0}) - 2p(n/2)$ for even n, and $\tilde{\mathbf{\Phi}}(n, \mathbf{v}) = \mathbf{\Phi}(n, \mathbf{v})$, $\tilde{\mathbf{\Phi}}(n) = \mathbf{\Phi}(n)$ in other cases; $\tilde{\mathbf{\Phi}}$ is the number of so-called classes of non-degenerate symbols.

Let

$$\mathbf{\Phi}_{\mathbf{0}}(t) := \sum_{m\geq 1} \mathbf{\Phi}(m, \mathbf{0})t^{2m}, \quad \mathbf{\Phi}_{\mathbf{w}}(t) := \sum_{m\geq 1} \mathbf{\Phi}(m, \mathbf{w})t^{2m}.$$

Similarly let

$$\tilde{\mathbf{\Phi}}_{\mathbf{0}}(t) := \sum_{m\geq 1} \tilde{\mathbf{\Phi}}(m,\mathbf{0})t^{2m}, \quad \tilde{\mathbf{\Phi}}_{\mathbf{w}}(t) := \sum_{m\geq 1} \tilde{\mathbf{\Phi}}(m,\mathbf{w})t^{2m}.$$

We have

(16.57) $$2+\mathbf{\Phi}_{\mathbf{0}}(t)+\mathbf{\Phi}_{\mathbf{w}}(t) = \tfrac{1}{2}\mathcal{P}_2{}^2(\mathcal{G}(t^2)+3\mathcal{G}(-t^2))$$

and

(16.58) $$2+\mathbf{\Phi}_{\mathbf{0}}(t)-\mathbf{\Phi}_{\mathbf{w}}(t) = 2\mathcal{P}_2{}^2\mathcal{G}(-t^2) = 2\mathcal{P}_4,$$

hence $\mathbf{\Phi}_{\mathbf{w}}(t) = \mathcal{P}_2{}^2(\mathcal{G}(t^2)-\mathcal{G}(-t^2))/4$. As $\tilde{\mathbf{\Phi}}_{\mathbf{0}}(t)+\tilde{\mathbf{\Phi}}_{\mathbf{w}}(t) = \mathcal{P}_2{}^2(\mathcal{G}(t^2)-\mathcal{G}(-t^2))/2$ and $\tilde{\mathbf{\Phi}}_{\mathbf{w}} = \mathbf{\Phi}_{\mathbf{w}}$, we have also

(16.59) $$\tilde{\mathbf{\Phi}}_{\mathbf{0}}(t) = \tilde{\mathbf{\Phi}}_{\mathbf{w}}(t) = \frac{1}{4}\mathcal{P}_2{}^2(\mathcal{G}(t^2)-\mathcal{G}(-t^2)).$$

16.8. Last computation, y_1, y_2, y_4

We shall have to consider all even dimensions $2n$, so we will write eventually $y_i(2n)$ $(i = 1, 2, 4, n \geq 1)$.

Consider a semi-simple element $s \in (\mathbf{H}^*)^F$ such that $s^g = z_1 s$ with $z_1 \in \mathrm{Z}(\mathbf{H}^*)$ for some $g \in \mathbf{H}^*$. As $\mathbf{H}^* = \mathrm{Z}(\mathbf{H}^*)[\mathbf{H}^*, \mathbf{H}^*]$, we may assume $g \in [\mathbf{H}^*, \mathbf{H}^*]$. But $s = zs_0$ with $z \in \mathrm{Z}(\mathbf{H}^*)$ and $s_0 \in [\mathbf{H}^*, \mathbf{H}^*]$. Therefore $s_0^g = z_1 s_0$, $z_1 = s_0^g s_0^{-1} \in [\mathbf{H}^*, \mathbf{H}^*]$. As z_1 is of order prime to the characteristic, for some power a of p one has $s_0^{g^a} = z_1 s_0$ and g^a is semi-simple, so assume g is semi-simple. But $F(s) = s$ implies $z_1^{-1}F(z_1) = s_0 F(s_0)^{-1} \in \mathrm{Z}([\mathbf{H}^*, \mathbf{H}^*])$. There exists $t \in \mathrm{C}_{[\mathbf{H}^*,\mathbf{H}^*]}(g)$ with $t^{-1}F(t) = s_0F(s_0)^{-1}$ (Lang's theorem in the connected group $\mathrm{C}_{[\mathbf{H}^*,\mathbf{H}^*]}(g)$). Then $F(ts_0) = ts_0$ and $(ts_0)^g = z_1(ts_0)$. The series $\mathcal{E}(\mathbf{H}^F, zs_0)$ and $\mathcal{E}(\mathbf{H}^F, ts_0)$ are in one-to-one correspondence with $\mathcal{E}(\mathrm{C}_{\mathbf{H}^{*F}}(s_0), 1)$, with action of g or of λ_{z_1} in (16.56).

Now if $z \in \mathrm{Z}(\mathbf{H}^*)$, then the conjugacy class of zts_0 is F-stable if and only if ts_0 is a conjugate of $z^{-1}F(z)ts_0$, i.e. $z \in \mathcal{L}^{-1}(\mathcal{A}'(ts_0))$, where $\mathcal{A}'(ts_0)$ is the stabilizer of the conjugacy class of ts_0 under translation by elements of $\mathrm{Z}(\mathbf{H}^*)$ (clearly, $\mathcal{A}'(ts_0) \subset \mathcal{A}^*$).

It follows that, for any subgroup A' of $\mathrm{Z}([\mathbf{H}^*, \mathbf{H}^*])$, the number of semi-simple conjugacy classes C of $\mathbf{H}^*$ such that $C = F(C) = A'C$ is $|\mathrm{Z}(\mathbf{H}^*)^F/\mathrm{Z}([\mathbf{H}^*, \mathbf{H}^*])^F|$ times the similar number for $[\mathbf{H}^*, \mathbf{H}^*]$. We may restrict the computation of the numbers y_j to the series defined by $s \in [\mathbf{H}^*, \mathbf{H}^*]^F$. Put $N = |\mathrm{Z}(\mathbf{H}^*)^F/\mathrm{Z}([\mathbf{H}^*, \mathbf{H}^*])^F|$.

The condition $\lambda_{-e} \otimes \chi = \chi$ for $\chi \in \mathrm{Irr}(\mathbf{H}^F)$. A semi-simple element s is conjugate to $-es$ in $\mathbf{H}^F$ if and only if the centralizer of $\pi(s)$ in the special group is not connected, hence if and only if $\pi(s)$ has eigenvalues 1 and -1 (Proposition 16.25). Then the conjugacy is induced by an element g whose image under π belongs to $\mathrm{SO}(V_1 \perp V_{-1}) \setminus \mathrm{SO}(V_1) \times \mathrm{SO}(V_{-1})$. By Proposition 15.17, g exchanges irreducible representations, associated with degenerate symbols, and fixes the others.

Let (μ, ψ_+, ψ_-) or (μ', ψ'_+, ψ'_-) be the parameter of the conjugacy class of $\pi(s) \in G_1$; see Proposition 16.49.

In case $\pi(s) \in \mathrm{SO}_{2n}(q)$, the number of fixed irreducible characters in $\mathcal{E}(\mathbf{H}^F, s)$ is $\tilde{\mathbf{\Phi}}(\mu(1)/2, \psi_+)\tilde{\mathbf{\Phi}}(\mu(-1)/2, \psi_-)\prod_{(f,\tilde{f})\subset\mathcal{F}_0} p(\mu(f))$ and is given by the generating function

$$\Sigma_{(\mu,\psi)}\tilde{\Phi}(\mu(1)/2, \psi_+)\tilde{\mathbf{\Phi}}(\mu(-1)/2, \psi_-)t^{\mu(1)+\mu(-1)}\Pi_{(f,\tilde{f})}p(\mu(f))t^{\delta(f)\mu(f)},$$

where $\delta(f) = |f|$ if $f = \tilde{f}$, or $\delta(f) = 2|f|$. The product on the right is $F_0^{(\Sigma)}(t)$, defined in §16.2. The classes of semi-simple elements with $\mu(-1)\mu(1) \neq 0$ are equally distributed between the two Witt types of forms (see Proposition 16.53). The generating function for the number of such characters is therefore $N((\tilde{\mathbf{\Phi}}_{\mathbf{0}} + \tilde{\mathbf{\Phi}}_{\mathbf{w}})^2)F_0^{(\Sigma)}/2$.

Consider now all the series $\mathcal{E}(\mathbf{H}^F, s)$ with $s \in \mathrm{Z}(\mathbf{H}^*)s_0$ and $\pi(s_0) \in (G_1 \setminus \mathrm{SO}_{2n,\mathbf{0}}(q))$. The number of fixed irreducible characters in $\mathcal{E}(\mathbf{H}^F, s)$ is $\tilde{\mathbf{\Phi}}(\mu'(1)/2, \psi'_+)\prod p(\mu'(f'))$ where the product on the right runs over the set of $(f', \tilde{f}', F(\tilde{f}'), F(\tilde{f}')) \subset \mathcal{F}_0$. By Lemma 16.7(ii), it is equal to $\mathcal{P}_2{}^2\mathcal{P}_4{}^{-1}F_0^{(\Sigma)}$. The generating function for the number of such characters is therefore $N(\tilde{\mathbf{\Phi}}_{\mathbf{0}}(t^2) + \tilde{\mathbf{\Phi}}_{\mathbf{w}}(t^2))\mathcal{P}_2\mathcal{P}_4{}^{-1}F_0^{(\Sigma)}$. By (16.59), the generating function $Y_{-e}(t) = 1 + \sum_{n\geq 1} y_{-e}(2n)t^{2n}$ for the number y_e of irreducible characters whose stabilizer in $\mathcal{A}$ contains λ_{-e} is

(16.60)
$$Y_{-e}(t) = \frac{N}{8}\mathcal{P}_2{}^4((\mathcal{G}(t^2) - \mathcal{G}(-t^2))^2 + 4\mathcal{G}(-t^2)(\mathcal{G}(t^4) - \mathcal{G}(-t^4)))F^{(\Sigma)}.$$

The contribution of $y_{-e}(2n)$ to $y_2(2n)$ is $y_{-e}(2n) - y_4(2n)$.

The number $y_{(\pm e)z_0,2}(4n)$ of $\chi \in \mathrm{Irr}(\mathbf{H}^F)$ such that $I_\chi = <(\pm e)z_0>$. Assume s is a conjugate of $z_0 s$ or of $(-e)z_0 s$. In the special group $\pi(s)$ and $-\pi(s)$ are conjugate. Let $\lambda = \lambda_z$ and $z \in \{z_0, (-e)z_0\}$.

We prove the following.

Proposition 16.61. *(i) The parameter (μ, ψ_+, ψ_-) defines a semi-simple conjugacy class of $\mathrm{SO}_{2n,\mathbf{v}}(q)$ containing both an element x and $(-1_V)x$ if and*

only if

(a) $\mu(\bar{f}) = \mu(f)$ *for all* $f \in \mathcal{F}$, $\psi_+ = \psi_-$ *and*

(b) n *is even or* $\mu(1) \neq 0$.

(ii) The parameter $(\mu', \psi'_+ = \psi'_-)$ *defines a semi-simple conjugacy class of* $(G_1 \setminus \mathrm{SO}_{2n,\mathbf{0}}(q))$ *containing both an element* x *and* $(-1_V)x$ *if and only if*

(a′) $\mu'(\bar{f}') = \mu'(f')$ *for all* $f' \in \mathcal{F}'$, $\psi'_+ = \psi'_-$ *and*

(b′) if n *is even, then* $\psi'_+ = 0$, *if* n *is odd then* $\mu'(1) \neq 0$ *and* $\psi'_+ = \mathbf{w}$.

Proof. Note first that:

1. we know that when n is odd $z_0^2 = -e$, so that if s is conjugate to $z_0 s$ in G_1, then s is conjugate to $(-e)s$, hence $\pi(s)$ has eigenvalues 1 and -1,

2. with our convention $\psi'_+ = \mathbf{0}$ when $\mu'(1) = 0$, (b′) may be written $\mu'(1) \equiv n \pmod{2}$.

If a semi-simple conjugacy class of the orthogonal group $\mathrm{O}_{2n,\mathbf{v}}(q)$ contains both x and $(-1_V)x$, then its parameter (μ, ψ_+, ψ_-) satisfies (a).

Conversely assume (a) is true for some (m, ψ_+, ψ_-). By Proposition 16.8 there exists a rational orthogonal transformation g of V such that $gxg^{-1} = (-1_V)x$. This equality defines g modulo the centralizer of x. If 1 is an eigenvalue of x, then -1 is an eigenvalue of x, so that the centralizer of x in the orthogonal group is not contained in the special group. Hence there are some g as above in the special group. Assume now that $\mu(1) = \mu(-1) = 0$. Then the centralizer of x is contained in the special group so that all g that satisfy the equality have equal determinant, and this is true for g rational or not. But $gxg^{-1} = (-1_V)x$ is equivalent to $g(V_\alpha) = V_{-\alpha}$ for all eigenvalues α and eigenspaces V_α of x. Let β be such that $\beta^4 = 1$ but $\beta \notin \{1, -1\}$. If $\alpha \notin \{\alpha^{-1}, -\alpha^{-1}\}$, i.e. $\alpha \notin \{-1, 1, \beta, -\beta\}$, then the exchange of $V_\alpha \oplus V_{\alpha^{-1}}$ and $V_{-\alpha} \oplus V_{-\alpha^{-1}}$ may be realized by an element of $\mathrm{SO}(V_\alpha \oplus V_{\alpha^{-1}} \oplus V_{-\alpha} \oplus V_{-\alpha^{-1}})$, where the direct sum of the four distinct eigenspaces is endowed with the restriction of the quadratic form. But if $\alpha \in \{\beta, -\beta\}$, then $g(V_\alpha) = V_{-\alpha}$ and $g(V_{-\alpha}) = V_\alpha$ is realized by some special g only if the dimension $\mu(\beta)$ (or $\mu(\{\beta, -\beta\})$ when $q \equiv 3 \pmod{4}$) of V_β is even. Clearly $n \equiv \mu(\beta) \pmod{2}$. (b) follows.

Applying (a) and (b) in $\mathrm{SO}_{2n,\mathbf{v}}(q^2)$, we obtain that if a conjugacy class of $(G_1 \setminus \mathrm{SO}_{2n,\mathbf{v}}(q))$ contains x and $(-1_V)x$ then its parameter (m', ψ') satisfies $\mu'(\bar{f}') = \mu'(f')$ for all $f' \in \mathcal{F}'$, $\psi'_+ = \psi'_-$ and $\mu'(1) \neq 0$ when n is odd. These conditions imply the existence of $g \in \mathrm{SO}_0(2n)$ such that $gxg^{-1} = (-1_v)x$. Furthermore, as $F(x) = -x$, $g^{-1}F(g) \in \mathrm{C}(x)$. As g is defined modulo $\mathrm{C}_{\mathrm{SO}}(x)$, there exists such a $g \in \mathrm{SO}(q)$ if and only if, for any such g, $g^{-1}F(g) \in [F, \mathrm{C}_{\mathrm{SO}}(x)]$. There is such a g in $\mathrm{SO}(q)$ at least when $\mathrm{C}_{\mathrm{SO}}(g)$ is connected.

Consider now the action of g on the subspace $V_1 \perp V_{-1}$ of dimension $2\mu'(1) \neq 0$. Assume first that $\mu'(1) = n$. If $\psi'_+ = 0$, then $V_1(q^2)$ admits an orthonormal basis $\{e_j\}_{1\le j\le \mu'(1)}$ with respect to the restriction of the form. Then $\{F(e_j)\}_{1\le j\le \mu'(1)}$ is an orthonormal basis of V_{-1} and we may take $g(e_j) = F(e_j)$, $g(F(e_j)) = e_j$ $(1 \le j \le \mu'(1))$ to define the restriction of g to $V_1 \perp V_{-1}$ as an element of $\mathrm{SO}(q)$. If $\psi'_+ = 1$, there exists a basis $\{e_j\}_{1\le j\le \mu'(1)}$ of V_1 such that the value of the form on $\Sigma_j x_j e_j$ $(x_j \in \mathbb{F}_{q^2})$ is $\gamma x_1^2 + \Sigma_{1<j} x_j^2$, where $-\gamma$ is not a square in $\mathbb{F}_{q^2}$. On the basis $\{F(e_j)\}_j$ of V_{-1}, the restriction of the form is $\gamma^q x_1^2 + \Sigma_{1<j} x_j^2$. One sees that a solution in $g \in \mathrm{SO}(q^2)$ for the equalities $g(V_1) = V_{-1}$, $g(V_{-1}) = V_1$ is $g(e_1) = \alpha F(e_1)$, $g(F(e_1)) = \alpha^{-1} e_1$, $g(e_j) = F(e_j)$, $g(F(e_j)) = e_j$ $(1 < j)$ with $\alpha^2\gamma^{q-1} = 1$. Then the restriction of $g^{-1}F(g)$ to V_1 has determinant $\alpha^{-(q+1)} = \gamma^{(q^2-1)/2} = -1$ because γ is not a square in $\mathbb{F}_{q^2}$. In the general case, decompose g into $g_1 g_0$, g_1 and g_0 acting respectively on the spaces $V_1 \perp V_{-1}$ and $(V_1 \perp V_{-1})^\perp$. When n is even, g_0 has to be special, so that g may be special only if g_1 is special, hence $\psi_+ = 0$. When n is odd g_0 has to be non-special, so that g may be special only if $\psi_+ = 1$. □

Let $f \in \mathcal{F}_0$ be such that $\mu(f) \neq 0$.

If $f \neq \bar{f}$, then g^* exchanges the isomorphic components in f and $\bar{f}$ of $\mathrm{C}_{\mathbf{H}^*}(s)^F$; each one has $p(\mu(f))$ unipotent irreducible characters. If $\tilde{f} \neq f = \bar{f}$, then $|f| \in 2\mathbb{N}$, and the action of g^*, up to an element of $\mathrm{C}_{\mathbf{H}^*}(s)^F$, is induced by $F^{|f|/2}$. Any element of $\mathcal{E}(\mathrm{C}_{\mathbf{H}^*}(s)^F_f, 1)$ is fixed, and $|\mathcal{E}(\mathrm{C}_{\mathbf{H}^*}(s)^F_f, 1)| = p(\mu(f))$.

When $(\pi(s)^2 - 1_V)$ is one-to-one, the number of λ_{z_0}-fixed or $\lambda_{(-e)z_0}$-fixed irreducible characters is given by the product $\prod p(\mu(f))$ on the sets $\{f, \tilde{f}, \bar{f}, \bar{\tilde{f}}\}$. By Lemma 16.6, it is the coefficient in degree $2n$ of the series $\mathcal{P}_2 F_0^{(\Sigma)}(t^2)$, or $\mathcal{P}_2{}^2\mathcal{P}_4{}^{-1}F_0^{(\Delta)}$ by (16.20). The even subseries of degrees divisible by 4 is $(\mathcal{P}_2 + \mathcal{P}_2^-)\mathcal{P}_2\mathcal{P}_4^{-1}F_0^{(\Delta)}/2$. By Proposition 16.53, the contribution to y_2 of series $\mathcal{E}(\mathbf{H}^F, s)$ without the eigenvalue 1 or -1 is given by the function $2N(\mathcal{P}_2 + \mathcal{P}_2^-)\mathcal{P}_2\mathcal{P}_4^{-1}F_0^{(\Delta)}$.

Consider now the two components of $\mathrm{C}_{\mathbf{H}^*}(s)^F$ for the eigenvalues 1 and -1, in case $\pi(s)$ has eigenvalues 1 and -1. Then s is conjugate to $(-e)z_0 s$ as well as to $z_0 s$. We may assume that the element $g^* \in (\mathbf{H}^*)^F$ that satisfies $g^* s g^{*-1} = z_0 s$ exchanges the two components. It defines a one-to-one map between the sets of irreducible unipotent characters, hence an involutive map on the set of symbols $\gamma\colon \Phi(\mu(1)/2, \psi_+) \to \Phi(\mu(1)/2, \psi_+) = \Phi(\mu(-1)/2, \psi_-)$. The fixed points are the pairs $(\Lambda, \gamma(\Lambda))$. But s and $(-e)s$ are conjugate and we have seen that non-degenerate symbols are Jordan parameters of fixed points of λ_{-e} in $\mathcal{E}(\mathbf{G}^F, s)$. Hence the stabilizer in $\mathcal{A}$ of $\chi \in \mathcal{E}(\mathbf{G}^F, s)$ is exactly $<\lambda_{z_0}>$ if and only if its Jordan parameter is $(\Lambda, \gamma(\Lambda))$ for some degenerate symbol Λ. Such a symbol exists only if $\psi'_+ = \mathbf{0}$; their number is $\boldsymbol{\Phi}(\mu(1)/2, \mathbf{0}) - \tilde{\boldsymbol{\Phi}}(\mu(1)/2, \mathbf{0})$, coefficient of degree $2\mu(1)$ of $2(\mathcal{P}_8 - 1)$ by (16.58). From what we have seen on the action of λ_{-e}, if $g_1^* s g_1^{*-1} = (-e)z_0 s$, the induced map $\gamma'\colon \Phi(\mu(1)/2, \psi_+) \to \Phi(\mu(1)/2, \psi_+)$ differs from γ on degenerate symbols, so that the stabilizer in

$\mathcal{A}$ of $\chi \in \mathcal{E}(\mathbf{G}^F, s)$ is exactly $<\lambda_{(-e)z_0}>$ if and only if its Jordan parameter is $(\Lambda, \gamma(\Lambda)')$ for some degenerate symbol Λ, where $\gamma(\Lambda)' = \gamma'(\Lambda)$ is the twin of $\gamma(\Lambda)$. Using Proposition 16.53, we obtain the total number of irreducible characters in the considered series with stabilizer $<\lambda_{z_0}>$ or $<\lambda_{(-e)z_0}>$; it is given by the function $2N(\mathcal{P}_8 - 1)\mathcal{P}_2(\mathcal{P}_2 + \mathcal{P}_2^-)\mathcal{P}_4{}^{-1} F_0^{(\Delta)}$.

Consider now the series defined by classes whose parameter is the parameter in $\mathrm{SO}_{2n,\mathbf{0}}(q^2)$ of a class of $(G_1 \setminus \mathrm{SO}_{2n,\mathbf{0}}(q))$. The parameter $(\mu', \psi'_\pm)$ has to satisfy $\mu'(f') = \mu'(\bar{f}')$. Thus μ' is constant on an orbit under the group B in Lemma 16.7 whose (ii) states that $\Pi_{\omega' \in \mathcal{F}'_0/<\mathcal{A}',F>} \mathcal{P}(t^{|\omega'|}) = \mathcal{P}_2{}^2\mathcal{P}_4{}^{-1} F_0^{(\Delta)}$. In case $\mu'(1) \neq 0$, there is only one component in $\mathrm{C}_{\mathbf{H}^*}(s)^F$ corresponding to eigenvalues 1 and -1, Jordan parameters of irreducible unipotent characters are elements of $\Phi(\mu'(1)/2, \psi'_+)$ and we have yet seen that only degenerate symbols are not fixed by λ_{-e}. By Proposition 16.53, the number of considered irreducible characters is given by the function $2N\mathcal{P}_8(\mathcal{P}_2 + \mathcal{P}_2^-)\mathcal{P}_2\mathcal{P}_4^{-1} F_0^{(\Delta)}$.

Hence, with $Y_{(\pm e)z_0,2}(t) := 1 + \sum_j y_{(\pm e)z_0,2}(4n)t^{4n}$,

$$Y_{(\pm e)z_0,2}(t) = 4N\mathcal{P}_2\mathcal{P}_4{}^{-1}\mathcal{P}_8(\mathcal{P}_2 + \mathcal{P}_2^-)F_0^{(\Delta)}. \tag{16.62}$$

Number of $\chi \in \mathrm{Irr}(\mathrm{H}^F)$ with $I_\chi = A$. The parameter (μ, ψ) or (μ', ψ') of s has to satisfy $\forall f \in \mathcal{F}$, $\mu(f) = \mu(\bar{f})$, $\mu(1) \neq 0$ and $\psi_+ = \psi_-$ or, $\forall f' \in \mathcal{F}'$, $\mu'(f') = \mu'(\bar{f}')$, $\psi'_+ = \psi'_- = \mathbf{0}$ and $\mu'(1) \neq 0$.

From the preceding discussion it follows that the number of A-fixed elements of $\mathcal{E}(\mathbf{H}^F, s)$ is then $\tilde{\mathbf{\Phi}}(\mu(1)/2, \psi_+)\Pi_f p(\mu(f))$, where f runs over a set of representatives of the orbits under $\mathcal{A} = <f \mapsto \bar{f}, f \mapsto \tilde{f}>$ on $\mathcal{F}_0$, or $\tilde{\mathbf{\Phi}}(\mu'(1)/2, 0)\Pi_f p(\mu(f))$, where f runs over a set of representatives of the orbits under $\mathcal{A}' = <f' \mapsto \bar{f}', f' \mapsto \tilde{f}', f' \mapsto F(f')>$ on $\mathcal{F}'_0$.

Let $Y_4(t) := 1 + \sum_{n \geq 1} y_4(2n)t^{2n}$. From Proposition 16.53 and Lemmas 16.6 and 16.7 (iii), we have

$$Y_4(t) = N(\tilde{\mathbf{\Phi}}_{\mathbf{0}}(t^2) + \tilde{\mathbf{\Phi}}_{\mathbf{w}}(t^2) + 2\tilde{\mathbf{\Phi}}_{\mathbf{0}}(t^2))\mathcal{P}_2{}^2\mathcal{P}_4{}^{-1} F_0^{(\Delta)}.$$

From (16.59), we have

$$Y_4(t) = N\mathcal{P}_2{}^2\mathcal{P}_4\big(\mathcal{G}(t^4) - \mathcal{G}(-t^4)\big) F_0^{(\Delta)}. \tag{16.63}$$

The equality $4|\mathrm{Irr}(K)| = y_1 + 4y_2 + 16y_4$. One has $y_1 + y_2 + y_4 = |\mathrm{Irr}(\mathbf{H}^F)|$. It is given by the coefficients of degree $4n$ in the series given in Proposition 16.55. The number $3y_2 + 15y_4$ in dimension $4n$ is the coefficient of degree $4n$ in $3Y_e + 3Y_{(\pm e)z_0,2} + 12Y_4$, by formulae (16.60), (16.62) and (16.63). On the other side, up to a multiplier N, $|\mathrm{Irr}(K)|$ is given by the coefficient of degree $4n$ in formula (16.43). The verification reduces to two facts: On the coefficient of $F_0^{(\Sigma)}$ it is $2\mathcal{G}(t^4)^2 = \mathcal{G}(t^2)^2 + \mathcal{G}(-t^2)^2$, on the coefficient of $F_0^{(\Delta)}$, by (16.2), (16.20) and (16.15), ${\mathcal{P}_4}^2 F_0^{(\Delta)}$ and ${\mathcal{P}_2}^2\mathcal{P}_4\mathcal{G}(-t^4)F_0^{(\Delta)}$ agree on degrees divisible by 4. □

Remark. The number of irreducible representations of $\mathbf{H}^F$ may be computed by Jordan decomposition, semi-simple conjugacy classes are described by Proposition 16.53. We may describe an irreducible representation by a parameter (m_0, ϕ_+, ϕ_-) or (m', ϕ'), where $m_0\colon \mathcal{F}_0 \times \mathbb{N}^* \to \mathbb{N}$ is subjected to the same conditions as the restriction of m to $\mathcal{F}_0 \times \mathbb{N}^*$, and ϕ_+, ϕ_-, ϕ' are symbols. Up to the factor N introduced in Proposition 16.53, the generating functions are

$$T = \tfrac{1}{2}\big((1+\Phi_0)^2 + \Phi_{\mathrm{w}}^2\big)\big(F_0^{(\Sigma)} + F_0^{(\Delta)}\big) + (1+\Phi_0)\Phi_{\mathrm{w}}\big(F^{(\Sigma)} - F_0^{(\Delta)}\big),$$

which give the number of all parameters defined on $\mathbb{F}_q$,

$$T_4 = \tfrac{1}{2}\big(F_0^{(\Sigma)} + F_0^{(\Delta)}\big),$$

which gives the number of parameters (m_0, ϕ_+, ϕ_-) associated to four conjugacy classes of semi-simple elements, and

$$T_1 = \tfrac{1}{2}\big(\Phi_0^2 + \Phi_{\mathrm{w}}^2\big)\big(F_0^{(\Sigma)} + F_0^{(\Delta)}\big) + \Phi_0\Phi_{\mathrm{w}}\big(F^{(\Sigma)} - F_0^{(\Delta)}\big),$$

which gives the number of parameters (m_0, ϕ_+, ϕ_-) associated to only one conjugacy class.

As for parameters (m_0', Φ') defined on $\mathbb{F}_{q^2}$, their total number $T'(t)$ is a sum of

$(1+\Phi_0(t^2))({\mathcal{P}_2}^2{\mathcal{P}_4}^{-1}F_0^{(\Sigma)} + {\mathcal{P}_8}^{-1}{\mathcal{P}_4}^2F_0^{(\Delta)})/2$ and

$\Phi_{\mathrm{w}}(t^2)({\mathcal{P}_2}^2{\mathcal{P}_4}^{-1}F_0^{(\Sigma)} - \quad {\mathcal{P}_8}^{-1}{\mathcal{P}_4}^2F_0^{(\Delta)})/2$,

$$\begin{aligned} T'(t) = {} & (1+\Phi_0(t^2)+\Phi_{\mathrm{w}}(t^2)){\mathcal{P}_2}^2{\mathcal{P}_4}^{-1}F_0^{(\Sigma)}/2 \\ & + (1+\Phi_0(t^2)-\Phi_{\mathrm{w}}(t^2)){\mathcal{P}_4}^2{\mathcal{P}_8}^{-1}F_0^{(\Delta)}/2 \end{aligned}$$

and the number of parameters (m_0', ϕ') associated to four conjugacy classes is given by the function

$$T_4'(t) = \tfrac{1}{2}\big({\mathcal{P}_2}^2{\mathcal{P}_4}^{-1}F_0^{(\Sigma)} + {\mathcal{P}_4}^2{\mathcal{P}_8}^{-1}F_0^{(\Delta)}\big).$$

Then $2T' + 2T_4' = {\mathcal{P}_2}^4\mathcal{G}(-t^2)(\mathcal{G}(t^4) + 3\mathcal{G}(-t^4))F_0^{(\Sigma)}/2 + 2{\mathcal{P}_4}^2F_0^{(\Delta)}$. We have $y_1 + y_2 + y_4 = N(2T + 2T_4 - T_1 + 2T' + 2T_4')$ and find the function given in Proposition 16.55, thanks to (16.57) and (16.58), is

$${\mathcal{P}_2}^4\left(\frac{(\mathcal{G}_2 + 3\mathcal{G}_{-2})^2}{8} + \frac{\mathcal{G}_{-2}(\mathcal{G}_4 + 3\mathcal{G}_{-4})}{2}\right)F_0^{(\Sigma)} + 4{\mathcal{P}_4}^2F_0^{(\Delta)}.$$

Exercises

1. Find a generating function for the number of conjugacy classes of finite general linear groups $\mathrm{GL}_n(q)$.
2. Find a generating function for the number of conjugacy classes of unitary groups $\mathrm{GL}(n, -q)$.

   Show that a conjugacy class of $\mathrm{GL}(n, -q)$ for some n is defined by an application $m' \colon \mathcal{F}' \to \mathbb{N}$ such that $m'(f') = m'(\tilde{f}')$, where $\mathcal{F}'$ is defined on $\mathbb{F}_{q^2}$. The corresponding generating function is $\prod_{j\geq 1}\left(\frac{1+t^j}{1-qt^j}\right) = \mathcal{P}_1{}^4\mathcal{P}_2{}^{-1}\Phi$.
3. Find a generating function for the number of conjugacy classes of symplectic groups $\mathrm{Sp}_{2n}(q)$.

   A conjugacy class of $\mathrm{Sp}_{2n}(q)$ for some n is defined by a triple (m, Ψ_+, Ψ_-) as in Proposition 16.10 with conditions $m(\pm 1, 2j+1) \in 2\mathbb{N}$ and $\Psi_+(j)$ (resp. $\Psi_-(j)$), defined for $j > 0$, is a Witt symbol on $\mathbb{F}_q$ in dimension $m(1, 2j)$ (resp. $m(-1, 2j)$). The corresponding generating function is $\prod_{j\geq 1}\left(\frac{(1+t^{2j})^4}{1-qt^{2j}}\right) = \mathcal{P}_2{}^6\mathcal{P}_4{}^{-4}F_0^{(\Sigma)}$.
4. Find generating functions for the number of conjugacy classes of the groups $\Omega_{n,\mathbf{v}}(q)$.
5. ([Lu77], 6.4). Let $(d_n^{\mathbf{0}})_{n\in\mathbb{N}}$, $(d_n^{\mathbf{w}})_{n\in\mathbb{N}}$ be two sequences of integers.

   Let V be an orthogonal $\mathbf{F}$-space, of even dimension $2n$, with a rational structure on $\mathbb{F}_q$, of Witt symbol $\mathbf{v} \in \{\mathbf{0}, \mathbf{w}\}$. The special Clifford group $\mathbf{H} = \mathrm{CL}(V)^\circ$ (§16.4) is in duality over $\mathbb{F}_q$ with the conformal group $\mathrm{CSO}(V)$ (§16.6). For any semi-simple element s of $\mathbf{H}^F$, $\pi(s) \in \mathrm{SO}(V)^F$ has a parameter (μ, ψ_+, ψ_-). Consider the orthogonal decomposition $V = V_1(\pi(s)) \perp V_{-1}(\pi(s)) \perp V^0(\pi(s))$ (two eigenspaces for 1, -1, and an orthogonal complement as before Proposition 16.8). Let $b^0(s)$ be the number of unipotent conjugacy classes in the centralizer of the restriction $\pi(s)\,|_{V^0(\mathbb{F}_q)}$ in $\mathrm{SO}(V^0)^F$. Put

$$b^{\mathbf{v}}(s) = d^{\psi_+}_{\mu(1)/2}\, d^{\psi_-}_{\mu(-1)/2}\, b^0(s).$$

Verify the identity

$$1 + \frac{1}{2(q-1)}\sum_{n\in\mathbb{N}^*}\left(\sum_{(s)} b^{\mathbf{v}}(s) + \sum_{(s)} b^{\mathbf{w}}(s)\right)t^{2n}$$
$$= \left(1 + \frac{1}{2}\sum_{n\in\mathbb{N}^*}(d_n^{\mathbf{0}} + d_n^{\mathbf{w}})t^{2n}\right)^2 F_0^{(\Sigma)}$$

where (s) runs over the sets of semi-simple conjugacy classes in the respective rational special Clifford groups.

With $d_n^{\mathbf{v}} = \Phi(n, \mathbf{v}) = |\mathcal{E}(\mathrm{SO}_{2n,\mathbf{v}}(q), 1)|$, according to the Jordan decomposition, one has $b^{\mathbf{v}}(s) = |\mathcal{E}(\mathrm{CSO}(V)^F, s)|$ and the preceding series gives, up to the divisor $2(q-1)$, the numbers of elements in $\mathrm{Irr}(\mathrm{CSO}_{2n,\mathbf{0}}(q)) \cup \mathrm{Irr}(\mathrm{CSO}_{2n,\mathbf{w}}(q))$. Verify the compatibility with Proposition 16.51.

6. On orthogonal groups in odd dimension (type $\mathbf{B}_n$), check the following results.

   The number of conjugacy classes of $\mathrm{SO}_{2n+1,\mathbf{v}}(q)$ ($\mathbf{v} \in \{\mathbf{1}, \mathbf{d}\}$) is given by the generating function $t\mathcal{P}_2{}^4\mathcal{G}_1(t^4)\mathcal{G}(t^4)F_0^{(\Sigma)}$, where $\mathcal{G}_1(t) = \sum_{j\in\mathbb{N}} t^{j(j+1)}$.

   Let $(c_n)_{n\in\mathbb{N}}$ be a sequence of integers.

   The special orthogonal group $\mathrm{SO}(2n+1)$ and the symplectic group $\mathrm{Sp}(2n)$ may be defined as dual groups over $\mathbf{F}$ and $\mathbb{F}_q$. Let V be a symplectic $\mathbf{F}$-space with a rational structure on $\mathbb{F}_q$. For any semi-simple element of $\mathrm{Sp}(V)^F$ consider as in Exercise 5 the decomposition $V = V_1(s) \perp V_{-1}(s) \perp V^0(s)$, let $b^0(s)$ be the number of unipotent conjugacy classes in the centralizer of the restriction $s\,|_{V^0(\mathbb{F}_q)}$ in $\mathrm{Sp}(V^0)^F$. Put

$$b(s) = c_{\mu(1)/2}c_{\mu(-1)/2}b^0(s)$$

   where $\mu(\pm 1)$ is the dimension of the eigenspace $V_{\pm 1}(s)$. One has an identity

$$\sum_{n\in\mathbb{N}} \left(\sum_{(s)} b(s)\right) t^{2n} = \left(\sum_{n\in\mathbb{N}} c_n t^{2n}\right)^2 F_0^{(\Sigma)}$$

   where (s) runs over the set of semi-simple conjugacy classes of $\mathrm{Sp}_{2n}(q)$.

   Let

$$\Phi(n) = \sum_{d>0} p_2(n - (d^2 - d))$$

   hence

$$\sum_{n\in\mathbb{N}} \Phi(n)t^n = \mathcal{P}_1{}^2\mathcal{G}_1\,.$$

   Lusztig proved

$$\Phi(n) = |\mathcal{E}(\mathrm{Sp}_{2n}(q), 1)| = |\mathcal{E}(\mathrm{SO}_{2n+1}(q), 1)|\,.$$

   With $c_n = \Phi(n)$, one has $b(s) = |\mathcal{E}(\mathrm{C}^\circ_{\mathrm{Sp}(V)}(s)^F, 1)| = |\mathcal{E}(\mathrm{SO}(2n+1)^F, s)|$ so that $\sum_{(s)} b(s)$ is $|\mathrm{Irr}(\mathrm{SO}_{2n+1}(q))|$.

   Indeed one has $\sum_{n\in\mathbb{N}} |\mathrm{Irr}(\mathrm{SO}_{2n+1}(q))| t^{2n+1} = t\mathcal{P}_2{}^4\mathcal{G}_1(t^4)\mathcal{G}(t^4)F_0^{(\Sigma)}$, thanks to the identity $\mathcal{G}(t^2)\mathcal{G}_1(t^2) = \mathcal{G}_1(t)^2$.

7. On the conformal symplectic group CSp(2n).

   The group of symplectic similitudes is in duality with the Clifford group of an orthogonal space of odd dimension $2n+1$. Arguing as in Exercises 5 and 6, show the following results.

Let $(d_n^{\mathbf{0}})_{n\in\mathbb{N}}$, $(d_n^{\mathbf{w}})_{n\in\mathbb{N}}$, $(c_n)_{n\in\mathbb{N}}$, define $b(s)$ for any semi-simple element s of the rational special Clifford group $\mathrm{CL}^\circ(2n+1)^F$, so that

$$\sum_{n\in\mathbb{N}}\left(\sum_{(s)} b(s)\right)t^{2n}$$
$$= (q-1)\Big(\sum_{n\in\mathbb{N}} c_n t^{2n}\Big)\Big(1+\frac{1}{2}\sum_{n\in\mathbb{N}^*}(d_n^{\mathbf{0}}+d_n^{\mathbf{w}})t^{2n}\Big)F_0^{(\Sigma)}$$

where (s) runs over the set of semi-simple conjugacy classes of $\mathrm{CL}^\circ(2n+1)^F$.

With $d_n^{\mathbf{v}} = \Phi(n,\mathbf{v})$ and $c_n = \boldsymbol{\Phi}(n)$, one obtains a generating function for the number of irreducible characters (or of conjugacy classes) of $\mathrm{CSp}_{2n}(q)$, i.e., up to a factor $(q-1)/4$, $\mathcal{P}_2{}^4\left(\mathcal{G}(t^2)+3\mathcal{G}(-t^2)\right)\mathcal{G}_1(t^2)F_0^{(\Sigma)}$.

Notes

Proofs of the results of §16.3 and generalizations, including characteristic 2, may be found in [Wall63]. For older references, see its introduction. A survey on conjugacy classes in reductive groups is given in [SpSt70].

In his fundamental work on the classification of representations of finite classical groups, Lusztig considers groups of rational points of reductive algebraic groups with connected center [Lu77]. His proof includes the verification that the number of exhibited irreducible representations is actually equal to the number of conjugacy classes. Corresponding generating functions, as in Proposition 16.51 above, are given there, including the characteristic 2 case (see Exercises 5, 6 and 7 for types in odd characteristic).

17

Standard isomorphisms for unipotent blocks

Let $\mathbf{G}$ be a connected reductive $\mathbf{F}$-group defined over $\mathbb{F}_q$, with Frobenius endomorphism $F: \mathbf{G} \to \mathbf{G}$. Let $\mathbf{G}^*$ be in duality with $\mathbf{G}$ (see Chapter 8).

Let ℓ be prime, different from the characteristic of $\mathbf{F}$. Let $(\mathcal{O}, K, k)$ be an ℓ-modular splitting system for $\mathbf{G}^F$. We recall from Chapter 9 that there is a product of ℓ-blocks $\mathcal{O}\mathbf{G}^F.b_\ell(\mathbf{G}^F, 1)$ whose set of irreducible characters is the union of rational series $\mathcal{E}(\mathbf{G}^F, s)$ for s ranging over ℓ-elements of $(\mathbf{G}^*)^F$. Recall that we call these blocks the "unipotent blocks", they are the ones not annihilated by at least one unipotent character (see Theorem 9.12). Since the unipotent characters have $\mathrm{Z}(\mathbf{G}^F)$ in their kernel, and since we have a bijection

$$\mathrm{Res}^{\mathbf{G}^F}_{[\mathbf{G},\mathbf{G}]^F} : \mathcal{E}(\mathbf{G}^F, 1) \to \mathcal{E}([\mathbf{G}, \mathbf{G}]^F, 1),$$

one may expect that the algebra $\mathcal{O}\mathbf{G}^F.b_\ell(\mathbf{G}^F, 1)$ depends essentially only on the type of $(\mathbf{G}, F)$. It is easily proved that the partitions of $\mathcal{E}(\mathbf{G}^F, 1)$ and $\mathcal{E}(\mathbf{G}^F_{\mathrm{ad}}, 1)$ induced by ℓ-blocks are the same (Theorem 17.1). To get an isomorphism of $\mathcal{O}$-algebras one must, however, take care of the whole of $\mathcal{E}_\ell(\mathbf{G}^F, 1)$ (not just unipotent characters). Under a stronger hypothesis on ℓ (namely, ℓ does not divide the order of $\mathrm{Z}(\mathbf{G}_{\mathrm{sc}})^F$), we prove the isomorphism of $\mathcal{O}$-algebras

$$\mathcal{O}\mathbf{G}^F.b_\ell(\mathbf{G}^F, 1) \cong \mathcal{O}\mathrm{Z}(\mathbf{G}^F)_\ell \otimes_{\mathcal{O}} \mathcal{O}\mathbf{G}^F_{\mathrm{ad}}.b_\ell(\mathbf{G}^F_{\mathrm{ad}}, 1)$$

(see Theorem 17.7). The proof uses the results of Chapter 15 (and therefore also Chapter 16); see Proposition 17.4 below.

In this chapter, $(\mathbf{G}, F)$ is a connected reductive $\mathbf{F}$-group defined over $\mathbb{F}_q$, and $\sigma: \mathbf{G} \to \widetilde{\mathbf{G}}$ is an embedding that satisfies (15.1(σ)). One has a commutative diagram

$$\begin{array}{ccc} \mathbf{G}^F & \xrightarrow{\sigma} & \widetilde{\mathbf{G}}^F \\ \downarrow \pi & (\mathbf{D}) & \downarrow \tilde{\pi} \\ \pi(\mathbf{G}^F) & \xrightarrow{j} & {\mathbf{G}_{\mathrm{ad}}}^F \end{array}$$

and the associated restriction maps Res_σ, Res_π, Res_j, $\mathrm{Res}_{\tilde{\pi}}$ going the other way around on the corresponding sets of central functions.

17.1. The set of unipotent blocks

A first result shows that $\mathbf{G} \to \mathbf{G}_{\mathrm{ad}}$ behaves well with respect to the decomposition into unipotent blocks (see Definition 9.13).

Theorem 17.1. *Let ℓ be a prime not dividing q. Then $\mathbf{G}^F$ and $\mathbf{G}_{\mathrm{ad}}^F$ have the same number of unipotent ℓ-blocks, and the map from $\mathbb{Z}\mathcal{E}_\ell(\mathbf{G}_{\mathrm{ad}}^F, 1)$ to $\mathbb{Z}\mathcal{E}_\ell(\mathbf{G}^F, 1)$ induced by Res_j preserves the orthogonal decomposition induced by ℓ-blocks.*

The proof requires two lemmas. Let G be a finite group.

The first lemma is about blocks of G and $G/Z(G)$ (see [NaTs89] §5.8).

Lemma 17.2. *If $Z \subseteq Z(G)$ and χ, $\chi' \in \mathrm{Irr}(G)$ are such that $Z \subseteq \mathrm{Ker}(\chi) \cap \mathrm{Ker}(\chi')$, then χ and χ' define the same ℓ-block of G if and only if they define the same ℓ-block as characters of G/Z.*

The next lemma is about normal subgroups (see [NaTs89] §5.5).

Lemma 17.3. *Let $A \lhd G$. Let b (resp. a) be an ℓ-block of G (resp. A) with $\chi_0 \in \mathrm{Irr}(G, b)$ such that $\mathrm{Res}^G_A \chi_0 \in \mathrm{Irr}(A, a)$.*

For all $\chi \in \mathrm{Irr}(G, b)$, $\mathrm{Res}^G_A \chi \in \mathbb{Z}\mathrm{Irr}(A, a)$ and each element of $\mathrm{Irr}(A, a)$ occurs in such a restriction.

Proof of Theorem 17.1. The map $\mathrm{Res}_{\tilde{\pi}}$ sends $\mathcal{E}_\ell(\mathbf{G}_{\mathrm{ad}}^F, 1)$ into $\mathrm{Irr}(\widetilde{\mathbf{G}}^F)$. By Proposition 15.9, it induces $\mathcal{E}(\mathbf{G}_{\mathrm{ad}}^F, 1) \xrightarrow{\sim} \mathcal{E}(\widetilde{\mathbf{G}}^F, 1)$. Lemma 17.2 tells us that it preserves the partition induced by ℓ-blocks, so, by Theorem 9.12, it sends $\mathcal{E}_\ell(\mathbf{G}_{\mathrm{ad}}^F, 1)$ into $\mathcal{E}_\ell(\widetilde{\mathbf{G}}^F, 1)$ preserving ℓ-blocks (one may also prove that it preserves generalized characters $R^{\mathbf{G}}_{\mathbf{T}}\theta$).

Let us now look at the restriction map $\mathrm{Res}_\sigma = \mathrm{Res}^{\widetilde{\mathbf{G}}^F}_{\mathbf{G}^F} : \mathbb{Z}\mathcal{E}_\ell(\widetilde{\mathbf{G}}^F, 1) \to \mathbb{Z}\mathcal{E}_\ell(\mathbf{G}^F, 1)$ and show that it preserves the orthogonal decomposition into blocks. We have $\mathrm{Res}_\sigma(\mathcal{E}_\ell(\widetilde{\mathbf{G}}^F, 1)) \subseteq \mathbb{Z}\mathcal{E}_\ell(\mathbf{G}^F, 1)$ by Proposition 15.6. Each ℓ-block of $\mathcal{E}_\ell(\mathbf{G}^F, 1)$ and $\mathcal{E}_\ell(\widetilde{\mathbf{G}}^F, 1)$ contains a unipotent character (Theorem 9.12(ii)) and one has $\mathrm{Res}_\sigma(\mathcal{E}(\widetilde{\mathbf{G}}^F, 1)) = \mathcal{E}(\mathbf{G}^F, 1)$ by Proposition 15.9. This implies that all ℓ-block idempotents b of $\mathbf{G}^F$ such that $\mathrm{Irr}(\mathbf{G}^F, b) \subseteq \mathcal{E}_\ell(\mathbf{G}^F, 1)$ are $\widetilde{\mathbf{G}}^F$-fixed. It now remains to show that such a block idempotent b of $\mathbf{G}^F$ cannot split as a sum of several block idempotents of $\widetilde{\mathbf{G}}^F$. Those blocks $\{b'\}$ of $\widetilde{\mathbf{G}}^F$ would all satisfy $\mathrm{Irr}(\widetilde{\mathbf{G}}^F, b') \cap \mathcal{E}(\widetilde{\mathbf{G}}^F, 1) \neq \emptyset$ (Theorem 9.12(ii)). So, to prove our claim, it suffices to check that, if χ_1, $\chi_2 \in \mathcal{E}(\widetilde{\mathbf{G}}^F, 1)$ and $b_{\mathbf{G}^F}(\mathrm{Res}_\sigma \chi_1) = b_{\mathbf{G}^F}(\mathrm{Res}_\sigma \chi_2)$, then $b_{\widetilde{\mathbf{G}}^F}(\chi_1) = b_{\widetilde{\mathbf{G}}^F}(\chi_2)$. Since $\mathrm{Res}_\sigma \chi_2 \in \mathrm{Irr}(\mathbf{G}^F, b_{\mathbf{G}^F}(\mathrm{Res}_\sigma \chi_1))$, there exists $\chi_3 \in \mathrm{Irr}(\widetilde{\mathbf{G}}^F, b_{\widetilde{\mathbf{G}}^F}(\chi_2))$ such that $\langle \mathrm{Res}_\sigma \chi_3, \mathrm{Res}_\sigma \chi_1 \rangle_{\mathbf{G}^F} \neq 0$ (Lemma 17.3 for $a = b_{\mathbf{G}^F}(\mathrm{Res}_\sigma \chi_1)$ and $b = b_{\widetilde{\mathbf{G}}^F}(\chi_2)$). But $\mathrm{Res}_\sigma \chi_1 \in \mathrm{Irr}(\mathbf{G}^F)$ and $\widetilde{\mathbf{G}}^F/\mathbf{G}^F$ is commutative, so Clifford theory implies that $\chi_3 = \lambda\chi_1$ where λ is a linear character of $\widetilde{\mathbf{G}}^F$ with $\lambda(\mathbf{G}^F) = 1$.

Since χ_1 is unipotent, there exists an F-stable maximal torus $\mathbf{T} \subseteq \widetilde{\mathbf{G}}$ such that $\langle \chi_1, \mathrm{R}_{\mathbf{T}}^{\widetilde{\mathbf{G}}}(1) \rangle_{\widetilde{\mathbf{G}}^F} \neq 0$. Then $\langle \lambda\chi_1, \mathrm{R}_{\mathbf{T}}^{\widetilde{\mathbf{G}}}(\mathrm{Res}_{\mathbf{T}^F}^{\widetilde{\mathbf{G}}^F} \lambda) \rangle_{\widetilde{\mathbf{G}}^F} = \langle \chi_1, \lambda^{-1}\mathrm{R}_{\mathbf{T}}^{\widetilde{\mathbf{G}}}(\mathrm{Res}_{\mathbf{T}^F}^{\widetilde{\mathbf{G}}^F} \lambda) \rangle_{\widetilde{\mathbf{G}}^F} = \langle \chi_1, \mathrm{R}_{\mathbf{T}}^{\widetilde{\mathbf{G}}}(1) \rangle_{\widetilde{\mathbf{G}}^F} \neq 0$ (use Proposition 9.6(iii)). We get $\langle \chi_3, \mathrm{R}_{\mathbf{T}}^{\widetilde{\mathbf{G}}}(\mathrm{Res}_{\mathbf{T}^F}^{\widetilde{\mathbf{G}}^F} \lambda) \rangle_{\widetilde{\mathbf{G}}^F} \neq 0$, but $\chi_3 \in \mathrm{Irr}(\widetilde{\mathbf{G}}^F, b_{\widetilde{\mathbf{G}}^F}(\chi_1)) \subseteq \mathcal{E}_\ell(\widetilde{\mathbf{G}}^F, 1)$ (see Definition 9.13), so the definition of Lusztig series implies that $\mathrm{Res}_{\mathbf{T}^F}^{\widetilde{\mathbf{G}}^F} \lambda$ is an ℓ-element of the group $\mathrm{Irr}(\mathbf{T}^F)$. But we have $\mathbf{T}^F\mathbf{G}^F = \widetilde{\mathbf{G}}^F$ (use Proposition 8.1 and the fact that $[\mathbf{T} \cap \mathbf{G}, F] = \mathbf{T} \cap \mathbf{G}$ since this is connected), so λ is an ℓ-element.

But then multiplication by λ preserves the partition of $\mathrm{Irr}(\widetilde{\mathbf{G}}^F)$ induced by ℓ-blocks. This can be seen as follows. The automorphism of the group algebra over $\mathcal{O}$ defined on group elements by $g \mapsto \lambda(g^{-1})g$, sends e_χ to $e_{\lambda\chi}$ (see Definition 9.1), but fixes the central idempotents since it is trivial mod. $J(\mathcal{O})$ and one may use the lifting of idempotents.

Then we have $b_{\widetilde{\mathbf{G}}^F}(\chi_1) = b_{\widetilde{\mathbf{G}}^F}(\chi_3)$, and hence $b_{\widetilde{\mathbf{G}}^F}(\chi_1) = b_{\widetilde{\mathbf{G}}^F}(\chi_2)$ as claimed. □

17.2. ℓ-series and non-connected center

We now show essentially that $\mathrm{Res}_{\mathbf{G}^F}^{\widetilde{\mathbf{G}}^F}$ bijects characters in rational series associated with semi-simple elements whose order is not involved in non-connexity of centers. This is where we use Theorem 15.11.

Proposition 17.4. *Let $\sigma : \mathbf{G} \to \widetilde{\mathbf{G}}$ be as above, $\tilde{\pi} : \widetilde{\mathbf{G}} \to \widetilde{\mathbf{G}}_{\mathrm{ad}} = \mathbf{G}_{\mathrm{ad}}$ and $\pi = \tilde{\pi} \circ \sigma$ the canonical epimorphisms, hence the inclusion $j : \pi(\mathbf{G}^F) \to \mathbf{G}_{\mathrm{ad}}^F$ and the diagram* (**D**)*. Dually, let $\sigma^* : (\widetilde{\mathbf{G}})^* \to \mathbf{G}^*$, $\pi^* : \mathbf{G}_{\mathrm{ad}}{}^* \to \mathbf{G}^*$, $(\tilde{\pi})^* : \mathbf{G}_{\mathrm{ad}}{}^* \to (\widetilde{\mathbf{G}})^*$ be dual morphisms such that $\pi^* = \sigma^* \circ (\tilde{\pi})^*$.*

(i) Let t be a semi-simple of $\mathbf{G}^{*F}$ whose order is prime to $|(\mathrm{Z}(\mathbf{G})/\mathrm{Z}^\circ(\mathbf{G}))_F|$.

Let $s_1 \in (\mathbf{G}_{\mathrm{ad}}{}^*)^F$ and $s \in (\widetilde{\mathbf{G}}^*)^F$ be such that $t = \sigma^*(s)$ and $s = (\tilde{\pi})^*(s_1)$. Then the commutative diagram (**D**) *gives rise by the associated restrictions to* *the following commutative diagram of bijections*

$$\begin{array}{ccc} \mathcal{E}(\mathbf{G}^F, t) & \xleftarrow{\mathrm{Res}_\sigma} & \mathcal{E}(\widetilde{\mathbf{G}}^F, s) \\ \big\uparrow \mathrm{Res}_\pi & & \big\uparrow \mathrm{Res}_{\widetilde{\pi}} \\ \mathcal{E}(\pi(\mathbf{G}^F), \pi(t)) & \xleftarrow{\mathrm{Res}_j} & \mathcal{E}(\mathbf{G}_{\mathrm{ad}}^F, s_1) \end{array}$$

where $\mathcal{E}(\pi(\mathbf{G}^F), \pi(t)) := \mathrm{Res}_j(\mathcal{E}(\mathbf{G}_{\mathrm{ad}}^F, s_1))$ *is a set of irreducible characters of* $\pi(\mathbf{G}^F)$.

(ii) π^* *induces a bijection from the set of conjugacy classes of elements of* $(\mathbf{G}_{\mathrm{ad}}{}^*)^F$ *with order prime to* $|(\mathrm{Z}(\mathbf{G})/\mathrm{Z}^\circ(\mathbf{G}))_F|$ *to the set of conjugacy classes of elements of* $(\mathbf{G}^*)^F$ *contained in the image of* π^* *and with order prime to* $|(\mathrm{Z}(\mathbf{G})/\mathrm{Z}^\circ(\mathbf{G}))_F|$.

Proof. (i) We first use the equality $t = \sigma^*(s)$. By Proposition 15.6, restrictions to $\mathbf{G}^F$ of elements of $\mathcal{E}(\widetilde{\mathbf{G}}^F, s)$ decompose in $\mathcal{E}(\mathbf{G}^F, t)$. Hence the group $\widetilde{\mathbf{G}}^F$, acting on the representations of its normal subgroup $\mathbf{G}^F$, leaves stable the subset $\mathcal{E}(\mathbf{G}^F, t)$ and this operation induces an operation of $Q := \mathbf{G}_{\mathrm{ad}}^F/\pi(\mathbf{G}^F)$ on $\mathcal{E}(\mathbf{G}^F, t)$. The cardinality of an orbit of Q on the series $\mathcal{E}(\mathbf{G}^F, s)$ is the order of some subgroup of $\mathrm{C}_{\mathbf{G}^*}(t)^F/\mathrm{C}^\circ_{\mathbf{G}^*}(t)^F$ (see Corollary 15.14). But the prime divisors of the order of $\mathrm{C}_{\mathbf{G}^*}(t)^F/\mathrm{C}^\circ_{\mathbf{G}^*}(t)^F$ divide the order of t (see Proposition 13.16(i)). Yet, by (15.4) and hypotheses on t, the order of Q is prime to the order of t. So the action of $\widetilde{\mathbf{G}}^F$ on $\mathcal{E}(\mathbf{G}^F, t)$ is trivial. On the other hand, by Theorem 15.11, we know that the restrictions to $\mathbf{G}^F$ of elements of $\mathcal{E}(\widetilde{\mathbf{G}}^F, s)$ are multiplicity free. By Clifford theory, $\mathrm{Res}_\sigma(\mathcal{E}(\widetilde{\mathbf{G}}^F, s)) \subseteq \mathcal{E}(\mathbf{G}^F, t)$.

If $\mu \in \mathrm{Irr}\big(\widetilde{\mathbf{G}}^F/\sigma(\mathbf{G}^F)\big)$ corresponds to $z \in \mathrm{Ker}(\sigma^*)^F$ by (15.2), then multiplication by μ induces a bijection from $\mathcal{E}(\widetilde{\mathbf{G}}^F, s)$ onto $\mathcal{E}(\widetilde{\mathbf{G}}^F, zs)$ (see Proposition 8.26). There is $s_0 \in \mathrm{Ker}(\sigma^*)^F.s$ whose order is prime to $|(\mathrm{Z}(\mathbf{G})/\mathrm{Z}^\circ(\mathbf{G}))_F|$ as is the order of t. If s_0 and zs_0 are conjugate in $(\widetilde{\mathbf{G}}^*)^F$, then z belongs to $[\widetilde{\mathbf{G}}^*, \widetilde{\mathbf{G}}^*] \cap \mathrm{Z}(\widetilde{\mathbf{G}}^*)^F$. But the order of z divides the order of s_0 hence is prime to $|[\widetilde{\mathbf{G}}^*, \widetilde{\mathbf{G}}^*] \cap \mathrm{Z}(\widetilde{\mathbf{G}}^*)^F|$ by (15.3). So $z = 1$. If $\mu \neq 1$, then $z \neq 1$ and $\mathcal{E}(\widetilde{\mathbf{G}}^F, s_0) \neq \mathcal{E}(\widetilde{\mathbf{G}}^F, zs_0)$. This shows that Res_σ injects $\mathcal{E}(\widetilde{\mathbf{G}}^F, s_0)$ into $\mathcal{E}(\mathbf{G}^F, t)$.

The elements of other series $\mathcal{E}(\widetilde{\mathbf{G}}^F, s')$ of irreducible characters of $\widetilde{\mathbf{G}}^F$ decompose by restriction to $\mathbf{G}^F$ on elements of $\mathcal{E}(\mathbf{G}^F, t)$ if and only if $\sigma^*(s')$ and t are conjugate; but if $\sigma^*(s') = \sigma^*(s_0)$, with $s', s_0 \in (\widetilde{\mathbf{G}}^*)^F$, then $s's_0{}^{-1} \in \mathrm{Ker}(\sigma^*)$ and $s's_0{}^{-1}$ defines $\zeta \in \mathrm{Irr}\big(\widetilde{\mathbf{G}}^F/\sigma(\mathbf{G}^F)\big)$. As tensor multiplication by ζ preserves restrictions to $\mathbf{G}^F$, one has $\mathrm{Res}_\sigma(\mathcal{E}(\widetilde{\mathbf{G}}^F, s')) = \mathrm{Res}_\sigma(\mathcal{E}(\widetilde{\mathbf{G}}^F, s_0))$, hence $\mathrm{Res}_\sigma(\mathcal{E}(\widetilde{\mathbf{G}}^F, s')) = \mathrm{Res}_\sigma(\mathcal{E}(\widetilde{\mathbf{G}}^F, s_0)) = \mathcal{E}(\mathbf{G}^F, t)$. So Res_σ is a bijection on any such series as claimed.

Suppose now $s = (\widetilde{\pi})^*(s_1)$, so that $t = \pi^*(s_1)$. As $\mathrm{Z}(\widetilde{\mathbf{G}})$ is connected, $(\widetilde{\pi})^*$ is an injection. Then $\mathrm{Z}(\widetilde{\mathbf{G}})^F$ is in the kernel of every element of $\mathcal{E}(\widetilde{\mathbf{G}}^F, s)$ so that restriction via $(\widetilde{\mathbf{G}}^F \to \mathbf{G}_{\mathrm{ad}}^F)$ induces a bijection $\mathrm{Res}_{\widetilde{\pi}}$ from $\mathcal{E}(\mathbf{G}_{\mathrm{ad}}^F, s_1)$ onto $\mathcal{E}(\widetilde{\mathbf{G}}^F, s)$. The restriction via $\widetilde{\pi} \circ \sigma = j \circ \pi$ sends an element of $\mathcal{E}(\mathbf{G}_{\mathrm{ad}}^F, s_1)$ onto an irreducible character of $\mathbf{G}^F$. As $\pi(\mathbf{G}^F)$ is a quotient group of $\mathbf{G}^F$ and a normal subgroup of $\mathbf{G}_{\mathrm{ad}}^F$, the restriction via j of an element of $\mathcal{E}(\mathbf{G}_{\mathrm{ad}}^F, s_1)$ is irreducible.

(ii) The algebraic group $\mathbf{G}_{\mathrm{ad}}{}^*$ is simply connected and the image of π^* is the derived group of $\mathbf{G}^*$. Thus π^* may be defined by restriction of the dual morphism of an embedding of $\mathbf{G}$ in a group with connected center, and (15.3) applies, the kernel C of π^* is isomorphic, with Frobenius action, to $\mathrm{Irr}(\mathrm{Z}(\mathbf{G})/\mathrm{Z}^\circ(\mathbf{G}))$. Consider the morphism $\tau \colon (\mathbf{G}_{\mathrm{ad}}{}^*)^F \to [\mathbf{G}^*, \mathbf{G}^*]^F$ induced by π^*. The kernel of τ is C^F and $|C^F| = |(\mathrm{Z}(\mathbf{G})/\mathrm{Z}^\circ(\mathbf{G}))_F|$. The image of τ is a normal subgroup of $[\mathbf{G}^*, \mathbf{G}^*]^F$ whose index divides $|C_F| = |C^F]$ (Proposition 8.1). Then τ induces a natural bijection between $(\mathbf{G}_{\mathrm{ad}}{}^*)^F$-conjugacy classes of elements whose order is prime to $|C^F|$ and the $(\mathbf{G}_{\mathrm{ad}}{}^*)^F/C^F$-conjugacy classes of their images.

Let t_1 be such a semi-simple element and $t = \pi^*(t_1) \in [\mathbf{G}^*, \mathbf{G}^*]^F$. Write $C(t)$ for $\mathrm{C}_{[\mathbf{G}^*,\mathbf{G}^*]}(t)$ and $C(t_1)$ for $\mathrm{C}^\circ_{\mathbf{G}_{\mathrm{ad}}{}^*}(t_1)$ The exponent of $C(t)/C(t)^\circ$ divides the order of t because the map $g \mapsto [g, t_1]$ defines a morphism from $(\pi^*)^{-1}(C(t))$ to C with kernel $C(t_1) = (\pi^*)^{-1}(C(t)^\circ)$ and $[g, t_1]^k = [g, t_1^k]$ for any integer k. By Theorem 15.13, $A(t)$ is trivial, i.e. $C(t)^F = (C(t)^\circ)^F$. The equality can be written as $C(t)^F = (C(t_1)/C)^F$. Proposition 8.1 applied to the quotient $\mathbf{G}_{\mathrm{ad}}{}^*/C$ with endomorphism F gives an isomorphism $[\mathbf{G}^*, \mathbf{G}^*]^F/\mathrm{Im}\,\tau \to C/[C, F]$. Proposition 8.1 applied to the quotient $C(t_1)/C$ with endomorphism F gives an isomorphism $C(t)^F/(C(t_1)^F/C^F) \to C/[C, F]$. But $C(t_1)^F/C^F = \pi^*(C(t_1)^F) = C(t) \cap \mathrm{Im}\,\tau$. Thus the conjugacy class of t in $\mathrm{Im}\,\tau$ is one conjugacy class in $[\mathbf{G}^*, \mathbf{G}^*]^F$. As $\mathbf{G}^* = \mathrm{Z}^\circ(\mathbf{G}^*)[\mathbf{G}^*, \mathbf{G}^*]$, we also have $\mathrm{C}_{\mathbf{G}^*}(t)^F = \mathrm{C}^\circ_{\mathbf{G}^*}(t)^F$. By [DiMi91] 3.25, the $\mathbf{G}^*$-conjugacy class of t, which intersects $[\mathbf{G}^*, \mathbf{G}^*]$ in one class, contains only one $\mathbf{G}^{*F}$-conjugacy class, and the same is true in $[\mathbf{G}^*, \mathbf{G}^*]$. □

Under stricter restriction we obtain isomorphisms between blocks (see Theorem 17.7).

Definition 17.5. *Let $\Pi(\mathbf{G}, F)$ be the set of primes not dividing $q.|\mathrm{Z}(\mathbf{G}_{\mathrm{sc}})^F|$ (see §8.1 and Table 13.11).*

Let us gather some information on $\Pi(\mathbf{G}, F)$-elements and the series they define in connection with the morphisms in diagram (**D**).

Lemma 17.6. *Let $(\mathbf{G}, F)$ be a connected reductive algebraic group defined over $\mathbb{F}_q$, let $\tau \colon \mathbf{G}_{\mathrm{sc}} \to [\mathbf{G}, \mathbf{G}]$ be a simply connected covering, A any subgroup of $\mathbf{G}^F$ containing $\tau(\mathbf{G}_{\mathrm{sc}}{}^F)$, $\pi \colon \mathbf{G} \to \mathbf{G}_{\mathrm{ad}}$ the natural epimorphism and $\pi^* \colon \mathbf{G}_{\mathrm{ad}}{}^* \to \mathbf{G}^*$ a morphism dual to π. Then the following hold.*

(i) The groups $\mathrm{Z}(\mathbf{G})^F \cap [\mathbf{G}, \mathbf{G}]$, $\mathbf{G}_{\mathrm{ad}}^F/\pi(\mathbf{G}^F)$ *and* $\mathbf{G}^F/\mathrm{Z}^\circ(\mathbf{G})^F A$ *are commutative* $\Pi(\mathbf{G}, F)'$*-groups (see Definition 17.5).*

(ii) If $\chi \in \mathcal{E}_\ell(\mathbf{G}^F, 1)$*, then* $\mathrm{Z}(\mathbf{G})_{\ell'}^F \subseteq \mathrm{Ker}(\chi)$*. If* s *is a* $\Pi(\mathbf{G}, F)$*-element of* $(\mathbf{G}^*)^F$ *and* $\chi \in \mathcal{E}(\mathbf{G}^F, s)$*, there exists a unique* $z \in \mathrm{Z}(\mathbf{G}^*)_{\Pi(\mathbf{G},F)}^F$ *corresponding by duality to* $\lambda \in \mathrm{Irr}(\mathbf{G}^F/\tau({\mathbf{G}_{\mathrm{sc}}}^F))$ *such that* $\mathrm{Z}(\mathbf{G})^F \subseteq \mathrm{Ker}(\lambda^{-1}\chi)$ *and* $\lambda^{-1}\chi \in \mathcal{E}(\mathbf{G}^F, z^{-1}s)$*.*

Proof. Note that $\Pi(\mathbf{G}, F) = \Pi(\mathbf{G}^*, F)$. We write Π for $\Pi(\mathbf{G}, F)$.

(i) As $[\mathbf{G}, \mathbf{G}] = \tau(\mathbf{G}_{\mathrm{sc}})$ and $\mathrm{Z}(\mathbf{G}) \cap [\mathbf{G}, \mathbf{G}] \subseteq \tau\big(\mathrm{Z}(\mathbf{G}_{\mathrm{sc}})\big)$, the group $\mathrm{Z}(\mathbf{G})^F \cap [\mathbf{G}, \mathbf{G}]$ is a group of F-fixed points on a section of $\mathrm{Z}(\mathbf{G}_{\mathrm{sc}})$. We know by (15.4) that $\mathbf{G}_{\mathrm{ad}}^F/\pi(\mathbf{G}^F)$ is isomorphic to the group $(\mathrm{Z}(\mathbf{G})/\mathrm{Z}^\circ(\mathbf{G}))_F$, whose order divides that of $\mathrm{Z}(\mathbf{G})^F \cap [\mathbf{G}, \mathbf{G}]$. Proposition 8.1(i) applied to the quotient morphism $\mathrm{Z}^\circ(\mathbf{G}) \times \mathbf{G}_{\mathrm{sc}} \to \mathbf{G}$, defined by inclusion and τ, shows that $\mathbf{G}^F/\mathrm{Z}^\circ(\mathbf{G})^F\tau({\mathbf{G}_{\mathrm{sc}}}^F)$ is a commutative Π'-group. So is $\mathbf{G}^F/\mathrm{Z}^\circ(\mathbf{G})^F A$.

(ii) When $\mathbf{T}$ and $\mathbf{T}^*$ are F-stable maximal tori of $\mathbf{G}$ and $\mathbf{G}^*$ respectively, a duality between $\mathbf{T}$ and $\mathbf{T}^*$ defines an isomorphism $(s \mapsto \hat{s})$ from $(\mathbf{T}^*)^F$ onto $\mathrm{Irr}(\mathbf{T}^F)$ (8.14). Let $z \in \mathrm{Z}(\mathbf{G})^F$ and $\chi \in \mathcal{E}(\mathbf{G}^F, s)$. With the notation of (and by) Theorem 8.17(ii), one has $\chi(z) = \langle \pi_z^{\mathbf{G}^F}, \chi\rangle_{\mathbf{G}^F}$, hence $\chi(z) = \varepsilon_{\mathbf{G}}|\mathbf{G}^F|_p^{-1}\sum_{\mathbf{T}}\varepsilon_{\mathbf{T}}\langle \mathrm{R}_{\mathbf{T}}^{\mathbf{G}}(\pi_z^{\mathbf{T}^F}), \chi\rangle_{\mathbf{G}^F}$. But $\pi_z^{\mathbf{T}^F} = \sum_{\tau\in\mathrm{Irr}(\mathbf{T}^F)}\theta(z)\theta$ and $\langle \mathrm{R}_{\mathbf{T}}^{\mathbf{G}}\theta, \chi\rangle_{\mathbf{G}^F} = 0$ if $\theta \neq \hat{s}$. One has

$$\chi(z) = \varepsilon_{\mathbf{G}}|\mathbf{G}^F|_p^{-1}\sum_{(\mathbf{T},\theta)\leftrightarrow(\mathbf{T},s)}\varepsilon_{\mathbf{T}}\hat{s}(z)\langle \mathrm{R}_{\mathbf{T}}^{\mathbf{G}}s, \chi\rangle_{\mathbf{G}^F}$$

and so $\chi(z) = \hat{s}(z)\chi(1)$. Clearly, if s is an ℓ-element, $\mathbf{T}_{\ell'}^F \subseteq \mathrm{Ker}(\hat{s})$ so that $\mathrm{Z}(\mathbf{G})_{\ell'}^F \subseteq \mathrm{Ker}(\chi)$.

More generally if s is a Π-element of $(\mathbf{G}^*)^F$ and $\chi \in \mathcal{E}(\mathbf{G}^F, s)$, then the kernel of χ contains the Hall Π'-subgroup of $\mathrm{Z}(\mathbf{G})^F$. By (i) there exists a unique $\lambda \in \mathrm{Irr}((\mathbf{G}^F/\theta({\mathbf{G}_{\mathrm{sc}}}^F))_\Pi)$ such that $\chi(g) = \lambda(g)\chi(1)$ on any Π-element g of $\mathrm{Z}(\mathbf{G})^F$. If λ corresponds to $z \in \mathrm{Z}(\mathbf{G}^*)_\Pi^F$ by (8.19) and $\chi \in \mathcal{E}(\mathbf{G}^F, s)$, then $\lambda\chi \in \mathcal{E}(\mathbf{G}^F, zs)$ (see Proposition 8.26). The existence of z and λ as in the second assertion of (ii) follows. □

17.3. A ring isomorphism

Here is the main result of this chapter.

Theorem 17.7. *Let* $\ell \in \Pi(\mathbf{G}, F)$ *and* $(\mathcal{O}, K, k)$ *be an* ℓ*-modular splitting system for* $\mathbf{G}^F$*.*

Then the following $\mathcal{O}$-algebras are isomorphic

$$\mathcal{O}\mathbf{G}^F.b_\ell(\mathbf{G}^F,1) \cong \mathcal{O}\mathrm{Z}(\mathbf{G})^F_\ell \otimes \mathcal{O}\mathbf{G}^F_{\mathrm{ad}}.b_\ell(\mathbf{G}^F_{\mathrm{ad}},1)$$

(see Definition 9.9).

The following statements are about general finite groups. The first is standard (see [NaTs89] §5.8) about central quotients. The second is about normal subgroups in a situation generalizing the restriction $\mathrm{Res}^{\widetilde{\mathbf{G}}^F}_{\mathbf{G}^F}$ needed in the proof of Theorem 17.7.

Let G be a finite group, ℓ be a prime and $(\mathcal{O}, K, k)$ be an ℓ-modular splitting system for G.

Proposition 17.8. *(i) Let $m\colon \mathcal{O}G \to \mathcal{O}G/Z$ be the reduction map mod. Z, for Z an ℓ-subgroup of $\mathrm{Z}(G)$. If B is an ℓ-block of G, then $m(B)$ is an ℓ-block of G/Z and $\mathrm{Irr}\big(G/Z, m(B)\big) = \{\mathrm{Res}_m\chi \mid \chi \in \mathrm{Irr}(G,B),\ \chi(Z) = \chi(1)\}$.*

(ii) Let $m'\colon \mathcal{O}G \to \mathcal{O}G/Z'$ be the reduction map mod. Z', for Z' an ℓ'-subgroup of $\mathrm{Z}(G)$. If B is an ℓ-block of G, then $m'(B) \neq \{0\}$ if and only if for all $\chi \in \mathrm{Irr}(G,B)$, $\chi(Z') = \chi(1)$. Moreover, if $m'(B) \neq \{0\}$, then $m'\colon \mathcal{O}G.B \to (\mathcal{O}G/Z').m'(B)$ is an isomorphism of algebras.

The following is a corollary of Theorem 9.18.

Proposition 17.9. *Let H be a subgroup of G, and b (resp. c) be a central idempotent of $\mathcal{O}G$ (resp. $\mathcal{O}H$). Assume that, for any $\chi \in \mathrm{Irr}(G,b)$, $\mathrm{Res}^G_H\chi \in \mathrm{Irr}(H,c)$ and Res^G_H induces a bijection $\mathrm{Irr}(G,b) \to \mathrm{Irr}(H,c)$. Then*

$$\mathcal{O}Hc \cong \mathcal{O}Gb$$

as $\mathcal{O}$-algebras.

Proof. Since simple KGb-modules are annihilated by $1-c$, we have $cb = bc = b$. Then Theorem 9.18 applies with $M = \mathcal{O}Gb$, $A = \mathcal{O}Hc$, $B = \mathcal{O}Gb$. We get that M and $M^\vee = \mathrm{Hom}_\mathcal{O}(M,\mathcal{O})$ (note that $M = {}_AB_B$ and $M^\vee = {}_BB_A$) induce inverse Morita equivalences between $A-\mathbf{mod}$ and $B-\mathbf{mod}$. Then $A \cong \mathrm{End}_B(M^\vee)^{\mathrm{opp}}$, i.e. $A \cong B$ (see also Exercise 9.6). □

Let us now finish the proof of Theorem 17.7. Let $Z := \mathrm{Z}(\mathbf{G})^F_\ell$, $Z' := \mathrm{Z}(\mathbf{G})^F_{\ell'}$ and $A := \mathrm{Z}(\mathbf{G})^F[\mathbf{G},\mathbf{G}]^F$.

By Theorem 9.12(i), Proposition 17.4(i), which applies to $[\mathbf{G},\mathbf{G}] \to \mathbf{G} \to \widetilde{\mathbf{G}}$, and Lemma 17.3, there exists a sum of ℓ-block idempotents of $\mathcal{O}A$, written e^A, such that the restriction from $\mathbf{G}^F$ to A induces a bijection from $\mathcal{E}_\ell(\mathbf{G}^F,1)$

onto $\mathrm{Irr}(A, e^A)$. By Proposition 17.9 one gets an isomorphism

$$\mathcal{O}\mathbf{G}^F.b_\ell(\mathbf{G}^F, 1) \cong \mathcal{O}Ae^A.$$

Furthermore, by Lemma 17.6(ii), Z' is in the kernel of every element of $\mathcal{E}_\ell(\mathbf{G}^F, 1)$ or $\mathrm{Irr}(A, e^A)$. By Proposition 17.8, $\pi(b_\ell(\mathbf{G}^F, 1))$ and $\pi(e^A)$ are sums of block idempotents of $\mathcal{O}\pi(\mathbf{G}^F)$ and $\mathcal{O}\pi(A)$ respectively. One gets as above an isomorphism inducing restriction on characters

$$\mathcal{O}\pi(\mathbf{G}^F)\pi(b_\ell(\mathbf{G}^F, 1)) \cong \mathcal{O}\pi(A)\pi(e^A).$$

By Proposition 17.8(ii), e^A has a natural image $e^{A/Z'}$ in $\mathcal{O}A/Z'$ such that $\mathcal{O}Ae^A \cong \mathcal{O}A/Z'e^{A/Z'}$. By Lemma 17.6(i), $\mathrm{Z} \cap [\mathbf{G}, \mathbf{G}] = 1$. Thus one has an isomorphism $A/Z' \cong \mathrm{Z} \times \pi(A)$ by a map which sends aZ' to $(z, \pi(a))$ where $z \in \mathrm{Z}$ is such that $az^{-1}Z' \cap [\mathbf{G}, \mathbf{G}] \neq \emptyset$. As $\pi(A) \cong (A/Z')/(ZZ'/Z')$, we get an isomorphism

$$\widetilde{\pi}\colon \mathcal{O}Ae^A \to \mathcal{O}\mathrm{Z} \otimes \mathcal{O}\pi(A)\pi(e^A).$$

Now, let us show that Res_j bijects $\mathcal{E}_\ell(\mathbf{G}_{\mathrm{ad}}^F, 1)$ and $\mathrm{Irr}(\pi(\mathbf{G}^F), \pi(b_\ell(\mathbf{G}^F, 1)))$. Again, by Proposition 17.9, this would imply that

$$\mathcal{O}\mathbf{G}_{\mathrm{ad}}^F.b_\ell(\mathbf{G}_{\mathrm{ad}}^F, 1) \cong \mathcal{O}\pi(\mathbf{G}^F)\pi(b_\ell(\mathbf{G}^F, 1)).$$

This would complete our proof.

By Proposition 17.4, if $\widetilde{s} \in (\widetilde{\mathbf{G}}^*)^F_\ell$, the series $\mathcal{E}(\mathbf{G}^F, i^*(\widetilde{s}))$ and $\mathcal{E}(\widetilde{\mathbf{G}}^F, \widetilde{s})$ are in bijection by Res_i. Since the kernel of i^* is connected, i^* restricts to a surjection between the groups of rational points. The group $\mathrm{Z}(\mathbf{G})^F$ (resp. $\mathrm{Z}(\widetilde{\mathbf{G}})^F$) is in the kernel of an element of $\mathcal{E}(\mathbf{G}^F, t)$ (resp. $\mathcal{E}(\widetilde{\mathbf{G}}^F, \widetilde{s})$) if and only if $t \in \pi^*(\mathbf{G}_{\mathrm{ad}}{}^*)$ (resp. $\widetilde{s} \in \widetilde{\pi}^*\big(\widetilde{\mathbf{G}_{\mathrm{ad}}{}^*}\big)$). If it is the case for $t := i^*(\widetilde{s})$ and $\widetilde{s}$, let $s_1 \in (\mathbf{G}_{\mathrm{ad}}{}^*)^F$ be such that $(\widetilde{\pi})^*(s_1) = \widetilde{s}$ and $\pi^*(s_1) = i^*(\widetilde{s})$. We obtain that Res_j is a bijection between $\mathcal{E}(\mathbf{G}_{\mathrm{ad}}^F, s_1)$ and $\mathrm{Res}_\pi(\mathcal{E}(\mathbf{G}^F, i^*(\widetilde{s})))$. This holds for every conjugacy class (s_1) of ℓ-elements of $(\mathbf{G}_{\mathrm{ad}}{}^*)^F$, so we get our claim by Proposition 17.4(ii).

Exercises

1. Let $\mathbf{G} \subseteq \widetilde{\mathbf{G}}$ be an embedding of connected reductive $\mathbf{F}$-groups defined over $\mathbb{F}_q$ satisfying Hypothesis 15.1. Let ℓ be a prime not dividing q. Let $\mathbf{G}^F \subseteq H \subseteq \widetilde{\mathbf{G}}^F$ be such that $H/\mathbf{G}^F$ is the Sylow ℓ-subgroup of the commutative group $\widetilde{\mathbf{G}}^F/\mathbf{G}^F$. Denote by $\mathcal{E}(H, \ell')$ the set of components of characters of type $\mathrm{Res}_H^{\widetilde{\mathbf{G}}^F}\chi$ for $\chi \in \mathcal{E}(\widetilde{\mathbf{G}}^F, \ell')$.

Show that $\mathrm{Res}^H_{\mathbf{G}^F}$ sends bijectively $\mathcal{E}(H, \ell')$ to $\mathcal{E}(\mathbf{G}^F, \ell')$ (use an argument similar to the proof of Proposition 17.4 (i)). Deduce that the elements of $\mathcal{E}(\mathbf{G}^F, \ell')$ are all fixed by H (see Theorem 14.6).

2. Assume the hypotheses of Theorem 17.7, whose notation is also used. Show the following, more precise, statement.

   There exists an isomorphism of algebras

$$\beta: \mathcal{O}\mathbf{G}^F.b_\ell(\mathbf{G}^F, 1) \to \mathcal{O}\mathrm{Z}(\mathbf{G})^F_\ell \otimes \mathcal{O}\mathbf{G}^F_{\mathrm{ad}}.b_\ell(\mathbf{G}^F_{\mathrm{ad}}, 1)$$

   satisfying the following:

   if $\chi \in \mathcal{E}_\ell(\mathbf{G}^F, 1)$, $\chi' \in \mathcal{E}_\ell(\mathbf{G}^F_{\mathrm{ad}}, 1)$, $\mu \in \mathrm{Irr}(\mathrm{Z}(\mathbf{G})^F_\ell)$ and $\chi = (\mu \otimes \chi') \circ \beta$, there exists a unique $\lambda \in \mathrm{Irr}(\mathbf{G}^F/\theta(\mathbf{G}_{\mathrm{sc}}{}^F))_\ell$, in duality with a unique $z \in \mathrm{Z}(\mathbf{G}^*)^F_\ell$, such that $\mathrm{Res}_{\mathrm{Z}(\mathbf{G})^F_\ell}(\chi/\chi(1)) = \mathrm{Res}_{\mathrm{Z}(\mathbf{G})^F_\ell}(\lambda) = \mu$ and $\mathrm{Res}_j \chi' = \mathrm{Res}_\pi \lambda^{-1}\chi \in \mathrm{Irr}(\pi(\mathbf{G}^F))$.

3. We use the notation of Proposition 17.9. Denote by $m_b: \mathcal{O}Hc \to \mathcal{O}Gb$ the morphism induced by right multiplication by b. Show that it is an isomorphism.

   Show that, if $\chi \in \mathrm{Irr}(G, b)$, then $\chi \circ m_b = \mathrm{Res}^G_H(\chi)$ and, if D is a defect group of $b_G(\chi)$, then $D \subseteq H$ and it is a defect group of $m_c^{-1}(b_G(\chi)) = b_H(\mathrm{Res}^G_H(\chi))$.

   In Theorem 17.7, show that if b is an ℓ-block of $\mathcal{O}\mathbf{G}^F.b_\ell(\mathbf{G}^F, 1)$ with defect group $D \subseteq \mathbf{G}^F$, then $\beta(b)$ is an ℓ-block of $\mathcal{O}(\mathbf{G}_{\mathrm{ad}})^F.b_\ell(\mathbf{G}^F_{\mathrm{ad}}, 1)$ with defect group $j(D/\mathrm{Z}(\mathbf{G})^F_\ell)$.

Note

See [CaEn93].

PART IV

Decomposition numbers and q-Schur algebras

We have seen in §5.4 the definition of decomposition matrices Dec(A) and a very elementary property of them in finite reductive groups. Let us take $(\mathbf{G}, F)$ a connected reductive group defined over $\mathbb{F}_q$, ℓ a prime not dividing q, $(\mathcal{O}, K, k)$ an ℓ-modular splitting system for $\mathbf{G}^F$, $\mathbf{B} \subseteq \mathbf{G}$ an F-stable Borel subgroup, and denote by B_1 the sum of unipotent blocks in $\mathcal{O}\mathbf{G}^F$ (see Definition 9.13). Then the decomposition matrix of $\mathcal{H} := \mathrm{End}_{\mathcal{O}\mathbf{G}^F}(\mathrm{Ind}_{\mathbf{B}^F}^{\mathbf{G}^F} \mathcal{O})$ is a submatrix of the decomposition matrix of B_1.

In general, Dec($\mathcal{H}$) does not have the same number of columns or rows as Dec(B_1) (Theorem 5.28). The rows of Dec($\mathcal{H}$) correspond to characters occurring in $\mathrm{Ind}_{\mathbf{B}^F}^{\mathbf{G}^F} K$, hence are unipotent, while the number of rows of Dec(B_1) is the number of characters in rational series corresponding with ℓ-elements (see Theorem 9.12).

Concerning columns, we know from Chapter 13 that, when $(\mathrm{Z}(\mathbf{G})/\mathrm{Z}^\circ(\mathbf{G}))^F$ is of order prime to ℓ, the number of columns of Dec(B_1) is $|\mathcal{E}(\mathbf{G}^F, 1)|$, the number of unipotent characters. In the case of $\mathbf{G}^F = \mathrm{GL}_n(\mathbb{F}_q)$, this is the number of partitions of n. For the columns of Dec($\mathcal{H}$), let us consider the case of $\mathbf{G}^F = \mathrm{GL}_n(\mathbb{F}_q)$ with $q \equiv 1$ mod. ℓ. Then $\mathcal{H} \otimes k \cong k\mathfrak{S}_n$, so the number of columns of Dec($\mathcal{H}$) equals the number of ℓ-regular partitions of n. So, in this case, $\mathcal{H}$ seems not big enough.

Starting from the Hecke algebra $\mathcal{H}$ of type A_{n-1} over $\mathcal{O}$, one introduces a new algebra, the so-called q-Schur algebra, defined as

$$\mathcal{S}_{\mathcal{O}}(n, q) := \mathrm{End}_{\mathcal{H}}(\Pi_V \mathcal{H} x_V),$$

where V ranges over the parabolic subgroups of the symmetric group and x_V is the sum of the corresponding basis elements of $\mathcal{H}$ in the usual presentation (see Chapter 3).

A fundamental result about q-Schur algebras is that Dec($\mathcal{S}_{\mathcal{O}}(n, q)$) is square (i.e. $\mathcal{S}_{\mathcal{O}}(n, q) \otimes K$ and $\mathcal{S}_{\mathcal{O}}(n, q) \otimes k$ have the same number of simple modules;

see [Mathas] 4.15). One proves, moreover, that it is lower triangular unipotent for suitable orderings of rows and columns, and a maximal square submatrix of $\mathrm{Dec}(B_1)$ for $\mathrm{GL}_n(\mathbb{F}_q)$.

Relating $\mathcal{O}\mathbf{G}^F$-modules to $\mathcal{H}$-modules requires us to consider $\mathcal{O}$-analogues of Gelfand–Graev and Steinberg $K\mathbf{G}^F$-modules. This analysis also provides a lot of information about simple $B_1 \otimes k$-modules and the partition induced on them by Harish-Chandra series (defined in Chapter 1).

When the rational type of $(\mathbf{G}, F)$ is no longer ("split") **A**, but still among ^{2}**A**, **B**, **C**, **D** or ^{2}**D**, a generalization of the above is possible under the condition that ℓ and the order of q mod. ℓ are odd (see Chapter 20). The latter essentially ensures that $\mathcal{H}$ is then similar in structure to Hecke algebras of type A (see §18.6).

18

Some integral Hecke algebras

In this chapter, we gather some preparatory material pertaining to Hecke algebras and related modules, mainly for Coxeter groups of types A, BC and D.

Our algebras are defined over a local ring $\mathcal{O}$. Assume $\mathcal{H} = \bigoplus_{w \in W} \mathcal{O} a_w$ is a Hecke algebra over $\mathcal{O}$ associated with a Coxeter group (W, S) and certain parameters taken in $\mathcal{O} \setminus J(\mathcal{O})$ (see Definition 3.4). If $I \subseteq S$, denote $x_I = \sum_{w \in W_I} a_w$. In the first and second sections, we give properties of the right ideals $x_I \mathcal{H}$ and determine the morphisms between them. The second section is more precisely about type A_{n-1}, i.e. Hecke algebras associated with the symmetric group $\mathfrak{S}_n$. Then those ideals can be indexed by partitions of n, and one may show several properties of the above morphism groups with regard to tensoring with K, the field of fractions of $\mathcal{O}$. If M denotes the product of those ideals $x_I \mathcal{H}$, then the q-Schur algebra is $\mathcal{S}_{\mathcal{O}}(n, q) := \mathrm{End}_{\mathcal{H}}(M)$ (though in the text we use another definition, clearly equivalent to the above; see Exercise 1). The elementary results of this section can also be interpreted as information about its decomposition matrix. Its triangular shape appears in a very straightforward fashion. In the next chapter, we shall prove that it is also a square (unipotent) matrix, and a maximal one in the unipotent block of the decomposition matrix of general linear groups.

Concerning Hecke algebras of type BC, our task is mainly to show how the issues may reduce to type A. Dipper–James have shown in [DipJa92] that, if the parameters $Q, q \in \mathcal{O} \setminus J(\mathcal{O})$ are such that $Q + q^j \notin J(\mathcal{O})$ for any $j \in \mathbb{Z}$, then the corresponding Hecke algebra $\mathcal{H}_{\mathcal{O}}(\mathrm{BC}_n)$ of type BC_n over $\mathcal{O}$ is Morita equivalent to a product of Hecke algebras associated with groups $\mathfrak{S}_m \times \mathfrak{S}_{n-m}$ $(m = 0, \dots, n)$. This is a result one should compare with the corresponding one for group algebras (where $Q = q = 1$). Namely, when 2 is invertible in $\mathcal{O}$, the group algebra $\mathcal{O}[W(\mathrm{BC}_n)]$ is Morita equivalent with $\Pi_{m=0}^{n} \mathcal{O}[\mathfrak{S}_m \times \mathfrak{S}_{n-m}]$, an easy consequence of the fact that $W(\mathrm{BC}_n) \cong (\mathbb{Z}/2\mathbb{Z})^n \rtimes \mathfrak{S}_n$. The proof of the statement about Hecke algebras requires some technicalities since the multiplication in $\mathcal{H}_{\mathcal{O}}(\mathrm{BC}_n)$ resembles the one in $W(\mathrm{BC}_n)$ only when lengths add.

Quite predictably, the case of type D_{2n} is even more technical, since in the case of group algebras, one finds that $\mathcal{O}[W(\mathrm{D}_{2n})]$ is Morita equivalent to

$$\left(\Pi_{m=0,\ m\neq n}^{2n}\mathcal{O}[\mathfrak{S}_m \times \mathfrak{S}_{2n-m}]\right) \times \mathcal{O}[(\mathfrak{S}_n \times \mathfrak{S}_n)\rtimes \mathbb{Z}/2\mathbb{Z}],$$

where the last term is obviously not the group algebra of a Coxeter group. In this case, we limit ourselves to showing just that the "unipotent decomposition matrix" phenomenon mentioned above is preserved by the kind of Clifford theory occurring between Hecke algebras of types BC and D. Our statement is a general one about restriction of a module from A to B when $A = B \oplus B\tau$ is an $\mathcal{O}$-free algebra, B a subalgebra and τ an invertible element of A such that $B\tau = \tau B$, $\tau^2 \in B$.

18.1. Hecke algebras and sign ideals

In this section, we denote by $\mathcal{O}$ a commutative ring.

Let (W, S) be a Coxeter system (see Chapter 2) such that W is finite. In this chapter we consider Hecke algebras such as those introduced in Definition 3.6 with the extra condition that the parameters q_s are invertible.

Definition 18.1. *Let $(q_s)_{s\in S}$ be a family of invertible elements of $\mathcal{O}$ such that $q_s = q_t$ when s and t are W-conjugate. One defines $\mathcal{H}_{\mathcal{O}}(W, (q_s))$ to be the $\mathcal{O}$-algebra with generators a_s ($s \in S$) obeying the relations $(a_s + 1)(a_s - q_s) = 0$ and $a_sa_ta_s \ldots = a_ta_sa_t \ldots$ ($|st|$ terms on each side) for all $s, t \in S$.*

Recall ([Hum90] §7.1, [GePf00] 4.4.6) that $\mathcal{H}_{\mathcal{O}}(W, (q_s))$ is then $\mathcal{O}$-free with a basis indexed by W, allowing us to write $\mathcal{H}_{\mathcal{O}}(W, (q_s)) = \bigoplus_{w\in W} \mathcal{O}a_w$. An alternative presentation using the a_w's is $a_wa_{w'} = a_{ww'}$ when $w, w' \in W$ satisfy $l(ww') = l(w) + l(w')$ and $a_wa_s = (q_s - 1)a_w + q_sa_{ws}$ when $s \in S$ and $l(ws) = l(w) - 1$.

Note that, if $I \subseteq S$, this allows us to consider $\mathcal{H}_{\mathcal{O}}(W_I, (q_s)_{s\in I})$ as the subalgebra of $\mathcal{H}_{\mathcal{O}}(W, (q_s)_{s\in S})$ corresponding with the subspace $\bigoplus_{w\in W_I} \mathcal{O}a_w$.

Notation 18.2. $q_w := q_{s_1}q_{s_2}\ldots q_{s_l}$ when $w = s_1s_2\ldots s_l$ is a reduced expression of w as a product of elements of S, does not depend on the reduced expression (see, for instance, [Bour68] IV.1 Proposition 5). If $I \subseteq S$, denote

$$x_I = \sum_{w \in W_I} a_w \quad \text{and} \quad y_I = \sum_{w \in W_I} (-1)^{l(w)} q_w^{-1} a_w.$$

Proposition 18.3. *$I \subseteq S$, $w \in W_I$.*

(i) $x_I a_w = a_w x_I = q_w x_I$,

(ii) $y_I a_w = a_w y_I = (-1)^{l(w)} y_I$,

(iii) If $J \subseteq I$, $x_I x_J = x_J x_I = (\sum_{w \in W_J} q_w) x_I$ and $y_I y_J = y_J y_I = (\sum_{w \in W_J} q_w^{-1}) y_I$.

Proof. It clearly suffices to check the case when $I = S$ and $w = s \in S$.

Separating the elements $w \in W$ such that $l(sw) = l(w) + 1$ (denoted by $D_{s,\emptyset}$, see §2.1) and the ones such that $l(sw) = l(w) - 1$, one gets $a_s x_S = \sum_{w \in D_{s,\emptyset}} a_{sw} + \sum_{w \in W \setminus D_{s,\emptyset}} q_s a_{sw} + (q_s - 1) a_w$. Using the bijection $w \mapsto sw$ between $D_{s,\emptyset}$ and its complement, the first and last terms rearrange as $q_s \sum_{w \in D_{s,\emptyset}} a_w$, whence the equality $a_s x_S = q_s x_S$. The equality $x_S a_s = q_s x_S$ is proved in the same way.

Using the same discussion, one has $y_I a_s = \sum_{w \in D_{\emptyset,s}} (-1)^{l(w)} q_w^{-1} a_{ws} + \sum_{w \in W \setminus D_{\emptyset,s}} (-1)^{l(w)} q_w^{-1} (q_s - 1) a_w + \sum_{w \in W \setminus D_{\emptyset,s}} (-1)^{l(w)} q_w^{-1} q_s a_{ws}$. We have $q_{ws} = q_w q_s$ when $w \in D_{\emptyset,s}$, $q_{ws} q_s = q_w$ otherwise. So our sum rearranges to give the sought equality $y_S a_s = q_s y_S$. The equality $a_s y_S = q_s y_S$ is proved in the same way.

The equalities in (iii) are straightforward from (i) and (ii). □

Proposition 18.4. *$I, J \subseteq S$, $w \in W$. If $y_I a_w x_J \neq 0$, then $W_I \cap {}^w W_J = 1$. The line $\mathcal{O} y_I a_w x_J$ only depends on the double coset $W_I w W_J$.*

Proof. (a) Assume $I \supseteq J = \{s\}$ and $w = 1$. Then Proposition 18.3(ii) gives $y_I x_s = y_I + y_I a_s = 0$.

(b) Assume $w \in D_{IJ}$, then $W_I{}^w \cap W_J = W_{I^w \cap J}$ by Theorem 2.6. If $W_I{}^w \cap W_J \neq 1$, then let $s \in I^w \cap J$. One has $a_w x_s = x_{s'} a_w$ where $s' = wsw^{-1} \in I$. One has $x_J = x_s x'$ where $x' = \sum_{w \in W_J \cap D_{s,\emptyset}} a_w$. Then $y_J a_w x_I = y_J a_w x_s x' = y_J x_{s'} a_w x'$, but this is 0 by the above case (a).

(c) For arbitrary w, write $w = w_1 w' w_2$ with $w_1 \in W_I$, $w_2 \in W_J$, $w' \in D_{IJ}$ and therefore lengths add. Then $y_I a_w x_J = y_I a_{w_1} a_{w'} a_{w_2} x_J = (-1)^{l(w_1)} q_{w_2} y_I a_{w'} x_J$ by Proposition 18.3(i–ii). This and case (b) give our Proposition. □

The property of symmetry of Hecke algebras, already encountered in Chapter 1, can be used to study the morphisms between the sign ideals $y_I.\mathcal{H}_{\mathcal{O}}(W, (q_s))$.

The following is proved in [GePf00] 8.1.1.

Proposition 18.5. *$\mathcal{H}_{\mathcal{O}}(W, (q_s))$ is a symmetric algebra (see Definition 1.19).*

We now assume that $\mathcal{O}$ is a complete discrete valuation ring.

Theorem 18.6. *We abbreviate $\mathcal{H} = \mathcal{H}_{\mathcal{O}}(W, (q_s))$ (see Definition 18.1). Let $I, J \subseteq S$. Then $\mathrm{Hom}_{\mathcal{H}}(y_I\mathcal{H}, y_J\mathcal{H}) = \bigoplus_{d\in D_{IJ}} \mathcal{O}\mu_d^{\mathcal{O}}$ where $\mu_d^{\mathcal{O}}$ is defined by $\mu_d^{\mathcal{O}}(h) = y_J a_d h$ for all $h \in y_I\mathcal{H}$.*

We need the following.

Lemma 18.7. *Let A be an $\mathcal{O}$-free finitely generated* **symmetric** *algebra. Let $V \subseteq A$ be an $\mathcal{O}$-pure right ideal and $t \in \mathrm{Hom}_A(V, A_A)$. Then there exists $a \in A$ such that $t(v) = av$ for all $v \in V$.*

Proof of Lemma 18.7. When M is a right A-module, denote $M^\vee = \mathrm{Hom}_{\mathcal{O}}(M, \mathcal{O})$. This is a left A-module and this defines by transposition a contravariant functor from **mod**$-A$ to $A-$**mod**. This functor preserves $\mathcal{O}$-free modules and, on those $\mathcal{O}$-free finitely generated A-modules, it is an involution. The regular module A_A is sent to a left module isomorphic with ${}_AA$ since A is symmetric.

In the situation of the lemma, since V is pure in A, the transpose of the inclusion gives a surjection $r: A^\vee \to V^\vee$ (restriction to V of the linear forms). By the projectivity of $A^\vee \cong {}_AA$, the transpose $t^\vee: A^\vee \cong {}_AA \to V^\vee$ factors as $t^\vee = r \circ \theta$ where $\theta: {}_AA \to {}_AA$ is in $A-$**mod**. Then $\widehat{t} := \theta^\vee$ extends t since $r^\vee: V \to A$ is the canonical injection. Since $\theta^\vee \in \mathrm{Hom}_A(A_A, A_A)$, it is a left multiplication. □

Lemma 18.8. *Denote $\mathcal{H}_I = \mathcal{H}_{\mathcal{O}}(W_I, (q_s)_{s\in I})$, identified with the subalgebra of $\mathcal{H}$ equal to $\oplus_{w\in W_I}\mathcal{O}a_w$. Then $y_I\mathcal{H} \cong y_I\mathcal{H}_I \otimes_{\mathcal{H}_I} \mathcal{H}$ by the obvious map.*

Proof of Lemma 18.8. As a left $\mathcal{H}_I$-module, we have $\mathcal{H} = \bigoplus_d \mathcal{H}_I a_d$ where the sum is over $d \in D_{I\emptyset}$. Since each a_w is invertible, we have $\mathcal{H}_I a_d \cong \mathcal{H}_I$ by the obvious map. However, $y_I\mathcal{H}_I = \mathcal{H}_I y_I = \mathcal{O}y_I$ by Proposition 18.3(ii), therefore $y_I\mathcal{H} = \bigoplus_d \mathcal{O}y_I a_d$ (sum over $d \in D_{I\emptyset}$ as above). Then $y_I\mathcal{H}_I \otimes_{\mathcal{H}_I} \mathcal{H} \cong \bigoplus_d y_I\mathcal{H}_I \otimes \mathcal{H}_I a_d$. It suffices to show that $y_I\mathcal{H}_I \otimes \mathcal{H}_I a_d \cong \mathcal{O}y_I a_d$ by the obvious map. Since a_d is invertible, one may assume $d = 1$. Then it is just the trivial isomorphism $y_I\mathcal{H}_I \otimes \mathcal{H}_I \cong y_I\mathcal{H}_I = \mathcal{O}y_I$. □

Proof of Theorem 18.6. By the adjunction between $\bigotimes_{\mathcal{H}_I}\mathcal{H}$ and $\mathrm{Res}^{\mathcal{H}}_{\mathcal{H}_I}$ (see [Ben91a] 2.8.6), we have

$$\mathrm{Hom}_{\mathcal{H}}(y_I\mathcal{H}_I \otimes_{\mathcal{H}_I} \mathcal{H}, y_J\mathcal{H}) \cong \mathrm{Hom}_{\mathcal{H}_I}\left(y_I\mathcal{H}_I, \mathrm{Res}^{\mathcal{H}}_{\mathcal{H}_I} y_J\mathcal{H}\right)$$

by the map restricting each $f: y_I\mathcal{H}_I \otimes_{\mathcal{H}_I} \mathcal{H} \to y_J\mathcal{H}$ to the subspace $y_I\mathcal{H}_I \otimes 1$. Through this isomorphism and the identification $y_I\mathcal{H} \cong y_I\mathcal{H}_I \otimes_{\mathcal{H}_I} \mathcal{H}$ of

Lemma 18.8, it now suffices to check that

$$\mathrm{Hom}_{\mathcal{H}_I}(y_I\mathcal{H}_I, \mathrm{Res}^{\mathcal{H}}_{\mathcal{H}_I} y_J\mathcal{H}) = \bigoplus_{d\in D_{IJ}} \mathcal{O}\mu'_d$$

where μ'_d is defined by $\mu'_d(h) = y_J a_d h$ for all $h \in y_I\mathcal{H}_I$.

Using the double coset decomposition $W = \cup_{d\in D_{IJ}} W_I d W_J$ with lengths adding (Proposition 2.4), we get ${}_{\mathcal{H}_J}\mathcal{H}_{\mathcal{H}_I} \cong \oplus_{d\in D_{IJ}} \mathcal{H}_J a_d \mathcal{H}_I$, so that $\mathrm{Res}^{\mathcal{H}}_{\mathcal{H}_I} y_J\mathcal{H} = \bigoplus_{d\in D_{IJ}} y_J a_d \mathcal{H}_I$ by Proposition 18.3(ii). Each summand is isomorphic with $\mathcal{H}_I$ by the evident map, again by the double coset decomposition above. Through this further identification, μ'_d becomes the injection $y_I\mathcal{H}_I \to \mathcal{H}_I$, and all we must check is that this injection generates $\mathrm{Hom}_{\mathcal{H}_I}(y_I\mathcal{H}_I, \mathcal{H}_I)$. Applying Lemma 18.7 to $A = \mathcal{H}_I$ and $V = y_I\mathcal{H}_I$, we get that any morphism $y_I\mathcal{H}_I \to \mathcal{H}_I$ is in the form $h \mapsto ah$ for $a \in \mathcal{H}_I$. This gives our claim since $\mathcal{H}_I y_I = y_I\mathcal{H}_I = \mathcal{O}y_I$ by Proposition 18.3(ii). □

The following shows that sign ideals $y_I\mathcal{H}$ have properties similar to those of permutation modules for group algebras (see [Ben91a] 5.5, [Thévenaz] §27). Compare the following with [Thévenaz] 27.11.

Corollary 18.9. *With the same notation as in Theorem 18.6, let $k = \mathcal{O}/J(\mathcal{O})$ be the residue field of $\mathcal{O}$. Let $M := \Pi_{I\subseteq S} y_I\mathcal{H}$ considered as a right $\mathcal{H}$-module. Then $\mathrm{End}_{k\otimes\mathcal{H}}(k\otimes M) \cong k\otimes \mathrm{End}_{\mathcal{H}}(M)$ by the functor $k\otimes_{\mathcal{O}} -$. In particular, $M_i \mapsto k\otimes M_i$ is a bijection between the (isomorphism types of) indecomposable direct summands of M and $k\otimes M$.*

Proof. Applying Theorem 18.6 for $\mathcal{O}$ and k, one finds $k\otimes_{\mathcal{O}} \mathrm{Hom}_{\mathcal{H}}(y_I\mathcal{H}, y_J\mathcal{H}) \cong \mathrm{Hom}_{k\otimes\mathcal{H}}(y_I k\otimes\mathcal{H}, y_J k\otimes\mathcal{H})$ by the $k\otimes_{\mathcal{O}} -$ functor, since $\mu^k_d = k\otimes\mu^{\mathcal{O}}_d$. This gives $\mathrm{End}_{k\otimes\mathcal{H}}(k\otimes M) \cong k\otimes\mathrm{End}_{\mathcal{H}}(M)$ by $k\otimes_{\mathcal{O}} -$.

Now $E := \mathrm{End}_{\mathcal{H}}(M)$ is a finitely generated, $\mathcal{O}$-free $\mathcal{O}$-algebra. So the conjugacy classes of its primitive idempotents are in bijection with those of $k\otimes E$ by the classical theorem on idempotent liftings (see [Ben91a] 1.9.4, [Thévenaz] 1.3.2). This gives our claim about indecomposable direct summands of M and $k\otimes M$ by the above decomposition and the relation with idempotents (see [Thévenaz] 1.1.16). □

18.2. Hecke algebras of type A

In this section, (W, S) is the symmetric group on n letters (a Coxeter group of type A_{n-1}; see Example 2.1(i)). We take $\mathcal{O}$ a commutative ring and $q \in \mathcal{O}^\times$ an invertible element. We denote by $\mathcal{H}_{\mathcal{O}} = \mathcal{H}_{\mathcal{O}}(\mathfrak{S}_n, q)$ the associated Hecke algebra over $\mathcal{O}$ (see Definition 18.1).

We shall give in this case some more specific properties of the right ideals $y_I\mathcal{H}$ (see Notation 18.2).

Let us fix some (classical) vocabulary (see also §5.2). A *partition* $\lambda \vdash n$ of n is a sequence $\lambda_1 \geq \lambda_2 \geq \ldots \geq \lambda_l$ such that $\lambda_1 + \lambda_2 + \cdots + \lambda_l = n$. The dual partition λ^* is defined by $\lambda_i^* = |\{j\ ;\ \lambda_j \geq i\}$ for $i = 1, \ldots, \lambda_1$. If $\mu \vdash n$ is the partition $\mu_1 \geq \mu_2 \geq \ldots \geq \mu_m$, one writes $\lambda \ll \mu$ if and only if $m \leq l$ and $\lambda_1 + \lambda_2 + \cdots + \lambda_i \leq \mu_1 + \mu_2 + \cdots + \mu_i$ for all $i \leq m$.

Denote $\mathbf{n} := \{1, \ldots, n\}$. For a finite set F, $\mathfrak{S}_F$ denotes the permutation group of F.

A partition Λ of $\mathbf{n}$ is a set of non-empty, pairwise disjoint subsets of $\mathbf{n}$ whose union is $\mathbf{n}$ (i.e. a quotient of $\mathbf{n}$ by an equivalence relation). The cardinalities of the elements of Λ define a unique partition of n called the **type** of Λ.

If $\Lambda = \{\Lambda_1, \ldots \Lambda_l\}$, one defines $\mathfrak{S}_\Lambda = \mathfrak{S}_{\Lambda_1} \times \cdots \times \mathfrak{S}_{\Lambda_l} \subseteq \mathfrak{S}_{\mathbf{n}}$. If Λ, Γ are two partitions of $\mathbf{n}$, one says they are **disjoint** if and only if $\mathfrak{S}_\Lambda \cap \mathfrak{S}_\Gamma = \{1\}$, i.e. no intersection of an element of Λ with an element of Γ has cardinality ≥ 2.

For the following we refer to [Gol93] 6.2 and 6.3 (see also [CuRe87] 75.13).

Theorem 18.10. *(i) Let λ, $\mu \vdash n$. There exist disjoint partitions Λ and Γ of $\mathbf{n}$ of types λ and μ, if and only if $\lambda \ll \mu^*$.*

(ii) If Λ is a partition of $\mathbf{n}$ of type λ, then $\mathfrak{S}_\Lambda$ is transitive on the set of partitions Γ of $\mathbf{n}$ that are disjoint of Λ and of type λ^*.

It is clear that the conjugates of parabolic subgroups W_I ($I \subseteq S$) are the $\mathfrak{S}_\Lambda$'s considered above. Similarly, one may define elements of the Hecke algebra associated with partitions $\lambda \vdash n$.

Definition 18.11. *If $\lambda \vdash n$, let I_λ be the subset of $S = \{s_1, s_2, \ldots, s_{n-1}\}$ where $s_i = (i, i+1)$ defined by $S \setminus I_\lambda = \{s_{\lambda_1}, s_{\lambda_1+\lambda_2}, \ldots, s_{\lambda_1+\cdots+\lambda_{l-1}}\}$. Let $\mathfrak{S}_\lambda$ be the subgroup of $\mathfrak{S}_n$ generated by I_λ. Let $x_\lambda = x_{I_\lambda}$, $y_\lambda = y_{I_\lambda}$ (see Notation 18.2).*

Theorem 18.12. *Let λ, $\mu \vdash n$.*

(i) If $y_\lambda \mathcal{H} x_\mu \neq 0$, then $\lambda \ll \mu^*$.

(ii) $y_\lambda \mathcal{H} x_{\lambda^*} = \mathcal{O} y_\lambda a_{w_\lambda} x_{\lambda^*} \neq 0$ for some $w_\lambda \in W$.

Proof. (i) By Proposition 18.4, we may have $y_\lambda a_w x_\mu \neq 0$ only if $W_{I_\lambda} \cap {}^w W_{I_\mu} = 1$. But, with our definition of I_λ, it is clear that $W_{I_\lambda} = \mathfrak{S}_{\Lambda_\lambda}$ where Λ_λ denotes the partition $\{\{1, 2, \ldots, \lambda_1\}, \{\lambda_1 + 1, \ldots, \lambda_1 + \lambda_2\}, \ldots\}$ which is of type λ. So Λ_λ and $w\Lambda_\mu$ are disjoint partitions of types λ and μ. Then $\lambda \ll \mu^*$ by Theorem 18.10(i).

(ii) Denote $\lambda = \lambda_1 \geq \lambda_2 \geq \ldots \geq \lambda_l$, $\lambda^* = \lambda'_1 \geq \lambda'_2 \geq \ldots \geq \lambda'_{l'}$ (note that $l' = \lambda_1$, $l = \lambda'_1$), $I = I_\lambda$, $I' = I_{\lambda'}$. Let $T = (t_{i,j})_{1 \leq i \leq l\ ,\ 1 \leq j \leq l'}$,

$T' = (t'_{i,j})_{1\le i\le l\ ,\ 1\le j\le l'}$ be defined by

$$T = \begin{bmatrix} 1 & 2 & \dots & & \lambda_1 \\ \lambda_1+1 & \dots & & \lambda_1+\lambda_2 & \\ \vdots & & & & \\ n-\lambda_l+1 & \dots & n & & \end{bmatrix}$$

$$T' = \begin{bmatrix} 1 & \lambda'_1+1 & \dots & n-\lambda'_{l'}+1 \\ 2 & \vdots & & \vdots \\ \vdots & & \dots & n \\ & \lambda'_1+\lambda'_2 & & \\ \lambda'_1 & & & \end{bmatrix}$$

where the ends of lines and columns are completed by zero (note that they are in the same places in T and T' thanks to the definition of λ^* from λ). Let Λ be the partition of the set $\mathbf{n}$ defined by the rows of T and Λ' by the columns of T'. Then $\mathfrak{S}_\Lambda = W_I$ and $\mathfrak{S}_{\Lambda'} = W_{I'}$.

Let $v \in \mathfrak{S}_\mathbf{n}$ be defined by $v(t_{ij}) = t'_{ij}$ when $t_{ij} \neq 0$. Then $v\Lambda$ is the partition defined by the rows of T'. It is therefore clear that $v\Lambda$ and Λ' are disjoint. Then $W_{I'} \cap {}^vW_I = 1$ and by Theorem 18.10(ii), $W_{I'}vW_I$ is the only double coset with this property. So $y_\lambda \mathcal{H} x_{\lambda^*} = y_I \mathcal{H} x_{I'} = \mathcal{O} y_I a_{v^{-1}} x_{I'}$ by Proposition 18.4 again. We prove the following below.

Lemma 18.13. $l(w'vw) = l(w') + l(v) + l(w)$ for any $w \in W_I$, $w' \in W_{I'}$.

This completes our proof of Theorem 18.12 since then $y_I a_{v^{-1}} x_{I'} = \sum_{w\in W_I\ ,\ w'\in W_{I'}} (-1)^{l(w)} q_w^{-1} a_{wv^{-1}w'}$ and this sum is a decomposition in the basis of the a_x ($x \in W$) since ${}^vW_I \cap W_{I'} = 1$. □

Proof of Lemma 18.13. Using the reflection representation and root system of $W = \mathfrak{S}_n$ (see Example 2.1(i)), one sees easily that D_I^{-1} is the set of $x \in W$ such that the rows in

$$x.T := \begin{bmatrix} x(1) & x(2) & \dots & & x(\lambda_1) \\ x(\lambda_1+1) & \dots & & x(\lambda_1+\lambda_2) & \\ \vdots & & & & \\ x(n-\lambda_l+1) & \dots & x(n) & & \end{bmatrix}$$

are increasing (except for zeros, of course).

Our lemma amounts to showing that, for all $w' \in W_{I'} = \mathfrak{S}_{\Lambda'}$, $w'v \in D_I^{-1}$ (see Proposition 2.3(i)). First $v.T$ is T', so we have to check the rows of $w'.T'$.

But $\mathfrak{S}_{\Lambda'}$ permutes only elements inside the sets $\{1, \ldots, \lambda'_1\}, \ldots$ making up Λ', so the non-zero elements of the columns of $w'.T'$ are obtained by permuting those of the same columns of T'. Any element of the ith column of T' is less than or equal to any non-zero element of the $(i+1)$th column of T', so $w'.T'$ has the same property. Thus our claim is proved. □

Theorem 18.14. *Assume that $\mathcal{O}$ is a complete discrete valuation ring (so that the Krull–Schmidt theorem holds for $\mathcal{O}$-free algebras of finite rank). Then there exists a family $(M^\lambda_\mathcal{O})_{\lambda\vdash n}$ of pairwise non-isomorphic indecomposable right $\mathcal{H}_\mathcal{O}(\mathfrak{S}_n, q)$-modules such that*

- *if $\lambda \vdash n$, then $y_{\lambda^*}\mathcal{H}_\mathcal{O}(\mathfrak{S}_n, q)$ has $M^\lambda_\mathcal{O}$ as a direct summand with multiplicity 1,*
- *if $\lambda, \mu \vdash n$ and $M^\mu_\mathcal{O}$ is isomorphic with a direct summand of $y_{\lambda^*}\mathcal{H}_\mathcal{O}(\mathfrak{S}_n, q)$, then $\mu \ll \lambda$.*

Theorem 18.15. *Assume that $\mathcal{O}$ is a complete discrete valuation ring with fraction field K. Assume that $\mathcal{H}_\mathcal{O}(\mathfrak{S}_n, q)\otimes K$ is split semi-simple. Then the M^λ_K $(\lambda \vdash n)$ are the simple $\mathcal{H}_K$-modules. Moreover, the multiplicity of M^μ_K in $M^\lambda_\mathcal{O}\otimes K$ is non-zero only if $\mu \ll \lambda$. It is 1 when $\lambda = \mu$.*

Proof of Theorems 18.14 and 18.15. Fix $\lambda \vdash n$. Take a decomposition $y_{\lambda^*}\mathcal{H}_\mathcal{O} = \bigoplus_i M_i$ with indecomposable right $\mathcal{H}_\mathcal{O}$-modules M_i. Then $y_{\lambda^*}\mathcal{H}_\mathcal{O}x_\lambda = \bigoplus_i M_i x_\lambda$ is a line by Theorem 18.12(ii), so there exists a unique i_λ such that $M_{i_\lambda}x_\lambda \neq 0$. Denote $M_{i_\lambda} = M^\lambda_\mathcal{O}$. Now, if $M^\mu_\mathcal{O}$ is a direct summand of $y_{\lambda^*}\mathcal{H}_\mathcal{O}$, then $y_{\lambda^*}\mathcal{H}_\mathcal{O}x_\mu \neq 0$ and therefore, by Theorem 18.12(i), $\lambda^* \ll \mu^*$. This is the same as $\mu \ll \lambda$ (use the equivalence in Theorem 18.10(i) or use [JaKe81] 1.4.11).

If $M^\lambda_\mathcal{O} \cong M^\mu_\mathcal{O}$, then by the above we have both $\lambda \ll \mu$ and $\mu \ll \lambda$, i.e. $\lambda = \mu$. This completes the proof of Theorem 18.14.

If $\mathcal{H}_K$ is semi-simple, then it is isomorphic with $K\mathfrak{S}_n$, the group algebra of $\mathfrak{S}_n$ (see Theorem 3.16, [CuRe87] 68.21, or [Cart85] 10.11.3). The M^λ_K's are non-isomorphic simple modules. Since their number is the number of λ's, which is also the number of conjugacy classes of $\mathfrak{S}_n$, i.e. the number of simple $K\mathfrak{S}_n$-modules, the M^λ_K are the simple $\mathcal{H}_K$-modules. Now, we have an injection $M^\lambda_\mathcal{O}\otimes K \supseteq M^\lambda_K$, since in $y_{\lambda^*}\mathcal{H}_\mathcal{O}\otimes K = y_{\lambda^*}\mathcal{H}_K$, the submodule $M^\lambda_\mathcal{O}\otimes K$ is not annihilated by x_λ while M^λ_K is the simple and isotypic component defined by this same condition. So $\mathrm{Hom}_{\mathcal{H}_K}(M^\lambda_K, M^\mu_\mathcal{O}\otimes K) \subseteq \mathrm{Hom}_{\mathcal{H}_K}(M^\lambda_\mathcal{O}\otimes K, M^\mu_\mathcal{O}\otimes K) = \mathrm{Hom}_{\mathcal{H}_K}(M^\lambda_\mathcal{O}, M^\mu_\mathcal{O})\otimes K$, which is 0 when $\mu \not\ll \lambda$. When $\lambda = \mu$, then $0 \neq \mathrm{Hom}_{\mathcal{H}_K}(M^\lambda_K, M^\lambda_\mathcal{O}\otimes K) \subseteq \mathrm{Hom}_{\mathcal{H}_K}(M^\lambda_K, y_{\lambda^*}\mathcal{H}_K) = K$ by Theorem 18.14. □

18.3. Hecke algebras of type $\mathbb{B}\mathbb{C}$; Hoefsmit's matrices and Jucys–Murphy elements

Definition 18.16. (see also Definition 3.6) *$\mathcal{H}_R(\mathbb{B}\mathbb{C}_n)$ is the algebra over $R = \mathbb{Z}[x, y, y^{-1}]$ (where x, y are indeterminates) defined by n generators $a_0, a_1, \dots, a_{n-1}$ satisfying the relations $(a_0 - x)(a_0 + 1) = 0$, $(a_i - y)(a_i + 1) = 0$ if $i \geq 1$, $a_0a_1a_0a_1 = a_1a_0a_1a_0$, $a_ia_j = a_ja_i$ if $|i - j| \geq 2$, $a_ia_{i+1}a_i = a_{i+1}a_ia_{i+1}$ if $n - 2 \geq i \geq 1$. If $K \supseteq R$, let $\mathcal{H}_K(\mathbb{B}\mathbb{C}_n) := \mathcal{H}_R(\mathbb{B}\mathbb{C}_n) \otimes_R K$.*

We denote $t_i = a_{i-1}a_{i-2}\dots a_0 \dots a_{i-2}a_{i-1}$ for $n \geq i \geq 1$.

Theorem 18.17. *As a $\mathbb{Q}(x, y)$-algebra, $\mathcal{H}_{\mathbb{Q}(x,y)}(\mathbb{B}\mathbb{C}_n)$ injects in a matrix algebra $\mathrm{Mat}_N(\mathbb{Q}(x, y))$ (N an integer ≥ 0) such that each t_i is sent to a diagonal matrix whose diagonal elements are taken in the following subset of R*

$$\{x, xy, xy^2, \dots, xy^{2n-2}, -1, -y, -y^2, \dots, -y^{2n-2}\}.$$

Proof. Denote $K = \mathbb{Q}(x, y)$. We essentially need to recall Hoefsmit's description of a set of representations of $\mathcal{H}_K(\mathbb{B}\mathbb{C}_n)$.

Recall that a Young diagram is a way to visualize a partition $\lambda_1 \geq \lambda_2 \geq \dots \lambda_m > 0$ as a series of m rows in an $m \times \lambda_1$-matrix with lengths $\lambda_1, \dots, \lambda_m$. The sum $\sum_i \lambda_i$ is called the size of the Young diagram. It is customary to represent it as lines of square empty boxes, hence the term diagram. A Young **tableau** is the same thing but with integers filling the boxes; the associated Young diagram is called its type. A Young tableau is said to be **standard** if each row and each column is increasing (so the integers are all distinct). Note that we admit the empty diagram and tableau $\emptyset$, associated with a partition of size 0. A pair of standard tableaux of size n is $(T^{(1)}, T^{(2)})$ where each $T^{(i)}$ is standard of type $D^{(i)}$, the sum of their sizes is n and the integers filling them are the ones in $\{1, \dots, n\}$.

Hoefsmit's construction associates with each such pair $(D^{(1)}, D^{(2)})$ of Young diagrams of total size n, a representation of $\mathcal{H}_K(\mathbb{B}\mathbb{C}_n)$ on $V_{(D^{(1)},D^{(2)})} = K\tau_1 \oplus \dots \oplus K\tau_f$ where the τ_i are the distinct Young standard tableaux of type $(D^{(1)}, D^{(2)})$ (see [Hoef74] 2, [GePf00] 10).

The action of $\mathcal{H}_K(\mathbb{B}\mathbb{C}_n)$ on the sum V of the $V_{(D^{(1)},D^{(2)})}$'s is faithful since $\mathcal{H}_K(\mathbb{B}\mathbb{C}_n)$ is semi-simple and the Hoefsmit representations give all the simple modules of this algebra (see [GePf00] 10.5.(ii), [Hoef74] 2.2.14).

The action of t_i $(1 \leq i \leq n)$ is given in [GePf00] 10.1.6, [Hoef74] 3.3.3. If $\tau = (T^{(1)}, T^{(2)})$ is a pair of Young tableaux of type $(D^{(1)}, D^{(2)})$, the action of t_i

on $\tau \in V_{(D^{(1)},D^{(2)})}$ is given by

$$t_i.\tau = y^{c-r+i-1}x\tau \quad \text{or} \quad t_i.\tau = -y^{c-r+i-1}\tau$$

according to whether i is found in $T^{(1)}$ or $T^{(2)}$, and where c and r denote the column and row labels of the box where it is found.

It is clear that $c - r \leq n - 1$. The fact that our tableaux are standard implies that $i \geq c + r - 1$ (just go from i to the origin $c = r = 1$ along the cth column, then along the first row). Then $c - r + i - 1 \geq 2c - 2 \geq 0$. So the t_i's do act on V diagonally with scalars among those given in the theorem. □

Corollary 18.18. *Let $\mathcal{H}_{\mathbb{Z}[y,y^{-1}]}(\mathfrak{S}_n, y)$ be the algebra generated by $a_1, \ldots, a_{n-1}$ satisfying $a_i^2 = (y-1)a_i + y$, $a_ia_{i+1}a_i = a_{i+1}a_ia_{i+1}$, $a_ja_i = a_ja_i$ when $|i-j| \geq 2$. Denote $\zeta_n = a_{n-1}a_{n-2}\ldots a_1a_1\ldots a_{n-2}a_{n-1}$. Then $(\Pi_{i=0}^{2n-2}(\zeta_n - y^i))^2 = 0$.*

Proof. Checking the relations defining $\mathcal{H}_R(\mathrm{BC}_n)$, it is easily checked that there is a morphism of $\mathbb{Z}[y, y^{-1}]$-algebras

$$\theta\colon \mathcal{H}_R(\mathrm{BC}_n) \to \mathcal{H}_{\mathbb{Z}[y,y^{-1}]}(\mathfrak{S}_n, y)$$

defined by $\theta(a_0) = \theta(x) = -1$, $\theta(a_i) = a_i$ for $i \geq 1$. Then $\zeta_n = -\theta(t_n)$.

Theorem 18.17 implies that $\Pi_{i=0}^{2n-2}(t_n - xy^i).\Pi_{i=0}^{2n-2}(t_n + y^i) = 0$. The image by θ gives the sought equality. □

Remark 18.19. We have a series of inclusions: $R(t_1, \ldots, t_n) \subseteq \mathcal{H}_R(\mathrm{BC}_n) \subseteq \mathcal{H}_K(\mathrm{BC}_n) \subseteq \mathrm{Mat}_N(K)$ where the first is also a subalgebra of R^N, hence commutative. Note that this representation does not imbed $\mathcal{H}_R(\mathrm{BC}_n)$ as a subalgebra of $\mathrm{Mat}_N(R)$ when $n \geq 2$ (see [Hoef74] §2.2, [GePf00] §10.1).

The case of the group algebra of $W(\mathrm{BC}_n)$ is deceptive at this point. The specialization for $x = y = 1$ sends $R(t_1, \ldots, t_n)$ to the group algebra of the normal subgroup of $W(\mathrm{BC}_n)$ of order 2^n. But $R(t_1, \ldots, t_n)$ itself is R-free with a rank much bigger when $n \geq 2$ (see [ArKo] 3.17). Through a specialization f giving a semi-simple algebra isomorphic with the group algebra of $W(\mathrm{BC}_n)$, $K(t_1, \ldots, t_n)$ is sent to a maximal commutative semi-simple subalgebra.

Remark 18.20. In order to get Corollary 18.18, one could in fact just look at representations of type $V_{(D^{(1)},\emptyset)}$ (notation of the proof of Theorem 18.17) where a_0 acts by the scalar x and whose sum gives a faithful matrix representation of $\mathcal{H}_R(\mathfrak{S}_n)$ (see [Hoef74] 2.3.1). We chose the more general Theorem 18.17 for the convenience of references.

18.4. Hecke algebras of type BC: some computations

Recall the basis $(a_w)_{w\in W}$. We sometimes write $a(w)$ when the expression for w is complicated.

Note that each a_i ($i \geq 1$) is invertible, so $a(w)$ is invertible when $w \in \mathfrak{S}_n$. If $\lambda = (\lambda_1, \lambda_2, \ldots)$ is a sequence of integers ≥ 1 whose sum is n, recall that $\mathfrak{S}_\lambda = \mathfrak{S}_{\lambda_1} \times \mathfrak{S}_{\{\lambda_1+1,\ldots,\lambda_1+\lambda_2\}} \times \cdots \subseteq \mathfrak{S}_n$.

The t_m's commute with each other, so we can set the following.

Definition 18.21. *Let $m \in \{0, 1, \ldots, n\}$. We denote $\pi_m = \Pi_{j=1}^m (t_j - xy^{j-1})$, $\tilde{\pi}_m = \Pi_{j=1}^m (t_j + y^{j-1})$.*

Let $w_m^{(n)} \in \mathfrak{S}_n$ be the permutation of $\{1, \ldots, n\}$ corresponding to addition of m mod. n, namely $w_m^{(n)} = (1, 2, \ldots, n)^m$.

Denote $v_m^{(n)} = \pi_m a(w_m^{(n)})\tilde{\pi}_{n-m}$.

We shall sometimes omit the superscript $^{(n)}$ when the dimension is clear from the context.

Here are the basics about the law of $\mathcal{H}_R(\mathrm{BC}_n)$.

Proposition 18.22. *(i) The t_i's commute.*

(ii) The family $t_{i_1} \ldots t_{i_l} a(w)$, for $1 \leq i_1 < \ldots < i_l \leq n$ and $w \in \mathfrak{S}_n$, is an R-basis of $\mathcal{H}_R(\mathrm{BC}_n)$.

(iii) a_i commutes with t_j if $j \neq i, i+1$.

(iv) a_i commutes with $\Pi_{j=1}^k (t_j - y^{j-1}\lambda)$ for $\lambda \in R$, $k \neq i$.

Proof. Denote by $s_0, s_1, \ldots, s_{n-1}$ the generators of the Coxeter group of type BC_n satisfying the same braid relations as the a_i's. Denote $s_i' = s_{i-1} \ldots s_0 \ldots s_{i-1}$. Then $a_i = a(s_i)$, $t_i = a(s_i')$.

(i) has already been seen.

(iii) Since $a_0 = t_1$, we may assume $i \geq 1$. By the braid relations, it is clear that a_i commutes with $a_0, a_1, \ldots a_{i-2}, a_{i+2}, \ldots, a_{n-1}$. This implies our claim as soon as we have proved that a_i commutes with $t_{i+2} = a_{i+1}a_i \ldots a_0 \ldots a_i a_{i+1}$. Using the braid relation between a_i and a_{i+1}, along with the fact that a_{i+1} commutes with $t_i = a_{i-1} \ldots a_0 \ldots a_{i-1}$, one gets $a_i t_{i+2} = a_i a_{i+1} a_i t_i a_i a_{i+1} = a_{i+1} a_i a_{i+1} t_i a_i a_{i+1} = a_{i+1} a_i t_i a_{i+1} a_i a_{i+1} = a_{i+1} a_i t_i a_i a_{i+1} a_i = t_{i+2} a_i$, i.e. our claim.

We now prove that

(iv′) a_i commutes with $yt_i + t_{i+1}$ and $t_i t_{i+1}$.

Using $t_{i+1} = a_i t_i a_i$ and the quadratic equation satisfied by a_i, the commutator $[a_i, t_{i+1}]$ equals $a_i^2 t_i a_i - a_i t_i a_i^2 = -y[a_i, t_i]$, whence the first part of (iv′). Similarly, and since $t_i t_{i+1} = t_{i+1} t_i$, we have $a_i t_i t_{i+1} - t_i t_{i+1} a_i = [a_i^2, t_i a_i t_i] =$

$(y-1)[a_i, t_ia_it_i] = (y-1)(t_{i+1}t_i - t_it_{i+1}) = 0$, whence the second part of (iv′).

(iv) follows easily from (iv′) since a_i commutes with $(t_i - y^{i-1}\lambda)(t_{i+1} - y^i\lambda)$, and with $(t_j - y^{j-1}\lambda)$ for each $j \neq i, i+1$.

(ii) Denote $\mathcal{H} = \sum Rt_{i_1}\dots t_{i_l}a(w)$. Since $\mathcal{H}_R(\mathrm{BC}_n)$ is R-free and $\mathcal{H}_K(\mathrm{BC}_n)$ has dimension $2^n n!$ for any field K containing R, it suffices to prove $\mathcal{H} = \mathcal{H}_R(\mathrm{BC}_n)$.

Since $\mathcal{H}_R(\mathrm{BC}_n)$ is generated by $a_0, a_1, \dots, a_{n-1}$, it suffices to check that $[a_i, t_{i_1}\dots t_{i_l}] \in \mathcal{H}$ for any sequence $1 \leq i_1 < \dots < i_l \leq n$. If $i = 0$, this is clear. If $i \geq 1$, using (iii), (iv′) above, and invertibility of y, one may assume that $i_1 = i = i_l$. Then we may write $a_it_i = t_{i+1}a_i^{-1} = t_{i+1}(y^{-1}a_i + y^{-1} - 1)$. □

Lemma 18.23. *Let $1 \leq m, m' \leq n$.*

(i) $\pi_m\mathcal{H}_R(\mathrm{BC}_n)\widetilde{\pi}_{m'} = 0$ when $m + m' > n$.

(ii) When $m < n$, $v_m^{(n)} = v_m^{(n-1)}.a(s_{n,n-m})(t_{n-m} + y^{n-m-1}) = (t_m - xy^{m-1})a(s_{m,n}).v_{m-1}^{(n-1)}$.

(iii) $\pi_m a_w\widetilde{\pi}_{n-m} = 0$ if $w \in \mathfrak{S}_n \setminus \mathfrak{S}_{m,n-m}w_m^{(n)}\mathfrak{S}_{n-m,m}$.

(iv) Let $\sigma\colon \mathcal{H}_R(\mathfrak{S}_{n-m,m}) \to \mathcal{H}_R(\mathfrak{S}_{m,n-m})$ be the isomorphism defined by $\sigma(a_i) = a_{w_m^{(n)}(i)}$ for $i \neq 0, m$ (note that $w_m^{(n)}(i) \equiv i + m$ mod. n). Then, for all $h \in \mathcal{H}_R(\mathfrak{S}_{n-m,m})$,

$$a(w_m^{(n)})h = \sigma(h)a(w_m^{(n)}), \quad a(w_{n-m}^{(n)})^{-1}h = \sigma(h)a(w_{n-m}^{(n)})^{-1},$$
$$\text{and} \quad v_m h = \sigma(h)v_m.$$

(v) $v_m\mathcal{H}_R(\mathrm{BC}_n) = v_m\mathcal{H}_R(\mathfrak{S}_n)$.

Proof. Each $w \in \mathfrak{S}_n$ such that $w(n) < n$ can be written (see Example 2.5)

(A) $w = w's_{n,w^{-1}(n)} = s_{w(n),n}w''$ with $w', w'' \in \mathfrak{S}_{n-1}$ and lengths added,

i.e. $l(w) = l(w') + n - w^{-1}(n) = n - w(n) + l(w'')$.

(i) Assume $m + m' > n$.

Since $\mathcal{H}_R(\mathrm{BC}_n) = R(t_1, \dots, t_n)\mathcal{H}_R(\mathfrak{S}_n)$ and $\pi_m, \widetilde{\pi}_{m'} \in R(t_1, \dots, t_n)$, it suffices to show that $\pi_m\mathcal{H}_R(\mathfrak{S}_n)\widetilde{\pi}_{m'} = 0$. We prove $\pi_m a_w\widetilde{\pi}_{m'} = 0$ for any $w \in \mathfrak{S}_n$ by induction on $l(w)$ and n.

If $w = 1$ (this also accounts for $n = 1$), one gets $\pi_m\widetilde{\pi}_{m'}$, which is a multiple of $\pi_1\widetilde{\pi}_1$. This is zero by the quadratic relation satisfied by $t_1 = a_1$. Similarly, if $m = n$, then π_m is central by Proposition 18.22(iv) and we get again $\pi_m\widetilde{\pi}_{m'} = 0$. The same is true if $m' = n$. So we assume $m, m' < n$.

If $w \in \mathfrak{S}_{n-1}$, the induction hypothesis on n gives the result. When $w \notin \mathfrak{S}_{n-1}$, (A) above gives $a_w = a_{w'}a_{n-1}\dots a_k$ for $w(k) = n$ and $w' \in \mathfrak{S}_{n-1}$. If $k \geq m'$, then $a_{n-1}, \dots, a_k$ commute with $\widetilde{\pi}_{m'-1}$ by Proposition 18.22(iv). Then

$\pi_m a_w \widetilde{\pi}_{m'} = (\pi_m a_{w'} \widetilde{\pi}_{m'-1}) a_{n-1} \dots a_k (t_{m'} + y^{m'-1})$ and the first parenthesis is 0 by the induction hypothesis on n. If $k < m'$, then a_k commutes with $\widetilde{\pi}_{m'}$ by Proposition 18.22(iv), so we get $\pi_m a_w \widetilde{\pi}_{m'} = \pi_m a_{w'} a_{n-1} \dots a_{k+1} \widetilde{\pi}_{m'} a_k = (\pi_m a(w' s_{n-1} \dots s_{k+1}) \widetilde{\pi}_{m'}) a_k$ and the parenthesis is 0 by the induction hypothesis on $l(w)$.

(ii) Applying (A) for $w = w_m^{(n)}$ (addition of m mod. n in $\{1, \dots, n\}$), we get $w^{-1}(n) = n - m$, $w(n) = m$, $w' = w_m^{(n-1)}$, and $w'' = w_{m-1}^{(n-1)}$.

In the Hecke algebra, this implies that

$$a\big(w_m^{(n)}\big) = a\big(w_m^{(n-1)}\big) a_{n-1} a_{n-2} \dots a_{n-m} = a_m a_{m+1} \dots a_{n-1} a\big(w_{m-1}^{(n-1)}\big).$$

Using the commutation of Proposition 18.22(iv), one gets $(t_m - xy^{m-1})$ $a(s_{m,n}).v_{m-1}^{(n-1)} = (t_m - xy^{m-1}) a(s_{m,n}).\pi_{m-1} a(w_{m-1}^{(n-1)}) \widetilde{\pi}_{n-m} = (t_m - xy^{m-1}) \pi_{m-1}$ $a(s_{m,n}) a(w_{m-1}^{(n-1)}) \widetilde{\pi}_{n-m} = \pi_m a(w_m^{(n)}) \widetilde{\pi}_{n-m} = v_m^{(n)}$. This gives (ii) since the first equality is proved in the same fashion.

(iii) Let $w \in \mathfrak{S}_n$ be such that $\pi_m a(w) \widetilde{\pi}_{n-m} \neq 0$. Let us write $w = w_1 w_2 w_3$ with $w_1 \in \mathfrak{S}_{m,n-m}$, $w_3 \in \mathfrak{S}_{n-m,m}$ and $l(w) = l(w_1) + l(w_2) + l(w_3)$ (see Proposition 2.4). Then π_m (resp. $\widetilde{\pi}_{n-m}$) commutes with $a(w_1)$ (resp. $a(w_3)$) by Proposition 18.22(iv). Then $\pi_m a(w_2) \widetilde{\pi}_{n-m} \neq 0$. So we may assume that w is such that $w \in D_{I,I'}$ where $W_I = \mathfrak{S}_{m,n-m}$ and $W_{I'} = \mathfrak{S}_{n-m,m}$ in $W = \mathfrak{S}_n$ (Proposition 2.4). We prove that $w = w_m^{(n)}$ by induction on n. Since $w \in D_{I,I'}$, each reduced expression of w begins with s_m, so (A) implies $w(n) = m$ and $w = s_{m,n} w'$ with $w' \in \mathfrak{S}_{n-1}$ ($w \in \mathfrak{S}_{n-1}$ is impossible by (i)). Then, using Proposition 18.22(iv) again, $\pi_m a(s_{m,n}) a(w') \widetilde{\pi}_{n-m} = (t_m - xy^{m-1}) \pi_{m-1} a(s_{m,n})$ $a(w') \widetilde{\pi}_{n-m} = (t_m - xy^{m-1}) a(s_{m,n}) (\pi_{m-1} a(w') \widetilde{\pi}_{n-m})$, so $\pi_{m-1} a(w') \widetilde{\pi}_{n-m} \neq 0$ (we assume $m \neq 0$, the case when $m = 0$ is clear). The induction gives $w' \in \mathfrak{S}_{m-1,n-m,1} w_{m-1}^{(n-1)} \mathfrak{S}_{n-m,m-1,1}$. Then $w \in s_{m,n} \mathfrak{S}_{m-1,n-m,1} w_{m-1}^{(n-1)} \mathfrak{S}_{n-m,m-1,1} \subseteq \mathfrak{S}_{m,n-m} s_{m,n} w_{m-1}^{(n-1)} \mathfrak{S}_{n-m,m-1,1}$. We have seen above that $s_{m,n} w_{m-1}^{(n-1)} = w_m^{(n)}$, thus our claim is proved.

(iv) Let $i \neq 0, n-m$. Using the geometric representation of $\mathfrak{S}_n$ one gets $w_m s_i = s_{w_m(i)} w_m$ and $s_i w_{n-m} = w_{n-m} s_{w_m(i)}$ with lengths adding and $w_m(i) \neq m$. Then $a(w_m) a_i = a(w_m s_i) = a_{w_m(i)} a(w_m)$ and $a_i a(w_{n-m}) = a(w_{n-m}) a_{w_m(i)}$. Moreover, $v_m a_i = \pi_m a(w_m) \widetilde{\pi}_{n-m} a_i = \pi_m a(w_m) a_i \widetilde{\pi}_{n-m} = \pi_m a_{w_m(i)} a(w_m) \widetilde{\pi}_{n-m} = a_{w_m(i)} \pi_m a(w_m) \widetilde{\pi}_{n-m}$ by Proposition 18.22(iv).

(v) Let us show first that

$$\text{(v}'\text{)} \qquad v_m t_i \in v_m \mathcal{H}_R(\mathfrak{S}_n)$$

for all i. We prove this by induction on i. Note that, by the quadratic relation satisfied by $a_0 = t_1$, one has $\pi_1 t_1 = -\pi_1$ and therefore $\pi_m t_1 = -\pi_m$. Similarly $\widetilde{\pi}_m t_1 = x \widetilde{\pi}_m$. This gives our claim for $i = 1$ according to $m = n$ or not. Assume (v′) for i and let us prove it for $i+1$ when $i \leq n-1$. One has $t_{i+1} = a_i t_i a_i$. Assume $i \neq n-m$. Then (iv) and the induction hypothesis imply $v_m t_{i+1} = a_{w_m(i)} v_m t_i a_{i+1} \in a_{w_m(i)} v_m \mathcal{H}_R(\mathfrak{S}_n) = v_m a_i \mathcal{H}_R(\mathfrak{S}_n)$, thus our claim. Assume $i = n-m$. By (i), we have $v_m(t_{i+1} + y^i) = \pi_m a(w_m)\widetilde{\pi}_i(t_{i+1} + y^i) = \pi_m a(w_m)\widetilde{\pi}_{n-m+1} = 0$. So $v_m t_{i+1} = -y^i v_m$ in that case, thus (v′) again.

To show (v), it suffices to prove that $v_m \mathcal{H}_R(\mathfrak{S}_n)$ is stable under right multiplication by $a_0 = t_1$. Let $w \in \mathfrak{S}_n$. By a clear variant of (A), $w = s_{w(1),1} w'$ where $w' = s_{1,w(1)} w \in \mathfrak{S}_{1,n-1} = <s_2, \ldots, s_{n-1}>$ and lengths add $l(w) = w(1) - 1 + l(w')$. Then $w' s_0 = s_0 w'$ with lengths adding, so we can write $v_m a(w) t_1 = v_m a(s_{w(1),1}) a(w' s_0) = v_m a(s_{w(1),1}) t_1 a(w') = v_m t_{w(1)} a(s_{w(1),1})^{-1} a(w') \in v_m \mathcal{H}_R(\mathfrak{S}_n)$ by (v′) above. □

Proposition 18.24. *(i)* $\pi_m \mathcal{H}_R(\mathfrak{S}_n)\widetilde{\pi}_{n-m} = v_m \mathcal{H}_R(\mathfrak{S}_{n-m,m}) = \mathcal{H}(\mathfrak{S}_{m,n-m}) v_m$.

(ii) The map

$$\mathcal{H}_R(\mathfrak{S}_n) \to v_m \mathcal{H}_R(\mathrm{BC}_n), \quad h \mapsto v_m h$$

is a bijection.

Proof. (i) The last equality comes from Lemma 18.23(iv). By Lemma 18.23(ii)–(iii) and Proposition 18.22(iv), we also have $\pi_m \mathcal{H}_R(\mathfrak{S}_n)\widetilde{\pi}_{n-m} = \pi_m \mathcal{H}_R(\mathfrak{S}_{m,n-m}) a(w_m) \mathcal{H}_R(\mathfrak{S}_{n-m,m})\widetilde{\pi}_{n-m} = \mathcal{H}_R(\mathfrak{S}_{m,n-m})\pi_m a(w_m)\widetilde{\pi}_{n-m}\mathcal{H}_R(\mathfrak{S}_{n-m,m}) = \mathcal{H}_R(\mathfrak{S}_{m,n-m}) v_m \mathcal{H}_R(\mathfrak{S}_{n-m,m})$, whence $\pi_m \mathcal{H}_R(\mathfrak{S}_n)\widetilde{\pi}_{n-m} = v_m \mathcal{H}_R(\mathfrak{S}_{n-m,m}) = \mathcal{H}_R(\mathfrak{S}_{m,n-m}) v_m$.

(ii) Lemma 18.23(v) gives the surjectivity. It remains to check injectivity.

By Proposition 18.22(ii), we have $\mathcal{H}_R(\mathrm{BC}_n) = t_1 t_2 \ldots t_n \mathcal{H}_R(\mathfrak{S}_n) \oplus \mathcal{H}'$ as an R-module, where $\mathcal{H}' = \bigoplus_{1 \leq i_1 < \ldots < i_l \leq n,\ l \neq n} R t_{i_1} \ldots t_{i_l} \mathcal{H}_R(\mathfrak{S}_n)$. Also $t_1 t_2 \ldots t_n \mathcal{H}_R(\mathfrak{S}_n) \cong \mathcal{H}_R(\mathfrak{S}_n)$ as R-module (left multiplication by $t_1 t_2 \ldots t_n$). So let

$$p_n \colon \mathcal{H}_R(\mathrm{BC}_n) \to \mathcal{H}_R(\mathfrak{S}_n)$$

be such that $t_1 t_2 \ldots t_n p_n(h)$ is the projection of $h \in \mathcal{H}_R(\mathrm{BC}_n)$ with kernel $\mathcal{H}'$. We prove

(ii′) $$p_n(v_m) = a(s_{1,m+1})^{-1} a(s_{2,m+2})^{-1} \ldots a(s_{n-m,n})^{-1}$$

by induction on n. If $m = n$, then $v_m = \pi_n$ and $p_n(\pi_n) = 1$ by Proposition 18.22(ii). Assume $m < n$ (then $n \neq 0$) and that our claim is proved for $v_m^{(n-1)} \in \mathcal{H}_R(\mathrm{BC}_{n-1})$. By Lemma 18.23(ii), one has $v_m^{(n)} = v_m^{(n-1)} a(s_{n,n-m})(t_{n-m} + y^{n-m-1}) = v_m^{(n-1)}(t_n a(s_{n-m,n})^{-1} + y^{n-m-1} a(s_{n,n-m})) = t_n v_m^{(n-1)} a(s_{n-m,n})^{-1} + y^{n-m-1} v_m^{(n-1)} a(s_{n,n-m})$ since t_n commutes with $\mathcal{H}_R(\mathrm{BC}_{n-1})$ by Proposition 18.22(iii). We also have $\mathcal{H}_R(\mathrm{BC}_{n-1}) \subseteq \mathcal{H}'$ and $p_n(t_n h) = p_{n-1}(h)$ for all $h \in$

$\mathcal{H}_R(\mathrm{BC}_{n-1})$ by Proposition 18.22(ii). Then $p_n(v_m^{(n)}) = p_{n-1}(v_m^{(n-1)})a(s_{n-m,n})^{-1}$. The induction hypothesis then gives (ii′).

Now, it is clear that, if $h \in \mathcal{H}_R(\mathfrak{S}_n)$, then $a(s_{n-m,n}) \dots a(s_{2,m+2})a(s_{1,m+1}) p_n(v_m h) = h$. This gives an inverse to our map $h \mapsto v_m h$. □

18.5. Hecke algebras of type BC: a Morita equivalence

In what follows we assume that $\mathcal{O}$ is a principal ideal domain and that Q, q are two elements of $\mathcal{O}$ such that the following hypothesis is satisfied

Hypothesis 18.25. *q, $\Pi_{i=0}^{n-1}(Q + q^i)$, and $\Pi_{i=0}^{n-1}(Qq^i + 1)$ are invertible in $\mathcal{O}$.*

Definition 18.26. *$\mathcal{H}_\mathcal{O}(\mathrm{BC}_n, Q, q)$ (most of the time abbreviated as $\mathcal{H}_\mathcal{O}(\mathrm{BC}_n)$) is defined as $\mathcal{H}_R(\mathrm{BC}_n) \otimes_R \mathcal{O}$ where the morphism $R = \mathbb{Z}[x, y, y^{-1}] \to \mathcal{O}$ is the one sending x to Q and y to q.*

Our main goal is to prove the following theorem.

Theorem 18.27. *There exist $\varepsilon_0, \varepsilon_1, \dots, \varepsilon_n$ in $\mathcal{H}_\mathcal{O}(\mathrm{BC}_n, Q, q)$ such that*

(i) the ε_m are orthogonal idempotents such that $\varepsilon_m \mathcal{H}_\mathcal{O}(\mathrm{BC}_n)\varepsilon_{m'} = 0$ when $m \neq m'$,

(ii) ε_m centralizes $\mathcal{H}_\mathcal{O}(\mathfrak{S}_{m,n-m}, q)$ and $\varepsilon_m \mathcal{H}_\mathcal{O}(\mathrm{BC}_n)\varepsilon_m = \varepsilon_m \mathcal{H}_\mathcal{O}(\mathfrak{S}_{m,n-m}, q) \cong \mathcal{H}_\mathcal{O}(\mathfrak{S}_{m,n-m}, q)$, as $\mathcal{O}$-algebras, by the natural map,

(iii) letting $\varepsilon = \varepsilon_0 + \dots + \varepsilon_n$, $\mathcal{H}_\mathcal{O}(\mathrm{BC}_n)$ is Morita equivalent to $\varepsilon \mathcal{H}_\mathcal{O}(\mathrm{BC}_n)\varepsilon$.

Proposition 18.28. *π_m and $\widetilde{\pi}_m$ are idempotents up to units of the center of $\mathcal{H}_\mathcal{O}(\mathfrak{S}_m)$.*

Proof. One may assume $m = n$. In $\mathcal{H}_R(\mathrm{BC}_n)$, π_n and $\widetilde{\pi}_n$ are central by Proposition 18.22(iv), and $\pi_n t_1 = -\pi_n$, $\widetilde{\pi}_n t_1 = y\widetilde{\pi}_n$ by the quadratic equation satisfied by $t_1 = a_0$. In $\mathcal{H}_R(\mathrm{BC}_n)$, one has $\pi_n t_i = \pi_n a_{i-1} \dots a_1 t_1 a_1 \dots a_{i-1} = a_{i-1} \dots a_1 \pi_n t_1 a_1 \dots a_{i-1} = -\pi_n a_{i-1} \dots a_1 a_1 \dots a_{i-1}$ and similarly $\widetilde{\pi}_n t_i = y\widetilde{\pi}_n a_{i-1} \dots a_1 a_1 \dots a_{i-1}$.

Then, in $\mathcal{H}_\mathcal{O}(\mathrm{BC}_n)$,

$$(\pi_n)^2 = \Pi_{i=1}^n \pi_n (t_i - Qq^{i-1}) = (-1)^n \pi_n \Pi_{i=1}^n (a(s_{i,1})a(s_{1,i}) + Qq^{i-1}).$$

It remains to show that $\Pi_{i=1}^n (a(s_{i,1})a(s_{1,i}) + Qq^{i-1})$ is a unit in the center of $\mathcal{H}_\mathcal{O}(\mathfrak{S}_n)$. This element is central since it is the image of $(-1)^n \pi_n$ under the epimorphism $\mathcal{H}_R(\mathrm{BC}_n) \to \mathcal{H}_\mathcal{O}(\mathfrak{S}_n)$ sending x to Q, y to q, a_0 to -1, a_i ($i \geq 1$) to the element denoted the same. To show that it is a unit, it suffices to check that, for $i = 1, \dots, n$, $A := a(s_{i,1})a(s_{1,i}) + Qq^{i-1}$ is a unit. Let

$P(X) \in \mathcal{O}[X]$ be the polynomial $\Pi_{j=0}^{2i-2}(X - Qq^{i-1} - q^j)^2$. Then $P(0)$ is a unit of $\mathcal{O}$ by Hypothesis 18.25. One has $P(A) = 0$ by Corollary 18.18. So this allows us to write $AA' = A'A = 1$ for A' a polynomial expression in A.

The same can be done for $\widetilde{\pi}_m$, using $(\widetilde{\pi}_n)^2 = \widetilde{\pi}_n \Pi_{i=1}^n (Qa(s_{i,1})a(a_{1,i}) + q^{i-1})$ and defining $\widetilde{P}(X) = \Pi_{j=0}^{2i-2}(X - q^{i-1} - Qq^j) \in \mathcal{O}[X]$. □

Theorem 18.29. *The regular right $\mathcal{H}_{\mathcal{O}}(\mathrm{BC}_n, Q, q)$-module is isomorphic with*

$$v_0\mathcal{H}_{\mathcal{O}}(\mathrm{BC}_n) \oplus \ldots \oplus (v_m\mathcal{H}_{\mathcal{O}}(\mathrm{BC}_n))^{\binom{n}{m}} \oplus \ldots \oplus v_n\mathcal{H}_{\mathcal{O}}(\mathrm{BC}_n).$$

Proof of Theorem 18.29. Denote $I_m^{(n')} = v_m^{(n')}\mathcal{H}_{\mathcal{O}}(\mathrm{BC}_n)$ for $0 \le m \le n' \le n$.

Lemma 18.30. *Let $0 \le m < n' \le n$.*
(i) $I_m^{(n')}$ is $\mathcal{O}$-free of rank $2^{n-n'}n!$.
(ii) If $m \neq 0$, then $I_{m-1}^{(n')}$ is an $\mathcal{O}$-pure submodule of $I_m^{(n')}$.

Proof of Lemma 18.30. (i) Proposition 18.24(ii) tells us that $v_m^{(n')}\mathcal{H}_{\mathcal{O}}(\mathrm{BC}_{n'})$ is $\mathcal{O}$-free of rank $|\mathfrak{S}_{n'}|$. Using the standard basis of $\mathcal{H}_{\mathcal{O}}(\mathrm{BC}_n)$ and the multiplication formula when lengths add, we see that $\mathcal{H}_{\mathcal{O}}(\mathrm{BC}_n)$ is a free left $\mathcal{H}_{\mathcal{O}}(\mathrm{BC}_{n'})$-module of basis the a_d's for d ranging over $D_{I,\emptyset}$ where $W(\mathrm{BC}_{n'}) = W(\mathrm{BC}_n)_I$. Its cardinality is $|W(\mathrm{BC}_n) : W(\mathrm{BC}_{n'})|$. Then $v_m^{(n')}\mathcal{H}_{\mathcal{O}}(\mathrm{BC}_n) \cong v_m^{(n')}\mathcal{H}_{\mathcal{O}}(\mathrm{BC}_{n'}) \otimes_{\mathcal{H}_{\mathcal{O}}(\mathrm{BC}_{n'})} \mathcal{H}_{\mathcal{O}}(\mathrm{BC}_n)$ is $\mathcal{O}$-free of rank $|\mathfrak{S}_{n'}|.|W(\mathrm{BC}_n) : W(\mathrm{BC}_{n'})|$. That is our claim.

(ii) The inclusion $v_m^{(n')}\mathcal{H}_{\mathcal{O}}(\mathrm{BC}_n) \subseteq v_m^{(n'-1)}\mathcal{H}_{\mathcal{O}}(\mathrm{BC}_n)$ takes place for any commutative ring, by Lemma 18.23(ii). The fact that the quotient is $\mathcal{O}$-free when $\mathcal{O}$ is a given principal ideal domain clearly follows from (i) since it is satisfied for any quotient field of $\mathcal{O}$. □

By Lemma 18.23(ii), left multiplication by $(t_m - Qq^{m-1})a(s_{m,n'})$ induces a surjection

$$\mu \colon I_{m-1}^{(n'-1)} \to I_m^{(n')}.$$

One has $I_{m-1}^{(n')} \subseteq \mathrm{Ker}(\mu)$ since $a(s_{m,n'})$ commutes with π_m (Proposition 18.22 (iv)) and therefore $(t_m - Qq^{m-1})a(s_{m,n'})v_{m-1}^{(n')} \in (t_m - Qq^{m-1})\pi_{m-1}\mathcal{H}\mathcal{O}(\mathrm{BC}_{n'})\widetilde{\pi}_{n'-m+1} = \pi_m\mathcal{H}\mathcal{O}(\mathrm{BC}_{n'})\widetilde{\pi}_{n'-m+1} = 0$ by Lemma 18.23(i).

By Lemma 18.30, $I_{m-1}^{(n')}$ is an $\mathcal{O}$-pure $\mathcal{O}$-submodule of $I_{m-1}^{(n'-1)}$ of corank the rank of $I_m^{(n')}$. Then $I_{m-1}^{(n')} = \mathrm{Ker}(\mu)$, thus giving an exact sequence of right $\mathcal{H}_{\mathcal{O}}(\mathrm{BC}_n)$-modules

$$\text{(E)} \qquad 0 \to I_{m-1}^{(n')} \to I_{m-1}^{(n'-1)} \to I_m^{(n')} \to 0$$

for any $1 \le m \le n' \le n$.

We now show that $I_m^{(n')}$ is projective as a right $\mathcal{H}_\mathcal{O}(\mathrm{BC}_n)$-module, for any $0 \leq m \leq n' \leq n$. When $m = 0$ (resp. $m = n'$), $I_m^{(n')} = \tilde{\pi}_{n'}\mathcal{H}_\mathcal{O}(\mathrm{BC}_n)$ (resp. $I_m^{(n')} = \pi_m\mathcal{H}_\mathcal{O}(\mathrm{BC}_n)$) and Proposition 18.28 gives projectivity. We now prove that $I_m^{(n')}$ is projective by induction on $n' - m$. If in the exact sequence (E) the third and fourth terms are projective, then the sequence splits and the second term is projective. This and the induction hypothesis give our claim.

Knowing that all terms in (E) are projective, we get

$$I_{m-1}^{(n'-1)} \cong I_{m-1}^{(n')} \oplus I_m^{(n')}$$

for any $1 \leq m \leq n' \leq n$. By iteration of the above from the case $m = n' = 1$, we get $I_0^{(0)} \cong \oplus_{k=0}^{n'}(I_k^{(n')})^{\binom{n'}{k}}$ for any $n' \leq n$. Taking $n' = n$, this gives the theorem since $I_0^{(0)} = \mathcal{H}_\mathcal{O}(\mathrm{BC}_n)$. □

We can now complete the proof of Theorem 18.27.

By Proposition 18.24(i) and (ii), there is a unique $z_m \in \mathcal{H}_\mathcal{O}(\mathfrak{S}_{n-m,m})$ such that

$$v_m a(w_{n-m})v_m = v_m z_m.$$

We are going to prove that z_m is invertible and that $\varepsilon_m := v_m z_m^{-1} a(w_{n-m})$ (for $m = 0, 1, \ldots, n$) satisfy the requirements of the theorem.

By Lemma 18.23(iv), it is clear that z_m is central in $\mathcal{H}_\mathcal{O}(\mathfrak{S}_{n-m,m})$.

Let us check that z_m is invertible. From Theorem 18.29, we see that the right ideal $v_m\mathcal{H}_\mathcal{O}(\mathrm{BC}_n)$ is projective, so it can be written $\varepsilon\mathcal{H}_\mathcal{O}(\mathrm{BC}_n)$ for an idempotent ε. Then $v_m\mathcal{H}_\mathcal{O}(\mathrm{BC}_n)v_m\mathcal{H}_\mathcal{O}(\mathrm{BC}_n) = v_m\mathcal{H}_\mathcal{O}(\mathrm{BC}_n)$ and therefore $v_m \in v_m\mathcal{H}_\mathcal{O}(\mathrm{BC}_n)v_m\mathcal{H}_\mathcal{O}(\mathrm{BC}_n)$. This can also be written $v_m \in v_m\mathcal{H}_\mathcal{O}(\mathfrak{S}_n)$ $v_m\mathcal{H}_\mathcal{O}(\mathfrak{S}_n) = v_m a(w_{n-m})v_m\mathcal{H}_\mathcal{O}(\mathfrak{S}_n) = v_m z_m\mathcal{H}_\mathcal{O}(\mathfrak{S}_n)$ by Lemma 18.23(v) and (ii). Using Proposition 18.24(ii), we get the existence of a right inverse of z_m in $\mathcal{H}_\mathcal{O}(\mathfrak{S}_n)$. Since $\mathcal{H}_\mathcal{O}(\mathfrak{S}_n)$ is $\mathcal{O}$-free, this right inverse is also a left inverse, and therefore is also one in $\mathcal{H}_\mathcal{O}(\mathfrak{S}_{n-m,m})$ (see also Exercise 14).

Let $\varepsilon_m := v_m a(w_{n-m})\sigma(z_m^{-1}) = v_m z_m^{-1} a(w_{n-m})$ (see Lemma 18.23(iv)). We have

$$\begin{aligned}(\varepsilon_m)^2 &= v_m a(w_{n-m})\sigma(z_m^{-1})v_m a(w_{n-m})\sigma(z_m^{-1})\\ &= v_m a(w_{n-m})v_m z_m^{-1} a(w_{n-m})\sigma(z_m^{-1})\\ &= v_m a(w_{n-m})\sigma(z_m^{-1}) = \varepsilon_m\end{aligned}$$

by Lemma 18.23(iv) and the definition of z_m. So ε_m is an idempotent. Moreover (i) follows from Lemma 18.23(i).

We have $\varepsilon_m\mathcal{H}_\mathcal{O}(\mathrm{BC}_n) = v_m\mathcal{H}_\mathcal{O}(\mathrm{BC}_n)$, so Theorem 18.29 implies that any projective indecomposable right $\mathcal{H}_\mathcal{O}(\mathrm{BC}_n)$-module is a summand of some

$\varepsilon_m \mathcal{H}_\mathcal{O}(\mathrm{BC}_n)$. This gives (iii) by the standard definition of Morita equivalences (see [Thévenaz] 1.9.9, [Ben91a] §2.2).

We have seen that z_m is central in $\mathcal{H}_\mathcal{O}(\mathfrak{S}_{n-m,m})$. Then ε_m centralizes $\mathcal{H}_\mathcal{O}(\mathfrak{S}_{m,n-m}) = \sigma\big(\mathcal{H}_\mathcal{O}(\mathfrak{S}_{n-m,m})\big)$ by Lemma 18.23(iv). The natural map $\mathcal{H}_\mathcal{O}(\mathfrak{S}_{m,n-m}) \to \varepsilon_m \mathcal{H}_\mathcal{O}(\mathfrak{S}_{m,n-m})$ sends h to $v_m z_m^{-1} a(w_{n-m})h$, so it is a bijection by Proposition 18.24(ii).

It remains to compute $\varepsilon_m \mathcal{H}_\mathcal{O}(\mathrm{BC}_n)\varepsilon_m$. We postpone the proof of the following lemma until after the present proof.

Lemma 18.31. $\tilde{\pi}_{n-m}\mathcal{H}_\mathcal{O}(\mathfrak{S}_n)\pi_m = \tilde{\pi}_{n-m}a(w_{n-m})\pi_m\mathcal{H}_\mathcal{O}(\mathfrak{S}_{m,n-m})$.

Using the above lemma and Lemma 18.23, we get

$$\begin{aligned}
\varepsilon_m \mathcal{H}_\mathcal{O}(\mathrm{BC}_n)\varepsilon_m &= v_m \mathcal{H}_\mathcal{O}(\mathrm{BC}_n) v_m z_m^{-1} a(w_{n-m})\\
&= v_m \mathcal{H}_\mathcal{O}(\mathfrak{S}_n) v_m z_m^{-1} a(w_{n-m})\\
&= \pi_m a(w_m)\tilde{\pi}_{n-m}\mathcal{H}_\mathcal{O}(\mathfrak{S}_n)\pi_m a(w_m)\tilde{\pi}_{n-m} z_m^{-1} a(w_{n-m})\\
&= \pi_m a(w_m)\tilde{\pi}_{n-m} a(w_{n-m})\pi_m \mathcal{H}_\mathcal{O}(\mathfrak{S}_{m,n-m})\\
&\quad \times a(w_m)\tilde{\pi}_{n-m} z_m^{-1} a(w_{n-m})\\
&= v_m a(w_{n-m})\pi_m a(w_m)\mathcal{H}_\mathcal{O}(\mathfrak{S}_{n-m,m})\tilde{\pi}_{n-m} z_m^{-1} a(w_{n-m})\\
&= v_m a(w_{n-m}) v_m z_m^{-1} a(w_{n-m})\mathcal{H}_\mathcal{O}(\mathfrak{S}_{m,n-m})\\
&= v_m a(w_{n-m})\mathcal{H}_\mathcal{O}(\mathfrak{S}_{m,n-m}) = \varepsilon_m \mathcal{H}_\mathcal{O}(\mathfrak{S}_{m,n-m}).
\end{aligned}$$

This completes the proof of Theorem 18.27. □

Proof of Lemma 18.31. From the defining relations of $\mathcal{H}_\mathcal{O}(\mathrm{BC}_n)$, it is easy to see that one may define an $\mathcal{O}$-linear anti-automorphism ι of $\mathcal{H}_\mathcal{O}(\mathrm{BC}_n)$ by $\iota(a_w) = a_{w^{-1}}$. One has $\iota(a(w_m)) = a(w_m^{-1}) = a(w_{n-m})$ and ι induces the identity on the commutative algebra generated by the t_i's. Then Proposition 18.24(i) gives $\tilde{\pi}_{n-m}\mathcal{H}_\mathcal{O}(\mathfrak{S}_n)\pi_m = \iota(\pi_m\mathcal{H}_\mathcal{O}(\mathfrak{S}_n)\tilde{\pi}_{n-m}) = \iota(\mathcal{H}_\mathcal{O}(\mathfrak{S}_{m,n-m})v_m) = \tilde{\pi}_{n-m}$ $a(w_{n-m})\pi_m\mathcal{H}_\mathcal{O}(\mathfrak{S}_{m,n-m})$. □

18.6. Cyclic Clifford theory and decomposition numbers

Assume that 2 is invertible in $\mathcal{O}$. Let A be an $\mathcal{O}$-free finitely generated $\mathcal{O}$-algebra.

We assume there is $\tau \in A^\times$ and a subalgebra B such that $A = B \oplus B\tau$, with $B\tau = \tau B$ and $\tau^2 \in B$. In other words, A is a $\mathbb{Z}/2\mathbb{Z}$-graded algebra in the sense of [CuRe87] §11.

Let

$$\theta\colon A \to A \text{ be defined by } \theta(b_1 + b_2\tau) = b_1 - b_2\tau$$

for $b_1, b_2 \in B$. This is clearly a ring automorphism.

Our aim is to prove the following (for decomposition matrices, we refer to Definition 5.25).

Theorem 18.32. *Let M be a θ-stable A-module (i.e. ${}^{\theta}M \cong M$). We assume that $(\mathcal{O}, K, k)$ is a splitting system for* $\mathrm{End}_A(M)$ *and* $\mathrm{End}_B(\mathrm{Res}_B^A M)$*, and that $A \otimes K$ and $B \otimes K$ are split semi-simple.*

If $\mathrm{Dec}_A(M)$ *is square lower unitriangular, then* $\mathrm{Dec}_B(\mathrm{Res}_B^A M)$ *is also square lower unitriangular.*

Remark 18.33. (i) From the hypothesis that A is a graded algebra over B, it is easy to see that $A \otimes K$ is semi-simple if and only if $B \otimes K$ is semi-simple (use [CuRe87] 11.16).

(ii) Theorem 18.32 is only about the indecomposable direct summands of M up to isomorphism (the multiplicities of those summands have no influence on the conclusion), so one might relax the hypothesis to assume only that θ permutes those indecomposable summands up to isomorphism. An equivalent hypothesis would be that $A \otimes_B \mathrm{Res}_B^A M$ is a multiple of M (use Lemma 18.34(iii) below).

The proof of Theorem 18.32 requires some lemmas and notation.

We denote by

$$\mathrm{Ind}_B^A\colon B\textbf{-mod} \to A\textbf{-mod}$$

the tensor product functor ${}_A A_B \otimes_B -$.

Lemma 18.34. *(i)* Res_B^A *and* Ind_B^A *are left and right adjoint to each other.*

(ii) If V is a (right) B-module, then $\mathrm{Res}_B^A \mathrm{Ind}_B^A V \cong V \oplus {}^{\tau}V$

(iii) If U is a (right) A-module, then $\mathrm{Ind}_B^A \mathrm{Res}_B^A U \cong U \oplus {}^{\theta}U$.

Proof. (i) This is a consequence of the fact that A is a $\mathbb{Z}/2\mathbb{Z}$-graded algebra (see [CuRe87] 11.13(i)).

(ii) and (iii). The restriction functor Res_B^A can be seen as the tensor functor ${}_B A_A \otimes_A -$. So (ii) and (iii) reduce to the following isomorphisms between bimodules

$${}_B A_A \otimes_A {}_A A_B \cong {}_B B_B \oplus {}_B B_B^{\tau}$$

and

$${}_A A_B \otimes_B {}_B A_A \cong {}_A A_A \oplus {}_A A_A^{\theta}$$

as bimodules. The first is easy since ${}_BA_A \otimes_A {}_AA_B \cong {}_BA_B$ by the evident map and $A = B \oplus B\tau$ as a B-bimodule.

For the second, let us define

$$\mu\colon {}_AA_B \otimes_B {}_BA_A \to {}_AA_A \oplus {}_AA_A^\theta \text{ by } h \otimes h' \mapsto (hh', h\theta(h')).$$

This is well defined since θ is the identity on B. Moreover, it is clearly a morphism for the bimodule structures. To check that it is onto, it suffices to note that its image clearly contains $(1, -1)$ and $(1, 1)$ while 2 is invertible in $\mathcal{O}$.

However, the $\mathcal{O}$-rank of ${}_AA_B \otimes_B {}_BA_A$ is $4.(B:\mathcal{O})$ since $A_B \cong (B_B)^2$ and ${}_BA \cong ({}_BB)^2$ by the graded structure. So μ is an isomorphism. □

Lemma 18.35. *Let U be an indecomposable A-module such that k is a splitting field for the semi-simple quotients of* $\mathrm{End}_A(U)$ *and* $\mathrm{End}_B(\mathrm{Res}_B^A U)$. *One (and obviously only one) of the following cases occurs.*

(i) $U \cong {}^\theta U$, $\mathrm{Res}_B^A U \cong V \oplus {}^\tau V$ *for some indecomposable B-module V such that $V \not\cong {}^\tau V$, and* $U \cong \mathrm{Ind}_B^A V$.

(ii) $U \not\cong {}^\theta U$, $\mathrm{Res}_B^A U = V$ *is indecomposable, and* $\mathrm{Ind}_B^A V \cong U \oplus {}^\theta U$.

Proof. By Lemma 18.34(iii), U is a direct summand of $\mathrm{Ind}_B^A\mathrm{Res}_B^A U$. By the Krull–Schmidt theorem, this implies that there is a direct summand V of $\mathrm{Res}_B^A U$ such that U is a direct summand of $\mathrm{Ind}_B^A V$. Then $\mathrm{Res}_B^A U$ is a direct summand of $\mathrm{Res}_B^A\mathrm{Ind}_B^A V = V \oplus^\tau V$, by Lemma 18.34(ii). Note that ${}^\tau V$ is a direct summand of $\mathrm{Res}_B^A {}^\tau U \cong \mathrm{Res}_B^A U$. Then two cases may occur:

(i′) $\mathrm{Res}_B^A U \cong V \oplus {}^\tau V$ with $V \not\cong {}^\tau V$,

(ii′) ${}^\tau V \cong V$ and $\mathrm{Res}_B^A U \cong V$ or $V \oplus V$.

In case (i′), U is a direct summand of $\mathrm{Ind}_B^A V$ but ranks coincide, so $U \cong \mathrm{Ind}_B^A V$. Then $U \cong {}^\theta U$ since θ is trivial on B. This implies the first case of our lemma.

Assume (ii′). Denote $E = \mathrm{End}_B(V)$. Let us study $\mathrm{End}_A(\mathrm{Ind}_B^A V)$. By a standard result (see [CuRe87] 11.14(iii)), this is $E \oplus E\hat{\tau}$ where $\hat{\tau}$ is a unit of $\mathrm{End}_A(\mathrm{Ind}_B^A V)$ and $\hat{\tau}^2 \in E$. (Namely $\hat{\tau}$ is the map defined by $\hat{\tau}(a \otimes v) = a\tau \otimes \phi(v)$, for $a \in A$, $v \in V$, and where $\phi \in \mathrm{End}_{\mathcal{O}}(V)$ induces an isomorphism $V \cong {}^\tau V$.) By the splitting hypothesis and the indecomposability of V, one has $E/J(E) = k$. But $J(E)[\hat{\tau}] := J(E) \oplus J(E)\hat{\tau}$ is clearly a two-sided ideal of $\mathrm{End}_A(\mathrm{Ind}_B^A V) = E \oplus E\hat{\tau}$, and it is nilpotent mod. $J(\mathcal{O})$. Since $\hat{\tau}^2$ is a unit of E, its class mod. $J(E)$ is $\lambda.1$ where $\lambda \in k^\times$. Now, we have $\mathrm{End}_A(\mathrm{Ind}_B^A V)/J(E)[\hat{\tau}] \cong k[X]/(X^2 - \lambda)$. Since $2 \neq 0$ in k, the ring $k[X]/(X^2 - \lambda)$ is a product of extension fields of k. By the splitting hypothesis, each of those fields must be k, so the maximal semi-simple quotient of $\mathrm{End}_A(\mathrm{Ind}_B^A V)$ is k^2. By the classical correspondence between simple modules for $\mathrm{End}_A(\mathrm{Ind}_B^A V)$ and the isomorphism types of summands in a decomposition of $\mathrm{Ind}_B^A V$ as a direct sum of indecomposable A-modules (see, for instance, [Ben91a] §1.4, [NaTs89] 1.14.7), the isomorphism $\mathrm{End}_A(\mathrm{Ind}_B^A V)/J\big(\mathrm{End}_A(\mathrm{Ind}_B^A V)\big) \cong k^2$ implies

$$\mathrm{Ind}_B^A V \cong U_1 \oplus U_2$$

where $U_1 \not\cong U_2$.

However, we have $V^m \cong \mathrm{Res}_B^A U$ for $m = 1$ or 2. Therefore $(\mathrm{Ind}_B^A V)^m \cong U \oplus {}^\theta U$ by Lemma 18.34(iii). Since U is indecomposable, the decomposition we have above for $\mathrm{Ind}_B^A V$ implies $m = 1$ and $U \not\cong {}^\theta U$. This is case (ii). □

Let us denote by $S_1, S_2, \ldots, S_f$ the simple submodules of $M \otimes K$, up to isomorphism. By the hypothesis on $\mathrm{Dec}_A(M)$, this list can be made in such a way that the indecomposable direct summands of M are (up to isomorphism) $M_1, \ldots, M_f$, and moreover

$$d_{ij} := \dim_K \mathrm{Hom}_{A\otimes K}(S_i, M_j \otimes K)$$

satisfies $d_{ii} = 1$ and $d_{ij} = 0$ for all $1 \leq i < j \leq f$.

Lemma 18.36. *Let $\rho \in \mathfrak{S}_f$ be defined by ${}^\theta S_i \cong S_{\rho(i)}$.*
Let $1 \leq i \leq f$.
(i) ${}^\theta M_i \cong M_{\rho(i)}$.
(ii) If $k \in \mathbb{Z}$ and $\rho^k(i) \neq i$, then $d_{i,\rho^k(i)} = 0$.

Proof. (i) First ${}^\theta M_i$ is a direct summand of M by hypothesis. So it is isomorphic to some $M_{\rho'(i)}$ for $1 \leq \rho'(i) \leq f$. This defines $\rho' \in \mathfrak{S}_f$.

We have $\mathrm{Hom}_{A\otimes K}({}^\theta X_1, {}^\theta X_2) = \mathrm{Hom}_{A\otimes K}(X_1, X_2)$ for any $A \otimes K$-modules X_1, X_2. Applied to $X_1 = S_i$, $X_2 = M_j \otimes K$, this gives

(D) $$d_{\rho(i),\rho'(j)} = d_{ij}$$

for any $1 \leq i, j \leq f$. Denoting by $E_v \in \mathrm{GL}_f(\mathbb{Q})$ the permutation matrix associated with $v \in \mathfrak{S}_f$, the equation (D) above also can also be written $\mathrm{Dec}_A(M)E_{\rho'} = E_\rho \mathrm{Dec}_A(M)$. Since $\mathrm{Dec}_A(M)$ is lower triangular, the Bruhat decomposition in $\mathrm{GL}_f(\mathbb{Q})$ implies $\rho = \rho'$.

(ii) Now let $\{i, \rho(i)\}$ be a non-trivial orbit under ρ. One may assume that $i < \rho(i)$. Then $d_{i,\rho(i)} = 0$. Otherwise, we have $d_{\rho^m(i),\rho^{m'}(i)} = d_{i,\rho^{m'-m}(i)}$ for all $m, m' \in \mathbb{Z}$, by (D) above. This implies it is zero when $\rho^m(i) \neq \rho^{m'}(i)$. This is (ii). □

Definition 18.37. *Let $\rho \in \mathfrak{S}_f$ be defined by ${}^\theta S_i \cong S_{\rho(i)}$. Let $\mathcal{I}_\rho \subseteq \{1, \ldots, f\} \times \{0, 1\}$ be the set of pairs (i, k) such that, if $\rho(i) \neq i$, then $k = 0$ and $i < \rho(i)$.*

For instance, taking $f = 5$ and $\rho = (15)(34) \in \mathfrak{S}_5$, one gets the following list (in lexicographic order) for $\mathcal{I}_\rho$, $(1, 0) < (2, 0) < (2, 1) < (3, 0)$.

By Lemma 18.35, if $\rho(i) \neq i$ we may denote $N_i = \mathrm{Res}_B^A M_i$ and this is an indecomposable direct summand of $\mathrm{Res}_B^A M$. When $\rho(i) = i$, one has $\mathrm{Res}_B^A M_i = N_i \oplus {}^\tau N_i$ where N_i is an indecomposable direct summand of $\mathrm{Res}_B^A M$.

By the same lemma and the fact that Res_B^A sends simple A-modules to semi-simple B-modules (see [CuRe87] 11.16(i)), we may denote $T_i = \mathrm{Res}_B^A S_i$ when $\rho(i) \neq i$, $\mathrm{Res}_B^A S_i = T_i \oplus {}^\tau T_i$ where T_i is a simple submodule of $\mathrm{Res}_B^A M \otimes K$.

Lemma 18.38. *The maps*

$$(i, k) \mapsto {}^{\tau^k} N_i \quad \text{and} \quad (i, k) \mapsto {}^{\tau^k} T_i$$

are bijections between $\mathcal{I}_\rho$ and the indecomposable direct summands of $\mathrm{Res}_B^A M$ *for the first map, the simple submodules of* $\mathrm{Res}_B^A M \otimes K$ *for the second. The multiplicities of the ${}^{\tau^k} T_i$'s in the ${}^{\tau^k} N_i \otimes K$ are as in the following table.*

	N_j *for* $j \neq \rho(j)$	${}^{\tau^l} N_j$ *for* $j = \rho(j)$
T_i *for* $i \neq \rho(i)$	$d_{ij} + d_{i\rho(j)}$	d_{ij}
${}^{\tau^k} T_i$ *for* $i = \rho(i)$	d_{ij}	$c_{ij}(k - l)$

where $c_{ij}(1) = c_{ij}(-1) = d_{ij} - c_{ij}(0)$.

Moreover, the choice of which summand in $\mathrm{Res}_B^A M_i$ *is called N_i for each ρ-fixed i can be made such that all values in the diagonal are* 1*'s.*

Proof. Two N_i's of the first kind ($\rho(i) \neq i$) can be isomorphic $N_i \cong N_j$ only if $M_i \cong^{\theta^k} M_j$ for some k (take Ind_B^A and apply Lemma 18.35(ii)), i.e. $i = \rho^k(j)$. Conversely, M_i and ${}^{\theta^k} M_i$ have the same Res_B^A.

An isomorphism between indecomposable modules of the second kind ${}^{\tau^k} N_i \cong {}^{\tau^l} N_j$, where i and j are ρ-fixed, can take place only if $i = j$ and $k = l$. One obtains $i = j$ by taking Ind_B^A and applying Lemma 18.35(i). Then Lemma 18.35(i) gives $k = l$.

The two kinds of indecomposable summands do not overlap because of the action of τ.

The same is done for the ${}^{\tau^k} T_i$'s. This gives all simple submodules of $\mathrm{Res}_B^A M \otimes K$ by adjunction and the fact that Ind_B^A brings simple modules to semi-simple ones as can be seen from the regular $B \otimes K$-module.

When $\rho(i) \neq i$, by adjunction (Lemma 18.34(i)), we have $\mathrm{Hom}_{B\otimes K}(T_i, V \otimes K) \cong \mathrm{Hom}_{A\otimes K}(S_i, \mathrm{Ind}_B^A V \otimes K)$ for any direct summand V of $\mathrm{Res}_B^A M$, so this can be computed using our knowledge of the $\mathrm{Ind}_B^A V \otimes K$'s (Lemma 18.35).

When $\rho(i) = i$ and $\rho(j) \neq j$, then $\mathrm{Hom}_{B\otimes K}({}^{\tau^k} T_i, N_j \otimes K) = \mathrm{Hom}_{B\otimes K}({}^{\tau^k} T_i, \mathrm{Res}_B^A M_j \otimes K) \cong \mathrm{Hom}_{A\otimes K}(\mathrm{Ind}_B^A {}^{\tau^k} T_i, M_j \otimes K) = \mathrm{Hom}_{A\otimes K}(S_i, M_j \otimes K)$ by adjunction and Lemma 18.35(i) for the statement $\mathrm{Ind}_B^A {}^{\tau^k} T_i \cong S_i$.

Concerning the $c_{ij}(k-l)$'s, note first that $\mathrm{Hom}_{B\otimes K}({}^{\tau} V_1, {}^{\tau} V_2) = \mathrm{Hom}_{B\otimes K}(V_1, V_2)$ for any $B \otimes K$-modules V_1, V_2. So the dimension of $\mathrm{Hom}_{B\otimes K}({}^{\tau^k} T_i, {}^{\tau^l} N_j \otimes K)$ depends only on i, j and the parity of $k - l$. For the remaining equation, $c_{ij}(-l) + c_{ij}(1-l)$ is the dimension of $\mathrm{Hom}_{B\otimes K}(T_i \oplus {}^{\tau} T_i, {}^{\tau^l} N_j \otimes K) = \mathrm{Hom}_{B\otimes K}(\mathrm{Res}_B^A S_i, {}^{\tau^l} N_j \otimes K) \cong \mathrm{Hom}_{A\otimes K}(S_i, M_j \otimes K)$ by adjunction and Lemma 18.35(i).

We now come to the last statement. When $\rho(i) \neq i$, then $d_{ii} + d_{i\rho(i)} = d_{ii} = 1$ by Lemma 18.36(ii) and hypothesis (ii) of the theorem. We now fix i such that $\rho(i) = i$. Then $c_{ii}(0) + c_{ii}(-1) = c_{ii}(0) + c_{ii}(1) = d_{ii} = 1$, so there is only one summand in the first sum which is 1 while the other is 0. This means that T_i occurs once in one of $\{N_i, {}^{\tau} N_i\}$ but not in the other. So if the choice of which summand of $\mathrm{Res}_B^A S_i$ is called T_i has been made, then one can choose which summand of $\mathrm{Res}_B^A M_i$ is called N_i in such a way that the multiplicity of T_i in it is 1. Then the diagonal of the table bears just 1's. □

We can now complete our proof of Theorem 18.32. The set $\mathcal{I}_\rho \subseteq \{1, \ldots, f\} \times \{0, 1\}$ inherits the lexicographic order of $\{0, 1\}$. Since we know that the diagonal contains only 1's, it just remains to check that the entry in the table of Lemma 18.38 is 0 when it corresponds to a relation $(i, k) < (j, l)$ in $\mathcal{I}_\rho$. We review the four possible cases.

- $\rho(i) \neq i$, $\rho(j) \neq j$ (hence $k = l = 0$ and $i < j$). Then the decomposition number is $d_{ij} + d_{i\rho(j)}$ and this is zero since $i < j < \rho(j)$ for all k (recall that $j < \rho(j)$ by Definition 18.37).
- $\rho(i) \neq i$, $\rho(j) = j$.
- $\rho(i) = i$, $\rho(j) \neq j$. In this case and the above, we have $i < j$ and the table in Lemma 18.38 gives our claim since $d_{ij} = 0$.
- $\rho(i) = i$, $\rho(j) = j$. If $i < j$, then $d_{i,j} = 0$ and the equation of Lemma 18.38 implies that $c_{ij}(-l) = c_{ij}(1-l) = 0$ for all l. If $i = j$ and $k < l$, then the same equation, along with the statement about the diagonal, implies that $c_{ii}(k-l) = 0$ for $(k, l) = (0, 1)$ or $(1, 0)$ (while $c_{ii}(0) = d_{ii} = 1$). □

Remark 18.39. The functors Ind_B^A and Res_B^A have quite symmetric roles in the proof of Theorem 18.32. One may obtain a theorem dual to Theorem 18.32, and a converse to it. See Exercise 15.

Exercises

1. Make a new definition of a Hecke algebra $\mathcal{H}_\mathcal{O}(W,(q_s),(q'_s))$ with two sets of parameters, the quadratic relation becoming $(a_s - q_s)(a_s - q'_s) = 0$. Define y_I and y'_I where the second becomes x_I upon specializing $q'_s = -1$. Use an automorphism $h \mapsto h'$ of $\mathcal{H}_\mathcal{O}(W,(q_s),(q'_s))$ to reduce the checkings of §18.1 by half.
2. Show that $a_s \mapsto (q_s - 1) - a_s = -(q_s a_s)^{-1}$ defines an involutory automorphism of the $\mathcal{O}$-algebra $\mathcal{H}_\mathcal{O}(W,(q_s)$ of Definition 18.1. Show that it exchanges x_I with y_I.

   Deduce from this that one may replace the y_λ's by x_λ's in Theorem 18.14.
3. Let (W, S) be a finite Coxeter group. Let $I, J \subseteq S$. Using the notation of §18.1, show that $x_I y_J \neq 0$ is equivalent to $I \cap J = \emptyset$. Show that, if W_I and W_J are conjugate subgroups of W, then y_I and y_J are conjugate in $\mathcal{H}_\mathcal{O}(W,(q_s)_{s\in S})$.
4. Show that, if $W = \mathfrak{S}_n$, $q = 1$ and K is algebraically closed of characteristic zero, then $y_\lambda \mathcal{H}_K \cong \mathrm{Ind}_{\mathfrak{S}_\lambda}^{\mathfrak{S}_n}\mathrm{sgn}$ and $x_\lambda \mathcal{H}_K \cong \mathrm{Ind}_{\mathfrak{S}_\lambda}^{\mathfrak{S}_n} 1$ (notation of §18.1) where sgn denotes the one-dimensional representation defined by the signature. Show that $\langle \mathrm{Ind}_{\mathfrak{S}_\lambda}^{\mathfrak{S}_n}\mathrm{sgn}, \mathrm{Ind}_{\mathfrak{S}_\mu}^{\mathfrak{S}_n} 1\rangle_{\mathfrak{S}_n} \neq 0$ implies $\lambda \ll \mu^*$.

   Deduce the fact about $\mathrm{Irr}(\mathfrak{S}_n)$ mentioned at the end of the proof of Theorem 18.15.
5. Let G be a finite group acting on a ring A by ring automorphisms. Show that there exists a ring denoted by $A \rtimes G$ whose underlying commutative group is $A^{(G)}$ (maps from G to A) and such that G and its action on A inject in the units of $A \rtimes G$.

   If $\mathcal{O}$ is a commutative ring and $H \rtimes G$ is a semi-direct product of finite groups, show that $\mathcal{O}[H \rtimes G] \cong (\mathcal{O}[H]) \rtimes G$.
6. Let $\mathcal{O}$ be a commutative ring, let A be the $\mathcal{O}$-agebra $\mathcal{O}^n$ (multiplication is defined componentwise), let G be a finite group action on A by $\mathcal{O}$-algebra automorphisms.
   (a) Show that $A \rtimes G$ is Morita equivalent to $\Pi_i \mathcal{O}[G_i]$ where i ranges over the G-classes of primitive idempotents of A and G_i denotes the centralizer of an element of i (if $\varepsilon \in A$ is a primitive idempotent, let $\bar{\varepsilon}$ denote the sum of elements of the orbit of ε under G; show that $A \rtimes G = A \rtimes G\bar{\varepsilon} \oplus A \rtimes G(1 - \bar{\varepsilon})$ is a decomposition as a direct product of rings and that $\varepsilon(A \rtimes G)\varepsilon \cong \mathcal{O}[G_\varepsilon]$).
   (b) Apply the above to get the representation theory of semi-direct products $H \rtimes G$ of finite groups, where H is commutative, in all characteristics not dividing $|H|$.
   (c) Case of $W(\mathrm{BC}_n)$ in odd characteristic.
   (d) Case of $W(\mathrm{D}_n)$ in odd characteristic.

7. (Ariki–Koike) Show that the subalgebra of $\mathcal{H}_R(\mathrm{BC}_2)$ generated by t_1 and t_2 equals $R \oplus Rt_1 \oplus Rt_2 \oplus Rt_1t_2 \oplus Rt'_1 \oplus Rt'_2$ where $t'_1 = (y-1)(t_1t_2 + xy)a_2$, $t'_2 = x(y-1)(xy - yt_1 - t_2 - y)a_2$.
8. Prove Proposition 18.22(i) and (iii) by using the geometric representation of $W(\mathrm{BC}_n)$ and the roots (see Example 2.1(ii)).
9. Prove that $\mathcal{H}_R(\mathrm{BC}_n)$ is not a subalgebra of $\Pi_i \mathrm{Mat}_{N_i}(R)$, where the N_i's are the dimensions of Hoefsmit's representations.
10. Show that the coefficient of v_m on $a(w_m)$ is $(-x)^m y^{\binom{m}{2}+\binom{n-m}{2}}$ and that it is the only component on $\mathcal{H}_R(\mathfrak{S}_n)$ with respect to the standard basis of the a_w's.
11. Show that $p_n(v_m) = a(w_{n-m})^{-1}$ (notation of the proof of Proposition 18.24(ii)).
12. Assume $\mathcal{O}$ is a complete discete valuation ring. Let A be a finitely generated $\mathcal{O}$-free algebra. Let $a \in A$, and assume aA is projective as a right module. Show that there is a unit $u \in A$ such that au is an idempotent.
13. Let $R' = \mathbb{Z}[x_1, x_2, y, y^{-1}]$. Define $\mathcal{H}_{R'}(\mathrm{BC}_n)$ by the same relations as $\mathcal{H}_R(\mathrm{BC}_n)$ but with the quadratic relation for a_0 being $(a_0 - x_1)(a_0 - x_2) = 0$. Define the t_i's, the π_i's and the $\widetilde{\pi}_i$'s of this algebra. Show that the π_i's and the $\widetilde{\pi}_i$'s are exchanged by the automorphism $h \to \widetilde{h}$ permuting x_1 and x_2. Simplify certain proofs in §18.1 by use of this automorphism.
14. $\mathcal{O}$ is a commutative ring, A an $\mathcal{O}$-free $\mathcal{O}$-algebra. If $a \in A$ is invertible on one side, show that it satisfies a polynomial equation $P(a) = 0$ with $P \in \mathcal{O}[t]$ and $P(0)$ invertible in $\mathcal{O}$ (work in the matrix algebra $\mathrm{End}_{\mathcal{O}}(A)$ and take P to be the characteristic polynomial of right multiplication by a). Deduce that a is invertible (on both sides) in the subalgebra it generates.
15. Show the results of §18.6 when $A = B \oplus B\tau \oplus \ldots \oplus B\tau^{r-1}$ is a $\mathbb{Z}/r\mathbb{Z}$-graded algebra over $\mathcal{O}$, where r is invertible in $\mathcal{O}$, where the triple $(\mathcal{O}, K, k)$ is assumed to be a splitting system for the endomorphism rings of all considered modules.

    Show that the implication of Theorem 18.32 is an equivalence.

    $\mathrm{Dec}(A)$ is square unitriangular if and only if $\mathrm{Dec}(B)$ is square unitriangular.
16. Find an example of a $\mathbb{Z}/2\mathbb{Z}$-graded algebra $A = B \oplus B\tau$ over a field of characteristic 2 and such that $A \otimes_B A$ is indecomposable as an A-bimodule.

Notes

The q-Schur algebra $\mathrm{End}_{\mathcal{H}(\mathfrak{S}_n)}(\bigoplus_{\lambda \vdash n} y_\lambda \mathcal{H}(\mathfrak{S}_n))$ was introduced by Dipper–James in [DipJa89]. This, and the many generalizations that followed, opened a new chapter of representation theory (see the books [Donk98a], [Mathas], and the references given there).

The Morita equivalence for type BC was proved by Dipper–James [DipJa92], using [DipJa86] and [DipJa87]. This was generalized by Du–Rui and Dipper–Mathas (see [DuRui00], [DipMa02]) to the context of Ariki–Koike algebras [ArKo]. Concerning the generalization of Hoefsmit construction to all types of Coxeter groups and also valuable historical remarks, see [Ram97].

The result in §18.6 is due to Genet; see [Gen03].

19

Decomposition numbers and q-Schur algebras: general linear groups

Let us recall the notion of decomposition matrices from §5.3 and §18.6 earlier. If $(\mathcal{O}, K, k)$ is an ℓ-modular splitting system for an $\mathcal{O}$-free finitely generated $\mathcal{O}$-algebra A, and if M is an $\mathcal{O}$-free finitely generated A-module, we defined $\mathrm{Dec}_A(M)$ as a matrix recording the multiplicities of the simple $A \otimes K$-modules in the various indecomposable summands of M. The classical ℓ-decomposition matrix $\mathrm{Dec}(\mathcal{O}G)$ of a finite group G is $\mathrm{Dec}_{\mathcal{O}G}({}_{\mathcal{O}G}\mathcal{O}G)$ (see Definition 5.25).

In the case of the symmetric group $\mathfrak{S}_n$, it is a well-known result that the ℓ-decomposition matrix can be written

$$\mathrm{Dec}(\mathcal{O}\mathfrak{S}_n) = \begin{pmatrix} 1 & 0 & 0 \\ * & \ddots & 0 \\ * & * & 1 \\ * & * & * \\ \vdots & \vdots & \vdots \\ * & * & * \end{pmatrix}$$

for a suitable ordering of the columns (see [JaKe81] 6.3.60).

Concerning finite reductive groups $\mathbf{G}^F$ with connected $\mathrm{Z}(\mathbf{G})$, we know from Theorem 14.4 that the unipotent characters are a basic set for the sum of unipotent ℓ-blocks $B_1 = \mathcal{O}\mathbf{G}^F.b_\ell(\mathbf{G}^F, 1)$ (see Definition 9.9). In other words, the lines of $\mathrm{Dec}(B_1)$ corresponding with $\mathcal{E}(\mathbf{G}^F, 1)$ define a square submatrix of determinant ± 1. We show the following theorem of Dipper–James; see [DipJa89]. For $G = \mathrm{GL}_n(\mathbb{F}_q)$, this submatrix is of the form $\mathrm{Dec}_{\mathcal{H}}(M)$ where $\mathcal{H}$ is the Hecke algebra associated with the symmetric group $\mathfrak{S}_n$ and parameter q (see Definition 18.1), and M is a direct product of ideals $M \cong \Pi_{\lambda \vdash n} y_\lambda \mathcal{H}$ (see Theorem 19.15 and Theorem 19.16). For $\lambda = (1, \ldots, 1)$, one gets (in a special case) the inclusion of decomposition matrices already mentioned in §5.4. One shows in addition that the lexicographic order on partitions λ makes the above decomposition matrix lower triangular as in the case of $\mathfrak{S}_n$.

The connection between $\mathrm{Dec}_{\mathcal{H}}(M)$ and $\mathrm{Dec}(\mathcal{O}\mathbf{G}^F)$ relies essentially on the symmetry of $\mathcal{H}$ as an $\mathcal{O}$-algebra and certain isomorphisms between Hom spaces related to the equivalence of §1.5. The q-Schur algebra is $\mathrm{End}_{\mathcal{H}}(M)$, an algebra whose number of simple modules is the same over k and K (see [Mathas] 4.15 or Corollary 19.17 below).

The upper triangular shape of the $\mathrm{Dec}(B_1)$ above allows us to determine the simple cuspidal kG-modules (see Chapter 1) pertaining to this block. They are of type $\mathrm{hd}(Y)$ where Y is the reduction mod. $J(\mathcal{O})$ of a version over $\mathcal{O}$ of the Steinberg module. Denote by e the order of q mod. ℓ. The existence of a simple cuspidal B_1-module is equivalent to $n = 1$ or $n_{\ell'} = e$ (Theorem 19.18). It is also possible to say which simple $kG.b_\ell(G, 1)$-modules are in the series defined by a cuspidal module for a standard Levi subgroup of $G = \mathrm{GL}_n(\mathbb{F}_q)$ (Dipper–Du, Geck–Hiss–Malle, see [DipDu93], [GeHiMa94]). This completes, for the product of the unipotent blocks of $\mathrm{GL}_n(\mathbb{F}_q)$, the program set in Chapter 1.

19.1. Hom functors and decomposition numbers

Let $\mathcal{O}$ be a local ring with field of fractions K.

Definition 19.1. *If V is an $\mathcal{O}$-free $\mathcal{O}$-module and $V' \subseteq V$ is a submodule, one denotes $\sqrt{V'} = V \cap (V' \otimes_{\mathcal{O}} K)$, an $\mathcal{O}$-submodule of $V \otimes_{\mathcal{O}} K$, i.e. $\sqrt{V'} = \{v \in V \mid J(\mathcal{O})^a v \subseteq V' \text{ for some } a \geq 0\}$.*

One says V' is $\mathcal{O}$-pure if and only if $\sqrt{V'} = V'$, i.e. V/V' is $\mathcal{O}$-free.

We assume in this section that ℓ is a prime and $(\mathcal{O}, K, k)$ is an ℓ-modular splitting system for a finite group G. Let Y be an $\mathcal{O}$-free $\mathcal{O}G$-module. Denote $E := \mathrm{End}_{\mathcal{O}G}(Y)$. Recall from §1.5

$$H_Y \colon \mathcal{O}G\text{--}\mathbf{mod} \to \mathbf{mod}\text{--}E$$

the functor $\mathrm{Hom}_{\mathcal{O}G}(Y, -)$. If M is an E-submodule of some $H_Y(V) = \mathrm{Hom}_{\mathcal{O}G}(Y, V)$, one denotes by $MY \subseteq V$ the $\mathcal{O}G$-submodule $M.Y := \sum_{m \in M} m(Y)$.

The main result is a version "over $\mathcal{O}$" of §1.5; see also Exercise 3.

Theorem 19.2. *Assume that E is symmetric (see Definition 1.19).*

(i) If J is an $\mathcal{O}$-pure right ideal of E, then $H_Y(\sqrt{JY})$, considered as a right ideal of $E = H_Y(Y)$, equals J.

(ii) If J_1, J_2 are $\mathcal{O}$-pure right ideals of E, then H_Y induces an isomorphism $\mathrm{Hom}_{\mathcal{O}G}(\sqrt{J_1 Y}, \sqrt{J_2 Y}) \xrightarrow{\sim} \mathrm{Hom}_E(J_1, J_2)$.

Proof. (i) One has clearly $J \subseteq H_Y(JY) \subseteq H_Y(\sqrt{JY})$. However, it is easy to see that $H_Y(\sqrt{JY}) = \sqrt{H_Y(JY)}$. So both J and $H_Y(\sqrt{JY})$ are pure, and it suffices to prove that $J \otimes_{\mathcal{O}} K \supseteq H_Y(\sqrt{JY}) \otimes_{\mathcal{O}} K$.

One has $H_Y(\sqrt{JY}) \otimes_{\mathcal{O}} K = \mathrm{Hom}_{\mathcal{O}G}(Y, \sqrt{JY}) \otimes K \subseteq \mathrm{Hom}_{KG}(Y \otimes K, (JY) \otimes K) \subseteq \mathrm{Hom}_{KG}(Y \otimes K, (J \otimes K)(Y \otimes K))$ but this is the image of $(J \otimes K)(Y \otimes K)$ by the Hom-functor associated to the KG-module $Y \otimes K$. The endomorphism algebra $\mathrm{End}_{KG}(Y \otimes K)$ is Frobenius (see Definition 1.19) since it is semi-simple, so $\mathrm{Hom}_{KG}(Y \otimes K, (J \otimes K)(Y \otimes K)) = J \otimes K$ by Lemma 1.28(iii).

(ii) The proof parallels the one of Theorem 1.25(i).

The functor H_Y provides a map $\mathrm{Hom}_{\mathcal{O}G}(\sqrt{J_1 Y}, \sqrt{J_2 Y}) \to \mathrm{Hom}_E(H_Y(\sqrt{J_1 Y}), H_Y(\sqrt{J_2 Y}))$ defined by $H_Y(f)$ being the composition on the left with f. By (i), it suffices to check that this is a bijection.

If $f \in \mathrm{Hom}_{\mathcal{O}G}(\sqrt{J_1 Y}, \sqrt{J_2 Y})$ satisfies $H_Y(f) = 0$, then $f(e(Y)) = 0$ for all $e \in H_Y(\sqrt{J_1 Y})$. Then $f(H_Y(\sqrt{J_1 Y}).Y) = 0$, i.e. $f(J_1.Y) = 0$ by (i). Thus clearly $f(\sqrt{J_1 Y}) = 0$, i.e. $f = 0$. So H_Y is injective.

We now prove that H_Y is onto. Let $h \in \mathrm{Hom}_E(J_1, J_2)$. By Lemma 18.7, there is $e \in E$ such that $h(i) = ei$ for all $i \in J_1$. Then $eJ_1 \subseteq J_2$ and therefore $e(J_1 Y) \subseteq J_2 Y$, $e(\sqrt{J_1 Y}) \subseteq \sqrt{J_2 Y}$. Taking $f \in \mathrm{Hom}_{\mathcal{O}G}(\sqrt{J_1 Y}, \sqrt{J_2 Y})$ to be the restriction of e, one has clearly $H_Y(f) = h$. □

Recall decomposition matrices (Definition 5.25) and basic sets of characters (Definition 14.3).

Proposition 19.3. *If $\mathcal{B}$ is a subset of* $\mathrm{Irr}(G, b)$ *for a central idempotent $b \in$* $\mathrm{Z}(\mathcal{O}G)$ *(see §5.1), the submatrix of* $\mathrm{Dec}(\mathcal{O}G)$ *corresponding to simple KG-modules in $\mathcal{B}$ and projective indecomposable $\mathcal{O}Gb$-modules is square invertible (over $\mathbb{Z}$) if and only if $\mathcal{B}$ is a basic set of characters for $\mathcal{O}Gb$.*

Proof. The lattice in $\mathrm{CF}(G, K)$ generated on $\mathbb{Z}$ by the $d^1\chi$ for $\chi \in \mathrm{Irr}(G)$ has a basis $\mathrm{IBr}(G)$ ("Brauer characters," i.e. central functions $G_{\ell'} \to \mathcal{O}$ obtained by lifting in $\mathcal{O}$ the roots of 1 in k, see [Ben91a] §5.3, [NaTs89] §3.6) which partitions along blocks of $\mathcal{O}G$ as $\mathrm{IBr}(G) = \bigcup_i \mathrm{IBr}(G, b_i)$. As a classical result, the decomposition matrix can be seen as giving in each row the coordinates in this basis of each $d^1\chi$ for $\chi \in \mathrm{Irr}(G)$ (see [NaTs89] 3.6.14 and its proof).

Then $\mathcal{B}$ is a basic set of characters for $\mathcal{O}Gb$ if and only if the corresponding $d^1\chi$'s generate the same lattice as $\mathrm{IBr}(G, b)$ and have the same cardinality. This is equivalent to the corresponding block of the decomposition matrix being invertible (over $\mathbb{Z}$). □

The following gathers some technical conditions related to Theorem 19.2. It will allow us to embed in $\mathrm{Dec}(\mathcal{O}G)$ some decomposition matrices $\mathrm{Dec}_{\mathcal{O}G}(X)$ for certain $\mathcal{O}G$-modules that differ from their projective covers by a somewhat canonical submodule.

Theorem 19.4. *Let $A = \mathcal{O}Gb$ for b a central idempotent of $\mathcal{O}G$. We assume we have a collection $(X_\sigma)_\sigma$ of A-modules and, for each σ a collection $(y_\lambda)_{\lambda\in\Lambda_\sigma}$ of elements of $\mathrm{End}_A(X_\sigma)$. We assume the following conditions are all satisfied.*

(a) Letting $\mathcal{B}_\sigma$ be the set of irreducible components of $X_\sigma \otimes K$, the union $\mathcal{B} := \cup_\sigma \mathcal{B}_\sigma$ is disjoint and a basic set of characters for A (see Definition 14.3).

(b) For each σ, the $\mathcal{O}$-algebra $\mathcal{H}_\sigma := \mathrm{End}_A(X_\sigma)$ is symmetric (see Definition 1.19) and there is some $\lambda \in \Lambda_\sigma$ such that $y_\lambda = 1$.

(c) For each σ and $\lambda \in \Lambda_\sigma$, the right ideal $y_\lambda\mathcal{H}_\sigma$ is $\mathcal{O}$-pure in $\mathcal{H}_\sigma$.

(d) For each σ and $\lambda \in \Lambda_\sigma$, there exists an exact sequence

$$0 \to \Omega_{\sigma,\lambda} \to P_{\sigma,\lambda} \to \sqrt{y_\lambda.X_\sigma} \to 0$$

*in $A-$**mod** where $P_{\sigma,\lambda}$ is projective, and $\Omega_{\sigma,\lambda} \otimes K$ has no irreducible component in $\mathcal{B}$.*

Then, denoting $S_\sigma = \mathrm{End}_{\mathcal{H}_\sigma}(\Pi_{\lambda\in\Lambda_\sigma} y_\lambda\mathcal{H}_\sigma)$ and choosing an ordering of the σ's, one has

$$\mathrm{Dec}(A) = \begin{pmatrix} D_0 & D_1 \\ D_0' & D_1' \end{pmatrix}$$

$$\text{where } D_0 = \begin{pmatrix} \mathrm{Dec}(S_{\sigma_1}) & 0 & \dots & 0 \\ 0 & \mathrm{Dec}(S_{\sigma_2}) & & \vdots \\ \vdots & & \ddots & 0 \\ 0 & \dots & 0 & \mathrm{Dec}(S_{\sigma_m}) \end{pmatrix},$$

and where the columns of D_1 correspond to the projective indecomposable A-modules not occurring in the projective covers of the $\sqrt{y_\lambda.X_\sigma}$'s.

Moreover, D_1 is empty if and only if each $\mathrm{Dec}(S_\sigma)$ has a number of columns greater than or equal to its number of rows. Then each $\mathrm{Dec}(S_\sigma)$, and therefore D_0, is a square matrix.

Proof. Let us fix σ. The hypotheses (b) and (c) allow us to apply Theorem 19.2(ii) to $Y = X_\sigma$, $E = \mathcal{H}_\sigma$, and the right ideals generated by the y_λ's. One gets an isomorphism

$$\mathrm{End}_A(\Pi_{\lambda\in\Lambda_\sigma}\sqrt{y_\lambda X_\sigma}) \cong S_\sigma$$

induced by the functor $H_{X_\sigma} := \mathrm{Hom}_A(X_\sigma, -)$. Using Proposition 5.27, one gets

$$\mathrm{Dec}(S_\sigma) = \mathrm{Dec}(\mathrm{End}_A(\Pi_\lambda \sqrt{y_\lambda X_\sigma})) = \mathrm{Dec}(\Pi_\lambda \sqrt{y_\lambda X_\sigma}).$$

We may assume that, in (d), $P_{\lambda,\sigma}$ is a projective cover of $\sqrt{y_\lambda X_\sigma}$ (since a projective cover and the corresponding quotient would be direct summands of $P_{\lambda,\sigma}$ and $\Omega_{\lambda,\sigma}$).

Since $1 = y_\lambda$ for some $\lambda \in \Lambda_\sigma$, the rows of $\mathrm{Dec}(\Pi_\lambda \sqrt{y_\lambda X_\sigma})$ are indexed by $\mathcal{B}_\sigma$. Choose a numbering of the σ's, then a numbering of $\mathrm{Irr}(A \otimes K)$ listing first the elements of $\mathcal{B}_{\sigma_1}, \mathcal{B}_{\sigma_2}, \ldots, \mathcal{B}_{\sigma_m}$, then $\mathrm{Irr}(A \otimes K) \setminus \mathcal{B}$. Choosing also a numbering of the projective A-modules beginning with the $P_{\lambda,\sigma}$'s, one gets the form stated once we have proved that $P_{\lambda,\sigma} \otimes K$ and $P_{\lambda',\sigma'} \otimes K$ have no element of $\mathcal{B}$ in common when $\sigma \neq \sigma'$. By (d), such an element would be in both $\sqrt{y_\lambda . X_\sigma} \otimes K$ and $\sqrt{y_{\lambda'} . X_{\sigma'}} \otimes K$, hence in both $X_\sigma \otimes K$ and $X_{\sigma'} \otimes K$. This is impossible since $\mathcal{B}_\sigma \cap \mathcal{B}_{\sigma'} = \emptyset$ by (a).

The matrix $\mathrm{Dec}(S_\sigma)$ has $|\mathcal{B}_\sigma|$ rows. If its number of columns is greater than or equal to $|\mathcal{B}_\sigma|$, then D_0 is square and therefore D_1 is empty since $(D_0 D_1)$ is square of dimension the number $|\mathcal{B}|$ of projective indecomposable A-modules, $\mathcal{B}$ being a basic set of characters for A (the number of projective indecomposable modules equals the number of simple $A \otimes k$-modules). Conversely, if D_1 is empty, we have an invertible square matrix (Proposition 19.3) which is block diagonal. It is easy to check that each block must be square. Then $\mathrm{Dec}(S_\sigma)$ must have exactly $|\mathcal{B}_\sigma|$ columns. □

19.2. Cuspidal simple modules and Gelfand–Graev lattices

Let $(\mathbf{G}, F)$ be a connected reductive $\mathbf{F}$-group defined over $\mathbb{F}_q$. We assume that $\mathrm{Z}(\mathbf{G})$ is connected. After the next proposition, we shall assume that $\mathbf{G}$ is the general linear group $\mathrm{GL}_n(\mathbf{F})$ with its usual definition over $\mathbb{F}_q$.

Let ℓ be a prime not dividing q. Let $(\mathcal{O}, K, k)$ be an ℓ-modular splitting system for $\mathbf{G}^F$. Recall $e_{\ell'}^{\mathbf{G}^F} \in K\mathbf{G}^F$ and $b_\ell(\mathbf{G}^F, 1) \in \mathcal{O}\mathbf{G}^F$ (see Definition 9.9 and Theorem 9.12).

Let $\mathbf{B}$ be an F-stable Borel subgroup of $\mathbf{G}$. Denote $U = R_u(\mathbf{B})^F$, a Sylow p-subgroup of $\mathbf{G}^F$. Let ψ be a **regular** linear character of U (see [DiMi91] 14.27). The same letter may denote a line over $\mathcal{O}$ affording $\psi: U \to \mathcal{O}^\times$. All regular linear characters of U are $\mathbf{B}^F$-conjugate ([DiMi91] 14.28), so that the $\mathcal{O}\mathbf{G}^F$-module $\mathrm{Ind}_U^{\mathbf{G}^F} \psi$ does not depend on the choice of ψ.

Definition 19.5. *Denote by* $\Gamma_{\mathbf{G}^F} := \mathrm{Ind}_U^{\mathbf{G}^F} \psi$. *Let* $\Gamma_{\mathbf{G}^F,1} = b_\ell(\mathbf{G}^F, 1).\Gamma_{\mathbf{G}^F}$ *(see Definition 9.4). Denote* $\mathrm{St}_{\mathbf{G}^F} = e_{\ell'}^{\mathbf{G}^F}.\Gamma_{\mathbf{G}^F,1}$, *an* $\mathcal{O}\mathbf{G}^F$*-module. We also denote* $\overline{\Gamma}_{\mathbf{G}^F,1} = \Gamma_{\mathbf{G}^F,1} \otimes_\mathcal{O} k$ *and* $\overline{\mathrm{St}}_{\mathbf{G}^F} = \mathrm{St}_{\mathbf{G}^F} \otimes_\mathcal{O} k$.

Proposition 19.6. *(i) The* $\Gamma_{\mathbf{G}^F,1}$ *are projective indecomposable modules.*

(ii) $\mathrm{St}_{\mathbf{G}^F}\otimes_{\mathcal{O}} K$ *is the simple* $K\mathbf{G}^F$*-module corresponding to the Steinberg character (see [DiMi91] §9). Moreover,* $\mathrm{hd}(\overline{\mathrm{St}}_{\mathbf{G}^F})$ *is simple with projective cover* $\overline{\Gamma}_{\mathbf{G}^F,1}$.

(iii) If L is a standard Levi subgroup of $\mathbf{G}^F$*, then* ${}^*\mathrm{R}_L^{\mathbf{G}^F}\mathrm{St}_{\mathbf{G}^F}=\mathrm{St}_L$.

Proof. Since $|U|$ is invertible in $\mathcal{O}$, one has $\Gamma_{\mathbf{G}^F}=\mathcal{O}\mathbf{G}^F.u_\psi$ where u_ψ is the idempotent $|U|^{-1}\sum_{u\in U}\psi(u^{-1})u$. So $\Gamma_{\mathbf{G}^F}$, and therefore $\Gamma_{\mathbf{G}^F,1}$ are projective. Now $\Gamma_{\mathbf{G}^F,1}$ is the projective cover of $\mathrm{St}_{\mathbf{G}^F}$ by Theorem 9.10. To show that $\Gamma_{\mathbf{G}^F,1}$ is indecomposable, it suffices to check that $\mathrm{St}_{\mathbf{G}^F}$ is. But $\mathrm{St}_{\mathbf{G}^F}\otimes K$ is simple since $\Gamma_{\mathbf{G}^F}\otimes K$ has exactly one component in each rational series (see [DiMi91] 14.47). We have (i). Moreover, [DiMi91] 14.47 tells us that the component of $\Gamma_{\mathbf{G}^F}\otimes K$ in $\mathcal{E}(\mathbf{G}^F,1)$ is the Steinberg character.

Since $\Gamma_{\mathbf{G}^F,1}\to\mathrm{St}_{\mathbf{G}^F}$ is a projective cover with indecomposable $\Gamma_{\mathbf{G}^F,1}$, $\overline{\Gamma}_{\mathbf{G}^F,1}$ is also indecomposable ([Thévenaz] 1.5.2) and $\overline{\Gamma}_{\mathbf{G}^F,1}\to\overline{\mathrm{St}}_{\mathbf{G}^F}$ is a projective cover, so $\mathrm{hd}(\overline{\mathrm{St}}_{\mathbf{G}^F})=\mathrm{hd}(\overline{\Gamma}_{\mathbf{G}^F,1})$ is simple. This is (ii).

(iii) By the definition of $\mathrm{St}_{\mathbf{G}^F}$ and Proposition 9.15, it suffices to check that ${}^*\mathrm{R}_L^{\mathbf{G}^F}\Gamma_{\mathbf{G}^F,1}=\Gamma_{L,1}$. Both sides are projective modules, so the equality may be checked on associated characters (see [Ben91a] 5.3.6). One has ${}^*\mathrm{R}_L^{\mathbf{G}^F}(\Gamma_{\mathbf{G}^F}\otimes K)=\Gamma_L\otimes K$ (see [DiMi91] 14.32). So the sought equality follows by Proposition 9.15. □

We now assume that $\mathbf{G}=\mathrm{GL}_n(\mathbf{F})$, F is the usual Frobenius map $(x_{ij})\mapsto(x_{ij}^q)$.

We denote $G=\mathbf{G}^F=\mathrm{GL}_n(\mathbb{F}_q)$.

Let $\mathbf{T}_1$ be the torus of diagonal matrices, $\mathbf{B}$ the Borel of upper triangular matrices in GL. The Weyl group $\mathrm{N}_{\mathbf{G}}(\mathbf{T}_1)/\mathbf{T}_1$ identifies with $\mathfrak{S}_n$ (permutation matrices). One parametrizes the $\mathbf{G}^F$-classes of F-stable maximal tori from $\mathbf{T}_1$ by conjugacy classes of $\mathfrak{S}_n$ (see §8.2). When $w\in\mathfrak{S}_n$, choose $\mathbf{T}_w$ in the corresponding class. When $f\in\mathrm{CF}(\mathfrak{S}_n,K)$, let

$$R_f^{(G)}:=(n!)^{-1}\sum_{w\in\mathfrak{S}_n}f(w)\mathrm{R}_{\mathbf{T}_w}^{\mathbf{G}}(1_{\mathbf{T}_w^F})\in\mathrm{CF}(G,K).$$

Then (see [DiMi91] §15.4) we have the following.

Theorem 19.7. *(i)* $f\mapsto R_f^{(G)}$ *is an isometry sending* $\mathrm{Irr}(\mathfrak{S}_n)$ *onto* $\mathcal{E}(G,1)$.

(ii) A unipotent character of $\mathrm{GL}_n(\mathbb{F}_q)$ *can be cuspidal only if* $n=1$.

The trivial representations of G and $\mathfrak{S}_n$ correspond. The sign representation of $\mathfrak{S}_n$ corresponds with the Steinberg character of G.

Recall the usual parametrization of $\mathrm{Irr}(\mathfrak{S}_n)$ by partitions $\lambda\vdash n$ (see [CuRe87] 75.19). The element of $\mathrm{Irr}(\mathfrak{S}_n)$ corresponding with $\lambda\vdash n$ is the only irreducible character present in both $\mathrm{Ind}_{\mathfrak{S}_\lambda}^{\mathfrak{S}_n} 1$ and $\mathrm{Ind}_{\mathfrak{S}_{\lambda^*}}^{\mathfrak{S}_n} \mathrm{sgn}$ (see §18.2 for the notation λ^*). This can be related with Theorem 18.14 above (see also Exercise 18.4). We then get a parametrization of the unipotent characters of $\mathrm{GL}_n(\mathbb{F}_q)$ by partitions of n.

Notation 19.8. Let $\lambda \mapsto \chi_\lambda$ be the parametrization of $\mathcal{E}(G, 1)$ by partitions $\lambda \vdash n$.

If $\lambda = (\lambda_1 \geq \lambda_2 \geq \ldots) \vdash n$, denote by $L_\lambda = L(\lambda) \cong \mathrm{GL}_{\lambda_1}(\mathbb{F}_q) \times \mathrm{GL}_{\lambda_2}(\mathbb{F}_q) \times \cdots$ the associated standard Levi subgroups of G.

Proposition 19.9. *Keep $G = \mathrm{GL}_n(\mathbb{F}_q)$. The matrix of inner products $(\langle \chi_\lambda, \mathrm{R}_{L(\mu^*)}^G(\Gamma_{L(\mu^*),1} \otimes K)\rangle_G)_{\lambda,\mu}$ is lower triangular unipotent, more precisely the element corresponding to (λ, μ) is zero unless $\lambda \ll \mu$ (see §18.2), and equal to 1 when $\lambda = \mu$.*

Proof. The proposition is about ordinary characters, so we denote the modules $\Gamma_{L(\mu^*),1}$, St_G etc. by their characters.

The unipotent component of the Gelfand–Graev character is the Steinberg character (see [DiMi91] 14.40 and 14.47(ii)), so $\mathrm{R}_L^G\Gamma_{L,1}$ has a component on unipotent characters equal to $\mathrm{R}_L^G(R_{\mathrm{sgn}}^{(L)})$. This in turn is $R^{(G)}_{\mathrm{Ind}_{W_L}^{\mathfrak{S}_n}(\mathrm{sgn})}$ ([DiMi91] 15.7). Through the isometry of Theorem 19.7 above, the claim now reduces to checking that the inner product of central functions on $\mathfrak{S}_n$ $\langle \chi_\lambda, \mathrm{Ind}_{\mathfrak{S}_{\mu^*}}^{\mathfrak{S}_n} \mathrm{sgn}\rangle_{\mathfrak{S}_n}$ is zero unless $\lambda \ll \mu$, and 1 if $\lambda = \mu$. This is classical (see [JaKe81] 2.1.10, [Gol93] 7.1) or an easy consequence of Theorem 18.15 (see Exercise 18.4). □

The following gives a very restrictive condition on cuspidal simple kG-modules. A more complete result will be obtained in Theorem 19.18. Recall that $\mathrm{St}_G = e_{\ell'}^G \Gamma_{G,1}$ (see Definition 19.5).

Lemma 19.10. *If M is a simple cuspidal $kG.b_\ell(G, 1)$-module, then $M \cong \mathrm{hd}(\overline{\mathrm{St}}_G)$.*

Proof. By Theorem 14.4, the $d^1\chi$'s for $\chi \in \mathcal{E}(G, \ell')$ generate over $\mathbb{Z}$ the group of characters of projective $\mathcal{O}G$-modules. So, by Brauer's second Main Theorem (which implies that d^1 and the projection on an ℓ-block commute) the $d^1\chi$ for unipotent χ's generate the group of characters of projective $\mathcal{O}G.b_\ell(G, 1)$-modules. By the unitriangularity property of Proposition 19.9, the $\mathrm{R}_{L(\lambda)}^G(\Gamma_{L(\lambda),1})$'s for varying λ are projective modules whose characters generate the same group as the projective indecomposable modules. This implies that every simple $kG.b_\ell(G, 1)$-module is in the head of some of the

$\mathrm{R}^G_{L(\lambda)}(\Gamma_{L(\lambda),1}) \otimes k = \mathrm{R}^G_{L(\lambda)}(\overline{\Gamma}_{L(\lambda),1})$'s. If S is a simple cuspidal kG-module, and $\mathrm{Hom}_{kG}(\mathrm{R}^G_{L(\lambda)}(\overline{\Gamma}_{L(\lambda),1}), S) \neq 0$ for some λ, then by adjunction and cuspidality of S, one has $L(\lambda) = G$. Proposition 19.6(i) and (ii) imply that $\mathrm{hd}(\overline{\Gamma}_{G,1}) = \mathrm{hd}(\overline{\mathrm{St}}_G) \cong S$. □

Theorem 19.11. *Let $B := \mathbf{B}^F$. Let $\tau = (B_p, B, k)$ be the associated cuspidal "triple" (see Notation 1.10).*

(i) $\mathrm{soc}(\overline{\mathrm{St}}_G)$ *is simple and is the only composition factor of* $\overline{\mathrm{St}}_G$ *in* $\mathcal{E}(kG, \tau)$ *(see Notation 1.30).*

(ii) There is a unique $\mathcal{O}$-pure submodule of $\mathrm{Ind}_B^G\mathcal{O}$ *with character that of* $\mathrm{St}_G \otimes K$ *(i.e. the Steinberg character of G). It is isomorphic with* St_G.

Proof. Let $T = \mathbf{T}_1^F$ be the diagonal torus of $G = \mathrm{GL}_n(\mathbb{F}_q)$.

Let us show first:

(i′) there is exactly one composition factor of $\overline{\mathrm{St}}_G$ *in* $\mathcal{E}(kG, \tau)$.

One has $\mathrm{Hom}_{kG}(\mathrm{R}_T^G\overline{1_T}, \overline{\mathrm{St}}_G) \cong \mathrm{Hom}_{kT}(\overline{1_T}, {}^*\mathrm{R}_T^G\overline{\mathrm{St}}_G) = \mathrm{Hom}_{kT}(\overline{1_T}, \overline{1_T}) \cong k$, by adjunction and Proposition 19.6(iii). This implies that $\overline{\mathrm{St}}_G$ has at least one composition factor in $\mathcal{E}(kG, \tau)$. Let kT denote the regular kT-module. Again, by adjunction and Proposition 19.6(iii), we also have $\mathrm{Hom}_{kG}(\mathrm{R}_T^G kT, \overline{\mathrm{St}}_G) \cong \mathrm{Hom}_{kT}(kT, {}^*\mathrm{R}_T^G\overline{\mathrm{St}}_G) = \mathrm{Hom}_{kT}(kT, \overline{1_T}) \cong k$. But $\mathrm{R}_T^G kT$ is a projective module. Then $\mathrm{R}_T^G kT$ has among its indecomposable summands all the projective covers of the elements of $\mathcal{E}(kG, \tau)$. So, the above equation $\mathrm{Hom}_{kG}(\mathrm{R}_T^G kT, \overline{\mathrm{St}}_G) \cong k$ actually gives (i′).

Assume now that a simple submodule S of $\overline{\mathrm{St}}_G$ is cuspidal. By Lemma 19.10, $S \cong \mathrm{hd}(\overline{\mathrm{St}}_G)$. Since $\overline{\Gamma}_{G,1}$ is a projective cover of $\overline{\mathrm{St}}_G$, the multiplicity of S in $\overline{\mathrm{St}}_G$ is given by $\mathrm{Hom}_{kG}(\overline{\Gamma}_{G,1}, \overline{\mathrm{St}}_G) = \mathrm{Hom}_{\mathcal{O}G}(\Gamma_{G,1}, \mathrm{St}_G) \otimes k$. Now $\mathrm{Hom}_{\mathcal{O}G}(\Gamma_{G,1}, \mathrm{St}_G)$ is a line since $\mathrm{Hom}_{KG}(\Gamma_{G,1} \otimes K, \mathrm{St}_G \otimes K)$ is a line as stated in the proof of Proposition 19.9 (combine [DiMi91] 14.40 and 14.47(ii)). This now implies that $\overline{\mathrm{St}}_G = S$. Therefore St_G is cuspidal since ${}^*\mathrm{R}$ functors, being multiplication with idempotents $e(V) \in \mathcal{O}G$ (see Notation 3.11 and Proposition 1.5(i)), commute with reduction mod. $J(\mathcal{O})$. By the form of cuspidal unipotent characters for this kind of group (see Theorem 19.7(ii)), we obtain that $G = T$. We get

(i″) If $G \neq T$, then $\overline{\mathrm{St}}_G$ *has no simple cuspidal submodule.*

Let us now show (i) by induction on the index of T in G. If $G = T$, then Theorem 18.12(ii) gives our claim.

Assume $G \neq T$. Then (i″) applies. Let S be a simple submodule of $\overline{\mathrm{St}}_G$. By (i′), it suffices to show $S \in \mathcal{E}(kG, \tau)$. Since S is non-cuspidal there is a proper Levi subgroup $L \subset G$ such that ${}^*\mathrm{R}_L^G S \neq 0$. This module is a submodule of the reduction mod. $J(\mathcal{O})$ of ${}^*\mathrm{R}_L^G\mathrm{St}_G = \mathrm{St}_L$ (Proposition 19.6(iii)). The induction hypothesis implies that $\mathrm{Hom}_{kL}(\mathrm{R}_T^L\overline{1_T}, {}^*\mathrm{R}_L^G S) \neq 0$. By adjunction and transitivity, this gives $\mathrm{Hom}_{kG}(\mathrm{R}_T^G\overline{1_T}, S) \neq 0$, i.e. $S \in \mathcal{E}(kG, \tau)$.

(ii) Again by Proposition 19.6(iii), $\mathrm{Hom}_{\mathcal{O}G}(\mathrm{St}_G, \mathrm{R}_T^G 1_T)$ is a line. Let f be a generating element over $\mathcal{O}$. The reduction mod $J(\mathcal{O})$, $\overline{f}: \overline{\mathrm{St}_G} \to \overline{\mathrm{R}_T^G 1_T} = \mathrm{R}_T^G\overline{1_T}$ is non-zero. From (i), we know that $\mathrm{soc}(\overline{\mathrm{St}_G})$ is simple and the only composition factor in $\mathcal{E}(kG, \tau)$. But all the composition factors of $\mathrm{soc}(\mathrm{R}_T^G\overline{1_T})$ are in $\mathcal{E}(kG, \tau)$ by Theorem 1.29. So $\overline{f}$ is injective. This means that $f(\mathrm{St}_G)$ is pure in $\mathrm{R}_T^G 1_T$, thus our claim holds up to uniqueness.

The uniqueness relies on the following easy lemma. □

Lemma 19.12. *Y is an $\mathcal{O}G$-module, $\chi \in \mathrm{Irr}(G)$ with multiplicity 1 in $Y \otimes K$. Then $Y \cap e_\chi Y$ (intersection taken in $Y \otimes K$) is the unique $\mathcal{O}$-pure submodule of Y with character χ.*

19.3. Simple modules and decomposition matrices for unipotent blocks

We keep $\mathbf{G} = \mathrm{GL}_n(\mathbf{F})$, $F: \mathbf{G} \to \mathbf{G}$ defined by $F((x_{ij})) = (x_{ij}^q)$, $G = \mathbf{G}^F = \mathrm{GL}_n(\mathbb{F}_q)$ and its usual BN-pair B, T, $W = \mathfrak{S}_n$. If $\lambda = (n_1, \ldots)$ is such that $\sum_i n_i = n$, we denote by $L(\lambda) = \mathrm{GL}_{n_1}(\mathbb{F}_q) \times \cdots$ the associated standard Levi subgroup. Recall that ℓ is a prime not dividing q and that $(\mathcal{O}, K, k)$ is an ℓ-modular splitting system for G.

Definition 19.13. *Let $X := \mathrm{Ind}_B^G \mathcal{O}$, $\mathcal{H} := \mathrm{End}_{\mathcal{O}G}(X)$, i.e. the Hecke algebra of type A_{n-1} and parameter q (see Theorem 3.3). If λ is a partition of n, let $y_\lambda \in \mathcal{H}$ be as in Definition 18.11.*

Proposition 19.14. *Let $\lambda \vdash n$.*

(i) $y_\lambda\mathcal{H}$ is $\mathcal{O}$-pure in $\mathcal{H}$.

(ii) The submodule $\sqrt{y_\lambda X}$ of X is isomorphic with $\mathrm{R}_{L(\lambda)}^G \mathrm{St}_{L(\lambda)}$, the latter having $\mathrm{R}_{L(\lambda)}^G \Gamma_{L(\lambda),1}$ as a projective cover (notation of Definition 19.5).

Proof. (i) Denote by $\mathfrak{S}_\lambda$ the subgroup of $\mathfrak{S}_n$ corresponding to the Weyl group of $L(\lambda)$ (see Definition 18.11). Let $\mathcal{H}_\lambda$ be the subalgebra of $\mathcal{H}$ corresponding to the basis elements a_w for $w \in \mathfrak{S}_\lambda$. It is clearly a commutative tensor product of Hecke algebras of type A. We have $y_\lambda \in \mathcal{H}_\lambda$ and $\mathcal{H}_\lambda$ is $\mathcal{O}$-pure in $\mathcal{H}$. So our claim reduces to the case of $\lambda = (n)$. Then $y_{(n)}\mathcal{H} = \mathcal{O}y_{(n)}$ by Proposition 18.3(ii). This implies our claim since the greatest common divisor of the coefficients of $y_{(n)}$ is 1.

(ii) Assume first that $\lambda = (n)$. We must prove that $\sqrt{y_{(n)}X} \cong \mathrm{St}_G$. By Theorem 19.11(ii), it suffices to check that $y_{(n)}\mathrm{Ind}_B^G K$ represents the Steinberg character St_G^K. If $\mu \vdash n$, we have $x_\mu \mathrm{Ind}_B^G K = \mathrm{R}_{L(\mu)}^G K$ since in $KG \otimes_{KB} K$, $a_w(1 \otimes 1) = \sum_{b \in BwB/B} b \otimes 1$ and therefore $x_\mu(1 \otimes 1) = \sum_{b \in P_\mu/B} b \otimes 1$ where $P_\mu = L(\mu)B$. However, since x_μ's and y_μ's are proportional to idempotents in $\mathcal{H} \otimes K$ (Proposition 18.3(iii)), $\mathrm{Hom}_{KG}(x_\mu \mathrm{Ind}_B^G K, y_{(n)} \mathrm{Ind}_B^G K) = y_{(n)}(\mathcal{H} \otimes K)x_\mu$ is $\neq 0$ only if $\mu = (1, \ldots, 1)$ where it is a line (Theorem 18.12). Then the character of $y_{(n)} \mathrm{Ind}_B^G K$ has the same scalar products with the $\mathrm{R}_{L(\mu)}^G K$'s as the Steinberg character since $\langle \mathrm{R}_{L(\mu)}^G 1, \mathrm{St}_G^K \rangle_G = \langle 1, \mathrm{St}_{L(\mu)}^K \rangle_{L(\mu)}$ (apply [DiMi91] §9). Then the character of $y_{(n)} \mathrm{Ind}_B^G K$ is the Steinberg character by the corresponding property of characters of $\mathfrak{S}_n$ with regard to induced characters $\mathrm{Ind}_{\mathfrak{S}_\mu}^{\mathfrak{S}_n} 1$. Another proof of the above may be given, using [CuRe87] 71.14.

For general $\lambda \vdash n$, let $L := L(\lambda)$. Denote $X_L := \mathrm{Ind}_{B \cap L}^L \mathcal{O}$. We know that $\sqrt{y_\lambda X_L} \cong \mathrm{St}_L$ by the above. But $\mathrm{R}_L^G(\sqrt{Y}) = \sqrt{\mathrm{R}_L^G Y}$ for any $Y \subseteq Y'$ (Y' an $\mathcal{O}L$-lattice). We then get $\sqrt{y_\lambda X_G} \cong \mathrm{R}_L^G \sqrt{y_L \mathrm{R}_T^L \mathcal{O}} \cong \mathrm{R}_L^G \mathrm{St}_L$ as claimed.

Concerning the projective cover, we have by definition $\mathrm{St}_L = e_{\ell'}^L \Gamma_{L,1}$, so $\mathrm{R}_L^G \mathrm{St}_L = e_{\ell'}^G \mathrm{R}_L^G \Gamma_{L,1}$ (Proposition 9.15) admits $\mathrm{R}_L^G \Gamma_{L,1}$ as a projective cover by Theorem 9.10. □

Theorem 19.15. *There is a parametrization of simple $kG.b_\ell(G,1)$-modules (up to isomorphism) by partitions $\lambda \vdash n$*

$$\lambda \mapsto Z_\lambda$$

where Z_λ is characterized by the fact that its projective cover is a direct summand of $\mathrm{R}_{L(\lambda^*)}^G \Gamma_{L(\lambda^*),1}$ but not of any $\mathrm{R}_{L(\mu^*)}^G \Gamma_{L(\mu^*),1}$ with $\mu \not\ll \lambda$.

Theorem 19.16. *With the above parametrization of simple $kG.b_\ell(G,1)$-modules, and the usual parametrization of unipotent characters (see Proposition 19.7(i)), one has*

$$\mathrm{Dec}(\mathcal{O}G.b_\ell(G,1)) = \begin{pmatrix} D \\ * \end{pmatrix},$$

where D is a square lower unitriangular matrix for the order relation $\ll$ on partitions (i.e. $D = (d_{\lambda\mu})$ with $d_{\lambda\mu} = 0$ if $\mu \not\ll \lambda$ and $d_{\lambda\lambda} = 1$). Moreover,

$$D = \mathrm{Dec}_{\mathcal{H}}(\Pi_{\lambda \vdash n} y_\lambda \mathcal{H})$$

(see Definition 19.13), the bijection between indecomposable direct summands of $\Pi_{\lambda \vdash n} y_\lambda \mathcal{H}$ and indecomposable projective $\mathcal{O}G.b_\ell(G,1)$-modules being induced by the functor $H_X = \mathrm{Hom}_{\mathcal{O}G}(X, -)$ where $X = \mathrm{Ind}_B^G \mathcal{O}$.

Proof of Theorems 19.15 and 19.16. We check first that Theorem 19.16 is true for some parametrization of the simple unipotent kG-modules. Denote

$S := \mathrm{End}_{\mathcal{H}}(\Pi_{\lambda\vdash n} y_\lambda \mathcal{H})$. We prove that, for some ordering of the rows and columns,

$$\mathrm{Dec}(\mathcal{O}G.b_\ell(G,1)) = \begin{pmatrix} \mathrm{Dec}(S) & D_1 \\ D_0' & D_1' \end{pmatrix}$$

by showing that Theorem 19.4 applies for a single $X_\sigma := X = \mathrm{Ind}_B^G \mathcal{O}$ and the family of y_λ's for $\lambda \vdash n$. We have to check that the four conditions of Theorem 19.4 are fulfilled.

Condition (a) is satisfied since the unipotent characters of $G = \mathrm{GL}_n(\mathbb{F}_q)$ are simply the components of $\mathrm{Ind}_B^G K$ (see Theorem 19.7(ii)) and they form a basic set of characters for $\mathcal{O}G.b_\ell(G,1)$ (Theorem 14.4).

To check condition (b) of Theorem 19.4, one may apply Theorem 1.20(ii), noting that Condition 1.17(b) is then trivial. Another proof consists in combining Theorem 3.3 and Proposition 18.5.

Condition (c) is Proposition 19.14(i) above.

Condition (d). Proposition 19.14(ii) gives $\sqrt{y_L \mathrm{R}_T^G 1_T} \cong \mathrm{R}_L^G \mathrm{St}_L$. Moreover, we have a projective cover $\mathrm{R}_L^G \Gamma_{L,1} \longrightarrow \mathrm{R}_L^G \mathrm{St}_L$ by Proposition 19.14(ii).

It remains to check that S has a square lower unitriangular decomposition matrix.

According to the last statement of Theorem 19.4, in order to show that $\mathrm{Dec}(S)$ is square, it suffices to show that D_1 is empty. Then, to show that D_1 is empty, it suffices to show that every projective indecomposable $\mathcal{O}G.b_\ell(G,1)$-module is among the direct summands of the $\mathrm{R}_L^G \Gamma_{L,1}$'s. As was already noted in the proof of Lemma 19.10, by the unitriangularity property of Proposition 19.9, the $\mathrm{R}_{L(\lambda)}^G(\Gamma_{L(\lambda),1})$ for varying λ are projective modules whose characters generate the same group as the projective indecomposable $\mathcal{O}G.b_\ell(G,1)$-modules. So any projective indecomposable module is a direct summand of some $\mathrm{R}_{L(\lambda)}^G \Gamma_{L(\lambda),1}$.

Let us show that $\mathrm{Dec}(S)$ is unitriangular. By Proposition 5.27, this matrix is the decomposition matrix of the right $\mathcal{H}$-module $M := \Pi_{\lambda\vdash n} y_\lambda \mathcal{H}$. We know that $\mathcal{H} \otimes_{\mathcal{O}} K = \mathrm{End}_{KG}(\mathrm{Ind}_B^G K)$ is semi-simple since KG is semi-simple. So Theorem 18.15 applies. Knowing that our decomposition matrix is square, of size the number of simple $\mathcal{H} \otimes K$-modules, i.e. the number of partitions of n, Theorem 18.14 implies that the $M_{\mathcal{O}}^\lambda$'s for $\lambda \vdash n$ are the only indecomposable direct summands of M, up to isomorphism. Now Theorem 18.15 gives our claim.

It remains to check Theorem 19.15 and that Theorem 19.16 can be stated with the parametrization defined in Theorem 19.15, the unitriangularity of the decomposition matrix corresponding to zeros at (λ, μ) when $\lambda \not\ll \mu$.

Let us recall from the above that every indecomposable summand of $M = \Pi_\lambda y_\lambda \mathcal{H}$ is isomorphic to some $M^\mu_\mathcal{O}$ ($\mu \vdash n$).

Note that since $H_X(\sqrt{y_{\lambda^*}X}) \cong y_{\lambda^*}\mathcal{H}$ (Theorem 19.2(i)) and $\sqrt{y_{\lambda^*}X} \cong \mathrm{R}^G_{L(\lambda^*)}\mathrm{St}_{L(\lambda^*)}$, the latter having $\mathrm{R}^G_{L(\lambda^*)}\Gamma_{L(\lambda^*)}$ as a projective cover (Proposition 19.14(ii)), there is an indecomposable direct summand $\mathcal{P}_\lambda$ of the projective module $\mathrm{R}^G_{L(\lambda^*)}\Gamma_{L(\lambda^*)}$ such that

$$H_X(e^G_{\ell'}\mathcal{P}_\lambda) \cong M^\lambda_\mathcal{O}$$

(see Theorem 18.14). Define $Z_\lambda := \mathrm{hd}(\mathcal{P}_\lambda)$. The $\mathcal{P}_\lambda$ are pairwise non-isomorphic by the same property of the $M^\lambda_\mathcal{O}$'s, so the Z_λ's have the same property. Their number is the number of partitions of n, i.e. the cardinality of $\mathcal{E}(G,1)$ (Theorem 19.7(i)), so they are all the simple $kG.b_\ell(G,1)$-modules by Theorem 14.4.

$\mathcal{P}_\lambda$ has multiplicity one in $\mathrm{R}^G_{L(\lambda^*)}\Gamma_{L(\lambda^*)}$ since $e^G_{\ell'}.\mathrm{R}^G_{L(\lambda^*)}\Gamma_{L(\lambda^*)} \cong \sqrt{y_{\lambda^*}X}$ and the same property is satisfied by its image $M^\lambda_\mathcal{O}$ in $y_{\lambda^*}\mathcal{H} = H_X(\sqrt{y_{\lambda^*}X})$ with H_X satisfying Theorem 19.2(ii). Moreover, if $e^G_{\ell'}\mathcal{P}_\lambda$ is a summand of $\sqrt{y_{\mu^*}X}$, then $M^\lambda_\mathcal{O}$ is a summand of $y_{\mu^*}\mathcal{H}$ and therefore $\mu \ll \lambda$ by Theorem 18.14. This gives Theorem 19.15: $\mathcal{P}(Z_\lambda)$ is now the only summand of the projective module $Y_\lambda := \mathrm{R}^G_{L(\lambda^*)}\Gamma_{L(\lambda^*),1}$ not occurring in any $\mathrm{R}^G_{L(\mu^*)}\Gamma_{L(\mu^*),1}$ since the other summands of Y_λ are $\mathcal{P}(Z_\nu)$'s with $\lambda \ll \nu$, $\lambda \neq \nu$.

The same uniqueness argument as above shows that χ_λ is characterized by its appearance in the character of $Y_\lambda \otimes K$ but in no $Y_\mu \otimes K$ for $\mu \not\ll \lambda$. But defining a simple KG-module as the summand of $X \otimes K$ whose image by $H_{X\otimes K}$ is M^λ_K would give a summand of $Y_\lambda \otimes K$ satisfying the same condition by Theorem 19.2 (in the trivial case of a semi-simple KG). So $H_{X\otimes K}(\chi_\lambda) \cong M^\lambda_K$ (in order to spare notation, we identify χ_λ with any KG-module having this character).

It remains to check that the multiplicity of χ_λ in the character of $\mathcal{P}(Z_\mu)$ is the multiplicity of M^λ_K in $M^\mu_\mathcal{O} \otimes K$. Using Theorem 19.2 and $H_{X\otimes K} = H_X \otimes K$ on summands of X, one has $\mathrm{Hom}_{KG}(\chi_\lambda, \mathcal{P}(Z_\mu)\otimes K) = \mathrm{Hom}_{KG}(\chi_\lambda, e_{\ell'}.\mathcal{P}(Z_\mu) \otimes K) \cong \mathrm{Hom}_{\mathcal{H}\otimes K}(M^\lambda_K, H_X(e_{\ell'}.\mathcal{P}(Z_\mu))\otimes K) = \mathrm{Hom}_{\mathcal{H}\otimes K}(M^\lambda_K, M^\mu_\mathcal{O}\otimes K)$. □

Corollary 19.17. *Let $\mathcal{H} = \bigoplus_{w\in\mathfrak{S}_n} \mathcal{O}a_w$ denote the Hecke algebra of type A_{n-1} over $\mathcal{O}$ and with parameter a power q of a prime invertible in $\mathcal{O}$ (see Definition 3.6 or Definition 18.1). If $\lambda = (\lambda_1, \ldots, \lambda_t)$ is a sequence of integers greater than or equal to 1 whose sum is n, we denote $\lambda \models n$ and let $\mathfrak{S}_\lambda = \mathfrak{S}_{\lambda_1} \times \mathfrak{S}_{\{\lambda_1+1,\ldots,\lambda_1+\lambda_2\}} \ldots$ (see also Definition 18.11, $y_\lambda := \sum_{w\in\mathfrak{S}_\lambda}(-q)^{-l(w)}a_w$). Then $\mathrm{End}_\mathcal{H}(\Pi_{\lambda\models n} y_\lambda\mathcal{H})$ has a square lower unitriangular decomposition matrix.*

Proof. By Proposition 5.27, our claim is about the decomposition matrix of the right $\mathcal{H}$-module $\Pi_{\lambda \models n} y_\lambda \mathcal{H}$. Theorem 19.16 above tells us that the direct summand $\Pi_{\lambda \vdash n} y_\lambda \mathcal{H}$ satisfies our claim. So it suffices to show that, if $\lambda \models n$ and $\tilde{\lambda} \vdash n$ are the same up to the order of terms, then $y_\lambda \mathcal{H} \cong y_{\tilde{\lambda}} \mathcal{H}$. One has clearly $\mathfrak{S}_\lambda = v \mathfrak{S}_{\tilde{\lambda}} v^{-1}$ for some $v \in \mathfrak{S}_n$. Taking v of minimal length, it is clear that, for all $w \in \mathfrak{S}_\lambda$, $\tilde{w} \in \mathfrak{S}_{\tilde{\lambda}}$, one has $l(wv) = l(w) + l(v)$, $l(v\tilde{w}) = l(\tilde{w}) + l(v)$ and therefore $l(v\tilde{w}v^{-1}) = l(\tilde{w})$. Then $a_v y_{\tilde{\lambda}} = \sum_{\tilde{w} \in \mathfrak{S}_{\tilde{\lambda}}} (-q)^{-l(\tilde{w})} a_{v\tilde{w}} = y_\lambda a_v$. On the other hand, any a_x ($x \in \mathfrak{S}_n$) is invertible since q is a unit in $\mathcal{O}$. So $y_{\tilde{\lambda}} \mathcal{H} \cong y_\lambda \mathcal{H}$ by $h \mapsto a_v h$. □

19.4. Modular Harish-Chandra series

We keep $\mathbf{G} = \mathrm{GL}_n(\mathbf{F})$, $G = \mathrm{GL}_n(\mathbb{F}_q)$ endowed with their usual BN-pairs, $(\mathcal{O}, K, k)$ an ℓ-modular splitting system for G. We recall the parametrization of simple $kG.b_\ell(G, 1)$-modules by partitions $\lambda \vdash n$ (see Theorem 19.15)

$$\lambda \mapsto Z_\lambda.$$

Theorem 19.18. *With the notation recalled above,* $\mathrm{hd}(\overline{\mathrm{St}_{G,1}}) = Z_{(1,\dots,1)}$ *is cuspidal if and only if $n = 1$ or $n = e\ell^m$ where e is the order of q mod. ℓ and $m \geq 0$. It is the only simple cuspidal $kG.b_\ell(G, 1)$-module. Its dimension is* $|G|q^{-\binom{n}{2}}(q^n - 1)^{-1}$

If $n > 1$, there exist a Coxeter torus $\mathbf{T} \subseteq \mathbf{G}$ (i.e. of type a cycle of order n in $\mathfrak{S}_n$ with regard to the diagonal torus of $\mathbf{G}$, see Example 13.4(i) and (iii)) and $\theta \in \mathrm{Hom}(\mathbf{T}^F, \mathcal{O}_\ell^\times)$ a character in general position (see [Cart85] p. 219) and of order a power of ℓ, such that the Brauer character of $Z_{(1,\dots,1)}$ is the restriction to $G_{\ell'}$ of the cuspidal irreducible character equal to $\varepsilon_\mathbf{G} \varepsilon_\mathbf{T} \mathrm{R}_\mathbf{T}^\mathbf{G}(\theta)$ (or equivalently, $Z_{(1,\dots,1)} \cong M \otimes_\mathcal{O} k$ where M is an $\mathcal{O}G$-lattice in the cuspidal KG-module affording the character $\varepsilon_\mathbf{G} \varepsilon_\mathbf{T} \mathrm{R}_\mathbf{T}^\mathbf{G}(\theta)$; see §8.3 for the notation $\varepsilon_\mathbf{G}$).

Definition 19.19. *Let $d \geq 2$, $n \geq 0$. Recall that the exponential notation for partitions $\lambda \vdash n$ is denoted by $(1^{(m_1)}, 2^{(m_2)}, \dots, i^{(m_i)}, \dots)$, meaning that i is repeated m_i times and $\sum_i i m_i = n$. A d-regular partition of n is any partition $(1^{(m_1)}, 2^{(m_2)}, \dots)$ such that all m_i's are less than d. We denote by $\pi(n)$ (resp. $\pi_d(n)$) the number of partitions (resp. d-regular partitions) of n. We set $\pi(0) = \pi_d(0) = 1$.*

If $c, d \geq 2$, $\lambda = (\lambda_1 \geq \lambda_2 \geq \dots) \vdash n$, let $\rho_{c,d}(\lambda) \vdash n$ be defined by $\lambda_i = \lambda_i^{(-1)} + c \sum_{j \geq 0} d^j \lambda_i^{(j)}$ with $0 \leq \lambda_i^{(-1)} \leq c - 1$, $0 \leq \lambda_i^{(j)} \leq d - 1$ and $\rho_{c,d}(\lambda) = (1^{(m_{-1})}, c^{(m_0)}, (cd)^{(m_1)}, (cd^2)^{(m_2)}, \dots)$ with $m_j = \sum_i \lambda_i^{(j)}$. We set $\rho_{1d} = \rho_{dd}$.

Note that $\rho_{c,d}(\lambda) = \lambda$ if and only if all parts of λ are in $\{1, c, cd, cd^2, cd^3, \dots\}$.

We recall that, when ℓ is a prime, $\pi_\ell(n)$ equals the number of conjugacy classes of elements of $\mathfrak{S}_n$ whose order is prime to ℓ (see [JaKe81] 6.1.2 or Exercise 4 below).

Theorem 19.20. *Let $L = L(\lambda)$ be the Levi subgroup of G associated with a partition $\lambda \vdash n$ where all parts are of the form $e\ell^i$ (m_i such parts) where $i \geq 0$ and e is the order of q mod. ℓ, or $= 1$ (m_{-1} such parts). Let Z_L be the unique simple $kL.b_\ell(L, 1)$-module that is cuspidal (see Theorem 19.18), giving rise to the cuspidal "triple" (L, Z_L) in the sense of Notation 1.10 (the parabolic subgroup is omitted since the induced module $\mathrm{R}_L^G Z_L$ does not depend on it; see Notation 3.11).*

(i) $\mathrm{End}_{kG}\mathrm{R}_L^G Z_L$ is isomorphic with $\mathcal{H}_k(\mathfrak{S}_{m_{-1}}, q.1_k) \otimes k[\mathfrak{S}_{m_0} \times \mathfrak{S}_{m_1} \times \cdots]$.

(ii) If $\mu \vdash n$, then $Z_\mu \in \mathcal{E}(kG, L, Z_L)$ if and only if $\lambda = \rho_{e,\ell}(\mu^*)$.

We are going to prove both theorems in the remainder of this section. We shall need to know the number of simple modules for Hecke algebras of type A (see [Mathas] 3.43).

Theorem 19.21. *Let $n \geq 1$, and let q be an integer prime to ℓ. Let d be the smallest integer such that $1 + q + \cdots + q^{d-1} \equiv 0$ mod. ℓ (i.e. $d = \ell$ if $q \equiv 1$ mod. ℓ, d is the order of q mod. ℓ otherwise). Then the number of simple $\mathcal{H}_k(\mathrm{A}_{n-1}, q)$-modules equals $\pi_d(n)$, the number of d-regular partitions of n (see Definition 19.19).*

Remark 19.22. A slightly different approach, not using Theorem 19.21 above and showing in an elementary fashion that only dimensions $e\ell^m$ give rise to unipotent cuspidal modules, is sketched in Exercises 6–9. Note that in case $q \equiv 1$ mod. ℓ, $\mathcal{H}_k(\mathrm{A}_{n-1}, q) = \mathcal{H}_k(\mathrm{A}_{n-1}, 1)$ is the group algebra $k\mathfrak{S}_n$ and Theorem 19.21 is easy (see Exercise 4 or [JaKe81] 6.1.12).

Proof of Theorems 19.18 and 19.20. We assume that $G = \mathrm{GL}_n(\mathbb{F}_q)$ has a cuspidal $kG.b_\ell(G, 1)$-module. We have seen that such a simple module has to be isomorphic with $\mathrm{hd}(\overline{\mathrm{St}}_G)$ (Lemma 19.10).

Viewing $\mathrm{GL}_n(\mathbb{F}_q)$ as the group of $\mathbb{F}_q$-linear endomorphisms of $\mathbb{F}_{q^n}$, letting $s \in \mathbb{F}_{q^n}$ be an element of multiplicative order $(q^n - 1)_\ell$, it is clear that, since ℓ divides $\phi_n(q)$, s is in no proper subfield of $\mathbb{F}_{q^n}$. So the element of $\mathrm{GL}_n(\mathbb{F}_q)$ it induces by multiplication is a regular element $s \in \mathbf{T}^*$ (i.e. $\mathrm{C}_{\mathbf{G}^*}(s) = \mathbf{T}^*$) in a Coxeter torus of $\mathbf{G}^* = \mathrm{GL}_n(\mathbf{F})$.

Let M be an $\mathcal{O}G$-lattice in the KG-module affording the irreducible character $\varepsilon_\mathbf{G}\varepsilon_\mathbf{T}\mathrm{R}_\mathbf{T}^\mathbf{G}\hat{s}$ (Theorem 8.27). Its rank is $|G|q^{-\binom{n}{2}}(q^n - 1)^{-1}$ by Theorem 8.16(ii). Since s is an ℓ-element, $b_\ell(G, 1)$ acts by the identity on $M \otimes K$, hence on M and $M \otimes k$. By the Mackey formula ([DiMi91] 11.13) and the fact that no $\mathbf{G}^F$-conjugate of $\mathbf{T}$ embeds in a proper standard Levi subgroup, $\varepsilon_{\mathbf{G}}\varepsilon_{\mathbf{T}}\mathrm{R}_{\mathbf{T}}^{\mathbf{G}}\hat{s}$ is cuspidal, so M is cuspidal. The same is true for $M \otimes k$ and all its composition factors since $^*\mathrm{R}$ and $\bigotimes_{\mathcal{O}} k$ commute. This implies at once that $kG.b_\ell(G, 1)$ has cuspidal simple modules.

Now Lemma 19.10 implies that, since it is cuspidal, it is a multiple of $\mathrm{hd}(\overline{\mathrm{St}}_G)$. Since $\overline{\Gamma}_{G,1}$ is a projective cover of $\mathrm{hd}(\overline{\mathrm{St}}_G)$ (see Proposition 19.6(ii)), the multiplicity equals the dimension of $\mathrm{Hom}_{kG}(\overline{\Gamma}_{G,1}, M \otimes_{\mathcal{O}} k)$, which is also the rank of $\mathrm{Hom}_{\mathcal{O}G}(\Gamma_{G,1}, M)$, i.e. the dimension of $\mathrm{Hom}_{KG}(\Gamma_{G,1} \otimes K, M \otimes K)$. This is 1 since the Gelfand–Graev character has no multiplicity (see [DiMi91] 14.47) while $M \otimes K$ is simple of character $\pm\mathrm{R}_{\mathbf{T}}^{\mathbf{G}}\hat{s}$. This implies that $\mathrm{hd}(\overline{\mathrm{St}}_G) = Z_{(1,\ldots,1)}$ is cuspidal when $n = 1$ or $e\ell^m$. It satisfies what is stated in Theorem 19.18.

We shall see below that the converse is true.

Structure of $\mathcal{H} := \mathrm{End}_{kG}\mathrm{R}_L^G Z_L$. Define M_n as in Theorem 19.18 (we add the index to recall the ambient dimension). We assume the above choice of s has been made once and for all for any dimension less than or equal to n. Write $L = L_\lambda$ with $\lambda = n_1 \geq n_2 \geq \ldots$, so $L \cong \mathrm{GL}_{n_1}(\mathbb{F}_q) \times \mathrm{GL}_{n_2}(\mathbb{F}_q) \times \cdots$, and define $M_L \cong M_{n_1} \otimes M_{n_2} \otimes \ldots$. Then $Z_L = M_L \otimes_{\mathcal{O}} k$ since there is a single cuspidal simple $kL.b_\ell(L, 1)$-module. By the uniqueness of M in dimension n, we have the following.

Lemma 19.23. *Let* $\lambda = (1^{(m_{-1})}, (e)^{(m_0)}, (el)^{(m_1)}, \ldots)$ *(see Definition 19.19), then* $\mathrm{N}_G(L, M_L) = \mathrm{N}_G(L, M_L \otimes K) = \mathrm{N}_G(L) \cong \Pi_i\big(\mathrm{GL}_{e\ell^i}(\mathbb{F}_q)\big) \wr \mathfrak{S}_{m_i}$ *and* M_L *extends to an* $\mathcal{O}\mathrm{N}_G(L)$*-lattice.*

One has $\mathcal{H} = \bigoplus_g \mathcal{O}a_{g,\tau,\tau}$ (see Definition 1.12) where $\tau = (L, Z_L)$ and g ranges over a representative system which, in the case of a finite group with a BN-pair, is the subgroup of the Weyl group $W(I_L, M_L) := \{w \in W\ ;\ wI_L = I_L \text{ and } {}^wM_L \cong M_L\}$ (see Theorem 2.27(iv)). We have an injective map from $\mathcal{H}$ to $\mathrm{End}_{kG}\mathrm{R}_L^G M \otimes k$ sending $a_{g,\tau,\tau}$ to the element labeled in the same fashion. So it yields an isomorphism of algebras $\mathcal{H} \otimes k \cong \mathrm{End}_{kG}\mathrm{R}_L^G M \otimes k$. So it suffices to show the following.

Lemma 19.24. $\mathrm{End}_{\mathcal{O}G}\mathrm{R}_L^G M$ *is isomorphic with* $\mathcal{H}_{\mathcal{O}}(\mathfrak{S}_{m_{-1}}, q) \otimes (\otimes_{i\geq 0}\mathcal{H}_{\mathcal{O}}(\mathfrak{S}_{m_i}, q^{e\ell^i}))$.

The same basis is used in Chapter 3 to describe $\mathcal{H}_L \otimes_{\mathcal{O}} K = \mathrm{End}_{KG}(X_L \otimes K)$. Note that in our case $W(I_L, M_L) = W^{I_L} = \{w \in W\ ;\ wI_L = I_L\}$ (see Definition 2.26). Then the group $C(I_L, M_L)$ is trivial. The basis elements, once normalized as in the proof of Theorem 3.16, satisfy certain relations involving a cocycle λ and coefficients $c_\alpha \in K$. It is clear from its definition in Theorem 3.16 that λ takes its values in $\mathcal{O}^\times$. By Remark 3.18, the cocycle is trivial.

By Proposition 1.23, it suffices to check the quadratic relation in the case $\lambda = (d, d)$, $n = 2d$, $d > 1$ (the case of $d = 1$ is Theorem 3.3). Denote by $P = VL$ the parabolic subgroup containing lower triangular matrices such that $L = L_{(d,d)}$. Remember that M has character $\pm \mathrm{R}_{\mathbf{T}}^{\mathbf{L}}\hat{s}$ where $s \in (\mathbf{T}^*)_\ell^F$ is a regular element in the dual of a Coxeter torus $\mathbf{T}$ of $\mathbf{L}$. Let us define the following matrices, where $g \in \mathrm{Mat}_d(\mathbb{F}_q)$, $h \in \mathrm{GL}_d(\mathbb{F}_q)$,

$$x := \begin{pmatrix} 0 & \mathrm{Id}_d \\ \mathrm{Id}_d & 0 \end{pmatrix}, \quad v_g := \begin{pmatrix} \mathrm{Id}_d & 0 \\ g & \mathrm{Id}_d \end{pmatrix} \in V, \quad t_h = \begin{pmatrix} -h^{-1} & 0 \\ 0 & h \end{pmatrix} \in L.$$

We know that x generates $W(L, M)$ (see above). So our claim about parameters is implied by the following.

Lemma 19.25. *$b := (-1)^{\ell-1} q^{\binom{d+1}{2}} a_{x,\tau,\tau}$ satisfies $(b+1)(b-q^d) = 0$.*

Proof. Denote $\varepsilon = (-1)^{\ell-1}$. The above equation is equivalent to $(a_{x,\tau,\tau})^2 = q^{-d^2} + \varepsilon q^{-\binom{d}{2}}(1-q^{-d})a_{x,\tau,\tau}$.

By Proposition 3.9(ii) (or the computations in the proof of Proposition 1.18), we have $(a_{x,\tau,\tau})^2 = q^{-d^2} + \beta a_{x,\tau,\tau}$ for some $\beta \in K$. So, taking $m \in M$, one must look at the projection of $(a_{x,\tau,\tau})^2(1 \otimes_P m)$ on $KPxP \otimes_P M$ as a direct summand of $KG \otimes_P M$, and check that it is $\varepsilon q^{-\binom{d}{2}}(1-q^{-d})a_{x,\tau,\tau}(1 \otimes m)$. We have $a_{x,\tau,\tau}(1 \otimes m) = e(V)x \otimes \theta(m)$, where $\theta \colon M \to M$ makes the kL-modules M and ${}^xM$ isomorphic (see Definition 1.12). Then $(a_{x,\tau,\tau})^2(1 \otimes m) = |V|^{-1}\sum_g e(V)xv_gx \otimes m$ where g ranges over $\mathrm{Mat}_d(\mathbb{F}_q)$. It is easily checked that a product xv_gx may be in $PxP = VxP$ only if g is invertible (look at when a product $v_{g'}xv_g$ can possibly be in P^x). In that case, $xv_gx = v_{g^{-1}}xt_gv_{g^{-1}}$ and $e(V)xv_gx \otimes m = e(V)x \otimes t_gm$ since V acts trivially on M. So our projection is $e(V)x \otimes m'$ where $m' = q^{-d^2}\sum_{g \in G'} t_gm$, the sum being over $g \in G' := \mathrm{GL}_d(q)$. Since this projection is expected to be $\varepsilon q^{-\binom{d}{2}}(1-q^{-d})a_{x,\tau,\tau}(1 \otimes m) = \varepsilon q^{-\binom{d}{2}}(1-q^{-d})e(V)x \otimes \theta(m)$, it remains to check that

$$\sum_{g \in G'} t_gm = \varepsilon q^{\binom{d}{2}}(q^d - 1)\theta(m).$$

We recall that $M = M_0 \otimes_K M_0$ where M_0 is a representation of $G' = \mathrm{GL}_d(\mathbb{F}_q)$ of character $\pm \mathrm{R}_{\mathbf{T}}^{\mathbf{G}'}\hat{s}$ for s a regular ℓ-element of the Coxeter torus $\mathbf{T}^*$, so that $\dim M_0 = |G'|q^{-\binom{d}{2}}(q^d-1)^{-1}$ (see Theorem 19.18). Note that $-\mathrm{Id}_d$ is a central element which is in $\mathbf{T}_\ell^F$ if and only if $\ell = 2$, so it acts on M_0 by ε. The elements of L can be written as (g_1, g_2) with $g_i \in G'$ and act by $(g_1, g_2).(m_1 \otimes m_2) = g_1.m_1 \otimes g_2.m_2$. We have $\theta(m_1 \otimes m_2) = m_2 \otimes m_1$. Let us take a K-basis of $M_0 = Ke_1 \oplus \ldots \oplus Ke_f$, where each $g \in G'$ acts by a matrix $(\mu_{i,j}(g))$. To check the equation above, we may take $m = e_i \otimes e_j$. By the definition of t_g, we have $\sum_{g\in G'} t_g.(e_i \otimes e_j) = \varepsilon \sum_{g\in G'} g^{-1}.e_i \otimes g.e_j = \varepsilon \sum_{g\in G',\, k,l} \mu_{ki}(g^{-1})\mu_{lj}(g)e_k \otimes e_l$. But, in K, $\sum_{g\in G'} \mu_{ki}(g^{-1})\mu_{lj}(g) = |G'|(\dim(M_0))^{-1}\delta_{kj}\delta_{il}$ by a well-known orthogonality relation (see [Serre77a] p. 27). So $\sum_{g\in G'} t_g.(e_i \otimes e_j) = \varepsilon q^{\binom{d}{2}}(q^d - 1)e_j \otimes e_i$ as expected. □

Here are some immediate properties of $\rho_{c,d}$ reductions (see Definition 19.19).

Lemma 19.26. $c, d \geq 2$, $n \geq 0$.

(i) Let $(m_{-1}, m_0, m_1, \ldots)$ be a sequence of integers ≥ 0 such that $n = m_{-1} + c\sum_{i\geq 0} d^i m_i$. Then $|\{\lambda \vdash n; \rho_{c,d}(\lambda) = (1^{(m_{-1})}, c^{(m_0)}, (cd)^{(m_1)}, (cd^2)^{(m_2)}, \ldots)\}| = \pi_c(m_{-1})\pi_d(m_0)\pi_d(m_1)\pi_d(m_2)\ldots$.

(ii) $\pi(n) = \sum \pi_c(m_{-1})\pi_d(m_0)\pi_d(m_1)\ldots$ where the sum is over all sequences $(m_{-1}, m_0, m_1, \ldots)$ of integers ≥ 0 such that $n = m_{-1} + c\sum_{i\geq 0} d^i m_i$.

Proof. (i) is easy from the definition of $\rho_{c,d}$ and the uniqueness of d-adic expansion.

(ii) is a consequence of (i) and of the partition of $\{\lambda \mid \lambda \vdash n\}$ induced by the map $\rho_{c,d}$. □

Lemma 19.27. *Assume the same hypotheses as above. Let $\lambda, \mu \vdash n$. Let $\subseteq_G$ denote inclusion up to G-conjugacy. If $L(\lambda) \subseteq_G L(\mu)$, then $L(\rho_{c,d}(\lambda)) \subseteq_G L(\rho_{c,d}(\mu)) \subseteq_G L(\mu)$.*

Proof. The relation $L(\lambda) \subseteq_G L(\mu)$ is clearly equivalent to the parts of μ being disjoint sums of parts of λ (argue on simple submodules of $(\mathbb{F}_q)^n$ seen as a module over $L(\mu)$). Now, $L(\rho_{c,d}(\mu)) \subseteq_G L(\mu)$ is clear from the definition of $\rho_{c,d}$. The other inclusion reduces to the case $\mu = (n)$. We prove it in the case $\mu = (n)$ by induction on n. Write $\lambda = (\lambda_1 \geq \lambda_2 \geq \ldots)$, $\lambda_i = m^i_{-1} + c(m^i_0 + m^i_1 d + m^i_2 d^2 + \cdots)$, $n = n_{-1} + c(n_0 + n_1 d + n_2 d^2 + \cdots)$, with n_{-1}, $m^i_{-1} \in [0, c-1]$, $m^i_0 + m^i_1 d + m^i_2 d^2 + \cdots$ and $n_0 + n_1 d + n_2 d^2 + \cdots$ being d-adic expansions. We have $n_{-1} = n = \sum_j \lambda_j \equiv \sum_j m^j_{-1}$ mod. c, so that $\sum_j m^j_{-1} \geq n_{-1}$. If $n_{-1} \neq 0$, one may then replace n with $n-1$, find some $m^j_{-1} \neq 0$ and replace the corresponding λ_j with $\lambda_j - 1$. The inclusion we obtain by induction in GL_{n-1} gives our claim. If $n_{-1} = 0$, then one replaces (c, d) with (d, d), each λ_j with $c^{-1}(\lambda_j - m^j_{-1})$ and n with $c^{-1}(n - \sum_j m^j_{-1})$. The induction gives our claim. □

In what follows, we abbreviate $\rho_{e,\ell} = \rho$. By what we know from Theorem 19.18, all cuspidal triples are of type $\tau_\lambda := (L(\lambda), \mathrm{hd}(\overline{\mathrm{St}}_{L(\lambda)}))$ for $\lambda \vdash n$ and $\lambda = \rho(\lambda)$. From the structure of $\mathrm{End}_{kG}\mathrm{R}^G_{L(\lambda)}Z(\lambda)$ that we have just seen, if

$\lambda = \rho(\lambda) = (1^{(m_{-1})}, e^{(m_0)}, (e\ell)^{(m_1)}, (e\ell^2)^{(m_2)}, \ldots)$, then $\mathrm{End}(\mathrm{R}^G_{L(\lambda)} Z_{L(\lambda)})$ is isomorphic with $\mathcal{H}_k(\mathfrak{S}_{m_{-1}}, q) \otimes k[\mathfrak{S}_{m_0} \times \mathfrak{S}_{m_1} \times \cdots]$ by Theorem 19.20(i).

We are now in a position to show that all cuspidal triples are of type $\tau_\lambda := (L(\lambda), \mathrm{hd}(\overline{\mathrm{St}}_{L(\lambda)}))$ for $\lambda \vdash n$ and $\lambda = \rho(\lambda)$. If $\lambda = \rho(\lambda) = (1^{(m_{-1})}, e^{(m_0)}, (e\ell)^{(m_1)}, (e\ell^2)^{(m_2)}, \ldots)$, then $\mathrm{End}(\mathrm{R}^G_{L(\lambda)} Z_{L(\lambda)})$ is isomorphic with $\mathcal{H}_k(\mathfrak{S}_{m_{-1}}, q) \otimes k[\mathfrak{S}_{m_0} \times \mathfrak{S}_{m_1} \times \cdots]$ by Lemma 19.24. Then its number of simple modules is

$$s(\lambda) := \pi_d(m_{-1})\pi_\ell(m_0)\pi_\ell(m_1)\ldots$$

by Theorem 19.21. This number $s(\lambda)$ is also the number of indecomposable summands (up to isomorphism) of $\mathrm{R}^G_{L(\lambda)} Z_{L(\lambda)}$, which is also the number of elements of $\mathcal{E}(G, \tau_\lambda)$ (Theorem 1.29). To show that there are no other cuspidal triples than the τ_λ, it therefore suffices to check that $\sum_{\lambda\vdash n\ ;\ \lambda=\rho(\lambda)} s(\lambda) = \pi(n)$. But we have $\pi(n) = \sum \pi_d(m_{-1})\pi_\ell(m_0)\pi_\ell(m_1)\ldots$ where the sum is over $(1^{(m_{-1})}, e^{(m_0)}, (e\ell)^{(m_1)}, (e\ell^2)^{(m_2)}, \ldots) \vdash n$ by Lemma 19.26(ii). When $d = e$, this is actually Lemma 19.26(ii). When $d \neq e$, then $e = 1$ and $d = \ell$, forcing $m_{-1} = 0$ by our conventions, and this is again the same identity.

This gives our claim.

Note that the above argument uses just the statement that the number of simple $\mathcal{H}_k(\mathfrak{S}_n, q)$-modules is greater than or equal to $\pi_d(n)$.

It remains to prove Theorem 19.20(ii). Denote $Z'_\lambda = Z_{\lambda^*}$. Let $\mathcal{C} = \{\mu \vdash n \mid \rho(\mu) = \mu\}$, $\mathcal{Z}_\nu = \{Z'_\lambda \mid \rho(\lambda) = \nu\}$. We must show that $\mathcal{Z}_\nu = \mathcal{E}(G, \tau_\nu)$ for all $\nu \in \mathcal{C}$.

We have seen above that if $\lambda, \mu \vdash n$, the relation $L_\lambda \subseteq_G L(\mu)$ is equivalent to μ being formed by disjoints sums of parts of λ; we denote by $\lambda \ll' \mu$ the corresponding order relation on $\mathcal{C}$.

To prove that $\mathcal{Z}_\nu = \mathcal{E}(G, \tau_\nu)$ for all $\nu \in \mathcal{C}$, it suffices to check

$$\bigcup_{\nu\in\mathcal{C}\ \nu\ll'\mu} \mathcal{Z}_\nu = \bigcup_{\nu\in\mathcal{C}\ \nu\ll'\mu} \mathcal{E}(G, \tau_\nu)$$

for all $\mu \in \mathcal{C}$ (disjoint unions on both sides).

Let $\nu = (1^{(m_{-1})}, e^{(m_0)}, (e\ell)^{(m_1)}, (e\ell^2)^{(m_2)}, \ldots) \in \mathcal{C}$ (with $m_{-1} = 0$ if $e = 1$). We have $|\mathcal{Z}_\nu| = \pi_e(m_{-1})\pi_\ell(m_0)\pi_\ell(m_1)\ldots$ (with π_1 constant equal to 1) by Lemma 19.26(i). But $|\mathcal{E}(G, \tau_\nu)|$ is the number of indecomposable summands of $\mathrm{R}^G_{L(\nu)} Z(\nu)$ (up to isomorphism), by Theorem 1.29. This is the number of simple modules for the endomorphism ring $\mathrm{End}_{kG}(\mathrm{R}^G_{L(\nu)} Z(\nu))$ by the correspondence between conjugacy classes of primitive idempotents and simple modules. By Theorem 19.20(i) and Theorem 19.21, this is $|\mathcal{E}(G, \tau_\nu)| = \pi_e(m_{-1})\pi_\ell(m_0)\pi_\ell(m_1)\ldots$. So we get $|\mathcal{Z}_\nu| = |\mathcal{E}(G, \tau_\nu)|$ for all $\nu \in \mathcal{C}$.

Now, it suffices to prove the inclusion

$$\bigcup_{\nu\in\mathcal{C}\ \nu\ll'\mu}\mathcal{Z}_\nu \subseteq \bigcup_{\nu\in\mathcal{C}\ \nu\ll'\mu}\mathcal{E}(G,\tau_\nu)$$

for all $\mu\in\mathcal{C}$. Let $Z'_\lambda\in\mathcal{Z}_\nu$ with $\nu\ll'\mu$ in $\mathcal{C}$, i.e. $\lambda\vdash n$ and $\rho(\lambda)=\nu$. We have $Z'_\lambda = Z_{\lambda^*}$, which is a quotient of $\mathrm{R}^G_{L(\lambda)}\Gamma_{L(\lambda)}\otimes k$ (see Theorem 19.15). Then $Z'_\lambda\in\mathcal{E}(G,\tau_\gamma)$ for a cuspidal triple such that $L(\gamma)\subseteq_G L(\lambda)$. This implies $\rho(\gamma)\ll'\rho(\lambda)$ by Lemma 19.27, i.e. $\gamma\ll'\nu$ and therefore $\gamma\ll'\mu$. So Z'_λ is in $\bigcup_{\nu\in\mathcal{C}\ \nu\ll'\mu}\mathcal{E}(G,\tau_\nu)$ as stated. □

Exercises

1. Assume the hypotheses of Definition 19.5. Let s be a semi-simple ℓ'-element of $(\mathbf{G}^*)^F$. Define $\Gamma_{\mathbf{G}^F,s} = b_\ell(\mathbf{G}^F,s).\Gamma_{\mathbf{G}^F}$, $\mathrm{St}_{\mathbf{G}^F,s} = e_{\ell'}^{\mathbf{G}^F}.\Gamma_{\mathbf{G}^F,s}$. Show an analogue of Proposition 19.6 for those modules. What about non-connected Z($\mathbf{G}$)?
2. Let $\mathcal{O}$, K be as in §19.1. Let A be an $\mathcal{O}$-free finitely generated $\mathcal{O}$-algebra such that $A\otimes K$ is semi-simple. Let $0\to\Omega\to P\to Y\to 0$ be an exact sequence between $\mathcal{O}$-free A-modules with projective P. Show the equivalence of the following three conditions.
   (1) Y_K and Ω_K have no simple component in common.
   (2) Ω, as a submodule of P, is stable under the action of $\mathrm{End}_A(P)$.
   (2′) $\Omega\otimes K$, as a submodule of $P\otimes K$, is stable under the action of $\mathrm{End}_{A\otimes K}(P\otimes K)$.

   When those conditions are satisfied, show that $\mathrm{End}_A(Y)$ is a quotient of $\mathrm{End}_A(P)$.
3. Let $\mathcal{O}$ a local ring with fraction field K.
   (a) Let E be an $\mathcal{O}$-free finitely generated **symmetric** algebra (see Definition 1.19). Let $I\subseteq E^n$ be an $\mathcal{O}$-pure right submodul and $t\in\mathrm{Hom}_E(I,(E_E)^m)$. Then show that there exists $\widehat{t}\in\mathrm{Hom}_E((E_E)^n,(E_E)^m)$ such that it coincides with t on I.

   Show Theorem 19.2 for a pure submodule of E^n, when E is symmetric.
   (b) Let A be an $\mathcal{O}$-free finitely generated $\mathcal{O}$-algebra. Denote by $A-$**mod** (resp. **mod**$-A$) the category of $\mathcal{O}$-free finitely generated left (resp. right) A-modules. Let Y be in $A-$**mod**. Assume $A\otimes K$ is semi-simple and $E:=\mathrm{End}_A(Y)$ is symmetric. Show that **mod**$-E$ is equivalent to the full subcategory $\mathbf{C}_Y$ (resp. $\mathbf{C}'_Y$) in $A-$**mod** consisting of modules $I.Y$ (resp. $\sqrt{I.Y}$) for $I\subseteq E^l$ $\mathcal{O}$-pure.

4. Let $d \geq 2$, denote by $\pi'_d(n)$ the number of partitions of n whose parts are all non-divisible by d (where conventionally $\pi'_d(0) = 1$). Show that the associated series is $\prod_{k\geq 1}\frac{1-t^{dk}}{1-t^k} = \prod\frac{1}{1-t^k}$ where the second product is over $k \geq 1$ not divisible by d. Show that the series associated with $\pi_d(n)$ is $\prod_{k\geq 1}(1 + t^k + t^{2k} + \cdots + t^{(d-1)k})$. Deduce that $\pi_d = \pi'_d$. Show that the number of conjugacy classes of ℓ'-elements of $\mathfrak{S}_n$ is $\pi'_\ell(n) = \pi_\ell(n)$. What about conjugacy classes of $\mathfrak{S}_n$ whose elements are of order not divisible by (or prime to) d, when d is no longer a prime?
5. Generalize Lemma 19.26 and Lemma 19.27 with $\rho_{c,d}$ being replaced by the following expansion process. Let $\mathbf{d}$ be a sequence $1 = d_0|d_1|d_2|d_3|\ldots$ of integers in $[1, \infty]$ and increasing for division. The $\mathbf{d}$-adic expansion of n is $n = n_0 + n_1d_1 + n_2d_2 + \cdots$ with $0 \leq n_id_i < d_{i+1}$. Define $\rho_{\mathbf{d}}(\lambda)$ for any composition $\lambda \models n$. Show Lemma 19.26. Show that $L(\lambda) \subseteq_G L(\mu)$ implies $L(\rho_{\mathbf{d}}(\lambda)) \subseteq_G L(\rho_{\mathbf{d}}(\mu)) \subseteq_G L(\mu)$.
6. For the language of Green vertices, sources and Green correspondence for modules, we refer to [Ben91a] §§3.10, 3.12. Let G be a finite group, k a field of characteristic $\ell \neq 0$ and containing a $|G|$th root of 1. Assume $G = U \rtimes L$, a semi-direct product where U is of order prime to ℓ.
   (a) Show that the inflation functor $\mathrm{Infl}_L^G : kL\mathbf{-mod} \to kG\mathbf{-mod}$ preserves Green vertices and sources.
   (b) If D is an ℓ-subgroup of G such that $\mathrm{N}_G(D) \subseteq L$, and M is an indecomposable kG-module of vertex D, show that U acts trivially on M (show that if f is the Green correspondence between kG-modules with vertex D and kL-modules with same vertex (see [Ben91a] §3.12, [Thévenaz] 20.8), then $f\big(\mathrm{Infl}_L^G f^{-1}(M)\big)$ is isomorphic to M).
7. Let G be a finite group with a strongly split BN-pair of characteristic p. Let ℓ be a prime $\neq p$. Let k be a field of characteristic $\ell \neq 0$ and containing a $|G|$th root of 1. Let $P = U_P \rtimes L$ be the Levi decomposition of a parabolic subgroup $P \neq G$. Assume $|G : P|$ is prime to ℓ. Let N be an indecomposable kG-module whose dimension is prime to ℓ.
   (a) Show that any Sylow ℓ-subgroup of G is a vertex of N (see [NaTs89] 4.7.5 or [Thévenaz] Exercise 23.2). Show that there is an indecomposable kP-module M such that N is a direct summand of $\mathrm{Ind}_P^G M$.
   (b) Show that the vertices of M are Sylow ℓ-subgroups of G.

We assume now that a Sylow ℓ-subgroup D of G satisfies $D.\mathrm{C}_G(D) \subseteq L$.

   (c) Show that $\mathrm{N}_P(D) \subseteq L$.
   (d) Show that no component of $\mathrm{hd}(N)$ is cuspidal (apply Exercise 6 to show that U acts trivially on M).

8. Let $G = \mathrm{GL}_n(\mathbb{F}_q)$, $n \geq 2$.
   (a) Show that, if it has a Sylow ℓ-subgroup D such that $DC_G(D)$ can't be included in any proper split Levi subgroup, then $n = el^a$ where e is the order of q mod. ℓ and a is an integer.
   (b) Use Exercises 6–7 to show that, if $\mathrm{hd}(\overline{\mathrm{St}}_{G,1})$ is cuspidal, then $n = e\ell^a$.
9. Use Exercise 8 to obtain a classification of cuspidal pairs for the unipotent block of $\mathrm{GL}_n(\mathbb{F}_q)$ at characteristic ℓ using "just" the description of certain cuspidal modules given at the beginning of the proof of Theorem 19.18 and Exercise 8 to show that there is no other.

   Use this classification to derive Theorem 19.21 in the case where q is an integer prime to ℓ (q being a power of a prime, e its order mod. ℓ, $s(m)$ denoting the number of simple $\mathcal{H}_k(\mathfrak{S}_m, q)$-modules, show that the partition of simple unipotent kG-modules into the Harish-Chandra series gives the equality $\pi(n) = \sum s(m_{-1})\pi_\ell(m_0)\pi_\ell(m_1)\ldots$ where the sum is over $(1^{(m_{-1})}, e^{(m_0)}, (e\ell)^{(m_1)}, (e\ell^2)^{(m_2)}, \ldots) \vdash n$, then use Lemma 19.26(ii) to get inductively $s(m) = \pi_e(m)$).
10. Show that the lexicographic order refines $\ll$ on partitions of n. Show that $\ll$ refines $\ll'$ (see the end of §19.4). Draw the diagram of $\ll$ and $\ll'$ for the least integers such that they are not total orderings (6 and 4).

Notes

Theorem 19.16, Theorem 19.18 and Theorem 19.20(i) are due to Dipper, see [Dip85a], [Dip85b], and also [Ja86]. Dipper–James' reinterpretation of Dipper's work gave rise mainly to the notion of q-Schur algebras [DipJa89] and to new theorems on modular representations of Hecke algebras, see [DipJa86], [DipJa87]. Our exposition follows [Ca98] and [Geck01]. A different approach is introduced in [BDK01], see also [Tak96].

The partition of simple modules of the unipotent blocks over k into Harish-Chandra series is due to Dipper–Du [DipDu97], but we have essentially followed the idea introduced in [GeHiMa94] §7. We thank G. Malle for having pointed it out to us. Exercises 6–8 are also an adaptation of [GeHiMa94] §7.

20

Decomposition numbers and q-Schur algebras: linear primes

In this chapter, we intend to prove theorems similar to Theorem 19.16 but for $G = \mathbf{G}^F$, where $\mathbf{G}$ is defined over $\mathbb{F}_q$ with associated Frobenius endomorphism F, and $(\mathbf{G}, F)$ is of rational type $\mathbf{A}$, ${}^2\mathbf{A}$, $\mathbf{B}$, $\mathbf{C}$, $\mathbf{D}$, or ${}^2\mathbf{D}$. Let ℓ be a prime not dividing $2q$ and such that q is of odd order mod. ℓ. Let $(\mathcal{O}, K, k)$ be an ℓ-modular splitting system for G. Gruber–Hiss have proved that $B_1 := \mathcal{O}G.b_\ell(G, 1)$, the product of the unipotent blocks of G (see Definition 9.4), has a decomposition matrix in the form

$$\mathrm{Dec}(B_1) = \begin{pmatrix} 1 & 0 & 0 \\ * & \ddots & 0 \\ * & * & 1 \\ * & * & * \\ \vdots & \vdots & \vdots \\ * & * & * \end{pmatrix}$$

(see [GeHi97]). This is done essentially by relating $\mathrm{Dec}(B_1)$ with the decomposition matrices $\mathrm{Dec}(\mathcal{S}_\mathcal{O}(n, q))$ for various q and n, where $\mathcal{S}_\mathcal{O}(n, q)$ is the q-Schur algebra obtained from the Hecke algebra of $\mathfrak{S}_n$ and parameter q (see the introduction to Chapter 18).

The process is close to the one used in the preceding chapter but, to use it, one must make the above strong restriction on ℓ. The term "linear" is to recall that in this case the process used in the case of $\mathrm{GL}_n(\mathbb{F}_q)$ applies.

When ℓ is linear, the unipotent cuspidal characters of standard Levi subgroups $L_I \subseteq G$ are in blocks of central defect, thus being characters of projective $\mathcal{O}[L_I/\mathrm{Z}(L_I)]$-modules Ψ_I. Then $\mathrm{Ind}_{P_I}^G \Psi_I$ and its endomorphism algebra are to replace the modules $\mathrm{Ind}_B^G \mathcal{O}$ and Hecke algebras $\mathcal{H}_\mathcal{O}(\mathfrak{S}_n, q)$ used in the case of $\mathrm{GL}_n(\mathbb{F}_q)$. The main difference is that the resulting Hecke algebra over $\mathcal{O}$ is now of type BC or D.

We then use the results gathered in Chapter 18 about these Hecke algebras.

20.1. Finite classical groups and linear primes

Theorem 20.1. *Let* $(\mathbf{G}, F)$ *be a connected reductive group defined over* $\mathbb{F}_q$. *Assume that the rational type of* $(\mathbf{G}, F)$ *only involves "classical" types* $\mathbf{A}$, ${}^2\mathbf{A}$, $\mathbf{B}$, $\mathbf{C}$, $\mathbf{D}$ *and* ${}^2\mathbf{D}$ *(see §8.1).*

Let ℓ *be a prime not dividing* q *and let* $(\mathcal{O}, K, k)$ *be an* ℓ*-modular splitting system for* $\mathbf{G}^F$.

Assume moreover that ℓ *and its order mod.* q *are odd, and* $\ell \in \Pi(\mathbf{G}, F)$ *(see Definition 17.5).*

Then, up to an ordering of the rows and columns,

$$\mathrm{Dec}(\mathcal{O}\mathbf{G}^F.b_\ell(\mathbf{G}^F, 1)) = \begin{pmatrix} 1 & 0 & 0 \\ * & \ddots & 0 \\ * & * & 1 \\ * & * & * \\ \vdots & \vdots & \vdots \\ * & * & * \end{pmatrix}$$

(i.e. there is a maximal square submatrix which is lower triangular unipotent). Moreover, the $|\mathcal{E}(\mathbf{G}^F, 1)|$ *first rows correspond to* $\mathcal{E}(\mathbf{G}^F, 1)$.

Remark 20.2. (1) When ℓ is odd and q is of odd order mod. ℓ, the condition $\ell \in \Pi(\mathbf{G}, F)$ is satisfied except possibly if $(\mathbf{G}, F)$ has rational types $\mathbf{A}$ (see Definition 17.5 and Table 13.11). Otherwise, the case of $\mathrm{GL}_n(\mathbb{F}_q)$ has been treated in Chapter 19 without any restriction on ℓ.

(2) Note that, by Bruhat decomposition for rational matrices, the unipotent triangular shape in Theorem 20.1 determines a unique ordering of columns from the ordering of lines. This translates into the fact that simple $kG.b_\ell(G, 1)$-modules are parametrized by $\mathcal{E}(G, 1)$ in a unique way (see also Remark 20.14).

In view of Theorem 17.7, it will be enough to prove Theorem 20.1 for some groups with connected center producing all the expected rational types ${}^2\mathbf{A}$, $\mathbf{B}$, $\mathbf{C}$, $\mathbf{D}$, and ${}^2\mathbf{D}$. We fix the notation below.

Definition 20.3. *If* $n \geq 1$ *is an integer, let* $G_n(q)$ *denote one of the following finite groups. Recall that* $\mathbf{F}$ *is an algebraic closure of* $\mathbb{F}_q$.

(1) The unitary group $\mathrm{GU}_n(\mathbb{F}_{q^2})$, *which is obtained as* $\mathbf{G}^F$ *from* $\mathbf{G} = \mathrm{GL}_n(\mathbf{F})$ *defined over* $\mathbb{F}_q$ *but is such that the associated Frobenius endomorphism induces an automorphism of order 2 of the root system (see §8.1).*

(2) The special orthogonal group $\mathrm{SO}_n(\mathbb{F}_q)$ *for odd* n*. This is obtained as* $\mathbf{G}^F$ *for* $\mathbf{G} = \mathrm{SO}_n(\mathbf{F})$ *and split Frobenius endomorphism.*

(3) The conformal symplectic group $\mathrm{CSp}_n(\mathbb{F}_q)$ *with even* n*. This is obtained as* $\mathbf{G}^F$ *for* $\mathbf{G}$ *the group generated by symplectic matrices and homotheties, possessing split Frobenius endomorphism.*

(4) The conformal special orthogonal group $\mathrm{CSO}_n^+(\mathbb{F}_q)$ *with even* n*. This is obtained as* $\mathbf{G}^F$ *for* $\mathbf{G}$ *the group generated by homotheties and matrices of determinant 1 that are orthogonal with respect to a non-degenerate symmetric form, endowed with split Frobenius endomorphism.*

(5) The conformal special orthogonal group $\mathrm{CSO}_n^-(\mathbb{F}_q)$ *with even* n*. This is obtained as* $\mathbf{G}^F$ *for* $\mathbf{G}$ *the group generated by homotheties and matrices of determinant 1 that are orthogonal with respect to a non-degenerate symmetric form, endowed with a Frobenius endomorphism inducing a symmetry of order 2 of the root system (see §8.1).*

Denote $\widetilde{q} = q^2$ *in case (1),* $\widetilde{q} = q$ *otherwise.*

Let $\mathbf{T}_0 \subseteq \mathbf{B}_0$ *be a maximal torus and a Borel subgroup, both* F*-stable, let* $T = \mathbf{T}_0^F$, $B = \mathbf{B}_0^F$, $N = \mathrm{N}_{\mathbf{G}}(\mathbf{T}_0)^F$, $S \subseteq W := N/T$ *be the associated BN-pair of the finite group* $\mathbf{G}^F$ *(see §8.1).*

We also denote by Δ *the basis of the reflection representation of* W *and occasionally represent it by a diagram (see Example 2.1).*

Note that the above Δ (in bijection with S) is related to the corresponding notion for the underlying group $\mathbf{G}$ but differs in cases (1) and (5).

Accordingly, note that (W, S) is of type $\mathrm{BC}_{[n/2]}$ in cases (1), (2), (3), (5), and of type $\mathrm{D}_{n/2}$ in case (4) (see §8.1 giving the correspondence between rational type and type of the associated finite BN-pair).

The hypotheses on ℓ will be used throughout the following.

Proposition 20.4. *Keep* $G = G_n(q)$ *one of the above. Assume* ℓ *is a prime not dividing* q*, odd and such that the order of* q *mod.* ℓ *is odd too.*

(i) If χ *is a unipotent cuspidal character of* G *(identified with a* KG*-module), then* $b_G(\chi)$ *has central defect group. Moreover,* $\chi = (e_{\ell'}^G.\Psi) \otimes K$ *(see Definition 9.9) where* Ψ *is a projective indecomposable* $\mathcal{O}G$*-module.*

(ii) If L_I $(I \subseteq \Delta)$ *has a cuspidal unipotent character and* $\emptyset \neq I \subseteq J \subseteq \Delta$*, then* $L_J \cong G_m(q) \times \mathrm{GL}_{\lambda_1}(\widetilde{q}) \times \mathrm{GL}_{\lambda_2}(\widetilde{q}) \cdots$ *where* $n - m = 2(\lambda_1 + \lambda_2 + \cdots)$*.*

(iii) Each standard Levi subgroup of $G_n(q)$ *has at most one cuspidal unipotent character.*

(iv) If Q *is a power of* q*, Hypothesis 18.25 is satisfied in any local ring whose residual field is of characteristic* ℓ*.*

Proof. (i) First note that, since χ is unipotent, it has $\mathrm{Z}(G)$ in its kernel. So it suffices to check that $\mathcal{O}G.b_G(\chi)$ has central defect groups, since then $\Psi = \mathcal{O}G.b_G(\chi)$ is such that $\mathrm{Z}(G)$ acts trivially on $e^G_{\ell'}.\Psi \otimes K$ while this block has only one irreducible character trivial on $\mathrm{Z}(G)$ (see Remark 5.6) hence equal to χ. Using again the fact that $\mathrm{Z}(G)$ is in the kernel of G, it suffices now to check that the integer $|G{:}\,\mathrm{Z}(G)|/\chi(1)$ is prime to ℓ.

Assume $G = G_n(q)$ is a unitary group. From [Lu84] p. 358 or [Cart85] §13.8, we know that $n = m(m+1)/2$ for some $m \geq 1$ and χ is a character such that $|G{:}\,\mathrm{Z}(G)|/\chi(1) = \Pi_{h=1}^{m}(q^{2h-1}+1)^{m-h+1}$ up to a power of q. This is prime to ℓ since $\ell \neq 2$ and the order of q mod. ℓ is odd.

Assume $G = G_n(q)$ is among the cases (2) to (5) of Definition 20.3. From [Lu84] p. 359 or [Cart85] §13.8, we know that $n = m(m+2)/4$ or $(m+1)^2/4$ for some integer $m \geq 1$ and χ is such that $|G{:}\,\mathrm{Z}(G)|/\chi(1) = \Pi_{h=1}^{m-1}(q^h+1)^{m-h}$ up to powers of q and 2. This is again prime to ℓ.

For an alternative approach to this question, see Exercise 22.7.

(ii) The rational type of a finite reductive group having unipotent cuspidal characters can't include $\mathbf{A}_m$ as a connected component for $m \geq 1$ (see [Cart85] §13.8, [Lu84] Appendix, or Theorem 19.7 above). So in the $(\mathbf{G}, F)$ we are considering, the type of a Levi subgroup $\mathbf{L}$ such that $\mathcal{E}(\mathbf{L}^F, 1)$ contains cuspidal characters, if non-empty, must be irreducible and contain the subsystem of rational type ${}^2\mathbf{A}_2$, $\mathbf{B}_2$, $\mathbf{C}_2$, $\mathbf{D}_4$, ${}^2\mathbf{D}_3$ in cases (1), (2), …, (5), respectively. This means $\mathbf{L}^F$ is L_I where $I \subseteq \Delta$ is non-empty and corresponds to the m first simple roots in type BC or D in the lists of Example 2.1(ii) and (iii).

Then L_I is a direct product as stated in (ii). To see that, one may assume that $\Delta \setminus I$ is a single root. In unitary or orthogonal groups these Levi subgroups consist of the matrices that can be written $\mathrm{diag}(x, y, w_0.{}^t\overline{x}^{-1}.w_0)$ for $x \in \mathrm{GL}_{n-m}(\mathbb{F}_q)$, $y = w_0.{}^t\overline{y}^{-1}.w_0 \in \mathrm{GL}_n(\mathbb{F}_{\tilde{q}})$, where $\lambda \mapsto \overline{\lambda}$ is an involution of $\mathbb{F}_{\tilde{q}}$ and w_0 is the permutation matrix reversing the ordering of the basis. In symplectic groups w_0 is to be replaced by an antisymmmetric matrix (see Exercise 2.6(d)).

In the associated conformal groups, the block diagonal matrices above are multiplied by scalar matrices but then $L_I \cong G_m(q) \times \mathrm{GL}_{n-m}(\mathbb{F}_{\tilde{q}})$ by the map sending $\mathrm{diag}(x, y, z)$ to (y, x), an inverse map being $(a, b) \mapsto (b, a, {}^ta.a.w_0.{}^tb^{-1}.w_0)$.

(iii) A finite reductive group of rational type $\mathbf{A}$, ${}^2\mathbf{A}$, $\mathbf{B}$, $\mathbf{C}$, $\mathbf{D}$ or ${}^2\mathbf{D}$ may not have more than one cuspidal unipotent character. This is due to the classification of Lusztig; see [Lu84] pp. 358–9, [Cart85] §13.8.

(iv) This is clear from the fact that q has odd order mod. ℓ. □

Definition 20.5. *Let $G = G_n(q)$ as in Definition 20.3 with split BN-pair $(B = UT, N)$ and set of simple roots Δ. Let Σ_G be the set of pairs $\sigma = (I_\sigma, \chi)$ where $I_\sigma \subseteq \Delta$ and $\chi \in \mathcal{E}(L_{I_\sigma}, 1)$ is cuspidal unipotent. Then (Proposition 20.4(ii)) $L_\sigma = G_m(q) \times T_\sigma$ where $m \leq n$ and $T_\sigma \subseteq T$ is isomorphic with $(\mathbb{F}_q^\times)^{\frac{n-m}{2}}$, and $\chi = \chi_m \otimes 1_{T_\sigma}$ where $\chi_m \in \mathcal{E}(G_m(q), 1)$ is cuspidal. Let Ψ_σ be the projective indecomposable $\mathcal{O}G_m(q)$-module corresponding to χ_m (see Proposition 20.4(i)).*

Let $M_\sigma := (e_{\ell'}.\Psi_\sigma) \times \mathcal{O}_{T_\sigma}$ (see Definition 9.9).

Let $X_\sigma := \mathrm{R}_{L_\sigma}^G M_\sigma$.

Let $\mathcal{H}_\sigma := \mathrm{End}_{\mathcal{O}G}(X_\sigma)$.

Define Λ_σ as the set of subsets $I' \subseteq \Delta$ such that I' is of type a sum of A's and $I' \subseteq (I_\sigma)^\perp$. Let I'_σ be the union of the elements of Λ_σ.

By Proposition 20.4, I_σ is connected, not of type A. Therefore, $I'_\sigma = \Delta$ when $I_\sigma = \emptyset$ and Δ is of type D; otherwise, $I'_\sigma \in \Lambda_\sigma$.

When M is an L_I-module, recall the notation $W(I, M) := \{w \in W \mid wI = I \text{ and } {}^wM \cong M\}$. By the uniqueness of Proposition 20.4(iii) and the definition of Ψ_σ from the cuspidal character $M_\sigma \otimes K$, we have the following (see Definition 2.26).

Lemma 20.6. $W(I_\sigma, M_\sigma) = W(I_\sigma, M_\sigma \otimes K) = W^{I_\sigma} = \{w \in W \mid wI_\sigma = I_\sigma\}$

We now start the proof of Theorem 20.1.

It suffices to take G to be one of the groups from Definition 20.3 since, by Theorem 17.7, the unipotent block is isomorphic with the unipotent block of a direct product of such groups and general linear groups, the case of which was treated in Theorem 19.16.

We apply Theorem 19.4 to the algebra $\mathcal{O}G.b_\ell(G, 1)$ (see Definition 9.4) for the collection of modules X_σ introduced in Definition 20.5.

Condition (a) of Theorem 19.4 is satisfied since the disjunction of the union $\cup_\sigma \mathcal{B}_\sigma$ is simply a consequence of Harish–Chandra theory for characters ([DiMi91] 6.4) and Lusztig's determination of cuspidal pairs (see Proposition 20.4(ii)). The union is the whole set of unipotent characters since Harish–Chandra theory preserves unipotence of characters, by Proposition 8.25. This produces a basic set of characters for $\mathcal{O}G.b_\ell(G, 1)$ by Theorem 14.4. □

20.2. Hecke algebras

The endomorphism algebra $\mathcal{H}_\sigma := \mathrm{End}_{\mathcal{O}G}(X_\sigma)$ can be described by means of Theorem 1.20, once we check Condition 1.17(b).

In order to define the elements $y_\lambda \in \mathcal{H}_\sigma$ of Theorem 19.4, we must analyze the law of $\mathcal{H}_\sigma$.

One has $\mathcal{H}_\sigma = \bigoplus_g a_{g,\tau,\tau}$ where $\tau = (P_{I_\sigma}, U_{I_\sigma}, M_\sigma)$ and g ranges over a representative system which, in the case of a finite group with a BN-pair, is the group $W(I_\sigma, M_\sigma) := \{w \in W \mid wI_\sigma = I_\sigma \text{ and } {}^wM_\sigma \cong M_\sigma\}$ (see Theorem 2.27(iv)). The same basis is used in Chapter 3 to describe $\mathcal{H}_\sigma \otimes_\mathcal{O} K = \mathrm{End}_{KG}(X_\sigma \otimes K)$. By Lemma 20.6, the group $C(I_\sigma, M_\sigma)$ is trivial. The basis elements, once normalized as in the proof of Theorem 3.16, satisfy certain relations involving a cocycle λ and coefficients $c_\alpha \in K$. It is clear from its definition in Theorem 3.16 that λ takes its values in $\mathcal{O}^\times$. In [Lu84] 8.5.12, it is shown that the c_α's are of the form $q^{n(\alpha)} - q^{-n(\alpha)}$ where $2n(\alpha)$ is an integer ≥ 1. (Note: this is where we use the fact that the center of $\mathbf{G}$ is connected, though this restriction could be lifted for a unipotent character.) Then $c_\alpha \neq 0$. This implies easily that the cocycle is cohomologous to 1 over K (see [Cart85] 10.8.4 where the condition $c_\alpha \neq 0$ is included in the definition of his group $R_{J,\phi}$, or [Lu84] 8.6), hence over $\mathcal{O}$.

We now give a presentation of $\mathcal{H}_\sigma$.

Assume $I_\sigma \neq \emptyset$ (the case of $I_\sigma = \emptyset$ is covered by Theorem 3.3). By Proposition 20.4(ii), (W_{I_σ}, I_σ) is a Coxeter group of the same type BC or D as (W, S) but in lower degree. So we are within the cases discussed in Example 2.28(ii) and (iii) with $|I_\sigma| \neq 1$ in type D (case (4)) since rational type $(\mathbf{A}_1, q)$ has no cuspidal unipotent character (see the proof of Proposition 20.4(ii)). Then W^{I_σ} is also of type BC or D. Note that the second possibility occurs only if $I_\sigma = \emptyset$ and Δ is of type D. When $I_\sigma \neq \Delta$, the simple roots of W^{I_σ} make a set $\Delta^{I_\sigma} = \{\delta_\sigma\}\dot\cup I'_\sigma$ (see Definition 20.5) where δ_σ is outside Δ

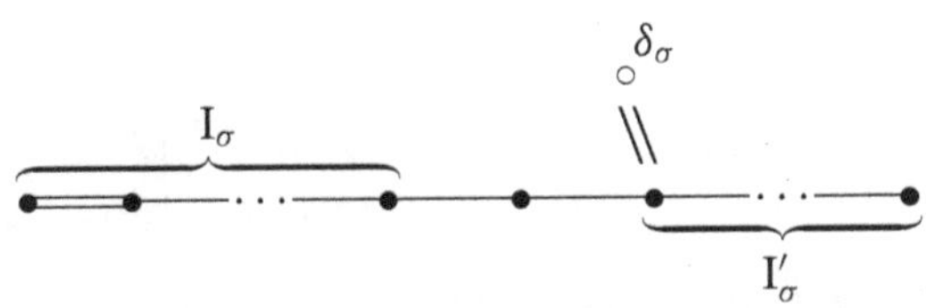

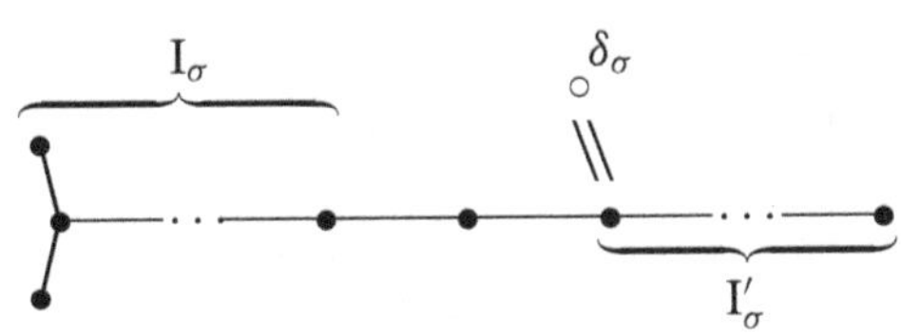

(see Example 2.28(ii) and (iii)). The generators of the Hecke algebra corresponding to elements of I'_σ are just images of the same in the subalgebra corresponding with $\mathrm{R}_{L_{I_\sigma}}^{L_{I_\sigma \cup I'_\sigma}}(M_\sigma)$ by Proposition 1.23. Since $M_\sigma = e_{\ell'}.\Psi_\sigma \times \mathcal{O}_{T_\sigma}$, then $\mathrm{R}_{L_{I_\sigma}}^{L_{I_\sigma\cup I'_\sigma}}(M_\sigma) = e_{\ell'}.\Psi_\sigma \times \mathrm{R}_{T_\sigma}^{L_{I'_\sigma}}(\mathcal{O}_{T_\sigma})$. The corresponding isomorphism for the endomorphism algebra implies that the law of this subalgebra is given by Theorem 3.3 with a parameter ind(s) constant equal to $\widetilde{q}$. By what we have recalled about the cocycle and the quadratic equation satisfied by the extra generator corresponding with the root δ_σ, we get the following generation,

$$\mathcal{H}_\sigma = \bigoplus_{w\in W^{I_\sigma}} \mathcal{O}b_w$$

where, abbreviating $b_w = b_\delta$ if $w = s_\delta$ for $\delta \in \Delta^{I_\sigma}$,

$$(b_\delta)^2 = (q(\delta) - 1)b_\delta + q(\delta)$$

where $q(\delta) = \widetilde{q}$ if $\delta \neq \delta_\sigma$, $q(\delta_\sigma)$ is a power of q,

$$b_w b_{w'} = b_{ww'}$$

when lengths add in the Coxeter group W^{I_σ}.

Definition 20.7. *Keep $\sigma \in \Sigma_G$ (see Definition 20.5). Recall that $I'_\sigma = \bigcup_{\lambda\in\Lambda_\sigma}\lambda$. We now define the following cases.*

Case (I). If $I_\sigma = \emptyset$ and Δ is of type D *(case (4) of Definition 20.3), then $I'_\sigma = \Delta$, and Λ_σ consists of all the subsets of Δ not including type* D_4.

Case (II). In other cases, I'_σ is of type A, and Λ_σ is just the set of subsets of I'_σ.

If $\lambda \in \Lambda_\sigma$, let

$$y_\lambda = \sum_{w\in(W^{I_\sigma})_\lambda} (-q)^{-l(w)} b_w$$

(note that in $W_{I'_\sigma}$ the length maps l of W^{I_σ} and of W coincide since they correspond to simple roots that are already in Δ).

With this definition of Λ_σ and y_λ, condition (b) of Theorem 19.4 is satisfied since $\mathcal{H}_\sigma$ is a symmetric algebra by Theorem 1.20(ii). Moreover, $y_\emptyset = 1$.

Let $\lambda \in \Lambda_\sigma$. Note that y_λ is in the subalgebra corresponding to $\mathrm{R}_{L_{I_\sigma}}^{L_{I_\sigma\cup I'_\sigma}}(M_\sigma)$, which we denote $\mathcal{H}_\mathcal{O}(I'_\sigma) = \oplus_{w\in W_{I'_\sigma}}\mathcal{O}b_w = \mathrm{End}_{\mathcal{O}L_{I_\sigma\cup I'_\sigma}}(\mathrm{R}_{L_{I_\sigma}}^{L_{I_\sigma\cup I'_\sigma}}(M_\sigma)) \cong \mathrm{End}_{\mathcal{O}L'_{I'_\sigma}}(\mathrm{R}_{T'_\sigma}^{L'_{I'_\sigma}}(\mathcal{O}))$ where $L'_{I'_\sigma}$ is the product of general linear groups described in Proposition 20.4(ii), and $T'_\sigma = T_\sigma \cap L'_{I'_\sigma}$ is its diagonal torus.

We now check condition (c) of Theorem 19.4. To show that $y_\lambda\mathcal{H}_\sigma$ is $\mathcal{O}$-pure in $\mathcal{H}_\sigma$, it suffices to find a subalgebra $\mathcal{H}' \subseteq \mathcal{H}_\sigma$ such that $\mathcal{H}_\sigma$ is free as (left) $\mathcal{H}'$-module, $y_\lambda \in \mathcal{H}'$, and $y_\lambda\mathcal{H}'$ is $\mathcal{O}$-pure in $\mathcal{H}'$.

Assume case (II) of Definition 20.7. We take $\mathcal{H}' := \mathcal{H}_\mathcal{O}(I'_\sigma)$ defined above. The inclusion $\mathcal{H}' \subseteq \mathcal{H}_\sigma$ corresponds to the inclusion of type A_{r-1} in type BC_r and we have an obvious analogue of Proposition 18.22(ii) with $a(w)$'s on the right. So $\mathcal{H}_\sigma$ is $\mathcal{H}'$-free. But Proposition 19.14(i) implies that $y_\lambda\mathcal{H}_\mathcal{O}(I'_\sigma)$ is $\mathcal{O}$-pure in $\mathcal{H}_\mathcal{O}(I'_\sigma)$.

Assume case (I) of Definition 20.7. Then $y_\lambda \in \mathcal{H}(I')$ where $\lambda \subseteq I' \subseteq \Delta$ and I' is of type A_{r-1} if Δ is of type D_r. Proposition 19.14(i) again implies that $y_\lambda\mathcal{H}_\mathcal{O}(I')$ is $\mathcal{O}$-pure in $\mathcal{H}_\mathcal{O}(I')$. As we will do more systematically in §20.4 below, we may embed $\mathcal{H}_\sigma = \mathcal{H}_\mathcal{O}(\mathrm{D}_r)$ in $\mathcal{H}_\mathcal{O}(\mathrm{BC}_r, 1, q)$ (see Definition 18.26), where $\mathcal{H}_\sigma$ is generated by $a_0a_1a_0$, $a_1, \ldots, a_{r-1}$ because of the corresponding embedding of Coxeter groups (see Example 2.1(iii)). Then $a_0^2 = 1$ and $\mathcal{H}_\sigma$ correspond with elements such that l is even in the description of Proposition 18.22(ii). Then $\mathcal{H}_\sigma$ is $\mathcal{H}(I'')$-free for $\mathcal{H}(I'') = \langle a_1, \ldots, a_{r-1}\rangle$ corresponding to a subsystem of type A_{r-1}. We may assume that $I'' = I'$ since conjugacy by a_0 would otherwise exchange them. Then $y_\lambda \in \mathcal{H}(I'')$ and we can conclude as in case (II).

We now check condition (d).

Assume case (II) (see Definition 20.7 above). Using the functoriality of the inclusion $\mathrm{End}_{\mathcal{O}L_{I_\sigma\cup I'_\sigma}}(\mathrm{R}_{L_{I_\sigma}}^{L_{I_\sigma\cup I'_\sigma}}(M_\sigma)) \subseteq \mathrm{End}_{\mathcal{O}G}(\mathrm{R}_{L_{I_\sigma}}^G(M_\sigma))$, we get $y_\lambda X_\sigma = y_\lambda \mathrm{R}_{L_{I_\sigma}}^G M_\sigma = \mathrm{R}_{L_{I_\sigma\cup I'_\sigma}}^G(y_\lambda \mathrm{R}_{L_{I_\sigma}}^{L_{I_\sigma\cup I'_\sigma}} M_\sigma) = \mathrm{R}_{L_{I_\sigma\cup I'_\sigma}}^G(e_{\ell'}.\Psi_\sigma \otimes y_\lambda.(\mathrm{R}_{T'_\sigma}^{L_{I'_\sigma}}\mathcal{O}))$, where in the last expression, y_λ is considered as an element of the Hecke algebra $\mathcal{H}_\mathcal{O}(I'_\sigma) = \mathrm{End}_{\mathcal{O}L_{I'_\sigma}}(\mathrm{R}_{T'_\sigma}^{L_{I'_\sigma}}\mathcal{O})$. By Proposition 19.14(ii) and Proposition 9.15, we have $\sqrt{y_\lambda \mathrm{R}_{T'_\sigma}^{L_{I'_\sigma}}\mathcal{O}} \cong \mathrm{R}_{L_\lambda}^{L_{I'_\sigma}} e_{\ell'}.\Gamma_{L_\lambda,1} \cong e_{\ell'}.\mathrm{R}_{L_\lambda}^{L_{I'_\sigma}}\Gamma_{L_\lambda,1}$. Then $\sqrt{y_\lambda \mathrm{R}_{L_{I_\sigma}}^{L_{I_\sigma\cup I'_\sigma}} M_\sigma} \cong e_{\ell'}.(\Psi_\sigma \times \sqrt{y_\lambda \mathrm{R}_{T'_\sigma}^{L_{I'_\sigma}}\mathcal{O}}) \cong e_{\ell'}.(\Psi_\sigma \times \mathrm{R}_{L_{I'_\sigma}}^{L_\lambda}\Gamma_{L_\lambda,1})$. Taking the images under $\mathrm{R}_{L_{I_\sigma\cup I'_\sigma}}^G$, we get

$$\sqrt{y_\lambda X_\sigma} \cong e_{\ell'}.\big(\mathrm{R}_{L_{I_\sigma\cup\lambda}}^G(\Psi_\sigma \times \Gamma_{L_\lambda,1})\big)$$

(since the Harish–Chandra induction commutes with $e_{\ell'}$ and $\sqrt{-}$; see Proposition 9.15). This type of module has a projective cover by the same without the idempotent $e_{\ell'}$, by Theorem 9.10. This gives condition (d) of Theorem 19.4 in this case, since a covering $P \to e_{\ell'}.P$ always satisfies it by the definition of $e_{\ell'}$.

Assume case (I) of Definition 20.7, i.e. $I_\sigma = \emptyset$ and Δ is of type D. The same reasoning as above (in a simpler situation) gives $\sqrt{y_\lambda X_\sigma} \cong e_{\ell'}.(\mathrm{R}_{L_\lambda}^G(\Gamma_{L_\lambda,1}))$ (in fact, one might use the same proof as in the case of GL; see the proof of Theorem 19.16). Then $e_{\ell'}.\sqrt{y_\lambda X_\sigma}$ has a projective cover by $\mathrm{R}_{L_\lambda}^G(\Gamma_{L_\lambda,1})$ and we get condition (d) of Theorem 19.4.

We now have all conditions of Theorem 19.4 for the σ's and the y_λ's of Definition 20.5 and Definition 20.7.

Theorem 19.4 tells us that $\mathrm{Dec}(\mathcal{O}G.b_\ell(G,1))$ can be written $\begin{pmatrix} D_0 & D_1 \\ D_0' & D_1' \end{pmatrix}$ where D_0 is the diagonal matrix made with the decomposition matrices D_σ of the $\mathcal{H}_\sigma$-modules $V_\sigma = \bigoplus_{\lambda\in\Lambda_\sigma} y_\lambda\mathcal{H}_\sigma$.

To get the claim of our Theorem 20.1, it now suffices to show that each D_σ has at least as many columns as rows (hence is square) and is lower triangular (see the last statement in Theorem 19.4).

We show this in the next two sections. The first corresponds to the case where all $\mathcal{H}_\sigma$'s are of type BC.

20.3. Type BC

Concerning Hecke algebras, let us give an analogue of Corollary 19.17 for type BC. Corollary 19.17 gives the following lemma about type A.

Let n, a be integers greater than or equal to 1. Let $\mathcal{O}$ be a complete discrete valuation ring with residue field of prime characteristic ℓ.

Lemma 20.8. *The q-Schur algebra* $\mathcal{S}_\mathcal{O}(n,q^a) := \mathrm{End}_{\mathcal{H}_\mathcal{O}(\mathfrak{S}_n,q^a)}(\bigoplus_{\lambda\vdash n} y_\lambda\mathcal{H}_\mathcal{O}(\mathfrak{S}_n,q^a))$ *has a square lower unitriangular decomposition matrix.*

The next result is about type BC. We use the notation of §18.3. Recall from Corollary 19.17 the notation $\lambda \models n$ to mean that λ is a finite sequence of integers greater than or equal to 1 whose sum is n. When moreover $1\leq m\leq n$, we denote $\lambda\models_m n$ when $\lambda=(\lambda',\lambda'')$ with $\lambda'\models m$, $\lambda''\models n-m$. Here is a consequence of Theorem 18.27 whose notation ε_m is also used.

Lemma 20.9. *Let q be a power of a prime $\neq\ell$, Q a power of q (possibly $Q=1$). Assume that ℓ and the order of q mod. ℓ are odd. When $\lambda\models n$, let* $y_\lambda=\sum_{w\in\mathfrak{S}_\lambda}(-q)^{-l(w)}a_w\in\mathcal{H}:=\mathcal{H}_\mathcal{O}(\mathrm{BC}_n,Q,q)$ *(see Definition 18.26). Then* $\bigoplus_{m,\lambda\models_m n}\varepsilon_m y_\lambda\mathcal{H}$, *as a right $\mathcal{H}$-module, has a decomposition matrix equal to the diagonal matrix with blocks* $\mathrm{Dec}(\mathcal{S}_\mathcal{O}(m,q)\otimes\mathcal{S}_\mathcal{O}(n-m,q))$ *(see Lemma 20.8) for $m=0,1,\ldots,n$ (we put $\mathcal{S}_\mathcal{O}(0,q)=\mathcal{S}_\mathcal{O}(1,q)=\mathcal{O}$). This makes a square lower unitriangular matrix.*

Proof. The assumptions on q and ℓ allow us to apply Theorem 18.27(iii) since, for all $i\in\mathbb{N}$, $(Qq^i+1)(q^i+Q)$ is not divisible by ℓ. Recall that $\varepsilon=\varepsilon_0+\varepsilon_1+\cdots+\varepsilon_n$. By the Morita equivalence of Theorem 18.27(iii), for each right $\mathcal{H}$-module M, its decomposition matrix as an $\mathcal{H}$-module is the same as the decomposition matrix of $M\epsilon$ as an $\varepsilon\mathcal{H}\varepsilon$-module (see [Thévenaz] Theorem 18.4(f) and Remark 18.9).

So we are left to compute the decomposition matrix of the $\varepsilon\mathcal{H}\varepsilon$-module $\bigoplus_{m,\lambda\models_m n}\varepsilon_m y_\lambda \mathcal{H}\varepsilon$. When $\lambda \models_m n$, then $y_\lambda \in \mathcal{H}(\mathfrak{S}_{m,n-m})$, so Theorem 18.27(i) implies $\varepsilon_m y_\lambda \mathcal{H}\varepsilon = y_\lambda \varepsilon_m \mathcal{H}\varepsilon_m$. The isomorphism of $\varepsilon_m\mathcal{H}\varepsilon_m$ with $\mathcal{H}(\mathfrak{S}_{m,n-m})$ (see Theorem 18.27(ii)) sends $y_\lambda\varepsilon_m\mathcal{H}\varepsilon_m = \varepsilon_m y_\lambda \mathcal{H}(\mathfrak{S}_{m,n-m})$ to $y_\lambda\mathcal{H}(\mathfrak{S}_{m,n-m})$. So, for a fixed m, the decomposition matrix of $\bigoplus_{\lambda\models_m n}\varepsilon_m y_\lambda\mathcal{H}$ as $\mathcal{H}$-module is that of $\bigoplus_{\lambda\models_m n} y_\lambda\mathcal{H}(\mathfrak{S}_{m,n-m})$ as an $\mathcal{H}(\mathfrak{S}_{m,n-m})$-module. Writing each $\lambda\models_m n$ as (λ_1,λ_2) with $\lambda_1\models m$, $\lambda_2\models n-m$, one gets $y_\lambda\mathcal{H}(\mathfrak{S}_{m,n-m}) = y_{\lambda_1}\mathcal{H}(\mathfrak{S}_m)\otimes y_{\lambda_2}\mathcal{H}(\mathfrak{S}_{n-m})$ as $\mathcal{H}(\mathfrak{S}_m)\otimes\mathcal{H}(\mathfrak{S}_{n-m})$-module, so the decomposition matrix of $\bigoplus_{\lambda\models_m n}\varepsilon_m y_\lambda\mathcal{H}$ is the Kronecker tensor product of matrices $\mathrm{Dec}(\mathcal{S}_\mathcal{O}(m,q))\otimes\mathrm{Dec}(\mathcal{S}_\mathcal{O}(n-m,q))$.

The summands of $\bigoplus_{m,\lambda\models_m n}\varepsilon_m y_\lambda\mathcal{H}$ for various m have no indecomposable summand in common since $\mathrm{Hom}_\mathcal{H}(\varepsilon_m\mathcal{H},\varepsilon_{m'}\mathcal{H})=0$ by Theorem 18.27(ii) when $m\neq m'$. The same occurs upon tensoring with K. So the decomposition matrix of $\bigoplus_{m,\lambda\models_m n}\varepsilon_m y_\lambda\mathcal{H}$ is the diagonal matrix stated in the lemma. □

Recall that, to complete the proof of Theorem 20.1, we must check that all decomposition matrices $\mathrm{Dec}_{\mathcal{H}_\sigma}(\bigoplus_{\lambda\in\Lambda_\sigma} y_\lambda\mathcal{H}_\sigma)$ are square.

In the remainder of this section, we assume that all $\mathcal{H}_\sigma$ are of type $\mathbb{BC}$.

In that case, the discussion before Definition 20.7 shows that any $\mathcal{H}_\sigma$ is of the type studied in Lemma 20.9. This lemma tells us that the right $\mathcal{H}_\sigma$-module $V'_\sigma = \bigoplus_{m,\lambda}\varepsilon_m y_\lambda\mathcal{H}_\sigma$ has a square lower unitriangular decomposition matrix. In the above sum, m ranges from 0 to $|I'_\sigma|+1$ and λ is a subset of I'_σ not containing the mth element in the enumeration of I'_σ given above (left to right in the pictures of §20.2). Since $V_\sigma = y_\lambda\mathcal{H}_\sigma = \varepsilon_m y_\lambda\mathcal{H}_\sigma\oplus(1-\varepsilon_m)y_\lambda\mathcal{H}_\sigma$, we know that V'_σ is a direct summand of a power of V_σ. So the number of columns of $\mathrm{Dec}(V_\sigma)$ is greater than or equal to the number of columns of $\mathrm{Dec}(V'_\sigma)$. As for the number of rows, one must look at the simple $\mathcal{H}_\sigma\otimes_\mathcal{O} K$ components of $V_\sigma\otimes K$ and $V'_\sigma\otimes_\mathcal{O} K$. We claim that they are in each case all the simple $\mathcal{H}_\sigma\otimes K$-modules. Since KG is semi-simple and $\mathcal{H}_\sigma\otimes K = \mathrm{End}_{\mathcal{O}G}(X_\sigma)\otimes K = \mathrm{End}_{KG}(X_\sigma\otimes K)$ is an endomorphism algebra of a KG-module, $\mathcal{H}_\sigma\otimes K$ is semi-simple. We have $V'_\sigma\supseteq\bigoplus_m\varepsilon_m\mathcal{H}_\sigma$ and therefore $V'_\sigma\otimes K\supseteq\bigoplus_m\varepsilon_m\mathcal{H}_\sigma\otimes K$. Since the Morita equivalence of Theorem 18.27 also holds upon tensoring with K, we see that any simple $\mathcal{H}_\sigma\otimes K$-module is present in $\bigoplus_m\varepsilon_m\mathcal{H}_\sigma\otimes K$, hence also in $V'_\sigma\otimes K$ and $V_\sigma\otimes K$. This gives our claim.

So we see that $\mathrm{Dec}(V'_\sigma)$ and $\mathrm{Dec}(V_\sigma)$ have the same number of rows. Then $\mathrm{Dec}(V_\sigma) = (\mathrm{Dec}(V'_\sigma), D')$ where D' is a matrix corresponding to the direct summands of V_σ not direct summands of V'_σ. Lemma 20.9 tells us that $\mathrm{Dec}(V'_\sigma)$ is square. Then Theorem 19.4 (last statement) tells us that each $\mathrm{Dec}(V_\sigma)$ is square and therefore equal to $\mathrm{Dec}(V'_\sigma)$. The unitriangularity property then also follows from that of $\mathrm{Dec}(V'_\sigma)$.

This completes our proof in that case. We extract from the above the following strengthened version of Lemma 20.9, which expresses the fact that the direct summands of V_σ are already in V'_σ.

Lemma 20.10. *Use the same notation and hypotheses as Lemma 20.9. Then* $\bigoplus_{\lambda\models n} y_\lambda\mathcal{H}$*, as a right* $\mathcal{H}$*-module, has a decomposition matrix equal to the diagonal matrix with blocks* $\mathrm{Dec}(\mathcal{S}_\mathcal{O}(m,q)\otimes\mathcal{S}_\mathcal{O}(n-m,q))$ *(see Lemma 20.8) for* $m=0,1,\ldots,n$ *(recall that* $\mathcal{S}_\mathcal{O}(1,q)=\mathcal{S}_\mathcal{O}(0,q)=\mathcal{O}$*). This makes a square lower unitriangular matrix.*

20.4. Type D

We assume now that not all $\mathcal{H}_\sigma$ are of type BC. As said before, one $\mathcal{H}_\sigma$ is of type D_r corresponding to $I_\sigma=\emptyset$, $M_\sigma=\mathcal{O}$. Note that the questions of W^{I_σ} (see Definition 2.26), cocycle, etc. discussed above are trivial in that case.

Denote by $\emptyset$ the corresponding σ, whence $\mathcal{H}_\emptyset=\mathcal{H}_\mathcal{O}(\mathrm{D}_r,q)$, $\Lambda_\emptyset$. Denote $V_\emptyset=\bigoplus_\lambda y_\lambda\mathcal{H}_\emptyset$ where the sum is over subsets of Δ involving no type D (i.e. products of types A for various ranks). We prove the following.

Lemma 20.11. *$V_\emptyset$ has a direct summand $V'_\emptyset$ such that* $\mathrm{Dec}_{\mathcal{H}_\emptyset}(V'_\emptyset)$ *is square unitriangular with size the number of simple* $\mathcal{H}_\emptyset\otimes K$*-modules.*

Then, just as in the above, the last statement of Theorem 19.4 will imply that $\mathrm{Dec}(V'_\sigma)=\mathrm{Dec}(V_\sigma)$ for all $\sigma\in\Sigma_G$ including $\sigma=\emptyset$.

It remains to prove Lemma 20.11.

Recall that $\mathcal{H}_\emptyset=\mathrm{End}_{\mathcal{O}G}(\mathrm{Ind}_B^G\mathcal{O})$ is $\mathcal{H}_\mathcal{O}(\mathrm{D}_r,q)$ (see Definition 18.1 and Theorem 3.3).

We abbreviate $\mathcal{H}_\emptyset=\mathcal{H}$.

Denote $\dot{\mathcal{H}}=\mathcal{H}_\mathcal{O}(\mathrm{BC}_r,1,q)$. Then Hypothesis 18.25 is satisfied.

The generator $a_0\in\dot{\mathcal{H}}$ (see Definition 18.16) satisfies $(a_0)^2=1$. Moreover, it is easy to check that $\mathcal{H}$ identifies with the subalgebra of $\dot{\mathcal{H}}$ generated by $a_0a_1a_0$, $a_1,a_2,\ldots,a_{r-1}$ (see [GePf00] 10.4.1, [Hoef74] §2.3). Those generators are permuted by conjugacy under a_0, so we have a decomposition $\dot{\mathcal{H}}=\mathcal{H}\oplus\mathcal{H}a_0$ inducing a structure of $\mathbb{Z}/2\mathbb{Z}$-graded algebra on $\dot{\mathcal{H}}$ (see §18.6). Recall the existence of $\theta\colon\dot{\mathcal{H}}\to\dot{\mathcal{H}}$, the automorphism equal to $-\mathrm{Id}$ on $\mathcal{H}a_0$, and fixing every element of $\mathcal{H}$. One denotes by $\tau\colon h\mapsto a_0ha_0$ the involutory automorphism of $\mathcal{H}$ induced by a_0. Note that, if $w\in W(\mathrm{D}_r)$, then $\tau(a_w)=a_{\tau^*(w)}$ where τ^* is the "flipping" automorphism of the Dynkin diagram (hence of the Weyl group) of type D_r fixing every simple root except two ($n\geq 4$).

Let $\dot{V}:=\bigoplus_{m,\lambda\models_m r}\varepsilon_m y_\lambda\dot{\mathcal{H}}$ as in Lemma 20.9 (with $Q=1$).

Lemma 20.12. *All the indecomposable direct summands of* $\operatorname{Res}^{\dot{\mathcal{H}}}_{\mathcal{H}}\dot{V}$ *are isomorphic with direct summands of* $V_\emptyset$.

Proof. Since the ε_m's are idempotents, $\bigoplus_{m,\lambda\models_m r}\varepsilon_m y_\lambda\dot{\mathcal{H}}$ is a direct summand of $\bigoplus_{m,\lambda\models_m r} y_\lambda\dot{\mathcal{H}}$. Then it suffices to check that any indecomposable direct summand of $\operatorname{Res}^{\dot{\mathcal{H}}}_{\mathcal{H}} y_\lambda\dot{\mathcal{H}}$ ($\lambda\models r$) is a direct summand of $V_\emptyset$. Since $\dot{\mathcal{H}}=\mathcal{H}\oplus a_0\mathcal{H}$ as a right $\mathcal{H}$-module, we have $\operatorname{Res}^{\dot{\mathcal{H}}}_{\mathcal{H}} y_\lambda\dot{\mathcal{H}}\cong y_\lambda\mathcal{H}\oplus\tau(y_\lambda)\mathcal{H}$.

Denote $\Delta=\{\delta'_0,\delta_1,\ldots,\delta_{r-1}\}$, so that $a_0a_1a_0=a_{s_0}$, $a_i=a_{s_i}$ $(i\geq 1)$, where s_i denotes the reflection associated with δ_i. If $\lambda=(\lambda_1,\ldots,\lambda_t)$, then $y_\lambda=y_I$ where $I=\Delta\setminus\{\delta'_0,\delta_{\lambda_1},\delta_{\lambda_1+\lambda_2},\ldots,\delta_{r-\lambda_t}\}$. Then $y_\lambda\mathcal{H}=y_I\mathcal{H}$ and $\tau(y_\lambda)\mathcal{H}=\tau(y_I)\mathcal{H}=y_{\tau^*(I)}\mathcal{H}$ are both summands of $V_\emptyset$ since both $\{\delta_1,\delta_2,\ldots,\delta_{r-1}\}$ and its image under τ^*, i.e. $\{\delta'_0,\delta_2,\ldots,\delta_{r-1}\}$, are of type A_{r-1}. □

A key fact, in order to apply the results of §18.6, is the following.

Lemma 20.13. θ *permutes the (isomorphism types) of the indecomposable direct summands of* $\dot{V}$.

Proof. The indecomposable summands of $\dot{V}$ are direct summands of the right modules $y_\lambda\dot{\mathcal{H}}$. Those are stable under θ since $y_\lambda\in\mathcal{H}$, so for each indecomposable direct summand V_i of $\dot{V}$, both V_i and V_i^θ are direct summands of $\bigoplus_I y_I\mathcal{H}$ where I ranges over the subsets of the distinguished generators of $W(\mathrm{BC}_r)$ (see Notation 18.2). By Corollary 18.9, it suffices to check that $k\otimes V_i^\theta$ is isomorphic with a direct summand of $k\otimes\dot{V}$. We have $k\otimes\dot{V}=\bigoplus_{m,\lambda\models_m r}\varepsilon_m y_\lambda k\otimes\dot{\mathcal{H}}$ where $k\otimes\dot{\mathcal{H}}=\mathcal{H}_k(\mathrm{BC}_r,1,\overline{q})$ (we denote by $\overline{q}$ the reduction of q mod. ℓ). Denote by e the order of q mod ℓ. Then $k\otimes\dot{\mathcal{H}}=k\otimes\ddot{\mathcal{H}}$ where $\ddot{\mathcal{H}}=\mathcal{H}_{\mathcal{O}}(\mathrm{BC}_r,q^e,q)$. Now Lemma 20.10 tells us that the indecomposable summands of $\ddot{V}:=\bigoplus_{m,\lambda\models_m r}\varepsilon_m y_\lambda k\otimes\ddot{\mathcal{H}}$ are the same as those of $\oplus_{\lambda\models r}y_\lambda\ddot{\mathcal{H}}$. The latter is clearly θ-stable since θ fixes every y_λ. So $k\otimes V_i^\theta$, which was a direct summand of $\bigoplus_{\lambda\models r}y_\lambda k\otimes\dot{\mathcal{H}}\cong\bigoplus_{\lambda\models r}y_\lambda k\otimes\ddot{\mathcal{H}}$, is now a summand of $k\otimes\ddot{V}\cong k\otimes\dot{V}$. Thus our claim is proved. □

Proof of Lemma 20.11. We now apply Theorem 18.32 with $A=\dot{\mathcal{H}}$, $B=\mathcal{H}$, $M=\dot{V}$. This is possible since $\operatorname{Dec}_{\dot{\mathcal{H}}}(\dot{V})$ is square unitriangular by Lemma 20.9, and the indecomposable direct summands of $\dot{V}$ are permuted by θ by Lemma 20.13 above. We get that $\operatorname{Dec}_{\mathcal{H}}(\operatorname{Res}^{\dot{\mathcal{H}}}_{\mathcal{H}}\dot{V})$ is square unitriangular.

Take now for $V'_\emptyset$ the sum of the non-isomorphic indecomposable direct summands of $\operatorname{Res}^{\dot{\mathcal{H}}}_{\mathcal{H}}\dot{V}$. Then $\operatorname{Dec}_{\mathcal{H}}(V'_\emptyset)=\operatorname{Dec}_{\mathcal{H}}(\operatorname{Res}^{\dot{\mathcal{H}}}_{\mathcal{H}}\dot{V})$. Lemma 20.12 tells us that $V'_\emptyset$ is a direct summand of $V_\emptyset$. This completes the proof of Lemma 20.11. □

Remark 20.14. The results of Theorem 18.32 can be made more precise here. By the theory of representations of $W(\mathrm{BC}_r)$ and its subgroup $W(\mathrm{D}_r)$ (see [GePf00] §10.4), the index set for the simple modules over K of Lemma 18.38 for $\mathcal{H}(\mathrm{BC}_r, 1, q)$ is the set of bi-partitions (λ_1, λ_2) where $\lambda_1 \vdash m$ and $\lambda_2 \vdash r - m$ for some $m = 0, 1, \ldots, r$. Moreover, the permutation of this index set defined in Definition 18.37 is in this case induced by the involution $(\lambda_1, \lambda_2) \mapsto (\lambda_2, \lambda_1)$. So the only multiplicities not determined by the case of type BC_r are for even r. Otherwise, i.e. in cases (1), (2), (3), (5) of Definition 20.3, one sees from Lemma 20.10 that the decomposition numbers for unipotent characters are determined by the case of $\mathrm{GL}_n(q)$. This also applies to case (4) for n not a multiple of 4, by Lemma 18.38.

Exercises

1. Show that the multiplicities in Theorem 18.15 are less than or equal to $n!$. Show that the integers in the square unitriangular submatrix of Theorem 20.1 corresponding to unipotent characters are bounded by the order of the Weyl group (use Remark 20.14). What about other decomposition numbers?
2. Show that, in the hypotheses of Theorem 20.1, the cuspidal simple $kG.b_\ell(G, 1)$-modules are in bijection with cuspidal unipotent characters $\chi \mapsto \psi_\chi$ where ψ_χ is the unique simple $kG.\overline{b}_G(\chi)$-module (see Proposition 20.4(i)).

Note

This is essentially based on [GruHi97].

PART V

Unipotent blocks and twisted induction

In this part, we relate the partition of unipotent characters induced by ℓ-blocks to the twisted induction map $R^{\mathbf{G}}_{\mathbf{L}\subseteq\mathbf{P}}$ (see Chapter 8) and draw some consequences. The local methods that are used in this part were introduced in §5.3. Take $(\mathcal{O}, K, k)$ an ℓ-modular splitting system for $\mathbf{G}^F$, take $\mathbf{B}_0 \supseteq \mathbf{T}_0$ some F-stable Borel subgroup and maximal torus in a connected reductive group $(\mathbf{G}, F)$ defined over $\mathbb{F}_q$ (ℓ does not divide q). In Chapter 5, we proved that, if $\mathbf{T}_0^F$ is the centralizer of its ℓ-elements, then the whole $\mathcal{O}\mathbf{G}^F$-module $\mathrm{Ind}_{\mathbf{B}_0^F}^{\mathbf{G}^F}\mathcal{O}$ is in the principal block of $\mathcal{O}\mathbf{G}^F$. That hypothesis on $\mathbf{T}_0^F$ is satisfied when $\mathbf{G}$ is defined over $\mathbb{F}_q$ (with F the corresponding Frobenius) and ℓ divides $q-1$.

Assume now that the multiplicative order of q mod. ℓ is some integer $e \geq 1$. We have seen in Chapter 12 the notion of an e-split Levi subgroup. In the present part, we show that, if $\mathbf{L}$ is an e-split Levi subgroup of $\mathbf{G}^F$ and $\zeta \in \mathcal{E}(\mathbf{L}^F, 1)$, then all components of $R^{\mathbf{G}}_{\mathbf{L}}\zeta$ are in a single $\mathrm{Irr}(\mathbf{G}^F, b)$ for b a block idempotent of $\mathcal{O}\mathbf{G}^F$ which only depends on the block of $\mathcal{O}\mathbf{L}^F$ defined by ζ. Thus, inducing unipotent characters from e-split Levi subgroups yields an "e-generalized" Harish-Chandra theory. We show that it defines a partition of $\mathcal{E}(\mathbf{G}^F, 1)$ which actually coincides with the one induced by ℓ-blocks. Each unipotent block is associated with a pair $(\mathbf{L}, \zeta)$ where $\mathbf{L}$ is e-split and $\zeta \in \mathcal{E}(\mathbf{L}^F, 1)$ can't be induced from a smaller e-split Levi subgroup. We also show that the finite reductive group $C^\circ_{\mathbf{G}}([\mathbf{L}, \mathbf{L}])^F$ concentrates most of the local structure of the unipotent block of $\mathcal{O}\mathbf{G}^F$ associated with $(\mathbf{L}, \zeta)$. We close this part by returning to principal blocks when ℓ divides $q-1$ (see §5.3). In this case, we check Broué's conjecture on blocks with commutative defect groups.

21

Local methods; twisted induction for blocks

Let $\mathbf{G}$ be a connected reductive $\mathbf{F}$-group defined over the finite field $\mathbb{F}_q$. Let $F: \mathbf{G} \to \mathbf{G}$ denote the associated Frobenius endomorphism (see A2.4 and A2.5). Let ℓ be a prime not dividing q. Let $(\mathcal{O}, K, k)$ denote an ℓ-modular splitting system for $\mathbf{G}^F$ and its subgroups.

In this chapter, we introduce a slight variant of ℓ-subpairs (see §5.1) in the case of finite groups $\mathbf{G}^F$. Instead of taking pairs (U, b) where b is a block idempotent of $\mathcal{O}C_{\mathbf{G}^F}(U)$, we take b a block idempotent of $\mathcal{O}C^\circ_{\mathbf{G}}(U)^F$ (at least when U is an ℓ-subgroup of $\mathbf{G}^F$ included in a torus). We show that there is a natural notion of inclusion for those pairs. It proves a handier setting to prove the following (Theorem 21.7).

Assume ℓ is good for $\mathbf{G}$ (see Definition 13.10). Let $\mathbf{L}$ be an F-stable Levi subgroup of $\mathbf{G}$ such that $\mathbf{L} = C^\circ_{\mathbf{G}}(Z(\mathbf{L})^F_\ell)$. Then for each unipotent ℓ-block B_{L} of $\mathbf{L}^F$, there exists a unipotent ℓ-block $R^{\mathbf{B}}_{\mathbf{L}} B_{\mathrm{L}}$ of $\mathbf{G}^F$ such that, if $\zeta \in \mathcal{E}(\mathbf{L}^F, 1) \cap \mathrm{Irr}(\mathbf{L}^F, B_L)$, then all irreducible components of $R^{\mathbf{G}}_{\mathbf{L}\subseteq\mathbf{P}}\zeta$ are in $\mathrm{Irr}(\mathbf{G}^F, R^{\mathbf{G}}_{\mathbf{L}} B_{\mathrm{L}})$.

We conclude the chapter with a determination of 2-blocks. Assume $\mathbf{G}$ involves only types **A**, **B**, **C** or **D**. Then the principal block of $\mathbf{G}^F$ is only one unipotent 2-block.

21.1. "Connected" subpairs in finite reductive groups

Let G be a finite group, ℓ a prime, and $(\mathcal{O}, K, k)$ an ℓ-modular splitting system for G. An important fact is the following (see [NaTs89] 3.6.22).

Proposition 21.1. *All block idempotents of $\mathcal{O}G$ are in the submodule of $\mathcal{O}G$ generated by $G_{\ell'}$.*

In the following, we define a mild generalization of the subpairs in the case when $G = \mathbf{G}^F$, where $\mathbf{G}$ is a connected reductive group defined over a finite field of characteristic $p \neq \ell$ and F is the associated Frobenius endomorphism of $\mathbf{G}$. For simplicity, we just consider (commutative) ℓ-subgroups of groups $\mathbf{T}^F$ where $\mathbf{T}$ is an F-stable torus of $\mathbf{G}$. Assume $(\mathcal{O}, K, k)$ is an ℓ-modular splitting system for $\mathbf{G}^F$ and its subgroups.

Definition–Proposition 21.2. *Let $V \subseteq U$ be commutative ℓ-subgroups of $\mathbf{G}^F$ such that $U \subseteq \mathrm{C}^\circ_{\mathbf{G}}(U)$. Let b_U be a block idempotent of $\mathcal{O}\mathrm{C}^\circ_{\mathbf{G}}(U)^F$. Then there is a unique block idempotent b_V of $\mathcal{O}\mathrm{C}^\circ_{\mathbf{G}}(V)^F$ satisfying the two equivalent conditions*

(a) $\mathrm{Br}_U(\overline{b_V}).\overline{b_U} \neq 0$, *and*

(b) $\mathrm{Br}_U(\overline{b_V}).\overline{b_U} = \overline{b_U}$.

Then one writes $(V, b_V) \lhd (U, b_U)$.

Proof. The group $\mathrm{C}^\circ_{\mathbf{G}}(U)$ is F-stable and reductive, so the hypothesis $U \subseteq \mathrm{C}^\circ_{\mathbf{G}}(U)$ is equivalent to the fact that there is an F-stable torus $\mathbf{T}$ such that $U \subseteq \mathbf{T}^F$. Then the same is satisfied by V and $U \subseteq \mathrm{C}^\circ_{\mathbf{G}}(V)$. Let b_V be some block idempotent of $\mathcal{O}\mathrm{C}^\circ_{\mathbf{G}}(V)^F$. Now b_V is fixed by U, so $\mathrm{Br}_U(\overline{b_V})$ makes sense and is in $k[\mathrm{C}_{\mathbf{G}^F}(U)]$. Moreover, b_V has non-zero coefficients only on ℓ'-elements (Proposition 21.1), so does $\mathrm{Br}_U(\overline{b_V})$ and therefore $\mathrm{Br}_U(\overline{b_V}) \in k[\mathrm{C}^\circ_{\mathbf{G}}(U)^F]$ since $\mathrm{C}^\circ_{\mathbf{G}}(U)^F \lhd \mathrm{C}_{\mathbf{G}^F}(U)$ with index a power of ℓ (see Proposition 13.16(i)). In fact, $\mathrm{Br}_U(\overline{b_V}) \in \mathrm{Z}(k[\mathrm{C}^\circ_{\mathbf{G}}(U)^F])$ since b_V is $\mathrm{C}^\circ_{\mathbf{G}}(V)^F$-fixed and $\mathrm{C}^\circ_{\mathbf{G}}(U)^F \subseteq \mathrm{C}^\circ_{\mathbf{G}}(V)^F$.

Now $\mathrm{Br}_U(\overline{b_V})$ (resp. $\overline{b_U}$) is an idempotent (resp. a primitive idempotent) in $\mathrm{Z}(k[\mathrm{C}^\circ_{\mathbf{G}}(U)^F])$ and the equivalence between (a) and (b) is clear. The existence of $\overline{b_V}$ satisfying (a) follows from $\mathrm{Br}_U(1).\overline{b_U} = \overline{b_U}$ and the decomposition of 1 in the sum of block idempotents. The uniqueness of $\overline{b_V}$ is due to the fact that, if $\overline{b'_V}$ is a block idempotent of $\mathrm{C}^\circ_{\mathbf{G}}(V)^F$ distinct from $\overline{b_V}$, then $\overline{b_V}$ and $\overline{b'_V}$, and therefore $\mathrm{Br}_U(\overline{b_V})$ and $\mathrm{Br}_U(\overline{b'_V})$, are orthogonal. □

Here are two basic properties of the "connected" subpairs introduced in Proposition 21.2. The proof is quite easy from the definition and the corresponding properties of ordinary subpair inclusion.

Proposition 21.3. *Let $(\mathbf{G}, F)$, $V \subseteq U$, b_U be as in Proposition 21.2. Let b_V be a block idempotent of $\mathcal{O}\mathrm{C}^\circ_{\mathbf{G}}(V)^F$. One has the following properties.*

(i) $(V, b_V) \lhd (U, b_U)$ in $\mathbf{G}$ if and only if $(\{1\}, b_V) \lhd (U, b_U)$ in $\mathrm{C}^\circ_{\mathbf{G}}(V)$.

(ii) The relation $\lhd$ is transitive on the above type of pairs.

21.2. Twisted induction for blocks

A very important property of the adjoint ${}^*\mathrm{R}^{\mathbf{G}}_{\mathbf{L}}$ of twisted induction in connection with ℓ-blocks is its commutation with decomposition maps d^x (see Definition 5.7). The following is to be compared with Proposition 5.23.

Theorem 21.4. *Let* $(\mathbf{G}, F)$, ℓ *be as before a connected reductive group defined over* $\mathbb{F}_q$, *and a prime* ℓ *not dividing* q. *Let* $\mathbf{P} = \mathbf{L} \rtimes R_u(\mathbf{P})$ *be a Levi decomposition of a parabolic subgroup of* $\mathbf{G}$ *such that* $F\mathbf{L} = \mathbf{L}$. *Let* $x \in \mathbf{G}^F_\ell$ *(so that* x *is semi-simple). Then* $\mathrm{C}^\circ_{\mathbf{P}}(x) = \mathrm{C}^\circ_{\mathbf{L}}(x) \rtimes \mathrm{C}^\circ_{R_u(\mathbf{P})}(x)$ *is a Levi decomposition of a parabolic subgroup of* $\mathrm{C}^\circ_{\mathbf{G}}(x)$ *and*

$$d^x \circ {}^*\mathrm{R}^{\mathbf{G}}_{\mathbf{L}\subseteq\mathbf{P}} = {}^*\mathrm{R}^{\mathrm{C}^\circ_{\mathbf{G}}(x)}_{\mathrm{C}^\circ_{\mathbf{L}}(x)\subseteq\mathrm{C}^\circ_{\mathbf{P}}(x)} \circ d^x$$

on $\mathrm{CF}(\mathbf{G}^F, K)$.

Proof. Note that d^x sends a central function on $\mathbf{G}^F$ to a central function on $\mathrm{C}_{\mathbf{G}^F}(x)$ vanishing outside $\mathrm{C}_{\mathbf{G}^F}(x)_{\ell'}$. But since $\mathrm{C}_{\mathbf{G}^F}(x)_{\ell'} \subseteq \mathrm{C}^\circ_{\mathbf{G}}(x)$ (Proposition 13.16(i)), the above ${}^*\mathrm{R}^{\mathrm{C}^\circ_{\mathbf{G}}(x)}_{\mathrm{C}^\circ_{\mathbf{L}}(x)\subseteq\mathrm{C}^\circ_{\mathbf{P}}(x)} \circ d^x$ actually makes sense.

Then our claim is an easy consequence of the character formula for ${}^*\mathrm{R}^{\mathbf{G}}_{\mathbf{L}\subseteq\mathbf{P}}$ maps (see [DiMi91] 12.5 or Theorem 8.16(i)). □

Let E be a set of integers greater than or equal to 1. Recall that a ϕ_E-subgroup is an F-stable torus of $\mathbf{G}$ whose polynomial order is a product of cyclotomic polynomials in $\{\phi_e \mid e \in E\}$. An E-split Levi subgroup is the centralizer $\mathrm{C}_{\mathbf{G}}(\mathbf{T})$ of a ϕ_E-subgroup $\mathbf{T}$.

Definition 21.5. *Let* $\mathbf{L}_i$ $(i = 1, 2)$ *be* F*-stable Levi subgroups of* $\mathbf{G}$. *Let* $\zeta_i \in \mathcal{E}(\mathbf{L}_i^F, 1)$. *We write* $(\mathbf{L}_1, \zeta_1) \geq (\mathbf{L}_2, \zeta_2)$ *whenever* $\mathbf{L}_1 \supseteq \mathbf{L}_2$ *and there is a parabolic subgroup* $\mathbf{P}_2 \subseteq \mathbf{L}_1$ *for which* $\mathbf{L}_2$ *is a Levi supplement and* $\langle \zeta_2, \mathrm{R}^{\mathbf{L}_1}_{\mathbf{L}_2\subseteq\mathbf{P}_2}\zeta_1\rangle_{\mathbf{L}_1^F} \neq 0$.

If $\emptyset \neq E \subseteq \{1, 2, 3, \ldots\}$ *and* $\mathbf{L}_i$ $(i = 1, 2)$ *are* E*-split Levi subgroups of* $\mathbf{G}$, *the above relation is written* $(\mathbf{L}_1, \zeta_1) \geq_E (\mathbf{L}_2, \zeta_2)$. *The character* ζ_1 *is said to be* E*-cuspidal if any relation* $(\mathbf{L}_1, \zeta_1) \geq_E (\mathbf{L}_2, \zeta_2)$ *implies* $\mathbf{L}_1 = \mathbf{L}_2$. *The pair* $(\mathbf{L}_1, \zeta_1)$ *is then called an* E*-cuspidal pair.*

We define $\gg_E$ *by transitive closure of* $\geq_E$ *above on pairs* $(\mathbf{L}, \zeta)$ *where* $\mathbf{L}$ *is an* E*-split Levi subgroup of* $\mathbf{G}$ *and* $\zeta \in \mathcal{E}(\mathbf{L}^F, 1)$.

Example 21.6. Let us look at the case of Example 13.4(ii), i.e. $\mathbf{G} = \mathrm{GL}_n(\mathbf{F})$, F being the usual Frobenius map $(x_{ij}) \mapsto (x_{ij}^q)$. Let us parametrize the $\mathbf{G}^F$-classes of F-stable maximal tori from the diagonal torus of $\mathbf{G}$ by conjugacy classes of $\mathfrak{S}_n$ (see §8.2). When $w \in \mathfrak{S}_n$, choose $\mathbf{T}_w$ in the corresponding class. When $f \in \mathrm{CF}(\mathfrak{S}_n, K)$, let

$$R^{(\mathbf{G})}_f := (n!)^{-1} \sum_{w\in\mathfrak{S}_n} f(w)\mathrm{R}^{\mathbf{G}}_{\mathbf{T}_w}(1_{\mathbf{T}_w^F}) \in \mathrm{CF}(\mathbf{G}^F, K).$$

Then $f \mapsto R^{(\mathbf{G})}_f$ is an isometry sending $\mathrm{Irr}(\mathfrak{S}_n)$ onto $\mathcal{E}(\mathbf{G}^F, 1)$ (see Theorem 19.7). Since $\mathrm{Irr}(\mathfrak{S}_n)$ is in turn parametrized by partitions of n (see

[CuRe87] 75.19) as stated in §5.3 and Exercise 18.4, we write $\chi_\lambda^{\mathbf{G}} \in \mathcal{E}(\mathbf{G}^F, 1)$ for the unipotent character of $\mathrm{GL}_n(q)$ associated with $\lambda \vdash n$.

Let $e \geq 1$. Let $\mathbf{L}^{(m)} \cong \mathrm{GL}_{n-me}(\mathbf{F}) \times (\mathbf{S}_{(e)})^m$ be the e-split Levi subgroup of $\mathrm{GL}_n(\mathbf{F})$ defined in Example 13.4(ii) when $me \leq n$. If $\lambda \vdash n$, we have

$$\text{(E)} \qquad {}^*\mathrm{R}^{\mathbf{G}}_{\mathbf{L}^{(1)}}(\chi_\lambda^{\mathbf{G}}) = \sum_\gamma \varepsilon(\lambda, \gamma)\chi_{\lambda*\gamma}^{\mathbf{L}^{(1)}}$$

where the sum is over e-hooks of λ, $\lambda * \gamma \vdash n - e$ and the signs $\varepsilon(\lambda, \gamma)$ are as defined in §5.2. Note that we have identified the unipotent characters of $(\mathbf{L}^{(1)})^F$ with those of $\mathrm{GL}_{n-e}(\mathbb{F}_q)$. Let us say briefly how one checks (E) (see also [DiMi91] 15.7). To compute ${}^*\mathrm{R}^{\mathbf{G}}_{\mathbf{L}^{(1)}} \circ \mathrm{R}^{\mathbf{G}}_{\mathbf{T}_w}$, one may use a Mackey formula similar to the one satisfied by Harish-Chandra induction since one of the two Levi subgroups involved is a torus (see [DiMi91] 11.13). Note that $\mathbf{L}^{(1)}$ contains a $\mathbf{G}^F$-conjugate of $\mathbf{T}_w$ only if w can be written, up to conjugacy in $\mathfrak{S}_n$, as $w'c$ where $w' \in \mathfrak{S}_{n-e}$ and $c = (n - e + 1, \ldots, n)$ is a cycle of order e. Then the Mackey formula gives ${}^*\mathrm{R}^{\mathbf{G}}_{\mathbf{L}^{(1)}}\mathrm{R}^{\mathbf{G}}_{\mathbf{T}_{w'c}}(1) = n!/(n - e)!\mathrm{R}^{\mathbf{G}}_{\mathbf{T}_{w'c}}(1)$. Then (E) above is a consequence of the Murnaghan–Nakayama formula about restrictions to $\mathfrak{S}_{n-e}.c$ of irreducible characters of $\mathfrak{S}_n$ (see Theorem 5.13).

Assume $\lambda \vdash n$ is of e-core $\kappa \vdash n - me$. Then an iteration of (E), along with the corresponding iteration of the Murnaghan–Nakayama formula (Theorem 5.15(iii)), gives

$$(\mathbf{G}, \chi_\lambda^{\mathbf{G}}) \gg_e (\mathbf{L}^{(m)}, \chi_\kappa^{\mathbf{L}^{(m)}}).$$

(Note that actually one may replace the above $\gg_e$ by $\geq$ (see Definition 21.5).)

Let us show that $\chi_\kappa^{\mathbf{L}^{(m)}}$ is e-cuspidal. We may assume that $m = 0$ and we need to show that ${}^*\mathrm{R}^{\mathbf{G}}_{\mathbf{L}}\chi_\kappa^{\mathbf{G}} = 0$ for any proper e-split Levi subgroup $\mathbf{L} \subseteq \mathbf{G}$. The above parametrization of unipotent characters shows that any unipotent character of $\mathbf{L}^F$ is a linear combination of $\mathrm{R}^{\mathbf{L}}_{\mathbf{T}}1$'s for $\mathbf{T}$ ranging over F-stable maximal tori of $\mathbf{L}$. So, by transitivity of twisted induction (and independence with regard to Borel subgroups in the case of tori; see Theorem 8.17(i)) our claim reduces to checking that ${}^*\mathrm{R}^{\mathbf{G}}_{\mathbf{T}}\chi_\kappa^{\mathbf{G}} = 0$ for any F-stable maximal torus of $\mathbf{G}$ that embeds in a proper e-split Levi subgroup of $\mathbf{G}$. By the description of maximal proper e-split Levi subgroups of $\mathrm{GL}_n(\mathbf{F})$ (see Example 13.4(ii)), a $\mathbf{T}$ as above would be a maximal torus in some $\mathbf{L}' = \mathrm{GL}_{n-em}(\mathbf{F}) \times \mathbf{S}_{(em)}$ for some $m \geq 1$. Applying (E) above and the fact that, if κ is an e-core, then it is an em-core for any $m \geq 1$ (see Theorem 5.15(i)), we get ${}^*\mathrm{R}^{\mathbf{G}}_{\mathbf{L}'}\chi_\kappa^{\mathbf{G}} = 0$ and therefore ${}^*\mathrm{R}^{\mathbf{G}}_{\mathbf{T}}\chi_\kappa^{\mathbf{G}} = 0$ as claimed.

Theorem 21.7. *Assume ℓ is good for $\mathbf{G}$. Denote by $E(q, \ell)$ the set of integers d such that ℓ divides $\phi_d(q)$. Let $\mathbf{L}$ be an F-stable Levi subgroup of $\mathbf{G}$ such that $\mathbf{L} = \mathrm{C}^\circ_{\mathbf{G}}(\mathrm{Z}(\mathbf{L})^F_\ell)$. Let $\zeta \in \mathcal{E}(\mathbf{L}^F, 1)$ and denote by $b_{\mathbf{L}^F}(\zeta)$ the ℓ-block idempotent* *of* $\mathbf{L}^F$ *not annihilated by* ζ *(see §5.1). There is a unique block idempotent* $b_{\mathbf{G}}$ *of* $\mathcal{O}\mathbf{G}^F$ *such that*

$$(\{1\}, b_{\mathbf{G}}) \lhd (\mathrm{Z}(\mathbf{L})^F_\ell, b_{\mathbf{L}^F}(\zeta))$$

in the sense of Proposition 21.2, and

$$\chi \in \mathrm{Irr}(\mathbf{G}^F, b_{\mathbf{G}})$$

for any $\chi \in \mathcal{E}(\mathbf{G}^F, 1)$ *such that* $(\mathbf{G}, \chi) \gg_{E(q,\ell)} (\mathbf{L}, \zeta)$.

Notation 21.8. Under the hypothesis of the above theorem on $(\mathbf{G}, F)$, ℓ, $\mathbf{L}$ and if $B_{\mathbf{L}}$ is a block of $\mathcal{O}\mathbf{L}^F$, one may define $\mathrm{R}^{\mathbf{G}}_{\mathbf{L}} B_{\mathbf{L}}$ as the block of $\mathcal{O}\mathbf{G}^F$ acting by 1 on all the irreducible components of the $\mathrm{R}^{\mathbf{G}}_{\mathbf{L}\subseteq\mathbf{P}}\zeta$'s for $\zeta \in \mathrm{Irr}(\mathbf{L}^F, B_{\mathbf{L}}) \cap \mathcal{E}(\mathbf{L}^F, 1)$ and $\mathbf{P}$ a parabolic subgroup of $\mathbf{G}$ containing $\mathbf{L}$ as a Levi subgroup.

Proof of Theorem 21.7. We abbreviate $E(q, \ell) = E$.

Assume that $\mathrm{Z}(\mathbf{G})$ is connected. The following proof is to be compared with that of Theorem 5.19.

Since the center of $\mathbf{G}$ is connected, the E-split Levi subgroups of $\mathbf{G}$ all satisfy $\mathbf{L} = \mathrm{C}^\circ_{\mathbf{G}}(\mathrm{Z}(\mathbf{L})^F_\ell)$ (see Proposition 13.19). So, by transitivity of the inclusion of connected subpairs (Proposition 21.3(ii)), one is left to prove that

$$(1, b_{\mathbf{G}^F}(\chi)) \lhd \left(\mathrm{Z}(\mathbf{L})^F_\ell, b_{\mathbf{L}^F}(\zeta)\right)$$

as long as $\chi \in \mathcal{E}(\mathbf{G}^F, 1)$, $\zeta \in \mathcal{E}(\mathbf{L}^F, 1)$, and $(\mathbf{G}, \chi) \geq (\mathbf{L}, \zeta)$ (see Definition 21.5).

We prove this by induction on dim $\mathbf{G}$ - dim $\mathbf{L}$.

If $\mathbf{L} = \mathbf{G}$, everything is clear. Assume $\mathbf{L} \neq \mathbf{G}$. Fix $b_{\mathbf{L}}$ a block idempotent of $\mathcal{O}\mathbf{L}^F.b_\ell(\mathbf{L}^F, 1)$ (see Definition 9.4).

Since $\mathrm{Z}(\mathbf{L})^F_\ell \not\subseteq \mathrm{Z}(\mathbf{G})$, we may choose $x \in \mathrm{Z}(\mathbf{L})^F_\ell \setminus \mathrm{Z}(\mathbf{G})$ and denote $\mathbf{C} = \mathrm{C}^\circ_{\mathbf{G}}(x)$, a Levi subgroup by Proposition 13.16(ii). Then $x \in \mathbf{L} \subseteq \mathbf{C}$ and therefore $\mathbf{C} = \mathrm{C}^\circ_{\mathbf{G}}(\mathrm{Z}(\mathbf{C})^F_\ell)$. We have $\mathbf{L} \subseteq \mathbf{C} \subseteq \mathbf{G}$ with $\mathbf{C} \neq \mathbf{G}$. By the induction hypothesis, there is an ℓ-block idempotent $b_{\mathbf{C}}$ of $\mathbf{C}^F$ such that, for any $\zeta \in \mathrm{Irr}(\mathbf{L}^F, b_{\mathbf{L}}) \cap \mathcal{E}(\mathbf{L}^F, 1)$ and any parabolic subgroup $\mathbf{Q} \subseteq \mathbf{C}$ having $\mathbf{L}$ as Levi subgroup, all the irreducible components of $\mathrm{R}^{\mathbf{C}}_{\mathbf{L}\subseteq\mathbf{Q}}\zeta$ are in $\mathrm{CF}(\mathbf{C}^F, b_{\mathbf{C}})$ and moreover $(\{1\}, b_{\mathbf{C}}) \lhd (\mathrm{Z}(\mathbf{L})^F_\ell, b_{\mathbf{L}})$ in $\mathbf{C}$ whenever $\mathbf{L} = \mathrm{C}^\circ_{\mathbf{C}}(\mathrm{Z}(\mathbf{L})^F_\ell)$ (which is the case when $\mathbf{L} = \mathrm{C}^\circ_{\mathbf{G}}(\mathrm{Z}(\mathbf{L})^F_\ell)$; see Proposition 13.13(ii)).

Let χ be an irreducible component of $\mathrm{R}^{\mathbf{G}}_{\mathbf{L}\subseteq\mathbf{P}}\zeta$ for some $\zeta \in \mathrm{Irr}(\mathbf{L}^F, b_{\mathbf{L}}) \cap \mathcal{E}(\mathbf{L}^F, 1)$ and $\mathbf{P}$ a parabolic subgroup with Levi subgroup $\mathbf{L}$.

When b is a block idempotent of some group algebra $\mathcal{O}G$, denote by $c \mapsto b.c$ the projection morphism $\mathrm{CF}(G, K) \to \mathrm{CF}(G, K, b)$.

Recall the decomposition map $d^x \colon \mathrm{CF}(G, K) \to \mathrm{CF}(\mathrm{C}_G(x), K)$.

Let $f = b_{\mathbf{L}}.d^1{}^*\mathrm{R}^{\mathbf{G}}_{\mathbf{L}\subseteq\mathbf{P}}\chi$ be the projection on $\mathrm{CF}(\mathbf{L}^F, K, b_{\mathbf{L}})$ of the restriction of ${}^*\mathrm{R}^{\mathbf{G}}_{\mathbf{L}\subseteq\mathbf{P}}\chi$ to ℓ'-elements of $\mathbf{L}^F$.

Lemma 21.9. *For all* $\zeta' \in \mathrm{Irr}(\mathbf{L}^F, b_{\mathbf{L}}) \cap \mathcal{E}(\mathbf{L}^F, 1)$, *one has* $\langle d^x\chi, \mathrm{R}^{\mathbf{C}}_{\mathbf{L}\subseteq\mathbf{P}\cap\mathbf{C}}\zeta'\rangle_{\mathbf{C}^F} = \langle f, d^1\zeta'\rangle_{\mathbf{L}^F}$.

Proof. First $\mathrm{C}_{\mathbf{G}}(x)^F_{\ell'} = \mathbf{C}^F_{\ell'}$ since $\mathrm{C}_{\mathbf{G}}(x)/\mathrm{C}^\circ_{\mathbf{G}}(x)$ is an ℓ-group. Then $d^x\chi$ is a central function on $\mathbf{C}^F$ and $\langle d^x\chi, \mathrm{R}^{\mathbf{C}}_{\mathbf{L}\subseteq\mathbf{P}\cap\mathbf{C}}\zeta'\rangle_{\mathbf{C}^F}$ makes sense.

Theorem 21.4 implies $\langle d^x\chi, \mathrm{R}^{\mathbf{C}}_{\mathbf{L}\subseteq\mathbf{P}\cap\mathbf{C}}\zeta'\rangle_{\mathbf{C}^F} = \langle d^x{}^*\mathrm{R}^{\mathbf{G}}_{\mathbf{L}\subseteq\mathbf{P}}\chi, \zeta'\rangle_{\mathbf{L}^F}$. But ${}^*\mathrm{R}^{\mathbf{G}}_{\mathbf{L}\subseteq\mathbf{P}}\chi \in \mathbb{Z}\mathcal{E}(\mathbf{L}^F, 1)$ by Proposition 8.25, and each unipotent character of $\mathbf{L}^F$ is trivial on x, so $\langle d^x{}^*\mathrm{R}^{\mathbf{G}}_{\mathbf{L}\subseteq\mathbf{P}}\chi, \zeta'\rangle_{\mathbf{L}^F} = \langle d^1{}^*\mathrm{R}^{\mathbf{G}}_{\mathbf{L}\subseteq\mathbf{P}}\chi, \zeta'\rangle_{\mathbf{L}^F} = \langle f, \zeta'\rangle_{\mathbf{L}^F} = \langle f, d^1\zeta'\rangle_{\mathbf{L}^F}$ since ζ' and $d^1\zeta'$ both equal their projections on $\mathrm{CF}(\mathbf{L}^F, K, b_{\mathbf{L}})$ (by Brauer's second Main Theorem, see Theorem 5.8, for $d^1\zeta'$). Combining the two equalities gives our lemma. □

Lemma 21.10. $b_{\mathbf{C}}.d^x(\chi) \neq 0$ *(which makes sense since* $\mathrm{C}_{\mathbf{G}^F}(x)_{\ell'} = \mathbf{C}^F_{\ell'}$*).*

Proof. By the hypothesis on χ, one may write ${}^*\mathrm{R}^{\mathbf{G}}_{\mathbf{L}\subseteq\mathbf{P}}\chi = m_1\zeta_1 + \cdots + m_\nu\zeta_\nu$ with distinct $\zeta_i \in \mathcal{E}(\mathbf{L}^F, 1)$ (see Proposition 8.25), the m_i's being non-zero integers and $\zeta = \zeta_1$. Now $b_{\mathbf{L}^F}(\zeta_i).d^1(\zeta_i) = d^1(\zeta_i)$ again by Brauer's second Main Theorem, so

$$f = b_{\mathbf{L}}.d^1({}^*\mathrm{R}^{\mathbf{G}}_{\mathbf{L}\subseteq\mathbf{P}}\chi) = m_1 d^1(\zeta_1) + \sum_{i\geq 2\,;\, b_{\mathbf{L}^F}(\zeta_i)=b_{\mathbf{L}}} m_i d^1(\zeta_i).$$

Since the center of $\mathbf{G}$ is connected, we may apply Theorem 14.4. So we know that the $d^1(\zeta_i)$'s for $1 \leq i \leq \nu$ are linearly independent, and therefore $b_{\mathbf{L}}.d^1({}^*\mathrm{R}^{\mathbf{G}}_{\mathbf{L}\subseteq\mathbf{P}}\chi) = f \neq 0$. One sees clearly that $f(x^{-1})$ is the complex conjugate of $f(x)$ for all $x \in \mathbf{L}^F$, so $\langle f, f\rangle_{\mathbf{L}^F} \neq 0$. Then there is some ζ_i in the expression for f above, let us call it ζ', such that $\zeta' \in \mathcal{E}(\mathbf{L}^F, \ell') \cap \mathrm{Irr}(\mathbf{L}^F, b_{\mathbf{L}})$ and $\langle f, d^1\zeta'\rangle_{\mathbf{L}^F} \neq 0$.

We may apply Lemma 21.9, so $\langle d^x\chi, \mathrm{R}^{\mathbf{C}}_{\mathbf{L}\subseteq\mathbf{P}\cap\mathbf{C}}\zeta'\rangle_{\mathbf{C}^F} \neq 0$. This implies that $b_{\mathbf{C}}.d^x\chi \neq 0$ since all the irreducible components of $\mathrm{R}^{\mathbf{C}}_{\mathbf{L}\subseteq\mathbf{P}\cap\mathbf{C}}\zeta'$ are in $b_{\mathbf{C}}$ by the induction hypothesis. This is our claim. □

Let us now consider the Brauer map $\mathrm{Br}_{<x>}$ in the group algebra $k[\mathbf{G}^F]$. Let us show the following.

Lemma 21.11. $\mathrm{Br}_{<x>}(\overline{b}_{\mathbf{G}^F}(\chi)).\overline{b_{\mathbf{C}}} = \overline{b_{\mathbf{C}}}$.

Proof. We have $\mathbf{C}^F = \mathrm{C}^\circ_{\mathbf{G}}(x)^F \lhd \mathrm{C}_{\mathbf{G}^F}(x)$ with index a power of ℓ. Let $b' = \sum_g g.b_{\mathbf{C}}.g^{-1}$, where g ranges over $\mathrm{C}_{\mathbf{G}^F}(x)$ mod the stabilizer of $b_{\mathbf{C}}$. The above is a sum of orthogonal idempotents and therefore b' is a central idempotent of $\mathcal{O}\mathrm{C}_{\mathbf{G}^F}(x)$ such that $b'.b_{\mathbf{C}} = b_{\mathbf{C}}.b' = b_{\mathbf{C}}$. Moreover b' is a block idempotent of $\mathcal{O}\mathrm{C}_{\mathbf{G}^F}(x)$, i.e. it is primitive in $\mathrm{Z}(\mathcal{O}\mathrm{C}_{\mathbf{G}^F}(x))$; see [NaTs89] 5.5.6 (an easy consequence of $C^\circ_{\mathbf{G}}(x)^F \lhd C_{\mathbf{G}^F}(x)$ with index a power of ℓ). So, to check Lemma 21.11, it suffices to check that $\mathrm{Br}_{<x>}(\overline{b}_{\mathbf{G}^F}(\chi)).\overline{b'} = \overline{b'}$ or equivalently the ordinary subpair inclusion $(\{1\}, b_{\mathbf{G}^F}(\chi)) \subseteq (<x>, b')$ in $\mathbf{G}^F$. By Brauer's second Main Theorem (Theorem 5.8), it suffices to check that $b'.d^x(\chi) \neq 0$. This is clearly a consequence of Lemma 21.10 and $b_{\mathbf{C}}.b' = b_{\mathbf{C}}$.

We have now checked Lemma 21.11 for χ an irreducible component of some $R^{\mathbf{G}}_{\mathbf{L}\subseteq\mathbf{P}}\zeta$ for $\zeta \in \mathrm{Irr}(\mathbf{L}^F, b_{\mathbf{L}}) \cap \mathcal{E}(\mathbf{L}^F, 1)$ and $\mathbf{P}$ a parabolic for $\mathbf{L}$ in $\mathbf{G}$. If χ' satisfies the same hypothesis as χ for other ζ and $\mathbf{P}$, then $\mathrm{Br}_{<x>}(\overline{b}_{\mathbf{G}^F}(\chi')).\overline{b_{\mathbf{C}}} = \overline{b_{\mathbf{C}}}$ so $\overline{b}_{\mathbf{G}^F}(\chi') = \overline{b}_{\mathbf{G}^F}(\chi)$ since otherwise $\overline{b}_{\mathbf{G}^F}(\chi')\overline{b}_{\mathbf{G}^F}(\chi) = 0$ and therefore $0 = \mathrm{Br}_{<x>}(\overline{b}_{\mathbf{G}^F}(\chi')\overline{b}_{\mathbf{G}^F}(\chi)).\overline{b_{\mathbf{C}}} = \mathrm{Br}_{<x>}(\overline{b}_{\mathbf{G}^F}(\chi'))\mathrm{Br}_{<x>}(\overline{b}_{\mathbf{G}^F}(\chi)).\overline{b_{\mathbf{C}}} = \mathrm{Br}_{<x>}(\overline{b}_{\mathbf{G}^F}(\chi')).\overline{b_{\mathbf{C}}} = \overline{b_{\mathbf{C}}}$, a contradiction. This yields the theorem when the center of $\mathbf{G}$ is connected.

Now, we no longer assume that $\mathrm{Z}(\mathbf{G})$ is connected. Let $\mathbf{G} \subseteq \widetilde{\mathbf{G}}$ be an embedding of $\mathbf{G}$ into a reductive group with connected center and such that $\widetilde{\mathbf{G}} = \mathbf{G}\mathrm{Z}(\widetilde{\mathbf{G}})$ (see §15.1).

Let $\widetilde{\mathbf{L}} = C_{\widetilde{\mathbf{G}}}(\mathrm{Z}^\circ(\mathbf{L})) = \mathrm{Z}(\widetilde{\mathbf{G}})\mathbf{L}$ be the Levi subgroup of $\widetilde{\mathbf{G}}$ containing $\mathbf{L}$. We have bijections $\mathbf{M} \mapsto \widetilde{\mathbf{M}} = \mathrm{Z}(\widetilde{\mathbf{G}})\mathbf{M}$ between Levi subgroups (resp. E-split Levi subgroups) of $\mathbf{G}$ and $\widetilde{\mathbf{G}}$. Moreover, $\mathrm{Res}^{\widetilde{\mathbf{M}}^F}_{\mathbf{M}^F}$ induces a bijection $\mathcal{E}(\widetilde{\mathbf{M}}^F, 1) \to \mathcal{E}(\mathbf{M}^F, 1)$ which commutes with twisted inductions (see Proposition 15.9).

Since E-split Levi subgroups correspond by the above (Proposition 13.7), a relation $(\mathbf{G}, \chi) \gg_E (\mathbf{L}, \zeta)$ as in the hypotheses of the theorem implies $(\widetilde{\mathbf{G}}, \widetilde{\chi}) \gg_E (\widetilde{\mathbf{L}}, \widetilde{\zeta})$ for unipotent characters such that $\chi = \mathrm{Res}^{\widetilde{\mathbf{G}}^F}_{\mathbf{G}^F}\widetilde{\chi}$ and $\zeta = \mathrm{Res}^{\widetilde{\mathbf{L}}^F}_{\mathbf{L}^F}\widetilde{\zeta}$.

Concerning the block idempotents, $b_{\mathbf{G}^F}(\chi)$ is the only primitive idempotent $b \in \mathrm{Z}(\mathcal{O}\mathbf{G}^F)$ such that $\chi(b) \neq 0$. Since $\chi(b) = \widetilde{\chi}(b)$, and since $b_{\mathbf{G}^F}(\chi) \in \mathrm{Z}(\mathcal{O}\widetilde{\mathbf{G}}^F)$ ($\chi = \mathrm{Res}^{\widetilde{\mathbf{G}}^F}_{\mathbf{G}^F}\widetilde{\chi}$ being $\widetilde{\mathbf{G}}^F$-stable), one has $b_{\mathbf{G}^F}(\chi).b_{\widetilde{\mathbf{G}}^F}(\widetilde{\chi}) = b_{\widetilde{\mathbf{G}}^F}(\widetilde{\chi})$. Similarly $b_{\mathbf{L}^F}(\zeta).b_{\widetilde{\mathbf{L}}^F}(\widetilde{\zeta}) = b_{\widetilde{\mathbf{L}}^F}(\widetilde{\zeta})$.

Denote $Z = \mathrm{Z}(\mathbf{L})^F_\ell$, $\widetilde{Z} = \mathrm{Z}(\widetilde{\mathbf{L}})^F_\ell$. Then $\widetilde{\mathbf{L}} = C^\circ_{\widetilde{\mathbf{G}}}(\widetilde{Z})$ and $\mathbf{L} = C^\circ_{\mathbf{G}}(Z)$. Since we have proved the theorem in the case of connected center, we have $\big(1, b_{\widetilde{\mathbf{G}}^F}(\widetilde{\chi})\big) \lhd \big(\widetilde{Z}, b_{\widetilde{\mathbf{L}}^F}(\widetilde{\zeta})\big)$ in $\widetilde{\mathbf{G}}^F$. Similarly, we have $\widetilde{\mathbf{L}} = C^\circ_{\widetilde{\mathbf{G}}}(\mathrm{Z}(\widetilde{\mathbf{L}})^F_\ell)$ by Proposition 13.19 . This means that $\mathrm{Br}_{\widetilde{Z}}\big(\overline{b}_{\widetilde{\mathbf{G}}^F}(\widetilde{\chi})\big).\overline{b}_{\widetilde{\mathbf{L}}^F}(\widetilde{\zeta}) = \overline{b}_{\widetilde{\mathbf{L}}^F}(\widetilde{\zeta})$ in $k\widetilde{\mathbf{L}}^F$. So

$$\text{(I)} \qquad \mathrm{Br}_{\widetilde{Z}}(\overline{b}_{\mathbf{G}^F}(\chi)).\overline{b}_{\widetilde{\mathbf{L}}^F}(\widetilde{\zeta}) = \overline{b}_{\widetilde{\mathbf{L}}^F}(\widetilde{\zeta}).$$

But $C_{\mathbf{G}^F}(Z)_{\ell'} = \mathbf{G}^F \cap C_{\widetilde{\mathbf{G}}^F}(\widetilde{Z})_{\ell'}$ since both equal $C^\circ_{\mathbf{G}}(Z)^F_{\ell'} = \mathbf{G}^F_{\ell'} \cap C^\circ_{\widetilde{\mathbf{G}}}(\widetilde{Z})$ (see Proposition 13.16(i)) and this is $\mathbf{L}^F_{\ell'}$. Then $\mathrm{Br}_{\widetilde{Z}}(\overline{b}_{\mathbf{G}^F}(\chi)) = \mathrm{Br}_Z(\overline{b}_{\mathbf{G}^F}(\chi))$ and the equation (I) above becomes $\mathrm{Br}_Z(\overline{b}_{\mathbf{G}^F}(\chi)).\overline{b}_{\widetilde{\mathbf{L}}^F}(\widetilde{\zeta}) = \overline{b}_{\widetilde{\mathbf{L}}^F}(\widetilde{\zeta})$. Now the fact that $b_{\widetilde{\mathbf{L}}^F}(\widetilde{\zeta}).b_{\mathbf{L}^F}(\zeta) = b_{\mathbf{L}^F}(\zeta)$ implies that $\mathrm{Br}_Z(\overline{b}_{\mathbf{G}^F}(\chi)).\overline{b}_{\mathbf{L}^F}(\zeta) \neq 0$. By Proposition 21.2, this gives the inclusion $(1, b_{\mathbf{G}^F}(\chi)) \lhd (Z, b_{\mathbf{L}^F}(\zeta))$.

This completes the proof of the theorem. □

Remark 21.12. If $\mathbf{L}$ is an $E_{q,\ell}$-split Levi subgroup of $\mathbf{G}$ and $B_{\mathbf{L}}$ is a unipotent ℓ-block of $\mathbf{L}^F$, it is clear from its characterization that $\mathrm{R}_{\mathbf{L}}^{\mathbf{G}} B_{\mathbf{L}}$ (see Notation 21.8) does not depend on the choice of the parabolic subgroup having $\mathbf{L}$ as Levi subgroup.

It is also clear that the transitivity $\mathrm{R}_{\mathbf{L}}^{\mathbf{G}} B_{\mathbf{L}} = \mathrm{R}_{\mathbf{M}}^{\mathbf{G}}(\mathrm{R}_{\mathbf{L}}^{\mathbf{M}} B_{\mathbf{L}})$ holds when $\mathbf{L} \subseteq \mathbf{M} \subseteq \mathbf{G}$ is an inclusion of $E_{q,\ell}$-split Levi subgroups.

While Theorem 21.7 defines a unipotent block from its unipotent characters, the following gives information on non-unipotent characters of a unipotent block.

Theorem 21.13. *Let $(\mathbf{G}, F)$ be a connected reductive group defined over $\mathbb{F}_q$. Let ℓ be a prime good for $\mathbf{G}$ and not dividing q. Let $\chi \in \mathcal{E}(\mathbf{G}^F, t)$, with $t \in (\mathbf{G}^*)^F_\ell$. Let $\mathbf{G}(t) \subseteq \mathbf{G}$ be a Levi subgroup in duality with $\mathrm{C}^\circ_{\mathbf{G}^*}(t)$, let $\mathbf{P}$ be a parabolic subgroup of $\mathbf{G}$ having $\mathbf{G}(t)$ as Levi subgroup. Then there is $\chi_t \in \mathcal{E}(\mathbf{G}(t)^F, 1)$ such that $\langle \chi, \mathrm{R}^{\mathbf{G}}_{\mathbf{G}(t)\subseteq\mathbf{P}}(\hat{t}\chi_t)\rangle_{\mathbf{G}^F} \neq 0$, where $\hat{t}$ is the linear character of $\mathbf{G}(t)^F$ associated with t by duality (see Proposition 8.26). For any such $(\mathbf{G}(t), \mathbf{P}, \chi_t)$ associated with χ, all the irreducible components of $\mathrm{R}^{\mathbf{G}}_{\mathbf{G}(t)\subseteq\mathbf{P}}\chi_t$ are in $\mathrm{Irr}\big(\mathbf{G}^F, b_{\mathbf{G}^F}(\chi)\big) \cap \mathcal{E}(\mathbf{G}^F, 1)$ (where $b_{\mathbf{G}^F}(\chi)$ denotes the ℓ-block idempotent of $\mathbf{G}^F$ not annihilated by χ; see §5.1).*

Proof. Since ℓ is good, $\mathrm{C}^\circ_{\mathbf{G}^*}(t)$ is a Levi subgroup of $\mathbf{G}^*$ by Proposition 13.16(ii). Applying Proposition 15.10, one gets the existence of some $\xi \in \mathcal{E}(\mathbf{G}(t)^F, t)$ such that $\langle \mathrm{R}^{\mathbf{G}}_{\mathbf{G}(t)\subseteq\mathbf{P}}\xi, \chi\rangle_{\mathbf{G}^F} \neq 0$. However, $\mathcal{E}(\mathbf{G}(t)^F, t) = \hat{t}\mathcal{E}(\mathbf{G}(t)^F, 1)$, since t is central in $\mathrm{C}^\circ_{\mathbf{G}^*}(t)$ (see Proposition 8.26), thus allowing us to write $\xi = \hat{t}\chi_t$ with $\chi_t \in \mathcal{E}(\mathbf{G}(t)^F, 1)$.

Let $b = b_{\mathbf{G}^F}(\chi)$. Assume now that $\chi \in \mathcal{E}(\mathbf{G}^F, t)$ is an irreducible component of $\mathrm{R}^{\mathbf{G}}_{\mathbf{G}(t)\subseteq\mathbf{P}}(\hat{t}\chi_t)$ for some $\chi_t \in \mathcal{E}(\mathbf{G}(t), 1)$. One must check that $\mathrm{R}^{\mathbf{G}}_{\mathbf{G}(t)\subseteq\mathbf{P}}(\chi_t) \in \mathbb{Z}[\mathrm{Irr}(\mathbf{G}^F, b)]$. First $\mathrm{C}^\circ_{\mathbf{G}^*}(t)$ is $E_{q,\ell}$-split by its definition and Proposition 13.19. Then $\mathbf{G}(t)$ is also $E_{q,\ell}$-split by Proposition 13.9. So Theorem 21.7 implies the existence of the block idempotent $\mathrm{R}^{\mathbf{G}}_{\mathbf{G}(t)} b_{\mathbf{G}(t)^F}(\chi_t)$ of $\mathbf{G}^F$ and it suffices to check that it equals b. By Brauer's second Main Theorem (Theorem 5.8), it suffices to check that $d^1(\mathrm{R}^{\mathbf{G}}_{\mathbf{G}(t)\subseteq\mathbf{P}}\chi_t)$ is not in the kernel of the projection $\mathrm{CF}(\mathbf{G}^F, K) \to \mathrm{CF}(\mathbf{G}^F, K, b)$. Using commutation between d^1 and twisted induction (Proposition 9.6(iii), or the adjoint of the equality in Theorem 21.4), one has $d^1(\mathrm{R}^{\mathbf{G}}_{\mathbf{G}(t)\subseteq\mathbf{P}}\chi_t) = \mathrm{R}^{\mathbf{G}}_{\mathbf{G}(t)\subseteq\mathbf{P}}(d^1\chi_t) = \mathrm{R}^{\mathbf{G}}_{\mathbf{G}(t)\subseteq\mathbf{P}} d^1(\hat{t}\chi_t) = d^1(\mathrm{R}^{\mathbf{G}}_{\mathbf{G}(t)\subseteq\mathbf{P}}\hat{t}\chi_t)$. By Proposition 15.10, one has $\varepsilon_{\mathbf{G}}\varepsilon_{\mathbf{G}(t)}\mathrm{R}^{\mathbf{G}}_{\mathbf{G}(t)\subseteq\mathbf{P}}\hat{t}\chi_t = \chi + \chi_2 + \cdots + \chi_v$ with $\chi_i \in \mathrm{Irr}(\mathbf{G}^F)$. Then $\varepsilon_{\mathbf{G}}\varepsilon_{\mathbf{G}(t)} d^1(\mathrm{R}^{\mathbf{G}}_{\mathbf{G}(t)\subseteq\mathbf{P}}\hat{t}\chi_t) = d^1\chi + \sum_i d^1\chi_i$. Its projection on $\mathrm{CF}(\mathbf{G}^F, K, b)$ is the subsum where one retains i such that $\chi_i \in \mathrm{Irr}(\mathbf{G}^F, b)$, by Brauer's second Main Theorem. So this is not equal to 0 by evaluation at 1, whence our claim. □

21.3. A bad prime

Here is a case where ℓ is bad, but, as it turns out, $\mathbf{G}$ is not.

Theorem 21.14. *Let* $(\mathbf{G}, F)$ *be a connected reductive group defined over* $\mathbb{F}_q$ *with odd* q. *Assume* $\mathbf{G}$ *only involves types* **A, B, C, D**. *Then* $\bigcup_s \mathcal{E}(G, s)$, *where* s *ranges over 2-elements of* $(\mathbf{G}^*)^F$, *is the set of irreducible characters of the principal 2-block of* $\mathbf{G}^F$.

Proof. By Theorem 9.12(ii), it suffices to show that all unipotent characters of $\mathbf{G}^F$ are in $\mathrm{Irr}(\mathbf{G}^F, B_0)$ for B_0 the principal 2-block of $\mathbf{G}^F$.

We use induction on $\dim(\mathbf{G})$. By Theorem 17.1, one may assume that $\mathbf{G} = \mathbf{G}_{\mathrm{ad}} \neq 1$. Let us introduce the following lemma; its proof is given after the proof of the theorem.

Lemma 21.15. *Assume the same hypotheses as above with* $\mathbf{G} = \mathbf{G}_{\mathrm{ad}} \neq 1$. *Let* $\chi \in \mathcal{E}(\mathbf{G}^F, 1)$. *Then there exists an* F*-stable maximal torus* $\mathbf{T}$ *of* $\mathbf{G}$ *such that* $\mathbf{T}_2^F \neq 1$ *and* ${}^*\mathrm{R}_{\mathbf{T}}^{\mathbf{G}}(\chi) \neq 0$.

Let $\chi \in \mathcal{E}(G, 1)$ and let $\mathbf{T}$ be as in the lemma. Let $x \in \mathbf{T}_2^F \setminus \{1\}$, then $\mathrm{C}_{\mathbf{G}}^\circ(x)$ is an F-stable reductive group (see Proposition 13.13(i)) and it satisfies the hypothesis of the theorem (see the description of types involved in [Bour68] VI. Ex.4.4). One has $\langle d^x\chi, \mathrm{R}_{\mathbf{T}}^{\mathrm{C}_{\mathbf{G}}^\circ(x)} 1_{\mathbf{T}^F}\rangle_{\mathrm{C}_{\mathbf{G}}^\circ(x)^F} = \langle d^{x*}\mathrm{R}_{\mathbf{T}}^{\mathbf{G}}(\chi), 1_{\mathbf{T}^F}\rangle_{\mathbf{T}^F}$ by Theorem 21.4. But ${}^*\mathrm{R}_{\mathbf{T}}^{\mathbf{G}}(\chi)$ is a non-zero combination of unipotent characters of $\mathbf{T}^F$, i.e. a multiple of $1_{\mathbf{T}^F}$, so $\langle d^x\chi, \mathrm{R}_{\mathbf{T}}^{\mathrm{C}_{\mathbf{G}}^\circ(x)} 1_{\mathbf{T}^F}\rangle_{\mathrm{C}_{\mathbf{G}}^\circ(x)^F} = \langle {}^*\mathrm{R}_{\mathbf{T}}^{\mathbf{G}}(\chi), 1_{\mathbf{T}^F}\rangle_{\mathbf{T}^F} \langle d^x 1_{\mathbf{T}^F}, 1_{\mathbf{T}^F}\rangle_{\mathbf{T}^F} = |\mathbf{T}_2^F|^{-1}\langle {}^*\mathrm{R}_{\mathbf{T}}^{\mathbf{G}}(\chi), 1_{\mathbf{T}^F}\rangle_{\mathbf{T}^F} \neq 0$. This implies that $d^x\chi$ has a non-zero projection on the principal 2-block $B_0(x)$ of $\mathcal{O}\mathrm{C}_{\mathbf{G}}^\circ(x)^F$, since by the induction hypothesis $\mathrm{R}_{\mathbf{T}^F}^{\mathrm{C}_{\mathbf{G}}^\circ(x)^F} 1_{\mathbf{T}^F}$ is in $\mathrm{CF}(\mathrm{C}_{\mathbf{G}}^\circ(x)^F, K, B_0(x))$. Writing $b_0(x)$ and $b_0'(x)$ as the block idempotents of the principal 2-blocks of $\mathcal{O}\mathrm{C}_{\mathbf{G}}^\circ(x)^F$ and $\mathcal{O}\mathrm{C}_{\mathbf{G}}(x)^F$ respectively, we have $d^x\chi(b_0(x)) \neq 0$. Knowing that $\mathrm{C}_{\mathbf{G}}(x)^F/\mathrm{C}_{\mathbf{G}}^\circ(x)^F$ is a 2-group, that 2-block idempotents can be written with elements of odd order only (Proposition 21.1), and that $b_0(x)$ is clearly $\mathrm{C}_{\mathbf{G}^F}(x)$-fixed since the trivial character is fixed, we get $b_0(x) = b_0'(x)$. So we may write $d^x\chi(b_0'(x)) \neq 0$. Now, Brauer's second and third Main Theorems (Theorem 5.8 and Theorem 5.10) imply that χ is in $\mathrm{Irr}(\mathbf{G}^F, B_0)$ as claimed. □

Proof of Lemma 21.15. Since the assertion is about $\mathbf{G}_{\mathrm{ad}}$, one may assume $\mathbf{G} = \mathbf{G}_{\mathrm{ad}}$ is simple and $(\mathbf{G}, F)$ is of rational type $(\mathbf{X}, \epsilon q)$ (up to changing q into a power).

Assume the type is **A**. Let $\mathbf{T}_0$ be a diagonal torus of $\mathbf{G}$, let $W = \mathrm{N}_{\mathbf{G}}(\mathbf{T}_0)/\mathbf{T}_0 \cong \mathfrak{S}_{n+1}$ for some $n \geq 1$. Then the unipotent characters of $\mathbf{G}^F$ are of the form $\pm R_\phi$

for $\phi \in \mathrm{Irr}(W)$ (see [DiMi91] 15.8 or Example 21.6 above for linear groups). One has $\langle \mathrm{R}_{\mathbf{T}_0}^{\mathbf{G}} 1_{\mathbf{T}_0^F}, R_\phi \rangle_{\mathbf{G}^F} = \phi(1)$ by [DiMi91] 15.5. This is not equal to 0 for all ϕ. However, $|\mathbf{T}_0^F| = (q-\varepsilon)^n$ which is even when $n \geq 1$.

Assume the type of $\mathbf{G}$ is $\mathbf{B}$, $\mathbf{C}$ or $\mathbf{D}$. Let $\mathbf{B}_0 \supseteq \mathbf{T}_0$ be an F-stable Borel subgroup and maximal torus respectively. Assume that the permutation F_0 of $X(\mathbf{T}_0)$ induced by F is Id or an involution; this excludes rational type ${}^3\mathbf{D}_4$. Then all maximal F-stable tori $\mathbf{T}$ are such that $\mathbf{T}^F$ is even, i.e., for every $w \in W := \mathrm{N}_{\mathbf{G}}(\mathbf{T}_0)/\mathbf{T}_0$, $\det(w - qF_0)$ is an even integer. To see this, note that in the presentation of root systems given in Example 2.1, w is represented by a matrix permutation with signs, relative to a basis $(e_1, \ldots, e_n)$ of $X(\mathbf{T}_0) \otimes \mathbb{R}$, while F_0 is either Id, $-w_0$ (where w_0 is the element of maximal length in W) or the reflection of vector e_1 (in the case of rational type ${}^2\mathbf{D}_{2n}$), i.e. in all cases a permutation matrix with signs. Then the characteristic polynomial of wF_0^{-1} is of type $(x^{a_1} - \varepsilon_1) \ldots (x^{a_r} - \varepsilon_r)$ with the ε_i's being signs. The value at q is even since $a_1 + \cdots + a_r = n \geq 1$. But we know that there is some F-stable maximal torus $\mathbf{T}$ such that $\langle \mathrm{R}_{\mathbf{T}}^{\mathbf{G}}(1_{\mathbf{T}^F}), \chi \rangle_{\mathbf{G}^F} \neq 0$ by the definition of unipotent characters.

The case of ${}^3\mathbf{D}_4$ is more difficult since there are cuspidal unipotent characters, and certain F-stable maximal tori $\mathbf{T}$ satisfy $\mathbf{T}_2^F = \{1\}$. The proof of the lemma must use the explicit determination of the scalar products $\langle \mathrm{R}_{\mathbf{T}}^{\mathbf{G}}(1_{\mathbf{T}^F}), \chi \rangle_{\mathbf{G}^F}$; see [Lu84] 4.23. In Exercise 7 below, we give a more elementary approach. □

Exercises

1. Let $G \lhd G'$ be finite groups. Let $(\mathcal{O}, K, k)$ be an ℓ-modular splitting system for G' and its subgroups. Let b, b' be block idempotents of $\mathcal{O}G$ and $\mathcal{O}G'$, respectively. One says that b' *covers* b if and only if $bb' \neq 0$. Show that this is equivalent to
   (i) $b'.\sum {}^g b = b'$, where the sum is over a representative system of G' mod. the stabilizer of b,
   (ii) for some $\chi' \in \mathrm{Irr}(G', b')$, $\mathrm{Res}_G^{G'} \chi'$ has a projection $\neq 0$ on $\mathrm{CF}(G, K, b)$,
   (iii) for all $\chi' \in \mathrm{Irr}(G', b')$, $\mathrm{Res}_G^{G'} \chi'$ has a projection $\neq 0$ on $\mathrm{CF}(G, K, b)$.

   If moreover $\mathrm{Br}_Q(\overline{b'}) \neq 0$ for some ℓ-subgroup $Q \subseteq G$, then $\mathrm{Br}_{{}^gQ}(\overline{b}) \neq 0$ for some $g \in G'$.
2. Assume the hypotheses of Proposition 21.3. Show that there is a unique block idempotent $\widehat{b_V}$ of $\mathcal{O}\mathrm{C}_{\mathbf{G}^F}(V)$ such that $b_V.\widehat{b_V} \neq 0$. One has $\widehat{b_V} = \sum_{t \in \mathrm{C}_{\mathbf{G}^F}(V)/I(b_V)} {}^t b_V$. Show that $(V, \widehat{b_V}) \subseteq (U, \widehat{b_U})$ (inclusion of ordinary subpairs) if and only if there is $c \in \mathrm{C}_{\mathbf{G}^F}(V)$ such that $(V, {}^c b_V) \lhd (U, b_U)$

(show first that if S is an ℓ-subgroup of $\mathbf{G}^F$ normalizing (V, b_V) and such that $\mathrm{Br}_S(b_V) \neq 0$, then S normalizes $\widehat{b_V}$ and $\mathrm{Br}_S(\widehat{b_V}) \neq 0$).

3. Build a generalization of "connected subpairs" (U, b_U) where the ℓ-subgroups $U \subseteq \mathbf{G}^F$ are not necessarily commutative. Prove an analogue of Theorem 5.3.
4. Show Theorem 21.7 with arbitrary ℓ-blocks instead of unipotent ℓ-blocks (replace $\mathcal{E}(\mathbf{L}^F, 1)$ with $\mathcal{E}(\mathbf{L}^F, \ell')$; see Definition 9.4).
5. Let G be a finite group, ℓ a prime, $(\mathcal{O}, K, k)$ be an ℓ-modular splitting system for G. If b is a sum of block idempotents of kG, denote by P_b^G the projection $\mathrm{CF}(G, K) \to \mathrm{CF}(G, K, b^*)$ where $b^* \in \mathcal{O}G$ is the block idempotent such that $\overline{b^*} = b$ (see §5.1).
   (a) Show that, if $x \in G_\ell$, then $d^x \circ P_b^G = P_{\mathrm{Br}_x(b)}^{\mathrm{C}_G(x)} \circ d^x$ (reduce to $\mathrm{Irr}(G, b)$ and use Brauer's second Main Theorem).
   (b) Conversely, if b_x is a sum of block idempotents of $k\mathrm{C}_G(x)$ and $P_{b_x}^{\mathrm{C}_G(x)}(\delta_1) = d^x \circ P_b^G(\delta_x)$ (where δ_y denotes the characteristic function of the G-conjugacy class of $y \in G$), show that $b_x = \mathrm{Br}_x(b)$ (compare $P_{b_x}^{\mathrm{C}_G(x)}$ and $P_{\mathrm{Br}_x(b)}^{\mathrm{C}_G(x)}$ at δ_1).
6. Use Exercise 5 to show that, under the hypotheses of §9.2, and if $x \in \mathbf{G}_\ell^F$, then $\mathrm{Br}_x\big(\overline{b}_\ell(\mathbf{G}^F, 1)\big) = \overline{b}_\ell(\mathrm{C}^\circ_{\mathbf{G}}(x)^F, 1)$ (note that δ_x is a uniform function, see Proposition 9.6(i)).

   What about $\mathrm{Br}_x(\overline{b}_\ell(\mathbf{G}^F, s))$ when $s \in (\mathbf{G}^*)^F$ is a semi-simple ℓ'-element (see §9.2)?
7. Use the above to show that Theorem 21.14 is equivalent to checking that, when $\mathbf{G} = \mathbf{G}_{\mathrm{ad}} \neq 1$, no $\chi \in \mathcal{E}(\mathbf{G}^F, 1)$ is of 2-defect zero.

   Show that if $\chi \in \mathcal{E}(\mathbf{G}_{\mathrm{ad}}^F, 1)$ is of 2-defect zero then it is cuspidal.

   Deduce Theorem 21.14 in the case of ${}^3\mathbf{D}_4$ from an inspection of degrees of cuspidal characters (see [Lu84] p. 373, [Cart85] §13.9). This applies also to the types mentioned in Theorem 21.14.
8. Assume the type of $\mathbf{G}$ is $\mathbf{B}$, $\mathbf{C}$ or $\mathbf{D}$ and q is odd. Show that the 2-blocks of $\mathbf{G}^F$ are the $B_2(\mathbf{G}^F, s)$ for $s \in \mathbf{G}^{*F}$, semi-simple of odd order. (Use Bonnafé–Rouquier's theorem (Theorem 10.1) or Exercise 9.5 about perfect isometries.)

Notes

We have used mainly [CaEn93] and [CaEn99a,b].

The property of $R_{\mathbf{L}}^{\mathbf{G}}$ relative to blocks (Theorem 21.7) is probably the shadow of some property of the complexes of $\mathbf{G}^F$-$\mathbf{L}^F$-bimodules defined by étale cohomology of varieties $\mathbf{Y}_{\mathbf{V}}$.

The ℓ-blocks of $\mathbf{G}^F$ for bad primes ℓ have been studied by Enguehard in [En00]. In particular, Theorem 21.7 still holds for ℓ bad. In fact, ℓ-blocks of $\mathbf{G}^F$ are in bijection with $\mathbf{G}^F$-classes of e-cuspidal pairs $(\mathbf{L}, \zeta)$ such that $\zeta(1)_\ell = |\mathbf{L}^F : \mathrm{Z}(\mathbf{L}^F)|_\ell$ (central defect, see Remark 5.6).

The result in Exercise 6 comes from [BrMi89].

22

Unipotent blocks and generalized Harish-Chandra theory

We take $\mathbf{G}$, F, q, ℓ, $(\mathcal{O}, K, k)$ as in the preceding chapter. Let e be the multiplicative order of q mod. ℓ. From Theorem 21.7 we know that, for any e-split Levi subgroup $\mathbf{L}$ of $\mathbf{G}$ and any $\zeta \in \mathcal{E}(\mathbf{L}^F, 1)$, all irreducible characters occurring in $\mathrm{R}_{\mathbf{L}}^{\mathbf{G}}\zeta$ correspond to a single ℓ-block of $\mathbf{G}^F$. So the partition of $\mathcal{E}(\mathbf{G}^F, 1)$ into ℓ-blocks seems to parallel a Harish-Chandra theory where twisted induction of characters of e-split Levi subgroups replaces the traditional Harish-Chandra induction (i.e. the case when $e = 1$). Recall the corresponding notion of e-cuspidality (Definition 21.5).

In the present chapter, we show that the twisted induction $\mathrm{R}_{\mathbf{L}}^{\mathbf{G}}$ for ℓ-blocks induces a bijection

$$(\mathbf{L}, \zeta) \mapsto \mathrm{R}_{\mathbf{L}}^{\mathbf{G}} B_{\mathbf{L}^F}(\zeta)$$

between unipotent blocks of $\mathbf{G}^F$ (see Definition 9.13) and $\mathbf{G}^F$-conjugacy classes of pairs $(\mathbf{L}, \zeta)$ where $\mathbf{L}$ is an e-split Levi subgroup of $\mathbf{G}$ and $\zeta \in \mathcal{E}(\mathbf{L}^F, 1)$ is e-cuspidal ([CaEn94]). We develop the case of $\mathrm{GL}_n(\mathbb{F}_q)$, thus giving the partition of its unipotent characters induced by ℓ-blocks (see Example 21.6 and Example 22.10).

The proof we give follows the local methods sketched in §5.2 and §5.3.

A first step consists in building a Sylow ℓ-subgroup of the finite reductive group $\mathrm{C}^\circ_{\mathbf{G}}([\mathbf{L}, \mathbf{L}])^F$. Sylow ℓ-subgroups of groups $\mathbf{G}^F$ (for ℓ a good prime) are made of two "layers" (see Exercise 6). First, one takes an F-stable maximal torus $\mathbf{T}$ such that its polynomial order $P_{\mathbf{T},F}$ contains the biggest power of ϕ_e dividing that of $P_{\mathbf{G},F}$. Then one takes a Sylow ℓ-subgroup W of $\mathrm{N}_{\mathbf{G}}(\mathbf{T})^F/\mathbf{T}^F = (\mathrm{N}_{\mathbf{G}}(\mathbf{T})/\mathbf{T})^F$. The Sylow ℓ-subgroup of $\mathbf{G}^F$ is then an extension of W by $\mathbf{T}_\ell^F$.

For subpairs, this allows us to build a maximal subpair (D, b_D) containing $(\{1\}, \mathrm{R}_{\mathbf{L}}^{\mathbf{G}} b_{\mathbf{L}^F}(\zeta))$ where D is a Sylow ℓ-subgroup of $\mathrm{C}^\circ_{\mathbf{G}}([\mathbf{L}, \mathbf{L}])^F$. This needs some technicalities (in particular a review of characteristic polynomials of elements of Weyl groups of exceptional types) due to the rather inaccurate information we have about e-cuspidal pairs $(\mathbf{L}, \zeta)$.

22.1. Local subgroups in finite reductive groups, ℓ-elements and tori

Concerning the values of cyclotomic polynomials on integers, recall that if ℓ is a prime and $q, d \geq 1$ are integers then ℓ divides $\phi_d(q)$ if and only if $q_\ell = 1$ and $d_{\ell'}$ is the order of q mod. ℓ.

Let $(\mathbf{G}, F)$ be a connected reductive group defined over $\mathbb{F}_q$.

The following technical condition will be useful. It defines a subset of the set $\Pi(\mathbf{G}, F)$ of Definition 17.5, essentially by removing bad primes. Let ℓ be a prime.

Condition 22.1. We consider the conjunction of the following conditions

(a) ℓ is good for $\mathbf{G}$ (see Definition 13.10),

(b) ℓ does not divide $2q.|\mathrm{Z}(\mathbf{G}_{\mathrm{sc}})^F|$,

(c) $\ell \neq 3$ if the rational type of $(\mathbf{G}, F)$ includes type ${}^3\mathbf{D}_4$.

Note that this condition only depends on the rational type of $(\mathbf{G}, F)$ and can be read off from Table 13.11.

The following will be needed to show that e-cuspidal unipotent characters define ℓ-blocks with central defect (at least for good ℓ's).

Theorem 22.2. *Let $(\mathbf{G}, F)$ be a connected reductive $\mathbf{F}$-group defined over $\mathbb{F}_q$. Let ℓ be a prime satisfying Condition 22.1. Let $E_{q,\ell} = \{e, e\ell, \ldots, e\ell^i, \ldots\}$ be the set of integers d such that ℓ divides $\phi_d(q)$ (see Theorem 21.7).*

Any proper $E_{q,\ell}$-split Levi subgroup of $\mathbf{G}$ is included in a proper e-split Levi subgroup of $\mathbf{G}$.

Lemma 22.3. *(i) The g.c.d. of the polynomial orders of the F-stable maximal tori of $\mathbf{G}$ is the polynomial order of $\mathrm{Z}^\circ(\mathbf{G})$.*

(ii) Let $d \geq 1$ be an integer and ℓ be a prime satisfying Condition 22.1. If $\mathbf{T}$ is an F-stable maximal torus of $\mathbf{G}$ such that $\mathbf{T}_{\phi_{d\ell}} \not\subseteq \mathrm{Z}(\mathbf{G})$, then $\mathbf{T}_{\phi_d} \not\subseteq \mathrm{Z}(\mathbf{G})$.

Proof of Lemma 22.3. The various $\mathbf{G}^F$-conjugacy classes of F-stable maximal tori in $(\mathbf{G}, F)$ are parametrized by F-conjugacy classes in the Weyl group $W := W(\mathbf{G}, \mathbf{T}_0)$ where $\mathbf{T}_0$ is some F-stable torus (see §8.2). If $\mathbf{T}$ is obtained from $\mathbf{T}_0$ by twisting with $w \in W$, and if F_0 is defined by $F = qF_0$ where F and F_0 act on $Y(\mathbf{T}_0) \otimes \mathbb{R}$, then $P_{(\mathbf{T},F)}$ is the characteristic polynomial of ${F_0}^{-1}w$ on $Y(\mathbf{T}_0) \otimes \mathbb{R}$ (see §13.1).

For every F-stable torus $\mathbf{S}$ of $\mathbf{G}$ which contains $\mathrm{Z}(\mathbf{G})$, one has $P_{(\mathbf{S}/\mathrm{Z}(\mathbf{G}),F)} = P_{(\mathbf{S},F)}/P_{(\mathrm{Z}^\circ(\mathbf{G}),F)}$. So, in view of the statements we want to prove, we may assume that $\mathbf{G} = \mathbf{G}_{\mathrm{ad}}$. Then it is a direct product of its components, so we may also assume that $\mathbf{G}$ is rationally irreducible.

(i) The substitution $(x \mapsto x^m)$ in a set of coprime polynomials cannot produce common divisors. So, by Proposition 13.6, we assume $\mathbf{G}$ is irreducible. If F_0 above is the identity, the polynomial orders of $\mathbf{T}_0$ and of a torus obtained from $\mathbf{T}_0$ by twisting by a Coxeter element of $W(\mathbf{G}, \mathbf{T}_0)$ are coprime (Coxeter elements have no fixed points in the standard representation; see, for instance, [Cart72b] 10.5.6). If $(\mathbf{G}, F)$ is of type $({}^2\mathbf{A}_n, q) = (\mathbf{A}_n, -q)$, $({}^2\mathbf{D}_{2k+1}, q) = (\mathbf{D}_{2k+1}, -q)$ or $({}^2\mathbf{E}_6, q) = (\mathbf{E}_6, -q)$ (see §13.1), the property is deduced from the case $F_0 = 1$ by the substitution $(x \mapsto -x)$; see Example 13.4(iv). If $(\mathbf{G}, F)$ is of type $({}^2\mathbf{D}_{2k}, q)$, the Coxeter tori have polynomial order $(x^{2k} + 1)$, prime to the polynomial order $(x + 1)(x - 1)^{2k-1}$ of the quasi-split tori. If $(\mathbf{G}, F)$ is of type $({}^3\mathbf{D}_4, q)$, the quasi-split tori have polynomial order $(x - 1)(x^3 - 1) = \phi_1{}^2\phi_3$, while the maximal torus of type a product of two non-commuting simple reflections is of polynomial order $x^4 - x^2 + 1 = \phi_{12}$.

(ii) The point is to check that, if $w \in W(\mathbf{G}, \mathbf{T}_0)$ and the characteristic polynomial of wF_0 vanishes at a primitive $d\ell$th root of unity, it vanishes also at a primitive dth root. If $(\mathbf{G}, F)$ is of type $({}^2\mathbf{A}_n, q) = (\mathbf{A}_n, -q)$, $({}^2\mathbf{D}_{2k+1}, q) = (\mathbf{D}_{2k+1}, -q)$ or $({}^2\mathbf{E}_6, q) = (\mathbf{E}_6, -q)$ (i.e. $F_0 = -w_0$), the property is deduced from the case $F_0 = 1$ since $\ell \neq 2$. For a $\mathbf{G}$ of type $(\mathbf{A}_n)^m$ and $F_0 = \mathrm{Id}$, the characteristic polynomial of w is of type $(y - 1)^{-1}(y^{m_1} - 1)\ldots(y^{m_t} - 1)$ for $y = x^m$ where m_i's are integers greater than or equal to 1 of sum $n + 1$. Then the claim is due to the fact that ℓ satisfies Condition 22.1(b) (see Table 13.11). For $(\mathbf{G}, F)$ of type a sum of $\mathbf{B}$'s, $\mathbf{C}$'s, $\mathbf{D}$'s, and ${}^2\mathbf{D}$'s one gets polynomials $(y^{m_1} - \varepsilon_1)\ldots(y^{m_t} - \varepsilon_t)$ where $y = x^m$ and ε_i's are signs (see the proof of Lemma 21.15). Then our claim comes from $\ell \neq 2$. There remain the rational types $\mathbf{E}_6$ for $\ell = 5$, $\mathbf{E}_7$ for $\ell = 5, 7$, and $\mathbf{E}_8$ for $\ell = 7$ (other ℓ's either are bad or do not divide the order of $W(\mathbf{G}, \mathbf{T}_0) \rtimes \langle F_0 \rangle$). One may also note that d has to be 1 or 2 (see, for instance, Exercise 13.3). A list of elements is given in a paper by Carter (see [Cart72a]) and it is easily checked that our claim holds. Another possibility is to use a computer and GAP program to test all relevant elements, i.e. elements of order 5 or 7 of the Coxeter group of type $\mathbf{E}_8$ and their centralizers. □

Proof of Theorem 22.2. Let $\mathbf{H}$ be a proper $E_{q,\ell}$-split Levi subgroup of $\mathbf{G}$. Let $d \in E_{q,\ell}$.

Suppose $\mathrm{Z}^\circ(\mathbf{H})_{\phi_{d\ell}} \not\subseteq \mathrm{Z}(\mathbf{G})$ and $\mathrm{Z}^\circ(\mathbf{H})_{\phi_d} \subseteq \mathrm{Z}(\mathbf{G})$. Applying Lemma 22.3(i) to $(\mathbf{H}, F)$, one finds an F-stable maximal torus $\mathbf{T} \subseteq \mathbf{H}$ such that $\mathbf{T}_{\phi_d} \subseteq \mathrm{Z}^\circ(\mathbf{H})$

and therefore $\mathbf{T}_{\phi_d} \subseteq \mathrm{Z}^\circ(\mathbf{G})$. Since $\mathbf{T}$ is an F-stable maximal torus of $\mathbf{G}$, Lemma 22.3(ii) yields $\mathbf{T}_{\phi_{d\ell}} \subseteq \mathrm{Z}(\mathbf{G})$. But $\mathrm{Z}^\circ(\mathbf{H}) \subseteq \mathbf{T}$ and this contradicts $\mathrm{Z}^\circ(\mathbf{H})_{\phi_{d\ell}} \not\subseteq \mathrm{Z}(\mathbf{G})$.

So we find that, if $\mathrm{Z}^\circ(\mathbf{H})_{\phi_{d\ell}} \not\subseteq \mathrm{Z}(\mathbf{G})$, then $\mathrm{Z}^\circ(\mathbf{H})_{\phi_d} \not\subseteq \mathrm{Z}(\mathbf{G})$.

The fact that $\mathbf{H}$ is a proper $E_{q,\ell}$-split Levi subgroup implies that $\mathrm{Z}^\circ(\mathbf{H})_{\phi_{e\ell^a}} \not\subseteq \mathrm{Z}(\mathbf{G})$ for some $a \geq 0$. Applying the above a times, one gets $\mathrm{Z}^\circ(\mathbf{H})_{\phi_e} \not\subseteq \mathrm{Z}(\mathbf{G})$. Then $\mathbf{H} \subseteq \mathrm{C}_{\mathbf{G}}(\mathrm{Z}^\circ(\mathbf{H})_{\phi_e})$, the latter being a proper e-split Levi subgroup. □

Let us show how to isolate the F-stable components $\mathbf{G}'_i$ of $[\mathbf{G},\mathbf{G}]$ whose rational type ensures that $\mathrm{Z}(\mathbf{G}'_i)^F$ is an ℓ'-group (see Table 13.11).

Definition 22.4. *Let $(\mathbf{G}, F)$ be a connected reductive $\mathbf{F}$-group defined over $\mathbb{F}_q$. Let us write $[\mathbf{G},\mathbf{G}] = \mathbf{G}'_1 \ldots \mathbf{G}'_v$ for the finer decomposition as a central product of F-stable non-commutative reductive subgroups (see §8.1).*

Let ℓ be a prime not dividing q and good for $\mathbf{G}$. Define $\mathbf{G_a}$ as the central product of the subgroups $\mathrm{Z}(\mathbf{G})\mathbf{G}'_i$ whose rational type (see §8.1) is of the form $(\mathrm{A}_n, \varepsilon q^m)$ with ℓ dividing $q^m - \varepsilon$. Let $\mathbf{G_b}$ be the central product of the components $\mathbf{G}'_i$ of $[\mathbf{G},\mathbf{G}]$ not included in $\mathbf{G_a}$. Then $\mathbf{G} = \mathbf{G_a}.\mathbf{G_b}$ (central product).

Proposition 22.5. *(i) $\mathrm{Z}(\mathbf{G_b})^F$ and $\mathbf{G}^F/\mathbf{G}_\mathbf{a}^F.\mathbf{G}_\mathbf{b}^F$ are commutative ℓ'-groups.*

(ii) If Y is an ℓ-subgroup of $\mathbf{G}^F$ such that $\mathrm{Z}(\mathrm{C}_{\mathbf{G}^F}(Y))_\ell \subseteq \mathrm{Z}(\mathbf{G})\mathbf{G_a}$, then $Y \subseteq \mathbf{G_a}$.

Proof. (i) First the surjection $\mathbf{G}_{\mathrm{sc}} \to [\mathbf{G},\mathbf{G}]$ (see §8.1) induces a surjection $\mathrm{Z}((\mathbf{G_b})_{\mathrm{sc}}) \to \mathrm{Z}(\mathbf{G_b})$ compatible with the action of F. By Table 13.11, $\mathrm{Z}((\mathbf{G_b})_{\mathrm{sc}})^F$ is ℓ'. Now Proposition 8.1(ii) (in the form "$|(G/Z)^f|$ divides $|G^f|$") implies that $\mathrm{Z}(\mathbf{G_b})^F$ is of order ℓ'. By Proposition 8.1(i), $\mathbf{G}^F/\mathbf{G}_\mathbf{a}^F.\mathbf{G}_\mathbf{b}^F$ is isomorphic to a section of $\mathrm{Z}(\mathbf{G_b})$ on which F acts trivially, so it is an ℓ'-group.

(ii) By (i) above, $Y \subseteq \mathbf{G}_\mathbf{a}^F.\mathbf{G}_\mathbf{b}^F$. Let $Y_\mathbf{a} = \{y \in (\mathbf{G_a})_\ell^F \mid y\mathbf{G}_\mathbf{b}^F \cap Y \neq \emptyset\}$ and $Y_\mathbf{b}$ be similarly defined. One has $\mathrm{C}_{\mathbf{G}^F}(Y) = \mathrm{C}_{\mathbf{G_a}}(Y_\mathbf{a})^F.\mathrm{C}_{\mathbf{G_b}}(Y_\mathbf{b})^F$ and $\mathrm{Z}(Y_\mathbf{b})$ is clearly central in it. The hypothesis then implies that $\mathrm{Z}(Y_\mathbf{b}) \subseteq \mathrm{Z}(\mathbf{G_b})^F$, but the latter is an ℓ'-group by (i). So $\mathrm{Z}(Y_\mathbf{b}) = \{1\}$, hence $Y_\mathbf{b} = \{1\}$ and $Y \subseteq Y_\mathbf{a} \subseteq \mathbf{G_a}$.

Here is a proposition about diagonal tori in $\mathbf{G} = \mathbf{G_a}$. □

Proposition 22.6. *Let $(\mathbf{G}, F)$ be a connected reductive $\mathbf{F}$-group over $\mathbb{F}_q$, let ℓ be an odd prime not dividing q, let e be the order of q mod. ℓ. Assume $\mathbf{G} = \mathbf{G_a}$ and let $\mathbf{T}$ be a diagonal torus of $\mathbf{G}$. Then*

(i) $\mathbf{T} = \mathrm{C}_{\mathbf{G}}(\mathbf{T}_{\phi_e}) = \mathrm{C}^\circ_{\mathbf{G}}(\mathbf{T}_\ell^F)$,

(ii) $\mathbf{T}^F = \mathrm{C}_{\mathbf{G}^F}(\mathbf{T}_\ell^F)$,

(iii) $|\mathbf{G}^F : \mathrm{N}_{\mathbf{G}}(\mathbf{T})^F|$ is prime to ℓ.

Proof. Let $\times_\omega(\mathbf{A}_{n_\omega}, \varepsilon_\omega q^{m_\omega})$ be the rational type of $(\mathbf{G}, F)$ (see §8.1). For the diagonal torus, one has $P_{(\mathbf{T},F)} = P_{(\mathrm{Z}^\circ(\mathbf{G}),F)}\Pi_\omega(x^{m_\omega} - \varepsilon_\omega)^{n_\omega}$. For each ω, ℓ divides $q^{m_\omega} - \varepsilon_\omega$, so ϕ_e divides $x^{m_\omega} - \varepsilon_\omega$.

(i) Let us use induction on dim $\mathbf{G}/\mathbf{T}$. If $\mathbf{G}$ is a torus, there is nothing to prove. Otherwise, when $q^m - \varepsilon \equiv 0$ (mod. ℓ) and $n \geq 1$, then ℓ divides $(q^m - \varepsilon)^n/(q^m - \varepsilon, n+1)$ (here we use the fact that $\ell \neq 2$). This implies $|\mathbf{T}_\ell^F| > |\mathrm{Z}(\mathbf{G})_\ell^F|$ provided $\mathbf{G}$ is not a torus, so $\mathbf{T}_\ell^F$ is not central in $\mathbf{G}$. Then $\mathrm{C}^\circ_\mathbf{G}(\mathbf{T}_\ell^F)$ is a proper Levi subgroup (Proposition 13.16(ii)) with diagonal torus $\mathbf{T}$ and one obtains $\mathbf{T} = \mathrm{C}^\circ_\mathbf{G}(\mathbf{T}_\ell^F)$ by induction. Similarly, $\mathbf{T}_{\phi_e} \not\subseteq \mathrm{Z}^\circ(\mathbf{G})$, and one gets $\mathbf{T} = \mathrm{C}_\mathbf{G}(\mathbf{T}_{\phi_e})$ by considering $\mathrm{C}_\mathbf{G}(\mathbf{T}_{\phi_e})$ and applying induction.

(ii) We prove that $\mathrm{C}_{\mathbf{G}^F}(\mathbf{T}_\ell^F) \subset \mathbf{T}$, again by induction on the dimension of $\mathbf{G}/\mathbf{T}$.

Proposition 8.1 and Lang's theorem imply that $\mathbf{G}^F = \mathbf{T}^F[\mathbf{G},\mathbf{G}]^F$. As $\mathrm{C}_{\mathbf{G}^F}(\mathbf{T}_\ell^F) \subset \mathrm{C}((\mathbf{T} \cap [\mathbf{G},\mathbf{G}])_\ell^F)$, one may assume that $\mathbf{G} = [\mathbf{G},\mathbf{G}]$. If $\mathbf{G} = \mathbf{G}_1\mathbf{G}_2$ is a central product of two F-stable components, then $\mathbf{T}_i := \mathbf{T} \cap \mathbf{G}_i$ is a diagonal torus in $\mathbf{G}_i$ $(i = 1, 2)$, one has $\mathbf{G}^F = \mathbf{T}^F\mathbf{G}_1^F\mathbf{G}_2^F$ and $\mathrm{C}_{\mathbf{G}^F}(\mathbf{T}_\ell^F) \subset \mathrm{C}_{\mathbf{G}^F}(\mathbf{T}_1^F) \cap \mathrm{C}_{\mathbf{G}^F}(\mathbf{T}_2^F)$. As $\mathrm{C}_{\mathbf{G}^F}(\mathbf{T}_i^F) = \mathbf{T}^F\mathrm{C}_{\mathbf{G}_i^F}(\mathbf{T}_i^F)\mathbf{G}_j^F$ $(\{i,j\} = \{1,2\})$, one may assume that $\mathbf{G} = [\mathbf{G},\mathbf{G}]$ is of rational type a single $(\mathbf{A}_n, \epsilon q^m)$. If $[\mathbf{G},\mathbf{G}]$ is simply connected, then $\mathrm{C}_\mathbf{G}(\mathbf{T}_\ell^F)$ is connected (Theorem 13.14) and (ii) reduces to (i).

Let $g \in \mathrm{C}_{\mathbf{G}^F}(\mathbf{T}_\ell^F)$; one has to check that $g \in \mathbf{T}$. Let $\tau\colon \mathbf{G}_{\mathrm{sc}} \to [\mathbf{G},\mathbf{G}]$ be a simply connected covering, whose kernel K is central in $\mathbf{G}_{\mathrm{sc}}$. Clearly $|\mathrm{Z}(\mathbf{G})^F|$ divides $|\mathrm{Z}(\mathbf{G}_{\mathrm{sc}})^F|$ (see the proof of Proposition 22.5(i)). Let $S = \tau^{-1}(\mathbf{T}_\ell^F)$, let $g' \in \mathbf{G}_{\mathrm{sc}}$ be such that $\tau(g') = g$. One has $g'^{-1}F(g') \in K$ and, for any $s \in S$, $s^{-1}F(s) \in K$, $[g', s] \in K$, so that $[g', s] \in K^F$. The map $(s \mapsto [g', s])$ defines a morphism from S to K^F; let Z be its kernel. By definition, $g' \in \mathrm{C}_{\mathbf{G}_{\mathrm{sc}}}(Z)$. Furthermore, $\mathrm{C}_{\mathbf{G}_{\mathrm{sc}}}(Z)$ is connected by Theorem 13.14. As $\tau(\mathrm{C}_{\mathbf{G}_{\mathrm{sc}}}(Z))$ is a connected subgroup with finite index in $\mathrm{C}_\mathbf{G}(\tau(Z))$, so one has $\tau(\mathrm{C}_{\mathbf{G}_{\mathrm{sc}}}(Z)) = \mathrm{C}^\circ_\mathbf{G}(\tau(Z))$, so that $g \in \mathrm{C}^\circ_\mathbf{G}(\tau(Z))^F$.

Assume first that $\tau(Z)$ is not central in $\mathbf{G}$. Then $\mathbf{H} := \mathrm{C}^\circ_\mathbf{G}(\tau(Z))$ is a proper Levi subgroup of $\mathbf{G}$. In this group $\mathbf{T}$ is a diagonal torus and the induction hypothesis applies, so one has $\mathbf{T} \supseteq \mathrm{C}_{\mathbf{H}^F}(\mathbf{T}_\ell^F)$, hence $g \in \mathbf{T}$ as claimed.

One has $|Z| \geq |S|/|K^F|$, hence $|\tau(Z)|_\ell = |Z/K|_\ell \geq |\mathbf{T}_\ell^F|/|K_\ell^F|$. If $\tau(Z)$ is central, then $|\mathbf{T}_\ell^F| \leq |\mathrm{Z}(\mathbf{G})_\ell^F|.|K_\ell^F|$, hence $(q^m - \epsilon)_\ell^n \leq |(\mathrm{Z}(\mathbf{G})/\mathrm{Z}^\circ(\mathbf{G}))_\ell^F|.|K_\ell^F| \leq |\mathrm{Z}(\mathbf{G}_{\mathrm{sc}})_\ell^F|^2$. Given two integers $n, r \geq 1$, it is easy to check that either ℓ divides $\ell^{rn}/(\ell^r, n+1)^2$, or $\ell = n + 1 = 3$. So we are reduced to $\mathbf{G}_{\mathrm{sc}} = \mathrm{SL}_3(\mathbf{F})$, where $\mathrm{Z}(\mathbf{G}_{\mathrm{sc}})$ is cyclic of order 3 and therefore $\mathbf{G}$ is either $\mathbf{G}_{\mathrm{sc}} = \mathrm{SL}_3(\mathbf{F})$ itself (already done) or $\mathrm{PGL}_3(\mathbf{F})$. Then our claim is easily checked in the corresponding finite group $\mathrm{PGL}_3(\mathbb{F}_q)$ or $\mathrm{PU}_3(\mathbb{F}_q)$.

(iii) Since $\mathbf{G}/\mathbf{T}$ and $\mathrm{N}_{\mathbf{G}}(\mathbf{T})/\mathbf{T}$ are independent of $\mathrm{Z}(\mathbf{G})$, it suffices to check that $|(\mathbf{G}/\mathbf{T})^F|_\ell = |(\mathrm{N}_{\mathbf{G}}(\mathbf{T})/\mathbf{T})^F|_\ell$ for $\mathbf{G}_{\mathrm{ad}}$ or for a product of linear groups. Then the cardinalities are easy to compute and the key property is that if Q is an integer $\equiv 1$ mod. ℓ, then $(Q^m - 1)_\ell = m_\ell(Q-1)_\ell$ for any $m \geq 1$. □

Proposition 22.7. *Let $(\mathbf{G}, F)$ be a connected reductive $\mathbf{F}$-group defined over $\mathbb{F}_q$. Let ℓ be an odd prime not dividing q. Let e be the order of q mod. ℓ. If $\mathbf{S}$ is a maximal ϕ_e-subgroup of $\mathbf{G}$, then $\mathrm{N}_{\mathbf{G}}(\mathbf{S})^F$ contains a Sylow ℓ-subgroup of $\mathbf{G}^F$.*

Proof. We have $\mathbf{S} = (\mathbf{S} \cap \mathbf{G_a}).(\mathbf{S} \cap \mathbf{G_b})$ (see Definition 22.4) with multiplicativity of polynomial orders (Proposition 13.2(i)). By Proposition 22.5(i), it suffices to check that $|\mathbf{G}_i^F : \mathrm{N}_{\mathbf{G}_i}(\mathbf{S} \cap \mathbf{G}_i)^F|$ is prime to ℓ for $\mathbf{G}_i = \mathbf{G_a}$ and $\mathbf{G_b}$.

For $\mathbf{G}_i = \mathbf{G_a}$, note that Proposition 22.6(i) implies that, for $\mathbf{T}$ a diagonal torus of $\mathbf{G_a}$, $\mathbf{T}_{\phi_e}$ is a maximal ϕ_e-subgroup. By conjugacy of maximal ϕ_e-subgroups (see Theorem 13.18), one may assume $\mathbf{S} = \mathbf{T}_{\phi_e}$. One then has $\mathrm{N}_{\mathbf{G}^F}(\mathbf{S}) = \mathrm{N}_{\mathbf{G}^F}(\mathbf{T})$, and our claim follows from Proposition 22.6(iii).

For $\mathbf{G}_i = \mathbf{G_b}$, ℓ does not divide $|(\mathrm{Z}(\mathbf{G})/\mathrm{Z}^\circ(\mathbf{G}))^F|$ (Proposition 22.5(i)), so $\mathrm{C}_{\mathbf{G}}(\mathbf{S})^F = \mathrm{C}_{\mathbf{G}^F}(\mathbf{S}_\ell^F)$ by Lemma 13.17(ii). Denote $\mathbf{L} := \mathrm{C}_{\mathbf{G}}(\mathbf{S})$. Let D be a Sylow ℓ-subgroup of $\mathrm{N}_{\mathbf{G}}(\mathbf{S})^F$ containing $\mathbf{S}_\ell^F$. If D is not a Sylow ℓ-subgroup of $\mathbf{G}^F$, there is an ℓ-subgroup $E \supseteq D$ with $E \neq D$. Since $E \neq 1$, there is $z \in \mathrm{Z}(E) \setminus \{1\}$. We have $z \in \mathrm{C}_{\mathbf{G}^F}(E) \subseteq \mathrm{C}_{\mathbf{G}^F}(\mathbf{S}_\ell^F) = \mathbf{L}^F$, hence $z \in \mathbf{L}$ and $[z, \mathbf{S}] = 1$. Let $\mathbf{M} = \mathrm{C}^\circ_{\mathbf{G}}(z)$. Then $\mathbf{G} \neq \mathbf{M} \supseteq E$ and $\mathbf{S}$ by Proposition 13.16(i). Then the induction hypothesis implies that $\mathrm{N}_{\mathbf{M}}(\mathbf{S})^F$ contains a Sylow ℓ-subgroup of $\mathbf{M}^F$. This implies that $|D| \geq |E|$. A contradiction. □

22.2. The theorem

We begin with basic properties of the groups $\mathrm{C}^\circ_{\mathbf{G}}([\mathbf{L}, \mathbf{L}])$.

Proposition 22.8. *Let $(\mathbf{G}, F)$ be a connected reductive $\mathbf{F}$-group defined over $\mathbb{F}_q$. Let $\mathbf{L}$ be a Levi subgroup of $\mathbf{G}$. Let $\mathbf{K}$ be a closed subgroup of $\mathbf{G}$ such that $[\mathbf{L}, \mathbf{L}] \subseteq \mathbf{K} \subseteq \mathbf{L}$. Let $d \geq 1$.*

(i) $\mathrm{C}^\circ_{\mathbf{G}}(\mathbf{K})$ is a reductive subgroup of $\mathbf{G}$, $\mathrm{Z}^\circ(\mathbf{L})$ is a maximal torus in it.

(ii) If $\mathbf{L}$ is d-split and $\mathbf{K}$ is F-stable, then $\mathrm{Z}^\circ(\mathbf{L})_{\phi_d}$ is a maximal ϕ_d-subgroup of $\mathrm{C}^\circ_{\mathbf{G}}(\mathbf{K})$.

(iii) If $\mathbf{L}$ and $\mathbf{M}$ are d-split Levi subgroups of $\mathbf{G}$ and $[\mathbf{L}, \mathbf{L}] \subseteq \mathbf{M}$, then there exists $c \in \mathrm{C}^\circ_{\mathbf{G}}(\mathbf{L} \cap \mathbf{M})^F$ such that ${}^c\mathbf{L} \subseteq \mathbf{M}$, with equality ${}^c\mathbf{L} = \mathbf{M}$ when $[\mathbf{L}, \mathbf{L}] = [\mathbf{M}, \mathbf{M}]$.

Proof. Let $\mathbf{J} = C^\circ_{\mathbf{G}}(\mathbf{K})$ and let $\mathbf{T}$ be a maximal torus in $\mathbf{L}$. Recall that any connected reductive group satisfies $\mathbf{G} = Z^\circ(\mathbf{G}).[\mathbf{G}, \mathbf{G}]$.

(i) Clearly $\mathbf{K} = [\mathbf{L}, \mathbf{L}].(\mathbf{K} \cap \mathbf{T})$, so $\mathbf{T}$ normalizes $\mathbf{K}$ and $\mathbf{J}$. For $\mathbf{J}$, the corresponding roots clearly form a symmetric set. So ([DiMi91] 1.14) $\mathbf{J}$ is a reductive subgroup of $\mathbf{G}$. One has $\mathbf{L} = C_{\mathbf{G}}(Z^\circ(\mathbf{L}))$, so $C^\circ_{\mathbf{J}}(Z^\circ(\mathbf{L})) \subseteq C^\circ_{\mathbf{L}}([\mathbf{L}, \mathbf{L}]) = Z^\circ(\mathbf{L})$, whence the maximality of $Z^\circ(\mathbf{L})$.

(ii) One has $\mathbf{L} = C_{\mathbf{G}}(Z^\circ(\mathbf{L})_{\phi_d})$, so $C_{\mathbf{J}}(Z^\circ(\mathbf{L})_{\phi_d}) \subseteq Z^\circ(\mathbf{L})$, whence the maximality of $Z^\circ(\mathbf{L})_{\phi_d}$ as ϕ_d-subgroup of $\mathbf{J}$.

(iii) Take $\mathbf{K} := \mathbf{L} \cap \mathbf{M}$, then $Z^\circ(\mathbf{M}) \subseteq \mathbf{J}$. By (ii) and conjugacy of maximal ϕ_d-subgroups (Theorem 13.18), there is a $c \in \mathbf{J}^F$ such that ${}^c Z^\circ(\mathbf{L})_{\phi_d} \supseteq Z^\circ(\mathbf{M})_{\phi_d}$. Taking centralizers in $\mathbf{G}$ yields ${}^c\mathbf{L} \subseteq \mathbf{M}$ since $\mathbf{L}$ and $\mathbf{M}$ are d-split. When $[\mathbf{L}, \mathbf{L}] = [\mathbf{M}, \mathbf{M}]$, ${}^c\mathbf{L} = \mathbf{M}$ by symmetry. □

Here is our main theorem on defect groups and ordinary characters of unipotent ℓ-blocks.

Theorem 22.9. *Let $(\mathbf{G}, F)$ be a connected reductive $\mathbf{F}$-group defined over $\mathbb{F}_q$. Let ℓ be a prime not dividing q. Assume ℓ is odd, good for $\mathbf{G}$ and $\ell \neq 3$ if ${}^3\mathbf{D}_4$ occurs in the rational type of $(\mathbf{G}, F)$. Let e be the multiplicative order of q* mod. *ℓ. The map*

$$(\mathbf{L}, \zeta) \mapsto R^{\mathbf{G}}_{\mathbf{L}} B_{\mathbf{L}^F}(\zeta)$$

(see Notation 21.8) induces a bijection between the unipotent ℓ-blocks of $\mathbf{G}^F$ and the $\mathbf{G}^F$-conjugacy classes of pairs $(\mathbf{L}, \zeta)$ where $\mathbf{L}$ is e-split and $\zeta \in \mathcal{E}(\mathbf{L}^F, 1)$ is e-cuspidal. Moreover,

(i) $\quad \mathrm{Irr}(\mathbf{G}^F, R^{\mathbf{G}}_{\mathbf{L}} B_{\mathbf{L}^F}(\zeta)) \cap \mathcal{E}(\mathbf{G}^F, 1) = \{\chi \in \mathrm{Irr}(\mathbf{G}^F) \mid (\mathbf{G}, \chi) \gg_e (\mathbf{L}, \zeta)\}$ *(see Definition 21.5),*

(ii) any Sylow ℓ-subgroup of $C^\circ_{\mathbf{G}}([\mathbf{L}, \mathbf{L}])^F$ is a defect group of $R^{\mathbf{G}}_{\mathbf{L}} B_{\mathbf{L}^F}(\zeta)$.

Example 22.10. Let us show the consequence on ℓ-blocks of $\mathrm{GL}_n(\mathbb{F}_q) = \mathbf{G}^F$ where $\mathbf{G} = \mathrm{GL}_n(\mathbf{F})$. The unipotent characters of $\mathrm{GL}_n(\mathbb{F}_q)$ are parametrized

$$\lambda \mapsto \chi_\lambda$$

by partitions $\lambda \vdash n$. Then $(\mathbf{G}, \chi_\lambda) \gg_e (\mathbf{L}^{(m)}, \chi_\kappa)$ whenever λ has e-core $\kappa \vdash n - me$ and $\mathbf{L}^{(m)} \cong \mathrm{GL}_{n-me}(\mathbf{F}) \times (\mathbf{S}_{(e)})^m$ where $\mathbf{S}_{(e)}$ is a Coxeter torus of $\mathrm{GL}_e(\mathbf{F})$. Moreover χ_κ is e-cuspidal (see Example 21.6).

Theorem 22.9 then implies that two unipotent characters χ_λ and χ_μ are in the same $\mathrm{Irr}(\mathbf{G}^F, B)$ for B an ℓ-block if and only if λ and μ have the same e-core $\kappa \vdash n - me$. Then a defect group of B is given by a Sylow ℓ-subgroup of $\mathrm{GL}_{me}(\mathbb{F}_q) = C_{\mathbf{G}}([\mathbf{L}^{(m)}, \mathbf{L}^{(m)}])^F$.

Remark 22.11. (1) It is known that $\geq_e$ is a transitive relation among pairs $(\mathbf{L}, \zeta)$ where $\mathbf{L}$ is an e-split Levi subgroup of $\mathbf{G}$ and $\zeta \in \mathcal{E}(\mathbf{L}^F, 1)$ (Broué–Michel–Malle; see the notes below).

(2) In the case of $e = 1$, the 1-split Levi subgroups $\mathbf{L}$ give rise to Levi subgroups $\mathbf{L}^F$ of the finite BN-pair in $\mathbf{G}^F$ (see §8.1), and $\mathrm{R}_\mathbf{L}^\mathbf{G}$ is then the Harish-Chandra induction. Theorem 22.9 tells us that the partition of unipotent characters induced by ℓ-blocks is the same as the one induced by Harish-Chandra series (note that $\geq_e$ is transitive by positivity of Harish-Chandra induction; see [DiMi91] §6). Under the hypotheses of Theorem 22.9 with $e = 1$, if $\mathbf{P} = \mathbf{L} \rtimes R_u(\mathbf{P})$ is a Levi decomposition of an F-stable parabolic subgroup and ζ is a cuspidal unipotent character of $\mathbf{L}^F$, then the components of $\mathrm{R}_\mathbf{L}^\mathbf{G}\zeta$ make the set of unipotent characters defined by a single ℓ-block of $\mathbf{G}^F$. This can be seen as a converse of Theorem 5.19.

22.3. Self-centralizing subpairs

The so-called self-centralizing subpairs are a basic tool to build maximal subpairs and defect groups. We recall some basic facts about this notion (see [Thévenaz] §41).

Fix G a finite group, ℓ a prime, and $(\mathcal{O}, K, k)$ an ℓ-modular splitting system for G.

Definition 22.12. *If B is an ℓ-block of G with central defect group, we define the* **canonical character** *of B as the only element of* $\mathrm{Irr}(G, B)$ *with* $\mathrm{Z}(G)$ *in its kernel (thus being identified with the character of a block of defect zero of* $G/\mathrm{Z}(G)$*; see Remark 5.6).*

An ℓ-subpair (Q, b) of G is called a **self-centralizing** *subpair if and only if b has defect group* $\mathrm{Z}(Q)$ *in* $\mathrm{C}_G(Q)$.

The following is fairly easy (see Theorem 5.3(iii) and [Thévenaz] 41.4).

Proposition 22.13. *If $(Q, b) \subseteq (Q', b')$ and (Q, b) is self-centralizing, then (Q', b') is self-centralizing and* $\mathrm{Z}(Q') \subseteq \mathrm{Z}(Q)$. *A self-centralizing subpair (Q, b) is maximal if and only if* $\mathrm{N}_G(Q, b)/Q\mathrm{C}_G(Q)$ *is ℓ'.*

Blocks with central defect are defined by their canonical character since this may be used to express the unit of the block (see Remark 5.6). So inclusion of self-centralizing subpairs relates well to scalar product of characters.

Proposition 22.14. *Let (Q, b) be a self-centralizing ℓ-subpair of G and let $\xi \in$* $\mathrm{Irr}(\mathrm{C}_G(Q), b)$ *be the canonical character of b. Let Q' be an ℓ-subgroup con-*

taining Q and $\xi' \in \mathrm{Irr}(\mathrm{C}_G(Q'))$. Then $(Q,b) \lhd (Q', b_{\mathrm{C}_G(Q')}(\xi'))$ with ξ' being the canonical character of $b(\xi')$ if and only if the following are both satisfied:

(i) Q' normalizes the pair (Q,ξ),

(ii) $\xi'(1)_\ell = |\mathrm{C}_G(Q'){:}\,\mathrm{Z}(Q)\cap \mathrm{C}_G(Q')|_\ell$ and $\langle \mathrm{Res}^{\mathrm{C}_G(Q)}_{\mathrm{C}_G(Q')}\xi, \xi'\rangle_{\mathrm{C}_G(Q')} \not\equiv 0$ mod. ℓ.

Then $(Q', b_{\mathrm{C}_G(Q')}(\xi'))$ is self-centralizing (see Definition 2.12).

Proof. Let us first note that in both sides of the equivalence we have $\mathrm{Z}(Q)\cap \mathrm{C}_G(Q') \subseteq \mathrm{Ker}(\xi')$: if ξ' is canonical, this is because $\mathrm{Z}(Q)\cap\mathrm{C}_G(Q') \subseteq \mathrm{Z}(Q') \subseteq \mathrm{Ker}(\xi')$; if (ii) holds, this is because $\mathrm{Z}(Q)\cap \mathrm{C}_G(Q')\subseteq \mathrm{Z}(Q)\subseteq \mathrm{Ker}(\xi)$ and the representation space of ξ' is a summand of the restriction to $\mathrm{C}_G(Q')$ of the representation space of ξ. As a first consequence $\xi'(1)^{-1}.|\mathrm{C}_G(Q'):\mathrm{Z}(Q)\cap\mathrm{C}_G(Q')|$ is an integer.

Now assume that $(Q, b(\xi))$ is self-centralizing with canonical ξ, Q' is an ℓ-subgroup normalizing (Q,ξ) and containing Q, $\xi'\in\mathrm{Irr}(\mathrm{C}_G(Q'))$. Let us check that $(Q,b(\xi))\lhd(Q',b(\xi'))$ if and only if $\sum_{g\in \mathrm{C}_G(Q)_{\ell'}}\xi(g^{-1})\xi'(g)\notin \xi'(1)J(\mathcal{O})$.

Denote by $\mathcal{B}'$ the set of block idempotents of $\mathcal{O}\mathrm{C}_G(Q')$ such that $\mathrm{Br}_{Q'}(\overline{b(\xi)}) = \sum_{b'\in\mathcal{B}'}\overline{b'}$. One has $(Q,b(\xi))\lhd(Q',b_{\mathrm{C}_G(Q')}(\xi'))$ if and only if $\overline{b}_{\mathrm{C}_G(Q')}(\xi')\in\mathcal{B}'$, i.e. $\xi'(\sum_{b'\in\mathcal{B}'}b') = \xi'(1)$ (otherwise it is zero). However, $b(\xi) = u\sum_{g\in\mathrm{C}_G(Q)_{\ell'}}\xi(g^{-1})g$ where $u = |\mathrm{C}_G(Q){:}\,\mathrm{Z}(Q)|^{-1}.\xi(1)$ is a unit in $\mathcal{O}$, by the fact that $b(\xi)$ has central defect $\mathrm{Z}(Q)$ (see Remark 5.6). So, one has $\overline{\sum_{b'\in\mathcal{B}'}b' - u\sum_{g\in\mathrm{C}_G(Q')_{\ell'}}\xi(g^{-1})g} = 0$, i.e. $\sum_{b'\in\mathcal{B}'}b' - u\sum_{g\in\mathrm{C}_G(Q')_{\ell'}}\xi(g^{-1})g\in J(\mathcal{O})\mathcal{O}G$. This difference is also in $\mathrm{Z}(\mathcal{O}\mathrm{C}_G(Q'))$, so $\sum_{b'\in\mathcal{B}'}b' - u\sum_{g\in\mathrm{C}_G(Q')_{\ell'}}\xi(g^{-1})g\in J(\mathcal{O})\mathrm{Z}(\mathcal{O}\mathrm{C}_G(Q'))$ since $\mathrm{Z}(\mathcal{O}\mathrm{C}_G(Q'))$ is clearly a pure submodule of $\mathcal{O}G$, being generated by sums of conjugacy classes of $\mathrm{C}_G(Q')$. Consequently, since central elements of $\mathcal{O}\mathrm{C}_G(Q')$ act by scalars on the representation space of ξ', one gets $\xi'(\sum_{b'\in\mathcal{B}'}b' - u\sum_{g\in\mathrm{C}_G(Q')_{\ell'}}\xi(g^{-1})g)\in J(\mathcal{O})\xi'(1)$. So $(Q,b(\xi))\lhd(Q',b(\xi'))$ if and only if $\sum_{g\in\mathrm{C}_G(Q')_{\ell'}}\xi(g^{-1})\xi'(g)\notin J(\mathcal{O})\xi'(1)$.

We obtain the claimed equivalence once we show that $\sum_{g\in\mathrm{C}_G(Q')_{\ell'}}\xi(g^{-1})\xi'(g) = |\mathrm{C}_G(Q'){:}\,\mathrm{Z}(Q)\cap\mathrm{C}_G(Q')|.\langle\mathrm{Res}^{\mathrm{C}_G(Q)}_{\mathrm{C}_G(Q')}\xi,\xi'\rangle_{\mathrm{C}_G(Q')}$. Using the fact that $\mathrm{Z}(Q)\cap\mathrm{C}_G(Q')$ is in the kernel of both ξ and ξ', one gets $\sum_{g\in\mathrm{C}_G(Q')_{\ell'}}\xi(g^{-1})\xi'(g) = |\mathrm{Z}(Q)\cap\mathrm{C}_G(Q')|^{-1}\sum_{g\in\mathrm{C}_G(Q')_{\ell'}.(\mathrm{Z}(Q)\cap\mathrm{C}_G(Q'))}\xi(g^{-1})\times\xi'(g)$. But this in turn equals $|\mathrm{Z}(Q)\cap\mathrm{C}_G(Q')|^{-1}\sum_{g\in\mathrm{C}_G(Q')}\xi(g^{-1})\xi'(g) = |\mathrm{C}_G(Q'){:}\,\mathrm{Z}(Q)\cap\mathrm{C}_G(Q')|.\langle\mathrm{Res}^{\mathrm{C}_G(Q)}_{\mathrm{C}_G(Q')}\xi,\xi'\rangle_{\mathrm{C}_G(Q')}$ since ξ, defining a block of $\mathrm{C}_G(Q)$ with defect $\mathrm{Z}(Q)$, is zero outside $\mathrm{C}_G(Q)_{\ell'}.\mathrm{Z}(Q)$ (see Remark 5.6). □

Remark 22.15. The condition (ii) above is satisfied when $\mathrm{Res}^{\mathrm{C}_G(Q)}_{\mathrm{C}_G(Q')}\xi$ is irreducible and equal to ξ'. The proof is as follows: (Q,b) being self-centralizing and ξ canonical in $\mathrm{Irr}(\mathrm{C}_G(Q), b)$, one has $\xi(1)_\ell = |\mathrm{C}_G(Q):\mathrm{Z}(Q)|_\ell$, so $\xi'(1)_\ell = \xi(1)_\ell \geq |\mathrm{C}_G(Q'):\mathrm{Z}(Q)\cap \mathrm{C}_G(Q')|_\ell$, but $\xi'(1)$ divides $|\mathrm{C}_G(Q'):\mathrm{Z}(Q)\cap \mathrm{C}_G(Q')|$ since $\mathrm{Z}(Q)\cap \mathrm{C}_G(Q')$ is in the kernel of $\xi' = \mathrm{Res}^{\mathrm{C}_G(Q)}_{\mathrm{C}_G(Q')}\xi$. This gives (ii).

22.4. The defect groups

We know that we have a "connected subpair" inclusion $(\{1\}, \mathrm{R}_{\mathbf{L}}^{\mathbf{G}} b_{\mathbf{G}^F}(\zeta)) \lhd (\mathrm{Z}(\mathbf{L})_\ell^F, b_{\mathbf{L}^F}(\zeta))$ (see Theorem 21.7). Our main task is to prove that $(\mathrm{Z}(\mathbf{L})_\ell^F, b_{\mathbf{L}^F}(\zeta))$ is an ordinary subpair and to associate a maximal subpair containing it.

Proposition 22.16. *Let $(\mathbf{G}, F)$ be a connected reductive $\mathbf{F}$-group defined over $\mathbb{F}_q$. Let ℓ be an odd prime, good for $\mathbf{G}$ and $\ell \neq 3$ if ${}^3\mathbf{D}_4$ is involved in $(\mathbf{G}, F)$. Let e be the order of q mod. ℓ. Assume that $\zeta \in \mathcal{E}(\mathbf{G}^F, 1)$ is e-cuspidal. Then $b_{\mathbf{G}^F}(\zeta)$ has a central defect group and ζ is the canonical character of $\mathcal{O}\mathbf{G}^F.b_{\mathbf{G}^F}(\zeta)$.*

Lemma 22.17. *Let $(\mathbf{G}, F)$ be a connected reductive group defined over $\mathbb{F}_q$. Let ℓ be an odd prime not dividing q and good for $\mathbf{G}$. Let $\mathbf{G} = \mathbf{G_a}.\mathbf{G_b}$ be the decomposition of Definition 22.4 relative to ℓ. Let e be the order of q mod. ℓ. Let $(\mathbf{L}, \zeta)$ be an e-cuspidal unipotent pair of $\mathbf{G}$. Then*

(i) $\mathbf{L}\cap\mathbf{G_a}$ is a diagonal torus of $\mathbf{G_a}$ (see Example 13.4(iii)),

(ii) $\mathrm{Z}(\mathbf{L})_\ell^F = \mathrm{Z}^\circ(\mathbf{L})_\ell^F$, $\mathbf{L} = \mathrm{C}^\circ_{\mathbf{G}}(\mathrm{Z}(\mathbf{L})_\ell^F)$ and $\mathbf{L}^F = \mathrm{C}_{\mathbf{G}^F}(\mathrm{Z}(\mathbf{L})_\ell^F)$.

Proof of Lemma 22.17. First $\mathbf{L} = (\mathbf{L}\cap\mathbf{G_a}).(\mathbf{L}\cap\mathbf{G_b})$ and $\mathbf{L}\cap\mathbf{G_a}$, $\mathbf{L}\cap\mathbf{G_b}$ are e-split Levi subgroups of $\mathbf{G_a}$, $\mathbf{G_b}$ with an e-cuspidal unipotent character, denoted $\zeta_{\mathbf{a}}$ for that of $(\mathbf{L}\cap\mathbf{G_a})^F$.

(i) Let $\mathbf{T}$ be a diagonal torus of $\mathbf{G_a}$. By Proposition 22.6(i) $\mathbf{T} = \mathrm{C}^\circ_{\mathbf{G}}(\mathbf{T}_{\phi_e})$, so $\mathbf{T}$ is e-split in $\mathbf{G_a}$ and $\mathbf{T}_{\phi_e}$ is a maximal ϕ_e-subgroup of $\mathbf{G_a}$ (Theorem 13.18). If one writes $\mathbf{L}\cap\mathbf{G_a} = \mathrm{C}_{\mathbf{G_a}}(\mathbf{S})$ for $\mathbf{S}$ a ϕ_e-subgroup of $\mathbf{G_a}$, then $\mathbf{S}$ is in a $\mathbf{G}_{\mathbf{a}}^F$-conjugate of $\mathbf{T}_{\phi_e}$ (Theorem 13.18 again), so one may assume $\mathbf{L}\cap\mathbf{G_a}\supseteq\mathbf{T}$. But now, using the parametrization of unipotent characters by irreducible characters of the Weyl group associated with a diagonal torus ([DiMi91] 15.8) along with the formula for twisted induction ([DiMi91] 15.5), one gets ${}^*\mathrm{R}_{\mathbf{T}}^{\mathbf{L}\cap\mathbf{G_a}}\zeta_{\mathbf{a}} \neq 0$ as in the proof of Lemma 21.15. This implies $\mathbf{T} = \mathbf{L}\cap\mathbf{G_a}$ since $\zeta_{\mathbf{a}}$ is e-cuspidal.

(ii) We have $\mathbf{L} = \mathbf{T}.\mathbf{M}$, where $\mathbf{T}$ is a diagonal torus of $\mathbf{G_a}$ and $\mathbf{M}$ is an e-split Levi subgroup of $\mathbf{G_b}$. Then $\mathrm{Z}(\mathbf{L}) = \mathbf{T}.\mathrm{Z}(\mathbf{M})$ and $\mathrm{Z}^\circ(\mathbf{L}) = \mathbf{T}.\mathrm{Z}^\circ(\mathbf{M})$. As $\mathbf{G_a}\cap\mathbf{G_b}$ is central and $(\mathbf{G_a}\cap\mathbf{G_b})^F$ is an ℓ'-group, Proposition 8.1 implies $\mathrm{Z}(\mathbf{L})_\ell^F = \mathbf{T}_\ell^F.\mathrm{Z}(\mathbf{M})_\ell^F$ and $\mathrm{Z}^\circ(\mathbf{L})_\ell^F = \mathbf{T}_\ell^F.\mathrm{Z}^\circ(\mathbf{M})_\ell^F$. By Proposition 13.12(ii), $\mathrm{Z}^\circ(\mathbf{M})_\ell^F = \mathrm{Z}(\mathbf{M})_\ell^F$, so $\mathrm{Z}^\circ(\mathbf{L})_\ell^F = \mathrm{Z}(\mathbf{L})_\ell^F$.

Now $\mathrm{C}_{\mathbf{G}}(\mathrm{Z}(\mathbf{L})^F_\ell) = \mathrm{C}_{\mathbf{G}_a}(\mathbf{T}^F_\ell).\mathrm{C}_{\mathbf{G}_b}(\mathrm{Z}(\mathbf{M})^F_\ell)$, so $\mathrm{C}^\circ_{\mathbf{G}}(\mathrm{Z}(\mathbf{L})^F_\ell) = \mathrm{C}^\circ_{\mathbf{G}_a}(\mathbf{T}^F_\ell).\mathrm{C}^\circ_{\mathbf{G}_b}(\mathrm{Z}(\mathbf{M})^F_\ell)$. Since ℓ does not divide $|(\mathrm{Z}(\mathbf{G}_b))^F|$, Proposition 13.19 in $\mathbf{G}_b$ and Proposition 22.6(i) in $\mathbf{G}_a$ yield $\mathrm{C}^\circ_{\mathbf{G}}(\mathrm{Z}(\mathbf{L})^F_\ell) = \mathbf{L}$. Furthermore, $\mathrm{C}_{\mathbf{G}}(\mathrm{Z}(\mathbf{L})^F_\ell)/\mathrm{C}^\circ_{\mathbf{G}}(\mathrm{Z}(\mathbf{L})^F_\ell)$ is isomorphic to $\mathrm{C}_{\mathbf{G}_a}(\mathbf{T}^F_\ell)/\mathbf{T}$. By Proposition 22.6(ii), F fixes only 1 in this quotient, so that $\mathrm{C}_{\mathbf{G}^F}(\mathrm{Z}(\mathbf{L})^F_\ell) = \mathbf{L}^F$. □

Proof of Proposition 22.16. Since $\mathrm{Z}(\mathbf{G}^F)$ is in the kernel of ζ, it suffices to check that if $h \in \mathbf{G}^F_\ell \setminus \mathrm{Z}(\mathbf{G}^F)$ then $d^{h,\mathbf{G}^F}\zeta = 0$ (Remark 5.6). By Lemma 22.17(i), $\mathbf{G}_a$ is a torus, so $\mathbf{G}_a = \mathrm{Z}^\circ(\mathbf{G})$. Since $h \in \mathbf{G}^F_a.\mathbf{G}^F_b$ (Proposition 22.5(i)) and $\mathbf{G}^F_a$ is in the kernel of ζ, one may assume $h \in \mathbf{G}^F_b$. So the problem lies entirely within $\mathbf{G}_b$. Assume $\mathbf{G} = \mathbf{G}_b$. Then ℓ satisfies Condition 22.1 by Table 13.11. The proper Levi subgroup $\mathrm{C}^\circ_{\mathbf{G}}(h)$ is $E_{q,\ell}$-split (Proposition 13.19). Then Theorem 22.2 implies that there exists a proper e-split Levi subgroup $\mathbf{L}$ such that $\mathrm{C}^\circ_{\mathbf{G}}(h) \subseteq \mathbf{L}$. Then Theorem 21.4 implies $d^{h,\mathbf{G}^F}\zeta = d^{h,\mathbf{L}^F}{}^*\mathrm{R}^{\mathbf{G}}_{\mathbf{L}}\zeta = 0$ since ζ is e-cuspidal. □

Proof of Theorem 22.9. First, let us fix an e-cuspidal pair $(\mathbf{L}, \zeta)$ of $\mathbf{G}$.

Abbreviate $\mathrm{R}^{\mathbf{G}}_{\mathbf{L}} b_{\mathbf{L}^F}(\zeta) = b_{\mathbf{G}^F}(\mathbf{L}, \zeta)$. Denote $Z := \mathrm{Z}^\circ(\mathbf{L})^F_\ell = \mathrm{Z}(\mathbf{L})^F_\ell$. Lemma 22.17(ii) also implies $\mathbf{L} = \mathrm{C}^\circ_{\mathbf{G}}(Z)$ and $\mathbf{L}^F = \mathrm{C}_{\mathbf{G}^F}(Z)$, so Theorem 21.7 actually gives a subpair inclusion

$$\big(\{1\}, \mathrm{R}^{\mathbf{G}}_{\mathbf{L}} b_{\mathbf{L}^F}(\zeta)\big) \lhd (Z, b_{\mathbf{L}^F}(\zeta))$$

in $\mathbf{G}^F$.

We now construct a maximal subpair. Denote $\mathbf{J} := \mathrm{C}^\circ_{\mathbf{G}}([\mathbf{L}, \mathbf{L}])$. By Proposition 22.8(ii), $\mathrm{Z}^\circ(\mathbf{L})_{\phi_e}$ is a maximal ϕ_e-subgroup in $(\mathbf{J}, F)$, so $\mathrm{N}_{\mathbf{J}^F}(\mathrm{Z}^\circ(\mathbf{L})_{\phi_e})$ contains a Sylow ℓ-subgroup D of $\mathbf{J}^F$ (Proposition 22.7). Since $\mathbf{L} = \mathrm{C}_{\mathbf{G}}(\mathrm{Z}^\circ(\mathbf{L})_{\phi_e}) = \mathrm{C}^\circ_{\mathbf{G}}(Z)$, one has $\mathrm{N}_{\mathbf{J}^F}(\mathrm{Z}^\circ(\mathbf{L})_{\phi_e}) = \mathrm{N}_{\mathbf{J}^F}(\mathbf{L}) = \mathrm{N}_{\mathbf{J}^F}(Z)$, so $Z \lhd D$.

Lemma 22.18. *$(Z, b_{\mathbf{L}^F}(\zeta))$ is a self-centralizing subpair and $(D, b_{\mathrm{C}_{\mathbf{G}^F}(D)}(\mathrm{Res}^{\mathbf{L}^F}_{\mathrm{C}_{\mathbf{G}^F}(D)}\zeta))$ is a maximal subpair with canonical character $\mathrm{Res}^{\mathbf{L}^F}_{\mathrm{C}_{\mathbf{G}^F}(D)}\zeta$. Moreover,*

$$\big(\{1\}, \mathrm{R}^{\mathbf{G}}_{\mathbf{L}} b_{\mathbf{L}^F}(\zeta)\big) \lhd (Z, b_{\mathbf{L}^F}(\zeta)) \lhd \big(D, b_{\mathrm{C}_{\mathbf{G}^F}(D)}\big(\mathrm{Res}^{\mathbf{L}^F}_{\mathrm{C}_{\mathbf{G}^F}(D)}\zeta\big)\big).$$

Proof of Lemma 22.18. By Proposition 22.16, $(Z, b_{\mathbf{L}^F}(\zeta))$ is self-centralizing.

One has $\mathrm{C}_{\mathbf{G}^F}(D) \subseteq \mathrm{C}_{\mathbf{G}^F}(Z) = \mathbf{L}^F$, so the restriction $\mathrm{Res}^{\mathbf{L}^F}_{\mathrm{C}_{\mathbf{G}^F}(D)}\zeta$ makes sense. However, $[\mathbf{L}, \mathbf{L}] \subseteq \mathrm{C}_{\mathbf{G}}(D)$ and $\mathrm{Res}^{\mathbf{L}^F}_{[\mathbf{L},\mathbf{L}]^F}\zeta$ is irreducible (Proposition 15.9), so $\mathrm{Res}^{\mathbf{L}^F}_{\mathrm{C}_{\mathbf{G}^F}(D)}\zeta \in \mathrm{Irr}(\mathrm{C}_{\mathbf{G}^F}(D))$. Denote $\zeta_D := \mathrm{Res}^{\mathbf{L}^F}_{\mathrm{C}_{\mathbf{G}^F}(D)}\zeta$.

We check the normal inclusion $(Z, b_{\mathbf{L}^F}(\zeta)) \lhd (D, b_{\mathrm{C}_{\mathbf{G}^F}(D)}(\zeta_D))$ by applying Proposition 22.14 and Remark 22.15. It just remains to verify that D normalizes

$(Z, b_{\mathbf{L}^F}(\zeta))$. If $g \in D$, then it normalizes Z and therefore $\mathbf{L}$. Moreover, ζ, $\zeta^g \in \mathcal{E}(\mathbf{L}^F, 1)$ and they have the same restriction to $[\mathbf{L}, \mathbf{L}]^F$ since g centralizes $[\mathbf{L}, \mathbf{L}]$. By Proposition 15.9, this implies $\zeta = \zeta^g$.

It remains to check that $(D, b_{\mathrm{C}_{\mathbf{G}^F}(D)}(\zeta_D))$ is maximal.

Assume $(D, b_{\mathrm{C}_{\mathbf{G}^F}(D)}(\zeta_D)) \subseteq (E, b_E)$ is an inclusion of subpairs in $\mathbf{G}^F$. We must prove $D = E$.

Lemma 22.19. *If $\mathrm{Z}(E) \subseteq \mathbf{G_a}$, then $E \subseteq \mathbf{G_a}$ and $\mathbf{L} = \mathbf{T_a}\mathbf{G_b}$ where $\mathbf{T_a}$ is a diagonal torus of $\mathbf{G_a}$. Otherwise, if $z \in \mathrm{Z}(E) \setminus \mathbf{G_a}$, then $\mathbf{G_a}\mathrm{C}^\circ_{\mathbf{G}}(z)$ is a proper Levi subgroup of $\mathbf{G}$ containing E, $\mathbf{L}$, and $\mathrm{C}_{\mathbf{G}^F}(z)$.*

Proof of Lemma 22.19. Assume that $\mathrm{Z}(E) \subseteq \mathbf{G_a}$. Since $\mathrm{Z}(E) = \mathrm{Z}(\mathrm{C}_{\mathbf{G}^F}(E))_\ell$ by Remark 5.6 and the fact that (E, b_E) is self-centralizing, then Proposition 22.5(ii) implies $E \subseteq \mathbf{G_a}$. Thus $Z \subseteq \mathbf{G_a}$ and therefore $\mathbf{L} = \mathbf{T_a}\mathbf{G_b}$ where $\mathbf{T_a}$ is a diagonal torus of $\mathbf{G_a}$ by Lemma 22.17.

Let now $z \in \mathrm{Z}(E) \setminus \mathbf{G_a}$ and let $\mathbf{H} := \mathbf{G_a}\mathrm{C}^\circ_{\mathbf{G}}(z)$. Then $\mathbf{H}$ is a Levi subgroup of $\mathbf{G}$ (Proposition 13.16(ii)) and $\mathbf{H} \neq \mathbf{G}$ since $\mathrm{Z}(\mathbf{G_b})^F$ is an ℓ'-group. Moreover, $z \in \mathrm{Z}^\circ(\mathbf{L})^F_\ell$, so $\mathbf{H} \supseteq \mathbf{L}$ by Lemma 22.17. To complete our proof, it clearly suffices to check $\mathrm{C}_{\mathbf{G}^F}(z) \subseteq \mathbf{H}$. The quotient $(\mathbf{G_a}\mathrm{C}_{\mathbf{G_b}}(z))^F/\mathbf{H}^F$ equals $(\mathbf{G_a}\mathrm{C}_{\mathbf{G_b}}(z)/\mathbf{G_a}\mathrm{C}^\circ_{\mathbf{G_b}}(z))^F$ by Lang's theorem and is in turn an isomorphic image of $(\mathrm{C}_{\mathbf{G_b}}(z)/\mathrm{C}^\circ_{\mathbf{G_b}}(z))^F$. But z acts on $\mathbf{G_b}$ as a rational ℓ-element of $\mathbf{G_b}$ since $z \in (\mathbf{G_a})^F(\mathbf{G_b})^F$ (Proposition 22.5(i)). So $(\mathrm{C}_{\mathbf{G_b}}(z)/\mathrm{C}^\circ_{\mathbf{G_b}}(z))^F = 1$ by Proposition 13.16(i). Then $(\mathbf{G_a}\mathrm{C}_{\mathbf{G_b}}(z))^F \subseteq \mathbf{H}^F$ and therefore $\mathrm{C}_{\mathbf{G}^F}(z) \subseteq \mathbf{H}$. □

We prove $D = E$ by induction on $\dim \mathbf{G}$.

If $\mathrm{Z}(E) \subseteq \mathbf{G_a}$, the above lemma implies that $E \subseteq \mathbf{G_a}$ and $[\mathbf{L}, \mathbf{L}] = \mathbf{G_b}$. Then D is a Sylow ℓ-subgroup of $\mathrm{C}^\circ_{\mathbf{G}}([\mathbf{L}, \mathbf{L}])^F = \mathbf{G}^F_{\mathbf{a}}$, and therefore $D = E$.

If $\mathrm{Z}(E) \not\subseteq \mathbf{G_a}$, let $z \in \mathrm{Z}(E) \setminus \mathbf{G_a}$ and define $\mathbf{H} := \mathbf{G_a}\mathrm{C}^\circ_{\mathbf{G}}(z)$ as in Lemma 22.19. Then $(\mathbf{H}, F)$ satisfies the hypotheses of the theorem, $(\mathbf{L}, \zeta)$ is an e-cuspidal pair of $(\mathbf{H}, F)$, and one has $(Z, b_{\mathbf{L}^F}(\zeta)) \subseteq (D, b_{\mathrm{C}_{\mathbf{G}^F}(D)}(\zeta_D)) \subseteq (E, b_E)$ in $\mathbf{H}^F$. Since $\mathbf{H} \neq \mathbf{G}$, the induction hypothesis implies $D = E$. □

This proves Theorem 22.9(ii).

The map of Theorem 22.9 is clearly onto since, for any $\chi \in \mathcal{E}(\mathbf{G}^F, 1)$, there is an e-split Levi subgroup $\mathbf{L}$ and e-cuspidal character $\zeta \in \mathrm{Irr}(\mathbf{L}^F)$ such that $(\mathbf{G}, \chi) \gg_e (\mathbf{L}, \zeta)$. Moreover, ζ has to be unipotent by Proposition 8.25.

In order to get Theorem 22.9, it just remains to prove that, if $(\mathbf{L}, \zeta)$ and $(\mathbf{L}', \zeta')$ are e-cuspidal and if $\mathrm{R}^{\mathbf{G}}_{\mathbf{L}} B_{\mathbf{L}^F}(\zeta) = \mathrm{R}^{\mathbf{G}}_{\mathbf{L}'} B_{\mathbf{L}'^F}(\zeta')$, then $(\mathbf{L}, \zeta)$ and $(\mathbf{L}', \zeta')$ are $\mathbf{G}^F$-conjugate. Suppose $\mathrm{R}^{\mathbf{G}}_{\mathbf{L}} B_{\mathbf{L}^F}(\zeta) = \mathrm{R}^{\mathbf{G}}_{\mathbf{L}'} B_{\mathbf{L}'^F}(\zeta')$ and construct maximal subpairs as in Lemma 22.18: $(\{1\}, \mathrm{R}^{\mathbf{G}}_{\mathbf{L}} b_{\mathbf{L}^F}(\zeta)) \lhd (Z, b_{\mathbf{L}^F}(\zeta)) \lhd (D, b) := (D, b_{\mathrm{C}_{\mathbf{G}^F}(D)}(\mathrm{Res}^{\mathbf{L}^F}_{\mathrm{C}_{\mathbf{G}^F}(D)}\zeta))$ and $(\{1\}, \mathrm{R}^{\mathbf{G}}_{\mathbf{L}'} b_{\mathbf{L}'^F}(\zeta')) \lhd (Z', b_{\mathbf{L}'^F}(\zeta')) \lhd$

$(D', b') := (D', b_{C_{G^F}(D')}(\mathrm{Res}^{L'^F}_{C_{G^F}(D')}\zeta'))$. By conjugacy of maximal subpairs (Theorem 5.3(ii)), one has $(D', b') = {}^g(D, b)$ for some $g \in \mathbf{G}^F$. One has $[\mathbf{L}, \mathbf{L}] \subseteq \mathrm{C}^\circ_{\mathbf{G}}(D) \subseteq \mathbf{L} = [\mathbf{L}, \mathbf{L}].\mathrm{Z}(\mathbf{L})$, so $[\mathbf{L}, \mathbf{L}] = [\mathrm{C}^\circ_{\mathbf{G}}(D), \mathrm{C}^\circ_{\mathbf{G}}(D)]$ and therefore $[\mathbf{L}', \mathbf{L}'] = {}^g[\mathbf{L}, \mathbf{L}]$. The equality $b' = {}^g b$ implies $\mathrm{Res}^{\mathbf{L}'^F}_{C_{\mathbf{G}^F}(D')}\zeta' = {}^g(\mathrm{Res}^{\mathbf{L}^F}_{C_{\mathbf{G}^F}(D)}\zeta)$ (canonical characters), whence

$$\text{(R)} \qquad \mathrm{Res}^{\mathbf{L}'^F}_{[\mathbf{L}',\mathbf{L}']^F}\zeta' = {}^g\big(\mathrm{Res}^{\mathbf{L}^F}_{[\mathbf{L},\mathbf{L}]^F}\zeta\big).$$

Both $\mathbf{L}'$ and ${}^g\mathbf{L}$ are e-split, so Proposition 22.8(iii) implies that there is $c \in \mathrm{C}^\circ_{\mathbf{G}}({}^g\mathbf{L} \cap \mathbf{L}')^F$ such that ${}^{cg}\mathbf{L} = \mathbf{L}'$. But c centralizes $[\mathbf{L}, \mathbf{L}] \subseteq {}^g\mathbf{L} \cap \mathbf{L}'$, so ${}^c({}^g(\mathrm{Res}^{\mathbf{L}^F}_{[\mathbf{L},\mathbf{L}]^F}\zeta)) = {}^g(\mathrm{Res}^{\mathbf{L}^F}_{[\mathbf{L},\mathbf{L}]^F}\zeta)$. Combined with (R) above, this implies that ${}^{cg}\zeta$ and ζ' are two unipotent characters of $\mathbf{L}'^F$ whose restrictions to $[\mathbf{L}', \mathbf{L}']^F$ coincide. So they are equal by Proposition 15.9, i.e. ${}^{cg}(\mathbf{L}, \zeta) = (\mathbf{L}', \zeta')$. This completes the proof of Theorem 22.9. □

Exercises

1. Let $(\mathbf{G}, F)$ be a connected reductive group defined over $\mathbb{F}_q$. Let $d \geq 1$ be an integer and ℓ a prime divisor of $\phi_d(q)$ satisfying Condition 22.1. Let $\mathbf{H}$ be an F-stable connected reductive subgroup containing a maximal torus of $\mathbf{G}$ and such that $\mathrm{Z}^\circ(\mathbf{H})_{\phi_{d\ell}} \not\subseteq \mathrm{Z}(\mathbf{G})$. Show that $\mathrm{Z}^\circ(\mathbf{H})_{\phi_d} \not\subseteq \mathrm{Z}(\mathbf{G})$.
2. Compare the local structure of $\mathbf{G}^F$ and $(\mathbf{G_a})^F \times (\mathbf{G_b})^F$.
3. We use the hypotheses and notation of Theorem 22.9. Show that $\mathrm{C}_{\mathbf{G}}([\mathbf{L}, \mathbf{L}])^F/\mathrm{C}^\circ_{\mathbf{G}}([\mathbf{L}, \mathbf{L}])^F$ is of order prime to ℓ.
4. Let $F: \mathbf{G} \to \mathbf{G}$ be the Frobenius map associated with the definition over $\mathbb{F}_q$ of a connected reductive $\mathbf{F}$-group. Let $\mathbf{T}$ be an F-stable maximal torus of $\mathbf{G}$. Show that there exists a subgroup $N \subseteq \mathrm{N}_{\mathbf{G}}(\mathbf{T})$ such that $F(N) = N$, $\mathbf{T}.N = \mathrm{N}_{\mathbf{G}}(\mathbf{T})$, and $\mathbf{T} \cap N$ is the subgroup of elements of $\mathbf{T}$ of order a power of 2.

   Hint: assume first that $\mathbf{T}_0 \subseteq \mathbf{B}_0$ (a maximal torus in a Borel subgroup) are F-stable. Construct N_0 by use of Theorem 7.11 and the permutation of basic roots induced by F. Then check the general case by writing $\mathbf{T} = {}^g\mathbf{T}_0$ with $g^{-1}F(g) \in n_0\mathbf{T}_0$ where $n_0 \in N_0$, and defining $N := {}^{gt}N_0$ where $t \in \mathbf{T}$ satisfies $n_0F(t)n_0^{-1}t^{-1} = n_0F(g^{-1})g$.
5. Let H be a group acting on a module V. One says that $h \in H$ is **quadratic** on V if and only if $[[v, h], h] = 0$ for all $v \in V$.
   (a) Show that if A is a commutative normal subgroup of a group G, then the following are equivalent.

(i) For any subgroup $H \subseteq G$ containing A, A is the unique maximal commutative normal subgroup of H.

(ii) In the action of G/A on A, only 1 is quadratic on A.

One then denotes $A \mathrel{\dot\triangleleft} G$.

(b) Let $\mathbf{T}$ be a maximal torus of a connected reductive $\mathbf{F}$-group $\mathbf{G}$. Show that $\mathbf{T} \mathrel{\dot\triangleleft} \mathrm{N}_{\mathbf{G}}(\mathbf{T})$.

(c) Let A be a $\mathbb{Z}$-module such that $6A = A$. Let $n \geq 1$, $H := \mathfrak{S}_n$ acting on $V := A^n$. Show that only $1 \in H$ is quadratic on $[V/V^H, H]$.

6. Let $\mathbf{G}$ be a connected reductive group over $\mathbf{F}$, defined over $\mathbb{F}_q$ with associated Frobenius endomorphism $F: \mathbf{G} \to \mathbf{G}$. Let ℓ be prime ≥ 5 and not dividing q. Let e be the order of q mod. ℓ. Let $\mathbf{S}$ be a maximal ϕ_e-subgroup of $\mathbf{G}$ (see Theorem 13.18). Let $\mathbf{L} := \mathrm{C}_{\mathbf{G}}(\mathbf{S})$. Let $\mathbf{T}$ be a maximal F-stable torus of $\mathbf{L}$. Let $W(\mathbf{L}, \mathbf{T})^\perp$ be the subgroup of $\mathrm{N}_{\mathbf{G}}(\mathbf{T})/\mathbf{T}$ generated by the reflections associated with roots orthogonal to all the roots corresponding to $\mathbf{L}$.

(a) Show that there exists $V \subseteq \mathrm{N}_{\mathbf{G}}(\mathbf{T})^F$ such that $V \cap \mathbf{T} = \{1\}$ and reduction mod. $\mathbf{T}$ makes V isomorphic with a Sylow ℓ-subgroup of $(W(\mathbf{L}, \mathbf{T})^\perp)^F$ (use Exercise 3).

(b) Show that the semi-direct product $\mathrm{Z}(\mathbf{L})^F_\ell \rtimes V$ is a Sylow ℓ-subgroup of $\mathbf{G}^F$ and $\mathrm{Z}(\mathbf{L})^F_\ell \mathrel{\dot\triangleleft} \mathrm{Z}(\mathbf{L})^F_\ell \rtimes V$. *Hint*: in case $\mathbf{G} = \mathbf{G_a}$, use Exercise 5(c) and Proposition 22.6. In case $\mathbf{G} = \mathbf{G_b}$, show first that $\mathbf{S} = \{1\}$ implies $|\mathbf{G}^F|_\ell = 1$. Then use induction (see Lemma 22.19).

7. Let $(\mathbf{G}, F)$ be a connected reductive $\mathbf{F}$-group defined over $\mathbb{F}_q$. Assume that its rational type only contains types ${}^2\mathbf{A}$, $\mathbf{B}$, $\mathbf{C}$, $\mathbf{D}$ and ${}^2\mathbf{D}$. Let ℓ be a prime not dividing $2q$ and belonging to $\Pi(\mathbf{G}, F)$ (see Definition 17.5). Recall the notation $E_{q,\ell}$ from Theorem 22.2. Show that any proper $E_{q,\ell}$-split Levi subgroup of $\mathbf{G}$ embeds in a proper 1-split Levi subgroup (analyze the proof of Theorem 22.2).

Deduce that any cuspidal unipotent character of $\mathbf{G}^F$ is $E_{q,\ell}$-cuspidal and therefore defines an ℓ-block of central defect (see Proposition 22.16).

Notes

Using Asai–Shoji's results on twisted induction of unipotent characters, Broué–Malle–Michel have shown that the relation $\geq_e$ is transitive and therefore defines an analogue of Harish-Chandra theory (see [BrMaMi93], and also [FoSr86] for classical groups). In the case of abelian defect the ℓ-block corresponding to the unipotent cuspidal pair $(\mathbf{L}, \zeta)$ has defect group $\mathrm{Z}(\mathbf{L})^F_\ell$. Those blocks are sometimes called "generic", as in [Cr95], since many invariants do not depend on q (a basic trait of twisted induction of unipotent characters). They seem related with generalized Weyl groups $N_{\mathbf{G}}(\mathbf{L})^F/\mathbf{L}^F$ and associated "cyclotomic" Hecke algebras, see [BrMa93], [Bro00].

For this chapter, we havc followed [CaEn94] and [CaEn99a]. Proposition 22.14 is a variation on a theorem of Brauer, [Br67] 6G. Exercises 4–6 are from [Ca94]; see also [Al65].

23

Local structure and ring structure of unipotent blocks

In the present chapter we give some further information on unipotent blocks described in the two preceding chapters.

Keep $(\mathbf{G}, F)$ a connected reductive $\mathbf{F}$-group defined over $\mathbb{F}_q$. Let ℓ be a prime not dividing q, and $(\mathcal{O}, K, k)$ be an ℓ-modular splitting system for $\mathbf{G}^F$. Let e be the multiplicative order of q mod. ℓ. Let $(\mathbf{L}, \zeta)$ be an e-cuspidal pair of $\mathbf{G}$ (see Definition 21.5). It defines an ℓ-block idempotent $b_{\mathbf{G}^F}(\mathbf{L}, \zeta) := \mathrm{R}_{\mathbf{L}}^{\mathbf{G}} b_{\mathbf{L}^F}(\zeta) \in \mathcal{O}\mathbf{G}^F$ (see Theorem 22.9).

We give below a description of the full set $\mathrm{Irr}(\mathbf{G}^F, b_{\mathbf{G}^F}(\mathbf{L}, \zeta))$ in terms of twisted induction (Theorem 23.2). The main ingredient is Theorem 21.13.

Recall that the local structure of a block $\mathcal{O}Gb$ of a finite group G is the datum of non-trivial (i.e. $\neq G$) centralizers and normalizers of subpairs containing $(\{1\}, b)$.

We show that the local structure of $\mathcal{O}\mathbf{G}^F b_{\mathbf{G}^F}(\mathbf{L}, \zeta)$ lies within a subgroup of the Weyl group extended by $\mathrm{C}^\circ_{\mathbf{G}}([\mathbf{L}, \mathbf{L}])^F$ (Theorem 23.8).

It is believed that the local structure of a finite group at a given prime ℓ should provide information on the ℓ-modular representations of the group (see [Al86]). This is generally a difficult problem. An instance is Alperin's weight conjecture (see §6.3). In the case of a block b with a maximal subpair (D, b_D) where D is commutative, the local structure is controlled by $\mathrm{N}_G(D)$. Broué's conjecture then asserts that the derived category $D^b(\mathcal{O}\mathbf{G}^F.b)$ is equivalent to $D^b(B_D)$ where B_D is a block of $\mathcal{O}\mathrm{N}_G(D)$. We check it in the case of principal blocks of $\mathbf{G}^F$ when ℓ divides $q-1$ (Theorem 23.12). Since, in this case, the twisted induction describing the irreducible characters of the unipotent blocks is then a Harish-Chandra induction defined by bimodules (not just complexes), it is not surprising that one gets a Morita equivalence instead of a derived equivalence.

23.1. Non-unipotent characters in unipotent blocks

Let $(\mathbf{G}, F)$ be a connected reductive $\mathbf{F}$-group defined over $\mathbb{F}_q$.

Definition 23.1. *If $\mathbf{L}_1$, $\mathbf{L}_2$ are F-stable Levi subgroups of $\mathbf{G}$ and $\chi_i \in \mathcal{E}(\mathbf{L}_i^F, 1)$, one writes $(\mathbf{L}_1, \chi_1) \sim (\mathbf{L}_2, \chi_2)$ if and only if $[\mathbf{L}_1, \mathbf{L}_1] = [\mathbf{L}_2, \mathbf{L}_2]$ and $\mathrm{Res}^{\mathbf{L}_1^F}_{[\mathbf{L}_1,\mathbf{L}_1]^F} \chi_1 = \mathrm{Res}^{\mathbf{L}_2^F}_{[\mathbf{L}_2,\mathbf{L}_2]^F} \chi_2$. Note that $\sim$ is an equivalence relation.*

Theorem 23.2. *Let ℓ be a prime not dividing q, good for $\mathbf{G}$, $\ell \neq 3$ if $(\mathbf{G}, F)$ has rational components of type ${}^3\mathbf{D}_4$. Let e be the order of q mod. ℓ. Let $(\mathbf{L}, \zeta)$ be a (unipotent) e-cuspidal pair (see Definition 21.5).*

Let $(\mathcal{O}, K, k)$ be an ℓ-modular splitting system for $\mathbf{G}^F$. Let $B_{\mathbf{G}^F}(\mathbf{L}, \zeta) := \mathrm{R}^{\mathbf{G}}_{\mathbf{L}}(B_{\mathbf{L}^F}(\zeta))$ be the block of $\mathcal{O}\mathbf{G}^F$ defined by $(\mathbf{L}, \zeta)$ (see Theorem 22.9). Let $(\mathbf{G}^*, F)$ be in duality with $(\mathbf{G}, F)$ (see §8.1).

With this notation, the elements of $\mathrm{Irr}(\mathbf{G}^F, B_{\mathbf{G}^F}(\mathbf{L}, \zeta))$ are the characters occurring in some $\mathrm{R}^{\mathbf{G}}_{\mathbf{G}(t)}(\hat{t}\chi_t)$, where $t \in (\mathbf{G}^*)^F_\ell$, $\mathbf{G}(t) \subseteq \mathbf{G}$ is an F-stable Levi subgroup in duality with $\mathrm{C}^\circ_{\mathbf{G}^*}(t)$, $\hat{t}$ denotes the associated linear character of $\mathbf{G}(t)^F$ (see Proposition 8.26), $\chi_t \in \mathcal{E}(\mathbf{G}(t)^F, 1)$, and $(\mathbf{G}(t), \chi_t) \gg_e (\mathbf{L}_t, \zeta_t)$ for a (unipotent) e-cuspidal pair $(\mathbf{L}_t, \zeta_t)$ of $\mathbf{G}(t)$ such that $(\mathbf{L}, \zeta) \sim (\mathbf{L}_t, \zeta_t)$ (this forces $[\mathbf{L}, \mathbf{L}] \subseteq \mathbf{G}(t)$).

We first relate the $\gg_e$ orderings for $\mathbf{G}$ and for its $E_{q,\ell}$-split Levi subgroups.

Proposition 23.3. *Let $(\mathbf{G}, F)$ and ℓ be as above. Let $\mathbf{H}$ be an $E_{q,\ell}$-split Levi subgroup of $\mathbf{G}$.*

(i) The condition $(\mathbf{L}, \zeta) \sim (\mathbf{L}_\mathbf{H}, \zeta_\mathbf{H})$ defines a unique bijection between the $\sim$-classes of (unipotent) e-cuspidal pairs $(\mathbf{L}, \zeta)$ in $\mathbf{G}$ such that $[\mathbf{L}, \mathbf{L}] \subseteq \mathbf{H}$, and the $\sim$-classes of (unipotent) e-cuspidal pairs $(\mathbf{L}_\mathbf{H}, \zeta_\mathbf{H})$ of $\mathbf{H}$.

(ii) Let $(\mathbf{L}, \zeta)$ (resp. $(\mathbf{L}_\mathbf{H}, \zeta_\mathbf{H})$) be a (unipotent) e-cuspidal pair of $\mathbf{G}$ (resp. $\mathbf{H}$) such that $(\mathbf{L}, \zeta) \sim (\mathbf{L}_\mathbf{H}, \zeta_\mathbf{H})$. If $\chi \in \mathcal{E}(\mathbf{G}^F, 1)$, $\chi_\mathbf{H} \in \mathcal{E}(\mathbf{H}^F, 1)$ satisfy $(\mathbf{H}, \chi_\mathbf{H}) \gg_e (\mathbf{L}_\mathbf{H}, \zeta_\mathbf{H})$ and $(\mathbf{G}, \chi) \geq (\mathbf{H}, \chi_\mathbf{H})$, then $(\mathbf{G}, \chi) \gg_e (\mathbf{L}, \zeta)$ (see Definition 21.5).

Proof of Proposition 23.3. The proof of (i) and (ii) is by induction on dim $\mathbf{G}$ − dim $\mathbf{H}$. If $\mathbf{G} = \mathbf{H}$, everything is clear. Another easy case is when $\mathbf{G} = \mathbf{G}_\mathbf{a}$ (Definition 22.4) and $\mathbf{H}$ is any F-stable Levi subgroup: then $\mathbf{H} = \mathbf{H}_\mathbf{a}$ and, by Lemma 22.17, there is only one conjugacy class of e-cuspidal pairs in $\mathbf{G}$ and $\mathbf{H}$, and it corresponds to diagonal tori and trivial character. Then (i) and (ii) are clear. From now on, we separate two cases.

If $\mathrm{Z}^\circ(\mathbf{H})_{\phi_E} \subseteq \mathbf{G}_\mathbf{a}$ (see Proposition 13.5), then $\mathbf{H} = (\mathbf{H} \cap \mathbf{G}_\mathbf{a}).\mathbf{G}_\mathbf{b}$ since $\mathbf{H}$ is E-split. But then (i) and (ii) are clear on each side since they reduce to the cases $\mathbf{G} = \mathbf{G}_\mathbf{a}$ and $\mathbf{G} = \mathbf{H}$.

Assume $Z^\circ(\mathbf{H})_{\phi_E} \not\subseteq \mathbf{G_a}$. Then, by Theorem 22.2, there is a proper e-split Levi subgroup $\mathbf{M} \subset \mathbf{G}$ such that $\mathbf{H} \subseteq \mathbf{M}$. By induction, (i) and (ii) are satisfied by $\mathbf{H}$ in $\mathbf{M}$. Then, if $(\mathbf{L_H}, \zeta_\mathbf{H})$ is an e-cuspidal pair of $\mathbf{H}$, there exists in $\mathbf{M}$ an e-cuspidal pair $(\mathbf{L}, \zeta)$ such that $(\mathbf{L}, \zeta) \sim (\mathbf{L_H}, \zeta_\mathbf{H})$. Conversely, let $(\mathbf{L}, \zeta)$ be e-cuspidal in $\mathbf{G}$ with $[\mathbf{L}, \mathbf{L}] \subseteq \mathbf{H}$. Then $[\mathbf{L}, \mathbf{L}] \subseteq \mathbf{M}$, so, by Proposition 22.8(iii), there is $c \in \mathrm{C}^\circ_\mathbf{G}([\mathbf{L}, \mathbf{L}])^F$ such that ${}^c\mathbf{L} \subseteq \mathbf{M}$. By induction hypothesis, there exists an e-cuspidal pair $(\mathbf{L_H}, \zeta_\mathbf{H})$ in $\mathbf{H}$ such that $(\mathbf{L_H}, \zeta_\mathbf{H}) \sim {}^c(\mathbf{L}, \zeta)$. But ${}^c(\mathbf{L}, \zeta) \sim (\mathbf{L}, \zeta)$ is clear, so we get (i).

Assume the hypotheses of (ii), take $c \in \mathrm{C}^\circ_\mathbf{G}([\mathbf{L}, \mathbf{L}])^F$ as above such that ${}^c\mathbf{L} \subseteq \mathbf{M}$. Then ${}^c(\mathbf{L}, \zeta) \sim (\mathbf{L_H}, \zeta_\mathbf{H})$. One has $\langle {}^*\mathrm{R}^\mathbf{G}_\mathbf{M}\chi, \mathrm{R}^\mathbf{M}_\mathbf{H}\chi_\mathbf{H}\rangle_{\mathbf{M}^F} = \langle \chi, \mathrm{R}^\mathbf{G}_\mathbf{H}\chi_\mathbf{H}\rangle_{\mathbf{G}^F} \neq 0$, so there exists $\chi_\mathbf{M} \in \mathcal{E}(\mathbf{M}^F, 1)$ such that $(\mathbf{M}, \chi_\mathbf{M}) \geq (\mathbf{H}, \chi_\mathbf{H})$ and $(\mathbf{G}, \chi) \geq (\mathbf{M}, \chi_\mathbf{M})$. The induction hypothesis then gives $(\mathbf{M}, \chi_\mathbf{M}) \gg_e {}^c(\mathbf{L}, \zeta)$. This implies $(\mathbf{G}, \chi) \gg_e {}^c(\mathbf{L}, \zeta)$, hence $(\mathbf{G}, \chi) \gg_e (\mathbf{L}, \zeta)$ as claimed. □

Proof of Theorem 23.2. By Theorem 9.12(i), every element of $\mathrm{Irr}(\mathbf{G}^F, B_{\mathbf{G}^F}(\mathbf{L}, \zeta))$ is in some $\mathcal{E}(\mathbf{G}^F, t)$ with $t \in (\mathbf{G}^*)^F_\ell$. Consider $\chi \in \mathcal{E}(\mathbf{G}^F, t)$. First $\mathrm{C}^\circ_{\mathbf{G}^*}(t)$ is an $E_{q,\ell}$-split Levi subgroup of $\mathbf{G}^*$ by Proposition 13.19, so one may denote by $\mathbf{G}(t)$ an $E_{q,\ell}$-split Levi subgroup of $\mathbf{G}$ in duality with it (Proposition 13.9) and by $\hat{t}$ the associated linear character of $\mathbf{G}(t)^F$ (see (8.19) or [DiMi91] 13.20). Let $(\mathbf{L}_t, \zeta_t)$ be an e-cuspidal pair of $\mathbf{G}(t)$ such that $(\mathbf{G}(t), \chi_t) \gg_e (\mathbf{L}_t, \zeta_t)$. Now Proposition 23.3(i) tells us that there exists a unipotent e-cuspidal pair $(\mathbf{L}', \zeta')$ of $\mathbf{G}$ such that $(\mathbf{L}_t, \zeta_t) \sim (\mathbf{L}', \zeta')$.

Assume $\chi \in \mathrm{Irr}(\mathbf{G}^F, B_{\mathbf{G}^F}(\mathbf{L}, \zeta))$. It suffices to check that $(\mathbf{L}, \zeta)$ and $(\mathbf{L}', \zeta')$ are $\mathbf{G}^F$-conjugate, or equivalently, by Theorem 22.9, that $B_{\mathbf{G}^F}(\mathbf{L}, \zeta) = B_{\mathbf{G}^F}(\mathbf{L}', \zeta')$.

By Theorem 21.13, $\mathrm{R}^\mathbf{G}_{\mathbf{G}(t)}\chi_t$ is in $\mathrm{CF}(\mathbf{G}^F, K, B_{\mathbf{G}^F}(\mathbf{L}, \zeta))$. So, there exists $\chi_1 \in \mathrm{Irr}(\mathbf{G}^F, B_{\mathbf{G}^F}(\mathbf{L}, \zeta))$ such that $(\mathbf{G}, \chi_1) \geq (\mathbf{G}(t), \chi_t)$. But then Proposition 23.3(ii) implies $(\mathbf{G}, \chi_1) \gg_e (\mathbf{L}', \zeta')$, hence $\chi_1 \in \mathrm{Irr}\big(\mathbf{G}^F, B_{\mathbf{G}^F}(\mathbf{L}', \zeta')\big)$ by Theorem 22.9. Then $B_{\mathbf{G}^F}(\mathbf{L}, \zeta) = B_{\mathbf{G}^F}(\mathbf{L}', \zeta')$ as claimed. □

The description of Theorem 23.2 simplifies a bit in the case of blocks with commutative defect groups.

Corollary 23.4. *Assume the same hypothesis as Theorem 23.2. Assume, moreover, that the defect groups of* $B_{\mathbf{G}^F}(\mathbf{L}, \zeta)$ *are commutative. Then the elements of* $\mathrm{Irr}(\mathbf{G}^F, B_{\mathbf{G}^F}(\mathbf{L}, \zeta))$ *are the irreducible components of the* $\mathrm{R}^\mathbf{G}_{\mathbf{G}(t)}(\hat{t}\chi_t)$ *for* $t \in (\mathbf{G}^*)^F_\ell$, $\mathbf{G}(t)$ *in duality with* $\mathrm{C}^\circ_{\mathbf{G}^*}(t)$, *and* $(\mathbf{G}(t), \chi_t) \gg_e (\mathbf{L}, \zeta)$.

Proof. Assume that $\mathbf{L} \subseteq \mathbf{G}(t)$. Then the relation $(\mathbf{L}_t, \zeta_t) \sim (\mathbf{L}, \zeta)$ of Theorem 23.2 holds between e-cuspidal pairs in $(\mathbf{G}(t), F)$, so they are $\mathbf{G}(t)^F$-conjugate by Proposition 23.3(i). Then Theorem 23.2 gives our claim.

It remains to prove that $\mathbf{G}(t) \supseteq \mathbf{L}^g$ for some $g \in \mathbf{G}^F$. We have $\mathbf{G}(t) \supseteq [\mathbf{L}, \mathbf{L}]$. As seen in the proof of Theorem 23.2, $\mathbf{G}(t)$ is an $E_{q,\ell}$-split Levi subgroup of $\mathbf{G}$. But the hypothesis on defect is equivalent to $\mathrm{Z}(\mathbf{L})^F_\ell = \mathrm{Z}^\circ(\mathbf{L})^F_\ell$ (see Lemma 22.17(ii)) being a Sylow ℓ-subgroup of $\mathrm{C}^\circ_{\mathbf{G}}([\mathbf{L}, \mathbf{L}])^F$ by Lemma 22.18. Arguing on the quotient of polynomial orders $P_{\mathrm{C}^\circ_{\mathbf{G}}([\mathbf{L},\mathbf{L}]),F} / P_{\mathrm{Z}^\circ(\mathbf{L}),F}$ (see Proposition 13.2(ii)), this implies that $\mathrm{Z}^\circ(\mathbf{L})$ contains a maximal ϕ_d-subgroup of $\mathrm{C}^\circ_{\mathbf{G}}([\mathbf{L}, \mathbf{L}])^F$ for any $d \in E_{q,\ell}$

Let us show that any $E_{q,\ell}$-split Levi subgroup $\mathbf{H} \subseteq \mathbf{G}$ such that $\mathbf{H} \supseteq [\mathbf{L}, \mathbf{L}]$ actually contains a $\mathbf{G}^F$-conjugate of $\mathbf{L}$. Using induction on $\dim(\mathbf{G}) - \dim(\mathbf{H})$, we may assume that $\mathbf{H} = \mathrm{C}_{\mathbf{G}}(\mathbf{Z})$ where $\mathbf{Z}$ is a ϕ_d-subgroup of $\mathbf{G}$ for some $d \in E_{q,\ell}$. Then $\mathbf{Z}$ is $\mathbf{G}^F$-conjugate with a subtorus of $\mathrm{Z}^\circ(\mathbf{L})$ by the above and Theorem 13.18, whence our claim is proved. □

23.2. Control subgroups

For the notion of control subgroups, we refer to [Thévenaz].

Definition 23.5. Let G be a finite group for which $(\mathcal{O}, K, k)$ is an ℓ-modular splitting system. Let b be a block idempotent of $\mathcal{O}G$. A subgroup $H \subseteq G$ is said to be a $\mathcal{O}Gb$-control subgroup if and only if there is a maximal subpair (D, b_D) containing $(\{1\}, b)$ such that, for any subpair $(P, b_P) \subseteq (D, b_D)$ and any $g \in G$ such that $(P, b_P)^g \subseteq (D, b_D)$, there is $c \in \mathrm{C}_G(P)$ such that $c^{-1}g \in H$.

For the following, see [Thévenaz] 48.6 and 49.5.(c′).

Theorem 23.6. *Assume the same hypotheses as above. A subgroup H is a $\mathcal{O}Gb$-control subgroup if and only if there is a maximal subpair (D, b_D) containing $(\{1\}, b)$ such that, for any self-centralizing subpair $(Y, b_Y) \subseteq (D, b_D)$ (see §22.3), one has $\mathrm{N}_G(Y, b_Y) \subseteq \mathrm{C}_G(Y).H$.*

Example 23.7. (1) Let $n \geq 1$ be an integer. The ℓ-blocks of the symmetric group $\mathfrak{S}_n$ are in bijection with e-cores of size $n - m\ell$ for $m \geq 0$

$$\kappa \mapsto B(\kappa)$$

(see Theorem 5.16). A defect group for $B(\kappa)$ is given by any Sylow ℓ-subgroup of $\mathfrak{S}_{m\ell}$. For any subgroup Y of $\mathfrak{S}_{m\ell}$, we have clearly $\mathrm{N}_{\mathfrak{S}_n}(Y) \subseteq \mathrm{C}_{\mathfrak{S}_n}(Y).\mathfrak{S}_{m\ell}$. This implies that $\mathfrak{S}_{m\ell}$ is a $B(\kappa)$-control subgroup.

(2) We have seen in Example 22.10 that, if $\mathbf{G}^F = \mathrm{GL}_n(\mathbb{F}_q)$, then its unipotent ℓ-blocks are defined by e-cores of size $n - me$ (where e is the order of q mod. ℓ, and $m \geq 0$). If κ is such an e-core, a defect group of the corresponding ℓ-block

is given by a Sylow ℓ-subgroup of $\mathrm{GL}_{me}(\mathbb{F}_q)$. Then it is clear that $\mathrm{GL}_{me}(\mathbb{F}_q)$ is a control subgroup for that ℓ-block.

For general finite reductive groups, we prove the following. See also Exercise 1.

Theorem 23.8. *Let $(\mathbf{G}, F)$ be a connected reductive group defined over $\mathbb{F}_q$. Let ℓ be a prime not dividing q. Assume ℓ is odd, good for $\mathbf{G}$ and $\ell \neq 3$ if ${}^3\mathbf{D}_4$ is involved in $(\mathbf{G}, F)$.*

Let $(\mathbf{L}, \zeta)$ be an e-cuspidal pair (see Definition 21.5) defining the ℓ-block $B_{\mathbf{G}^F}(\mathbf{L}, \zeta) := \mathrm{R}_{\mathbf{L}}^{\mathbf{G}} B_{\mathbf{L}^F}(\zeta)$ of $\mathbf{G}^F$ (see Theorem 22.9).

Let H be a subgroup of $\mathbf{G}^F$ such that $\mathrm{C}^\circ_{\mathbf{G}}([\mathbf{L}, \mathbf{L}])^F \subseteq H$ and H covers the quotient $\mathrm{N}_{\mathbf{G}^F}([\mathbf{L}, \mathbf{L}], \mathrm{Res}^{\mathbf{L}^F}_{[\mathbf{L},\mathbf{L}]^F}\zeta)/[\mathbf{L}, \mathbf{L}]^F$. Then H is a $B_{\mathbf{G}^F}(\mathbf{L}, \zeta)$-control subgroup of $\mathbf{G}^F$.

Lemma 23.9. *If $\mathbf{G} = \mathbf{G_a}$ relative to a prime ℓ (Definition 22.4) and $Y \subseteq \mathbf{G}^F$ is a nilpotent group of semi-simple elements, then $\mathbf{H} := \mathrm{C}^\circ_{\mathbf{G}}(Y)$ is reductive, F-stable, and $\mathbf{H} = \mathbf{H_a}$.*

Proof of Lemma 23.9. If $\pi : \mathbf{G} \to \mathbf{G}_{\mathrm{ad}}$ is the natural epimorphism, then $\pi(\mathrm{C}^\circ_{\mathbf{G}}(Y)) = \pi(\mathrm{C}^\circ_{\mathbf{G}}(\mathrm{Z}(\mathbf{G})Y)) = \mathrm{C}^\circ_{\mathbf{G}_{\mathrm{ad}}}(\pi(Y))$ because of the standard description of connected centralizers in terms of roots (see Proposition 13.13(i)), and the type of $\mathrm{C}^\circ_{\mathbf{G}}(Y)$ is determined by $Y' := \pi(Y) \subseteq {\mathbf{G}_{\mathrm{ad}}}^F$. Now, let π': $\mathbf{G}' \to \mathbf{G}_{\mathrm{ad}}$ where $\mathbf{G}'$ is the direct product of general linear groups GL with the same rational type as $\mathbf{G}$ and corresponding reduction modulo its (connected) center. Then $\mathrm{C}^\circ_{\mathbf{G}_{\mathrm{ad}}}(Y') = \pi'\big(\mathrm{C}^\circ_{\mathbf{G}'}(Y'')\big)$ where $Y'' = (\pi')^{-1}(Y')^F$ since $(\pi')^{-1}(Y') = Y''\mathrm{Z}(\mathbf{G}')$ by connectedness of $\mathrm{Z}(\mathbf{G}')$ and Lang's theorem.

So, to prove the last assertion, assume $\mathbf{G} = \mathbf{G_a}$ is a direct product of general linear groups. Then $\mathrm{C}^\circ_{\mathbf{G}}(Y)$ is the direct product of the centralizers of the projections of Y. We assume $\mathbf{G} \cong \mathrm{GL}(n)^c$ has irreducible rational type $(\mathbf{A}_n, \varepsilon q^c)$ with $\varepsilon q^c \equiv 1$ (mod. ℓ). Using the fact that Y has a semi-simple representation in the underlying nc-dimensional space and F permutes the isotypic components, one sees that $\mathrm{C}^\circ_{\mathbf{G}}(Y)$ is a product of general linear groups of type $\times_i(\mathbf{A}_{m_i}, (\varepsilon q^c)^{c_i})$ where i ranges over the orbits of F^c acting on the simple representations of Y involved, $(m_i + 1)$ is the multiplicity of the corresponding simple representation and c_i the cardinality of the F^c-orbit. It is then clear that $\mathrm{C}^\circ_{\mathbf{G}}(Y) = \mathrm{C}^\circ_{\mathbf{G}}(Y)_{\mathbf{a}}$. □

Proof of Theorem 23.8. By Lemma 22.18, there is a maximal subpair (D, b_D) containing $(\{1\}, b_{\mathbf{G}^F}(\mathbf{L}, \zeta))$ and such that D is a Sylow ℓ-subgroup of $\mathrm{C}^\circ_{\mathbf{G}}([\mathbf{L}, \mathbf{L}])^F$, b_D is the block defined by the character $\mathrm{Res}^{\mathbf{L}^F}_{\mathrm{C}_{\mathbf{G}^F}(D)}\zeta$ and $\big(\mathrm{Z}^\circ(\mathbf{L})^F_\ell, b_{\mathbf{L}^F}(\zeta)\big) \lhd (D, b_D)$.

The following is about self-centralizing subpairs of (D, b_D). We use the terminology of §22.3.

Lemma 23.10. *Let (Y, b_Y) be a self-centralizing subpair of $\mathbf{G}^F$ with canonical character $\chi \in \mathrm{Irr}(\mathrm{C}_{\mathbf{G}^F}(Y), b_Y)$ and assume $(Y, b_Y) \subseteq (D, b_D)$. Then $\mathrm{C}^\circ_{\mathbf{G}}(Y)$ is reductive, $\mathrm{C}^\circ_{\mathbf{G}}(Y)_{\mathbf{b}} = [\mathbf{L}, \mathbf{L}]$, and $\mathrm{Res}^{\mathrm{C}_{\mathbf{G}}(Y)^F}_{[\mathbf{L},\mathbf{L}]^F} \chi = \mathrm{Res}^{\mathbf{L}^F}_{[\mathbf{L},\mathbf{L}]^F} \zeta$.*

Proof of Lemma 23.10. Let us prove the lemma by induction on $\dim \mathbf{G}$. If $\mathbf{G}$ is a torus, then $D = Y = \mathbf{G}^F_\ell$ and our claim is clear. In the general case, note that, since (Y, b_Y) and $(\mathrm{Z}^\circ(\mathbf{L})^F_\ell, b_{\mathbf{L}^F}(\zeta))$ are self-centralizing and included in (D, b_D), then $\mathrm{Z}(Y) \cap \mathrm{Z}^\circ(\mathbf{L})^F_\ell \supseteq \mathrm{Z}(D)$ (Proposition 22.13).

Assume that $\mathrm{Z}(D) \not\subseteq \mathbf{G}_{\mathbf{a}}$. Let $z \in \mathrm{Z}(D) \setminus \mathbf{G}_{\mathbf{a}}$ and let $\mathbf{H} := \mathbf{G}_{\mathbf{a}}\mathrm{C}^\circ_{\mathbf{G}}(z)$. Then Lemma 22.19 for $E = D$ implies that the inclusion $(Y, b_Y) \subseteq (D, b_D)$ actually holds in $\mathbf{H}^F$. The induction hypothesis then gives our claim.

Assume that $\mathrm{Z}(D) \subseteq \mathbf{G}_{\mathbf{a}}$. Then, again by Lemma 22.19, $D \subseteq \mathbf{G}_{\mathbf{a}}$ and $\mathbf{L} = \mathbf{T}_{\mathbf{a}}\mathbf{G}_{\mathbf{b}}$ where $\mathbf{T}_{\mathbf{a}}$ is a diagonal torus of $\mathbf{G}_{\mathbf{a}}$. Since $Y \subseteq D$, then $\mathrm{C}^\circ_{\mathbf{G}}(Y) = \mathrm{C}^\circ_{\mathbf{G}_{\mathbf{a}}}(Y)\mathbf{G}_{\mathbf{b}}$. Lemma 23.9 then implies that this is reductive, and $\mathrm{C}^\circ_{\mathbf{G}}(Y)_{\mathbf{b}} = \mathbf{G}_{\mathbf{b}} = [\mathbf{L}, \mathbf{L}]$. By Proposition 15.9, $\zeta \in \mathcal{E}(\mathbf{L}^F, 1)$ is the unipotent character whose restriction to $\mathbf{G}^F_{\mathbf{b}}$ is a certain e-cuspidal $\zeta_{\mathbf{b}} \in \mathcal{E}(\mathbf{G}^F_{\mathbf{b}}, 1)$. Note that $\zeta_{\mathbf{b}}$ defines a block of $\mathbf{G}^F_{\mathbf{b}}$ with central defect, by Proposition 22.16. Note also that $\zeta_{\mathbf{b}}$, considered as a character of $\mathbf{G}^F_{\mathbf{b}}/(\mathbf{G}_{\mathbf{a}} \cap \mathbf{G}_{\mathbf{b}})^F$, extends to a unipotent character $\tilde{\zeta}_{\mathbf{b}}$ of $(\mathbf{G}_{\mathbf{b}}/(\mathbf{G}_{\mathbf{a}} \cap \mathbf{G}_{\mathbf{b}}))^F$. Now, $(\mathbf{G}_{\mathbf{b}}/(\mathbf{G}_{\mathbf{a}} \cap \mathbf{G}_{\mathbf{b}}))^F = (\mathbf{G}/\mathbf{G}_{\mathbf{a}})^F = \mathbf{G}^F/\mathbf{G}^F_{\mathbf{a}}$ by connectedness of $\mathbf{G}_{\mathbf{a}}$ and Lang's theorem, so $\tilde{\zeta}_{\mathbf{b}}$ defines a unique $\tilde{\zeta} \in \mathrm{Irr}(\mathbf{G}^F)$ having $\mathbf{G}^F_{\mathbf{a}}$ in its kernel. Applied to $\mathbf{L}$ instead of $\mathbf{G}$, this construction clearly defines an element of $\mathcal{E}(\mathbf{L}^F, 1)$ which coincides with $\zeta_{\mathbf{b}}$ on $\mathbf{G}^F_{\mathbf{b}}$, so this is ζ. Thus $\zeta = \mathrm{Res}^{\mathbf{G}^F}_{\mathbf{L}^F} \tilde{\zeta}$ with $\tilde{\zeta} \in \mathrm{Irr}(\mathbf{G}^F)$.

We now apply Proposition 5.29 with $G = \mathbf{G}^F$, $H = {\mathbf{G}_{\mathbf{b}}}^F$, $\rho = \tilde{\zeta}$. One obtains $(Y, b_{\mathrm{C}_{\mathbf{G}^F}(Y)}(\mathrm{Res}^{\mathbf{G}^F}_{\mathrm{C}_{\mathbf{G}^F}(Y)}\tilde{\zeta})) \subseteq (D, b_{\mathrm{C}_{\mathbf{G}^F}(D)}(\mathrm{Res}^{\mathbf{G}^F}_{\mathrm{C}_{\mathbf{G}^F}(D)}\tilde{\zeta})) = (D, b_D)$, hence by Theorem 5.3(i) $b_Y = b_{\mathrm{C}_{\mathbf{G}^F}(Y)}(\mathrm{Res}^{\mathbf{G}^F}_{\mathrm{C}_{\mathbf{G}^F}(Y)}\tilde{\zeta})$. But $\tilde{\zeta}$ has $\mathrm{Z}(Y)$ in its kernel since $Y \subseteq \mathbf{G}^F_{\mathbf{a}}$. Then $\mathrm{Res}^{\mathbf{G}^F}_{\mathrm{C}_{\mathbf{G}^F}(Y)}\tilde{\zeta}$ is the canonical character of b_Y, so $\chi = \mathrm{Res}^{\mathbf{G}^F}_{\mathrm{C}_{\mathbf{G}^F}(Y)}\tilde{\zeta}$. Restricting further to $[\mathbf{L}, \mathbf{L}]^F$ completes our proof. □

We now complete the proof of Theorem 23.8. By Theorem 23.6, it suffices to check that $\mathrm{N}_{\mathbf{G}^F}(Y, b_Y) \subseteq H.\mathrm{C}_{\mathbf{G}^F}(Y)$ for any self-centralizing (Y, b_Y) included in (D, b_D).

By Lemma 23.10, $\mathrm{C}^\circ_{\mathbf{G}}(Y)_{\mathbf{b}} = [\mathbf{L}, \mathbf{L}]$ and the restriction to $[\mathbf{L}, \mathbf{L}]^F$ of the canonical character of b_Y equals $\mathrm{Res}^{\mathbf{L}^F}_{[\mathbf{L},\mathbf{L}]^F}\zeta$. Then $\mathrm{N}_{\mathbf{G}^F}(Y, b_Y)$, which acts by algebraic automorphisms of $\mathrm{C}^\circ_{\mathbf{G}}(Y)$ commuting with F, normalizes the pair $([\mathbf{L}, \mathbf{L}], \mathrm{Res}^{\mathbf{L}^F}_{[\mathbf{L},\mathbf{L}]^F}\zeta)$. So $\mathrm{N}_{\mathbf{G}^F}(Y, b_Y) \subseteq H.[\mathbf{L}, \mathbf{L}]^F \subseteq H.\mathrm{C}_{\mathbf{G}^F}(Y)$ as claimed. □

23.3. $(q-1)$-blocks and abelian defect conjecture

When the defect groups of an ℓ-block are commutative, then the normalizer of one of them is a control subgroup in the sense of the preceding section (apply, for instance, Theorem 23.6). In this case, Broué's conjecture (see [KLRZ98] §6.3.3, §9.2.4 and the notes below) postulates that this control subgroup concentrates most of the ring structure of the block considered.

Let G be a finite group, ℓ be a prime and $(\mathcal{O}, K, k)$ be an ℓ-modular splitting system for G. Let $b \in \mathrm{Z}(\mathcal{O}G)$ be an ℓ-block idempotent. Let (D, b_D) be a maximal subpair in G containing $(\{1\}, b)$. Let $b'_D = \sum_{x \in \mathrm{N}_G(D)/\mathrm{N}_G(D,b_D)} {}^x b_D \in \mathrm{Z}(\mathcal{O}\mathrm{N}_G(D))$ be the block idempotent of $\mathrm{N}_G(D)$ such that $b'_D b_D \neq 0$.

Conjecture 23.11. *(Broué) If D is a commutative group, then there exists an equivalence of the derived categories (see A1.12)*

$$D^b(\mathcal{O}Gb) \xrightarrow{\sim} D^b(\mathcal{O}\mathrm{N}_G(D)b'_D).$$

Note that when $\mathcal{O}Gb$ is the principal block (see Definition 5.9), then $\mathcal{O}\mathrm{C}_G(D)b_D$ is the principal block (Brauer's third Main Theorem, see Theorem 5.10), so that $b_D = b'_D$ and $\mathcal{O}\mathrm{N}_G(D)b'_D$ is the principal block of $\mathrm{N}_G(D)$.

We check Conjecture 23.11 in the case of principal blocks of finite reductive groups $\mathbf{G}^F$ over $\mathbb{F}_q$ in a case (namely, ℓ divides $q-1$) where the above module categories themselves are (Morita) equivalent.

Theorem 23.12. *Let $(\mathbf{G}, F)$ be a connected reductive $\mathbf{F}$-group defined over $\mathbb{F}_q$. Let ℓ be a prime* dividing $q-1$, *odd, good for* $\mathbf{G}$ *and $\ell \neq 3$ when rational type ${}^3\mathbf{D}_4$ is involved in the type of $(\mathbf{G}, F)$.*

Assume that a Sylow ℓ-subgroup $D \subseteq \mathbf{G}^F$ is commutative. Let $(\mathcal{O}, K, k)$ be an ℓ-modular splitting system for $\mathbf{G}^F$. Then the principal blocks of $\mathcal{O}\mathbf{G}^F$ and $\mathcal{O}\mathrm{N}_{\mathbf{G}^F}(D)$ are Morita equivalent.

Apart from the description of ordinary characters of unipotent blocks (see Theorem 23.2 and Corollary 23.4 above), the proof essentially relies on the following.

Theorem 23.13. *Let $(G, B = TU, N, S)$ be a finite group endowed with a split BN-pair of characteristic p (see Definition 2.20). Let ℓ be a prime such that $|N/T|_\ell = 1$ and $|B : B \cap B^s| \equiv 1$ mod. ℓ for any $s \in S$. Let $\mathcal{O}$ be a complete discrete valuation ring such that $\ell \in J(\mathcal{O})$. Then*

$$\mathcal{O}(N/T_{\ell'}) \cong \mathrm{End}_{\mathcal{O}G}\left(\mathrm{Ind}_{UT_{\ell'}}^G \mathcal{O}\right)$$

by an isomorphism which associates with any $t \in T_\ell$ the endomorphism of $\mathrm{Ind}_{UT_{\ell'}}^G \mathcal{O}$ *sending* $1 \otimes 1$ *to* $t^{-1} \otimes 1$.

If, moreover, $\mathrm{C}_N(T_\ell) = T$*, then* $\mathcal{O}(N/T_{\ell'})$ *is a block.*

Proof of Theorem 23.13. Note that $T = T_\ell \times T_{\ell'}$. The last statement is due to the fact that, if $\mathrm{C}_N(T_\ell) = T$, then $\mathrm{C}_{N/T_{\ell'}}(T_\ell) = T_\ell$, which implies that $N/T_{\ell'}$ has only one ℓ-block (see [Ben91a] 6.2.2).

Denote $W := N/T$. Under the hypothesis on $|W|$, the extension $1 \to T_\ell \to N/T_{\ell'} \to W \to 1$ splits by the Schur–Zassenhaus theorem on group cohomology (see, for instance, [Asch86] 18.1). So $N/T_{\ell'} = T_\ell \rtimes W$ for the usual action of W on T. The proof of the theorem will consist in reducing to a similar situation in $\mathrm{Aut}_{\mathcal{O}G}(\mathrm{Ind}_{UT_{\ell'}}^G \mathcal{O})$.

If V is a subgroup of a finite group H, then $\mathrm{End}_{\mathbb{Z}H}(\mathrm{Ind}_V^H \mathbb{Z}) \cong \mathbb{Z}[V\backslash H] \otimes_{\mathbb{Z}H} \mathbb{Z}[H/V] \cong \mathbb{Z}[V\backslash H/V]$ by the map sending the double coset VhV to the morphism

$$a_h^{(V)} \colon \mathrm{Ind}_V^H \mathbb{Z} \to \mathrm{Ind}_V^H \mathbb{Z}$$

defined by $a_h^{(V)}(1 \otimes 1) = \sum_{v \in V/V\cap V^h} vh^{-1} \otimes 1$. Note that, if R is any commutative ring, then $\mathrm{End}_{RH}(\mathrm{Ind}_V^H R) = \mathrm{End}_{\mathbb{Z}H}(\mathrm{Ind}_V^H \mathbb{Z}) \otimes_{\mathbb{Z}} R$.

Let $V \subseteq V'$ be subgroups of H, then we have a surjection of $\mathbb{Z}H$-modules $\mathrm{Ind}_V^H \mathbb{Z} \to \mathrm{Ind}_{V'}^H \mathbb{Z}$ sending $1 \otimes_V 1$ to $1 \otimes_{V'} 1$. Its kernel is stable under $\mathrm{End}_{\mathbb{Z}H}(\mathrm{Ind}_V^H \mathbb{Z})$ whenever $|V'\backslash H/V| = |V'\backslash H/V'|$ by an easy computation of scalar products of characters (see the proof of Theorem 5.28). In that case, the map $a_h^{(V)} \mapsto a_h^{(V')}$ is a ring morphism

$$\mathrm{End}_{RH}\big(\mathrm{Ind}_V^H R\big) \to \mathrm{End}_{RH}\big(\mathrm{Ind}_{V'}^H R\big).$$

Take now $H = G$, $V = UT_{\ell'}$, and $V' = B$. Denote $A := \mathrm{End}_{\mathcal{O}G}\mathrm{Ind}_V^G \mathcal{O}$. By Bruhat decomposition, we have $V\backslash G/V \cong N/T_{\ell'}$ and $B\backslash G/V \cong B\backslash G/B \cong W$. Then the above gives $A = \bigoplus_{n \in N/T_{\ell'}} \mathcal{O}a_n^{(V)}$. From the definition of the $a_n^{(V)}$'s, it is easily checked that

$$a_t^{(V)} a_n^{(V)} = a_{tn}^{(V)} = a_n^{(V)} a_{t^n}^{(V)} \tag{23.14}$$

for any $t \in T/T_{\ell'}$ and $n \in N/T_{\ell'}$. This implies that the submodule $\bigoplus_{t \in T_\ell} \mathcal{O}a_t^{(V)}$ is a subalgebra of A isomorphic to (and identified with) $\mathcal{O}T_\ell$. One has $J(\mathcal{O}T_\ell) = \{\sum_{t\in T_\ell} \lambda_t t \mid \sum_t \lambda_t \in J(\mathcal{O})\}$ and $J(\mathcal{O}T_\ell).A = A.J(\mathcal{O}T_\ell) = \{\sum_{n \in N/T_{\ell'}} \lambda_n a_n^{(V)} \mid \sum_{t \in T_\ell} \lambda_{tn} \in J(\mathcal{O})$ for all $n \in N/T_{\ell'}\}$ by (23.14) above.

Denote $k = \mathcal{O}/J(\mathcal{O})$. Let now

$$A \otimes k = \mathrm{End}_{kG}\big(\mathrm{Ind}_V^G k\big) \longrightarrow \mathrm{End}_{kG}\big(\mathrm{Ind}_B^G k\big)$$

be the map sending $a_n^{(V)}$ to $a_n^{(B)}$. The ring $\mathrm{End}_{kG}(\mathrm{Ind}_B^G k)$ is the Hecke algebra denoted by $\mathcal{H}_k(G, B)$ in Definition 3.4. It is $\cong kW$ by the map $a_n^{(B)} \mapsto nT \in W$ since $|B : B \cap B^s| = 1$ in k for any $s \in S$. Then the map above gives a morphism of $\mathcal{O}$-algebras

$$A \xrightarrow{\rho} kW$$

defined by $\rho(a_n^{(V)}) = nT \in W$. Its kernel is $J(\mathcal{O}T_\ell).A = A.J(\mathcal{O}T_\ell)$. This is clearly nilpotent mod. $J(\mathcal{O})$, so it is in $J(A)$. However, kW is semi-simple since ℓ does not divide $|W|$. So

(23.15) $$J(A) = J(\mathcal{O}T_\ell).A$$

and the exact sequence associated with ρ can be written

(23.16) $$0 \to J(A) \to A \xrightarrow{\rho} kW \to 0.$$

Since $1 + J(A) \subseteq A^\times$, this gives an exact sequence of groups

$$1 \to 1 + J(A) \to A^\times \to (kW)^\times \to 1.$$

Note that the $a_n^{(V)}$'s ($n \in N/T_{\ell'}$) are invertible since their classes mod. $J(A)$ are. Let Γ be the subgroup of $A^\times$ generated by the $a_n^{(V)}$'s for $n \in N/T_{\ell'}$. By (23.14), Γ normalizes T_ℓ and the restriction of ρ gives a map $\Gamma \to W$ indicating how the elements of Γ act on T_ℓ. So we get an exact sequence

$$1 \to \Gamma \cap (1 + J(A)) \to \Gamma \to W \to 1$$

where $\Gamma \cap (1 + J(A))$ acts trivially on T_ℓ.

If $\mathcal{O}$ is finite, then $1 + J(A)$ is a finite ℓ-group, so (23.16) splits by the Schur–Zassenhaus theorem.

In the general case, one may consider on A the $J(A)$-adic topology associated with the distance $(a, b) \mapsto 2^{-\nu(a-b)}$ where $\nu \colon A \to \mathbb{N}$ is defined by $\nu(J(A)^m \setminus J(A)^{m+1}) = m$. Then ρ is clearly continuous for the discrete topology on kW. This implies that $A^\times = \rho^{-1}((kW)^\times)$ and $1 + J(A) = \rho^{-1}(1)$ are closed in A. Moreover, $x \mapsto x^{-1}$ is continuous on $A^\times$. The groups $\mathrm{C}_{1+J(A)}(T_\ell)$ and $\mathrm{C}_{A^\times}(T_\ell)$ are then closed and normalized by Γ. So (23.16) induces the exact sequence

(23.17) $$1 \to \mathrm{C}_{1+J(A)}(T_\ell) \to \mathrm{C}_{1+J(A)}(T_\ell).\Gamma \to W \to 1$$

giving the way $\mathrm{C}_{1+J(A)}(T_\ell).\Gamma \subseteq \mathrm{N}_{A^\times}(T_\ell)$ acts on T_ℓ. Note that (23.17) implies that $\mathrm{C}_{1+J(A)}(T_\ell).\Gamma$ is closed since $\mathrm{C}_{1+J(A)}(T_\ell)$ is of finite index in it.

The proof of the following is an easy adaptation of [Thévenaz] 45.6 and its proof.

Lemma 23.18. *Let A be an $\mathcal{O}$-free $\mathcal{O}$-algebra of finite rank. Assume that $X \subseteq A^\times$ is a subgroup of $A^\times$ and that X is closed for the $J(A)$-adic topology in A. Assume $X/X \cap \big(1 + J(A)\big)$ is a finite ℓ'-group. Then the group morphism $X \to X/X \cap \big(1 + J(A)\big)$ splits.*

This implies that (23.17) splits. We now have a subgroup $W' \subseteq \mathrm{C}_{1+J(A)}(T_\ell).\Gamma$, isomorphic to W by ρ, and whose action on T_ℓ is that of W. Then $T_\ell.W'$ is a semi-direct product in $A^\times$ isomorphic to $T_\ell \rtimes W$ (for the usual action of W on T).

Denote by M the $\mathcal{O}$-submodule generated by $T_\ell.W'$ in A. It is an $\mathcal{O}T_\ell$-submodule such that $\rho(M) = kW$. This can also be written as $M + J(A) = M + J(\mathcal{O}T_\ell)A = A$ by (23.15). Applying the Nakayama lemma to the $\mathcal{O}T_\ell$-module A, this implies that $M = A$. Since $|T_\ell.W'| = |N : T_{\ell'}|$ is the rank of A, we now have that $T_\ell.W'$ is an $\mathcal{O}$-basis of A and therefore

$$A \cong \mathcal{O}(T_\ell \rtimes W).$$

This gives our claim since $N/T_{\ell'} \cong T_\ell \rtimes W$ as recalled at the beginning of the proof. □

Proof of Theorem 23.12. Let $\mathbf{T} \subseteq \mathbf{B}$ be a maximal torus and Borel subgroup of $\mathbf{G}$, both F-stable (see Theorem 7.1(iii)). Then $\mathbf{T}$ is a 1-split Levi subgroup of $\mathbf{G}$ (see §13.1), so $T = \mathbf{T}^F$, $B = \mathbf{B}^F$, $U = R_u(\mathbf{B})^F$, and $N = \mathrm{N}_{\mathbf{G}}(\mathbf{T})^F$ may be used to define the split BN-pair of $\mathbf{G}^F$. We may also apply Theorem 22.9 with $(\mathbf{L}, \zeta) = (\mathbf{T}, 1)$, which is clearly 1-cuspidal. The associated ℓ-block of $\mathbf{G}^F$ is the principal block since $(\mathbf{G}, 1) \geq (\mathbf{T}, 1)$ (see Definition 21.5; recall that $\mathrm{R}_{\mathbf{T}}^{\mathbf{G}}$ is here just the usual Harish-Chandra induction).

Since the ℓ-subgroups of $\mathbf{G}^F$ are commutative, the defect of the principal block is $\mathbf{T}_\ell^F$ by Lemma 22.18. This means that $\mathbf{T}_\ell^F$ is a Sylow ℓ-subgroup of $\mathbf{G}^F$. Then $\mathrm{N}_{\mathbf{G}}(\mathbf{T})^F/\mathbf{T}^F$ is ℓ'. Note that $\mathrm{N}_{\mathbf{G}^F}(T_\ell) = \mathrm{N}_{\mathbf{G}}(\mathbf{T})^F$ since $\mathbf{T} = \mathrm{C}^\circ_{\mathbf{G}}(T_\ell)$ by Lemma 22.17(ii).

The hypothesis of Theorem 23.13 on B is satisfied since $|B : B \cap B^n|$ ($n \in N$) is a product of orders of intersections $U \cap U^s$ ($s \in N/T$) which are powers of q, being cardinalities of the sets of points over $\mathbb{F}_q$ in affine spaces defined over $\mathbb{F}_q$ (see [DiMi91] 10.11(ii), but a more elementary argument may be given using root subgroups). We finally also have $\mathrm{C}_N(T_\ell) = T$, again by Lemma 22.17(ii).

Let $M := \mathrm{Ind}_{UT_{\ell'}}^G \mathcal{O}$, $A = \mathrm{End}_{\mathcal{O}G}(M)$. We consider M as a $\mathcal{O}G$-A^{opp}-bimodule. Then M is a projective $\mathcal{O}G$-module since $UT_{\ell'}$ is of order invertible in $\mathcal{O}$. We have seen (Theorem 23.13) that $A \cong \mathcal{O}(N/T_{\ell'})$ by an isomorphism sending $t \in T_\ell$ to the endomorphism a_t of M such that $a_t(1 \otimes 1) = t^{-1} \otimes 1$. Since N/T is an ℓ'-group, an $\mathcal{O}(N/T_{\ell'})$-module is projective if and only if it is so once restricted to $\mathcal{O}T_\ell$. This will be the case for M since M is $\mathcal{O}G$-projective and the action of the a_t's ($t \in T_\ell$) is the restriction of $\mathcal{O}G$-action to T_ℓ. So M is bi-projective.

Denote by $B_0 \subseteq \mathcal{O}\mathbf{G}^F$ the principal block.

We know that M is a bi-projective B_0-A-bimodule. To show that it induces a Morita equivalence between B_0 and A, by Theorem 9.18, it suffices to show that $M \otimes K$ induces an equivalence between $B_0 \otimes K$ and $A \otimes K$. But since $A \otimes K = \mathrm{End}_{B_0 \otimes K}(M \otimes K)$ and $B_0 \otimes K$ is semi-simple, it suffices to show that the simple components of $M \otimes K$ are all the simple $B_0 \otimes K$-modules. □

Lemma 23.19. *An element of* $\mathrm{Irr}(\mathbf{G}^F)$ *occurs in* $M \otimes K$ *if and only if it is in* $\mathrm{Irr}(\mathbf{G}^F, B_0)$.

Proof. The character of $M \otimes K = \mathrm{Ind}_{UT_{\ell'}}^{G} K$ is $\mathrm{R}_\mathbf{T}^\mathbf{G}(\mathrm{Ind}_{\mathbf{T}_{\ell'}^F}^{\mathbf{T}^F} 1) = \sum_\theta \mathrm{R}_\mathbf{T}^\mathbf{G}\theta$ where the sum is over ℓ-elements of $\mathrm{Irr}(\mathbf{T}^F)$. Corollary 23.4 gives us that $\mathrm{Irr}(\mathbf{G}^F, \mathrm{R}_\mathbf{T}^\mathbf{G} b_{\mathbf{T}^F}(1))$ is the set of components of generalized characters $\mathrm{R}_{\mathbf{G}(t)}^\mathbf{G}(\hat{t}\chi_t)$ where $(\mathbf{G}(t), \chi_t) \gg_1 (\mathbf{T}, 1)$. But $\gg_1$ is just $\geq$ (see Definition 21.5) by the evident "positivity" of Harish-Chandra induction, and the components of the $\mathrm{R}_{\mathbf{G}(t)}^\mathbf{G}(\hat{t}\chi_t)$'s are the components of the $\mathrm{R}_{\mathbf{G}(t)}^\mathbf{G}\mathrm{R}_\mathbf{T}^{\mathbf{G}(t)}(\hat{t}) = \mathrm{R}_\mathbf{T}^\mathbf{G}(\hat{t})$ since $\varepsilon_\mathbf{G}\varepsilon_{\mathbf{G}(t)}\mathrm{R}_{\mathbf{G}(t)}^\mathbf{G}$ sends $\mathcal{E}(\mathbf{G}(t)^F, t)$ to positive combinations of elements of $\mathcal{E}(\mathbf{G}^F, t)$ (Proposition 15.10), whence Lemma 23.19. □

Remark 23.20. Several finiteness conjectures assert that, once an ℓ-group D is given, the possible ℓ-blocks of finite groups having D as defect group should be taken in a finite list of possible "types" (see [Thévenaz] 38.5 for Puig's conjecture on source algebras). Donovan's conjecture asserts that there is only a finite number of Morita equivalence classes of ℓ-blocks $B \subseteq \mathcal{O}G$ of finite groups $G \supseteq D$ admitting D as a defect group (see [Al86]). Theorem 23.13 clearly implies it for the (few) blocks it considers. This is because those blocks are Morita equivalent to group algebras $\mathcal{O}(D \rtimes N')$ where N' is a *subgroup* of $\mathrm{Aut}(D)$.

Exercises

1. We use the notation of Theorem 23.8.
   (a) Show that $[\mathbf{L}, \mathbf{L}].\mathrm{C}_\mathbf{G}^\circ([\mathbf{L}, \mathbf{L}])$ is a connected reductive $\mathbf{F}$-group containing a maximal torus of $\mathbf{G}$ (use Proposition 22.8).
   (b) Show that $[\mathbf{L}, \mathbf{L}]^F.\mathrm{C}_\mathbf{G}^\circ([\mathbf{L}, \mathbf{L}])^F$ is transitive on 1-split maximal tori of $[\mathbf{L}, \mathbf{L}].\mathrm{C}_\mathbf{G}^\circ([\mathbf{L}, \mathbf{L}])$. Let $\mathbf{T}$ be such a maximal torus of $[\mathbf{L}, \mathbf{L}].\mathrm{C}_\mathbf{G}^\circ([\mathbf{L}, \mathbf{L}])$.
   (c) Show that $\mathrm{N}_\mathbf{G}([\mathbf{L}, \mathbf{L}])^F \subseteq \mathrm{N}_\mathbf{G}(\mathbf{T})^F.[\mathbf{L}, \mathbf{L}]^F.\mathrm{C}_\mathbf{G}^\circ([\mathbf{L}, \mathbf{L}])^F$.

(d) Deduce from the above that the control subgroup H of Theorem 23.8 can be taken such that $C^\circ_{\mathbf{G}}([\mathbf{L},\mathbf{L}])^F \subseteq H \subseteq N_{\mathbf{G}}(\mathbf{T},[\mathbf{L},\mathbf{L}])^F.C^\circ_{\mathbf{G}}([\mathbf{L},\mathbf{L}])^F$.

Notes

The description of non-unipotent characters of unipotent blocks is taken from [CaEn94], a generalization of the corresponding results of [FoSr82] and [FoSr89] for classical groups, and [BrMi93] for abelian defect.

Theorem 23.8 implies that the "Brauer category" (see [Thévenaz] §47) of $R^{\mathbf{G}}_{\mathbf{L}} B_{\mathbf{L}^F}(\zeta)$ is equivalent to that of the principal block of a finite reductive group suitably extended by a group of diagram automorphisms. In [En91] is shown the existence of perfect isometries (see Exercise 9.5) in similar cases for the linear and unitary groups. See [Jost96], [HiKe00] for checkings of Donovan's conjecture.

For Broué's abelian defect conjecture (and relation with Alperin's weight conjecture), see [Rou01], [Rick01]. See the web page

http://www.maths.bris.ac.uk/~majcr/adgc/adgc.html

for the current state of the conjecture. Puig has given a determination of "source algebras" (see [Thévenaz] §18) in a general case which implies Broué's conjecture for groups $\mathbf{G}^F$ when ℓ divides $q-1$, see [Pu90]. For arbitrary ℓ not dividing q, [Bro94] and [Bro95] give more precise conjectures about cohomology of Deligne–Lusztig varieties. See also [Rou02] §4.2.

Appendices

The following three appendices are an attempt to expound many classical results of use in the book, especially around Grothendieck's algebraic geometry. In particular, we tried to give all the necessary definitions so that the statements can be understood. The proofs are in the references we indicate. Some proofs are included for a couple of more special facts (see A2.10 and A3.17) in order to avoid too many direct references to [EGA] or [SGA].

While we have given the fundamental notions, some important theorems are omitted in order to keep this exposition to a reasonable size. So we recommend the basic treatises [Hart], [Weibel], [Milne80], and some more pedagogical texts such as [GelMan94], [Danil96].

Appendix 1

Derived categories and derived functors

We borrow from [McLane97] §VIII, [KaSch98] §1.2, [GelMan94] §§1–5, [Weibel], [Bour80], [KLRZ98] §2.

We use the basic language of categories and functors (full subcategories, natural transformations). In what follows, functors are covariant.

A1.1. Abelian categories

In a category $\mathcal{C}$, we write $X, Y \in \mathcal{C}$ to mean that X, Y are objects of $\mathcal{C}$, and denote by $\mathrm{Hom}_{\mathcal{C}}(X, Y)$ the corresponding set of morphisms.

An additive category is defined by the existence of a zero object, the fact that morphism sets are additive groups for which compositions of morphisms are linear and the existence of finite sums (denoted below as direct sums by $\oplus$). An example is $A-\mathbf{Mod}$, the category of A-modules for A a ring. A basic tool with modules is the existence of kernels and cokernels for any given morphism $X \to Y$ of modules.

In an additive category $\mathcal{A}$, a kernel of a map $X \xrightarrow{f} Y$ is defined as a map $K \xrightarrow{i} X$ such that $f \circ i = 0$ and (K, i) is "final" for this property, i.e. for any $K' \xrightarrow{i'} X$ such that $f \circ i' = 0$, there is a unique $K' \xrightarrow{g} K$ such that $i' = i \circ g$. It is unique up to isomorphism; one writes $\mathrm{Ker}(f) \to X$. One may also define cokernels $Y \to \mathrm{Coker}(f)$ in a formal way.

An additive category $\mathcal{A}$ is said to be abelian when kernels and cokernels exist for any morphism in the category, and left (resp. right) cancellable arrows are kernels (resp. cokernels).

When $X \xrightarrow{f} Y$ is a morphism in an abelian category, one may also define its image $\mathrm{Im}(f) \to Y$ (as the kernel of $Y \to \mathrm{Coker}(f)$) and co-image. Inclusions of objects may be defined as kernels, the corresponding quotients being their co-images.

A basic property in abelian categories is that isomorphisms are characterized by having kernel and cokernel both isomorphic to the zero object, or equivalently by being cancellable on both sides.

Functors between abelian categories are assumed to be additive. One may easily define notions of short exact sequences, projective (resp. injective) objects in $\mathcal{A}$, and also right (resp. left) exact functors between two abelian categories.

Note also that certain embedding theorems (see [GelMan94] 2.2.14.1) allow us to identify "small" abelian categories with subcategories of module categories.

A1.2. Complexes and standard constructions

Let $\mathcal{A}$ be an abelian category. One may define the category $C(\mathcal{A})$ of (cochain) complexes with objects the sequences

$$\dots X^{i-1} \xrightarrow{\partial^{i-1}} X^i \xrightarrow{\partial^i} X^{i+1} \dots$$

such that $\partial^i \partial^{i-1} = 0$ for all i. The morphisms are sequences of maps $f^i : X^i \to B^i$ such that $f^i \partial^{i-1} = \partial^i f^{i-1}$ for all i. One considers the full subcategories $C^+(\mathcal{A})$ (resp. $C^-(\mathcal{A})$, resp. $C^b(\mathcal{A})$) of complexes such that the X^i above are 0 for $-i$ (resp. i, resp. i or $-i$) sufficiently big. All are abelian categories with kernels and cokernels defined componentwise from the same in $\mathcal{A}$.

One defines shift operations as follows. If n is an integer and $X = (X^i, \partial_X^i)$ is a complex, one defines $X[n]$ by $X[n]^i = X^{i+n}$ and $\partial_{X[n]} = (-1)^n \partial_X$. If X is an object of $\mathcal{A}$, one uses the notation $X[n]$ to denote the complex with all terms equal to 0 except the $(-n)$th taken to be X.

A1.3. The mapping cone

The mapping cone of a morphism $f : X \to Y$ in $C(\mathcal{A})$ is defined as the following object $\mathrm{Cone}(f)$ of $C(\mathcal{A})$. One defines $\mathrm{Cone}(f)^i = X^{i+1} \oplus Y^i$ and

$$\partial^i_{\mathrm{Cone}(f)} = \begin{pmatrix} -\partial_X^{i+1} & 0 \\ f^{i+1} & \partial_Y^i \end{pmatrix}$$

(where the matrix stands for the appropriate combination of projections and products of maps in $\mathcal{A}$). One easily defines an exact sequence

$$(T) \qquad 0 \to Y \to \mathrm{Cone}(f) \to X[1] \to 0.$$

A1.4. Homology

Let $\mathcal{A}$ be an abelian category. The homology of a complex $X \in C(\mathcal{A})$ is defined as the sequence of objects $\mathrm{H}^i(X) \in \mathcal{A}$ $(i \in \mathbb{Z})$ in the following way. Since the composition

$$X^{i-1} \xrightarrow{\partial^{i-1}} X^i \xrightarrow{\partial^i} X^{i+1}$$

equals 0, we get $\mathrm{Im}(\partial^{i-1}) \to \mathrm{Ker}(\partial^i)$, which is a kernel, and we define $\mathrm{H}^i(X) := \mathrm{Ker}(\partial^i)/\mathrm{Im}(\partial^{i-1})$.

One may also consider this sequence of objects of $\mathcal{A}$ as a single object $\mathrm{H}(X)$ of $C(\mathcal{A})$ with $\partial_{\mathrm{H}(X)} = 0$. This defines a functor $X \mapsto \mathrm{H}(X)$ (on complexes), $f \mapsto \mathrm{H}(f)$ (on morphisms), from $C(\mathcal{A})$ to itself.

An element of $C(\mathcal{A})$ is said to be acyclic if and only if $\mathrm{H}(X) = 0$. A morphism $f\colon X \to X'$ in $C(\mathcal{A})$ is called a **quasi-isomorphism** if and only if $\mathrm{H}(f)\colon \mathrm{H}(X) \to \mathrm{H}(X')$ is an isomorphism. This is equivalent to $\mathrm{Cone}(f)$ being acyclic.

Note that, in A–**Mod**, one has $\mathrm{H}^i(X) = \mathrm{Ker}(\partial^i)/\partial^{i-1}(X^{i-1})$.

A1.5. The homotopic category

A morphism $f\colon X \to Y$ in $C(\mathcal{A})$ is said to be **null homotopic** if and only if there exists $s\colon X[1] \to Y$ such that $f^i = s^i\partial_X^i + \partial_Y^{i-1}s^{i-1}$ for all i. One may clearly factor out null homotopic morphisms in each morphism group and still have a composition of the corresponding classes. The resulting category $K(\mathcal{A})$ is called the **homotopic category**. Its objects are the same as in $C(\mathcal{A})$, only morphism groups differ. If the identity of a complex is null homotopic, one says that the complex is null homotopic; this is equivalent to being isomorphic to 0 in $K(\mathcal{A})$ (and implies acyclicity). "Conversely," a morphism $f\colon X \to Y$ in $C(\mathcal{A})$ gives an isomorphism in $K(\mathcal{A})$ if and only if $\mathrm{Cone}(f)$ is null homotopic.

One defines $K^b(\mathcal{A})$ and $K^+(\mathcal{A})$ by selecting the corresponding objects in $K(\mathcal{A})$ (see A1.2). These categories are additive but generally not abelian. Note that the above does not use the fact that $\mathcal{A}$ is abelian; it can be done for any additive category. Note also that any additive functor $F\colon \mathcal{A} \to \mathcal{A}'$ induces $K(F)\colon K(\mathcal{A}) \to K(\mathcal{A}')$, and that, if $\mathcal{A}$ is a full subcategory of $\mathcal{A}'$, then $K(\mathcal{A})$ is a full subcategory of $K(\mathcal{A}')$.

Homotopy has the following homological interpretation. If $X, Y \in C(\mathcal{A})$, define $\mathrm{Homgr}(X, Y) \in C(\mathbb{Z}$–**Mod**$)$ by

$$\mathrm{Homgr}(X, Y)^i = \bigoplus_{j\in\mathbb{Z}} \mathrm{Hom}_{\mathcal{A}}(X^j, Y^{j+i})$$

with differential defined as

$$f \mapsto \partial^i_{X,Y}(f) = f \circ \partial_X^{j-1} + (-1)^i \partial_Y^{j+i} \circ f$$

on $\mathrm{Hom}_{\mathcal{A}}(X^j, Y^{j+i})$. Then it can easily be seen that, inside $\bigoplus_j \mathrm{Hom}_{\mathcal{A}}(X^j, Y^{j+i})$, we have $\mathrm{Ker}(\partial^i_{X,Y}) = \mathrm{Hom}_{C(\mathcal{A})}(X, Y[i])$, where $\mathrm{Im}(\partial^i_{X,Y})$ is the subspace corresponding to null homotopic morphisms. We get

$$\mathrm{H}^i(\mathrm{Homgr}(X, Y)) = \mathrm{Hom}_{K(\mathcal{A})}(X, Y[i]).$$

A1.6. Derived categories

The derived category is a category $D(\mathcal{A})$ (additive but not necessarily abelian) with a functor $\delta\colon C(\mathcal{A}) \to D(\mathcal{A})$ such that any quasi-isomorphism is sent to an isomorphism and $(D(\mathcal{A}), \delta)$ is "initial" for this property. Note then that the homology functor factors as $H\colon D(\mathcal{A}) \to C(\mathcal{A})$.

This problem of "localization" with regard to a class of morphisms (here the quasi-isomorphisms) is solved formally in the following way. One keeps the same objects as $C(\mathcal{A})$ while morphisms are now chains $s'_1 f_1 s'_2 f_2 \ldots s'_m f_m$ where $f_i\colon X_i \to Y_i$ are morphisms in $C(\mathcal{A})$ and s'_i are symbols associated with quasi-isomorphisms $s_i\colon Y_i \to X_{i-1}$, and one makes chains equivalent according to the rule $s_i s'_i = \mathrm{Id}_{X_{i-1}}$, $s'_i s_i = \mathrm{Id}_{Y_i}$.

Since the functor $C(\mathcal{A}) \to D(\mathcal{A})$ has to factor as $C(\mathcal{A}) \to K(\mathcal{A}) \to D(\mathcal{A})$, one may start the construction with $K(\mathcal{A})$. This has the following advantage.

Let ε be the empty symbol, or $+$, or b. For any morphism $f\colon X \to Y$ and any quasi-isomorphism $s\colon X \to X'$, resp. $t\colon Y' \to Y$, in $C^\varepsilon(\mathcal{A})$, there is a commutative diagram in $K^\varepsilon(\mathcal{A})$

$$\begin{array}{ccc} X & \xrightarrow{f} & Y \\ \downarrow{\scriptstyle s} & & \downarrow \\ X' & \longrightarrow & Y' \end{array} \qquad \text{resp.} \qquad \begin{array}{ccc} X & \xrightarrow{f} & Y \\ \uparrow & & \uparrow{\scriptstyle t} \\ X' & \longrightarrow & Y' \end{array}$$

with vertical maps quasi-isomorphisms. This allows us to make the above construction of $D^\varepsilon(\mathcal{A})$ from $K^\varepsilon(\mathcal{A})$, taking formal expressions with $m = 1$, i.e. "roofs"

$$Y \xleftarrow{\ t\ } Y' \xrightarrow{\ f\ } X$$

to represent the elements of $\mathrm{Hom}_{D^\varepsilon(\mathcal{A})}(X, Y)$.

Note that the 0 object in $D^{\varepsilon}(\mathcal{A})$ corresponds with images of acyclic complexes.

Assume now that the abelian category $\mathcal{A}$ has enough injective objects, i.e. any object X admits an exact sequence $0 \to X \to I$ where I is injective. Then, injective resolutions allow us to prove that the functor $K^+(\mathcal{A}) \to D^+(\mathcal{A})$ restricts to an equivalence of categories $K^+(\mathbf{inj}_{\mathcal{A}}) \xrightarrow{\sim} D^+(\mathcal{A})$ where $\mathbf{inj}_{\mathcal{A}}$ is the additive full subcategory of $\mathcal{A}$ of injective objects.

More generally, if $\mathcal{I}$ is an additive subcategory of $\mathcal{A}$ such that, for any object X of $\mathcal{A}$, there is an exact sequence $0 \to X \to I$ with I in $\mathcal{I}$, then the functor $K^+(\mathcal{I}) \to D^+(\mathcal{A})$ is full and its "kernel" $\mathcal{N}$ is given by acyclic objects in $C^+(\mathcal{I})$:

$$K^+(\mathcal{I})/\mathcal{N} \cong D^+(\mathcal{A}),$$

the quotient notation meaning that we localize $K^+(\mathcal{I})$ by the morphisms $X \to Y$ embedding into a distinguished triangle $X \to Y \to Z \to X[1]$ in $K^+(\mathcal{I})$ with Z acyclic.

A1.7. Cones and distinguished triangles

The category $D^{\varepsilon}(\mathcal{A})$ (where $\varepsilon = b, +$ or empty) is not abelian in general. Short exact sequences are replaced by the notion of **distinguished triangles**. A triangle is any sequence of maps $X \to Y \to Z \to X[1]$ in $C^{\varepsilon}(\mathcal{A})$. If $X \xrightarrow{f} Y$ is a morphism in $C^{\varepsilon}(\mathcal{A})$, the exact sequence (T) of A1.3 allows to define a triangle

$$X \xrightarrow{f} Y \longrightarrow \mathrm{Cone}(f) \longrightarrow X[1]$$

in $C^{\varepsilon}(\mathcal{A})$. Any triangle is called **distinguished** if it is isomorphic in $K^{\varepsilon}(\mathcal{A})$ with one of the form above. In $D^{\varepsilon}(\mathcal{A})$, we take the images of the ones in $K^{\varepsilon}(\mathcal{A})$.

The distinguished triangles in $K^{\varepsilon}(\mathcal{A})$ or $D^{\varepsilon}(\mathcal{A})$ have many properties leading them to be considered as a reasonable substitute for exact sequences (made into axioms, they define **triangulated categories**, but we shall avoid this abstract notion). Note that each exact sequence $0 \to X \xrightarrow{f} Y \to Z \to 0$ in $\mathcal{A}$ yields a distinguished triangle $X[0] \to Y[0] \to Z[0] \to X[1]$ in $D^b(\mathcal{A})$ since $\mathrm{Cone}(f)$ is quasi-isomorphic to $Z[0]$.

If $\mathcal{A}$ and $\mathcal{B}$ are abelian categories, a functor $D^{\varepsilon}(\mathcal{A}) \to D^{\varepsilon}(\mathcal{B})$ is called **exact** if it preserves distinguished triangles.

If $\mathcal{S}$ is a set of objects of $C^{\varepsilon}(\mathcal{A})$, $K^{\varepsilon}(\mathcal{A})$, or $D^{\varepsilon}(\mathcal{A})$, we define the **subcategory generated by** $\mathcal{S}$, $<\mathcal{S}> \subseteq D^{\varepsilon}(\mathcal{A})$, as the smallest full subcategory containing the image of $\mathcal{S}$ in $D^{\varepsilon}(\mathcal{A})$, and stable under shifts, direct sums and distinguished triangles (and therefore direct summands), the last condition meaning that, if $X \to Y \to Z \to X[1]$ is a distinguished triangle and two of the three first objects are in $<\mathcal{S}>$, then the third also is. It is easily checked that the objects X of $\mathcal{A}$, considered as complexes $X[0]$, generate $D^b(\mathcal{A})$.

A1.8. Derived functors

Let $F: \mathcal{A} \to \mathcal{B}$ be a left-exact functor between abelian categories. Denote by $K^+(F): K^+(\mathcal{A}) \to K^+(\mathcal{B})$ the induced functor and by $Q_A: K^+(\mathcal{A}) \to D^+(\mathcal{A})$ the "localization" functor built with $D^+(\mathcal{A})$. One would like to build an *exact* functor $D^+(F): D^+(\mathcal{A}) \to D^+(\mathcal{B})$ such that there is a natural transformation of functors $\eta: Q_B \circ K^+(F) \to D^+(F) \circ Q_A$ and $(D^+(F), \eta)$ is "initial" for this property.

Let $\mathcal{I}$ be a class of objects of $\mathcal{A}$. We call it F-**injective** if any object X admits an exact sequence $0 \to X \to I$ with I in $\mathcal{I}$, and, for any exact sequence $0 \to I^1 \to I^2 \to I^3 \to 0$ in $\mathcal{A}$ with $I^1, I^2 \in \mathcal{I}$, we have $I^3 \in \mathcal{I}$ and the sequence $0 \to F(I^1) \to F(I^2) \to F(I^3) \to 0$ is exact in $\mathcal{B}$. If $\mathcal{A}$ has enough injective objects, this class will do.

If there is an F-injective class of objects of $\mathcal{A}$, then $D^+(\mathcal{A})$ can be defined by use of the construction of $K^+(\mathcal{I})/\mathcal{N} \cong D^+(\mathcal{A})$ recalled at the end of A1.6. Then $D^+(F)$ is defined by $K^+(F)$ on $K^+(\mathcal{I})$.

The functor $D^+(F): D^+(\mathcal{A}) \to D^+(\mathcal{B})$ is called the **right derived functor** associated with the left-exact functor $F: \mathcal{A} \to \mathcal{B}$. The classical notation is

$$\mathrm{R}F: D^+(\mathcal{A}) \to D^+(\mathcal{B}).$$

Composing $\mathrm{R}F$ with the homology functors $\mathrm{H}^i: D^+(\mathcal{B}) \to \mathcal{B}$, one gets the functors classically called ith right derived functors of F and denoted by $\mathrm{R}^i F$.

By the explicit construction of right derived functors, if $F_1 \to F_2$ is a natural transformation of functors $F_i: \mathcal{A} \to \mathcal{B}$ between abelian categories and $\mathcal{I}$ is both F_1-injective and F_2-injective in $\mathcal{A}$, one gets a natural transformation $\mathrm{R}F_1 \to \mathrm{R}F_2$ of functors $D^+(\mathcal{A}) \to D^+(\mathcal{B})$.

One may also define a notion of left derived functor $\mathrm{L}G: D^-(\mathcal{A}) \to D^-(\mathcal{B})$ called the **left derived functor** associated with the right-exact functor $G: \mathcal{A} \to \mathcal{B}$, injectivity being replaced by projectivity (this can be deduced from the right-handed theory applied to the opposite categories of $\mathcal{A}$ and $\mathcal{B}$).

A1.9. Composition of derived functors

The following is easy and concentrates part of what is also known as spectral sequences.

Let $F_1: \mathcal{A}_1 \to \mathcal{A}_2$, $F_2: \mathcal{A}_2 \to \mathcal{A}_3$, be additive left-exact functors between abelian categories. Assume there is an F_1-injective class $\mathcal{I}_1$ of objects of $\mathcal{A}_1$ and an F_2-injective class $\mathcal{I}_2$ of objects of $\mathcal{A}_2$ such that $F_1(\mathcal{I}_1) \subseteq \mathcal{I}_2$. Then $\mathrm{R}(F_2 \circ F_1) \cong \mathrm{R}F_2 \circ \mathrm{R}F_1$ (see [KaSch98] 1.8.7, [GelMan94] 4.4.15).

A1.10. Exact sequences of functors

Let $0 \to F_1 \to F_2 \to F_3 \to 0$ be an exact sequence of additive left-exact functors $F_i: \mathcal{A} \to \mathcal{B}$ between abelian categories (i.e. $F_i \to F_{i+1}$ are natural transformations such that the induced sequence $0 \to F_1(X) \to F_2(X) \to F_3(X) \to 0$ is exact for any object X of $\mathcal{A}$). Assume there is a class $\mathcal{I}$ of objects of $\mathcal{A}$ which is F_i-injective for each $i = 1, 2, 3$. Then there is a natural transformation $\mathrm{R}F_3 \to \mathrm{R}F_1[1]$ such that, for any object X in $D^+(\mathcal{A})$, the sequence $\mathrm{R}F_1(X) \to \mathrm{R}F_2(X) \to \mathrm{R}F_3(X)$ can be completed to form a distinguished triangle $\mathrm{R}F_1(X) \to \mathrm{R}F_2(X) \to \mathrm{R}F_3(X) \to \mathrm{R}F_1(X)[1]$ (see [KaSch98] 1.8.8). In short, an exact sequence of functors $0 \to F_1 \to F_2 \to F_3 \to 0$ is transformed into a distinguished triangle $\mathrm{R}F_1 \to \mathrm{R}F_2 \to \mathrm{R}F_3 \to \mathrm{R}F_1[1]$.

A1.11. Bi-functors

If $\mathcal{A}$ is a category, it is natural to consider $(X, Y) \mapsto \mathrm{Hom}_{\mathcal{A}}(X, Y)$ as a functor from $\mathcal{A}^{\mathrm{opp}} \times \mathcal{A}$ to the category of sets. The notion of bi-functor (see [KaSch98] §1.10) is devised from this model. It is defined on objects as $F: \mathcal{A} \times \mathcal{A}' \to \mathcal{A}''$ for three categories such that for X (resp. X') an object of $\mathcal{A}$ (resp. $\mathcal{A}'$), we have functors $F(X, -): \mathcal{A}' \to \mathcal{A}''$ (resp. $F(-, X'): \mathcal{A} \to \mathcal{A}''$) satisfying the compatibility condition $F(f, Y') \circ F(X, f') = F(Y, f') \circ F(f, X')$.

We assume now that $\mathcal{A}$, $\mathcal{A}'$ and $\mathcal{A}''$ are abelian categories.

We recall the notion of double complex and associated total complex (an example has been seen in A1.5 above: $\mathrm{Homgr}_{\mathcal{A}}$ complexes). A double complex is an object $X = ((X^{n,m}, \partial^{n,m})_{m\in\mathbb{Z}}), \phi^{n,m})_{n\in\mathbb{Z}}$ of $C(C(\mathcal{A}))$. If for each $n \in \mathbb{Z}$ there is a only a finite number of k such that $X^{k,n-k} \neq 0$, then one defines the total complex $t(X)$ associated with X, as $t(X)^n = \bigoplus_k X^{k,n-k}$ with $\partial^n_{t(X)}$ being $\partial^{k,n-k} \oplus (-1)^k \phi^{k,n-k}$ on the summand $X^{k,n-k}$.

Starting with an additive bi-functor F, left-exact with respect to each variable, the above construction of total complexes allows us to construct $F: C^+(\mathcal{A}) \times C^+(\mathcal{A}') \to C^+(\mathcal{A}'')$. It clearly induces $K^+(F): K^+(\mathcal{A}) \times K^+(\mathcal{A}') \to K^+(\mathcal{A}'')$ and one may ask for a reasonable notion of right derived bi-functor

$$\mathrm{R}F: D^+(\mathcal{A}) \times D^+(\mathcal{A}') \to D^+(\mathcal{A}'').$$

This bi-functor should be exact (i.e. preserve distinguished triangles) with respect to each variable; there should be a natural transformation $\eta: Q_{\mathcal{A}''} \circ K^+(F) \to \mathrm{R}F \circ (Q_{\mathcal{A}} \times Q_{\mathcal{A}'})$, and $(\mathrm{R}F, \eta)$ should be initial for those properties.

It is not difficult to check that, if $\mathcal{A}$ and $\mathcal{A}'$ have enough injective objects, then F admits derived functors with respect to each variable $\mathrm{R}_{\mathcal{A}}F(-, X')$ and $\mathrm{R}_{\mathcal{A}'}F(X, -)$ (see A1.8) for $X \in K^+(\mathcal{A})$, $X' \in K^+(\mathcal{A}')$. It can be defined by $\mathrm{R}_{\mathcal{A}}F(X, X') = K^+(F)(X, X')$ when $X \in K^+(\mathbf{inj}_{\mathcal{A}})$. Moreover $\mathrm{R}F$ also exists and

$$\mathrm{R}F \cong \mathrm{R}_{\mathcal{A}}F \cong \mathrm{R}_{\mathcal{A}'}F$$

(see [KaSch98] §1.10).

Let us return to the case of the bi-functor

$$\mathrm{Hom}_{\mathcal{A}}: \mathcal{A}^{\mathrm{opp}} \times \mathcal{A} \to \mathbb{Z}\text{-}\mathbf{Mod}.$$

When $\mathcal{A}$ is abelian, this is left-exact. Assume $\mathcal{A}$ has enough injective (or enough projective) objects. Using the bi-complex $\mathrm{Homgr}_{\mathcal{A}}$ introduced in A1.5, it is easy to check that

$$\mathrm{H}^i(\mathrm{RHom}_{\mathcal{A}}(C, C')) = \mathrm{Hom}_{D(\mathcal{A})}(C, C'[i])$$

for any $C \in D^+(\mathcal{A}^{\mathrm{opp}})$, $C' \in D^+(\mathcal{A})$, $i \in \mathbb{Z}$.

A1.12. Module categories

Let A be a ring. One denotes by $A\text{-}\mathbf{Mod}$ the category of left A-modules, by $A\text{-}\mathbf{mod}$ its full subcategory corresponding with finitely generated modules. Free modules are projective, so both $A\text{-}\mathbf{Mod}$ and $A\text{-}\mathbf{mod}$ have enough projective objects.

One uses the abbreviation $D^b(A) = D^b(A\text{-}\mathbf{mod})$.

It is customary to call **perfect** the (generally bounded) complexes of finitely generated projective A-modules.

The tensor product is a bi-functor $A^{\mathrm{opp}}\text{-}\mathbf{mod} \times A\text{-}\mathbf{mod} \to \mathbb{Z}\text{-}\mathbf{mod}$ which is right-exact while $A\text{-}\mathbf{mod}$ and $A^{\mathrm{opp}}\text{-}\mathbf{mod}$ have enough projective modules. This allows us to define

$$-\overset{\mathrm{L}}{\otimes}_A-: D^-(A^{\mathrm{opp}}\text{-}\mathbf{mod}) \times D^-(A\text{-}\mathbf{mod}) \to D^-(\mathbb{Z}\text{-}\mathbf{mod}).$$

If X or Y is perfect, $X\overset{\mathrm{L}}{\otimes}_A Y$ is easily defined from the obvious bi-complex $X \otimes_A Y$.

Assume for the remainder of this section that A is a finite-dimensional k-algebra for k a field. Then k-duality $M \mapsto M^*$ induces an exact functor

$A-\mathbf{mod} \to (A^{\mathrm{opp}}-\mathbf{mod})^{\mathrm{opp}}$. One has an isomorphism of bi-functors giving on objects $M^* \otimes_A N \cong \mathrm{Hom}_A(M, N)$ for $M, N \in A-\mathbf{mod}$ and therefore

$$M^* \overset{\mathrm{L}}{\otimes}_A N \cong \mathrm{RHom}_A(M, N)$$

on $D^b(A) \times D^b(A)$.

It can be easily checked that the simple A-modules generate $D^b(A)$.

When C is in $A-\mathbf{mod}$, denote by $\chi_C: A \to k$ the associated "character" defined by $\chi_C(a)$ being the trace of the action of a on C as k-vector space. This extends to $C^b(A-\mathbf{mod})$ by the formula $\chi_C = \sum_i (-1)^i \chi_{C^i}$ and this only depends on the image of C in $D^b(A)$ since $\chi_C = \chi_{\mathrm{H}(C)}$. This is sometimes called the **Lefschetz character** of C.

In the exercises below we give some properties of perfect complexes in relation to quotients $A \to A/I$ (I a two-sided ideal of A) or when A is symmetric.

A1.13. Sheaves on topological spaces

The theory of sheaves of commutative groups on a topological space is a model for many adaptations, specifically (in our case) schemes, coherent sheaves, and sheaves for the étale "topology" on schemes.

Let X be a topological space. One may identify it with the category X_{open} whose objects are open subsets of X and morphisms are inverse inclusions, i.e. $\mathrm{Hom}_X(U, U') = \{\to\}$ (a single element) if $U' \subseteq U$, $\mathrm{Hom}_X(U, U') = \emptyset$ otherwise. Note that any continuous map $f: X \to X'$ induces a functor $f^*_{\mathrm{open}}: X'_{\mathrm{open}} \to X_{\mathrm{open}}$ defined on objects by $U \mapsto f^{-1}(U)$.

If $\mathcal{C}$ is a category equal to $A-\mathbf{Mod}$ or **Sets**, a **presheaf** $\mathcal{F}$ on X with values in $\mathcal{C}$ is any functor

$$\mathcal{F}: X_{\mathrm{open}} \to \mathcal{C}.$$

This consists of a family of objects $(\mathcal{F}(U))_U$, also denoted by $\Gamma(\mathcal{F}, U)$, indexed by open subsets of X and "restriction" morphisms $\rho_{U,U'}: \mathcal{F}(U) \to \mathcal{F}(U')$ for every inclusion of open subsets $U' \subseteq U$. Elements of $\mathcal{F}(U)$ are often called **sections** of $\mathcal{F}$ over U; if $s \in \mathcal{F}(U)$, one often denotes $\rho_{U,U'}(s) = s_{|U'}$. Of course, a functor $X_{\mathrm{open}} \to \mathcal{C}$ may be composed with any functor $\mathcal{C} \to \mathcal{C}'$. Presheaves on X make a category $PSh_{\mathcal{C}}(X)$.

We abbreviate $PSh_{A-\mathbf{Mod}}$ as PSh_A. It is abelian, with kernels and cokernels defined in $A-\mathbf{Mod}$ at each $U \in X_{\mathrm{open}}$.

In **Sets** or any module category $A-\mathbf{Mod}$, arbitrary inductive limits exist. If $x \in X$, we call the **stalk** of $\mathcal{F}$ at x the limit $\mathcal{F}_x := \varinjlim \mathcal{F}(U)$ taken over $U \in X_{\mathrm{open}}$ such that $x \in U$. For any such U, one denotes by $s \mapsto s_x$ the map

$\mathcal{F}(U) \to \mathcal{F}_x$. Taking the stalk at a given x is an exact functor, sometimes denoted by $u_x \colon PSh_{\mathcal{C}}(X) \to \mathcal{C}$.

Sheaves are presheaves $\mathcal{F}$ satisfying one further condition: if $U = \bigcup_i U_i$ with each $U_i \in X_{\text{open}}$ and $s_i \in \mathcal{F}(U_i)$ is a family of sections such that $s_{i|U_i \cap U_j} = s_{j|U_i \cap U_j}$ for each pair i, j, then there is a *unique* $s \in \mathcal{F}(U)$ such that $s_{|U_i} = s_i$ for each i. The uniqueness above forces $\mathcal{F}(\emptyset) = 0$ if $\mathcal{C} = A-$**Mod**, $\mathcal{F}(\emptyset) = \emptyset$ if $\mathcal{C} =$ **Sets**. When $U \subseteq X$ is open, then the restriction of $\mathcal{F}$ to $U_{\text{open}} \subseteq X_{\text{open}}$ is again a sheaf, denoted by $\mathcal{F}_{|U}$.

Sheaves make a full subcategory $Sh_{\mathcal{C}}(X)$ in $PSh_{\mathcal{C}}(X)$. The forgetful functor $Sh_{\mathcal{C}}(X) \to PSh_{\mathcal{C}}(X)$ has a left adjoint called the **sheafification** functor. This is denoted as $\mathcal{F} \mapsto \mathcal{F}^+$ and constructed as follows. If $U \subseteq X$ is open, let $\mathcal{F}^+(U)$ be the set of maps $s \colon U \to \coprod_{x \in U} \mathcal{F}_x$ such that any $x \in U$ has a neighborhood $V \subseteq U$ and a section $t \in \mathcal{F}(V)$ such that $s(y) = t_y$ for all $y \in V$ (thus $s(x) \in \mathcal{F}_x$ for all x). The natural map $\mathcal{F} \to \mathcal{F}^+$ in $PSh(X)$, induces isomorphisms $\mathcal{F}_x \xrightarrow{\sim} \mathcal{F}_x^+$ on stalks.

If $\mathcal{F}$ is a sheaf on X and $U \subseteq X$ is open, one calls the elements of $\mathcal{F}(U)$ the "sections of $\mathcal{F}$ over U" (compare with the construction above). One may also use the notation $\mathcal{F}(U) = \Gamma(U, \mathcal{F})$ when one has to emphasize the functoriality with respect to $\mathcal{F}$ in $Sh_{\mathcal{C}}(X)$.

When M is an object of $\mathcal{C}$, one defines the constant presheaf in $PSh_{\mathcal{C}}(X)$ by $U \mapsto M$ for any open $U \subseteq X$. The sheafification is denoted by M_X and called the **constant sheaf** of stalk M. From the construction of $\mathcal{F}^+$ above, one sees that, if X is locally connected, then $S_X(U) = S \times \pi_0(U)$ if $S \in \mathcal{C} =$ **Sets**, resp. $M_X(U) = M^{\pi_0(U)}$ if $M \in \mathcal{C} = A-$**Mod** where π_0 denotes the set of connected components.

Let $f \colon X \to X'$ be a continuous map between topological spaces. One defines the **direct image** functor

$$f_* \colon PSh_{\mathcal{C}}(X) \to PSh_{\mathcal{C}}(X')$$

by $f_*\mathcal{F}(U') = \mathcal{F}(f^{-1}(U'))$. This preserves sheaves. The **inverse image**

$$f^* \colon Sh_{\mathcal{C}}(X') \to Sh_{\mathcal{C}}(X)$$

is the sheafification of the presheaf inverse image $f^\bullet \colon PSh_{\mathcal{C}}(X') \to PSh_{\mathcal{C}}(X)$ defined by $f^\bullet\mathcal{F}'(U) = \lim\limits_{\to} \mathcal{F}'(U')$ where U' ranges over the neighborhoods of $f(U)$ in X'. Then $f^*\mathcal{F}' := (f^\bullet\mathcal{F}')^+$. Those functors are adjoint on the categories of sheaves, i.e.

$$\mathrm{Hom}_{Sh_A(X')}(\mathcal{F}', f_*\mathcal{F}) \cong \mathrm{Hom}_{Sh_A(X)}(f^*\mathcal{F}', \mathcal{F})$$

as bi-functors.

In the case when $X' = \{\bullet\}$ (a single element), $Sh_A(\{\bullet\})$ identifies with $A-$**Mod** and there is a single continuous map $\sigma: X \to \{\bullet\}$. The functor σ_* identifies with $\Gamma(X, -)$. If M is an A-module, $\sigma^* M = M_X$, the associated constant sheaf. The maps $\{\bullet\} \to X$ are in bijection with elements of X, $x \mapsto \sigma_x$. Then the stalk functor $\mathcal{F} \mapsto \mathcal{F}_x$ coincides with σ_x^*.

A1.14. Locally constant sheaves and the fundamental group

A sheaf $\mathcal{F}$ is said to be **locally constant** if and only if every $x \in X$ has a neighborhood U such that $\mathcal{F}_{|U}$ is constant. Locally constant sheaves are also called local coefficient systems. They make a subcategory $LCS_{\mathcal{C}}(X)$ of $Sh_{\mathcal{C}}(X)$.

The existence of non-constant locally constant sheaves on connected spaces is related to simple connectedness in the following way. Assume X is pathwise connected and every element has a simply connected neighborhood. A **covering** of X is a continuous map $p: X' \to X$ such that any $x \in X$ has a neighborhood V such that $p^{-1}(V)$ is a disjoint union of open subsets all homeomorphic to V by p. Coverings (Y, p) of X make a category where morphisms are denoted by Hom_X. Fix $x_0 \in X$ and denote by $\pi_1(X, x_0)$ the associated fundamental group (homotopy classes of loops based at x_0). For any covering $p: Y \to X$, this group acts on $p^{-1}(x_0)$, thus defining a functor from coverings of X to $\pi_1(X, x_0)$-sets. Moreover, there is a covering $\widetilde{X} \to X$ such that this functor is isomorphic to $\mathrm{Hom}_X(\widetilde{X}, -)$. Denote now by $\mathbf{cov}(X)$ the category of coverings $p: Y \to X$ such that $p^{-1}(x_0)$ has a finite number of connected components, and by $\pi_1(X, x_0) -$ **sets** the category of finite $\pi_1(X, x_0)$-sets acted on continuously. What we have seen implies that they are equivalent. Let $LCS_f(X)$ be the category of locally constant sheaves of sets with finite stalks. It is inserted in the above equivalence as follows

$$\mathbf{cov}(X) \xrightarrow{\sim} LCS_f(X) \xrightarrow{\sim} \pi_1(X, x_0) - \mathbf{sets}$$

where the first arrow sends $p: Y \to X$ to $p_*(S_Y)$ (S a set with a single element) and the second is $\mathcal{F} \mapsto \mathcal{F}_{x_0}$.

An example is the unit circle $\mathbb{P}^1 = \{z \in \mathbb{C} \mid |z| = 1\}$. If $n \geq 2$ is an integer, let $e^{(n)}: \mathbb{P}^1 \to \mathbb{P}^1$ be defined by $z \mapsto z^n$. This is a finite covering corresponding to the quotient $\mathbb{Z}/n\mathbb{Z}$ of the fundamental group $\pi_1(\mathbb{P}^1) \cong \mathbb{Z}$. The sheaf $(e^{(n)})_* S_{\mathbb{P}^1}$ on $\mathbb{P}^1$ is locally constant but not constant.

A1.15. Derived operations on sheaves

The category of sheaves $Sh_A(X)$ is abelian: kernels are the same as in $PSh_A(X)$ (compute at the level of sections), while cokernels are the sheaves associated with presheaf cokernels. A sequence $\mathcal{F}^1 \to \mathcal{F}^2 \to \mathcal{F}^3$ is exact in $Sh_A(X)$ if and only if each sequence of stalks $\mathcal{F}^1_x \to \mathcal{F}^2_x \to \mathcal{F}^3_x$ is exact.

Inverse image functors f^* are exact (since $(f^*\mathcal{F}')_x = \mathcal{F}'_{f(x)}$ for any $x \in X$). Direct image functors f_* are left-exact. One has $(f \circ g)^* = g^* \circ f^*$ and $(f \circ g)_* = f_* \circ g_*$ when $X \xrightarrow{g} Y \xrightarrow{f} Z$ is a composition of continuous maps.

One has two bi-functors on $Sh_A(X)$ with values in $Sh_{\mathbb{Z}}(X)$ induced by Hom_A and $\otimes_A$ on $A-\mathbf{Mod}$. They are defined as follows. If $\mathcal{F}$, $\mathcal{G}$ are in $Sh_A(X)$, one defines $\mathcal{H}om_A(\mathcal{F}, \mathcal{G})$ by

$$U \mapsto \mathrm{Hom}_A(\mathcal{F}(U), \mathcal{G}(U))$$

(this is a sheaf); $\mathcal{H}om_A$ is left-exact with respect to each argument.

Let $\mathcal{F} \otimes_A \mathcal{G}$ be the sheafification of

$$U \mapsto \mathcal{F}(U) \otimes_A \mathcal{G}(U).$$

On stalks this corresponds to $- \otimes_A -$ on modules, so this bi-functor is right-exact.

The abelian category $Sh_A(X)$ has enough injective objects, a consequence of the fact that $A-\mathbf{Mod}$ has enough injective objects. If $\mathcal{F}$ is a sheaf on X and $\mathcal{F}_x \to I_x$ is the inclusion of $\mathcal{F}_x$ into an injective A-module, one has an adjunction map $\mathcal{F} \to \prod_{x\in X}(\sigma_x)_*\sigma_x^*\mathcal{F} = \prod_{x\in X}(\sigma_x)_*\mathcal{F}_x$ which can be composed with $\prod_{x\in X}(\sigma_x)_*\mathcal{F}_x \to \prod_{x\in X}(\sigma_x)_*I_x$. This is the first step of a construction known as "Godement resolution." The derived category $D^+(Sh_A(X))$ is denoted by $D^+_A(X)$ or simply $D^+(X)$.

Direct images f_* preserve injective sheaves (being left-adjoint to f^* which is exact) and are left-exact, so one may define

$$\mathrm{R}f_*\colon D^+_A(X) \to D^+_A(Y)$$

for any continuous map $f\colon X \to Y$, and one has

$$\mathrm{R}(f \circ g)_* = \mathrm{R}f_* \circ \mathrm{R}g_*$$

for any composition $W \xrightarrow{g} X \xrightarrow{f} Y$ of continuous maps (see A1.9).

In the case of $\sigma_X\colon X \to \{\bullet\}$, we write $\mathrm{R}\Gamma(X, \mathcal{F}) \in D^+(A-\mathbf{Mod})$. Its homology is called the homology of $\mathcal{F}$; one denotes

$$\mathrm{H}^i(\mathrm{R}\Gamma(X, \mathcal{F})) = \mathrm{H}^i(X, \mathcal{F}).$$

The above implies a natural isomorphism

$$\mathrm{R}\Gamma(X, \mathcal{F}) \cong \mathrm{R}\Gamma(Y, f_*\mathcal{F})$$

for any continuous map $f\colon X \to Y$, since $\sigma_Y \circ f = \sigma_X$.

Exercises

1. Let $\mathcal{A}$ be an abelian category. Let C be a complex of objects of $\mathcal{A}$. Assume $\mathrm{H}^i(X) = 0$ for all $i \neq 0$. Show that X is quasi-isomorphic to $\mathrm{H}(X)$. Generalize with an interval $\subseteq \mathbb{Z}$ instead of $\{0\}$, and truncation instead of $\mathrm{H}(X)$.
2. Let $\mathcal{A}$ be an abelian category. If X is an object of $\mathcal{A}$ and $n \in \mathbb{Z}$, denote by $X_{[n,n+1]} \in C^b(\mathcal{A})$ the complex defined by $X^n_{[n,n+1]} = X^{n+1}_{[n,n+1]} = X$, $\partial^n = \mathrm{Id}_X$, and all other $X^i_{[n,n+1]} = 0$ (i.e. the mapping cone of the identity morphism of $X[-n]$).
   (a) Show that $C \in C^b(\mathcal{A})$ is null homotopic if and only if it is a direct sum of complexes of the type above. Generalize to $C^+(\mathcal{A})$.
   (b) Let $C \in C^b(\mathbf{proj}_{\mathcal{A}})$ be a bounded complex of projective objects of $\mathcal{A}$. Assume $\mathrm{H}^i(C) = 0$ for $i \geq n_0$. Show that $C \cong C_0 \oplus C_1$ where C_0 is null homotopic and $C_1^i = 0$ for $i \geq n_0$.
3. Let $\mathcal{A}$ be an abelian category.
   (a) Let C be an object of $C^b(\mathcal{A})$ with $C^i = 0$ for $i \notin [m, m']$. Let $f\colon C \to C^m[-m]$ be defined by Id at degree m. Show that $\mathrm{Cone}(f) \cong (C^m)_{[m-1,m]} \oplus C^{>m}[1]$ where $(C^m)_{[m-1,m]}$ is as defined in Exercise 2 and $C^{>m}$ coincides with C on degrees strictly greater than m and is zero elsewhere.
   (b) Same hypothesis as in the question above. Let $g\colon C^{m'}[-m'] \to C$ be defined by Id at degree m'. Show that $\mathrm{Cone}(g) \cong (C^{m'})_{[m'-1,m']} \oplus C^{<m'}$.
   (c) Show that any term of a bounded acyclic complex is in the subcategory of $D^b(\mathcal{A})$ generated by the other terms.
   (d) If A is an artinian ring (or a finite-dimensional algebra over a field), show that $D^b(A\text{−}\mathbf{mod})$ is generated by the simple A-modules.
4. Let Λ be a principal ideal domain, and A be a Λ-algebra, Λ-free of finite rank. If $a \in A$, denote by $[a.]$ the left translation by a, i.e. $[a.](b) = ab$. Assume there is a Λ-linear form $f\colon A \to \Lambda$ such that $a \mapsto f \circ [a.]$ is a Λ-isomorphism between A and $\mathrm{Hom}_\Lambda(A, \Lambda)$ (for instance, $A = \Lambda G$ is the group algebra of a finite group G).

Denote by $M \mapsto {}^\vee M = \mathrm{Hom}_\Lambda(M, \Lambda)$ the Λ-duality functor from $A-\mathbf{mod}$ to $\mathbf{mod}-A$.

(a) Show that ${}^\vee M$ is projective when M is.

Let C be an object of $C^b(A-\mathbf{proj})$ and $m \le m'$ be integers such that $\mathrm{H}^i(C)$ is Λ-free for all $i \in [m, m']$ and $\mathrm{H}^i(C) = \{0\}$ for all $i \notin [m, m']$.

(b) Show that $\mathrm{Ker}({}^\vee\partial^i) = (\partial^i(C^i))^\perp$ and $\mathrm{Ker}(\partial^i)^\perp = {}^\vee\partial^i({}^\vee C^i)$ for any i.

(c) Show that $C \cong C_0 \oplus C_1$ in $C^b(A)$, where C_0 is null homotopic and $C_1^i = 0$ for all $i \notin [m, m']$ (apply Exercise 2.2 to C and ${}^\vee C$).

5. Let A be a ring and I a two-sided ideal of A such that $II = \{0\}$. Let $\rho\colon A/I-\mathbf{mod} \to A/I-\mathbf{mod}$ denote the functor $M \mapsto M/IM$.

(a) Show that ρ induces a functor $A-\mathbf{proj} \to A/I-\mathbf{proj}$ which is onto on objects and morphisms (lift idempotents in $\mathrm{Mat}_n(A) \to \mathrm{Mat}_n(A/I)$).

(b) Show that a map in $P \xrightarrow{f} Q$ in $A-\mathbf{proj}$ is a direct injection if and only if $\rho(P) \xrightarrow{\rho(f)} \rho(Q)$ is.

(c) Show that, if C_0 is in $C^b(A/I-\mathbf{proj})$ and null homotopic, then there is C in $C^b(A-\mathbf{proj})$, null homotopic, and such that $C_0 = \rho(C)$. If, moreover, C' is in $C^b(A-\mathbf{proj})$, any map $C_0 \to \rho(C')$ is of the form $\rho(C \to C')$.

(d) Show that, if C_1, C_2 are in $C^b(A-\mathbf{proj})$ and $\rho(C_1) \xrightarrow{x} \rho(C_2)$ is null homotopic then $x = \rho(y)$ where $C_1 \xrightarrow{y} C_2$ is null homotopic.

6. We keep the same hypotheses as in Exercise 5. Let C (resp. C_0) be an object of $C^b(A-\mathbf{proj})$ (resp. $C^b(A/I-\mathbf{proj})$) such that $\rho(C) \cong C_0$ in $D^b(A/I)$.

We want to show that there is an object C' of $C^b(A-\mathbf{proj})$ such that $\rho(C) \cong C_0$ and $C \cong C'$ in $D^b(A)$.

(a) Show that we may assume that there is a quasi-isomorphism $\rho(C) \xrightarrow{f} C_0$ in $C^b(A/I-\mathbf{proj})$ with each f^i onto (use Exercise 2), thus giving

$$0 \to R_0 \to \rho(C) \to C_0 \to 0$$

exact in $C^b(A/I-\mathbf{proj})$ with R_0 acyclic.

(b) Applying Exercise 5.3 to $R_0 \to \rho(C)$ above, define a quotient C/R in $C^b(A-\mathbf{proj})$ such that $\rho(C/R) \cong C_0$ (use Exercise 5.2) and C is quasi-isomorphic to C/R.

Notes

For a more complete introduction, emphasizing the rôle of unbounded complexes, see [Ke98]. See Houzel's introduction to [KaSch98] for a historical account of sheaf theory, from Leray's ideas up to $f^!$ functors.

(. . .) vers l'âge de douze ans, j'étais interné au camp de concentration de Rieucros (près de Mende). C'est là que j'ai appris, par une détenue, Maria, qui me donnait des leçons particulières bénévoles, la définition du cercle. Celle-ci m'avait impressionné par sa simplicité et son évidence, alors que la propriété de "rotondité parfaite" du cercle m'apparaissait auparavant comme une réalité mystérieuse au-delà des mots. C'est à ce moment, je crois, que j'ai entrevu pour la première fois (sans bien sûr me le formuler en ces termes) la puissance créatrice d'une "bonne" définition mathématique, d'une formulation *qui décrit l'essence.*

Alexandre Grothendieck, *Esquisse d'un programme*, 1984

Appendix 2

Varieties and schemes

In this appendix, we recall some basic results about varieties and schemes that are useful in this book. The main references for the subject are [Hart], [Milne98], [Mum70], [Mum88], [Kempf] (varieties, schemes), [CaSeMcD] (Lie groups), [Borel], [Springer] (algebraic groups), [Jantzen] (vector bundles) and [Danil94]. We indicate references for precise statements only when they seemed difficult to find. We begin with varieties over an algebraically closed field and properties of morphisms. Then we recall the basic results on algebraic groups. Schemes are introduced in A2.7. Working with schemes has many advantages (see the official list in [Mum88] p. 92, [Hart] pp. 58–9), even for studying varieties. This will be apparent in the notion and computation of étale cohomology (see Appendix 3). It is also necessary for the consistency of references in view of the quasi-affinity criterion of A2.10.

A2.1. Affine F-varieties

Let $\mathbf{F}$ be an algebraically closed field.

Affine $\mathbf{F}$-varieties are defined as sets of zeroes in $\mathbf{F}^n$ of elements of the polynomial ring $\mathbf{F}[t_1, \ldots, t_n]$ ("equations"). By Hilbert's "Nullstellensatz" the elements of such a set are in bijection with the set of maximal ideals $\mathrm{Max}(A)$ where $A = \mathbf{F}[t_1, \ldots, t_n]/I$ for I a reduced ideal (i.e. I contains any polynomial P that would satisfy $P^n \in I$ for some $n \geq 1$, i.e. $\sqrt{I} = I$). Note that $\mathrm{Max}(0) = \emptyset$.

This allows us to call **affine F-varieties** any $V = \mathrm{Max}(A)$ for A a commutative finitely generated $\mathbf{F}$-algebra which is reduced (nilradical = 0). One then denotes $A = \mathbf{F}[V]$ and calls the latter the **ring of regular functions on** V since it is isomorphic to the ring of maps on V obtained by restrictions of elements of $\mathbf{F}[t_1, \ldots, t_n]$ when A is written as a reduced quotient of $\mathbf{F}[t_1, \ldots, t_n]$.

The **Zariski topology** on V is the one generated by subsets of V of the form $V_f := \{x \in \mathrm{Max}(A) = V \mid f \notin x\}$ for some $f \in A = \mathbf{F}[V]$. In general this topology is not Hausdorff. In the case of $\mathbf{F}^n$ itself, the non-empty open subsets are dense. When $n = 1$ they are complements of finite sets.

One calls V_f the principal open subset associated with f; it is clearly in bijection with $\mathrm{Max}(A_f)$ where A_f is the localization $S^{-1}A$ for $S = \{f^n \mid n \geq 1\}$. The complement $V \setminus V_f$ is in bijection with $\mathrm{Max}(A/fA)$. If $x \in V = \mathrm{Max}(A)$, one denotes by A_x the localization $S^{-1}A$ where $S = A \setminus x$. It is a local ring with residual field $\mathbf{F}$.

A2.2. Locally ringed spaces and F-varieties

Putting together all the above bijections (and explaining the word "local") is achieved by defining a "structure" sheaf of rings $\mathcal{O}_V$ on V, for the Zariski topology. If U is open in V, one defines $\mathcal{O}_V(U)$ as the ring of maps $f: U \to \mathbf{F}$ such that for all $x \in U$ there is a neighbourhood $U' \ni x$ such that $f = g/h$ on U' where $g, h \in A$ and $h(U') \not\ni 0$. This is a sheaf for restrictions defined in the obvious way. These are ring morphisms (hence the term "sheaf of rings"). Stalks $\mathcal{O}_{V,x}$ are the A_x's defined above.

The affine variety $V = \mathrm{Max}(A)$ endowed with its structure sheaf $\mathcal{O}_V$ is what is called a **locally ringed space**: a topological space X endowed with a sheaf of commutative rings $\mathcal{O}_X$ such that restriction morphisms are ring morphisms and stalks are local rings. A morphism $(X, \mathcal{O}) \to (X', \mathcal{O}')$ of locally ringed spaces is any pair $(f, f^\sharp)$ where $f: X \to X'$ is continuous and $f^\sharp: \mathcal{O}' \to f_*\mathcal{O}$ is a morphism of sheaves on X' such that each $f^\sharp(U'): \mathcal{O}'(U') \to f_*\mathcal{O}(U') = \mathcal{O}(f^{-1}(U'))$ is a commutative ring morphism and each induced map $f_x^\sharp: \mathcal{O}'_{f(x)} \to (f_*\mathcal{O})_{f(x)} \to \mathcal{O}_x$ on stalks is such that $(f_x^\sharp)^{-1}(J(\mathcal{O}_x)) = J(\mathcal{O}'_{f(x)})$. Imposing further that all rings are $\mathbf{F}$-algebras and morphisms are $\mathbf{F}$-linear, one may even speak of locally $\mathbf{F}$-ringed spaces. Another way of doing that is to impose that each locally ringed space $(X, \mathcal{O})$ is endowed with a "structural" morphism $\sigma_X: X \to \mathrm{Max}(\mathbf{F})$ of locally ringed spaces, and morphisms $f: X \to X'$ must satisfy $\sigma_{X'} \circ f = \sigma_X$. One then speaks of **spaces "over F**."

In the case of affine $\mathbf{F}$-varieties, it is easily checked that morphisms $V \to V'$ coincide with $\mathbf{F}$-linear ring morphisms $\mathbf{F}[V'] \to \mathbf{F}[V]$, or more explicitly, considering V, resp. V', as a closed subset of $\mathbf{F}^n$, resp. $\mathbf{F}^{n'}$, with the restrictions to V of the polynomial maps $\mathbf{F}^n \to \mathbf{F}^{n'}$ that send V into V'.

One defines $\mathbf{F}$-prevarieties as locally ringed spaces $(X, \mathcal{O})$ over $\mathbf{F}$ admitting a finite covering $X = V_1 \cup \ldots \cup V_m$ by open subsets such that each $(V_i, \mathcal{O}_{|V_i})$

is isomorphic (as locally ringed space over **F**) to an affine **F**-variety. This makes a category. By reduction to affine varieties, it can be checked that open subsets and closed subsets of **F**-prevarieties are **F**-prevarieties (see [Hart] I.4.3).

It can also be checked that **F**-prevarieties are noetherian and quasi-compact (the same definition as compact but without the Hausdorff axiom). Any **F**-prevariety is a finite union of **irreducible** closed subsets.

Apart from affine **F**-varieties, an important example is that of projective spaces $\mathbb{P}^n_{\mathbf{F}}$ (lines of $\mathbf{F}^{n+1}$) where the V_i's are the subsets $\{\mathbf{F}.(x_1, \ldots, x_{n+1}) \mid x_i \neq 0\}$ in bijection with $\mathbf{F}^n$.

A basic property is that products exist in the category of **F**-prevarieties. The problem reduces to affine **F**-varieties where the solution is given by $\mathrm{Max}(A) \times \mathrm{Max}(B) := \mathrm{Max}(A \otimes_{\mathbf{F}} B)$. The basic example is $\mathbb{A}^n_{\mathbf{F}}$. Note that the Zariski topology on a product is generally not the product topology.

Arbitrary **F-varieties** are defined as prevarieties V such that the diagonal of $V \times V$ is closed. This is the case for affine **F**-prevarieties, projective spaces and any locally closed subset of an **F**-variety ("**subvariety**"). One denotes by $\mathbb{A}^n_{\mathbf{F}}$ the **F**-variety associated with $\mathbf{F}^n$, called the **affine space**. When $n \geq 2$, the open subvariety $\mathbb{A}^n_{\mathbf{F}} \setminus \{0\}$ is not affine (see [Kempf] 1.6.1). One calls **quasi-affine**, resp. **quasi-projective**, any **F**-variety isomorphic to a locally closed subvariety of some $\mathbb{A}^n_{\mathbf{F}}$, resp. $\mathbb{P}^n_{\mathbf{F}}$. Closed subvarieties of some $\mathbb{P}^n_{\mathbf{F}}$ are called **projective F-varieties**.

Note that our varieties are reduced in the sense that all rings defined by the structure sheaf are reduced. This constraint is traditional but not essential for many theorems (see below the notion of scheme over **F**).

For any affine variety $V = \mathrm{Max}(A)$, and any **F**-variety X there is a natural isomorphism

$$\mathrm{Hom}_{\mathbf{F}-Var}(X, V) \xrightarrow{\sim} \mathrm{Hom}_{\mathbf{F}-alg}(A, \mathcal{O}_X(X)).$$

An **F**-variety V is called **complete** if and only if, for any **F**-variety V', the projection $V' \times V \to V'$ is a closed map (the word *compact* is also used, referring to the corresponding notion for Hausdorff topological spaces). Closed subvarieties of complete varieties are complete. Projective varieties are complete.

The **dimension** $\dim_{\mathbf{F}}(V)$ of an **F**-variety V is the biggest d such that there is a chain of distinct non-empty irreducible closed subsets $V_0 \subset V_1 \subset \ldots \subset V_d$, i.e. the maximal dimension of its irreducible components. When V is irreducible, $\mathbf{F}[V]$ has no divisor of zero and the dimension of V is the transcendence degree of the field of quotients of $\mathbf{F}[V]$ over **F**. Dimension is additive with respect to product. The non-empty open subvarieties of $\mathbb{A}^n$ and $\mathbb{P}^n$ are of dimension n.

A2.3. Tangent sheaf, smoothness

Let $(X, \mathcal{O}_X)$ be an **F**-variety. The **tangent sheaf** $\mathcal{T}X$ on X is defined by $\mathcal{T}X(U) = \mathrm{Hom}(\mathrm{Max}(\mathbf{F}[t]/t^2), U)$ (morphisms of locally ringed spaces over **F**), clearly a functor on open subsets $U \subseteq X$. An element of $\mathcal{T}X(U)$ is the datum of some $u \in U$ along with a local morphism $\mathcal{O}_{X,u} \to \mathbf{F}[t]/t^2$. Thus the stalk at $x \in X$ is

$$\mathcal{T}X_x = \mathrm{Hom}_{\mathbf{F}}(J(\mathcal{O}_{X,x})/J(\mathcal{O}_{X,x})^2, \mathbf{F}).$$

If $f: X \to Y$ is a morphism of **F**-varieties, since $f^\sharp$ induces morphisms of local rings $\mathcal{O}_{Y,f(x)} \to \mathcal{O}_{X,x}$, we get morphisms of **F**-modules

$$\mathcal{T}f_x : \mathcal{T}X_x \to \mathcal{T}Y_{f(x)}.$$

They satisfy the properties expected from derivatives. One has $(\mathcal{T}\mathrm{Id}_X)_x = \mathrm{Id}_{\mathcal{T}X_x}$ and $\mathcal{T}(g \circ f)_x = (\mathcal{T}g)_{f(x)} \circ \mathcal{T}f_x$ (the "chain rule") when $Y \xrightarrow{f} X \xrightarrow{g} Z$ are morphisms of **F**-varieties. If $X' \subseteq X$ is a closed **F**-subvariety of X and $i: X' \to X$ denotes the associated closed immersion, then $\mathcal{T}i_x : \mathcal{T}X'_x \to \mathcal{T}X_x$ is injective for all $x \in X'$. We have $\mathcal{T}(X \times Y)_{(x,y)} = \mathcal{T}X_x \times \mathcal{T}Y_y$ for any $(x, y) \in X \times Y$.

If X is an affine **F**-variety (or an affine neighborhood of $x \in X$), i.e. a closed subvariety of $\mathbb{A}^n_{\mathbf{F}}$, $\mathcal{T}X_x$ is defined in $\mathbf{F}^n$ by the partial derivatives at x of the defining equations of $X \subseteq \mathbb{A}^n_{\mathbf{F}}$.

A point $x \in X$ is called **smooth** if and only if $\dim_{\mathbf{F}}(U) = \dim_{\mathbf{F}}(\mathcal{T}X_x)$ for some neighborhood U of x in X. The **F**-variety is called **smooth** if and only if all its points are smooth. In a smooth variety, irreducible components are connected components (i.e. they are open).

A **regular** variety is a variety such that, for all $x \in X$, $\mathcal{O}_{X,x}$ is an integrally closed (hence integral) ring. A smooth variety over **F** is regular (see [Atiyah–Macdonald] 11.23).

Two subspaces $X, Y \subseteq Z$ of a finite-dimensional **F**-space Z are said to intersect transversally if and only if $X + Y = Z$ or $X \cap Y = \{0\}$. Two closed subvarieties X, Y of an **F**-variety Z are said to intersect transversally at $z \in X \cap Y$ if and only if $\mathcal{T}X_z$ and $\mathcal{T}Y_z$ intersect transversally in $\mathcal{T}Z_z$. The subvarieties X, Y are said to **intersect transversally** (without z being specified) if the above holds for any $z \in X \cap Y$. When, moreover, X, Y, and Z are smooth, $X \cap Y$ is smooth.

A related notion is that of **smooth divisor with normal crossings**. Let X be a smooth connected **F**-variety of dimension $d \geq 1$. A smooth divisor is a finite union $D_1 \cup D_2 \cup \ldots \cup D_m$ of smooth connected subvarieties $D_i \subseteq X$ of dimension $d - 1$. It is said to have normal crossings if and only if for any $x \in D$

the hyperplanes $(\mathcal{T}D_i)_x$ (for i such that $x \in D_i$) are in general position in $\mathcal{T}X_x$ (i.e. have linearly independent equations).

A2.4. Linear algebraic groups and reductive groups

Linear algebraic groups, or **F-groups**, are defined as affine **F**-varieties with a group structure such that multiplication and inversion are morphisms for the structure of **F**-variety. This coincides in fact with (Zariski) closed subgroups of $\mathrm{GL}_n(\mathbf{F})$. Morphisms of algebraic **F**-groups are group morphisms that are at the same time morphisms of **F**-varieties. If $f : \mathbf{H} \to \mathbf{G}$ is a morphism, then both $\mathrm{Ker}(f)$ and $\mathrm{Im}(f)$ are closed subgroups, and we have

$$\dim\ \mathbf{G} = \dim\ \mathrm{Ker}(f) + \dim\ \mathrm{Im}(f).$$

Basic examples of linear algebraic groups are $\mathbb{A}^1_{\mathbf{F}}$ for the additive law, denoted by $\mathbb{G}_{\mathrm{a}}$, and $\mathbf{F}^\times$ for the multiplicative law, denoted by $\mathbb{G}_{\mathrm{m}}$.

In what follows, **G** is a linear algebraic group.

Algebraic groups are smooth **F**-varieties, with all irreducible components of the same dimension. One denotes by $\mathbf{G}^\circ$ the connected component of **G** containing the unit element; this is a closed normal subgroup. The irreducible components of **G** coincide with connected components and are the cosets of $\mathbf{G}/\mathbf{G}^\circ$. In order to lighten certain notation, we write $\mathrm{Z}^\circ(\mathbf{G}) := \mathrm{Z}(\mathbf{G})^\circ$ and $\mathrm{C}^\circ_{\mathbf{G}}(g) := \mathrm{C}_{\mathbf{G}}(g)^\circ$ for centers and centralizers.

An **action** of **G** on an **F**-variety **X** is a group action such that the associated map $\mathbf{G} \times \mathbf{X} \to \mathbf{X}$ is a morphism of **F**-varieties. The orbits $\mathbf{G}.x$ ($x \in \mathbf{X}$) are locally closed subvarieties of **X**.

We have a similar notion of linear representations (by action on vector **F**-spaces). For instance **G** acts on $\mathbf{F}[\mathbf{G}]$ by composition of regular functions with left translations, thus giving

$$\mathbf{G} \to \mathrm{GL}_{\mathbf{F}}(\mathbf{F}[\mathbf{G}]);\quad g \mapsto \lambda_g.$$

One has in **G** a **Jordan decomposition** $g = g_u g_s = g_s g_u$ where λ_{g_s} (resp. λ_{g_u}) is the semi-simple (resp. unipotent) part of λ_g in $\mathrm{GL}(\mathbf{F}[\mathbf{G}])$. This is preserved by any linear representation. The set of semi-simple elements of **G** is denoted by $\mathbf{G}_{\mathrm{ss}}$.

Note that, when **F** is the algebraic closure of a finite field, any element of **G** is of finite order and its Jordan decomposition coincides with its decomposition into p-part and p'-part (where p is the characteristic of **F**).

The tangent space at the unit element, $\mathcal{T}\mathbf{G}_1$, is a Lie algebra; we omit the subscript and write $\mathcal{T}\mathbf{G}$ or $\mathrm{Lie}(\mathbf{G})$. The differential of the multiplication

$\mathbf{G} \times \mathbf{G} \to \mathbf{G}$ is the addition $\mathrm{Lie}(\mathbf{G}) \times \mathrm{Lie}(\mathbf{G}) \to \mathrm{Lie}(\mathbf{G})$. The differential of inversion is $-\mathrm{Id}_{\mathrm{Lie}(\mathbf{G})}$.

A **torus** is any closed subgroup $\mathbf{T} \subseteq \mathbf{G}$ isomorphic to some finite product $\mathbb{G}_\mathrm{m} \times \cdots \times \mathbb{G}_\mathrm{m}$ (then a commutative divisible group where all elements are semi-simple). One then denotes $X(\mathbf{T}) := \mathrm{Hom}(\mathbf{T}, \mathbb{G}_\mathrm{m})$, a free $\mathbb{Z}$-module $\cong \mathbb{Z}^n$ if $\mathbf{T} \cong (\mathbb{G}_\mathrm{m})^n$. If V is a finite-dimensional $\mathbf{F}$-vector space and $\mathbf{T} \to \mathrm{GL}_\mathbf{F}(V)$ is a rational linear representation of a torus, one has $V = \bigoplus_{\alpha \in X(\mathbf{T})} V_\alpha$ where $V_\alpha := \{v \in V \mid t.v = \alpha(t)v \text{ for all } t \in \mathbf{T}\}$.

A **Borel subgroup** of $\mathbf{G}$ is any maximal connected solvable subgroup of $\mathbf{G}$. They are all $\mathbf{G}$-conjugates, so are maximal tori. The **unipotent radical** $R_u(\mathbf{G})$ is the biggest closed connected normal subgroup of $\mathbf{G}$ whose elements are all unipotent. One says $\mathbf{G}$ is a **reductive group** if and only if $R_u(\mathbf{G}) = \{1\}$.

Let $\mathbf{T} \subseteq \mathbf{G}$ be a maximal torus in a linear algebraic $\mathbf{F}$-group. Then the action of $\mathbf{T}$ on $\mathrm{Lie}(\mathbf{G})$ gives $\mathrm{Lie}(\mathbf{G}) = \mathrm{Lie}(\mathbf{G})^\mathbf{T} \oplus \bigoplus_{\alpha \in \Phi(\mathbf{G},\mathbf{T})} \mathrm{Lie}(\mathbf{G})_\alpha$ where $\Phi(\mathbf{G}, \mathbf{T})$ is the set of $\alpha \in X(\mathbf{T})$ such that $\alpha \neq 0$ and $\mathrm{Lie}(\mathbf{G})_\alpha \neq 0$. They are called the **roots** of $\mathbf{G}$ relative to $\mathbf{T}$. Assume now that $\mathbf{G}$ is reductive. Then we have $\mathrm{Lie}(\mathbf{G})^\mathbf{T} = \mathrm{Lie}(\mathbf{T})$ and each $\mathrm{Lie}(\mathbf{G})_\alpha$ for $\alpha \in \Phi(\mathbf{G}, \mathbf{T})$ is a line. The set $\Phi(\mathbf{G}, \mathbf{T})$ can be endowed naturally with a structure of crystallographic root system in $X(\mathbf{T}) \otimes_\mathbb{Z} \mathbb{R} \cong \mathbb{R}^{\dim(\mathbf{T})}$ whose associated reflection group is isomorphic to the "**Weyl group**" $W(\mathbf{G}, \mathbf{T}) := \mathrm{N}_\mathbf{G}(\mathbf{T})/\mathbf{T}$. If $\mathbf{B}$ is a Borel subgroup containing $\mathbf{T}$, then $\mathbf{B} = \mathbf{U} \rtimes \mathbf{T}$ where $\mathbf{U} = R_u(\mathbf{B})$, and $\Phi(\mathbf{B}, \mathbf{T})$ contains a unique basis Δ of the root system $\Phi(\mathbf{G}, \mathbf{T})$. The pair $(\mathbf{B}, \mathrm{N}_\mathbf{G}(\mathbf{T}))$ endows $\mathbf{G}$ with the BN-pair, or Tits system, satisfying $\mathbf{B} \cap \mathrm{N}_\mathbf{G}(\mathbf{T}) = \mathbf{T}$. The associated **length function** in $W(\mathbf{G}, \mathbf{T})$ is denoted by l. Each line $\mathrm{Lie}(\mathbf{G})_\alpha$ can also be written $\mathrm{Lie}(\mathbf{G})_\alpha = \mathrm{Lie}(\mathbf{X}_\alpha)$ where $\{\mathbf{X}_\alpha \mid \alpha \in \Phi(\mathbf{B}, \mathbf{T})\}$ is the set of minimal $\mathbf{T}$-stable non-trivial subgroups of $\mathbf{U}$ (**root subgroups**, all isomorphic to $\mathbb{G}_\mathrm{a}$). The parabolic subgroups of $\mathbf{G}$ containing $\mathbf{B}$ can be written as $\mathbf{P}_I = \mathbf{U}_I \rtimes \mathbf{L}_I$ (**Levi decomposition**) for $I \subseteq B$, with $\mathbf{U}_I = R_u(\mathbf{P}_I)$ and $\mathrm{Lie}(\mathbf{P}_I) = \mathrm{Lie}(\mathbf{U}_I) \oplus \mathrm{Lie}(\mathbf{L}_I)$ where $\mathrm{Lie}(\mathbf{L}_I) = \mathrm{Lie}(\mathbf{T}) \oplus \bigoplus_{\alpha \in \Phi_I} \mathrm{Lie}(\mathbf{G})_\alpha$, and $\mathrm{Lie}(\mathbf{U}_I) = \bigoplus_{\alpha \in \Phi(\mathbf{G},\mathbf{T}) \setminus \Phi_I} \mathrm{Lie}(\mathbf{G})_\alpha$ for $\Phi_I = \Phi(\mathbf{G}, \mathbf{T}) \cap \mathbb{R}I$. The "**Levi subgroup**" $\mathbf{L}_I$ is generated by $\mathbf{T}$ and the $\mathbf{X}_\alpha$'s such that $\alpha \in \Phi_I$; it is reductive. One also uses the term "parabolic subgroup" (and "Levi decomposition") for any $\mathbf{G}$-conjugate of the above.

The connected reductive groups (over $\mathbf{F}$) are classified by what is often called a **root datum** $(X, \Phi, Y, \Phi^\vee)$. This quadruple consists of two free abelian groups of finite rank, in duality over $\mathbb{Z}$ by $\langle -, - \rangle \colon X \times Y \to \mathbb{Z}$, along with subsets $\Phi \subseteq X$, $\Phi^\vee \subseteq Y$ in bijection $\Phi \to \Phi^\vee$ by some $\alpha \mapsto \alpha^\vee$ such that $\langle \alpha, \alpha^\vee \rangle = 2$ for all $\alpha \in \Phi$, and Φ (resp. $\Phi^\vee$) is a **root system** of $\mathbb{R}\Phi \subseteq X \otimes_\mathbb{Z} \mathbb{R}$ (resp. of $\mathbb{R}\Phi^\vee \subseteq Y \otimes_\mathbb{Z} \mathbb{R}$) for a scalar product on $\mathbb{R}\Phi$ (resp. $\mathbb{R}\Phi^\vee$) such that $x \mapsto x - \langle x, \alpha^\vee \rangle \alpha$ (resp. $y \mapsto y - \langle \alpha, y \rangle$) is the orthogonal reflection associated with α (resp. $\alpha^\vee$). The root datum $(X, \Phi, Y, \Phi^\vee)$ associated with a reductive group $\mathbf{G}$ and maximal torus $\mathbf{T}$ is such that $X = X(\mathbf{T})$, $\Phi = \Phi(\mathbf{G}, \mathbf{T})$, and $Y = \mathrm{Hom}(\mathbb{G}_{\mathrm{m}}, \mathbf{T})$ for some maximal torus $\mathbf{T}$.

Conversely, the connected reductive group associated with a root datum may be presented by generators and relations in a way quite similar to the case of reductive Lie algebras. Among the relations we recall the following. Let $\mathbf{T} \subseteq \mathbf{B}$ be a maximal torus and Borel subgroup, let Δ be the associated basis of the root system $\Phi(\mathbf{G}, \mathbf{T})$. Then there is a set $(n_\delta)_{\delta\in\Delta}$ of elements of $\mathrm{N}_{\mathbf{G}}(\mathbf{T})$ such that $n_\delta\mathbf{T} \in W(\mathbf{G}, \mathbf{T})$ is the reflection associated with δ, $n_\delta \in \mathbf{X}_\delta\mathbf{X}_{-\delta}\mathbf{X}_\delta$ and for any pair $\delta, \delta' \in \Delta$, denoting by $m_{\delta,\delta'}$ the order of the product of the corresponding two reflections, we have

$$n_\delta n_{\delta'} n_\delta \ldots = n_{\delta'} n_\delta n_{\delta'} \ldots$$

with $m_{\delta,\delta'}$ terms on each side (see [Springer] 9.3.2).

A2.5. Rational structures on affine varieties

We take $\mathbf{F}$ to be an algebraic closure of the finite field with q elements $\mathbb{F}_q$. A closed subvariety of $\mathbb{A}^n_{\mathbf{F}}$ is said to be defined over $\mathbb{F}_q$ if and only if it is the zero set of some subset $I \subseteq \mathbb{F}_q[x_1, \ldots, x_n]$. Then it is stable under the "Frobenius endomorphism" $F\colon \mathbb{A}^n_{\mathbf{F}} \to \mathbb{A}^n_{\mathbf{F}}$ raising coordinates to the qth power. More precisely, F induces a bijection $V \to V$ since $P(F(a)) = P(a)^q$ for any $a \in \mathbb{F}^n$ and $P \in \mathbb{F}_q[x_1, \ldots, x_n]$. The set V^F of fixed points is finite, $V = \bigcup_{n\geq 1} V^{F^n}$, and $(\mathcal{T}F)_x = 0$ for all $x \in V$.

A more intrinsic definition is as follows. An affine $\mathbf{F}$-variety $\mathbf{X}$ is defined over $\mathbb{F}_q$ if and only if its algebra of rational functions satisfies $\mathbf{F}[\mathbf{X}] = A_0 \otimes_{\mathbb{F}_q} \mathbf{F}$ where A_0 is a $\mathbb{F}_q$-algebra. Then the Frobenius endomorphism $F\colon \mathbf{X} \to \mathbf{X}$ is defined by its comorphism $F^\sharp \in \mathrm{End}(A_0 \otimes_{\mathbb{F}_q} \mathbf{F})$ being $a \otimes \lambda \mapsto a^q \otimes \lambda$ ($a \in A_0$, $\lambda \in \mathbf{F}$). As a kind of converse, we have that a closed subvariety of $\mathbf{X}$ is defined over $\mathbb{F}_q$ whenever it is F-stable (see [DiMi91] 3.3(iii)).

In the case where $\mathbf{X}$ is a linear algebraic group, any element is of finite order (being in some $\mathbf{X}^{F^n}$) and the Jordan decomposition coincides with the decomposition into p-part and p'-part (where p is the characteristic of $\mathbf{F}$).

A2.6. Morphisms and quotients

Let $f\colon Y \to X$ be a morphism of $\mathbf{F}$-varieties. It is said to be **quasi-finite** if and only if $f^{-1}(x)$ is a finite set for all $x \in X$. It is said to be **finite** if and only if for all open affine subvarieties $U \subseteq X$, $f^{-1}(U)$ is affine and $\mathbf{F}[f^{-1}(U)]$ is finitely generated as an $\mathbf{F}[U]$-*module* (recall $f^\sharp(U)\colon \mathbf{F}[U] = \mathcal{O}_X(U) \to \mathbf{F}[f^{-1}(U)] = \mathcal{O}_Y(f^{-1}(U))$). Finite morphisms are quasi-finite and closed. Conversely, we have **Zariski's main theorem** (in the form due to Grothendieck) which asserts that any quasi-finite morphism $Y \to X$ factors as $Y \hookrightarrow Y' \to X$ where the first map is an open immersion and the second is a finite morphism ([Milne80] I.1.8).

A morphism $f: Y \to X$ is said to be **dominant** if and only if, for each irreducible component $Y_i \subseteq Y$, the closure $\overline{f(Y_i)}$ is an irreducible component of X and $\overline{f(Y)} = X$. A dominant morphism between irreducible varieties induces a field extension $f^\sharp: \mathbf{F}(X) \hookrightarrow \mathbf{F}(Y)$. Then, such a dominant morphism f is said to be **separable** if and only if the field extension is separable. A morphism $f: Y \to X$ between irreducible **F**-varieties is separable if and only if there is a smooth point $y \in Y$ such that $f(y)$ is smooth and $\mathcal{T} f_y: \mathcal{T} Y_y \to \mathcal{T} X_{f(y)}$ is onto.

An **F**-variety is said to be **normal** if, for any $x \in X$, the ring $\mathcal{O}_{X,x}$ is an integral domain, noetherian and integrally closed. For any **F**-variety X, there is a normal **F**-variety $\widetilde{X}$ and a finite dominant morphism $\widetilde{X} \to X$. A "minimal" such $\widetilde{X} \to X$ exists and is called a normalization of X (see [Hart] Ex II.3.8 and [Miya94] 4.23).

The theorem of **purity of branch locus** (Zariski–Nagata) implies that, if $f: Y \to X$ is a finite dominant morphism between connected **F**-varieties with normal Y and smooth X, then the $y \in Y$ such that $\mathcal{T} f_y: \mathcal{T} Y_y \to \mathcal{T} X_{f(y)}$ is not an isomorphism form a closed subvariety of Y with dimension strictly less than $\dim(Y)$ (a complete proof is in [AltKlei70] VI.6.8; see also [Danil94] 3.1.3, [Milne80] I.3.7e).

Assume G is an algebraic **F**-group acting on Y. Note that G can be any finite group endowed with its structure of 0-dimensional **F**-variety.

One says that the morphism $f: Y \to X$ is an **orbit map** if and only if it induces a bijection between X and the (set-theoretic) quotient Y/G, i.e. f is onto and $f^{-1}(f(y)) = G.y$ for all $y \in Y$ (and therefore the G-orbits $G.y$ are closed, not just locally closed, subsets of Y). One says that $f: Y \to X$ is a **G-quotient** if and only if, for any morphism $Y \to Z$ of **F**-varieties which is constant on G-orbits, there is a unique morphism $X \to Z$ such that $Y \to Z$ factors as $Y \xrightarrow{f} X \to Z$. It is said to be locally trivial if and only if X is covered by open sets U_i such that each restriction of f, $f^{-1}(U_i) \to U_i$, admits a section morphism.

If $f: Y \to X$ is a dominant orbit map and X is irreducible, then G acts transitively on the irreducible components of Y, and the G-orbits all have dimension $\dim(Y) - \dim(X)$. If, moreover, X is smooth, then f is open ([Borel] 6.4).

If $f: Y \to X$ is a separable orbit map and both Y and X are smooth, then $Y \to X$ is a G-quotient ([Borel] 6.6). Note that, in the case when $G = \{1\}$, this yields a characterization of isomorphisms $Y \to X$ (see also [Har92] §14).

If G is a finite group acting on a quasi-projective variety Y, then there is a finite morphism $Y \to X$ satisfying the above (see [Har92] §10, [Mum70] §7, [SGA.1] V.1.8, or combine [Borel] 6.15 with [Serre88] §III.12). One denotes

$X = Y/G$. It is given locally by $\mathrm{Max}(A)/G = \mathrm{Max}(A^G)$ whenever $\mathrm{Max}(A)$ is an affine G-stable open subvariety of X.

If $\mathbf{H}$ is a closed subgroup of an algebraic (affine) $\mathbf{F}$-group $\mathbf{G}$, then the quotient $\mathbf{G}/\mathbf{H}$ is endowed with a structure of smooth quasi-projective $\mathbf{F}$-variety such that $\mathbf{G} \to \mathbf{G}/\mathbf{H}$ is a morphism of $\mathbf{F}$-varieties. If, moreover, $\mathbf{H} \lhd \mathbf{G}$, then $\mathbf{G}/\mathbf{H}$ has the structure of an algebraic (affine) $\mathbf{F}$-group with $\mathbf{F}[\mathbf{G}/\mathbf{H}] = \mathbf{F}[\mathbf{G}]^{\mathbf{H}}$.

Assume $\mathbf{G}$ is reductive. Let $\mathbf{T} \subseteq \mathbf{B} \subseteq \mathbf{P}$ be a maximal torus, a Borel subgroup and a parabolic subgroup, respectively. Denote by $S \subseteq W(\mathbf{G}, \mathbf{T})$ the set of simple reflections associated with $\mathbf{B}$, and recall

$$l \colon W(\mathbf{G}, \mathbf{T}) \to \mathbb{N},$$

the length function associated with S (see [Springer] §8.2). The quotient $\mathbf{G}/\mathbf{P}$ is a projective variety. When $w \in W(\mathbf{G}, \mathbf{T})$, denote $O(w) := \mathbf{B}w\mathbf{B}/\mathbf{B} \cong \mathbf{B}/\mathbf{B} \cap {}^w\mathbf{B} \cong \mathbb{A}_{\mathbf{F}}^{l(w)}$. The $O(w)$'s are locally closed, disjoint, and cover $\mathbf{G}/\mathbf{B}$. The Zariski closure of $O(w)$ ("Sçhubert variety") is the union of the $O(w')$ for $w' \leq w$, where $\leq$ denotes the Bruhat order in $W(\mathbf{G}, \mathbf{T})$ associated with S (see [Springer] 8.5.4, [Jantzen] II.13.7). If $w_0 \in W(\mathbf{G}, \mathbf{T})$ denotes the element of maximal length, the associated $O(w_0)$ is a dense open subvariety of $\mathbf{G}/\mathbf{B}$. The translates of $\mathbf{B}w_0\mathbf{B}$ allow us to show that $\mathbf{G}/R_u(\mathbf{B}) \to \mathbf{G}/\mathbf{B}$ and $\mathbf{G} \to \mathbf{G}/R_u(\mathbf{B})$ are locally trivial.

A2.7. Schemes

Let A be any commutative ring. Denote by $\mathrm{Spec}(A)$ its set of prime ideals. When $x \in \mathrm{Spec}(A)$, recall that A_x denotes the localization $(A \setminus x)^{-1}A$.

The **affine scheme** associated with A is the locally ringed space $\mathrm{Spec}(A)$ with the same definition of open subsets as in A2.1 and structure sheaf $\mathcal{O}$ defined by $\mathcal{O}(U)$ being the ring of maps $f \colon U \to \coprod_{x \in \mathrm{Spec}(A)} A_x$ such that, for all $u_0 \in U$, there is a neighborhood V of u_0 in U, and $a \in A$, $b \in A \setminus \bigcup_{x \in V} x$ such that $f(u) = a/b \in A_u$ for all $u \in V$. This is the sheafification of the presheaf $\mathcal{O}'$ defined by $\mathcal{O}'(U) = \{a/b \mid a \in A,\ b \in A \setminus \bigcup_{x \in U} x\}$. Note that $\mathcal{O}_x = A_x$ for all $x \in \mathrm{Spec}(A)$. General schemes $(X, \mathcal{O}_X)$ are defined by glueing affine schemes, just as $\mathbf{F}$-prevarieties are defined from affine $\mathbf{F}$-varieties (see A2.2, and note that the separation axiom is not required).

Noetherian schemes are defined as schemes obtained by glueing together a finite number of affine schemes associated with noetherian rings.

For any scheme $(X, \mathcal{O}_X)$, we have

$$\mathrm{Hom}(X, \mathrm{Spec}\, A) \cong \mathrm{Hom}(A, \mathcal{O}_X(X))$$

(use comorphisms).

We have a fully faithful functor $V \mapsto t(V)$ from the category of $\mathbf{F}$-varieties to the category of schemes. It is given by defining a sheaf of rings on the set $t(V)$ of irreducible closed non-empty subvarieties of V. Most notions defined for varieties (see A2.2 and A2.3) can be defined for schemes, especially schemes over $\mathbf{F}$, i.e. schemes X endowed with a morphism $X \to \mathrm{Spec}(\mathbf{F})$.

A point $x \in X$ is called **closed** if and only if $\{x\}$ is closed. In affine schemes, closed points correspond to $\mathrm{Max}(A)$ as subset of $\mathrm{Spec}(A)$. The map $x \mapsto \overline{\{x\}}$ is a bijection between X and the set of irreducible closed non-empty subsets of X. One calls x the **generic point** of $\overline{\{x\}}$, and its **dimension** is defined as the dimension of $\overline{\{x\}}$. In the affine case $X = \mathrm{Spec}(A)$, $\overline{\{x\}} = \mathrm{Spec}(A_x)$ is the set of $y \in \mathrm{Spec}(A)$ containing x. More generally, a **geometric point** of a scheme X is any morphism $\mathrm{Spec}(\Omega) \to X$ where Ω is a separably closed field extension of $\mathbf{F}$. Such a map amounts to the choice of its image $\{x\}$ along with an extension of the field of quotients of global sections on $\overline{\{x\}}$, $\mathcal{O}_{X,x}/J(\mathcal{O}_{X,x}) \to \Omega$.

An open subset of a scheme clearly inherits the structure of a scheme by restriction of the structure sheaf, whence the notion of **open immersion**. For closed subsets, the matter is a little more complicated since there may be several scheme structures on each (think of the various ideals of a polynomial ring giving rise to the same set of zeroes). One defines a **closed immersion** as a scheme morphism $i: Y \to X$ such that i is a homeomorphism of Y with $i(Y) = \overline{i(Y)}$ and $i^\sharp: \mathcal{O}_X \to i_*\mathcal{O}_Y$ is a surjection. A locally closed immersion, generally just called an **immersion**, is the composite of both types of immersions.

Finite products exist in the category of schemes. More generally, given a prescheme S, one defines the category of **schemes over** S, or S**-schemes**, with objects the scheme maps $X \to S$, and where morphisms are defined as commutative triangles. Products exist in this category. Given $X \to S$ and $X' \to S$, one denotes the product by $X \times_S X' \to S$ and calls it the **fibered product** of X and X' over S (endowed with its "projections" $X \times_S X' \to X$ and $X \times_S X' \to X'$). The operation consisting of changing $f: X \to S$ into the projection $X \times_S X' \to X'$ is called the **base change** (the base S of f becoming X' by use of $X' \to S$).

Fibered products are defined locally by $\mathrm{Spec}(A) \times_{\mathrm{Spec}(R)} \mathrm{Spec}(B) = \mathrm{Spec}(A \otimes_R B)$. When $X \xrightarrow{f} S$, $X' \xrightarrow{f'} S$ are maps between varieties, the fibered product $X \times_S X'$ identifies with the pull-back $\{(x, x') \mid f(x) = f'(x')\}$ viewed as a closed subvariety of $X \times X'$. Note that, in case f is an immersion $X \subseteq S$, then $X \times_S X'$ identifies with $f'^{-1}(X)$.

Let $f: X \to Y$ be a morphism of schemes. One calls it **separated** if the associated map $X \to X \times_Y X$ is a closed immersion. One calls it of **finite type** if and only if Y can be covered by affine open subschemes $Y_i = \mathrm{Spec}(B_i)$ such that $f^{-1}(Y_i)$ is in turn covered by a finite number of affine open subschemes

Spec($A_{i,j}$) where each $A_{i,j}$ is finitely generated as a B_i-*algebra*. One calls it **proper** if and only if it is separated, of finite type and, for any morphism $Y' \to Y$, the projection $X \times_Y Y' \to Y'$ obtained by base change is closed. For instance, any morphism $X \to Y$ of **F**-varieties with X complete (see A2.2), is proper. Finite morphisms are proper.

A property of morphisms immediately gives rise to the corresponding notion for schemes with a given base (for instance, schemes over **F**) by imposing it on the structure morphism, i.e. $\sigma_X: X \to \mathrm{Spec}(\mathbf{F})$ in the case of schemes over **F**.

A2.8. Coherent sheaves

(see [Hart] §II.5 and §II.7, [Kempf] 5, [Danil96] §2.1.1)

In the following, $(X, \mathcal{O}_X)$ is a scheme of finite type over **F**. An $\mathcal{O}_X$**-module**, or **sheaf over** X, is a sheaf $\mathcal{M}$ on the underlying topological space of X such that, for any open $V \subseteq U \subseteq X$, $\mathcal{M}(U)$ is an $\mathcal{O}_X(U)$-module, and the restriction maps $\mathcal{M}(U) \to \mathcal{M}(V)$ are group morphisms compatible with the restriction maps $\mathcal{O}_X(U) \to \mathcal{O}_X(V)$. Then $\mathcal{M}_x$ is an $\mathcal{O}_{X,x}$-module for each point $x \in X$. If U is an open subscheme of X, then $\mathcal{M}_{|U}$ is a sheaf over U for the structure sheaf $\mathcal{O}_{X|U}$.

Let $\mathcal{M}, \mathcal{M}'$ be two sheaves over X. We denote by $\mathrm{Hom}_{\mathcal{O}_X}(\mathcal{M}, \mathcal{M}')$ the set of sheaf morphisms (see also A1.13) such that each induced map $\mathcal{M}(U) \to \mathcal{M}'(U)$ is $\mathcal{O}_X(U)$-linear. With this definition of morphisms, sheaves over X make a category.

One defines $\mathcal{H}om(\mathcal{M}, \mathcal{M}')$ as the sheaf

$$U \mapsto \mathrm{Hom}_{\mathcal{O}_{X|U}}(\mathcal{M}(U), \mathcal{M}'(U)).$$

One defines $\mathcal{M} \otimes_{\mathcal{O}_X} \mathcal{M}'$ as the sheaf associated with the presheaf $U \mapsto \mathcal{M}(U) \otimes_{\mathcal{O}_X(U)} \mathcal{M}'(U)$ with evident restriction maps (see A1.15). One defines the dual as $\mathcal{M}^\vee = \mathcal{H}om(\mathcal{M}, \mathcal{O}_X)$.

Direct and inverse images are defined through a slight adaptation of the classical case (see A1.13). Let $f: Y \to X$, let $\mathcal{M}$, resp. $\mathcal{N}$, be a sheaf on X, resp. Y. One defines $f_*\mathcal{N}$ by noting that each $f_*\mathcal{N}(U) = \mathcal{N}(f^{-1}(U))$ is a $(f_*\mathcal{O}_Y)(U)$-module and that we have a ring morphism $f^\sharp: \mathcal{O}_X \to f_*\mathcal{O}_Y$, so we get an action of $\mathcal{O}_X(U)$ on each $f_*\mathcal{N}(U)$. Similarly, it is easy to consider the inverse image of $\mathcal{M}$ under f in $Sh_{\mathbb{Z}}(Y)$ as a $f^*\mathcal{O}_X$-module. One makes it into an $\mathcal{O}_Y$-module by tensor product (see above) knowing that $\mathcal{O}_X \to f_*\mathcal{O}_Y$ induces $f^*\mathcal{O}_X \to \mathcal{O}_Y$ by adjunction of f_* and f^*.

A prototype of $\mathcal{O}_X$-modules is as follows. Assume A is a finitely generated commutative $\mathbf{F}$-algebra and $X = \mathrm{Spec}(A)$ is the associated affine scheme. Let M be a finitely generated A-module. One defines the $\mathcal{O}_X$-module $\widetilde{M}$ by $\widetilde{M}(U) = \mathcal{O}_X(U) \otimes_A M$ where $A = \mathcal{O}_X(X)$ acts on $\mathcal{O}_X(U)$ through the restriction map $\mathcal{O}_X(X) \to \mathcal{O}_X(U)$. More explicitly in this case, if $f \in A \setminus (0)$ and $X_f := \{x \in X \mid f_x \notin J(\mathcal{O}_{X,x})\}$ is the associated open subset, then $\mathcal{O}_X(X_f) = A[f^{-1}]$ and $\widetilde{M}(X_f) = A[f^{-1}] \otimes_A M$. One calls a **coherent sheaf** on X, any $\mathcal{O}_X$-module $\mathcal{M}$ such that $\mathcal{M}(U)$ is of the type just described for any affine open subscheme $U \subseteq X$ (see [Hart] §II.5, [Danil96] §2.1.1). Coherent sheaves over X make an abelian category.

A sheaf $\mathcal{M}$ over X is said to be **generated by its global sections** if and only if, for every $x \in X$, the image of the restriction map $\mathcal{M}(X) \to \mathcal{M}_x$ generates $\mathcal{M}_x$ as an $\mathcal{O}_{X,x}$-module. If X is affine, $X = \mathrm{Spec}(A)$, and M is an A-module, then $\widetilde{M}_x = A_x \otimes_A M$ for any $x \in X$ and therefore $\widetilde{M}$ is generated by its global sections.

A sheaf $\mathcal{M}$ over X is said to be **invertible** if and only if it is locally isomorphic to $\mathcal{O}_X$, i.e. there is a covering of X by open sets U such that $\mathcal{M}_{|U} \cong \mathcal{O}_{X|U}$ as $\mathcal{O}_{X|U}$-module. This is equivalent to $\mathcal{M} \otimes \mathcal{M}^\vee \cong \mathcal{O}_X$. One says that $\mathcal{M}$ is **ample** if and only if it is invertible, and, for every coherent $\mathcal{O}_X$-module $\mathcal{M}'$, $\mathcal{M}' \otimes_{\mathcal{O}_X} \mathcal{M}^{\otimes n}$ is generated by its global sections as long as n is big enough.

If $i\colon X' \to X$ is an immersion and $\mathcal{M}$ is an ample sheaf; on X, then $i^*\mathcal{M}$ is ample (easy for open immersions, while for closed immersions one may use the notion of *very ample* sheaf; see [Hart] II.7.6).

A2.9. Vector bundles

([Jantzen] I.5 and II.4)

Keep $\mathbf{F}$ an algebraically closed field. Let $\mathbf{G}$ be an $\mathbf{F}$-group acting freely on an $\mathbf{F}$-variety X. Assume that the quotient variety $X/\mathbf{G}$ exists (see A2.2).

There is a functor

$$M \to \mathcal{L}_{X/\mathbf{G}}(M)$$

associating a coherent $\mathcal{O}_{X/\mathbf{G}}$-module with each finite-dimensional $\mathbf{F}$-vector space M endowed with a rational $\mathbf{G}$-action. If U is an affine $\mathbf{G}$-stable open subset of X, then $\mathcal{L}_{X/\mathbf{G}}(M)(U/\mathbf{G}) = (M \otimes \mathbf{F}[U])^{\mathbf{G}}$ where the action of $\mathbf{G}$ on the tensor product is diagonal ([Jantzen] I.5.8). One has $\mathcal{L}_{X/\mathbf{G}}(\mathbf{F}) = \mathcal{O}_{X/\mathbf{G}}$ ([Jantzen] I.5.10(1)). More properties hold when in addition the quotient $X \xrightarrow{\pi} X/\mathbf{G}$ is locally trivial (see A2.6). Then $\mathcal{L}_{X/\mathbf{G}}(M)^\vee \cong \mathcal{L}_{X/\mathbf{G}}(M^\vee)$ and $\mathcal{L}_{X/\mathbf{G}}$ commutes with tensor products ([Jantzen] II.4.1).

Assume $\mathbf{G}$ (resp. $\mathbf{G}'$) is an $\mathbf{F}$-group acting freely on the right on the $\mathbf{F}$-variety X (resp. X'), such that the quotient exists and is locally trivial. Let $\alpha\colon \mathbf{G}' \to \mathbf{G}$

be an injective morphism, let $\varphi: X' \to X$ be a morphism compatible with α, i.e. $\varphi(x'g') = \varphi(x')\alpha(g')$ for all $x' \in X', g' \in \mathbf{G}'$. Then φ induces $\bar{\varphi}: X'/\mathbf{G}' \to X/\mathbf{G}$ such that the following commutes

$$\begin{array}{ccc} X' & \xrightarrow{\varphi} & X \\ \downarrow & & \downarrow \\ X'/\mathbf{G}' & \xrightarrow{\bar{\varphi}} & X/\mathbf{G} \end{array}$$

(where vertical maps are quotients) and we have

$$\text{(I)} \qquad \bar{\varphi}^*\mathcal{L}_{X/\mathbf{G}}(M) \cong \mathcal{L}_{X'/\mathbf{G}'}(M^\alpha)$$

for any finite-dimensional rational representation space M of $\mathbf{G}$ over $\mathbf{F}$ (see [Jantzen] I.5.17 and Remark).

The above is related to another notion (see [Jantzen] II.5.16). Assume that $\mathbf{G}$ acts rationally on a finite-dimensional $\mathbf{F}$-vector space M. Then $X_M := (X \times M)/\mathbf{G}$ (for the diagonal action) is called the associated **vector bundle**. It is a scheme over $X/\mathbf{G}$. This is related to the $\mathcal{L}_{X/\mathbf{G}}(M)$ construction of coherent sheaves by noting that, for any open immersion $U \to X/\mathbf{G}$, $\mathrm{Hom}_{X/\mathbf{G}}(U, X_M)$ ("sections over U" of $X_M \to X/\mathbf{G}$) is an $\mathcal{O}_{X/\mathbf{G}}(U)$-module that identifies with $\mathcal{L}_{X/\mathbf{G}}(M)(U)$ (see [Hart] Ex. II.5.18).

A2.10. A criterion of quasi-affinity

Recalling the bijection $\mathrm{Hom}(Y, \mathrm{Spec}A) \to \mathrm{Hom}(A, \mathcal{O}_Y(Y))$, the identity of $A := \mathcal{O}_Y(Y)$ induces $u_Y: Y \to \mathrm{Spec}(\mathcal{O}_Y(Y))$, defined by $y \mapsto u_Y(y) = \{f \in \mathcal{O}_Y(Y) \mid f_y \in J(\mathcal{O}_{Y,y})\} \in \mathrm{Spec}(\mathcal{O}_Y(Y))$. When Y is affine, u_Y is clearly an isomorphism.

Let $(X, \mathcal{O}_X)$ be a noetherian scheme (see A2.7). Saying that $\mathcal{O}_X$ itself is ample amounts to saying that every coherent $\mathcal{O}_X$-module is generated by its global sections. The following shows that this is equivalent to quasi-affinity of X.

Theorem A2.11. *Let $(X, \mathcal{O}_X)$ be a noetherian scheme of finite type over* $\mathbf{F}$. *The following are equivalent*

(a) X is quasi-affine,

(b) $\mathcal{O}_X$ is ample

(c) $u_X: X \to \mathrm{Spec}(\mathcal{O}_X(X))$ is an open immersion.

Proof of Theorem A2.11. (c) implies (a) trivially.

(a) implies (b), see A2.8.

(b) implies (c). Assume that $\mathcal{O}_X$ is ample. Denote $A := \mathcal{O}_X(X)$. When $f \in A$, recall $X_f := \{x \in X \mid f_x \in J(\mathcal{O}_{X,x})\}$ the associated principal open subscheme of X.

Let $\mathbf{a} = \{f \in A \mid X_f \text{ is affine}\}$.

Lemma A2.12. $X = \bigcup_{f \in \mathbf{a}} X_f$.

Proof of Lemma A2.12. Let us prove first that the X_f's for $f \in A$ provide a base of neighborhoods for any $x_0 \in X$. Let $x_0 \in X$ and let $U \subseteq X$ be open with $x_0 \in U$. Denote $Y = X \setminus U$ and $i: Y \to X$, the closed immersion. We have an exact sequence $0 \to \mathcal{J} \to \mathcal{O}_X \to i_*\mathcal{O}_Y$ where the kernel is a coherent $\mathcal{O}_X$-module. At any open $U' \subseteq U$, we have $i_*\mathcal{O}_Y(U') = \mathcal{O}_Y(U' \cap Y) = 0$ since $U \cap Y = \emptyset$. Then $\mathcal{J}(U') = \mathcal{O}_X(U')$, and therefore $\mathcal{J}_{x_0} = \mathcal{O}_{X,x_0}$. Condition (b) implies that any coherent sheaf over X is generated by its global sections. Applied to $\mathcal{J}$, we get that there is $f \in \mathcal{J}(X) \subseteq \mathcal{O}_X(X)$ such that $f_{x_0} \notin J(\mathcal{O}_{X,x_0})$, i.e. $x_0 \in X_f$. It remains to check that $X_f \subseteq U$. Let $x \in X$ with $f_x \notin J(\mathcal{O}_{X,x})$. We have $\mathcal{J}_x \neq 0$ since $f \in \mathcal{J}(X)$. But this is possible only if $x \in U$, by the definition of $\mathcal{J}$ (see also the proof of [Hart] 5.9 which shows that Y is the "support" of $\mathcal{J}$).

Now, U above may be taken to be affine. Recall that there is $f \in A$ such that $X_f \subseteq U$. By the fundamental property of restriction maps, we have $X_f = \{x \in U \mid (f_{|U})_x \notin J(\mathcal{O}_{U,x})\} = U_f$ which is affine as U (see A2.1). This completes the proof of the lemma. □

We now finish the proof of Theorem A2.11. We apply Exercise 1 (see [Hart] Ex 2.17(a)) to $Y = \mathrm{Spec}(A)$, and $\phi = u_X$. For $f \in \mathbf{a}$, the affine scheme $\mathrm{Spec}(A_f)$ is identified with a principal open subscheme of $\mathrm{Spec}(A)$, the one consisting of prime ideals not containing f. To apply Exercise 1 to this collection of open subschemes of $\mathrm{Spec}(A)$, in view of the above lemma, it suffices to show that $u_X^{-1}(\mathrm{Spec}(A_f)) = X_f$ and that u_X induces an isomorphism $X_f \to \mathrm{Spec}(A_f)$. Whenever $x \in X$, $f \in A$, it is clear from the definition of $u_X(x)$ that

$$u_X(x) \in \mathrm{Spec}(A_f) \Leftrightarrow f \notin u_X(x) \Leftrightarrow x \in X_f.$$

So $u_X^{-1}(\mathrm{Spec}(A_f)) = X_f$. If $x \in X_f$, we have $\mathcal{O}_{X,x} = \mathcal{O}_{X_f,x}$ since X_f is open. Then $u_X(x) = u_{X_f}(x)$. When, moreover, $f \in \mathbf{a}$, i.e. X_f is affine, we have $\mathcal{O}_X(X_f) = A_f$ and $u_{X_f}: X_f \to \mathrm{Spec}(A_f)$ is an isomorphism. Thus our claim is proved. □

Quasi-affinity is preserved by quotients under finite group actions.

Corollary A2.13. *Let G be a finite group acting rationally on a quasi-projective $\mathbf{F}$-variety (so that Y/G can be considered as an $\mathbf{F}$-variety; see A2.6). Then Y is quasi-affine (resp. affine) if and only if Y/G is quasi-affine (resp. affine).*

Proof. Recall that the quotient map $Y \to Y/G$ is finite (A2.8), hence closed and open. A finite map is by definition an affine map, so the case when Y or Y/G is affine is clear.

If Y is quasi-affine, then Theorem A2.11 implies that u_Y is an open immersion. If we consider the composition

$$Y \xrightarrow{u_Y} \operatorname{Spec}(\mathbf{F}[Y]) \longrightarrow \operatorname{Spec}(\mathbf{F}[Y]^G)$$

where the second map is the quotient by G, the image of this composite is an open subvariety of $\operatorname{Spec}(\mathbf{F}[Y]^G)$, so Y/G is quasi-affine.

Assume conversely that we have an open immersion $Y/G \to X$ with X affine. Then the composition $Y \to Y/G \to X$ is a quasi-finite map to which we may apply Zariski's main theorem (see A2.6) to obtain a factorization $Y \xrightarrow{j} Y' \to X$ where j is an open immersion and $Y' \to X$ is a finite map. Then Y' is affine. So Y is quasi-affine. □

Exercise

1. Let $\phi: X \to Y$ be a morphism of schemes. Assume there are open subsets $U_i \subseteq Y$ such that $X = \bigcup_i \phi^{-1}(U_i)$, and, for all i, ϕ induces an isomorphism $\phi^{-1}(U_i) \to U_i$. Show that ϕ is an open immersion.

Notes

Theorem A2.11 is due to Grothendieck (see [EGA] II.5.1.2).

Algebraic geometry is a mixture of the ideas of two Mediterranean cultures. It is a superposition of the Arab science of the lightning calculation of the solutions of equations over the Greek art of position and shape. This tapestry was woven on European soil and is still being refined under the influence of international fashion.

[Kempf] p. ix.

Appendix 3

Etale cohomology

In this appendix we gather most of the results that are necessary for the purpose of the book. The outcome is a mix of fundamental notions, deep theorems (base change for proper morphisms, Künneth formula, Poincaré–Verdier duality, etc.), and useful (sometimes elementary) remarks.

Etale cohomology was introduced by Grothendieck and his team in [SGA] (especially [SGA.4], [SGA.4$\frac{1}{2}$], [SGA.5]). The other references are [Milne80], [FrKi88], [Tamme].

In what follows, $\mathbf{F}$ is an algebraically closed field.

A3.1. The étale topology

We consider schemes over $\mathbf{F}$, mainly noetherian of finite type unless otherwise specified. Morphisms are understood in the following sense.

A **flat** morphism $f: Y \to X$ between $\mathbf{F}$-schemes is any morphism such that, for any $y \in Y$, the induced map $\mathcal{O}_{X,f(y)} \to \mathcal{O}_{Y,y}$ makes $\mathcal{O}_{Y,y}$ into a flat $\mathcal{O}_{X,f(y)}$-module. Flat morphisms are open. A morphism $f: Y \to X$ is said to be **étale** if and only if it is flat and $\mathcal{O}_{Y,y}/J(\mathcal{O}_{X,f(y)})\mathcal{O}_{Y,y}$ is a finite separable extension of $\mathcal{O}_{X,f(y)}/J(\mathcal{O}_{X,f(y)}$ for all $y \in Y$.

Any open immersion is clearly étale. The notion of étale morphism is preserved by composition and (arbitrary) base change (see A2.7).

Let us fix X a scheme over $\mathbf{F}$. The **étale topology** $X_{\text{ét}}$ is the category of étale maps $U \to X$ of $\mathbf{F}$-schemes. The morphisms from $U \to X$ to $U' \to X$ are morphisms $U \to U'$ of X-schemes. If $U \to X$ is étale, then we clearly have a functor $U_{\text{ét}} \to X_{\text{ét}}$.

A final object is $X \to X$ ("identity" map), again abbreviated as X.

The objects of $X_{\text{ét}}$ are to be considered as "opens" of a generalized topology. The usual (Zariski) open subsets $U \subseteq X$ give rise to elements of $X_{\text{ét}}$ by considering the associated open immersion. The intersection of open subsets in ordinary topology is to be replaced with fibred product (see A2.7), i.e. $U \to X$ and $U' \to X$ give rise to $U \times_X U' \to X$. Similarly, if $f: X \to Y$ is a morphism, and $U \to Y$ is in $Y_{\text{ét}}$, one may define its inverse image by f as $U \times_X Y \to X$ obtained by base change.

A **neighborhood** of a geometric point $\text{Spec}(\Omega) \to X$ is any étale $U \to X$ endowed with a morphism of X-schemes $\text{Spec}(\Omega) \to U$.

A **covering** of $U \to X$ is a family of étale $(U_i \to X)_i$ whose images cover U.

A basic example is the following. Let K be a field, then $\text{Spec}(K)_{\text{ét}}$ is the category of maps $\coprod_i \text{Spec}(K_i) \to \text{Spec}(K)$ where each $\text{Spec}(K_i) \to \text{Spec}(K)$ is the map associated with a finite separable extension K_i/K.

A3.2. Sheaves for the étale topology

Let X be a scheme and A be a ring. In analogy with classical sheaf theory (see A1.13), a **presheaf** on $X_{\text{ét}}$ is defined as a functor

$$\mathcal{F}: X_{\text{ét}} \to A\mathbf{-Mod}.$$

When $U \to X$ is in $X_{\text{ét}}$, we abbreviate $\mathcal{F}(U \to X)$ as $\mathcal{F}(U)$, or use the sheaf-theoretic notation $\Gamma(U, \mathcal{F})$, especially in the case of $\Gamma(X, \mathcal{F})$ which in turn may be seen as a functor with respect to $\mathcal{F}$.

If $U \to X$ is étale, then we may define the restriction $\mathcal{F}_{|U}$ as a presheaf on $U_{\text{ét}}$ obtained by composing $\mathcal{F}$ with the functor $U_{\text{ét}} \to X_{\text{ét}}$ (see A3.1). A presheaf $\mathcal{F}$ is called a **sheaf** if and only if it satisfies the classical property with regard to coverings, where intersections are defined as in A3.1 and restrictions are defined as above. The corresponding category is denoted by $Sh_A(X_{\text{ét}})$ where morphisms are defined as in the classical way. It is abelian and has enough injective objects.

When K is a field, $Sh_{\mathbb{Z}}((\text{Spec}K)_{\text{ét}}) \cong \mathbb{Z}G\mathbf{-Mod}$ where G is the Galois group of the extension K_{sep}/K, K_{sep} denoting a separable closure of K.

Many theorems require that A is a finite ring where the characteristic of $\mathbf{F}$ is invertible (if $\neq 0$), and sheaves take their values in finitely generated (i.e. finite) modules; these are sometimes called **torsion sheaves**.

Sheafification $\mathcal{F} \to \mathcal{F}^+$ may be defined by an adaptation of the classical procedure (see §A1.13).

If $\mathcal{F}$ is a sheaf on $X_{\text{ét}}$ and $\overline{x}: \text{Spec}(\Omega) \to X$ is a geometric point, one defines the **stalk** $\mathcal{F}_{\overline{x}}$ as the inductive limit $\varinjlim \mathcal{F}(U)$ taken over the étale neighborhoods $U \to X$ of $\overline{x}$. This yields an exact functor $\mathcal{F} \to \mathcal{F}_{\overline{x}}$ ($\overline{x}$ is fixed).

The following is a very useful theorem. A sequence $\mathcal{F}^1 \to \mathcal{F}^2 \to \mathcal{F}^3$ in $Sh_A(X)$ is exact if and only if $\mathcal{F}^1_{\overline{x}} \to \mathcal{F}^2_{\overline{x}} \to \mathcal{F}^3_{\overline{x}}$ is exact for any geometric point $\overline{x}$ of X (see [Tamme] II.5.6, [Milne80] II.2.15). If X is the scheme corresponding to an **F**-variety, then only closed points $\mathrm{Spec}(\mathbf{F}) \to X$ need be checked (see [Milne80] II.2.17).

If M is an A-module, one defines the **constant sheaf** $M_X \in Sh_A(X_{\text{ét}})$ by $M_X(U) = M^{\pi_0(U)}$ for $U \to X$ in $X_{\text{ét}}$ (compare A1.13). A sheaf is called **locally constant** if and only if there is an étale covering $(U_i \to X)_i$ such that each restriction $\mathcal{F}_{|U_i}$ is a constant sheaf. If X is irreducible, a sheaf $\mathcal{F}$ in $Sh_{\mathbb{Z}}(X_{\text{ét}})$ is called **constructible** if and only if it has finite stalks, and there is an open subscheme $U \subseteq X$ such that $\mathcal{F}_{|U}$ is locally constant.

A3.3. Basic operations on sheaves

If $\mathcal{F}_1, \mathcal{F}_2$ are objects of $Sh_A(X_{\text{ét}})$, one defines $\mathcal{H}om(\mathcal{F}_1, \mathcal{F}_2)$ and $\mathcal{F}_1 \otimes_A \mathcal{F}_2$, two objects of $Sh_{\mathbb{Z}}(X_{\text{ét}})$, as in the classical topological case (see A1.15). They are constant (resp. locally constant) when both $\mathcal{F}_1$ and $\mathcal{F}_2$ are. If A is commutative, they may be considered as objects of $Sh_A(X_{\text{ét}})$, along with the "dual" $\mathcal{F}_1^\vee := \mathcal{H}om_A(\mathcal{F}_1, A_X)$.

Let $f: X \to Y$ be a morphism between **F**-schemes.

The **direct image** functor $\mathcal{F} \mapsto f_*\mathcal{F}$ from $Sh(X_{\text{ét}})$ to $Sh(Y_{\text{ét}})$ is defined by $(f_*\mathcal{F})(U \to Y) = \mathcal{F}(U \times_Y X \to X)$ whenever $U \to Y$ is in $Y_{\text{ét}}$. The **inverse image** functor $\mathcal{F}' \mapsto f^*\mathcal{F}'$ from $Sh(Y_{\text{ét}})$ to $Sh(X_{\text{ét}})$ is obtained by $f^*\mathcal{F}' = \mathcal{G}^+$ where $\mathcal{G}$ is the presheaf on X defined by $\mathcal{G}(V \xrightarrow{e} X) = \varinjlim \mathcal{F}'(U \to Y)$ where the limit is over commutative diagrams

$$\begin{array}{ccc} V & \longrightarrow & U \\ \downarrow{\scriptstyle e} & & \downarrow \\ X & \xrightarrow{f} & Y \end{array}$$

(compare A1.13).

For instance, if $\overline{x}: \mathrm{Spec}(\Omega) \to X$ is a geometric point, the stalk functor $\mathcal{F} \mapsto \mathcal{F}_{\overline{x}}$ (see A3.2) coincides with inverse image by $\overline{x}$. Similarly, noting that $Sh_A(\mathrm{Spec}(\mathbf{F})_{\text{ét}}) = A\text{–}\mathbf{Mod}$, the constant sheaf of stalk M in $A\text{–}\mathbf{Mod}$ is written as $M_X = \sigma_X^* M$ for $\sigma_X: X \to \mathrm{Spec}(\mathbf{F})$ the structure morphism.

A sheaf $\mathcal{F}$ on $X_{\text{ét}}$ is said to be **constructible** if and only if, for any closed immersion $i: Z \to X$ with Z irreducible, $i^*\mathcal{F}$ is constructible (see A3.2 above).

We have $(f \circ g)_* = f_* \circ g_*$ and $(f \circ g)^* = g^* \circ f^*$ whenever $f \circ g$ makes sense. Moreover f^* is right-adjoint to f_*. The direct image functor f_* is left-exact and the inverse image f^* is exact ([Milne80] II.2.6). This implies that f_* preserves injective objects of $Sh_A(X_{\text{ét}})$ ([Milne80] III.1.2(b)). If f is a finite morphism (see A2.6), then f_* is also right-exact ([Milne80] II.3.6). Concerning stalks for inverse images, one has $f^*\mathcal{F}_{f\circ\overline{x}} = \mathcal{F}_{\overline{x}}$ for any $f: X \to Y$ and any geometric point $\overline{x}$: Spec$(\Omega) \to X$.

Let $j: U \to X$ be an open immersion. One defines **extension by zero**, denoted by $j_!$, from sheaves on $U_{\text{ét}}$ to sheaves on $X_{\text{ét}}$, as follows. If $\mathcal{F} \in Sh(U_{\text{ét}})$, let $\mathcal{F}_!$ be the presheaf on X, defined at each $\phi: V \to X$ in $X_{\text{ét}}$ by $\mathcal{F}_!(V) = \mathcal{F}(V)$ if $\phi(V) \subseteq j(U)$, $\mathcal{F}_!(V) = 0$ otherwise. Define the extension by zero as $j_!\mathcal{F} = (\mathcal{F}_!)^+$. Then $j_!$ is left-adjoint to the functor j^* of "restriction to U." The functor $j_!$ preserves constructibility. Concerning stalks, one has $(j_!\mathcal{F})_{\overline{x}} = \mathcal{F}_{\overline{x}}$ if the image of $\overline{x}$ is in $j(U)$, $(j_!\mathcal{F})_{\overline{x}} = 0$ otherwise. Then $j_!$ is clearly exact. One also easily defines a natural transformation $j_! \to j_*$.

A3.4. Homology and derived functors

The category $Sh_A(X_{\text{ét}})$ being abelian, we may define its derived category $D(Sh_A(X_{\text{ét}}))$ (see A1.6). One defines $D^+_A(X)$ (resp. $D^b_A(X)$) as the full subcategory corresponding to complexes such that each cohomology sheaf in $Sh_A(X_{\text{ét}})$ is *constructible* and all are zero below a certain degree (resp. below and above certain degrees).

Let $f: X \to Y$ be a morphism of $\mathbf{F}$-schemes. Since $Sh_A(X_{\text{ét}})$ has enough injectives, the left-exact functor $f_*: Sh_A(X_{\text{ét}}) \to Sh_A(Y_{\text{ét}})$ gives rise to a right derived functor $Rf_*: D^+(Sh_A(X_{\text{ét}})) \to D^+(Sh_A(Y_{\text{ét}}))$. This functor preserves complexes whose cohomology sheaves are constructible and this subcategory has enough injectives, so we get

$$Rf_*: D^+_A(X) \to D^+_A(Y).$$

The above also preserves injectives, so we have

$$R(f \circ g)_* = Rf_* \circ Rg_*$$

whenever $f \circ g$ makes sense (see [Milne80] III.1.18). A special case is when $f = \sigma_X: X \to \text{Spec}(\mathbf{F})$ is the structure morphism and g is a morphism of $\mathbf{F}$-schemes. Then f_* identifies with the global section functor $\Gamma(X, -)$ (see A3.2) through the identification of sheaves on Spec($\mathbf{F}$) with abelian groups. The corresponding right derived functor $\mathrm{R}\Gamma(X, -)$: $D^+_A(X) \to D^+(A\text{-}\mathbf{Mod})$ gives rise to the **cohomology groups** $\mathrm{H}^i(X, \mathcal{F})$.

The composition formula above gives, for any $g: X \to X'$,

$$\mathrm{R}\Gamma(X, \mathcal{F}) \cong \mathrm{R}\Gamma(X', Rg_*\mathcal{F})$$

in a functorial way since $\sigma_{X'} \circ g = \sigma_X$.

A3.5. Base change for a proper morphism

(see [Milne80] VI.2.3, [Milne98] 17.7, 17.10)

Let $\pi: Y \to X$, $f: Z \to X$ be morphisms of **F**-schemes. Let

$$\begin{array}{ccc} Z \times_X Y & \xrightarrow{f'} & Y \\ \downarrow{\scriptstyle \pi'} & & \downarrow{\scriptstyle \pi} \\ Z & \xrightarrow{f} & X \end{array}$$

be the associated fibered product (see A2.7). From exactness of inverse images and adjunction between direct and inverse images, it is easy to define a natural morphism $f^*(\mathrm{R}\pi_*\mathcal{F}) \to \mathrm{R}\pi'_*(f'^*\mathcal{F})$ for any $\mathcal{F} \in Sh_A(X_{\text{ét}})$ ("base change morphism").

When π is proper (for instance, a closed immersion or any morphism between projective varieties, see A2.7) and $\mathcal{F}$ is a torsion sheaf, then the above morphism $f^*(\mathrm{R}\pi_*\mathcal{F}) \to \mathrm{R}\pi'_*(f'^*\mathcal{F})$ is an isomorphism.

A3.6. Homology and direct images with compact support

A morphism $f: X \to Y$ is said to be **compactifiable** if and only if it decomposes as $X \xrightarrow{j} \overline{X} \xrightarrow{\overline{f}} Y$ where j is an open immersion and $\overline{f}$ is a proper morphism. For the structure morphism $X \to \mathrm{Spec}(\mathbf{F})$, this corresponds to embedding X as an open subscheme of a complete scheme; this is possible for schemes associated with quasi-projective varieties.

The direct image with compact support is denoted by $\mathrm{R_c}f_*$ and defined by

$$\mathrm{R_c}f_* := \mathrm{R}\overline{f}_* \circ j_!$$

on torsion sheaves. This preserves cohomologically constructible sheaves and does not depend on the chosen compactification $f = \overline{f} \circ j$. So we get a well-defined functor

$$\mathrm{R_c}f_*: D_A^+(X) \to D_A^+(Y)$$

(see A3.4) for any finite ring A. Note that it is not stated that $\mathrm{R_c}f_*$ is the right derived functor of a functor $Sh_A(X_{\text{ét}}) \to Sh_A(Y_{\text{ét}})$.

When we have a composition $f \circ g$ with both f, g, and $f \circ g$ compactifiable, then $\mathrm{R_c}(f \circ g)_* = \mathrm{R_c}f_* \circ \mathrm{R_c}g_*$.

The case of a structure morphism $\sigma^X: X \to \mathrm{Spec}(\mathbf{F})$ allows us to define **cohomology with compact support**. If $\mathcal{F}$ is a sheaf on $X_{\text{ét}}$ and X is a quasi-projective **F**-variety, one denotes by $\mathrm{H}^i_c(X, \mathcal{F})$ the ith cohomology group of $\mathrm{R_c}\Gamma(X, \mathcal{F}) = \mathrm{R_c}\sigma^X_* \mathcal{F}$. If $g: X \to X'$ is compactifiable, one has similarly

$$\mathrm{R_c}\Gamma(X, \mathcal{F}) \cong \mathrm{R_c}\Gamma(X', \mathrm{R_c}g_*\mathcal{F}).$$

One often finds the notation $\mathrm{R}f_!$ and $\mathrm{R}\Gamma_c$ in the literature instead of $\mathrm{R_c}f_*$ and $\mathrm{R_c}\Gamma$ above.

When direct images are replaced by direct images with compact support (see A3.5), the base change theorem holds unconditonally, i.e. (notation of A3.5) π and f being any compactifiable morphisms, one has a natural map $f^*(\mathrm{R_c}\pi_*\mathcal{F}) \to \mathrm{R_c}\pi'_*(f'^*\mathcal{F})$ which is an isomorphism ([SGA.4] XVII.5.2, [SGA.4$\frac{1}{2}$] Arcata.IV.5).

A3.7. Finiteness of cohomology

Let $\pi: X \to Y$ be a proper morphism and $\mathcal{F}$ be a constructible sheaf (see A3.2) on X. Then $\mathrm{R}\pi_*\mathcal{F}$ is cohomologically constructible. As a consequence, if $\mathcal{F}$ is a constructible sheaf on a quasi-projective variety X, each $\mathrm{H}^i_c(X, \mathcal{F})$ is finite.

Assume X is a quasi-projective **F**-variety of dimension d. If $\mathcal{F}$ is a torsion sheaf with torsion prime to the characteristic of **F**, then $\mathrm{R}\Gamma(X, \mathcal{F})$ has a representative which has trivial terms in degrees outside $[0, 2d]$. One may even replace $2d$ by d whenever X is affine (see [Milne80] VI.1.1 and [Milne80] VI.7.2).

A3.8. Coefficients

Let us first introduce the notion of ℓ-**adic** sheaf cohomology. Let X be an **F**-variety, let ℓ be a prime, generally different from the characteristic of **F**. Let Λ be the ring of integers over $\mathbb{Z}_\ell$ in a finite extension of $\mathbb{Q}_\ell$. An ℓ-adic sheaf (over Λ) is a projective system $\mathcal{F} = (\mathcal{F}_{n+1} \to \mathcal{F}_n)_{n \geq 1}$ where $\mathcal{F}_n$ is an object of $Sh_{\Lambda/J(\Lambda)^n}(X_{\text{ét}})$ and each $\mathcal{F}_{n+1} \to \mathcal{F}_n$ induces an isomorphism $\mathcal{F}_{n+1} \otimes \Lambda/J(\Lambda)^n \cong \mathcal{F}_n$. Thus one could define a category (distinct from $Sh_\Lambda(X_{\text{ét}})$) with the basic operations (direct and inverse images, including stalks, $\mathcal{H}om$ and $\otimes$) and notions (locally constant sheaves) defined componentwise. One defines $\mathrm{H}^i(X, \mathcal{F})$, resp. $\mathrm{H}^i_c(X, \mathcal{F})$, as the corresponding projective limit (over $n \geq 1$) of the groups $\mathrm{H}^i(X, \mathcal{F}_n)$, resp. $\varprojlim \mathrm{H}^i_c(X, \mathcal{F}_n)$. These are Λ-modules. The constant ℓ-adic sheaf is defined by $\mathcal{F}_n$ being the constant sheaf $(\Lambda/J(\Lambda)^n)_X$, whence the notation $\mathrm{H}^i(X, \Lambda)$ and $\mathrm{H}^i_c(X, \Lambda)$.

Let $A \subseteq B$ be an inclusion of rings. Assume that B is a flat right A-module.

We have functors Res^B_A: B−**mod** $\to A$−**mod** and ${}_BB_A \otimes_A -$, both exact. They induce functors between the corresponding categories of sheaves on $X_{\text{ét}}$, and they are exact on them too, as can be seen on stalks. The classical adjunction between them on the module categories (see [Ben91a] 2.8.2) implies the same for the functors between the categories of sheaves on $X_{\text{ét}}$. We therefore get that

$$\text{Res}_A^B: Sh_B(X_{\text{ét}}) \to Sh_A(X_{\text{ét}})$$

is exact and preserves injectives. Then

$$\text{Res}_A^B(\text{R}\Gamma(X_{\text{ét}}, \mathcal{F})) \cong \text{R}\Gamma\big(X_{\text{ét}}, \text{Res}_A^B \mathcal{F}\big)$$

for any sheaf $\mathcal{F}$ on $X_{\text{ét}}$ with values in $B-$**mod**.

Since Res_A^B also clearly commutes with direct and inverse images, and also with $j_!$ (where j is an open immersion), the above implies the same with compact support

$$\text{Res}_A^B(\text{R}_\text{c}\Gamma(X_{\text{ét}}, \mathcal{F})) \cong \text{R}_\text{c}\Gamma\big(X_{\text{ét}}, \text{Res}_A^B \mathcal{F}\big).$$

The above implies that most theorems about $D_A^+(X)$ reduce to checking over $\mathbb{Z}$ or $\mathbb{Z}/n\mathbb{Z}$.

It may also be applied to use the theorems about ℓ-adic cohomology where $A = \mathbb{Z}/\ell^n\mathbb{Z}$ or $\mathbb{Z}_\ell$ in a framework where $\mathbb{Z}_\ell$ is replaced with Λ, the integral closure of $\mathbb{Z}_\ell$ in a finite field extension of $\mathbb{Q}_\ell$, and $\mathbb{Z}/\ell^n\mathbb{Z}$ with $\Lambda/J(\Lambda)^n$. One may also use the above in the case when A is commutative and $B = A[G]$ where G is a finite group acting on $\mathcal{F}$.

We know that a complex of sheaves is acyclic if and only if the corresponding complex on stalks is acyclic (see A3.2). The universal coefficient formula (see [Bour80] p. 98, [Weibel] 3.6.2 or Exercise 4.5) on module categories then implies that, if A is a principal ideal domain and $\mathcal{F}$ is an object of $D^b(Sh_A(X_{\text{ét}}))$, then $\mathcal{F} = 0$ if and only if $\mathcal{F} \otimes k = 0$ for all quotient fields $k = A/\mathfrak{M}$.

A3.9. The “open-closed” situation

([Tamme] II.8, [Milne80] II.3)

Let U be an open subscheme of the $\mathbf{F}$-scheme X. Let $j: U \to X$ and $i: X \setminus U \to X$ be the corresponding open and closed immersions (respectively). Let A be a ring. Consider the following as functors on $Sh_A(X_{\text{ét}})$, $Sh_A(U_{\text{ét}})$, and $Sh_A(X \setminus U_{\text{ét}})$. Then $j_!$, j^*, and i^* are exact, i_* is faithful and exact.

The natural transformations $i^*i_* \to \text{Id}$ and $j^*j_* \to \text{Id}$ are isomorphisms.

The natural transformation $j_! \to j_*$ (see A.3.3) composed with j^* yields exact sequences for any $\mathcal{F}$ in $Sh_A(X_{\text{ét}})$

$$0 \to j_!j^*\mathcal{F} \to \mathcal{F} \to i_*i^*\mathcal{F} \to 0.$$

This has the following consequence on cohomology, known as the "open-closed" long exact sequence (see [Milne80] III.1.30)

$$\ldots \to \mathrm{H}_{\mathrm{c}}^{m}(U) \to \mathrm{H}_{\mathrm{c}}^{m}(X) \to \mathrm{H}_{\mathrm{c}}^{m}(X \setminus U) \to \mathrm{H}_{\mathrm{c}}^{m+1}(U) \to \ldots$$

where one takes cohomology of a constant sheaf.

A3.10. Higher direct images and stalks

Let us recall the Hensel condition. A commutative local ring A is called **henselian** if and only if, for all monic $P \in A[x]$ such that its reduction mod $J(A)$ factors as $\overline{P} = Q'R'$ in $A/J(A)[x]$ with Q', R' relatively prime, there are $Q, R \in A[x]$ such that $\overline{Q} = Q'$, $\overline{R} = R'$ and $P = QR$. A is called **strictly henselian** if, moreover, $A/J(A)$ is separably closed. Each commutative local ring A admits a strict henselization $A \to A^{\mathrm{sh}}$, i.e. an initial object among local embeddings $A \to B$ where B is strictly henselian. If $\overline{x}\colon \mathrm{Spec}(\Omega) \to X$ is a geometric point of an $\mathbf{F}$-scheme and $x \in X$ is its image, one defines its **stalk** at $\overline{x}$ as $\mathcal{O}_{X,\overline{x}} := (\mathcal{O}_{X,x})^{\mathrm{sh}}$. This is the limit of $\mathcal{O}_U(U)$ where $U \to X$ ranges over étale neighborhoods of $\overline{x}$ ([Tamme] II.6.2). Note that $\overline{x}$ induces $\overline{x}^{\mathrm{sh}}\colon \mathrm{Spec}(\mathcal{O}_{X,\overline{x}}) \to X$.

Let $f\colon Y \to X$ be a morphism between noetherian $\mathbf{F}$-schemes. Let $\overline{x}\colon \mathrm{Spec}(\Omega) \to X$ be a geometric point with image $x \in X$ and such that Ω is the separable closure of $\mathcal{O}_{X,x}/J(\mathcal{O}_{X,x})$ (see A2.7). Let

$$\begin{array}{ccc} \overline{Y} := Y \times_X \mathrm{Spec}(\mathcal{O}_{X,\overline{x}}) & \xrightarrow{\ \xi\ } & Y \\ \downarrow & & \downarrow f \\ \mathrm{Spec}(\mathcal{O}_{X,\overline{x}}) & \xrightarrow{\ \overline{x}^{\mathrm{sh}}\ } & X \end{array}$$

be the associated fibered product. Then $(\mathrm{R}f_*\mathcal{F})_{\overline{x}} \cong \mathrm{R}\Gamma(\overline{Y}, \xi^*\mathcal{F})$ (see [Tamme] II.6.4.1, [Milne80] III.1.15, [SGA.4] VIII.5.2).

A3.11. Projection and Künneth formulae

Let X, Y be $\mathbf{F}$-schemes. Let A be a finite ring. For any $\mathcal{F}$ in $Sh_A(X_{\text{ét}})$, there exists an exact sequence $\mathcal{F}' \to \mathcal{F} \to 0$ in $Sh_A(X_{\text{ét}})$, where the stalks of $\mathcal{F}'$ are flat A-modules, hence $\mathcal{F}'$ is "flat" with respect to the tensor product in $Sh_A(X_{\text{ét}})$ defined in A3.3 (see [Milne80] VI.8.3). This allows us to define the left derived bi-functor $-\overset{\mathrm{L}}{\otimes}-$.

Let $\pi: Y \to X$ be a compactifiable morphism, $\mathcal{F}$ and $\mathcal{G}$ objects of $D^b_A(X)$, $D^b_A(Y)$ respectively. The **projection formula** is

$$(\mathrm{R_c}\pi_*\mathcal{G})\overset{\mathrm{L}}{\otimes}_A\mathcal{F} \cong \mathrm{R_c}\pi_*\big(\mathcal{G}\overset{\mathrm{L}}{\otimes}_A\pi^*\mathcal{F}\big)$$

in $D^b_{\mathbb{Z}}(X)$ (see [Milne80] VI.8.14).

When $\mathcal{F}$ is a locally constant sheaf, one has the same formula with R's instead of $\mathrm{R_c}$'s (easily deduced from the case $\mathcal{F} = A_X$).

The **Künneth formula** is a generalization of the projection formula. Keep the above notation and hypotheses, let S be another **F**-scheme and let

$$\begin{array}{ccc} X\times_S Y & \xrightarrow{f_X} & Y \\ \downarrow{\scriptstyle g_Y} & & \downarrow{\scriptstyle g} \\ X & \xrightarrow{f} & S \end{array}$$

be a fibred product of **F**-schemes, then

$$\mathrm{R_c}f_*\mathcal{F}\overset{\mathrm{L}}{\otimes}_A\mathrm{R_c}g_*\mathcal{G} \cong \mathrm{R_c}h_*\big(f_X^*\mathcal{F}\overset{\mathrm{L}}{\otimes}_A g_Y^*\mathcal{G}\big)$$

where $h = f\circ g_Y = g\circ f_X$ (see [Milne80] VI.8.14). The case $S = X$ gives back the projection formula.

A3.12. Poincaré–Verdier duality and twisted inverse images

The theorem known as Poincaré–Verdier duality gives the relation between the ordinary cohomology and cohomology with compact support for an étale sheaf, when the underlying variety is smooth.

Let X be a smooth quasi-projective **F**-variety with all connected components of the same dimension d. Let $\Lambda = \mathbb{Z}/n\mathbb{Z}$ where $n \geq 1$ is an integer.

Let $M \to M^\vee = \mathrm{Hom}_\Lambda(M, \Lambda)$ be the Λ-duality functor applied to $\Lambda-$**mod** (and therefore $C^b(\Lambda-\mathbf{mod})$ or $D^b(\Lambda-\mathbf{mod})$, up to changing the degrees into their opposite). On $Sh_\Lambda(X_{\text{ét}})$, we also have a similar functor sending $\mathcal{F}$ to $\mathcal{F}^\vee := \mathcal{H}om(\mathcal{F}, \Lambda_X)$.

Let $\mathcal{F}$ in $Sh(X_{\text{ét}})$ be locally constant with finite stalks. Then

$$\mathrm{R_c}\Gamma(X, \mathcal{F})^\vee \cong \mathrm{R}\Gamma(X, \mathcal{F}^\vee)[2d].$$

See [Milne80] VI.11.1, [SGA.4] XVIII.3.2.6.1.

As a generalization of the above, one may define a **twisted inverse image** functor, i.e. a right-adjoint $f^!$ to the functor $\mathrm{R_c} f_*$ (see [KaSch98] §III for the ordinary topological case and [SGA.4] §XVIII for the étale case; see also [Milne80] VI.11).

Let $f: X \to Y$ be a compactifiable morphism between $\mathbf{F}$-schemes (not necessarily smooth). There exists $f^! : D^b(Y) \to D^b(X)$ such that

$$\mathrm{R} f_* \mathrm{R}\mathcal{H}om(\mathcal{F}, f^!\mathcal{G}) \cong \mathrm{R}\mathcal{H}om(\mathrm{R_c} f_* \mathcal{F}, \mathcal{G})$$

for any objects $\mathcal{F}, \mathcal{G}$ of $D^b_\Lambda(X)$ and $D^b_\Lambda(Y)$ respectively (see [Milne80] VI.11.10, and, for a proof, [SGA.4] §XVIII 3.1.10 plus later arguments in [SGA.4$\frac{1}{2}$] "Dualité" §4).

One has

$$(f \circ g)^! = g^! \circ f^!$$

whenever the three functors make sense.

Applying the above to the structure morphism $\sigma_X: X \to \mathrm{Spec}(\mathbf{F})$ of an $\mathbf{F}$-variety (not necessarily smooth), one gets the following.

Define $\mathcal{D}: D^b_\Lambda(X) \to D^b_\Lambda(X)$ by $\mathcal{D}(\mathcal{F}) = \mathrm{R}\mathcal{H}om(\mathcal{F}, \sigma_X^! \Lambda)$. The above adjunction property of $f^!$ gives

$$\mathrm{R}\Gamma(X, \mathcal{D}(\mathcal{F})) \cong \mathrm{R_c}\Gamma(X, \mathcal{F})^\vee.$$

The main statement of Poincaré–Verdier duality is implied by the fact that, if moreover X is smooth with all connected components of dimension d, then $\sigma_X^! \Lambda \cong \Lambda_X[2d]$.

A3.13. Purity

Let X be an $\mathbf{F}$-variety, $Y \subseteq X$ a closed subvariety and $U = X \setminus Y$ the open complement. Denote by $U \xrightarrow{j} X \xleftarrow{i} Y$ the associated immersions. Assume X (resp. Y) is smooth with all connected components of the same dimension d (resp. $d-1$). Let $\mathcal{F}$ be a locally constant sheaf of $\mathbb{Z}/n\mathbb{Z}$-modules on $X_{\text{ét}}$, with $n \geq 1$ prime to the characteristic of $\mathbf{F}$. Then $\mathrm{R} j_* j^* \mathcal{F}$ may be represented by a complex

$$0 \to \mathcal{F} \to \mathcal{F}' \to 0 \ldots$$

concentrated in degrees 0 and 1 where $i^*\mathcal{F}'$ is locally isomorphic with $i^*\mathcal{F}$ (see [Milne80] VI.5.1).

A basic ingredient in the above is the computation of étale cohomology of affine spaces. One has $\mathrm{R}\Gamma(\mathbb{A}^1_{\mathbf{F}}, \mathbb{Z}/n\mathbb{Z}) \cong \mathbb{Z}/n\mathbb{Z}[0]$ (see [Milne80] VI.4.18).

A3.14. Finite group actions and constant sheaves

Let Y be a quasi-projective variety over $\mathbf{F}$ with a finite group G acting on it, thus defining a quotient map $\pi: Y \to Y/G = X$ (see A2.6).

Let Λ be a commutative ring and M be an Λ-module. The associated constant sheaf M_Y on Y then defines an object $\pi_*^G(M_Y)$ of $Sh_{\Lambda[G]}(X_{\text{ét}})$ as follows. If $U \to X$ is in $X_{\text{ét}}$, then $\pi_*(M_Y)_U$ is by definition $M^{\pi_0(Y \times_X U)}$ (see A3.2 and A3.3). But the action of G on Y clearly stabilizes the fiber product as a subset of $Y \times U$. So G permutes its connected components and this clearly extends to restriction morphisms, thus endowing $\pi_*(M_Y)$ with a structure of sheaf of $\Lambda[G]$-modules (which gives back $\pi_*(M_Y)$ by the forgetful functor $\Lambda G-\mathbf{Mod} \to \Lambda-\mathbf{Mod}$). The above construction corresponds to the "functoriality" of the constant sheaf with respect to X since it corresponds to the composition of the first line

$$\begin{array}{ccccc} Y \times_X U & \longrightarrow & Y & \xrightarrow{g} & Y \\ \pi_U \downarrow & & \pi \downarrow & & \pi \downarrow \\ U & \longrightarrow & X & \xrightarrow{\mathrm{Id}} & X \end{array}$$

(see also [Srinivasan] p. 51). Note that the above π_U is a G-quotient for the action of G mentioned before.

A consequence is also that $\mathrm{R}\Gamma(Y, M_Y) = \mathrm{R}\Gamma(X, \pi_* M_Y)$ and $\mathrm{R_c}\Gamma(Y, M_Y) = \mathrm{R_c}\Gamma(X, \pi_* M_Y)$ can be considered as objects of $D^b(\Lambda G-\mathbf{mod})$ giving back the usual cohomology Λ-modules by the restriction functor (see A3.8).

Concerning stalks, let $\overline{x}$: Spec($\mathbf{F}$) $\to X$ be a geometric point of X of image the closed point $x \in X = Y/G$. We have

$$(\pi_*^G M_Y)_{\overline{x}} \cong M^x$$

as a ΛG-submodule of Λ^Y (see [Tamme] II.6.4.2, [Milne80] II.3.5.(c) or A3.10 above).

A3.15. Finite group actions and projectivity

Let X be the scheme associated with a (quasi-projective) variety over an algebraically closed field $\mathbf{F}$. Let A be a finite ring (non-commutative) whose characteristic is invertible in $\mathbf{F}$. Let $\mathcal{F}$ be a constructible sheaf of A-modules on $X_{\text{ét}}$ such that, for every closed point $x \in X$, the stalk $\mathcal{F}_x$ is projective. Then $\mathrm{R}\Gamma(X, \mathcal{F})$ may be represented by an object of $C^b(A-\mathbf{proj})$. The proof (see [Srinivasan] p. 67) uses finiteness (see A3.7) and Godement resolutions built from stalks at closed points. One may also proceed as in [Milne80] VI.8.15, noting that the projection formula always holds for $\mathrm{R}\Gamma$ (see also [SGA.4] XVII.5.2). Consequently $\mathrm{R_c}\Gamma(X, \mathcal{F})$ may also be represented by an object of $C^b(A-\mathbf{proj})$, since for $j: X \to X'$ an open immersion, $j_!\mathcal{F}$ has the same non-trivial stalks as $\mathcal{F}$.

Assume the action of a finite group G on a quasi-projective $\mathbf{F}$-variety Y is given. Denote by $\pi: Y \to X = Y/G$ the associated quotient. Let Λ be a finite commutative ring of characteristic not divisible by that of $\mathbf{F}$, so that $\mathrm{R}\Gamma(X, \Lambda)$ and $\mathrm{R_c}\Gamma(X, \Lambda)$ can be considered as objects of $D^b(\Lambda[G]-\mathbf{mod})$ (see A3.14). If one has the additional hypothesis that all isotropy subgroups $G_y := \{g \in G \mid gy = y\}$ of closed points $y \in Y$ are of order invertible in Λ, then $\mathrm{R}\Gamma(X, \Lambda)$ and $\mathrm{R_c}\Gamma(X, \Lambda)$ can be represented by objects of $C^b(\Lambda[G]-\mathbf{proj})$. To check this, one uses the above along with A3.14 to check that the stalks of $\pi_*^G\Lambda$ at closed points are of type $\Lambda[G/G_y]$, hence projective (see also [Srinivasan] 6.4 and proof, [SGA.4] XVII and [SGA.4$\frac{1}{2}$] pp. 97–8).

We point out another consequence. Keep $Y \xrightarrow{\pi} X = Y/G$ the quotient of an $\mathbf{F}$-variety by a finite group such that the stabilizers of closed points are of order invertible in Λ. Let $\sigma: X \to S$ be a morphism of $\mathbf{F}$-varieties. Considering $\pi_*^G\Lambda_Y$ as in $Sh_{\Lambda G}(X_{\text{ét}})$ (see A3.14), we have

$$\text{(1)} \qquad \mathrm{R}\sigma_*(\pi_*^G\Lambda_Y)\overset{\mathrm{L}}{\otimes}_{\Lambda G}\Lambda_S \cong \mathrm{R}\sigma_*(\pi_*^G\Lambda_Y) \otimes_{\Lambda G} \Lambda_S \cong \mathrm{R}\sigma_*(\Lambda_X).$$

In particular $\mathrm{R}\Gamma(X, \Lambda) \cong \mathrm{R}\Gamma(Y, \Lambda)_G$ (the latter denoting the co-invariants in the action of G on $\mathrm{R}\Gamma(Y, \Lambda)$). The same holds for cohomology with compact support (see A3.6).

By the projection formula (see A3.11) and the fact that $\pi_*^G\Lambda_Y$ has projective stalks (see A3.14 and use the fact that stabilizers are of invertible order), the proof of (1) reduces to the isomorphism $\pi_*^G\Lambda_Y \otimes_{\Lambda G} \Lambda_X \cong \Lambda_X$. It is easy to define an "augmentation" map $\pi_*^G\Lambda_Y \otimes_{\Lambda G} \Lambda_X \longrightarrow \Lambda_X$. That it is an isomorphism is checked on stalks at closed points of $X = Y/G$, using A3.14 again.

Taking $S = \mathrm{Spec}(\mathbf{F})$, we get the statement about cohomology. As for cohomology with compact support, one may do the same as above with $\mathrm{R_c}$ instead of R (see A3.11).

A3.16. Locally constant sheaves and the fundamental group

(see [Milne80] I.5, [Milne98] 3, [FrKi88] AI, [Murre], [SGA.1] V, [BLR] 9)

Let X be a connected $\mathbf{F}$-scheme. One defines the category of **coverings** of X as the full subcategory of schemes over X (see A2.7) consisting of maps $Y \to X$ that are finite and étale (hence closed, open, and therefore onto). One uses the notation $\mathrm{Hom}_X(Y, Y')$ and $\mathrm{Aut}_X(Y)$ for morphism sets and automorphism groups in the category of schemes over X. If $\overline{x}$: Spec(Ω) $\to X$ is a geometric point of X, one denotes by $Y \mapsto Y(\overline{x})$ the functor associating $\mathrm{Hom}_X(\mathrm{Spec}(\Omega), Y)$ with each covering of X. The **degree** of a covering $Y \to X$ with connected Y is the cardinality of $Y(\overline{x})$ when $\overline{x}$: Spec(Ω) $\to X$ is a geometric point of X such that $Y(\overline{x})$ is non-empty. One calls $Y \to X$ a **Galois covering** if and only if, Y being connected, its degree equals the cardinality of $\mathrm{Aut}_X(Y)$. If X is an **F**-variety, then $G := \mathrm{Aut}_X(Y)$ acts freely on Y and $Y \to X$ is a G-quotient in the sense of A2.6. Conversely, if Y is a quasi-projective **F**-variety and G is a finite group acting freely on it, the associated quotient $Y \to Y/G$ is a Galois covering (see [Milne80] I.5.4, [Jantzen] I.5.7). More generally, if **G** is an algebraic **F**-group and $Y \to X$ is a **G**-quotient (see A2.6), it is called a **G-torsor** if it is locally trivial for the étale topology on X (see [Milne80] III.4.1).

If a connected **F**-scheme X and a geometric point $\overline{x}$: Spec(Ω) $\to X$ are fixed, the **fundamental group** $\pi_1(X, \overline{x})$ is defined as follows. The pairs (Y, α) where $Y \to X$ is a Galois covering and $\alpha \in Y(\overline{x})$ form a category that essentially dominates the connected coverings of X and one defines $\pi_1(X, \overline{x})$ as the limit of $\mathrm{Aut}_X(Y)$'s over pairs (Y, α) (Grothendieck's theory of fundamental groups for "Galois categories", see [SGA.1] V or [Murre]). The group $\pi_1(X, \overline{x})$ is endowed with the limit of the (discrete) topologies of the $\mathrm{Aut}_X(Y)$'s. If one restricts the limit to the Galois coverings whose automorphism groups are of order invertible in **F**, one obtains $\pi_1^{\mathrm{t}}(X, \overline{x})$, a quotient of $\pi_1(X, \overline{x})$ called the **tame** fundamental group.

For each (Y, α), we have an exact sequence of groups $1 \to \pi_1(Y, \alpha) \to \pi_1(X, \overline{x}) \to \mathrm{Aut}_X(Y) \to 1$. More generally, $(X, \overline{x}) \mapsto \pi_1(X, \overline{x})$ is a covariant functor. The fundamental group is also unique in the sense that any other geometric point $\overline{x'}$ would satisfy $\pi_1(X, \overline{x}) \cong \pi_1(X, \overline{x'})$. So we may sometimes omit $\overline{x}$.

One has $\pi_1(\mathbb{P}^1_{\mathbf{F}}) = 1$, and $\pi_1^{\mathrm{t}}(\mathbb{A}^1_{\mathbf{F}}) = 1$. If p denotes the characteristic of **F**, $\pi_1^{\mathrm{t}}(\mathbb{G}_{\mathrm{m}})$ is the closure of $(\mathbb{Q}/\mathbb{Z})_{p'}$ with regard to finite quotients.

As in the topological case (see A1.14), the functor $Y \mapsto Y(\overline{x})$ induces an equivalence between the category of coverings of X and the category of finite continuous $\pi_1(X, \overline{x})$-sets. An inverse functor consists of forming $(Y \times E)/G$ whenever G is a finite group, $Y \to X$ is a G-torsor and E is a finite G-set.

The functor $\mathcal{F} \mapsto \mathcal{F}_{\overline{x}}$ induces an equivalence between the category of locally constant constructible sheaves on X and the category of finite continuous $\pi_1(X, \overline{x})$-modules.

Let G be a finite group and $Y \xrightarrow{\pi} X$ be a G-torsor. Assume $|G|$ is prime to the characteristic p of **F**. Let $\Lambda = \mathbb{Z}/n\mathbb{Z}$ where n is prime to p. Let M be a finite ΛG-module, hence a finite $\pi_1(X)$-module. Then it is easy to check that $\pi_*^G \Lambda_Y \otimes_{\Lambda G} M_X$ is the sheaf corresponding to M through the above equivalence.

A3.17. Tame ramification along a divisor with normal crossings

(See [Milne80] I.5, [GroMur71], [BLR] 11)

Let us define (tame) ramification in a special context adapted to our needs. Let $Y, \overline{X}$ be smooth $\mathbf{F}$-varieties. Assume $\overline{X} = X \cup D$ (a disjoint union) where D is a smooth divisor with normal crossings (see A2.3).

We are interested in describing when locally constant sheaves on $X_{\text{ét}}$ can extend into locally constant sheaves on $\overline{X}_{\text{ét}}$. By A3.16, this is clearly related to an associated Galois covering $Y \to X$ extending to a Galois covering $Y' \to \overline{X}$. We describe how this leads to the construction of Grothendieck-Murre's $\pi_1^D(\overline{X})$ groups (see [GroMur71] §2).

Let $f: Y \to \overline{X}$ be finite, normal and étale between irreducible schemes over $\mathbf{F}$. Let d be one of the points of codimension 1 in D, generic point of an irreducible component D_d of D (see A2.7), then $\mathcal{O}_{\overline{X},d}$ is a discrete valuation ring and the ramification along D is the ramification of the extension $\mathbf{F}(Y)/\mathbf{F}(\overline{X})$ with respect to $\mathcal{O}_{\overline{X},d}$. Thus f is said to be **tamely ramified** with respect to D when $\mathbf{F}(Y)/\mathbf{F}(\overline{X})$ is tamely ramified with respect to the various $\mathcal{O}_{\overline{X},d}$, or "tamely ramified over the d's." Locally for the étale topology, one may replace the triple $(\mathcal{O}_{\overline{X},d} \to \mathbf{F}(\overline{X}) \to \mathbf{F}(Y))$ by $(\mathcal{O}_{\overline{X},\bar{d}} \to \mathbf{F}(\overline{X})_{\text{sep}} \to \mathbf{F}(Y)_{\text{sep}})$. Furthermore theses maps are given by the "fiber maps" of schemes $Y \times_{\overline{X}} \operatorname{Spec}(\mathcal{O}_{\overline{X},\bar{d}}) \to \operatorname{Spec}(\mathcal{O}_{\overline{X},\bar{d}})$. Finally the ramification with respect to D is the ramification of $Y \times_{\overline{X}} \operatorname{Spec}(\mathcal{O}_{\overline{X},\bar{d}}) \to \operatorname{Spec}(\mathcal{O}_{\overline{X},\bar{d}})$ with respect to the closed point of $\operatorname{Spec}(\mathcal{O}_{\overline{X},\bar{d}})$ ([GroMur 71] Lemmas 2.2.8, 2.2.10). It is clearly a local invariant. Up to an étale base change, a tamely ramified covering is a Kummer extension (i.e. essentially an extension by roots of some sections; see [GroMur71] §2.3, [Milne80] 5.2(e)).

By a general construction similar to the definition of $\pi_1(X)$ groups (see A3.16), given a geometric point ξ of $\overline{X}$ with image in X, there exists a profinite group $\pi_1^D(\overline{X}, \xi)$, the "tame fundamental group with respect to D and with base point ξ," such that the category of finite sets on which $\pi_1^D(\overline{X}, \xi)$ acts continuously is equivalent to a category of coverings $Y' \to \overline{X}$ that are tamely ramified with respect to D ([GroMur71] §2.4).

To a Galois covering $Y \to \overline{X}$ with group G there corresponds, from the theory of the ordinary fundamental group, the fiber functor on Y, i.e. $\operatorname{Hom}_{\overline{X}}(\xi, Y)$, acted on by G, hence by $\pi_1(X, \xi)$, of which G is a quotient. The similar construction of $\pi_1^D(\overline{X}, \xi)$ gives natural continuous surjective morphisms

$$\pi_1(X, \xi) \to \pi_1^D(\overline{X}, \xi) \quad \text{and} \quad \pi_1^D(\overline{X}, \xi) \to \pi_1(\overline{X}, \xi).$$

Assuming that G is a p'-group, all ramifications involved are tame and $\pi_1(X, \xi) \to G$ factors through $\pi_1^D(\overline{X}, \xi)$. The covering is unramified if and only if $\pi_1(X, \xi) \to G$ factors through the composed map $\pi_1(X, \xi) \to \pi_1(\overline{X}, \xi)$, and then the Galois covering extends to $\overline{X}$.

If D is irreducible of codimension 1, with generic point d, then, locally for the étale topology, the smooth pair $(X, \overline{X})$ is isomorphic to a standard affine pair $(\mathbb{A}^n_{\mathbf{F}}, \mathbb{A}^{n+1}_{\mathbf{F}})$ (all schemes being **F**-schemes) in the sense that there is a Zariski open neighborhood V of d and an étale map $\phi\colon V \to \mathbb{A}^{n+1}_{\mathbf{F}})$ that sends d to the generic point d' of $\mathbb{A}^n_{\mathbf{F}}$. The ramification of Y over d is that of $\phi_*(Y_{|V\cap X})$ over d'. That means that D is locally defined by an equation $z = 0$, where z is one of the $(n+1)$ variables z_j of $\mathbb{A}^{n+1}_{\mathbf{F}} = \mathrm{Spec}(\mathbf{F}[z_1, \ldots, z_{n+1}])$ and (writing $\mathbf{F}[z_1, \ldots, z_{n+1}] = \mathbf{F}[z_1, \ldots, z_n][z_{n+1}]$) $\mathcal{O}_{\overline{X},\bar{d}}$ is isomorphic to $\mathbb{A}^n_{\mathbf{F}} \times_{\mathrm{Spec}(\mathbf{F})} \mathbb{A}^{\mathrm{sh}}_{\mathbf{F}}$ where $\mathbb{A}^{\mathrm{sh}}_{\mathbf{F}} = \mathrm{Spec}(\mathbf{F}[z]^{\mathrm{sh}})$ (see A3.10). One may forget the first n variables (or replace $\mathbf{F}$ by $\mathbb{A}^n_{\mathbf{F}}$) and consider a map $x\colon \mathbb{A}^{\mathrm{sh}}_{\mathbf{F}} \to \overline{X}$ such that

- the closed point (z) of $\mathbb{A}^{\mathrm{sh}}_{\mathbf{F}}$ is mapped onto d so that $\delta = x \circ \delta'$, where δ (resp. δ') is a geometric point of $\overline{X}$ with image d (resp. of $\mathbb{A}^{\mathrm{sh}}_{\mathbf{F}}$ with image the closed point),
- the inverse image of D in $\mathbb{A}^{\mathrm{sh}}_{\mathbf{F}}$ is the reduced scheme defined by $z = 0$.

Finally the ramification with respect to D is given by $Y \times_{\overline{X}} \mathbb{A}^{\mathrm{sh}}_{\mathbf{F}} \to \mathbb{A}^{\mathrm{sh}}_{\mathbf{F}}$ with respect to the closed point of $\mathbb{A}^{\mathrm{sh}}_{\mathbf{F}}$.

Assume that D is no longer irreducible. One has $\pi_1^D(\overline{X}, \xi) = \prod_e \pi_1^{D_e}(\overline{X}, \xi)$, the product being over the irreducible components of D, where e is the generic point of its irreducible component $D_e := \overline{\{e\}}$. Put $D'_d = D_d \setminus \bigcup_{e \neq d} D_e$. Let $\overline{X}_d \to \overline{X}$ be the open subscheme whose support is $X \cup D'_d$, let d' be the generic point of D'_d in $\overline{X}_d$, hence $d' \mapsto d$. One obtains a triple $(\mathcal{O}_{\overline{X}_d, d'} \to \mathbf{F}(\overline{X}_d) \to \mathbf{F}(Y \times_{\overline{X}_d} X))$ that gives the ramification along D_d.

A3.18. Tame ramification and direct images

We keep the notation of the preceding section in relation to $\overline{\mathbf{X}} = \mathbf{X} \sqcup D$ where $\overline{\mathbf{X}}$ is a smooth **F**-variety and D is a smooth divisor with normal crossings. Let $\mathcal{F}$ be a locally constant constructible sheaf on X with no p-torsion. One says that $\mathcal{F}$ **ramifies** along D_d if and only if there is no locally constant sheaf $\overline{\mathcal{F}}$ on $\overline{X}_d$ such that $\mathcal{F} = j^*\overline{\mathcal{F}}$, where j is the open immersion $X \to \overline{X}_d$.

Recall ([Milne80] V.1.1, [SGA.4] IX) the existence of an equivalence between the category of locally constant sheaves with finite stalks on X and the category of finite étale schemes over X. The covering $Y \to X$ defines $\mathcal{F}_Y$ by $\mathcal{F}_Y(U) = \mathrm{Hom}_X(U, Y)$ so that $(\mathcal{F}_Y)_{\bar{x}} \cong \mathrm{Hom}_X(\mathrm{Spec}(\mathcal{O}_{X,\bar{x}}), Y)$. Then $\mathcal{F}_Y$

extends over $\overline{X}$ in a locally constant finite sheaf if and only if Y extends over $\overline{X}$ in a finite étale covering. One has $(j^*\mathcal{F}_Y)_{\bar{d}} = \mathrm{Hom}_X(\mathrm{Spec}(\mathcal{O}_{\overline{X},\bar{d}}) \times_{\overline{X}} Y, Y)$ (limit, on étale neighborhoods U of $\bar{d}$, of $\mathrm{Hom}_X(U \times_{\overline{X}} Y, Y)$).

Assuming D is irreducible of codimension 1, the obstruction to the extension is the ramification of Y with respect to D which is the ramification of $Y \times_{\overline{X}} \mathbb{A}^{\mathrm{sh}}_{\mathbf{F}} \to \mathbb{A}^{\mathrm{sh}}_{\mathbf{F}}$ with respect to the closed point of $\mathbb{A}^{\mathrm{sh}}_{\mathbf{F}}$. If $Y \times_{\overline{X}} \mathbb{A}^{\mathrm{sh}}_{\mathbf{F}} \to \mathbb{A}^{\mathrm{sh}}_{\mathbf{F}}$ is not ramified, as an étale finite covering of $\mathbb{A}^{\mathrm{sh}}_{\mathbf{F}}$, it is trivial. Hence $\mathcal{F}_Y$ extends to a constant sheaf locally around the generic point of D, so it extends to $\overline{X}$ in a locally constant sheaf. The non-ramification of $\mathcal{F}_Y$ along D is equivalent to the non-ramification of Y along D (see Theorem A3.19 (ii) below) .

Assume now that $\mathcal{F}$ is a locally constant sheaf of k-vector spaces with k a finite field of characteristic ℓ, and that $\mathcal{F}$ is associated with a linear character $\pi_1(X) \to k^\times$.

Theorem A3.19. *Let $i\colon D'_d \to \overline{X}_d$, $j\colon X \to \overline{X}_d$ be immersions.*
The following are equivalent.

(i) $\mathcal{F}$ ramifies along D_d.

(ii) $(j_*\mathcal{F})_d = 0$.

(iii) $i^* j_*\mathcal{F} = 0$.

(iv) $i^*(\mathrm{R}j_*\mathcal{F}) = 0$.

If the above are not satisfied, then $j_*\mathcal{F}$ is locally constant, $\mathcal{F} = j^* j_*\mathcal{F}$, $\mathrm{R}^1 j_*\mathcal{F} = i_* i^* j_*\mathcal{F}$, and $\mathrm{R}^q j_*\mathcal{F} = 0$ for $q \geq 2$.

Let us give an indication of the proof. Replacing $(X, \overline{X})$ with $(X, \overline{X}_d)$, and in view of the statements to prove, one may assume that $D = D_d$ is irreducible.

(ii) implies (i). If $\mathcal{F} = j^*\overline{\mathcal{F}}$, then $j_*\mathcal{F} = \overline{\mathcal{F}}$ by purity (A3.13). The fact that $j_*\mathcal{F}$ is locally constant implies that its stalk $\overline{\mathcal{F}}_x$ $(x \in \overline{X}$) only depends on the connected component of $\overline{X}$ containing x ([Milne80] V.1.10(a)). But $(j_*\mathcal{F})_{j(x)} = \mathcal{F}_x$ whenever $x \in X$. Moreover $\mathcal{F}_x \neq 0$. Since $j(X)$ meets any non-empty open subset of $\overline{X}$, this implies our claim.

(i) implies (iii). Let $Y \xrightarrow{\pi} X$ be a T-torsor where $T = \pi_1(X)/\theta^{-1}(1)$ is the finite quotient of $\pi_1(X)$ such that θ is a faithful representation of T. We must prove that $(j_*\mathcal{F}_\theta)_x = 0$ for all $x \in D$.

Let $Y \subseteq \overline{Y}$ be the normalization of $\overline{X}$ along Y (see A2.6), i.e. a normal **F**-variety $\overline{Y}$ minimal for the property that there is a finite morphism $\overline{\pi}\colon \overline{Y} \to \overline{X}$ such that $\overline{\pi}_{|Y} = j \circ \pi$. On T-stable affine open subschemes of Y, the construction of $\overline{Y}$ is as follows. We have $Y \to X$, which corresponds to the inclusion $A \supseteq A^T$, and $X \to \overline{X}$ to $A^T \supseteq A'$. Then $\overline{Y}$ corresponds with the integral closure B of A^T in A, but then $A' = B^T$ since it is integrally closed ($\overline{X}$ is smooth). From this, one sees that T acts on $\overline{Y}$ and the morphism $\overline{Y} \to \overline{X}$ is a T-quotient.

Similarly it is easy to see from this description that the closed points of $\overline{Y}$, where $\overline{\pi}$ does not induce an isomorphism between tangent spaces, are the $y \in \overline{Y}$ such that $|T.y| < |T|$. They are all in $\overline{\pi}^{-1}(D)$. Using the theorem of Zariski–Nagata on purity of branch locus (see A2.6), one sees that these are exactly the elements of $\overline{\pi}^{-1}(D)$. So we are left to prove that $(j_*\mathcal{F}_\theta)_x = 0$ as long as $|\overline{\pi}^{-1}(x)| < |T|$.

Recall the commutative square

$$\begin{array}{ccc} Y & \xrightarrow{j'} & \overline{Y} \\ \downarrow{\scriptstyle \pi} & & \downarrow{\scriptstyle \overline{\pi}} \\ X & \xrightarrow{j} & \overline{X} \end{array}$$

We have $\mathcal{F}_\theta = \pi_* k_Y \otimes_{kT} \Theta_X$ where Θ denotes the one-dimensional kT-module corresponding to θ. Then $\Theta_X = j^*\Theta_{\overline{X}}$ and the projection formula allows us to write $j_*\mathcal{F}_\theta = j_*\pi_*k_Y \otimes_{kT} \Theta_{\overline{X}}$.

Let us show that $j'_*k_Y = k_{\overline{Y}}$. We must take a connected $U \to \overline{Y}$ in $\overline{Y}_{\text{ét}}$ and check that $U \times_{\overline{Y}} Y$ is connected. Since $\overline{Y}$ is normal, U is also normal (apply [Milne80] I.3.17(b)), hence irreducible. But $U \times_{\overline{Y}} Y \to U$ is an open immersion since $Y \to \overline{Y}$ is, so $U \times_{\overline{Y}} Y$ is irreducible, hence connected.

Using $j\pi = \overline{\pi}j'$, this allows us to write $j_*\pi_*k_Y(U) \cong k^{\pi_0(U\times_{\overline{X}}Y)} \cong \pi'_*k_{\overline{Y}}(U)$ not just as k-modules but also as kT-modules, the action of T being on the right side of $U \times_{\overline{X}} Y$.

We now have $(j_*\mathcal{F}_\theta)_x = (\overline{\pi}_*k_{\overline{Y}})_x \otimes_{kT} \Theta$. This is zero when the stabilizer T_x of x in T is not equal to $\{1\}$ since then $(\overline{\pi}_*k_{\overline{Y}})_x \cong \mathrm{Ind}_{T_x}^T k$ (see A3.14) while Θ has no invariant under T_x.

(iii) is equivalent to (iv) (see also [SGA.4$\frac{1}{2}$] p. 180). Let $x \in D$. We must show that, if $(j_*\mathcal{F})_x = 0$, then $(\mathrm{R}j_*\mathcal{F})_x = 0$. Since D is a smooth divisor with normal crossings in a smooth variety over $\mathbf{F}$, there is a neighborhood V of x and an isomorphism $V \to \mathbb{A}^n_{\mathbf{F}}$ sending $D \cap V$ to $(\mathbb{G}_{\mathrm{m}})^n$. We are reduced to the case of a locally constant sheaf $\mathcal{F}$ on $(\mathbb{G}_{\mathrm{m}})^n$ such that $(j_*\mathcal{F})_x = 0$ for some $x \in \mathbb{A}^n_{\mathbf{F}} \setminus (\mathbb{G}_{\mathrm{m}})^n$ and $j\colon (\mathbb{G}_{\mathrm{m}})^n \to \mathbb{A}^n_{\mathbf{F}}$. We must show that $(\mathrm{R}j_*\mathcal{F})_x = 0$. By the Künneth formula, it suffices to treat the case of $n = 1$. Then we must show that $(\mathrm{R}j_*\mathcal{F})_0 = 0$ when $(j_*\mathcal{F})_0 = 0$. This last condition means that $\mathcal{F}$ is associated with a (finite) $\pi_1(\mathbb{G}_{\mathrm{m}})$-module without fixed points $\neq 0$ (use the equivalence (i)–(iii) we have proved). Then $\mathcal{F}^\vee$ satisfies the same. Now $(\mathrm{R}j_*\mathcal{F})_0 = \mathrm{R}\Gamma(\mathbb{G}_{\mathrm{m}} \times_{\mathbb{G}_{\mathrm{a}}} \mathbb{A}^{\mathrm{sh}}_{\mathbf{F}}, \mathcal{F}_0)$ by §A3.10, where $\mathcal{F}_0$ is the restriction of $\mathcal{F}$ to $\mathbb{G}_{\mathrm{m}} \times_{\mathbb{G}_{\mathrm{a}}} \mathbb{A}^{\mathrm{sh}}_{\mathbf{F}} \to \mathbb{G}_{\mathrm{m}}$. Both $\mathcal{F}$ and $\mathcal{F}^\vee$ satisfy the hypothesis $\mathrm{H}^0(\mathbb{G}_{\mathrm{m}} \times_{\mathbb{G}_{\mathrm{a}}} \mathbb{A}^{\mathrm{sh}}_{\mathbf{F}}, \mathcal{F}_0) = 0$. In this case of a curve, a strengthened version of A3.12, "local duality," applies (see [SGA.5] I.5.1, [SGA.4$\frac{1}{2}$] Dualité 1.3, or see [Milne80] V.2.2(a) and its proof). This implies

$H^1(\mathbb{G}_m \times_{\mathbb{G}_a} \mathbb{A}_{\mathbf{F}}^{sh}, \mathcal{F}_0) = 0$. The other H^m are 0 by dimension and affinity (see A3.7).

The last statements are obtained by purity (see A3.13).

Exercise

1. Show the commutative diagram

$$\begin{array}{ccc} Rf_! R\mathcal{H}om(\mathcal{F}, f^!\mathcal{F}') & \longrightarrow & R\mathcal{H}om(Rf_*\mathcal{F}, \mathcal{F}') \\ \downarrow & & \downarrow \\ Rf_* R\mathcal{H}om(\mathcal{F}, f^!\mathcal{F}') & \xrightarrow{\sim} & R\mathcal{H}om(Rf_!\mathcal{F}, \mathcal{F}') \end{array}$$

Notes

See Dieudonné's introduction to [FrKi88] for an account of how étale cohomology emerged as a solution to Weil's conjectures.

Concerning A3.15, a more general result about action of finite groups and étale cohomology is in [Rou02].

Abhyankhar's conjecture on (non-tame) fundamental groups of curves was solved rather recently by Raynaud (for $\pi_1(\mathbb{A}^1_{\mathbf{F}})$) and Harbater. See the contributions by Gille, Chambert-Loir, and Saïdi in the volume [BLR].

> *Le signe = placé entre deux groupes de symboles désignant des objets d'une catégorie signifiera parfois (par abus de notations) que ces objets sont canoniquement isomorphes. La catégorie et l'isomorphisme canonique devront en principe avoir été définis au préalable.*
>
> Pierre Deligne, [SGA.4] XVII

References

[Al65] Alperin, J. L., Large abelian subgroups of p-groups, *Trans. Amer. Math. Soc.*, **117** (1965), 10–20.

[Al86] Alperin, J. L., *Local Representation Theory*, Cambridge University Press, 1986.

[Al87] Alperin, J. L., Weights for finite groups, *in Proc. Symp. Pure Math.*, **47** I (1987), 369–79.

[AlFo90] Alperin, J. L. and Fong, P., Weights in symmetric and general linear groups, *J. Algebra,* **131** (1990), 2–22.

[AltKlei70] Altman, A. and Kleiman, S., *Introduction to Grothendieck Duality Theory*, Lecture Notes in Mathematics **146**, Springer, 1970.

[An93] An, J. B., 2-weights for classical groups. *J. reine angew. Math.,* **439** (1993), 159–204.

[An98] An, J. B., Dade's conjecture for 2-blocks of symmetric groups. *Osaka J. Math.,* **35** (1998), 417–37.

[An01] An, J. B., Dade's invariant conjecture for general linear and unitary groups in non-defining characteristics. *Trans. Amer. Math. Soc.,* **353** (2001), 365–90.

[And98] Andrews, G. E., *The Theory of Partitions*, Cambridge University Press, 1998.

[Ara98] Arabia, A., Objets quasi-projectifs (appendix to a paper by M.-F. Vignéras), *Sel. Math.*, **4** (1998), 612–23.

[ArKo] Ariki, S. and Koike, K., A Hecke algebra of $\mathbb{Z}/r\mathbb{Z} \wr \mathfrak{S}_n$ and construction of its irreducible representations, *Adv. Math.*, **106** (1994), 216–43.

[Artin] Artin, E., *Geometric Algebra*, Wiley, 1957.

[As84a] Asai, T., Unipotent class functions of split special orthogonal groups SO_{2n}^{+} over finite fields, *Comm. Algebra*, **12** (1984), 517–615.

[As84b] Asai, T., The unipotent class functions on the symplectic groups and the odd orthogonal groups over finite fields, *Comm. Algebra*, **12** (1984), 617–45.

[Asch86] Aschbacher, M., *Finite Group Theory*, Cambridge University Press, 1986.

[Atiyah–Macdonald] Atiyah, M. and Macdonald, I., *Introduction to Commutative Algebra*, Addison–Wesley, 1969.

[Aus74] Auslander, M. Representation theory of Artin algebras, *Comm. Algebra*, **1(3)** (1974), 177–268.

[BDK01] Brundan, J., Dipper, R. and Kleshchev, A., *Quantum Linear Groups and Representations of* $\mathrm{GL}_n(F_q)$, Memoirs of the American Mathematical Society, **149** (2001), no. 706, 112 pp.

[Ben91a] Benson, D. J., *Representations and Cohomology I: Basic Representation Theory of finite Groups and associative Algebras*, Cambridge University Press, 1991.

[Ben91b] Benson, D. J., *Representations and Cohomology II: Cohomology of Groups and Modules*, Cambridge University Press, 1991.

[BeOl97] Bessenrodt, C. and Olsson, J. B., The 2-blocks of the covering groups of the symmetric groups, *Adv. Math.*, **129**-2 (1997), 261–300.

[BiLa00] Billey, S. and Lakshmibai, V., *Singular Loci of Schubert Varieties*, Birkhäuser, 2000.

[BLR] Bost, J. B., Loeser, F. and Raynaud, M. (eds), *Courbes semi-stables et groupe fondamental en géométrie algébrique* (Proceedings of a Conference held in Luminy, 1998), *Progress in Mathematics*, **187**, Birkhäuser, 2000.

[Bo00] Bonnafé, C., Opérateur de torsion dans $\mathrm{SL}_n(q)$ et $\mathrm{SU}_n(q)$, *Bull. Soc. Math. France,* **128** (2000), 309–45.

[BoRo03] Bonnafé, C. and Rouquier, R., Variétés de Deligne–Lusztig et catégories dérivées, *Publ. Math. I.H.E.S.*, **97** (2003), 1–59.

[BoRo04] Bonnafé, C. and Rouquier, R. Sequel to [BoRo03], *work in progress.*

[Borel] Borel, A., *Linear Algebraic Groups*, Springer, 1969.

[BoTi] Borel, A. and Tits, J., Groupes réductifs, *Publ. Math. I.H.E.S.*, **27**, 1965.

[Bouc96] Bouc, S., Foncteurs d'ensembles munis d'une double action, *J. Algebra*, **183** (1996), 664–736.

[Bour59] Bourbaki, N., *Algèbre*, chapitre 9, Hermann, 1959.

[Bour68] Bourbaki, N., *Groupes et algèbres de Lie*, chapitres 4,5,6, Hermann, 1968.

[Bour80] Bourbaki, N., *Algèbre*, chapitre 10, Masson, 1980.

[Br47] Brauer, R., On a conjecture by Nakayama, *Trans Roy. Acad. Canada.* Sect.III.(3) **41** (1947), 11–19; also in *Collected Papers.* Vol. I., MIT Press, 1980.

[Br67] Brauer, R., On blocks and sections in finite groups I, *Amer. J. Math.*, **89** (1967), 1115–36.

[BrMa92] Broué, M. and Malle, G. Théorèmes de Sylow génériques pour les groupes réductifs sur les corps finis, *Math. Ann.*, **292** (1992), 241–62.

[BrMa93] Broué, M. and Malle, G. Zyklotomische Heckealgebren, *Astérisque* **212** (1993), 119–89.

[BrMaMi93] Broué, M., Malle, G. and Michel, J. Generic blocks of finite reductive groups, *Astérisque*, **212** (1993), 7–92.

[BrMaRo98] Broué, M., Malle, G. and Rouquier, R. Complex reflection groups, braid groups, Hecke algebras *J. reine angew. Math.*, **500** (1998), 127–90.

[BrMi89] Broué, M. and Michel, J., Blocs et séries de Lusztig dans un groupe réductif fini, *J. reine angew. Math.*, **395** (1989), 56–67.

[BrMi93] Broué, M. and Michel, J. Blocs à groupes de défaut abéliens des groupes réductifs finis, *Astérisque*, **212** (1993), 93–117.

[Bro86] Broué, M., Les ℓ-blocs des groupes $GL(n,q)$ et $U(n,q^2)$ et leurs structures locales, *Astérisque*, **133–4** (1986), 159–88.

[Bro88] Broué, M., Blocs, isométries parfaites, catégories dérivées, *C.R. Acad. Sci. Paris*, **307 Série I** (1988), 13–18.

[Bro90a] Broué, M., Isométries parfaites, types de blocs, catégories dérivées, in *Représentations linéaires des groupes finis* (ed. M. Cabanes), *Astérisque,* **181–2** (1990), 61–92.

[Bro90b] Broué, M., Isométries de caractères et équivalence de Morita ou dérivées, *Publ. Math. I.H.E.S.*, **71** (1990), 45–63.

[Bro94] Broué, M., Equivalences of blocks of group algebras, in *Finite-dimensional Algebras and Related Topics* (Proceedings of a Conference held in Ottawa, ON, 1992), *NATO Adv. Sci. Inst. Ser. C Math. Phys. Sci.*, **424**, Kluwer, 1994, pp. 1–26.

[Bro95] Broué, M., Rickard equivalences and block theory, in *Groups '93 Galway/St. Andrews*, Vol. 1 (Proceedings of a Conference held in Galway, 1993), London Math. Soc. Lecture Note Ser., **211**, Cambridge University Press, 1995, pp. 58–79.

[Bro00] Broué, M., Reflection groups, braid groups, Hecke algebras, finite reductive groups, *Current Developments in Mathematics*, (2000), 1–107.

[Brown] Brown, K. S., *Buildings*, Springer, 1989.

[Ca88] Cabanes, M., Brauer morphism between modular Hecke algebras, *J. Algebra,* **115-1** (1988), 1–31.

[Ca90] Cabanes, M., A criterion of complete reducibility and some applications, in *Représentations linéaires des groupes finis* (ed. M. Cabanes), *Astérisque,* **181–2** (1990), 93–112.

[Ca94] Cabanes, M., Unicité du sous-groupe abélien distingué maximal dans certains sous-groupes de Sylow, *C. R. Acad. Sci. Paris*, **318** Série I (1994), 889–94.

[Ca98] Cabanes, M., Algèbres de Hecke comme algèbres symétriques et théorème de Dipper, *C. R. Acad. Sci. Paris,* **327** Série I (1998), 531–6.

[CaEn93] Cabanes, M. and Enguehard, M., Unipotent blocks of finite reductive groups of a given type, *Math. Z.*, **213** (1993), 479–90.

[CaEn94] Cabanes, M. and Enguehard, M., On unipotent blocks and their ordinary characters, *Invent. Math.*, **117** (1994), 149–64.

[CaEn99a] Cabanes, M. and Enguehard, M., On blocks of finite reductive groups and twisted induction, *Adv. Math.*, **145** (1999), 189–229.

[CaEn99b] Cabanes, M. and Enguehard, M., On fusion in unipotent blocks, *Bull. Lond. Math. Soc.*, **31** (1999), 143–8.

[CaRi01] Cabanes, M. and Rickard, J., Alvis–Curtis duality as an equivalence of derived categories, in *Modular Representation Theory of Finite Groups* (ed. M. J. Collins, B. J. Parshall, and L. L. Scott), de Gruyter, 2001, pp. 157–74.

[Cart72a] Carter, R. W., Conjugacy classes in the Weyl group, *Compositio Math.*, **25** (1972), 1–59.

[Cart72b] Carter, R. W., *Simple Groups of Lie Type*, Wiley, 1972.

[Cart85] Carter, R. W., *Finite Groups of Lie Type: Conjugacy Classes and Complex Characters*, Wiley, 1985.

[Cart86] Carter, R. W., Representation theory of the 0-Hecke algebra, *J. Algebra*, **104** (1986), 89–103.

[CaSeMcD] Carter, R. W., Segal, G. and Mac Donald, I., *Lectures on Lie Groups and Lie Algebras*, LMSST **32**, Cambridge University Press, 1995.

[Ch55] Chevalley, C., Sur certains groups simples, *Tohoku Math. J.*, **7** (1955), 14–66.

[Cr95] Cartier, P., La théorie des blocs et les groupes génériques, *Séminaire Bourbaki*, Vol. 1993/94, *Astérisque*, **227** (1995), 171–208.

[Cu80a] Curtis, C. W., Truncation and duality in the character ring of a finite group of Lie type, *J. Algebra*, **62** (1980), 320–32.

[Cu80b] Curtis, C. W., *Homology Representations of Finite Groups*, Lecture Notes in Mathematics **832** Springer, 1980, pp. 177–94.

[CuLe85] Curtis, C. W. and Lehrer, G., Generic chain complexes and finite Coxeter groups, *J. reine angew. Math.*, **363** (1985), 146–73.

[CuRe87] Curtis, C. W and Reiner, I., *Methods of Representation Theory*, Wiley, 1987.

[Da71] Dade, E. C., Character theory pertaining to finite simple groups. *Finite Simple Groups* (Proceedings of an Instructional Conference held in Oxford, 1969), Academic Press, 1971, pp. 249–327.

[Da92] Dade, E. C. Counting characters in blocks I. *Invent. Math.*, **109** (1992), 187–210.

[Da99] Dade, E. C. Another way to count characters. *J. reine angew. Math.*, **510** (1999), 1–55.

[Danil94] Danilov, V. I., *Algebraic Varieties and Schemes*, in *Encyclopedia of Mathematical Sciences* **23**, Springer, 1994.

[Danil96] Danilov, V. I., *Cohomology of Algebraic Varieties*, in *Encyclopedia of Mathematical Sciences* **35**, Springer, 1996.

[DeLu76] Deligne, P. and Lusztig, G., Representations of reductive groups over finite fields, *Ann. Math.*, **103** (1976), 103–61.

[DeLu82] Deligne, P. and Lusztig, G., Duality for representations of a reductive group over a finite field, *J. Algebra*, **74** (1982), 284–91.

[DiMi90] Digne, F. and Michel, J., On Lusztig's parametrization of characters of finite groups of Lie type, in *Représentations linéaires des groupes finis*, (ed. M. Cabanes), *Astérisque,* **181–2** (1990), 113–56.

[DiMi91] Digne, F. and Michel, J., *Representations of Finite Groups of Lie Type*, Cambridge University Press, 1991.

[Dip80] Dipper, R. Vertices of irreducible representations of finite Chevalley groups in the describing characteristic, *Math. Z.*, **175** (1980), 143–59.

[Dip83] Dipper, R., On irreducible modules of twisted groups of Lie Type, *J. Algebra*, **81** (1983), 370–89.

[Dip85a] Dipper, R., On the decomposition numbers of the finite general linear groups, *Trans. Amer. Math. Soc.*, **290** (1985), 315–44.

[Dip85b] Dipper, R., On the decomposition numbers of the finite general linear groups II, *Trans. Amer. Math. Soc.*, **292** (1985), 123–33.

[Dip90] Dipper, R., On quotients of Hom-functors and representations of finite general linear groups I, *J. Algebra*, **130** (1990), 235–59.

[DipDu93] Dipper, R. and Du, J., Harish-Chandra vertices, *J. reine angew. Math.*, **437** (1993), 101–30.

[DipDu97] Dipper, R. and Du, J., Harish-Chandra vertices and Steinberg's tensor product theorems for finite general linear groups, *Proc. Lond. Math. Soc.*, **75**-3 (1997), 559–99.

[DipJa86] Dipper, R. and James, G. D., Representations of Hecke algebras of general linear groups, *Proc. Lond. Math. Soc.*, **52** (1986), 20–52.

[DipJa87] Dipper, R. and James, G. D., Blocks and idempotents of Hecke algebras of general linear groups, *Proc. Lond. Math. Soc.*, **54** (1987), 57–82.

[DipJa89] Dipper, R. and James, G. D., The q-Schur algebra, *Proc. Lond. Math. Soc.*, **59** (1989), 23–50.

[DipJa92] Dipper, R. and James, G. D., Representations of Hecke algebras of type B_n, *J. Algebra*, **146** (1992), 454–81.

[DipMa02] Dipper, R. and Mathas, A., Morita equivalences of Ariki–Koike algebras, *Math. Z.*, **240**-3 (2002), 579–610.

[Donk98a] Donkin, S., *The q-Schur Algebra*, London Mathematical Society Lecture Note Series, **253**, Cambridge University Press, 1998.

[Donk98b] Donkin, S. An introduction to the Lusztig conjecture, in *Representations of Reductive Groups*, Publ. Newton Inst., Cambridge University Press, 1998, pp. 173–87.

[DuRui00] Du, J. and Rui, H., Ariki–Koike algebras with semi-simple bottoms, *Math. Z.*, **234** (2000), 807–30.

[EGA] Grothendieck, A. and Dieudonné, J., *Éléments de géométrie algébrique*, II. Étude globale élémentaire de quelques classes de morphismes. *Publ. Math. Inst. Hautes Études Sci.*, **8**, 1961.

[En91] Enguehard, M., Isométries parfaites entre blocs des groupes linéaires ou unitaires, *Math. Z.*, **206** (1991), 1–24.

[En00] Enguehard, M. Sur les ℓ-blocs unipotents des groupes réductifs finis quand ℓ est mauvais, *J. Algebra*, **230** (2000), 334–77.

[FoSe73] Fong, P. and Seitz, G. M. Groups with a (B,N)-pair of rank 2, *Invent. Math.*, **21** (1973), 1–57; **24** (1973), 191–239.

[FoSr82] Fong, P. and Srinivasan, B., The blocks of finite general and unitary groups, *Invent. Math.*, **69** (1982), 109–53.

[FoSr86] Fong, P. and Srinivasan, B., Generalized Harish-Chandra theory for unipotent characters of finite classical groups, *J. Algebra*, **104** (1986), 301–9.

[FoSr89] Fong, P. and Srinivasan, B., The blocks of finite classical groups, *J. reine angew. Math.*, **396** (1989), 122–91.

[FrKi88] Freitag, E. and Kiehl, R., *Etale Cohomology and Weil Conjectures*, Springer, 1988.

[Geck93a] Geck, M., Basic sets of Brauer characters of finite groups of Lie Type II, *J. Lond. Math. Soc.*, **47** (1993), 255–68.

[Geck93b] Geck, M., A note on Harish-Chandra induction, *Manuscripta Math.*, **80**-4 (1993), 393–401.

[Geck01] Geck, M., Modular Harish-Chandra series, Hecke algebras and (generalized) q-Schur algebras, in *Modular Representation Theory of Finite Groups* (ed. M. J. Collins, B. J. Parshall, and L. L. Scott), de Gruyter, 2001, pp. 1–66.

[GeHi91] Geck, M. and Hiss, G., Basic sets of Brauer characters of finite groups of Lie type, *J. reine angew. Math.*, **418** (1991), 173–88.

[GeHi97] Geck, M. and Hiss, G., Modular representations of finite groups of Lie type in non-defining characteristic, in *Finite Reductive Groups* (ed. M. Cabanes), *Prog. Math.*, **141**, Birkhäuser, 1997, pp. 195–249.

[GeHiMa94] Geck, M., Hiss, G., and Malle, G., Cuspidal unipotent Brauer characters, *J. Algebra*, **168** (1994), 182–220.

[GeHiMa96] Geck, M., Hiss, G., and Malle, G., Towards a classification of the irreducible representations in non-describing characteristic of a finite group of Lie type, *Math. Z.*, **221** (1996), 353–86.

[GelMan94] Gelfand, S.I. and Manin, Y. I., *Homological Algebra*, EMS **38**, Springer, 1994 (Russian edition 1989).

[Gen02] Genet, G., On the commutator formula of a split BN-pair, *Pac. J. Math.*, **207** (2002), 177–81.

[Gen03] Genet, G. On stable decomposition matrices, *submitted* (2003).

[GePf00] Geck, M. and Pfeiffer, G, *Characters of Finite Coxeter Groups and Iwahori-Hecke Algebras*, Oxford University Press, 2000.

[God58] Godement, R., *Topologie algébrique et théorie des faisceaux*, Hermann, 1958.

[Gol93] Goldschmidt, D. M., *Group Characters, Symmetric Functions, and the Hecke Algebra*, University Lecture Series **4**, American Mathematical Society, 1993.

[Gre78] Green, J. A., On a theorem of Sawada, *J. Lond. Math. Soc.*, **18** (1978), 247–52.

[GroMur71] Grothendieck, A. and Murre, J., *The Tame Fundamental Group of a Formal Neighbourhood of a Divisor with Normal Crossings*, Lecture Notes in Mathematics **20**, Springer, 1971.

[GruHi97] Gruber, J. and Hiss, G., Decomposition numbers of finite classical groups for linear primes, *J. reine angew. Math.*, **455** (1997), 141–82.

[Haa86] Haastert, B., Die Quasiaffinität der Deligne–Lusztig Varietäten, *J. Algebra,* **102** (1986), 186–93.

[HaCh70] Harish-Chandra, Eisenstein series over finite fields, *Functional Analysis and Related Fields*, Springer, 1970, pp. 76–88.

[Har92] Harris, J., *Algebraic Geometry, GTM*, **133**, Springer, 1992.

[Hart] Hartshorne, R., *Algebraic Geometry, GTM*, **52**, Springer, 1977.

[Hi90] Hiss, G., Zerlegungszahlen endlicher Gruppen vom Lie-Typ in nicht-definierender Charakteristik, *Habilitationsschrift, RWTH Aachen*, 1990.

[Hi93] Hiss, G., Harish-Chandra series of Brauer characters in a finite group with a split BN-pair, *J. Lond. Math. Soc.*, **48** (1993), 219–28.

[HiKe00] Hiss, G. and Kessar, R., Scopes reduction and Morita equivalence classes of blocks in finite classical groups, *J. Algebra*, **230** (2000), 378–423.

[Hoef74] Hoefsmit, P. N., Representations of Hecke Algebras of finite Groups with BN-pair of classical type, PhD Thesis, University of British Columbia, Vancouver, 1974.

[How80] Howlett, B., Normalizers of parabolic subgroups of reflection groups, *J. Lond. Math. Soc.,* **21** (1980), 62–80.

[HowLeh80] Howlett, B. and Lehrer, G., Induced cuspidal representations and generalized Hecke rings, *Invent. Math.*, **58** (1980), 37–64.

[HowLeh94] Howlett, B. and Lehrer, G., On Harish-Chandra induction for modules of Levi subgroups, *J. Algebra*, **165** (1994), 172–83.

[Hum81] Humphreys, J. E., *Linear Algebraic Groups*, Springer, 1981.

[Hum90] Humphreys, J. E., *Coxeter Groups and Reflection Groups*, Cambridge University Press, 1990.

[Ja86] James, G. D., Irreducible representations of Hecke algebras of general linear groups, *Proc. Lond. Math. Soc.*, **52** (1986), 236–68.

[JaKe81] James, G. and Kerber, A., *The Representation Theory of the Symmetric Group*, in *Encyclopedia of Mathematics* **16**, Addison-Wesley, 1981.

[Jantzen] Jantzen, J. C., *Representations of Algebraic Groups*, Academic Press, 1987.

[Jost96] Jost, T., Morita equivalence for blocks of finite general linear groups, *Manuscripta Math.*, **91** (1996), 121–44.

[KaSch98] Kashiwara, M. and Schapira, P., *Sheaf Cohomology*, Springer, 1998.

[Ke98] Keller, B., Introduction to abelian and derived categories, in *Representations of Reductive Groups*, Publ. Newton Inst., Cambridge University Press, 1998, pp. 41–61.

[Kempf] Kempf, G. R., *Algebraic Varieties*, Cambridge University Press, 1993.

[KLRZ98] König, S. and Zimmermann, A., *Derived Equivalences for Group Rings*, Lecture Notes in Mathematics **1685**, Springer, 1998.

[KnRo89] Knörr, R. and Robinson, G. R., Some remarks on a conjecture of Alperin. *J. Lond. Math. Soc.* **39** (2) (1989), 48–60.

[KrThi99] Krob, D. and Thibon, J. Y., Noncommutative symmetric functions V. A degenerate version of $U_q(\mathrm{gl}_N)$, *Int. J. Algebra Comput.*, **9** (1999), 405–30.

[La56] Lang, S., Algebraic groups over finite fields, *Amer. J. Math.*, **78** (1956), 555–563.

[LT92] Lehrer, G. and Thévenaz, J., Sur la conjecture d'Alperin pour les groupes réductifs finis, *C. R. Acad. Sci. Paris*, **314** Série I (1992), 1347–51.

[Lu76a] Lusztig, G., On the finiteness of the number of unipotent classes, *Invent. Math.*, **34** (1976), 201–13.

[Lu76b] Lusztig, G., Coxeter orbits and eigenspaces of Frobenius, *Invent. Math.*, **38** (1976), 101–59.

[Lu77] Lusztig, G., Representations of finite classical groups, *Invent. Math.*, **43** (1977), 125–75.

[Lu84] Lusztig, G., *Characters of Reductive Groups over a Finite Field*, Ann. Math. Studies, **107**, Princeton University Press, 1984.

[Lu88] Lusztig, G., On the representations of reductive groups with disconnected centre, *Astérisque*, **168** (1988), 157–66.

[Lu90] Lusztig, G., Green functions and character sheaves, *Ann. Math.*, **131** (1990), 355–408.

[Mathas] Mathas, A., *Iwahori-Hecke Algebras and Schur Algebras of the Symmetric Group*, University Lecture Series **15**, American Mathematical Society, 1999.

[McLane63] Mac Lane, S., *Homology*, Springer, 1963.

[McLane97] Mac Lane, S., *Categories for the Working Mathematician*, Springer, 1997.

[Milne80] Milne, J. S., *Etale Cohomology*, Princeton University Press, 1980.

[Milne98] Milne, J. S., *Lectures on Etale Cohomology*, available at www.milne.org, 1998.

[Miya94] Miyanishi, M. , *Algebraic Geometry,* Translations of Math. Monographs, **136**, American Mathematical Society, 1994.

[Mum70] Mumford, D., *Abelian Varieties*, Oxford University Press, 1970.

[Mum88] Mumford, D., *The Red Book of Varieties and Schemes*, Lecture Notes in Mathematics **1358**, Springer, 1988.

[Murre] Murre, J., *Lectures on an Introduction to Grothendieck's Theory of the Fundamental Group*, Lecture Notes, Tata Institute for Fundamental Research, Bombay, 1967.

[NaTs89] Nagao, H. and Tsushima, Y., *Representations of Finite Groups*, Academic Press, 1989.

[No79] Norton, P. N., 0-Hecke algebras, *J. Austr. Math. Soc.*, **27** (1979), 337–57.

[Pe00] Peterfalvi, T., *Character Theory for the Odd Order Theorem*, LMS Lecture Note Series **272**, Cambridge University Press, 2000.

[Pu87] Puig, L., The Nakayama conjectures and the Brauer pairs, *Publications Mathématiques de l'Université Paris 7*, **25** (1987).

[Pu90] Puig, L., Algèbres de source de certains blocs des groupes de Chevalley, in *Représentations linéaires des groupes finis* (ed. M. Cabanes), *Astérisque,* **181–2** (1990), pp. 93–117.

[Ram97] Ram, A., Seminormal representations of Weyl groups and Iwahori–Hecke algebras, *Proc. Lond. Math. Soc.*, **75-1** (1997), 99–133.

[Ri69] Richen, F., Modular representations of split BN-pairs, *Trans. Amer. Math. Soc.*, **140** (1969), 435–460.

[Rick95] Rickard, J., Finite group actions and étale cohomology, *Publ. Math. I.H.E.S.*, **80** (1995), 81–94.

[Rick96] Rickard, J., Splendid equivalences: derived categories and permutation modules, *Proc. Lond. Math. Soc.*, **72** (1996), 331–58.

[Rick98] Rickard, J., An introduction to intersection cohomology, in *Representations of Reductive Groups,* Publ. Newton Inst., Cambridge University Press, 1998, pp. 151–71.

[Rick01] Rickard, J., The abelian defect group conjecture: some recent progress, in *Algebra and Representation Theory* (Proceedings of a conference held in Constanta, 2000), *NATO Sci. Ser. II Math. Phys. Chem.,* **28**, Kluwer, 2001, pp. 273–99.

[Rob98] Robinson, G. R., Further consequences of conjectures like Alperin's, *J. Group Theory,* **1**, (1998) 131–41.

[Rou01] Rouquier, R., Block theory via stable and Rickard equivalences, in *Modular Representation Theory of Finite Groups* (ed. M. J. Collins, B. J. Parshall, and L. L. Scott), de Gruyter, 2001, pp. 101–46.

[Rou02] Rouquier, R., Complexes de chaînes étales et courbes de Deligne–Lusztig, *J. Algebra*, **257** (2002), 482–508.

[Serre77a] Serre, J. P., *Linear Representations of Finite Groups*, GTM **42**, Springer, 1977.

[Serre77b] Serre, J. P., *Cours d'arithmétique*, Presses Universitaires de France, 1977.

[Serre88] Serre, J. P., *Algebraic Groups and Class Fields*, GTM **117**, Springer, 1988.

[SGA.1] Grothendieck, A., *Séminaire de géométrie algébrique du Bois-Marie 1960–61 (SGA.1), Revêtements étales et groupe fondamental*, Lecture Notes in Mathematics **224,** Springer, 1971.

[SGA.4] Artin, M., Grothendieck, A., and Verdier, J. L., *Séminaire de géométrie algébrique du Bois-Marie 1963–64 (SGA.4), Théorie des Topos et Cohomologie Étale des Schémas*, Lecture Notes in Mathematics **269, 270, 305,** Springer, 1972–1973.

[SGA.4$\frac{1}{2}$] Deligne, P., *Séminaire de géométrie algébrique du Bois-Marie (SGA.4$\frac{1}{2}$)*, Lecture Notes in Mathematics **569**, Springer, 1977.

[SGA.5] Grothendieck, A., *Séminaire de géométrie algébrique du Bois-Marie 1965–66 (SGA.5), Cohomologie ℓ-adique et Fonctions L*, Lecture Notes in Mathematics **589**, Springer, 1977.

[Sho85] Shoji, T., Some generalization of Asai's result for classical groups, in *Algebraic Groups and Related Topics* (Proceedings of a Conference held in Kyoto/Nagoya, 1983), *Adv. Stud. Pure Math.,* **6** (1985), 207–29.

[Sho87] Shoji, T., Shintani descent for exceptional groups over a finite field, *J. Fac. Sci. Univ. Tokyo Sect. IA Math.*, **34** (1987), 599–653.

[Sho97] Shoji, T., Unipotent characters of finite classical groups, in *Finite Reductive Groups* (ed. M. Cabanes), *Prog. Math.*, **141**, Birkhäuser, 1997, 373–413.

[Sm82] Smith, S. D., Irreducible modules and parabolic subgroups, *J. Algebra*, **75** (1982), 286–9.

[So95] Solomon, R., On finite simple groups and their classification, *Notices Am. Math. Soc.*, **42-2** (1995), 231–39.

[Spanier] Spanier, E., *Algebraic Topology*, McGraw-Hill, 1966.

[Sp70] Springer, T., Cusp forms for finite groups, in A. Borel *et al.*, *Seminar on Algebraic Groups and Related Finite Groups*, Lecture Notes in Mathematics **131**, Springer, 1970.

[Sp74] Springer, T., Regular elements in finite reflection groups, *Inv. Math.*, **25** (1974), 159–98.

[Springer] Springer, T., *Linear Algebraic Groups*, Second edition, Birkhäuser, 1999.

[SpSt70] Springer, T. and Steinberg, R., Conjugacy classes, in *Seminar on Algebraic Groups and Related Finite Groups* (ed. A. Borel *et al.*), Lecture Notes in Mathematics **131**, Springer, 1970.

[Srinivasan] Srinivasan, B., *Representations of Finite Chevalley Groups*, Lecture Notes in Mathematics **764** Springer, 1979.

[Stein68a] Steinberg, R., *Lectures on Chevalley groups*, Yale University Press, 1968.

[Stein68b] Steinberg, R., *Endomorphisms of Linear Algebraic Groups*, American Mathematical Society Memoirs, **80**, 1968.

[Tak96] Takeuchi, M., The group ring of $\mathrm{GL}_n(q)$ and the q-Schur algebra, *J. Math. Soc. Japan*, **48**-2 (1996), 259–74.

[Tamme] Tamme, G., *Introduction to étale cohomology*, Springer, 1994.

[Thévenaz] Thévenaz, J., *G-Algebras and Modular Representation Theory*, Oxford University Press, 1995.

[Tin79] Tinberg, N., Some indecomposable modules of groups with a BN-pair, *J. Algebra*, **61** (1979), 508–26.

[Tin80] Tinberg, N., Modular representations of finite groups with unsaturated split BN-pair, *Can. J. Math*, **32-3** (1980), 714–33.

[Tits] Tits, J., *Buildings of Spherical Type and Finite BN-pairs,* Lecture Notes in Mathematics **386**, Springer, 1974.

[Wall63] Wall, G. E., On the conjugacy classes in the unitary, symplectic and orthogonal groups, *J. Austr. Math. Soc.*, **3** (1963), 1–62.

[Weibel] Weibel, C., *An Introduction to Homological Algebra*, Cambridge University Press, 1994.

[Za62] Zassenhaus, H. On the spinor norm, *Arch. Math.*, **13** (1962), 434–51.

Index

Other notations

Finir par la vue des deux bonshommes penchés sur leur pupitre, et copiant.

Gustave Flaubert, *Bouvard et Pécuchet.*

For EU product safety concerns, contact us at Calle de José Abascal, 56–1°, 28003 Madrid, Spain or eugpsr@cambridge.org.

www.ingramcontent.com/pod-product-compliance
Lightning Source LLC
LaVergne TN
LVHW101454200926
848483LV00001B/128

* 9 7 8 0 5 2 1 8 2 5 1 7 7 *